COGNITIVE RADIO COMMUNICATIONS AND NETWORKING

COGNITIVE RADIO COMMUNICATIONS AND NETWORKING

PRINCIPLES AND PRACTICE

Robert C. Qiu and **Zhen Hu**
Tennessee Technological University, USA

Husheng Li
University of Tennessee at Knoxville, USA

Michael C. Wicks
University of Dayton Research Institute, USA

A John Wiley & Sons, Ltd., Publication

Library of Congress Cataloging-in-Publication Data

Qiu, Robert C.
  Cognitive radio communications and networking : principles and practice / Robert C. Qiu ... [et al.].
     p. cm.
  Includes bibliographical references and index.
  ISBN 978-0-470-97209-0 (cloth)
 1. Cognitive radio networks. I. Hu, Zhen. II. Li, Husheng, 1975- III. Title.
  TK5103.4815.Q58 2012
  621.382′1—dc23

                                                                                    2012009304

A catalogue record for this book is available from the British Library.

ISBN: 9780470972090

Set in 10/12pt Times by Laserwords Private Limited, Chennai, India.
Printed and bound in Malaysia by Vivar Printing Sdn Bhd

1 2012

To my family Lily Liman Li, Michelle, David and Jackie
—Robert C. Qiu

To my wife, Qingying Meng
—Zhen Hu

To my wife Min Duan and my son Siyi
—Husheng Li

Our families
—Michael C. Wicks

Contents

Preface

The idea of writing this book began at least five years ago when the first author taught a one first-year graduate course, on communications/wireless communications. After this course, some students pursued advanced topics such as convex optimization prior to their PhD research. MS students wanted to know more about the field before they began to design wireless systems. The first author taught such advanced courses regularly, and part of these materials provided the starting point for this book. After this book project began, additional authors were added so that we could meet with our deadlines and before the topics become outdated. Another title of this book could be Advanced Wireless Communications.

The most difficult part was to decide what to exclude. The wireless industry is still expanding rapidly after two decades of growth. The first author studied the second generation (2G) system—CDMA and GSM—during his university days. Now, 3G (WCDMA) and 4G (LTE) systems are available. Each system has its central concept and demands unique analytical skills. Generally professors find that their most significant responsibilities are to teach students the most difficult mathematical tools required to analyze and design fundamental system concepts. For example, in a GSM (TDMA) system, the equalizers are central to the system. For a CDMA system, a RAKE receiver is central (as is power control). For an LTE system, a multiple-input, multiple-output (MIMO) system combined with an orthogonal frequency division multiplexing (OFDM) is central.

This approach is adopted in our book. We cover the system concepts that are central to the next generation cognitive radio network (CRN). We claim that the following three analytical tools are central to the CRN: (1) large random matrices; (2) convex optimization; (3) game theory. The unified view is the so-called "Big Data"—high-dimensional data processing. Due to the unique nature of cognitive radio, we have an unparalleled challenge—having too much data at our disposal. In today's digital age, making sense of the data in real-time is central not only to major players like Facebook, Google and Amazon, but also to our telecommunication vendors. To successfully solve the Big Data problem however, there are still many hurdles. For one thing, the current tools are inadequate. Scientist and engineers with the skills to analyze the data are scarce. Future ECE students must learn the analytical skills obtained from studying Big Data. In addition to traditional fields, this book contains results from multi disciplinary fields: machine learning, financial engineering, statistics, quantum computing, etc. Social networking and the Smart Grid command more resources. Researchers must become more cost-conscious. Investments in other fields mentioned above can reduce the costs of solving these problems. Abstract mathematical connections are the best starting point toward this goal.

This justifies our belief in teaching students the most difficult analytical skills that are not readily obtained after leaving schools. By studying this book, practical engineers will understand system concepts, and may make connections with other fields. Peer researchers can use this book as a reference.

Compared with previous systems, the CRN contains radios that are highly programmable; their modulation waveforms are changing rapidly and their frequencies are agile; their radio frequency (RF) front-ends are wideband (up to several GHz). In addition to the highly programmable nature of their physical layer functions, a CRN radio senses the spectrum at an unprecedentedly low signal-to-noise-ratio (SNR) (e.g., -21 dB required by the FCC). To support this fundamental spectrum sensing function, the system allocates computing resources with the ultimate goal of real-time operations. From another viewpoint, this radio is a powerful sensor with almost unlimited computing and networking capabilities. Through the combination of these two views, communications and sensing are merged into one function that transmits, receives, and processes programmable modulated waveforms. Real-time distributed computing is embedded in these two functions.

It is believed that we lack a coherent network theory that is valid for numerous applications. Rather, the state-of-the-art network is designed for special needs; when a new need arises, the network must be redesigned. Costs are wasteful due to the lack of a network theory. The cognitive radio poses unique challenges in networking.

Wireless technology is proliferating rapidly; the vision of pervasive wireless computing, communication, sensing and control offers the promise of many societal and individual benefits. Cognitive radios, through dynamic spectrum access, offer the promise of being a disruptive technology. Cognitive radios are fully programmable wireless devices that can (1) sense their environment and (2) dynamically adapt their transmission waveform, channel access method, spectrum use and networking protocols. It is anticipated that cognitive radio technology will become a general-purpose programmable radio that will serve as a universal platform for wireless system development, as microprocessors have served a similar role for computation. There is, however, a big gap between having a flexible cognitive radio, effectively a building block, and the large-scale deployment of cognitive radio networks that dynamically optimize spectrum use. Testbeds are critical but totally ignored since the materials become outdated when the book is published. We want to focus on the materials that can last.

One goal is aimed toward a large-scale cognitive radio network; in particular, we need to study novel cognitive algorithms using quantum information and machine learning techniques, to integrate FPGA, CPU and graphics processing unit (GPU) technology into state-of-the-art radio platforms, and to deploy these networks as testbeds in the real-world university environment. Our applications range from communications to radar/sensing and Smart Grid technologies. Cognitive radio networking/sensing for unmanned aerial vehicles (UAVs) is also very interesting and challenging due to its high mobility. Synchronization is critical. UAVs can be replaced with robots.

One task will pursue a new initiative of CRN as sensors and explore the vision of a dual-use sensing/communication system based on CRN. The motivation is to push the convergence of sensing and communication systems into a unified cognitive networking system. CRN is a cyber-physical system with the integrated capabilities of control, communications, and computing.

Due to the embedded function of cooperative spectrum sensing in CRN, rich information about the radio environment may be obtained. This information *unique to CRN* can be exploited to detect, indicate, recognize, or track the target or intruder in the covered area of a CRN. The data for this kind of information system are intrinsically high-dimensional and random. Hence, we can employ quantum detection, quantum state estimation, and quantum information theory in our new initiative using CRN as sensors. In this way, the sensing capability of CRN can be explored together with great improvement in performance.

Very often one views a cognitive radio as two fundamental functions: (1) spectrum sensing; (2) spectrum-aware resources allocation. In this second function, convex optimization plays a central role. Optimization stems from human instinct. We always like to do something in the best way. Optimization theory gives us a way to realize this kind of human instinct. With the enhancement of computing capability, optimization theory, especially convex optimization, is a powerful signal processing tool to handle Big Data. If the data mining problem can be formulated as a convex optimization problem, the global optimum can be achieved. There is no doubt about the results or performances. However, there is still a challenge to make optimization algorithms scalable on the data sets of millions or trillions of elements. Thus, more effort is needed to explore optimization theory before we gain the benefit of it.

A collection of nodes are studied. These nodes, in analogy with human beings, can both collaborate and compete. Game theory captures the fundamental role of competition for resources. Of course, many algorithms in game theory can be formulated as convex optimization problems. For the games in CRN, we have provided plenty of working knowledge of generic games such that the readers can begin the research without reading specific books on game theory. Several typical examples in CRN are given to illustrate how to use game theory to analyze cognitive radio. Moreover, many unique concerns of games in cognitive radio are explained in order to motivate new research directions.

We will explain the networking issues in CRN in a layer by layer manner. Only challenges specific to CRN are explained in order to distinguish from traditional communication networks. We hope that the corresponding chapter not only explains the state-of-the-art of CRN, but also motivates new ideas in the design of CRN.

The overall picture of this book is presented in Figure P.1. Novel applications of the CRN include:

1. The Smart Grid; Security is a challenge.
2. Wireless networking for for unmanned aerial vehicles. Synchronization is a challenge.
3. Cloud computing is integrated with the CRN.
4. The CRN is used as distributed sensing.

Chapter 1 overviews the book. Twelve chapters are included.

Chapter 2 presents basic techniques for spectrum sensing. These techniques can be implemented in today's systems. Energy detection is the basis. The second-order statistics based detection is important. Features extracted using singular value decomposition (SVD) are also used. Cyclostationary detection is treated for completeness.

Chapter 3 is the core of spectrum sensing. It is also a stepping-stone to understand the algorithms of Chapter 4 that are believed to be new. The generalized likelihood ratio test (GLRT) is the culmination of the development of the whole Chapter 3. We focus on three

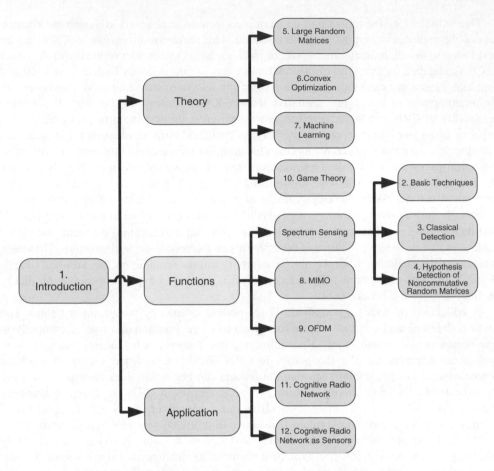

Figure P.1 Connections of different chapters in this book.

major analytical tools: (1) multivariate normal statistics; (2) sample covariance matrix that is a random matrix; (3) the GLRT. This chapter also prepares us for Chapter 5 (large random matrices).

Chapter 4 deals with noncommunicative random matrices and their detection. The nature of this chapter is exploratory. It connects us with some latest literature in quantum computing, applied linear algebra and machine learning. The basic mathematical objects are random matrices—matrix-valued random variables that are elements in an algebraic space such as C^* algebra. This chapter is designed not only for practical significance, but also for conceptual significance. These concepts are basic when we deal with Big Data in machine learning. A great number of random matrices are processed. The new algorithms achieve much better performance, compared with the classical algorithms that are treated in Chapter 3.

Chapter 5 is a long chapter. It was, however, not even included in the initial writing. In the last stage of the book project, we have reached the insight into its fundamental significance under our unified view of Big Data. In Chapter 4, we already established that the basic mathematical objects are the covariance matrices and their associated sample covariance matrices. We can asymptotically estimate the former from the data. Recently, the trend is to use the nonasymptotic sample covariance matrices instead. The data is huge but is not infinite. The central difficulty arises from the randomness of a sample covariance matrix that uses finite data samples. When a large collection of those sample covariance matrices are studied, the so-called random matrix theory is needed. Also, for detection, quantum information is needed. Under this context, Chapter 4 is connected when there quantum detection is used. Large random matrices were used to wireless communication as early as 1999 by Tse and Verdu to study CDMA systems. Later, they were used to study MIMO systems. They are especially critical to our vision of merging communications with sensing. Large random matrices are ideal mathematical objects to collect the intrinsic (quantum) information is a large network of cognitive nodes that are able to sense, compute, and reason. To study this collection of large random matrices, the so-called random matrix theory is needed. How to apply this theory in the large sensing network is clear from this chapter (see also Chapter 12). How to apply this theory for across-layer applications such as routing, physical layer optimization, etc. remains elusive at the time of writing. Compressive sensing is another fundamental concept that is only applied in this chapter. A comprehensive treatment is beyond our scope here. Its relevance is pointed out for emphasis. Compressive sensing exploits the structure of sparsity of the physical signals; large random matrices exploits the structure of random entries. Somehow it is believed that two theories must be combined together. We have only touched the surface of this issue and further research is still required.

Chapter 6 will give some background information about optimization theory. Optimization stems from human instinct. We always like to do something in the best way. Relying on mathematics, this human instinct can be written down. Convex optimization is a subfield of optimization theory. The strength of convex optimization is if a local minimum exists, then it is a global minimum. Hence, if the practical problem can be formulated as a convex optimization problem, then global optimum can be obtained. That is the reason why convex optimization has recently become popular. Linear programming, quadratic programming, geometric programming, Lagrange duality, optimization algorithm, robust optimization, and multi objective optimization will be covered. Some examples will be presented to show the beauty and benefit of optimization theory.

Chapter 7 will give some background information about machine learning. Machine learning can make the system intelligent. In order to give readers the whole picture of machine learning, almost all the topics related to machine learning will be covered, which include unsupervised learning, supervised learning, semi supervised learning, transductive inference, transfer learning, active learning, reinforcement learning, kernel-based learning, dimensionality reduction, ensemble learning, meta learning, Kalman filtering, particle filtering, collaborative filtering, Bayesian network, and so on. Machine learning will be the basic engine for cognitive radio network.

Chapter 8 will present MIMO transmission technique. MIMO in wireless communication exploits multiple antennas at both the transmitter and the receiver to improve the performance of wireless communication without additional radio bandwidth. Array gain, diversity gain, and multiplexing gain can be achieved. Space time coding, multi-user MIMO, MIMO network, and so on will be covered. MIMO can explore the spatial radio resources to support spectrum access and spectrum sharing in cognitive radio network.

Chapter 9 will present OFDM transmission technique. OFDM is a technique of digital data transmission based on multi carrier modulation. The critical issues in OFDM system including OFDM implementation, synchronization, channel estimation, peak power problem, adaptive transmission, spectrum shaping, OFDMA, and so on will be discussed. Spectrum access and spectrum sharing can also be well supported by OFDM in cognitive radio network.

Chapter 10 is devoted to the application of game theory in cognitive radio. There exist competition and collaboration in the spectrum, thus resulting in various games in cognitive radio. In this book, we will provide a brief introduction to game theory and then apply it to several typical types of games in cognitive radio.

Chapter 11 provides a systematic introduction to the design issues of networking with cognitive radio. We will explain the algorithms and protocols in various layers of cognitive radio networks. In particular, we will address the unique challenges brought by the mechanism of cognitive radio. We will also discuss the complex network phenomenon in cognitive radio networks.

Chapter 12 will describe a new initiative of cognitive radio network as sensors. This vision tries to explore a dual use sensing/communication system based on cognitive radio network. Cognitive radio network is a cyber-physical system with the integrated capabilities of control, communication, and computing. Cognitive radio network can provide an information superhighway and a strong backbone for the next generation intelligence, surveillance, and reconnaissance. Open issues together with the potential applications in cognitive radio network as sensors will be under investigation.

The author Qiu wants to thank his PhD graduates for their help in proof-reading: Jason Bonier, Shujie Hou, Xia Li, Feng Lin, and Changchun Zhang at TTU, especially Changchun Zhang for drawing numerous figures. Qiu and Hu want to thank their colleagues at TTU: Kenneth Currie, Nan Terry Guo, P. K. Rajan for many years' help. Qiu and Hu want to acknowledge their program director Dr. Santanu K. Das at Office of Naval Research (ONR) who supported their research contained in this book. This work is funded by National Science Foundation through two grants (ECCS-0901420 and ECCS-0821658), and Office of Naval Research through two grants (N00010-10-1-0810 and N00014-11-1-0006). The authors want to thank our editor Mark Hammond for his interest in this book and his encouragement during the whole process of the book development. The authors have received daily help from other editors: initially Sophia Travis and Sarah Tilley; later Susan Barclay.

For more information, please visit the companion website—www.wiley.com/go/qiu/cogradio.

1

Introduction

1.1 Vision: "Big Data"

"Big Data" [1] refers to datasets whose size is beyond the ability of typical database software tools to capture, store, manage, and analyze.

There is a convergence of communications, sensing and computing towards the objective of achieving some control. In particular, cloud computing is promising. Sensors become cheaper. A network becomes bigger. In particular, powered by Internet protocols, the Smart Grid—a huge network, much bigger than the traditional networks—becomes an "energy Internet."

Communications are becoming more and more like "backbones" for a number of applications. Sensing is a seamless ingredient in the future Internet of Things. In particular, it is the data acquisition mechanism to support the vision of "Big Data." Computing will become a commodity that is affordable by the common needs of everyday applications.

The economy is becoming a "digital economy," meaning that the jobs are more and more related to "soft power." This does not necessarily imply software programming. Rather, it implies that more and more job functions will be finished by a smart system which is driven by sophisticated mathematics. While job functions become more and more "soft," the needs for analytical analysis become more demanding. As a result, analytical skills, which are avoided by most of us at first sight, will be most useful in the lifelong education of a typical graduate student. Most often, our students know how to do their programing if they know the right mathematics. This is the central problem or dilemma. Analytical machinery is like our games of sports. Unless we practice with dedication, we will not become good players.

The book aims to focus on fundamentals, in particular, mathematical machinery. We primarily cover topics that are critical to cognitive radio network but hard to master without big efforts.

Cognitive Radio Communications and Networking: Principles and Practice, First Edition.
Robert C. Qiu, Zhen Hu, Husheng Li and Michael C. Wicks.
© 2012 John Wiley & Sons, Ltd. Published 2012 by John Wiley & Sons, Ltd.

1.2   Cognitive Radio: System Concepts

Radio spectrum is one of the most scarce and valuable resources, like real estate, in modern society. Competition for these scare resources is the basic drive for the telecommunication industry.

In the most general sense, cognitive radio takes advantage of Moore's law to capitalize on the computational power of the semiconductor industry [2]. When information is accessible in the digital domain, the force behind this novel radio is computationally intelligent algorithms. Machine learning and artificial intelligence have become the new frontier toward this vision—the analogy of robotics. Converting information from the analog domain to the digital domain plays a central role in this vision: revolutionary compressed sensing is, therefore, critical to expanding the territory of this new system paradigm. The agile, software defined radios that can perform according to algorithms are basic building blocks. When each node is computationally intelligent, wireless networking faces a novel revolution. At the system level, functions such as cognitive radio, cognitive radar and anti-jamming (even electronic warfare) have no fundamental difference and are unified into a single framework that requires interdisciplinary knowledge. Radar and communications should be unified since both require dynamic spectrum access (DSA)—the bottleneck. Spectrum agile/cognitive radio is a new paradigm in wireless communications—a special application of the above general radio.

Cognitive radio [3] takes advantage of the waveform programmable hardware platform, that is, so-called software-defined radio. Signal processing and machine learning are the core of the whole radio, called cognitive core (engine). In its fundamental nature, cognitive radio is a "mathematically-intensive" radio. It is policy based. The policy can be reasoned through the cognitive engine. In some sense, the whole book is focused on the fundamentals that are responsible for the cognitive engine. Here, our radio stands for a generalized sense. The radios can be used for communication networks, or sensor networks. So-called cognitive radar [4] is even included in this sense [2]. Our whole book can be viewed as a detailed spelling-out of Haykin's vision [3, 4]. Similar to Haykin, our style is mathematical in its nature. At the time of writing, the IEEE 802.22 standard on cognitive radio [5] was just released in July 2011. This book can be viewed as the mathematical justification for some critical system concepts, such as spectrum sensing (random matrices being the unifying theme), radio resource allocation (enabled by the convex optimization engine), and game theory (understanding the competition and collaboration of radio nodes in networking).

1.3   Spectrum Sensing Interface and Data Structures

Dynamic spectrum sharing in time and space is a fundamental system building block. An intelligent wireless communication system will estimate or predict spectrum availability and channel capacity, and adaptively reconfigure itself to maximum resource utilization while addressing interference mitigation [6]. Cognitive radio [3] is an attempt in this direction. It takes advantage of the waveform programmable hardware platform, that is, so-called software-defined radio.

The interface and data structures are significant in the context of system concepts. For example, we adopt the view of IEEE 1900.6 [6], as shown in Figure 1.1. Let us define some basic terms:

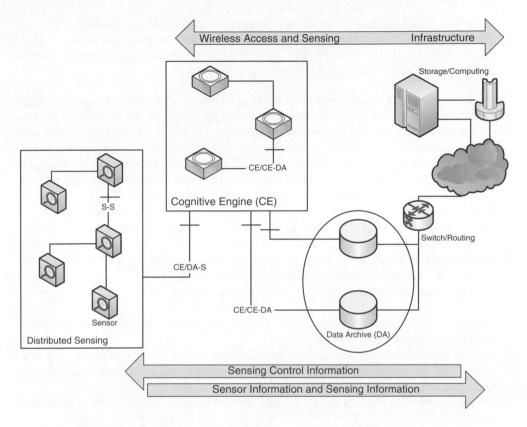

Figure 1.1 Sample topology of an IEEE 1900.6 distributed RF sensing system [6].

1. *Sensors*. The sensors are sometimes standalone or can form a small network of collaborating sensors that are inferring information about the available spectrum.
2. *Data archive*. The sensors talk to a data archive (DA), which can be considered a database where sensed information about spectrum occupancy is stored and provided.
3. *Cognitive engine*. A cognitive engine (CE) is an entity utilizing cognitive capabilities, including awareness, reasoning, solution making, and optimization for adaptive radio control and implementation of spectrum access policy. This CE is analogous with the human brain [3].
4. *Interface*. We need an interface that sensors utilize to talk to each other; so do CEs and DAs. It is necessary to change information between sensors, DAs and CEs, in order to disseminate spectrum availability and reduce interference to incumbent spectrum users.
5. *Distributed sensing*. In distributed scenarios, CEs and DAs must interface with communications devices; hence, generic but focused interface definitions are required.
6. *IEEE 1900.6*. The IEEE 1900.6 develops the interface and data structures that enable information flow among the various entities.
7. *Spectrum sensing*. Spectrum sensing is a core technology for DSA networks; it has recently been more and more intended not only as a stand-alone and real-time

technology, but also a necessary tool to constantly update the geolocalized spectrum map. Spectrum sensing is enabled by distributed mobile or fixed cognitive devices; this architecture allows the devices to monitor the spectrum occupancy and the overall level of interference with high precision and timeliness.

Spectrum sensing is fundamental to a cognitive radio. In some sense, a cognitive radio includes two parts: (1) spectrum sensing; (2) the radio resources are "cognitively" allocated using the available sensed spectrum information. In future evolved schemes, every "object" connected to the Internet could provide sensing features. This approach is oriented toward both the Internet of Things (IoT) and green radio communication paradigms [6]. The approach is also to create dynamic wide-area maps of spectrum usage that are being rapidly updated to optimize the overall electromagnetic emission and global interference. In this context, the jointly merging the notion of the cognitive radio network and the Smart Grid is relevant. The latter is a huge network of power grids (many sensors, mobile or fixed). The size of the network is many times bigger than the usual wireless communications network. The idea of this merging was explored (for the first time in the proposal of R. Qiu to the office of naval research (ONR)) [7].

Sensing related information basically consists of four categories:

1. *Sensing information* denotes any measurement information that can be obtained from a spectrum sensor.
2. *Sensing control* denotes any information required to describe the status or configuration, and to control or configure the data acquisition and RF sensing process of a spectrum sensor.
3. *Sensor information* denotes the parameters used to describe the capabilities of a spectrum sensor.
4. *Regulatory requirements* are unique to the application area of DSA by CRs.

1.4 Mathematical Machinery

1.4.1 Convex Optimization

Optimization stems from human instinct. We always want to do things in the best way. Relying on mathematics, this human instinct can be written down in terms of mathematical optimization. Practical problems can be formulated as optimization problems with objective functions, constraint functions, and optimization variables. Mathematical optimization attempts to minimize or maximize the objective function by systematically selecting the values of optimization variables from a specific set defined by the constraint functions.

Convex optimization is a subfield of mathematical optimization, which investigates the problem of minimizing convex objective function based on a compact convex set. The strength of convex optimization is if a local minimum exists, then it is a global minimum. Hence, if a practical problem can be formulated as a convex optimization problem, then global optimum can be obtained. That is one reason why convex optimization has recently become popular.

The other reason for the popularity of convex optimization is that convex optimization can be solved by the cutting plane method, ellipsoid method, subgradient method, or interior point method. Thus the interior point method, which was originally developed to solve linear programming problems, can also be used to solve convex optimization problems [8]. By taking advantage of the interior point method, convex optimization problems can be solved efficiently [8].

Convex optimization includes the well-known linear programming, second order cone programming (SOCP), semidefinite programming (SDP), geometric programming, and so on. Convex optimization is a powerful signal processing tool which can be exploited anywhere, for example, system control, machine learning, operation research, logistics, finance, management, telecommunication, and so on, due to the prevalence of convex optimization problems in practice [8].

Besides convex optimization, mathematical optimization also includes integer programming, combinatorial programming, nonlinear programming, fractional programming, stochastic programming, robust programming, multi-objective optimization, and so on.

Unfortunately, there are still a large amount of nonconvex optimization problems in the real-world. Relaxation is the common way to address the nonconvex optimization issues. The nonconvex optimization problem is relaxed to the convex optimization problem. Based on the global optimum to the convex optimization problem, we can find the sub-optimal solution to the original nonconvex optimization problem. The second strategy to deal with the nonconvex optimization problems makes use of stochastic methods. Stochastic methods exploit random variables to get the solution to the optimization problem. Stochastic methods do not need to explore the structures of objective functions and constraints. Stochastic methods include simulated annealing, stochastic hill climbing, genetic algorithm, ant colony optimization, particle swarm optimization, and so on.

When we enjoy the beauty and benefit of mathematical optimization, we cannot forget the contributors and the important researchers in mathematical optimization. Joseph Louis Lagrange found a way to identify optima. Carl Friedrich Gauss and Isaac Newton gave iterative methods to search for an optimum. In 1939, Leonid Kantorovich published an article "Mathematical Methods of Organizing and Planning Production," introducing the concept and theory of linear programming. Then George Bernard Dantzig developed simplex method for linear programming in 1947 and John von Neumann invented Duality Theorem for linear programming in the same year. Von Neumann's algorithm can be considered as the first interior-point method of linear programming. In 1984, a new polynomial-time interior-point method for linear programming was introduced by Narendra Karmarkar. Yurii Nesterov and Arkadi Nemirovski published a book *Interior-Point Polynomial Algorithms in Convex Programming* in 1994. Generally, the interior-point method is faster than the simplex method for the large-scale optimization problem. Besides, David Luenberger, Stephen P. Boyd, Yinyu Ye, Lieven Vandenberghe, Dimitri P. Bertsekas, and so on also made obvious contributions to mathematical optimization.

Mathematical optimization, especially convex optimization, has already greatly improved the performance of the current telecommunication system. For the next generation wireless communication system, that is, cognitive radio network, mathematical optimization will play a critical role. Cognitive radio network opens another stage for the show of mathematical optimization. Optimization will be the core of the cognitive engine. We can see the beauties of mathematical optimization in spectrum sensing, spectrum sharing, coding and decoding, waveform diversity, beamforming, radio resource management, cross-layer design, and security for cognitive radio network.

1.4.2 Game Theory

Game theory is an important analysis tool in cognitive radio. Essentially, a game involves multiple players, each of which makes individual decision and maximizes its own reward. Since the reward of each players is dependent on the actions of other players, the player must take the possible response of other players into account. All players will be satisfied at the equilibrium point, at which any individual deviation from the strategy only incurs reward loss. A natural question may arise, that is, why game theory is needed in cognitive radio?

The essential reason for the necessity of game theory is the existence of conflict or collaboration in cognitive radio. Some examples are given below:

- PUE attack: Primary user emulation (PUE) attack is a serious threat to the cognitive radio network, in which the attacker pretends to be a primary user and sends interference signals to scare secondary users away. Then, the secondary users need to evade the PUE attack. If there are multiple channels to choose, the secondary users need to make decisions on the channel use while the attacker needs to decide which channel to jam (if it is unable to jam all channels), thus forming a game.
- Channel synchronization: The control channel is of key importance in cognitive radio. Two secondary users need to convey control messages through the control channel. If the control channel is also in the unlicensed band, it is subject to the interruption of primary users. Hence, two secondary users need to collaborate to find a new control channel if the current one is no longer available. Such a collaboration is also a game.
- Suspicious collaborator: Collaborative spectrum sensing can improve the performance of spectrum sensing. However, the reports from a collaborator could be false if the collaborator is actually a malicious one. Hence, the honest secondary user needs to make a decision on whether trust the collaborator or not. Meanwhile, the attacker also needs to decide what type of report to share with the honest secondary user such that it can simultaneously spoof the honest user and disguise its goal.

The above examples concern zero-sum games, general sum games, Bayesian games and stochastic games. In this book, we will explain how to analyze a game, particularly the computation of Nash equilibrium, and apply the game theory to the above examples.

1.4.3 "Big Data" Modeled as Large Random Matrices

It turns out that random matrices are the unifying theme since "big data" can be modeled as large random matrices. With data acquisition and storage now easy, today's statisticians often encounter datasets for which the sample size, n, and the number of variables, p, are both large [9]: in the hundreds, thousands, millions and even billions in situations such as web search problems. This phenomenon is so-called "big data." The analysis of these datasets using classical methods of multivariate statistical analysis requires some care. In the context of wireless communications, networks become more and more dense. Spectrum sensing in cognitive radio collects much bigger datasets than the traditional multiple input multiple output (MIMO)-orthogonal frequency-division multiplexing (OFDM), and code division multiple access (CDMA) systems. For example, for a duration of 4.85

Table 1.1 Analogy of sensors and particles

Particles	Sensors	Random Matrices
Total energy	Information	Degrees of freedom
Energy levels		Eigenvalues

milliseconds, a data record (digital TV) consisting of more than 10^5 sample points is available for data processing. We can divide this long data record into vectors consisting of only p sample points. A number of sensors n can cooperate for spectrum sensing. The analogy of sensors and particles is shown in Table 1.1. Alternatively, we can view $n \cdot p = 10^5$ as using only one sensor to record a long data record. Thus we have $p = 100$ and $n = 1,000$ for the current example. In this example, both n and p are large and in the same order of magnitude.

Let X_{ij} be i.i.d. standard normal variables of $p \times n$ matrix $\mathbf{X}$

$$\mathbf{X} = \begin{bmatrix} X_{11} & X_{12} & \cdots & X_{1n} \\ X_{21} & X_{22} & \cdots & X_{2n} \\ \vdots & \vdots & \vdots & \vdots \\ X_{p1} & X_{p2} & \cdots & X_{pn} \end{bmatrix}_{p \times n}. \tag{1.1}$$

The sample covariance matrix is defined as

$$\mathbf{S}_n = \left(\frac{1}{n} \sum_{k=1}^{n} X_{ik} X_{jk} \right)_{i,j=1}^{p} = \frac{1}{n} \mathbf{X} \mathbf{X}^H, \tag{1.2}$$

where n vector samples of a p-dimensional zero-mean random vector with the population (or true covariance) matrix $\mathbf{I}$ and H stands for conjugate transpose (Hermitian) of a matrix.

The classical limit theorem is no longer suitable for dealing with large dimensional data analysis. The classical methods make an implicit assumption that p is fixed and n is growing infinitely large,

$$p \text{ fixed}, n \to \infty. \tag{1.3}$$

This asymptotic assumption (1.3) was consistent with the practice of statistics when these ideas were developed, since investigation of datasets with a large number of variables was very difficult. A better theoretical framework—that is, large p—for modern datasets, however, is the assumption of the so-called "large n, large p" asymptotics

$$p \to \infty, n \to \infty, \text{ but } \frac{p}{n} \to c > 0, \tag{1.4}$$

where c is a positive constant.

There is a large body of work concerned with the limiting behavior of the eigenvalues of a sample covariance matrix $\mathbf{S}_n$ when p and n both go to ∞ (1.4). A fundamental result is the Marchenko-Pastur equation, which relates the asymptotic behavior of the eigenvalues of the sample covariance matrix to that of the population covariance in the "large n, large p" asymptotic setting. We must change points of view: *from vectors to measures*.

One of the first problems to tackle is to find a mathematically efficient way to express the limit of a vector whose size grows to ∞. (Recall that there are p eigenvalues to estimate in our problem and p goes to ∞.) A fairly natural way to do so is to associate to any vector a probability measure. More explicitly, suppose we have a vector $(y_1, \ldots, y_p)$ in $\mathbb{R}^p$. We can associate to it the following measure:

$$dG_p(x) = \frac{1}{p} \sum_{i=1}^{p} \delta_{y_i}(x),$$

where δ_x is the Dirac delta function at x. G_p is thus a measure with p point masses of equal weight, one at each of the coordinates of the vector. The change of focus from vector to measure implies a change of focus in the notion of convergence—weak convergence of probability measure.

Following [10], we divide available techniques into three categories: (1) Moment approach; (2) Stieltjes transform; (3) Free probability. Applications for these basic techniques will be covered.

The Stieltjes transform is a convenient and very powerful tool in the study of the convergence of spectral distribution of matrices (or operators), just as the characteristic function of a probability distribution is a powerful tool for central limit theorems. More important, there is a simple connection between the Stieltjes transform of the spectral distribution of a matrix and its eigenvalues. By definition, the Stieltjes transform of a measure G on $\mathbb{R}$ is defined as

$$m_G(z) = \int \frac{1}{x - z} dG(x) \text{ for } z \in \mathbb{C}^+,$$

where $\mathbb{C}^+ \triangleq \mathbb{C} \cap \{z : \text{Im}(z) > 0\}$ is the set of complex numbers with strictly positive imaginary part. The Stieltjes transform is sometimes referred to as Cauchy or Abel-Stieltjes transform. Good references on Stieltjes transforms include [11] and [12].

The remarkable phenomenon is that the spectral distribution of the sample covariance matrix is asymptotically nonrandom. Furthermore, it is fully characterized by the true population spectral distribution, through the Marchenko-Pastur equation. The knowledge of the limiting distribution of the eigenvalues in the population, Σ, fully characterizes the limiting behavior of the eigenvalues of the sample covariance matrix, $\mathbf{S}$.

In the market for wireless communications, an excellent book by Couillet and Debbah (2011) [12] has just appeared, in addition to Tulino and Verdu (2004) [13]. Our aim in this book is to introduce the relevance of random matrix theory in the context of cognitive radio, in particular spectrum sensing. Our treatment is more practical than in those two books. Although some theorems are also compiled in our book, no proofs are given. We emphasize how to apply the theory through a large number of examples. It is our belief that future engineers must be familiar with random matrix methods since "big data" is the dominant theme across layers of the wireless network.

"One of the useful features, especially of the large dimensional random matrix theory approach, is its ability to predict the behavior of the empirical eigenvalue distribution of products and sums of matrices. The results are striking in terms of accuracy compared to simulations with reasonable matrix sizes." [12]

"Indeed, engineering education programs of the twentieth century were mostly focused on the Fourier transform theory due to the omnipresence of frequency spectrum. The twenty-first century engineers know by now that space is the next frontier due to the omnipresence of spatial modes, which refocuses the program towards a Stietjes transform theory." [12]

In the eyes of engineers, Bai and Silverstein (2010) [14], Hiai and Petz (2000) [11] and Forrester (2010) [15] are most readable among the mathematical literature. Anderson (2010) is also accessible [16] and Girko (1998) is comprehensive [17]. One excellent survey [10] is a good starting point for the massive literature. It is still the best survey. Two surveys [18] and [19] are very readable.

In the early 1980s, major contributions on the existence of the limiting spectral distribution (LSD) were made. In recent years, research on random matrix theory has turned toward second-order limiting theorems, such as the central limit theorem for linear spectral statistics, the limiting distributions of spectral spacings, and extreme eigenvalues.

Many applied problems require an estimate of a covariance matrix and/or of its inverse, where the matrix dimension is large compared to the sample size [20]. In such situations, the usual estimator, the sample covariance matrix, is known to perform poorly. When the matrix dimension p is larger than the number of observations available, the sample covariance matrix is not even invertible. When the ratio p/n is less than one but not negligible, the sample covariance matrix is invertible but numerically ill-conditioned, which means that inverting it amplifies estimation error dramatically. For large p, it is difficult to find enough observations to make p/n negligible, and therefore it is important to develop a well-conditioned estimator for large-dimensional covariance matrices such as in [20].

1.4.3.1 Why is Random Matrix Theory So Successful?

Random matrix theory is very successful in nuclear physics [21]. Here are several reasons:

1. **Flexibility**. It allows us to build in extra global symmetries, such as time reversal, spin, chiral symmetry, etc., treating several matrices—while maintaining its exact resolvability for all correlation functions of eigenvalues.
2. **Universality**. Random matrix theory can often be used as the simplest, solvable mode that captures the essential degrees of freedom of the theory. The role of the normal distribution in the classical limit theorem is played by the distributions arising in random matrix theory (Tracy-Widom distribution, sine distribution, . . .) in noncommutative settings that may or may not involve random matrices.
3. **Predictivity**. The scale or physical coupling can be extracted very efficiently by fitting data to random matrix theory's predictions.
4. **Rich mathematical structure**. This comes from the many facets of the large-n limit. The multiple connections of random matrix theory to various areas of mathematics make it an ideal bridge between otherwise almost unrelated fields (probability and analysis, algebra, algebraic geometry, differential systems, combinatorics). More generally, these developed techniques are fluent enough to be applied to other branches of sciences.

1.5 Sample Covariance Matrix

The study of sample covariance matrix is fundamental in multivariate analysis. With contemporary data, the matrix is often large, with number of variables comparable to sample size (so-called "big data") [22]. In this setting, relatively little is known about the distribution of the largest eigenvalue, or principal component variance. A surprise of the random matrix theory, the domain of mathematical physics and probability, is that the results seem to give useful information about principal components for quite small values of n and p.

Let $\mathbf{X}$, defined in (1.1), be an $p \times n$ data matrix. Typically, one thinks of n observations or cases $\mathbf{x}_i$ of a p-dimensional column vector which has covariance matrix $\mathbf{\Sigma}$. For definiteness, assume that rows $\mathbf{x}_i$ are independent Gaussian $\mathcal{N}(0, \mathbf{\Sigma})$. In particular, the mean has been subtracted out. If we also do not worry about dividing by n, we can call $\mathbf{X}\mathbf{X}^H$ a sample covariance matrix defined in (1.2). Under Gaussian assumption, $\mathbf{X}\mathbf{X}^H$ is said to have a Wishart distribution $\mathcal{W}(n, \mathbf{\Sigma})$. If $\mathbf{\Sigma} = \mathbf{I}$, the "null" case, we call it a white Wishart, in analogy with time series setting where a white spectrum is one with the same variance at all frequencies.

Large sample work in multivariate analysis has traditionally assumed that n/p, the number of observations per variable, is large. Today, it is common for p to be large or even huge, and so n/p may be moderate to small and in extreme cases less than one.

The eigenvalue and eigenvector decomposition of the sample covariance matrix

$$\mathbf{S} = \frac{1}{n}\mathbf{X}\mathbf{X}^H = \mathbf{U}\mathbf{L}\mathbf{U}^H = \sum_i l_i \mathbf{u}_i \mathbf{u}_i^H,$$

with eigenvalues in the diagonal matrix $\mathbf{L}$ and orthogonal eigenvectors collected as columns of $\mathbf{U}$. There is a corresponding population (or true) covariance matrix

$$\mathbf{\Sigma} = \mathbf{\Upsilon}\mathbf{\Lambda}\mathbf{\Upsilon}^H,$$

with eigenvalues λ_i and orthogonal eigenvectors collected as columns of $\mathbf{\Upsilon}$.

A basic phenomenon is that the same eigenvalues l_i are more spread out than the population λ_i. This effect is strongest in the null case when all population eigenvalues are the same.

Data matrices with complex Gaussian entries are of interest in statistics, signal processing and wireless communications. Suppose that $\mathbf{X} = (X_{ij})_{p \times n}$ with

$$ReX_{ij}, ImX_{ij} \sim \mathcal{N}(0, \frac{1}{2}),$$

all independently of one another. The matrix $\mathbf{S} = \mathbf{X}\mathbf{X}^H$ has the complex Wishart distribution, and its (real) eigenvalues are ordered $l_1 > \cdots > l_p$.

Define μ_{np} and σ_{np} as

$$\mu_{np} = \left(\sqrt{n} + \sqrt{p}\right)^2,$$

$$\sigma_{np} = \left(\sqrt{n} + \sqrt{p}\right)\left(\frac{1}{\sqrt{n}} + \frac{1}{\sqrt{p}}\right)^{1/3}.$$

Assume that $n = n(p)$ increases with p so that both μ_{np} and σ_{np} are increasing in p.

Theorem 1.1 (Johansson (2000) [23]) *Under the forementioned conditions, if $n/p \to$ $c \geq 1$, then*

$$\frac{l_1 - \mu_{np}}{\sigma_{np}} \xrightarrow{\mathcal{D}} W_2 \sim F_2,$$

where $\mathcal{D}$ stands for convergence in distribution.

The center and scale are essentially the same as the real case, but the limit distribution is

$$F_2(s) = \exp\left(-\int_s^\infty (x-s)\, q^2(x) dx\right),$$

where q is still the Painleve II function defined as

$$q''(x) = xq(x) + 2q^3(x),$$

$$q(x) \sim Ai(x) \quad \text{as} \quad x \to +\infty$$

and $Ai(x)$ denotes the Airy function. This distribution was found by Tracy and Widom [24, 25] as the limiting law of the largest eigenvalue of an p by n Gaussian symmetric matrix (Wigner matrix).

Simulations show the approximation to be informative for n and p as small as 5.

1.6   Large Sample Covariance Matrices of Spiked Population Models

A spiked population model, in which all the population eigenvalues are one except for a few fixed eigenvalues, has been extensively studied [26, 27]. In many examples, a few eigenvalues of the sample covariance matrix are separated from the rest of the eigenvalues, the latter being packed together as in the support of the Marchenko-Pastur density. Examples are so common in speech recognition, mathematical finance, wireless communications, physics of mixture, and data analysis and statistical learning.

The simplest non-null case would be the population covariance Σ is a finite rank perturbation of a multiple of the identity matrix $\mathbf{I}$. In other words, we say

$$\mathcal{H}_0 : \Sigma = \mathbf{I},$$

$$\mathcal{H}_1 : \Sigma = \Delta + \mathbf{I}, \Delta = \text{finite rank}$$

As mentioned in the above, Johnstone (2001) [22] derived the asymptotic distribution for the largest sample eigenvalue under the setting of an identity matrix $\mathbf{I}$ under Gaussianity. Soshnikov (2002) proved the distributional limits under weaker assumptions, in addition to deriving distributional limits of the k-th largest eigenvalue, for fixed but arbitrary k [28].

A few of the sample eigenvalues under $\mathcal{H}_1$ have limiting behavior that is different from $\mathcal{H}_0$ when the covariance is identity matrix $\mathbf{I}$.

A crucial aspect is the discovery of a phase transition phenomenon. Simply put, if the non-unit eigenvalues are close to one, then their sample versions will behave in roughly the same way as if the true covariance were the identity. However, when the true eigenvalues are larger than $1 + \sqrt{n/p}$, the sample eigenvalues have a different asymptotic property. The eigenvectors also undergo a phase transition. By performing a natural decomposition of the sample eigenvectors into "signal" and "noise" parts, it is shown that when $l_i > 1 + \sqrt{n/p}$, the "signal" part of the eigenvectors is asymptotically normal [27].

1.7   Random Matrices and Noncommutative Random Variables

Random matrices are noncommutative random variables [11], with respect to the expectation

$$\tau_N(\mathbf{H}) = \frac{1}{N} \sum_{i=1}^{N} \mathbb{E}(\mathbf{H_{ii}}),$$

for an $N \times N$ random matrix $\mathbf{H}$, where $\mathbb{E}$ represents the expectation of a classical random variable. It is a form of the Wigner theorem that

$$\tau_N\left(\mathbf{H}^{2k}(N)\right) \to \frac{1}{k+1} \binom{2k}{k}, N \to \infty$$

if the $N \times N$ real symmetric random matrix $\mathbf{H}(N)$ has independent identical Gaussian entries $\mathcal{N}(0, 1/N)$ so that

$$\tau_N\left(\mathbf{H}^2(N)\right) = 1.$$

The semicircle law is the limiting eigenvalue distribution density of $\mathbf{H}(N)$. It is also the limiting law of the free central limit. The reason why this is so was made clear by Voiculescu. Let

$$\mathbf{X}_1(N), \mathbf{X}_2(N), \ldots, \mathbf{X}_N(N),$$

be independent random matrices with the same distribution as $\mathbf{X}(N)$. It follows from the properties of Gaussians that the distribution of the random matrix

$$\frac{\mathbf{X}_1(N) + \mathbf{X}_2(N) + \ldots + \mathbf{X}_N(N)}{\sqrt{N}}$$

is the same as $\mathbf{X}(N)$. The convergence in moments to the semicircle law is understood in the sense that

$$\mathbf{X}_1(N), \mathbf{X}_2(N), \ldots, \mathbf{X}_N(N)$$

are in free relation. The conditions for the free relation include

$$\tau_N\left(\left[\mathbf{X}_1^k(N) - \tau_N^k(\mathbf{X}_1(N))\right]\right) \tau_N\left(\left[\mathbf{X}_2^l(N) - \tau_N^l(\mathbf{X}_2(N))\right]\right) = 0,$$

which is equivalently expressed as

$$\tau_N\left(\left[\mathbf{X}_1^k(N)\mathbf{X}_2^l(N)\right]\right) = \tau_N\left(\mathbf{X}_1^k(N)\right) \tau_N\left(\mathbf{X}_1^l(N)\right).$$

Independent symmetric Gaussian matrices and independent Haar distributed unitary matrices are asymptotically free. The notion of asymptotic freeness may serve as a bridge connecting random matrix theory with free probability theory.

1.8   Principal Component Analysis

Every 20–30 years, principal component analysis (PCA) is reinvented with slight revision. It has many different names. We model the communication signal or noise as random field. The Karhunen-Loeve decomposition (KLD) is also known as PCA, the Proper Orthogonal Decomposition (POD), and Empirical Orthogonal Function (EOF). Its kernel version, that is, Kernel PCA, is very popular. We apply PCA to spectrum sensing.

PCA is a standard tool for dimensionality reduction. PCA finds orthogonal directions with maximal variance of the data and allows its low-dimensional representation by linear projections onto these directions. This dimensionality reduction is a typical pre-processing setup. A spiked covariance model [29–32] implies that the underlying data is low-dimensional but each sample is corrupted by additive Gaussian noise.

1.9   Generalized Likelihood Ratio Test (GLRT)

The GLRT is the culmination of the theoretical development for spectrum sensing. Its kernel version, Kernel GLRT, performs well, in contrast to Kernel PCA.

Both GLRT and PCA (its kernel version Kernel PCA) use sample covariance matrices as their starting points. As a result, large-dimensional random matrices are natural objects of mathematics to study.

1.10   Bregman Divergence for Matrix Nearness

When dealing with random matrices, we still need some measure of distance between them. Matrix nearness problems ask for the distance from a given matrix to the nearest matrix with a certain property. The use of a Bregman divergence in place of a matrix norm is, for example, proposed by Dhillon and Tropp (2007) [33]. Bregman divergence is equivalent to quantum information [34, p. 203]. Let $\mathcal{C}$ is a convex set in a Banach space. For a smooth functional $\Psi : \mathcal{C} \to \mathbb{R}$,

$$D_{\Psi}\left(\mathbf{X}, \mathbf{Y}\right) \triangleq \Psi\left(\mathbf{X}\right) - \Psi\left(\mathbf{Y}\right) - \lim_{t \to +0} t^{-1}\left(\Psi\left(\mathbf{Y} + t\left(\mathbf{X} - \mathbf{Y}\right)\right) - \Psi\left(\mathbf{Y}\right)\right)$$

is called the Bregman divergence of $\mathbf{X}, \mathbf{Y} \in \mathcal{C}$. Now let $\mathcal{C}$ be the set of density matrices and let

$$\Psi\left(\rho\right) = \text{Tr } \rho \log \rho.$$

A density matrix is a positive definite matrix whose trace equals one. It can be shown that the Bregman divergence is the quantum relative entropy which is the basis for measuring quantum information. Problems of Bregman divergence can be formulated in terms of convex optimization. The semicircle law, free (matrix-valued) random variables, and quantum entropy are related [11], when we deal with "big data."

Functions of matrices are often needed in studying many problems in this book, for example, in spectrum sensing. The Matrix Function Toolbox contains MATLAB implementations to calculate functions of matrices [35]. It is available from http://www.maths.manchester.ac.uk/~higham/mftoolbox/

2

Spectrum Sensing: Basic Techniques

2.1 Challenges

Spectrum sensing in a cognitive radio is practically challenging, as shown in Table 2.2 [36, 37].

2.2 Energy Detection: No Prior Information about Deterministic or Stochastic Signal

Energy detection is the simplest spectrum sensing technique. It is a blind technique in that no prior information about the signal is required. It simply treats the primary signal as noise and decides on the presence or absence of the primary signal based on the energy of the observed signal. It does not involve complicated signal processing and has low complexity. In practice, energy detection is especially suitable for wideband spectrum sensing. The simultaneous sensing of multiple subbands can be realized by scanning the received wideband signal.

Two stages of sensing are desirable. The first stage uses the simplest energy detection. The second stage uses advanced techniques.

We follow [38] and [39–42] for the development below. Although the process is for band-pass, in general, one can still deal with its low-pass equivalent form and eventually translate it back to its band-pass type [43]. Besides, it has been verified [38] that both low-pass and band-pass processes are equivalent from the decision statistics perspective which is our main concern. Therefore, for convenience, we only address the problem for a low-pass process, following [41, 42].

Cognitive Radio Communications and Networking: Principles and Practice, First Edition.
Robert C. Qiu, Zhen Hu, Husheng Li and Michael C. Wicks.
© 2012 John Wiley & Sons, Ltd. Published 2012 by John Wiley & Sons, Ltd.

Table 2.1   Receiver parameters for 802.22 WRAN

Parameter	Analog TV	Digital TV	Wireless microphone
Bandwidth	6 MHz	6 MHz	200 kHz
Probability of detection	0.9	0.9	0.9
Probability of false alarm	0.1	0.1	0.1
Channel detection time	$\leq 2s$	$\leq 2s$	$\leq 2s$
Incumbent detection threshold	-94 dBm	-116 dBm	-106 dBm
SNR[a]	1 dB	-21 dB	-12 dB

[a]Receiver noise figure of 11 dB is assumed in IEEE 802.11 Working Group.

Table 2.2   Challenges for spectrum sensing

Practical challenges	Consequences	Comments
Very strict sensing requirements	See Table 2.1	To avoid "hidden node" problem
Unknown propagation channel and nonsynchronization	Make coherent detection unreliable	To relieve the primary user from the burden
Noise/interference uncertainty	Very difficult to estimate their power	Change with time and location

2.2.1   Detection in White Noise: Lowpass Case

The detection is a test of the following hypotheses:

1. $\mathcal{H}_0$: The input is noise alone:
   (a) $y(t) = n(t)$
   (b) $E[n(t)] = 0$
   (c) noise spectral density $= N_0$ (two-sided)
   (d) noise bandwidth $=$ W Hz
2. $\mathcal{H}_1$: The input is signal plus noise
   (a) $y(t) = s(t) + n(t)$
   (b) $E[s(t) + n(t)] = s(t)$

The output of the integrator is denoted by Y. We concentrate on a particular interval, say, $(0, T)$, and take the test statistic as Y or any quantity monotonic with Y. We shall find it convenient to express the false alarm and detection probabilities using the related quantity

$$\tilde{Y} = \frac{1}{N_0} \int_0^T y^2(t)dt. \tag{2.1}$$

The choice of T as the sampling instant is a matter of convenience; any interval of duration will serve.

It is known that a sample function, of duration T, of a process which has a bandwidth W (negligible outside this band) is described approximately by a set of sample values $2TW$ in number. In the case of low-pass processes, the values are obtained by sampling the processes at times $1/2W$ apart. In the case of relatively narrowband band-pass process, the values are obtained from the in-phase and quadrature modulation components sampled at times $1/W$ apart.

With an appropriate choice of time origin, we may express each sample of noise as [44]

$$n(t) = \sum_{i=-\infty}^{\infty} n_i \text{sinc}(2Wt - i) \tag{2.2}$$

where $\text{sinc}(x) = \sin \pi x / \pi x$, and

$$n_i = n\left(\frac{i}{2W}\right). \tag{2.3}$$

Clearly, each n_i is a Gaussian random variable with zero mean and with the same variance σ_i^2, which is the variance of $n(t)$; that is,

$$\sigma_i^2 = 2N_0 W, \quad \text{all } i. \tag{2.4}$$

Using the fact that

$$\int_{-\infty}^{\infty} \text{sinc}(2Wt - i)\text{sinc}(2Wt - k)dt = \tfrac{1}{2W}, \quad i = k \tag{2.5}$$
$$= 0, \quad i \neq k,$$

we may write

$$\int_{-\infty}^{\infty} n^2(t)dt = \frac{1}{2W} \sum_{i=-\infty}^{\infty} n_i^2. \tag{2.6}$$

Over the interval $(0, T)$, $n(t)$ may be approximated by a finite sum of $2TW$ terms, as follows:

$$n(t) = \sum_{i=1}^{2TW} n_i \text{sinc}(2Wt - i), \quad 0 < t < T. \tag{2.7}$$

Similarly, the energy in a sample of duration T is approximated by $2TW$ terms of the right side of (2.6):

$$\int_0^T n^2(t)dt = \frac{1}{2W} \sum_{i=1}^{2TW} n_i^2. \tag{2.8}$$

More rigor can be achieved by using the Karhunen-Loeve expansion (also called transform). Equation (2.8) may be considered as an approximation, valid for large T, after substituting (2.7) into the left-hand side of (2.8), or by using (2.39) and the statement in Section 2.2.5 to justify taking only $2TW$ terms of (2.6).

We can see that (2.8) is $N_0\tilde{Y}$, with $\tilde{Y}$ here being the test statistic under hypothesis $\mathcal{H}_0$. Let us write

$$\frac{n_i}{\sqrt{2WN_0}} = \xi_i, \tag{2.9}$$

$$\tilde{Y} = \sum_{i=1}^{2TW} \xi_i^2. \tag{2.10}$$

Thus, $\tilde{Y}$ is the sum of the squares of $2TW$ Gaussian random variables, each with zero mean and unit variance. $\tilde{Y}$ is said to have a chi-square distribution with $2TW$ degrees of freedom, for which extensive tables exist [45–47].

Now consider the case $\mathcal{H}_1$ where the input $y(t)$ has the signal $s(t)$ present. The segment of signal duration T may be represented by a finite sum of $2TW$ terms,

$$s(t) = \sum_{i=1}^{2TW} s_i \ \text{sinc}(2Wt - i), \quad 0 < t < T, \tag{2.11}$$

where

$$s_i = s(i/2W). \tag{2.12}$$

By following the reasoning above, we can approximate the signal energy in the interval $(0, T)$ by

$$\int_0^T s^2(t)dt = \frac{1}{2W} \sum_{i=1}^{2TW} s_i^2. \tag{2.13}$$

Define the coefficients by

$$\beta_i = s_i/\sqrt{2WN_0}. \tag{2.14}$$

Then

$$\frac{1}{N_0} \int_0^T s^2(t)dt = \sum_{i=1}^{2TW} \beta_i^2. \tag{2.15}$$

Using (2.11) and (2.2), the total input $y(t)$ with the signal present can be expressed as:

$$y(t) = \sum_{i=1}^{2TW} (\xi_i + s_i)\text{sinc}(2Wt - i). \tag{2.16}$$

2.2.2 Time-Domain Representation of the Decision Statistic

The energy of $y(t)$ in the interval $(0, T)$ is approximated by

$$\int_0^T y^2(t)dt = \frac{1}{2W} \sum_{i=1}^{2TW} (n_i + s_i)^2. \tag{2.17}$$

Under $\mathcal{H}_1$, the test statistic $\tilde{Y}$ is

$$\tilde{Y} = \frac{1}{N_0} \int_0^T y^2(t)dt = \sum_{i=1}^{2TW} (\xi_i + \beta_i)^2. \tag{2.18}$$

The sum in (2.18) is said to have a noncentral chi-square distribution with $2TW$ degrees of freedom and a noncentral parameter γ given by

$$\gamma = \sum_{i=1}^{2TW} \beta_i^2 = \frac{1}{N_0} \int_0^T s^2(t)dt \equiv \frac{E_s}{N_0}, \tag{2.19}$$

where γ, the ratio of signal energy to noise spectral density, provides a convenient definition of signal-to-noise ratio (SNR).

2.2.3 Spectral Representation of the Decision Statistic

The spectrum component on each spectrum subband of interest is obtained from the fast Fourier transform (FFT) of the sampled received signal. The test statistic of the energy detection, within M consecutive segments, is obtained as the observed energy summation,

$$Y = \begin{cases} \sum_{m=1}^{M} |W(m)|^2, & \mathcal{H}_0 \\ \sum_{m=1}^{M} |S(m) + W(m)|^2, & \mathcal{H}_1 \end{cases} \tag{2.20}$$

where $S(m)$ and $W(m)$ denote the spectral components of the received primary signal and the white noise on the subband of interest in the m-th segment, respectively. Interference is ignored in (2.20), to simplify analysis. The decision of the energy detection regarding the subband of interest is given by

$$\hat{\theta} = \begin{cases} \mathcal{H}_0, & Y > \lambda \\ \mathcal{H}_1 & Y < \lambda \end{cases}, \tag{2.21}$$

where the threshold λ is chosen to satisfy a target false-alarm probability.

Without loss of generality, we assume the noise $W(m)$ is white complex Gaussian with zero mean and variance two. The SNR of the received primary signal, within M segments, is defined as

$$\gamma = \frac{1}{2M} \sum_{m=1}^{M} |S(m)|^2. \tag{2.22}$$

The statistic of the energy detection Y follows a central chi-square distribution with $2M$ degrees of freedom under $\mathcal{H}_0$. Under $\mathcal{H}_1$, the Y follows a noncentral chi-square distribution with $2M$ degrees of freedom and a noncentrality parameter

$$\mu = \sum_{m=1}^{M} |S(m)|^2 = 2M\gamma. \tag{2.23}$$

In other words,

$$f_Y(y) \sim \begin{cases} \chi_{2M}^2, & \mathcal{H}_0 \\ \chi_{2M}^2(\mu), & \mathcal{H}_1 \end{cases} \tag{2.24}$$

where $f_Y(Y)$ denotes the probability density function (PDF) of Y, and χ_{2M}^2 and $\chi_{2M}^2(\mu)$ denote a central and noncentral chi-square distribution, respectively.

The PDF of Y can then be written as

$$f_Y(y) = \begin{cases} \frac{1}{2^u \Gamma(u)} y^{u-1} e^{-\frac{y}{2}}, & \mathcal{H}_0 \\ 2\left(\frac{y}{2\gamma}\right)^{\frac{u-1}{2}} e^{-\frac{2\gamma+y}{2}} I_{u-1}\left(\sqrt{2\gamma y}\right), & \mathcal{H}_1 \end{cases} \tag{2.25}$$

where $\Gamma(\cdot)$ is the gamma function and $I_\nu(\cdot)$ is the $\nu-$order modified Bessel function of the first kind [45, 48].

2.2.4 Detection and False Alarm Probabilities over AWGN Channels

The probability of detection and false-alarm can be defined as

$$P_D = P(Y > \lambda | \mathcal{H}_1) \tag{2.26}$$

$$P_F = P(Y > \lambda | \mathcal{H}_0), \tag{2.27}$$

where λ is the decision threshold. Using (2.25) to evaluate (2.27) yields the exact closed form expression

$$P_F = \frac{\Gamma(M, \frac{\lambda}{2})}{\Gamma(M)}, \tag{2.28}$$

where $\Gamma(\cdot, \cdot)$ is the upper incomplete gamma function [45, 48].

Given the target false-alarm probability, the threshold λ can be uniquely determined, using (2.28). Once λ is determined, the detection probability can be obtained by

$$
\begin{aligned}
P_D &= \int_0^{+\infty} P(Y > \lambda | \mathcal{H}_1, \mu) f_\mu(\mu) d\mu \\
&= \int_0^{+\infty} Q_M(\sqrt{\mu}, \sqrt{\lambda}) f_\mu(\mu) d\mu,
\end{aligned}
\tag{2.29}
$$

where

$$
\begin{aligned}
Q_M(a, b) &= e^{-(a^2 + b^2)/2} \sum_{n=0}^{\infty} \left(\frac{a}{b}\right)^n I_n(ab) \\
&= \int_b^{\infty} x \exp\left[-\frac{a^2 + x^2}{b^2}\right] I_0(x) dx
\end{aligned}
\tag{2.30}
$$

is the generalized Marcum Q-function and the PDF of μ. Making use of [43, Equation (2.1–124)], the cumulative distribution function (CDF) of Y can be evaluated (for an even number of degrees of freedom which is $2u$ in our case) in a closed form as

$$
F_Y(y) = 1 - Q_u(\sqrt{2\gamma}, \sqrt{y}),
\tag{2.31}
$$

where $Q_u(a, b)$ is the generalized Marcum Q-function [49]. Hence,

$$
P_D = Q_u(\sqrt{2\gamma}, \sqrt{\lambda}).
\tag{2.32}
$$

2.2.5 Expansion of Random Process in Orthonormal Series with Uncorrelated Coefficients: The Karhunen-Loeve Expansion

Representation of random process is the foundation for signal processing. Stationary and nonstationary processes require different treatment, as shown in Table 2.3.

The Karhunen-Loeve expansion [50–55] is used to show that $2TW$ terms suffice to approximate the energy in a finite duration sample of a band-limited process with a flat power density spectrum. This demonstration is more rigorous than that using the sampling theorem. This result is especially useful for ultra-wideband (UWB) systems. For

Table 2.3 Mathematical representation for random process

Random process	Stationary process	Nonstationary process
Continuous-time	Fourier Transform	Karhunen-Loeve Transform
Discrete-time	Discrete Fourier Transform (DFT)	Discrete Karhunen-Loeve Transform (DKLT)
Fast algorithms	Fast Fourier Transform (FFT)	Singular Value Decomposition (SVD)
Algorithms complexity	$O(N \log_2(N))$	$O(N^3)^a$

[a] Fast algorithms.

a narrowband example, $W = 1\,$kHz and $T = 5\,$ms, thus $2TW = 10$. For a UWB example, $W = 1\,$GHz and $T = 5\,$ns, thus $2TW = 10$. Rigorous treatment of signal detection is given in [52, 53]. The $2TW$ theorem [56–66] is critical to estimation and detection, optics, quantum mechanics, laser modes, etc.—to name a few [61]. For transient, UWB signals, nonstationary random processes are met: Fourier analysis is insufficient. Van Tree (1968) [54] gives a very readable treatment of this problem.

Consider a zero-mean, wide-sense stationary, Gaussian random process $n(t)$ with a flat power density spectrum extending over the frequency interval $(-W, W)$. Let its autocorrelation function $R(\tau)$ be given by

$$R(\tau) = \text{sinc}(2W\tau), \tag{2.33}$$

where $\text{sinc}(x) = \sin(\pi x)/\pi x$. The process $n(t)$ may be represented in the interval $(0, T)$ by the expansion of orthonormal functions $\phi_i(t)$:

$$n(t) = \sum_{i=1}^{\infty} \lambda_i \phi_i(t), \tag{2.34}$$

where λ_i is given by

$$\lambda_i = \int_0^T n(t)\phi_i(t)dt, \tag{2.35}$$

and the $\phi_i(t)$ are the eigenfunctions of the integral equation

$$\int_0^T R(t - \tau)\phi_i(\tau)d\tau = \kappa_i \phi_i(t), \tag{2.36}$$

where κ_i are the eigenvalues of the equation. The expansion coefficients λ_i are uncorrelated: statistically independent Gaussian random variables. It is in this case that the expansion finds its most important application [54]. The form of (2.36) is reminiscent of the matrix equation

$$\lambda \varphi = R_n \varphi, \tag{2.37}$$

where R_n is a symmetric, nonnegative definite matrix. This is the case when the discrete-time solution of (2.36) is attempted.

The number of terms in (2.34) which constitute a sufficiently good approximation with a finite number of terms depends on how rapidly the eigenvalues decrease in value after a certain index. The eigenvalues of (2.36) are the prolate spheroidal wave functions considered in [61–64, 66]. The cited sources show that the eigenvalues drop off rapidly after $2TW$ terms (except for $TW = 1$). Table 2.4 illustrates this rapid drop-off [54]. Therefore, we approximate (2.34) as

$$n(t) \approx \sum_{i=1}^{2TW} \lambda_i \phi_i(t). \tag{2.38}$$

The approximation (2.38) is more satisfactory than (2.7) based on sampling functions, because the rapidity of drop-off of the terms can be judged by how rapidly the eigenvalues λ_i drop off after $2TW$ terms.

Table 2.4 Eigenvalues for a bandlimited spectrum

$2TW = 2.55$	$2TW = 5.10$
$\lambda_1 = 0.996 \dfrac{P}{2W}$	$\lambda_1 = 1.0 \dfrac{P}{2W}$
$\lambda_2 = 0.912 \dfrac{P}{2W}$	$\lambda_2 = 0.999 \dfrac{P}{2W}$
$\lambda_3 = 0.519 \dfrac{P}{2W}$	$\lambda_3 = 0.997 \dfrac{P}{2W}$
$\lambda_4 = 0.110 \dfrac{P}{2W}$	$\lambda_4 = 0.961 \dfrac{P}{2W}$
$\lambda_5 = 0.009 \dfrac{P}{2W}$	$\lambda_5 = 0.748 \dfrac{P}{2W}$
$\lambda_6 = 0.0004 \dfrac{P}{2W}$	$\lambda_6 = 0.321 \dfrac{P}{2W}$
	$\lambda_7 = 0.061 \dfrac{P}{2W}$
	$\lambda_8 = 0.006 \dfrac{P}{2W}$
	$\lambda_9 = 0.0004 \dfrac{P}{2W}$

Since the $\phi_i(t)$ are orthonormal, the energy of $n(t)$ in the interval $(0, T)$ is, using (2.38),

$$\int_0^T n^2(t)dt \simeq \sum_{i=1}^{2TW} \lambda_i^2. \tag{2.39}$$

Since the process is Gaussian, then the λ_i are Gaussian. The variance of λ_i is κ_i and these are nearly the same for $i \leq 2TW$. Thus, the energy in the finite duration sample of $n(t)$ is the sum of $2TW$ squares of zero mean Gaussian variates all having the same variance. With the appropriate normalization, we are led to the chi-square distribution.

Definition [54]. A *Gaussian* white noise is a Gaussian process whose covariance function is $\sigma^2\delta(t-u)$. It may be decomposed over the interval $[0, T]$ by using *any* set of orthonormal functions $\phi_i(t)$. The coefficients along *each* coordinate function are statistically independent Gaussian variables with equal variance σ^2.

2.3 Spectrum Sensing Exploiting Second-Order Statistics

2.3.1 Signal Detection Formulation

There are two different frameworks regarding how to formulate spectrum sensing: (1) Signal Detection; (2) Signal Classification.

The problem is to decide whether the primary signal—deterministic or random process—is present or not from the observed signals. It can be formulated as the following two hypotheses:

$$y(t) = \begin{cases} i(t) + w(t), & \mathcal{H}_0 \\ x(t) + i(t) + w(t), & \mathcal{H}_1 \end{cases} \qquad (2.40)$$

where $y(t)$ is the received signal at the CR user, $x(t) = s(t) * h(t)$ with $s(t)$ the primary signal and $h(t)$ the channel impulse response, $i(t)$ is interference, and $w(t)$ is the additive Gaussian noise (AWGN). In (2.40), $\mathcal{H}_0$ and $\mathcal{H}_1$ denote the hypotheses corresponding to the absence and presence of the primary signal, respectively. Thus from the observation $y(t)$, the CR needs to decide between $\mathcal{H}_0$ and $\mathcal{H}_1$. The assumption is that the signal $x(t)$ is independent of the noise $n(t)$ and interference $i(t)$.

When the signal waveform is deterministically known exactly, the sensing filter is matched to the waveform of the signal. A more realistic picture is that the signal is a stochastic signal with second order statistics that will be exploited for detection.

2.3.2   Wide-Sense Stationary Stochastic Process: Continuous-Time

Due to unknown propagation and nonsynchronization, coherent detection is infeasible. A good model is that the received signal $x(t)$ in (2.40) is a stochastic process—wide-sense stationary (WSS) or not, but independent of the noise $w(t)$ and the interference $i(t)$. The noise $w(t)$ and the interference $i(t)$ are also independent. As a result, it follows that

$$R_{yy}(\tau) = \begin{cases} R_{ii}(\tau) + R_{ww}(\tau), & \mathcal{H}_0 \\ R_{xx}(\tau) + R_{ii}(\tau) + R_{ww}(\tau), & \mathcal{H}_1 \end{cases} \qquad (2.41)$$

The covariance function $R_{ff}(\tau)$ is defined as $R_{ff}(\tau) = \int_{-\infty}^{\infty} f(t)f(t+\tau)dt$. To gain insight, neglecting the interference leads to

$$R_{yy}(\tau) = \begin{cases} R_{ww}(\tau), & \mathcal{H}_0 \\ R_{xx}(\tau) + R_{ww}(\tau), & \mathcal{H}_1 \end{cases} \qquad (2.42)$$

For the white Gaussian noise, $R_{ww}(\tau) = \frac{N_0}{2}\delta(\tau)$ where N_0 is the two-sided power spectrum density (in a unit of watts per Hz). Or,

$$R_{yy}(\tau) = \begin{cases} \frac{N_0}{2}\delta(\tau), & \mathcal{H}_0 \\ R_{xx}(\tau) + \frac{N_0}{2}\delta(\tau), & \mathcal{H}_1. \end{cases} \qquad (2.43)$$

In the spectrum domain, it follows that

$$S_{yy}(f) = \begin{cases} \frac{N_0}{2}, & \mathcal{H}_0 \\ S_{xx}(f) + \frac{N_0}{2}, & \mathcal{H}_1 \end{cases} \qquad (2.44)$$

where $S_{yy}(f)$ and $S_{xx}(f)$ are the Fourier transform of $R_{yy}(\tau)$ and $R_{xx}(\tau)$. Unfortunately, for low SNR case, $S_{xx}(f)$ is much smaller than that of noise floor $N_0/2$. Practically, we cannot do spectrum sensing by visualizing the spectrum shape $S_{xx}(f)$—this is the most powerful approach in most times.

2.3.3 Nonstationary Stochastic Process: Continuous-Time

For the model of (2.40) $y(t) = x(t) + n(t)$, $0 \le t \le T$ where $n(t)$ is white Gaussian. Here the $x(t)$ can be a nonstationary stochastic process. Then it follows that [54, p. 201, Equation (143)]

$$C_y(t, s) = \frac{N_0}{2}\delta(t - s) + C_x(t, s). \tag{2.45}$$

The Karhunen-Loeve expansion gives [54, p. 181, Equation (50)]

$$C_x(t, s) = \sum_{i=1}^{\infty} \lambda_i \phi_i(t)\phi_i(s), \quad 0 \le t, s \le 0. \tag{2.46}$$

The Gaussian process implies that [54, p. 198, Equation (128)]

$$\delta(t - s) = \sum_{i=1}^{\infty} \phi_i(t)\phi_i(s), \quad 0 \le t, s \le T. \tag{2.47}$$

Combining (2.45), (2.46), and (2.47) yields

$$C_y(t, s) = \sum_{i=1}^{\infty} \left(\frac{N_0}{2} + \lambda_i\right)\phi_i(t)\phi_i(s), \quad 0 \le t, s \le T, \tag{2.48}$$

where the white Gaussian noise uniformly disturbs the eigenvalues across all the degrees of freedom. When $x(t)$ has the bandlimited spectrum

$$S_x(\omega) = \begin{cases} \frac{P}{2W}, & |f| \le W \\ 0, & |f| > W \end{cases}, \tag{2.49}$$

it follows [54, p. 192] that

$$C_x(t, s) = P\frac{\sin 2\pi W(t - s)}{2\pi W(t - s)}. \tag{2.50}$$

The covariance of the $y(t)$ is

$$\begin{aligned} C_y(t, s) &= \frac{N_0}{2}\delta(t - s) + P\frac{\sin 2\pi W(t-s)}{2\pi W(t-s)} \\ &= \sum_{i=1}^{2TW+1} \left(\frac{N_0}{2} + \lambda_i\right)\phi_i(t)\phi_i(s), \quad 0 \le t, s \le T, \end{aligned} \tag{2.51}$$

where the eigenfunctions and the eigenvalues are given by [54, p. 192]

$$\lambda\phi(t) = \int_{-T/2}^{T/2} P\frac{\sin 2\pi W(t - s)}{2\pi W(t - s)}\phi(s)ds. \tag{2.52}$$

When $2TW = 2.55$ and $2TW = 5.1$, Table 2.4 gives the eigenvalues.
Example $2TW = 2.55$.

Considering the first two eigenvalues: $\lambda_1 = 0.996\frac{P}{2W}$ and $\lambda_2 = 0.912\frac{P}{2W}$. Equation (2.51) becomes

$$C_y(t, s) = \left(\frac{N_0}{2} + 0.996\frac{P}{2W}\right)\phi_1(t)\phi_1(s) + \left(\frac{N_0}{2} + 0.992\frac{P}{2W}\right)\phi_2(t)\phi_2(s)$$

$$+ \sum_{i=3}^{2TW+1}\left(\frac{N_0}{2} + \lambda_i\right)\phi_i(t)\phi_i(s). \tag{2.53}$$

For the first term, the SNR defined as $\gamma_0 = \frac{P}{WN_0}$, which is as low as -21 dB. The first term in (2.53) becomes $(1 + 0.996\gamma_0)\frac{N_0}{2}\phi_1(t)\phi_1(s)$ or, approximately, $\frac{N_0}{2}\phi_1(t)\phi_1(s)$ since $\gamma_0 \cong 0.01$. Similarly, the second term in (2.53) becomes $(1 + 0.992\gamma_0)\frac{N_0}{2}\phi_2(t)\phi_2(s)$ or $\frac{N_0}{2}\phi_2(t)\phi_2(s)$.

We have three practical challenges: (1) SNR γ_0 is as low as -21 dB; (2) the signal power P is changing over time; (3) the noise power is σ_n^2. There is uncertain noise power, implying that σ_n^2 is a function of time.

As a result of the above reasons, the SNR $\gamma = \frac{P}{2W\sigma_n^2}$ is uncertain over time. So the SNR is not the best performance criterion sometimes. Ideally, the criterion should be invariant to the SNR terms in (2.53). The normalized correlation coefficient, fortunately, satisfies this condition. The normalized correlation coefficient is defined as

$$\rho(f, g) = \frac{\int_{-\infty}^{\infty} f(t)g(t)dt}{\sqrt{\int_{-\infty}^{\infty} f^2(t)dt}\sqrt{\int_{-\infty}^{\infty} g^2(t)dt}}, \tag{2.54}$$

which is $0 \le |\rho| \le 1$ and invariant to rotation and dilation of the functions $f(t)$ and $g(t)$. For the case of low SNR in (2.51), by neglecting the signal eigenvalue term λ_i and replacing $N_0/2$ with the uncertain noise variance σ_n^2 it follows that

$$C_y(t, s) \cong \sigma_n^2\left[\phi_1(t)\phi_1(s) + \phi_2(t)\phi_2(s) + \sum_{i=3}^{2TW+1}\phi_i(t)\phi_i(s)\right]$$

$$= \sigma_n^2\left[\sum_{i=1}^{2TW+1}\phi_i(t)\phi_i(s)\right] \tag{2.55}$$

$$\cong \sigma_n^2 C_x(t, s), \qquad \text{if } \lambda_i \cong 1 \quad \text{for } 1 \le i \le 2TW + 1$$

where the third line of the equation is valid for some special cases. According to (2.55), the $C_x(t, s)$ can be used as a feature for similarity measurement (defined in (2.54)) that is independent of the noise power term σ_n^2. This feature extraction has low computational cost since no eigenfunction is explicitly required.

In sum, the sensing filter measures the similarity of the received random signal, relative to the first eigenfunction (first feature) and the second eigenfunction (second feature) of the covariance function $C_x(t, s)$ defined in (2.40). Calculation of low-order eigenfunctions may be more accurate numerically. We can first extract the eigenfunctions of the covariance function $C_y(t, s)$ as the features. Then the similarity function of (2.54) is used for classification. When attempting recognition, the unclassified image (or waveform vector) is compared in turn to all of the database images, returning a vector of matching scores (one per feature) computed through normalized cross-correlation defined as (2.54). The unknown person is then classified as the one giving the highest cumulative score [67].

2.3.3.1 Flat Fading Signal

Let us consider the model of (2.40): $y(t) = x(t) + n(t)$, where $x(t)$ is a flat fading signal. According to (2.55), the $C_x(t, s)$ can be used as the feature for spectrum sensing. Fortunately, for a flat fading signal, or narrowband fading model [68], the $C_x(t, s)$ can be obtained in closed-form. We can gain insight into the problem by going through this exercise.

Following [68], the transmitted signal is an unmodulated carrier

$$s(t) = \text{Re}\{e^{j(2\pi f_c t + \phi_0)}\} = \cos(2\pi f_c t + \phi_0), \tag{2.56}$$

where f_c is the central frequency of the carrier with random phase offset ϕ_0.

For the narrowband flat fading channel, the received signal becomes

$$x(t) = \text{Re}\left\{ \left[\sum_{n=0}^{N(t)} \alpha_n(t) e^{-j\phi_n(t)} \right] e^{j2\pi f_c t} \right\}$$

$$= x_I(t) \cos 2\pi f_c t - x_Q(t) \sin 2\pi f_c t, \tag{2.57}$$

where the in-phase and quadrature components are given by

$$x_I(t) = \sum_{n=0}^{N(t)} \alpha_n(t) \cos\phi_n(t),$$

$$x_Q(t) = \sum_{n=0}^{N(t)} \alpha_n(t) \sin\phi_n(t), \tag{2.58}$$

where there are $N(t)$ components, each of which includes the amplitude $\alpha_n(t)$ and the phase $\phi_n(t)$. They are random. If $N(t)$ is large, the central limit theorem can be revoked to argue that $\alpha_n(t)$ and $\phi_n(t)$ are independent for different components in order to approximate $x_I(t)$ and $x_Q(t)$ as jointly Gaussian random processes. As a result, the autocorrelation and cross-correlation of the in-phase and quadrature received signal components: $x_I(t)$ and $x_Q(t)$ can be derived, following the Gaussian approximation.

The following properties can be derived [68]:

1. $x_I(t)$ and $x_Q(t)$, respectively, a zero-mean Gaussian process.
2. $x_I(t)$ and $x_Q(t)$ are, respectively, a WSS random process.
3. $x_I(t)$ and $x_Q(t)$ are uncorrelated, that is,

$$\mathbf{E}\left[x_I(t)x_Q(t)\right] = 0. \tag{2.59}$$

4. The received signal $x(t) = x_I(t) \cos 2\pi f_c t - x_Q(t) \sin 2\pi f_c t$ is also WSS with autocorrelation

$$R_x(\tau) = E[x(t)x(t+\tau)] = R_{x_I}(\tau) \cos(2\pi f_c \tau). \tag{2.60}$$

Here the autocorrelation functions of $x_I(t)$ and $x_Q(t)$ are equal:

$$R_{x_I}(\tau) = R_{x_Q}(\tau) = P_x J_0(2\pi f_D \tau), \tag{2.61}$$

where P_x is the power of the total received power, f_D is the Doppler frequency, and

$$J_0(x) = \frac{1}{\pi} \int_0^\pi e^{-jx\cos\theta} d\theta \tag{2.62}$$

is the Bessel function of zeroth order.

5. The power spectral densities (PSDs) of $x_I(t)$ and $x_Q(t)$—denoted by $S_{x_I}(f)$ and $S_{x_Q}(f)$, respectively—are obtained by taking the Fourier transform of their respective autocorrelation functions relative to the delay parameter τ:

$$S_{x_I}(f) = S_{x_Q}(f) = \begin{cases} \frac{2P_x}{\pi f_D} \frac{1}{\sqrt{1-(f/f_D)^2}}, & |f| \leq f_D \\ 0, & \text{else} \end{cases} . \tag{2.63}$$

The PSD of the received flat fading signal $x(t)$ is

$$S_x(f) = 0.25[S_{x_I}(f - f_c) + S_{x_I}(f + f_c)] = \begin{cases} \frac{P_x}{2\pi f_D} \frac{1}{\sqrt{1-(|f-f_c|/f_D)^2}}, & |f - f_c| \leq f_D \\ 0, & \text{else} \end{cases} . \tag{2.64}$$

It follows from (2.64) that the flat fading signal can be modeled : to pass two independent white Gaussian noise sources with PSD $N_0/2$ through lowpass filter with a frequency response $H(f)$ that satisfies

$$S_{x_I}(f) = S_{x_Q}(f) = \frac{N_0}{2}|H(f)|^2. \tag{2.65}$$

The filter outputs corresponds to the in-phase and quadrature components of the narrowband fading process with PSDs $S_{x_I}(f)$ and $S_{x_Q}(f)$.

Let us go back to our problem of spectrum sensing using the second order statistics. Since the fading signal is zero-mean WSS, only the second order statistics are sufficient for its characterization. Since it is zero-mean, the covariance function is identical to the autocorrelation function. It follows by inserting (2.61) into (2.60) that

$$R_x(\tau) = E[x(t)x(t + \tau)] = P_x J_0(2\pi f_D \tau)\cos(2\pi f_c \tau), \tag{2.66}$$

which, from (2.55), leads to

$$C_y(\tau) \cong \sigma_n^2 C_x(\tau) = \sigma_n^2 P_x J_0(2\pi f_D \tau)\cos(2\pi f_c \tau), \tag{2.67}$$

which requires a priori knowledge of f_c and $f_D = v/\lambda_c$ where v is the velocity of the mobile and λ_c is the wavelength of the carrier wave. If the similarity is used as the classification criterion, the uncertain noise power σ_n^2 and the (uncertain) total power of the received flat fading signal P_x are not required. In practice, the real challenge is to know the mobile velocity v which can be searched from the window $v_{min} \leq v \leq v_{max}$. Algorithm steps: (1) measure the autocorrelation function of the flat fading signal plus white Gaussian noise $C_y(\tau)$; (2) measure the similarity between $C_y(\tau)$ and $J_0(2\pi f_D \tau)\cos(2\pi f_c \tau)$; (3) If the similarity is above some pre-set threshold ρ_0, we assign hypothesis $\mathcal{H}_1$. Otherwise, we assign $\mathcal{H}_0$. Note that the threshold ρ_0 must be learned in advance.

2.3.4   Spectrum Correlation-Based Spectrum Sensing for WSS Stochastic Signal: Heuristic Approach

In general, there are three signal detection approaches for spectrum sensing: (1) energy detection, (2) matched filter (coherent detection), (3) feature detection. If only the local noise power is known, the energy detection is optimal [69]. If a deterministic pattern (e.g., pilot, preamble, or training sequence) of the primary signal is known, then the optimal detector usually applies a matched filtering structure to maximize the probability of detection. Depending on the available a priori information about the primary signal, one may choose different approaches. At very low SNR, the energy detection suffers from noise uncertainty, while the matched filter faces the problem of lost synchronization. Cyclostationary detection exploits the periodicity in the modulated schemes but requires high computational complexity. Covariance matrix based spectrum sensing can be viewed as the discrete-time formulation of second-order statistics in the time domain.

Results in Sections 2.3.2 and 2.3.3 give us insight into using second-order statistics, although continuous time has been used there. These classical results are still the foundation of our departure. Here, discrete-time second-order statistics are formulated in the spectrum domain. The connection of this section with Sections 2.3.2 and 2.3.3 will be pointed out later. In this section we follow [70] for the exposition of the theory. The flavor is practical.

The basic strategy is (1) to correlate the periodgram of the received signal with the selected spectrum features. For example, a particular TV transmission scheme can be selected as a feature that is constant during the transmissions. Then, (2) the correlation is examined for decision-making.

The discrete-time form of the problem (2.40) can be modeled as the l-th time instant

$$\begin{aligned}
\mathcal{H}_0 &: y(l) = n(l), \quad l = 0, 1, 2, \ldots, \\
\mathcal{H}_1 &: y(l) = x(l) + n(l), \quad l = 0, 1, 2, \ldots,
\end{aligned} \tag{2.68}$$

where $y(l)$ is the received signal by a second user, $x(l)$ is the transmitted incumbent signal, and $n(l)$ is the complex, zero-mean additive white Gaussian noise (AWGN), that is, $n(l) \sim \mathcal{CN}(0, \sigma_n^2)$. As in problem (2.40), the signal and the noise are assumed to be independent. Accordingly, the PSD of the received signal $S_Y(\omega)$ can be written as

$$\begin{aligned}
\mathcal{H}_0 &: S_Y(\omega) = \sigma_n, \\
\mathcal{H}_1 &: S_Y(\omega) = S_X(\omega) + \sigma_n, \quad 0 \leq \omega \leq 2\pi,
\end{aligned} \tag{2.69}$$

where $S_X(\omega)$ is the PSD function of the transmitted primary signal. Our task is to distinguish between $\mathcal{H}_0$ and $\mathcal{H}_1$, by exploiting the unique spectral signature exhibited in $S_X(\omega)$.

For WSS, the autocorrelation function and its Fourier transform, that is, the PSD, are good statistics to study. Due to the independence of the $x(t)$ from the $n(t)$, it follows that

$$\begin{aligned}
\mathcal{H}_0 &: \int_0^{2\pi} S_Y(\omega) S_X(\omega) d\omega = \sigma_n \int_0^{2\pi} S_X(\omega) d\omega, \\
\mathcal{H}_1 &: \int_0^{2\pi} S_Y(\omega) S_X(\omega) d\omega = \int_0^{2\pi} S_X(\omega) S_X(\omega) d\omega + \sigma_n \int_0^{2\pi} S_X(\omega) d\omega,
\end{aligned} \tag{2.70}$$

It is natural to define the test T_{SC} for the spectral correlation

$$T_{SC} = \int_0^{2\pi} S_Y(\omega) S_X(\omega) d\omega, \qquad (2.71)$$

and compare T_{SC} with the threshold

$$\gamma = \sigma_n \int_0^{2\pi} S_X(\omega) d\omega,$$

where $\int_0^{2\pi} S_X(\omega) d\omega$ is the average power of discrete-time random variable X [71]. In other words, we have

$$T_{SC} = \int_0^{2\pi} S_Y(\omega) S_X(\omega) d\omega \underset{\mathcal{H}_0}{\overset{\mathcal{H}_1}{\gtrless}} \sigma_n \int_0^{2\pi} S_X(\omega) d\omega = \gamma, \qquad (2.72)$$

where the threshold is a function of the noise PSD and the average power of the stochastic signal.

In Sections 2.3.2 and 2.3.3, we argue that the autocorrelation function is a good feature for classification. The results there are also valid for the cases when $x(t)$ is both WSS and nonstationary processes. The result in this section is valid for WSS only. But the discrete-time is explicitly considered.

2.3.4.1 Estimating the Power Spectrum Density

Practically, an estimate of $S_X(\omega)$ must be obtained. The original or "classical" methods are based directly on the Fourier transform. This approach is preferred here for two good reasons: (1) the fast algorithm (FFT computation engine) can be used; (2) the spectrum estimator must be valid for the low SNR region (e.g., -20 dB). The so-called "modern" methods for spectrum estimation [72]—with other terms *model-based*, *parametric*, *data adaptive* and *high resolution*—seem be out of reach to our low SNR region. The most recent "subspace methods" are connected intimately with the spectrum sensing approaches that are discussed in this chapter. As a result, these methods are not treated as "spectral estimators," rather as the spectrum sensing methods. See standard texts [73, 74] for details.

The classical methods of spectral estimation are based on the Fourier transform of the data sequence or its correlation. In spite of all the developments in newer, more "modern" techniques, classical methods are often the favorite when the data sequence is long and stationary. These methods are straightforward to apply and make no assumptions (other than stationarity) about the observed data sequence (i.e., the methods are *nonparametric*) [72]. The PSD is defined as the Fourier transform of the autocorrelation function. Since there are simple methods for estimating the correlation function—see [72, Chapter 6], it seems natural to estimate the PSD by using the estimated correlation function.

Suppose that the total number of data samples is N. The biased sample autocorrelation function is defined by

$$\hat{R}_x[l] = \frac{1}{N} \sum_{n=0}^{N-1-l} x[n+l]x^*[n]; \quad 0 \leq l \leq N, \tag{2.73}$$

where $\hat{R}_x[l] = \hat{R}_x^*[-l]$ for $0 \leq l$. This estimate is asymptotically unbiased and consistent [72, p. 586]. The expected value of the estimate is given by

$$E\{\hat{R}_x[l]\} = \frac{N - |l|}{N} R_x[l] \tag{2.74}$$

and its variance decreases as $1/N$, for small values of lag. Here $R_x[l]$ is the discrete-time autocorrelation function. It seems reasonable to define a spectral estimate as

$$\hat{S}_x(e^{j\omega}) = \sum_{l=-N+1}^{N-1} \hat{R}_x[l]e^{-j\omega l} ; \quad L < N. \tag{2.75}$$

This estimate for the PSD is known as the correlogram. It is typically used with large N and relatively small values of L (say $L \leq 10\% N$).

Now assume that the maximum lag l is taken to be equal to $N - 1$. We have

$$\hat{S}_x(e^{j\omega}) = \sum_{l=-N+1}^{N-1} \hat{R}_x[l]e^{-j\omega l} = \frac{1}{N}|X(e^{j\omega})|^2, \tag{2.76}$$

where

$$X(e^{j\omega}) = \sum_{n=0}^{N-1} x[n]e^{-j\omega n} \tag{2.77}$$

is the discrete-time Fourier transform of the data sequence. This estimate is called the *periodogram*. The N-point DFT approximation of the spectrum is denoted by

$$S_X^{(N)}(k) = \hat{S}_x(e^{j\omega})|_{\omega=2\pi k/N}, \quad k = 0, 1, \ldots, N - 1. \tag{2.78}$$

2.3.4.2   Spectral Correlation Using the Estimated Spectrum

As done in Sections 2.3.2 and 2.3.3 for continuous time, we assume that the (N-point sampled) PSD of the signal, $S_X^{(N)}(k)$ defined by (2.78) is known a priori at the receiver. We perform the following test:

$$T_N = \frac{1}{N} \sum_{k=0}^{N-1} S_Y^{(N)}(k) S_X^{(N)}(k) \underset{\mathcal{H}_0}{\overset{\mathcal{H}_1}{\gtrless}} \gamma, \tag{2.79}$$

where γ is the decision threshold.

Under hypothesis $\mathcal{H}_I$, we have

$$\mathbf{E}\,[T_{N,0}] = \frac{1}{N}\sigma_n^2 \sum_{k=0}^{N-1} S_X^{(N)}(k) = \sigma_n^2 P_x, \qquad (2.80)$$

where

$$P_x = \frac{1}{N} \sum_{k=0}^{N-1} S_X^{(N)}(k), \qquad (2.81)$$

is the average power transmitted across the whole bandwidth. Similarly, we have

$$\begin{aligned}
\mathbf{E}\,[T_{N,1}] &= \frac{1}{N} \sum_{k=0}^{N-1} \mathbf{E}\,[S_Y^{(N)}(k)] S_X^{(N)}(k) \\
&\approx \sigma_n^2 P_x + \frac{1}{N} \sum_{k=0}^{N-1} [S_X^{(N)}(k)]^2,
\end{aligned} \qquad (2.82)$$

where we have used the fact that the periodogram is an asymptotically unbiased estimate of the PSD [75, 76]. Here, we can use the difference between $\mathbf{E}[T_{N,1}]$ and $\mathbf{E}[T_{N,0}]$ to determine the detection performance.

2.3.5 Likelihood Ratio Test of Discrete-Time WSS Stochastic Signal

Here we mainly follow [69, 70, 77]. Considering a sensing interval of N samples, we can express the received signal and the transmitted signal in vector form

$$\mathbf{y} = [y(0), y(1), \ldots, y(n-1)]^T,$$

$$\mathbf{x} = [x(0), x(1), \ldots, x(n-1)]^T.$$

Some wireless signals experience propagation along multiple paths; it may be reasonable to approximately model them as being a second-order stationary zero-mean Gaussian stochastic process, as derived in Section 2.3.3 for flat fading signal. Formally,

$$\mathbf{x} \sim \mathcal{CN}(\mathbf{0}, \mathbf{R_x}) \qquad (2.83)$$

where $\mathbf{R_x} = \mathbf{E}(\mathbf{x}\mathbf{x}^T)$ is the covariance matrix. Equivalently, Equation (2.69) can be expressed as

$$\begin{aligned}
\mathcal{H}_0 &: \mathcal{CN}(\mathbf{0}, \mathbf{R_x}) \\
\mathcal{H}_1 &: \mathcal{CN}(\mathbf{0}, \mathbf{R_x} + \sigma_n^2\mathbf{I}),
\end{aligned} \qquad (2.84)$$

where $\mathbf{I}$ is the identity matrix.

The Neyman-Pearson theorem states that the binary hypothesis test uses the likelihood ratio

$$L(\mathbf{y}) = \frac{p(\mathbf{y}|\mathcal{H}_1)}{p(\mathbf{y}|\mathcal{H}_0)} \underset{\mathcal{H}_0}{\overset{\mathcal{H}_1}{\gtrless}} \gamma, \qquad (2.85)$$

where $\mathbf{y}$ is the observation. This test maximizes the probability of detection for a given probability of false alarm. For WSS Gaussian random processes, it follows that

$$L(\mathbf{y}) = \frac{p(\mathbf{y}|\mathcal{H}_1)}{p(\mathbf{y}|\mathcal{H}_0)} = \frac{\frac{1}{(2\pi)^{N/2}\det^{1/2}(\mathbf{R}_x+\sigma_n^2\mathbf{I})} \exp\left[-\frac{1}{2}\mathbf{y}^T(\mathbf{R}_x + \sigma_n^2\mathbf{I})^{-1}\mathbf{y}\right]}{\frac{1}{(2\pi\sigma_n^2)^{N/2}} \exp\left[-\frac{1}{2\sigma_n^2}\mathbf{y}^T\mathbf{y}\right]} \overset{\mathcal{H}_1}{\underset{\mathcal{H}_0}{\gtrless}} \gamma. \qquad (2.86)$$

The logarithm of the likelihood ratio is given by [78]

$$\log L(\mathbf{y}) = 2N\log\sigma_n - \log\det(\mathbf{R}_x + \sigma_n^2\mathbf{I}) - \mathbf{y}^T[(\mathbf{R}_x + \sigma_n^2\mathbf{I})^{-1} + \sigma_n^{-2}\mathbf{I}]\,\mathbf{y}. \qquad (2.87)$$

The constant terms can be absorbed into the threshold T_{LRT}. The optimal detection scheme in the sense of the Neyman-Pearson criterion requires only the logarithm likelihood ratio test (LRT) in the quadrature form

$$T_{LRT} = \mathbf{y}^T[\sigma_n^{-2}\mathbf{I} - g(\mathbf{R}_x + \sigma_n^2\mathbf{I})^{-1}]\,\mathbf{y} \overset{\mathcal{H}_1}{\underset{\mathcal{H}_0}{\gtrless}} \gamma'. \qquad (2.88)$$

2.3.5.1 Estimator-Correlator Structure

Using the matrix inversion lemma

$$(\mathbf{A} + \mathbf{BCD})^{-1} = \mathbf{A}^{-1} - \mathbf{A}^{-1}\mathbf{B}(\mathbf{D}\mathbf{A}^{-1}\mathbf{B} + \mathbf{C}^{-1})^{-1}\mathbf{D}\mathbf{A}^{-1}, \qquad (2.89)$$

we have upon using $\mathbf{A} = \sigma_n^2\mathbf{I}$, $\mathbf{B} = \mathbf{D} = \mathbf{I}$, $\mathbf{C} = \mathbf{R}_x$

$$(\mathbf{R}_x + \sigma_n^2\mathbf{I})^{-1} = \frac{1}{\sigma_n^2}\mathbf{I} - \frac{1}{\sigma_n^4}\left(\frac{1}{\sigma_n^2}\mathbf{I} + \mathbf{R}_x^{-1}\right)^{-1}, \qquad (2.90)$$

so that

$$T_{LRT} = \mathbf{y}^T\frac{1}{\sigma_n^2}\left(\frac{1}{\sigma_n^2}\mathbf{I} + \mathbf{R}_x^{-1}\right)^{-1}\mathbf{y} = \frac{1}{\sigma_n^2}\mathbf{y}^T\hat{\mathbf{x}}, \qquad (2.91)$$

where

$$\hat{\mathbf{x}} = \frac{1}{\sigma_n^2}\left(\frac{1}{\sigma_n^2}\mathbf{I} + \mathbf{R}_x^{-1}\right)^{-1}\mathbf{y}.$$

which can be rewritten as

$$\begin{aligned} \hat{\mathbf{x}} &= \frac{1}{\sigma_n^2}\left(\frac{1}{\sigma_n^2}\mathbf{I} + \mathbf{R}_x^{-1}\right)^{-1}\mathbf{y} = \frac{1}{\sigma_n^2}\left(\frac{1}{\sigma_n^2}(\mathbf{R}_x + \sigma_n^2\mathbf{I})\mathbf{R}_x^{-1}\right)^{-1}\mathbf{y} \\ &= \mathbf{R}_x(\mathbf{R}_x + \sigma_n^2\mathbf{I})^{-1}\mathbf{y}, \end{aligned} \qquad (2.92)$$

and can be viewed as the minimum mean square error (MMSE) estimation of $\mathbf{x}$.
Hence we decide $\mathcal{H}_1$ if

$$T_{LRT} = \sigma_n^2\mathbf{y}^T\hat{\mathbf{x}} > \gamma' \qquad (2.93)$$

or

$$T_{LRT} = \sigma_n^2 \sum_{n=0}^{N-1} y(n)\hat{x}(n). \tag{2.94}$$

2.3.5.2 White Gaussian Signal Assumption

If we assume that $\mathbf{x} \sim \mathcal{CN}(0, E_s\mathbf{I})$ or $\mathbf{R_x} = E_s\mathbf{I}$ in (2.84), and that $\{x[n]\}_{n=0,1,...,N-1}$ is an independent sequence, then the estimator-correlator structure (2.91) yields the test statistic:

$$\frac{E_s}{E_s + 2\sigma_n^2} \sum_{n=0}^{N-1} \|x[n]\|^2, \tag{2.95}$$

which is the equivalent to the energy detector. Thus the optimal detector is to decide $\mathcal{H}_1$ when

$$T_{ED} = \sum_{n=0}^{N-1} \|x[n]\|^2 > \gamma. \tag{2.96}$$

2.3.5.3 Low Signal-to-Noise Ratio

It follows from the Taylor series expansion that

$$\begin{aligned}
(\mathbf{R_x} + \sigma_n^2\mathbf{I})^{-1} &= (\mathbf{I} + \sigma_n^{-2}\mathbf{R_x})^{-1}\sigma_n^{-2} \\
&= (\mathbf{I} - \sigma_n^{-2}\mathbf{R_x} + \sigma_n^{-4}\mathbf{R_x}^2 - \cdots)\sigma_n^{-2}.
\end{aligned} \tag{2.97}$$

Here, we have used the infinite series [79, p. 705]

$$(\mathbf{I} + \mathbf{A})^{-1} = \mathbf{I} - \mathbf{A} + \mathbf{A}^2 - \mathbf{A}^3 + \mathbf{A}^4 + \cdots \tag{2.98}$$

for all real or complex matrices $\mathbf{A}$ of $n \times n$ such that

$$\text{sprad}(\mathbf{A}) < 1, \tag{2.99}$$

where

$$\text{sprad}(\mathbf{A}) \triangleq \max\{|\lambda| : \lambda \in \text{spec}(\mathbf{A})\},$$

saying that the matrix eigenvalue λ belongs to the spectrum $\text{spec}(\mathbf{A})$ of matrix $\mathbf{A}$. The convergence of the series (2.97) is assured if the maximum eigenvalues of $\sigma_n^{-2}\mathbf{R_x}$ are less than unity, as required by (2.99). This condition always holds in low SNR region where

$$\det^{1/N}(\mathbf{A}) \ll \sigma_n^2, \tag{2.100}$$

from which equation, by retaining the first two leading terms, (2.97) can be approximated as

$$(\mathbf{R_x} + \sigma_n^2\mathbf{I})^{-1} \approx (\mathbf{I} - \sigma_n^{-2}\mathbf{R_x})\sigma_n^{-2}. \tag{2.101}$$

By using (2.101), (2.88) is put into a more convenient form

$$T_{TRT} = \mathbf{y}^T[\sigma_n^{-2}\mathbf{I} - (\mathbf{R_x} + \sigma_n^2\mathbf{I})^{-1}]\mathbf{y} \approx \sigma_n^{-4}\mathbf{y}^T\mathbf{R_x}\mathbf{y} \overset{\mathcal{H}_1}{\underset{\mathcal{H}_0}{\gtrless}} \gamma'. \tag{2.102}$$

Or, we have

$$T_{LRT,N} \approx \frac{1}{N}\mathbf{y}^T\mathbf{R_x}\mathbf{y} \overset{\mathcal{H}_1}{\underset{\mathcal{H}_0}{\gtrless}} \gamma_{LRT} \tag{2.103}$$

and $\gamma_{LRT} = \sigma_n^4\gamma'/N$, which depends on the noise power σ_n^4. This noise dependence is challenging in practice.

2.3.6 Asymptotic Equivalence between Spectrum Correlation and Likelihood Ratio Test

Let us show the asymptotic equivalence between spectrum correlation and likelihood ratio detection at low SNR region. Consider a sequence of optimal LRT detectors as defined in (2.104):

$$T_{LRT,N} \approx \frac{1}{N}\mathbf{y}^T\mathbf{R_x}\mathbf{y} \overset{\mathcal{H}_1}{\underset{\mathcal{H}_0}{\gtrless}} \gamma_{LRT}, \quad N = 1, 2, \ldots. \tag{2.104}$$

Similarly, we define a sequence of spectral correlation detectors as

$$T_N = \frac{1}{N}\sum_{k=0}^{N-1} S_Y^{(N)}(k)S_X^{(N)}(k), \quad N = 1, 2, \ldots. \tag{2.105}$$

Notice that the LRT detectors are working in the time domain while the spectrum correlation detectors are in the frequency domain.

The sequence of spectrum correlation detector $\{T_N\}$ defined in (2.105) are, at very low SNR, asymptotically equivalent to the sequence of optimal LRT detectors $\{T_{LRT,N}\}$ defined as (2.104), that is,

$$\lim_{N\to\infty} |T_{LRT,N} - T_N| = 0. \tag{2.106}$$

The proof of (2.107) is in order. From the definition of two test statistics, it follows that

$$\lim_{N\to\infty} |T_{LRT,N} - T_N| = \lim_{N\to\infty} \frac{1}{N}|\mathbf{y}^*\mathbf{R_x}\mathbf{y} - \mathbf{y}^* W_N^* \Lambda W_N \mathbf{y}|, \tag{2.107}$$

where

$$\Lambda = \begin{pmatrix} S_X^{(N)}(0) & \cdots & 0 \\ \vdots & \ddots & \vdots \\ 0 & \cdots & S_X^{(N)}(N-1) \end{pmatrix} \tag{2.108}$$

is a diagonal matrix with the PSD of the incumbent signal in the diagonal, and $\mathbf{W}_N$ is the discrete-time Fourier transform (DFT) matrix defined as

$$
\mathbf{W}_N = \begin{pmatrix}
1 & 1 & 1 & & 1 \\
1 & w_N & w_N^2 & \cdots & w_N^{N-1} \\
1 & w_N^2 & w_N^4 & & w_N^{2(N-1)} \\
\vdots & & & \ddots & \vdots \\
1 & w_N^{N-1} & w_N^{2(N-1)} & \cdots & w_N^{(N-1)(N-1)}
\end{pmatrix},
\tag{2.109}
$$

where $w_N = e^{-j2\pi/N}$ being a primitive n-th root of unity. As a result,

$$
\lim_{N\to\infty} |T_{LRT,N} - T_N| = \lim_{N\to\infty} \frac{1}{N} |\mathbf{y}^*(\mathbf{R_x} - \mathbf{W}_N^* \Lambda \mathbf{W}_N)\mathbf{y}| = \lim_{N\to\infty} \frac{1}{N} |\mathbf{y}^*(\mathbf{R_x} - \mathbf{C}_N)\mathbf{y}|,
\tag{2.110}
$$

where $\mathbf{C}_N \overset{\Delta}{=} \mathbf{W}_N^* \Lambda \mathbf{W}_N$ is the circular matrix. As shown in [80], the Toeplitz matrix $\mathbf{R_x}$ is asymptotically equivalent to the circurlar matrix $\mathbf{C}_N$ since the weak norm (Hilbert-Schmidt norm) of $\mathbf{R_x} - \mathbf{C}_N$ goes to zero, that is,

$$
\lim_{N\to\infty} \|\mathbf{R_x} - \mathbf{C}_N\| = 0.
\tag{2.111}
$$

Thus, (2.106) follows from (2.107).

2.3.7 Likelihood Ratio Test of Continuous-Time Stochastic Signals in Noise: Selin's Approach

2.3.7.1 Derivation of the Likelihood Ratio

We present an approach that is originally due to Selin (1965) [81, Chapter 8].

The modulated signal with the center frequency f_c is

$$
s(t) = \text{Re}[S(t)e^{j2\pi f_c t}],
$$

transmitted through the multipath fading channel. The received stochastic signal is

$$
x(t) = \text{Re}[X(t)e^{j2\pi f_c t}],
$$

and

$$
E[X(t)X^*(u)] = 2R_X(t,u).
$$

The noise process is expressed as

$$
n(t) = \text{Re}[N(t)e^{j2\pi f_c t}].
$$

Following Selin [81], we consider the white Gaussian noise with flat one-sided PSD $2N_0$. The methods of this section apply to the case in which the PSD of the noise is not flat.

Defining $Y(t)$ as the complex envelope representation of the received waveform $y(t)$, that is,

$$y(t) = \text{Re}[Y(t)e^{j2\pi f_c t}],$$

the test may be expressed as

$$\mathcal{H}_0 : Y(t) = N(t),$$
$$\mathcal{H}_1 : Y(t) = X(t) + N(t).$$

The present test can be thought as a test for the covariance function of $Y(t)$:

$$\mathcal{H}_0 : E[Y(t)Y^*(u)] = 2N_0\delta(t-u),$$
$$\mathcal{H}_1 : E[Y(t)Y^*(u)] = 2R_X(t,u) + 2N_0\delta(t-u). \tag{2.112}$$

If the Gaussian envelope of the signal process experiences some deterministic modulation, the signal process is nonstationary.

The Karhunen-Loeve expansion is a good theoretical tool for the purpose of representing the likelihood ratio. In practice, numerical calculation can be performed in MATLAB.

We seek to represent the stochastic signal

$$Y(t) = \sum_{k=1}^{\infty} y_k \phi_k(t).$$

We hope the $\phi_k(t)$ satisfies the following: (1) deterministic functions of time; (2) the $\phi_k(t)$ are orthonormal for convenience

$$\int_0^T \phi_k(t)\phi_l^*(t)dt = \delta_{kl};$$

(3) The random coefficients should be normalized and uncorrelated, that is,

$$E[x_k x_l^*] = \delta_{kl};$$

(4) If $Y(t)$ is Gaussian, then the $\{y_k\}$ should also be Gaussian. Fortunately the Karhunen-Loeve expansion provides these properties.

Taking the ratio and then letting K approach infinity, we have

$$L[Y(t)] = \exp\left[\sum_{k=1}^{\infty} \frac{\lambda_k |y_k|^2}{4N_0(\lambda_k + N_0)}\right] \prod_{k=1}^{\infty} \left(\frac{1}{1 + \frac{\lambda_k}{N_0}}\right)^{1/2}. \tag{2.113}$$

The product term converges provided that

$$\sum_{k=1}^{\infty} (\lambda_k / N_0)$$

converges, in other words, provided that the signal-to-noise ratio is finite. The test statistic is

$$U(Y) = \sum_{k=1}^{\infty} \frac{\lambda_k^2}{\lambda_k + N_0} |y_k|^2. \tag{2.114}$$

2.3.7.2 Probabilities of Error

If $\mathcal{H}_0$ is true,

$$E_0[U] = E\left[\sum_{k=1}^{\infty} \frac{\lambda_k}{\lambda_k + N_0}|n_k|^2\right]$$

$$= 2N_0\left[\sum_{k=1}^{\infty} \frac{\lambda_k}{\lambda_k + N_0}\right] \tag{2.115}$$

$$\mathrm{Var}_0[U] = E\left[\sum_{k=1}^{\infty} \frac{\lambda_k}{\lambda_k + N_0}(|n_k|^2 - 2N_0)\right]$$

$$= 8(N_0)^2\sum_{k=1}^{\infty} \left(\frac{\lambda_k}{\lambda_k + N_0}\right)^2 \tag{2.116}$$

$$= 2\sum_{k=1}^{\infty} \left(\frac{2N_0\lambda_k}{\lambda_k + N_0}\right)^2.$$

If $\mathcal{H}_1$ is true,

$$E_1[U] = E\left[\sum_{k=1}^{\infty} \frac{\lambda_k}{\lambda_k + N_0}|x_k + n_k|^2\right]$$

$$= \sum_{k=1}^{\infty} \frac{\lambda_k}{\lambda_k + N_0}(2N_0 + \lambda_k) \tag{2.117}$$

$$= \sum_{k=1}^{\infty} 2\lambda_k$$

$$\mathrm{Var}_1[U] = E\left[\sum_{k=1}^{\infty} \frac{\lambda_k}{\lambda_k + N_0}[|n_k + x_k|^2 - 2(N_0 + \lambda_k)]\right]^2$$

$$= 2\sum_{k=1}^{\infty} (2\lambda_k)^2. \tag{2.118}$$

By the central limit theorem, if

$$\sum_{k=1}^{\infty} \lambda_k^2/K \ll N_0,$$

U is approximately normal.

For very weak signals in low SNR region,

$$\sum_{k=1}^{\infty} \left(\frac{N_0\lambda_k}{\lambda_k + N_0}\right)^2 \cong \sum_{k=1}^{\infty} \lambda_k^2, \tag{2.119}$$

and $\mathrm{Var}_0(U) \cong \mathrm{Var}_1(U)$. The signal detection probability depends only on the signal-to-noise (power) ratio d which is given by

$$
\begin{aligned}
d &\cong \frac{[E_1(U) - E_0(U)]^2}{\mathrm{Var}(U)} \\
&= \frac{\left[\sum\limits_{k=1}^{\infty} 2\lambda_k - \sum\limits_{k=1}^{\infty} \frac{2N_0\lambda_k}{\lambda_k + N_0}\right]^2}{8 \sum\limits_{k=1}^{\infty} \lambda_k^2} \\
&= \frac{\left[\sum\limits_{k=1}^{\infty} \frac{\lambda_k^2}{\lambda_k + N_0}\right]^2}{2 \sum\limits_{k=1}^{\infty} \lambda_k^2} \\
&\cong \frac{1}{2N_0^2} \sum\limits_{k=1}^{\infty} \lambda_k^2.
\end{aligned}
\tag{2.120}
$$

This is approximately equal to

$$
T \frac{1}{N_0^2} \int_{-\infty}^{\infty} |S_x(f)|^2 df,
\tag{2.121}
$$

where $S_x(f)$ is the PSD of the signal process if this process is stationary. (2.121) is essentially identical to the spectrum correlation rule in Section 2.3.4: comparing with (2.121) and (2.79). In deriving (2.121), we have used the following

$$
\begin{aligned}
\sum_{k=1}^{\infty} \lambda_k^2 &= \sum_k \lambda_k \sum_l \lambda_l \int_0^T \phi_k(t)\phi_l^*(t)dt \int_0^T \phi_k^*(u)\phi_l(u)du \\
&= 2 \int_0^T dt \int_0^T du |R_x(t-u)|^2 \\
&= 2 \int_0^T dt \int_t^{T+t} d\tau |R_x(\tau)|^2 \\
&= 2 \int_0^T dt \int_{-\infty}^{\infty} |S_x(f)|^2 df, \text{ for large } T, \\
&= 2T \int_{-\infty}^{\infty} |S_x(f)|^2 df
\end{aligned}
\tag{2.122}
$$

2.4 Statistical Pattern Recognition: Exploiting Prior Information about Signal through Machine Learning

2.4.1 Karhunen-Loeve Decomposition for Continuous-Time Stochastic Signal

We model the communication signal or noise as random field. KLD is also known as PCA, POD, and EOF. We follow [82, 83] for an exposition of the underlying model for turbulence in fluids—a subject of great scientific and technological importance, and yet one of the least understood. Like turbulence, radio signals involve the interaction of many degrees of freedom over broad ranges of spatial and temporal scales.

The POD is statistically based, and permits the extraction, from the electromagnetic field, of spatial and temporal structures (coherent structures) judged essential. The POD is a procedure for extracting a modal decomposition from an ensemble of signals. Its power lies in the mathematical properties that suggest it is the preferred basis. The existence of coherent structures, which contain most of the energy, suggests the drastic reduction in dimension. A suitable modal decomposition retains only these structures and appeals to averaging or modeling to account for the incoherent fluctuations.

Suppose we have an ensemble $\{\mathbf{u}^k\}$ of observations (experimental measurements or numerical simulations) of a turbulent velocity field or an electromagnetic field. We assume that each $\{\mathbf{u}^k\}$ belongs to an inner product (Hilbert) space X. Our goal is to obtain an orthogonal basis φ_j for X, so that almost every member of the ensemble can be decomposed relative to the φ_j:

$$\mathbf{u} = \sum_{j=0}^{\infty} a_j \varphi_j, \tag{2.123}$$

where the a_j are suitable modal coefficients. There is no prior reason to distinguish between space and time in the definition and derivation of the empirical basis functions, but we ultimately want a dynamic model for the coherent structures. We seek the spatial vector-valued functions φ_j, and subsequently determine the time-dependent scalar modal coefficients:

$$\mathbf{u}(\mathbf{x}, t) = \sum_{j} a_j(t) \varphi_j(\mathbf{x}). \tag{2.124}$$

Central to the POD is the concept of averaging operation $< \cdot >$. The operation of $< \cdot >$ may simply be thought as the average over a number of separate experiments, or, if we assume ergodicity, as a time average over the ensemble of observations obtained at different instants during a single experimental run. We restrict ourselves to the space of functions X which are square integrable, or, in physical terms, fields with finite energy on this interval. We need the inner product

$$(f, g) = \int_X f(x) g(x) dx,$$

and a norm

$$\|f\| = (f, f)^{1/2}.$$

2.4.1.1 Derivation of Empirical Functions

We start with an ensemble of observation $\{\mathbf{u}\}$, and ask which single (deterministic) element is most similar to the members of $\{\mathbf{u}\}$ on average? Mathematically, the notion of "most similar" corresponds to seeking an element φ such that

$$\max_{\varphi \in X} < |(\mathbf{u}, \varphi)|^2 > /(\varphi, \varphi), \tag{2.125}$$

where $| \cdot |$ denotes the modulus. In other words, we find the member of the φ which maximizes the (normalized) inner product with the field $\{\mathbf{u}\}$, which is most nearly parallel in function space. This is a classical problem in the calculus of variations. This can be reformulated in terms of the calculus of variations, with a functional for the constrained variational problem

$$J[\varphi] = < |(\mathbf{u}, \varphi)|^2 > -\lambda(||\varphi||^2 - 1), \tag{2.126}$$

where $||\varphi||^2 = (\varphi, \varphi)$ is the L^2-norm. A necessary condition for extrema is the vanishing of the functional derivative for all variations $\varphi + \varepsilon\psi \in X$:

$$\frac{d}{d\varepsilon}J[\varphi + \varepsilon\psi]|_{\varepsilon=0} = 0. \tag{2.127}$$

Some algebra, together with the fact that $\psi(\mathbf{x})$ is an arbitrary variation, shows that the condition of (2.127) reduces to

$$\int_\Omega \underbrace{< \mathbf{u}(\mathbf{x}, t)\mathbf{u}^*(\mathbf{x}', t) >}_{R(\mathbf{x},\mathbf{x}')}\varphi(\mathbf{x}')d\mathbf{x}' = \lambda\varphi(\mathbf{x}). \tag{2.128}$$

Here $\mathbf{x} \in \Omega$, where Ω denotes the spatial domain of the experiment. This is a Fredholm integral equation of the second kind whose kernel is the averaged autocorrelation tensor $R(\mathbf{x}, \mathbf{x}') = < \mathbf{u}(\mathbf{x}, t)\mathbf{u}^*(\mathbf{x}', t) >$, which we may rewrite as the operator equation $\mathbf{R}\varphi = \lambda\varphi$. The optimal basis is called empirical eigenfunctions, since the basis is derived from the ensemble of observations $\mathbf{u}^k$. The operator $\mathbf{R}$ is clearly self-adjoint, and also compact, so that Hilbert-Schmidt theory assures us that there is a countable infinity of eigenvalues $\{\lambda_j\}$ and eigenfunctions $\{\varphi_j\}$. Without loss of generality, for the solutions of (2.128), we can normalize so that $||\varphi_j|| = 1$ and re-order the eigenvalues so that $\lambda_j \geq \lambda_{j+1}$. By the first N eigenvalues (resp. eigenfunctions) we mean $\lambda_1, \lambda_2, \ldots, \lambda_N$ (resp. $\varphi_1, \varphi_2, \ldots, \varphi_N$). Note that the positive semidefiniteness of R implies that $\lambda_j > 0$. As a result, this representation provides a diagonal decomposition of the autocorrelation function

$$R(\mathbf{x}, \mathbf{x}') = \sum_{j=1}^{\infty}\lambda_j\varphi_j(\mathbf{x})\varphi_j^*(\mathbf{x}'). \tag{2.129}$$

It is these empirical functions that we use in the model decomposition (2.124) above. The diagonal representation (2.129) of the two-point correlation tensor ensures that the modal amplitudes are uncorrelated:

$$< a_i a_j^* > = \delta_{ij}\lambda_j. \tag{2.130}$$

In practice, only the eigenfunctions with strictly positive values are of interest. Those spatial structures have finite energy on average. Let us define the span S of these φ_j

$$S = \left\{\sum a_j\varphi_j | \lambda_j > 0, \sum |a_j|^2 < \infty\right\}. \tag{2.131}$$

What is the nature of the span S? Which functions can be reproduced by convergent linear combinations of these empirical eigenfunctions? It turns out that almost every member of the original ensemble $\{\mathbf{u}^k\}$ belongs to S! The span of the eigenfunctions is exactly the span of all the realizations of $\mathbf{u}(\mathbf{x})$, with the exception of a set of measure zero.

2.4.1.2 Optimality

Suppose we have an ensemble of members of $\mathbf{u}(\mathbf{x},t)$, decomposed in terms of an arbitrary orthonormal basis ψ_j,

$$\mathbf{u}(\mathbf{x}, t) = \sum_j b_j(t)\psi_j(\mathbf{x}). \tag{2.132}$$

Using the orthonormality of the ψ_j, the average energy is given by

$$\int_\Omega < \mathbf{u}(\mathbf{x}, t)\mathbf{u}^*(\mathbf{x}, t) > d\mathbf{x} = \sum_j < b_j(t)b_j^*(t) > . \tag{2.133}$$

For the particular case of the POD decomposition, the energy in the j-th mode is λ_j, as claimed by (2.130).

The optimality is stated as follows: For any N, the energy in the first N modes in a proper orthogonal decomposition is at least as great as that in any other N-dimensional projections:

$$\|\mathbf{u}_N\|^2 = \sum_j < a_j(t)a_j^*(t) > = \sum_{j=1}^N \lambda_j \geq \sum_{j=1}^N < b_j(t)b_j^*(t) > . \tag{2.134}$$

This follows from the general linear self-adjoint operators: the sum of the first N eigenvalues of $\mathbf{R}$ is greater than or equal to the sum of the diagonal terms in any N-dimensional projection of $\mathbf{R}$. Equation (2.134) states that, among all *linear* decompositions, the POD is the most efficient for modeling or reconstructing a signal $\mathbf{u}(\mathbf{x}, t)$, in the sense of capturing, on average, the most energy possible for a projection on a given number of modes. This observation motivates the use of the POD for low-dimensional modeling of coherent structures—dimensionality reduction. The rate of decay of the λ_j gives the indication of how fast finite-dimensional representations converge on average, and hence how well specific truncations might capture these structures.

2.5 Feature Template Matching

From pattern recognition, the eigenvectors are considered as features. We define the leading eigenvector as signal feature because for nonwhite WSS signal it is most robust against noise and stable over time [84]. The leading eigenvector is determined by the direction with the largest signal energy [84].

Assume we have 2×1 random vectors $\mathbf{x}_{s+n} = \mathbf{x}_s + \mathbf{x}_n$, where $\mathbf{x}_s$ is vectorized sine sequence and $\mathbf{x}_n$ is the vectorized WGN sequence. SNR is set to 0 dB. There are 1000 samples for each random vector in Figure 2.1. Now we use eigen-decomposition to set the new X axes for each random vector samples such that λ_1 is strongest along the corresponding new X axes. It can be seen that new X axes for $\mathbf{x}_s$ (SNR $= \infty$ dB) and $\mathbf{x}_{s+n}$ (SNR $= 0$ dB) are almost the same. X axes for $\mathbf{x}_n$ (SNR $= -\infty$), however, is rotated with some random angle. This is because WGN has almost the same energy distributed in every direction. New X axes for noise will be random and unpredictable but the direction for the signal is very robust.

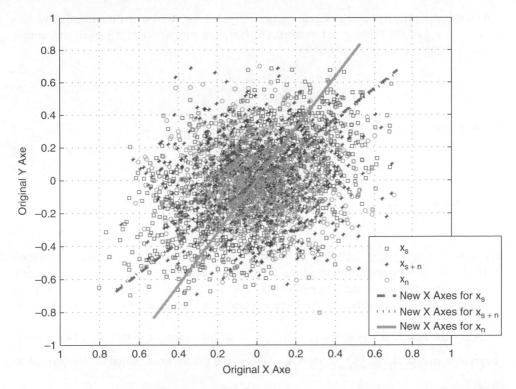

Figure 2.1 The leading eigenvector is determined by the direction with largest signal energy [84].

Based on leading eigenvector, feature template matching is explored for spectrum sensing. The secondary user receives the signal $y(t)$. Based on the received signal, there are two hypotheses: one is that the primary user is present $\mathcal{H}_1$, another one is the primary user is absent $\mathcal{H}_0$. In practice, spectrum sensing involves detecting whether the primary user is present or not from discrete samples of $y(t)$.

$$y(n) = \begin{cases} w(n) & \mathcal{H}_0 \\ x(n) + w(n) & \mathcal{H}_1, \end{cases} \qquad (2.135)$$

in which $x(n)$ are samples of the primary user's signal and $w(n)$ are samples of zero mean white Gaussian noise. In general, the algorithms of spectrum sensing aim at maximizing corresponding detection rate at a fixed false alarm rate with low computational complexity. The detection rate P_d and false alarm rate P_f are defined as

$$\begin{aligned} P_d &= prob(\text{detect } \mathcal{H}_1 | y(n) = x(n) + w(n)) \\ P_f &= prob(\text{detect } \mathcal{H}_1 | y(n) = w(n)), \end{aligned} \qquad (2.136)$$

in which *prob* represents probability.

The hypothesis detection is

$$\begin{aligned} \mathcal{H}_0 &: \mathbf{y} = \mathbf{w} \\ \mathcal{H}_1 &: \mathbf{y} = \mathbf{x} + \mathbf{w}. \end{aligned} \qquad (2.137)$$

Assume the primary user's signal is perfectly known. Given d-dimensional vectors $\mathbf{x}_1, \mathbf{x}_2, \cdots, \mathbf{x}_M$ of the training set constructed from the primary user's signal, the sample covariance matrix can be obtained by

$$\mathbf{R}_x = \frac{1}{M} \sum_{i=1}^{M} \mathbf{x}_i \mathbf{x}_i^T, \tag{2.138}$$

which assumes that the sample mean is zero,

$$\mathbf{u} = \frac{1}{M} \sum_{i=1}^{M} \mathbf{x}_i = \mathbf{0}. \tag{2.139}$$

The leading eigenvector of $\mathbf{R}_x$ can be extracted by eigen-decomposition of $\mathbf{R}_x$,

$$\mathbf{R}_x = \mathbf{V} \mathbf{\Lambda} \mathbf{V}^T, \tag{2.140}$$

where $\mathbf{\Lambda} = diag(\lambda_1, \lambda_2, \ldots, \lambda_d)$ is a diagonal matrix. $\lambda_i, i = 1, 2, \cdots, d$ are eigenvalues of $\mathbf{R}_x$. $\mathbf{V}$ is an orthonormal matrix, the columns of which $\mathbf{v}_1, \mathbf{v}_2, \cdots, \mathbf{v}_d$ are the eigenvectors corresponding to the eigenvalues $\lambda_i, i = 1, 2, \cdots, d$. For simplicity, take $\mathbf{v}_1$ as the eigenvector corresponding to the largest eigenvalue. The leading eigenvector $\mathbf{v}_1$ is the template of PCA.

For the measurement vectors $\mathbf{y}_i, i = 1, 2, \cdots, M$, the leading eigenvector of the sample covariance matrix $\mathbf{R}_y = \frac{1}{M} \sum_{i=1}^{M} \mathbf{y}_i \mathbf{y}_i^T$ is $\tilde{\mathbf{v}}_1$. Hence, the presence of signal is determined by [84, 85],

$$\rho = \max_{l=0,1,\ldots,d} \left| \sum_{k=1}^{d} \mathbf{v}_1[k] \tilde{\mathbf{v}}_1[k + l] \right| > T_{pca}, \tag{2.141}$$

where T_{pca} is the threshold value for PCA method, and ρ is the similarity between $\tilde{\mathbf{v}}_1$ and template $\mathbf{v}_1$ which is measured by cross-correlation. T_{pca} is assigned to arrive a desired false alarm rate. The detection with leading eigenvector under the framework of PCA is simply called PCA detection.

A nonlinear version of PCA—kernel PCA [86]—has been proposed based on the classical PCA approach. Kernel function is employed by kernel PCA to implicitly map the data into a higher dimensional feature space, in which PCA is assumed to work better than in the original space. By introducing the kernel function, the mapping φ need not be explicitly known which can obtain better performance without increasing much computational complexity.

The training set $\mathbf{x}_i, i = 1, 2, \cdots, M$ and received set $\mathbf{y}_i, i = 1, 2, \cdots, M$ in kernel PCA are obtained the same way as with the PCA framework.

The training set in the feature space are $\varphi(\mathbf{x}_1), \varphi(\mathbf{x}_2), \ldots, \varphi(\mathbf{x}_M)$ which are assumed to have zero mean, for example, $\frac{1}{M} \sum_{i=1}^{M} \varphi(\mathbf{x}_i) = \mathbf{0}$. The sample covariance matrix of $\varphi(\mathbf{x}_i)$ is

$$\mathbf{R}_{\varphi(x)} = \frac{1}{M} \sum_{i=1}^{M} \varphi(\mathbf{x}_i) \varphi(\mathbf{x}_i)^T. \tag{2.142}$$

Similarly, the sample covariance matrix of $\varphi(\mathbf{y}_i)$ is

$$\mathbf{R}_{\varphi(y)} = \frac{1}{M} \sum_{i=1}^{M} \varphi(\mathbf{y}_i)\varphi(\mathbf{y}_i)^T. \tag{2.143}$$

The detection algorithm with leading eigenvector under the framework of kernel PCA is summarized here as follows [85]:

1. Choose a kernel function k. Given the training set of the primary user's signal $\mathbf{x}_1, \mathbf{x}_2, \cdots, \mathbf{x}_M$, the kernel matrix is $\mathbf{K} = (\mathrm{k}(\mathbf{x}_i, \mathbf{x}_j))_{ij}$. $\mathbf{K}$ is positive semidefinite. Eigen-decomposition of $\mathbf{K}$ to obtain the leading eigenvector $\boldsymbol{\beta}_1$.
2. The received vectors are $\mathbf{y}_1, \mathbf{y}_2, \cdots, \mathbf{y}_M$. Based on the chosen kernel function, the kernel matrix $\tilde{\mathbf{K}} = (\mathrm{k}(\mathbf{y}_i, \mathbf{y}_j))_{ij}$ is obtained. The leading eigenvector $\tilde{\boldsymbol{\beta}}_1$ is also obtained by eigen-decomposition of $\tilde{\mathbf{K}}$.
3. The leading eigenvectors for $\mathbf{R}_{\varphi(x)}$ and $\mathbf{R}_{\varphi(y)}$ can be expressed as

$$\begin{aligned} \mathbf{v}_1^f &= (\varphi(\mathbf{x}_1), \varphi(\mathbf{x}_2), \dots, \varphi(\mathbf{x}_M))\boldsymbol{\beta}_1, \\ \tilde{\mathbf{v}}_1^f &= (\varphi(\mathbf{y}_1), \varphi(\mathbf{y}_2), \dots, \varphi(\mathbf{y}_M))\tilde{\boldsymbol{\beta}}_1. \end{aligned} \tag{2.144}$$

4. Normalize $\mathbf{v}_1^f$ and $\tilde{\mathbf{v}}_1^f$ to scale $\boldsymbol{\beta}_1$ and $\tilde{\boldsymbol{\beta}}_1$.
5. The similarity between $\mathbf{v}_1^f$ and $\tilde{\mathbf{v}}_1^f$ is

$$\rho = \boldsymbol{\beta}_1^T \mathbf{K}^t \tilde{\boldsymbol{\beta}}_1. \tag{2.145}$$

6. Determine the presence or absence of primary signal by evaluating $\rho > T_{kpca}$ or not. ρ is derived as

$$\begin{aligned} \rho &= \; < \mathbf{v}_1^f, \tilde{\mathbf{v}}_1^f > \; = \; < \sum_{i=1}^{M} \beta_i \varphi(\mathbf{x}_i), \sum_{j=1}^{M} \tilde{\beta}_i \varphi(\mathbf{y}_i) > \\[2mm] &= \{(\varphi(\mathbf{x}_1), \varphi(\mathbf{x}_2), ..., \varphi(\mathbf{x}_M))\boldsymbol{\beta}_1\}^T \cdot \\ &\qquad \{(\varphi(\mathbf{y}_1), \varphi(\mathbf{y}_2), ..., \varphi(\mathbf{y}_M))\tilde{\boldsymbol{\beta}}_1\} \\[2mm] &= \boldsymbol{\beta}_1^T \begin{pmatrix} \varphi(\mathbf{x}_1)^T \\ \varphi(\mathbf{x}_2)^T \\ \vdots \\ \varphi(\mathbf{x}_M)^T \end{pmatrix} (\varphi(\mathbf{y}_1), \varphi(\mathbf{y}_2), ..., \varphi(\mathbf{y}_M))\tilde{\boldsymbol{\beta}}_1 \\[2mm] &= \boldsymbol{\beta}_1^T \begin{pmatrix} \mathrm{k}(\mathbf{x}_1, \mathbf{y}_1), \mathrm{k}(\mathbf{x}_1, \mathbf{y}_2), ..., \mathrm{k}(\mathbf{x}_1, \mathbf{y}_M) \\ \mathrm{k}(\mathbf{x}_2, \mathbf{y}_1), \mathrm{k}(\mathbf{x}_2, \mathbf{y}_2), ..., \mathrm{k}(\mathbf{x}_2, \mathbf{y}_M) \\ ...... \\ \mathrm{k}(\mathbf{x}_M, \mathbf{y}_1), \mathrm{k}(\mathbf{x}_M, \mathbf{y}_2), ..., \mathrm{k}(\mathbf{x}_M, \mathbf{y}_M) \end{pmatrix} \tilde{\boldsymbol{\beta}}_1 \\[2mm] &= \boldsymbol{\beta}_1^T \mathbf{K}^t \tilde{\boldsymbol{\beta}}_1. \end{aligned} \tag{2.146}$$

$\mathbf{K}^t$ is the kernel matrix between $\varphi(\mathbf{x}_i)$ and $\varphi(\mathbf{y}_j)$. A measure of similarity between $\mathbf{v}_1^f$ and $\tilde{\mathbf{v}}_1^f$ has been obtained without giving $\mathbf{v}_1^f$ and $\tilde{\mathbf{v}}_1^f$ based on (2.146).

T_{kpca} is the threshold value for kernel PCA algorithm. The detection with leading eigenvector under the framework of kernel PCA is simply called kernel PCA detection.

DTV signal [87] captured in Washington D.C. will be employed to the experiment of spectrum sensing in this section. The first segment of DTV signal with $L = 500$ is taken as the samples of the primary user's signal $x(n)$.

First, the similarities of leading eigenvectors of the sample covariance matrix between first segment and other segments of DTV signal will be tested under the frameworks of PCA and kernel PCA. The DTV signal with length 10^5 is obtained and divided into 200 segments with the length of each segment 500. Similarities of leading eigenvectors derived by PCA and kernel PCA between the first segment and the rest 199 segments are shown in Figure 2.2. The result shows that the similarities are very high between leading eigenvectors of different segments' DTV signal (which are all above 0.94), on the other hand, kernel PCA is more stable than PCA.

The detection rates varied by SNR for kernel PCA and PCA compared with estimation-correlator (EC) and maximum minimum eigenvalue (MME) with $P_f = 10\%$ are shown in Figure 2.3 for 1000 experiments.

Experimental results show that kernel methods are 4 dB better than the corresponding linear methods. Kernel methods can compete with EC method.

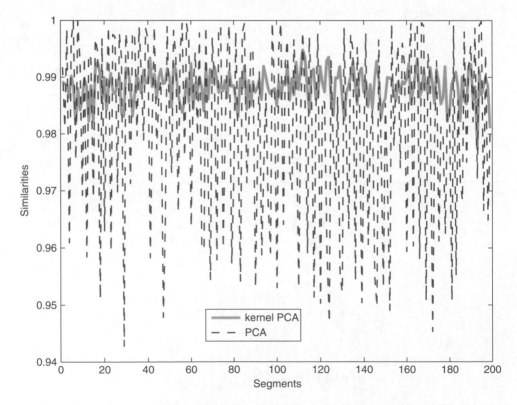

Figure 2.2 Similarities of leading eigenvectors derived by PCA and kernel PCA between the first segment and other 199 segments [85].

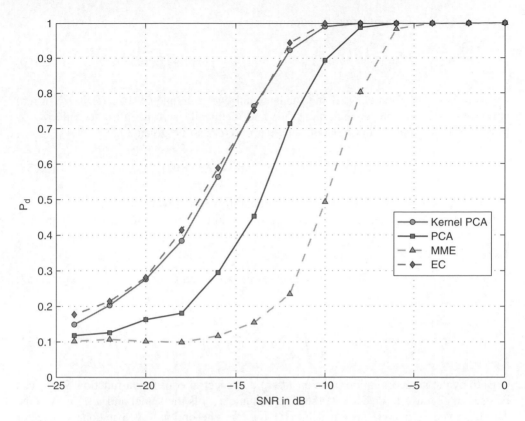

Figure 2.3 The detection rates for kernel PCA and PCA compared with EC and MME with $P_f = 10\%$ for DTV signal [85].

2.6 Cyclostationary Detection

Generally, noise in the communication system can be treated as wide-sense stationary process. Wide-sense stationary has time invariant autocorrelation function. Mathematically speaking, if $x(t)$ is a wide-sense stationary process, the autocorrelation function of $x(t)$ is

$$R_x(t, \tau) = E\{x(t)x^*(t - \tau)\} \qquad (2.147)$$

and

$$R_x(t, \tau) = R_x(\tau), \forall t. \qquad (2.148)$$

Generally, manmade signals are not wide-sense stationary. Some of them are cyclostationary [88]. A cyclostationary process is a signal exhibiting statistical properties which vary cyclically with time [89]. Hence, if $x(t)$ is a cyclostationary process, then

$$R_x(t, \tau) = R_x(t + T_0, \tau), \qquad (2.149)$$

where T_0 is the period in t not in τ.

Define cyclic autocorrelation function for $x(t)$ as

$$R_x^\alpha(\tau) = \lim_{T \to \infty} \frac{1}{T} \int_{-\frac{T}{2}}^{\frac{T}{2}} x\left(t + \frac{\tau}{2}\right) x^*\left(t - \frac{\tau}{2}\right) \exp(-j2\pi\alpha t)\, dt, \qquad (2.150)$$

where α is cyclic frequency. If the fundamental cyclic frequency of $x(t)$ is α_0, $R_x^\alpha(\tau)$ is nonzero only for integer multiples of α_0 and identically zero for all other values of α [88, 90, 91]. And spectral correlation function of $x(t)$ can be given as

$$S_x^\alpha(f) = \int_{-\infty}^{\infty} R_x^\alpha(\tau) \exp(-j2\pi f \tau)\, d\tau, \qquad (2.151)$$

where PSD is a special case of spectral correlation function when α is zero.

In practice, $S_x^\alpha(f)$ can also be calculated based on the two following steps [92]:

1.

$$X_T(t, f) = \int_{t-\frac{T}{2}}^{t+\frac{T}{2}} x(v) \exp(-j2\pi f v)\, dv. \qquad (2.152)$$

2.

$$S_x^\alpha(f) = \lim_{\Delta \to \infty} \lim_{T \to \infty} \frac{1}{\Delta} \frac{1}{T} \int_{-\frac{\Delta}{2}}^{\frac{\Delta}{2}} X_T\left(t, f + \frac{\alpha}{2}\right) X_T^*\left(t, f - \frac{\alpha}{2}\right) dt. \qquad (2.153)$$

Both cyclic autocorrelation function $R_x^\alpha(\tau)$ and spectral correlation function $S_x^\alpha(f)$ can be used as features to detect $x(t)$ [88, 92]. Assume $x(t)$ is the signal and $w(t)$ is AWGN. The observed signal $y(t)$ is equal to $x(t) + w(t)$. The optimal cyclostationary detector based on spectral correlation function can be [92–94],

$$z = \sum_\alpha \int S_x^{\alpha*}(f) S_y^\alpha(f)\, df. \qquad (2.154)$$

A novel approach to signal classification using spectral correlation and neural networks has been presented in [95]. α-profile is used as the feature in neural network for signal classification. The signal types under investigation include BPSK, QPSK, FSK, MSK, and AM [95]. α-profile is defined as [95],

$$\text{profile}(\alpha) = \max_f\{C_x^\alpha(f)\}, \qquad (2.155)$$

where $C_x^\alpha(f)$ is the spectral coherence function of $x(t)$ [95],

$$C_x^\alpha(f) = \frac{S_x^\alpha(f)}{\left(S_x^0\left(f + \frac{\alpha}{2}\right) S_x^{0*}\left(f - \frac{\alpha}{2}\right)\right)^{\frac{1}{2}}}. \qquad (2.156)$$

Similarly, signal classification based on spectral correlation analysis and SVM in cognitive radio has been presented in [96].

A low-complexity cyclostationary based spectrum sensing for UWB and WiMAX coexistence has been proposed in [97]. The cyclostationary property of WiMAX signals because of the cyclic prefix is used [97]. Cooperative cyclostationary spectrum sensing in cognitive radios has been discussed in [94, 98]. Cooperative spectrum sensing can improve the performance by the multiuser diversity [94]. The cyclostationary detector requires long detection time to obtain the feature, which causes inefficient spectrum utilization [99]. In order to address this issue, the sequential detection framework is applied together with the cyclostationary detector [99].

In an OFDM based cognitive radio system, cyclostationary signatures, which may be intentionally embedded in the communication signals, can be used to address a number of issues related to synchronization, blind channel identification, spectrum sharing and network coordination [100–102]. Hence, we can design cyclostationary signatures and the corresponding spectral correlation estimators for various situations and applications.

Besides, the blind source separation problem with the assumption that the source signals are cyclostationary has been studied in [103]. MMSE reconstruction for generalized undersampling of cyclostationary signals has been presented in [104]. Signal-selective direction of arrival (DOA) tracking for wideband cyclostationary sources has been discussed in [105]. Time difference of arrival (TDOA) and Doppler estimation for cyclostationary signals based on multicycle frequencies has been considered in [106].

3

Classical Detection

3.1 Formalism of Quantum Information

Fundamental Fact: The noise lies in a high-dimensional space; the signal, by contrast, lies in a much lower-dimensional space.

If a random matrix $\mathbf{A}$ has i.i.d rows $\mathbf{A}_i$, then $\mathbf{A}^*\mathbf{A} = \sum_i \mathbf{A}_i \mathbf{A}_i{}^T$ where $\mathbf{A}^*$ is the adjoint matrix of $\mathbf{A}$. We often study $\mathbf{A}$ through the $n \times n$ symmetric, positive semidefinite matrix, the matrix $\mathbf{A}^*\mathbf{A}$. The eigenvalues of $|\mathbf{A}| = \sqrt{\mathbf{A}^*\mathbf{A}}$ are therefore nonnegative real numbers.

An immediate application of random matrices is the fundamental problem of *estimating covariance matrices* of high-dimensional distributions [107]. The analysis of the row-independent models can be interpreted as a study of sample covariance matrices. For a general distribution in $\mathbb{R}^n$, its covariance can be estimated from a sample size of $N = O(n \log n)$ drawn from the distribution. For sub-Gaussian distributions, we have an even better bound $N = O(n)$. For low-dimensional distributions, much fewer samples are needed: if a distribution lies close to a subspace of dimension r in $\mathbb{R}^n$, then a sample of size $N = O(r \log n)$ is sufficient for covariance estimation.

There are deep results in random matrix theory. The main motivation of this subsection is to exploit the existing results in this field to better guide the estimate of covariance matrices, using nonasymptotic results [107].

3.2 Hypothesis Detection for Collaborative Sensing

The density operator (matrix) ρ is the basic building block. An operator ρ satisfies: (1) (**Trace condition**) ρ has trace equal to one, that is, $\mathrm{Tr}\rho = 1$; (2) (**Positivity**) ρ is a positive operator, that is, $\rho \geq 0$. Abusing terminology, we will use the term "positive" for "positive semide finite (denoted $A \geq 0$)." Covariance matrices satisfy the two necessary conditions. We say $\mathbf{A} > 0$ when $\mathbf{A}$ is a positive definite matrix; $\mathbf{B} > \mathbf{A}$ means that $\mathbf{B} - \mathbf{A}$ is a positive definite matrix. Similarly, we say this for nonnegative definite matrix for $\mathbf{B} - \mathbf{A} \geq 0$. The hypothesis test problems can be written as

$$\begin{aligned} \mathcal{H}_0 &: \mathbf{A} = \mathbf{R}_n \\ \mathcal{H}_1 &: \mathbf{B} = \mathbf{R}_S + \mathbf{R}_n, \end{aligned} \tag{3.1}$$

Cognitive Radio Communications and Networking: Principles and Practice, First Edition.
Robert C. Qiu, Zhen Hu, Husheng Li and Michael C. Wicks.
© 2012 John Wiley & Sons, Ltd. Published 2012 by John Wiley & Sons, Ltd.

where $\mathbf{R}_n$ is the covariance matrix of the noise and $\mathbf{R}_S$ that of the signal. The signal is assumed to be uncorrelated with the noise. From (3.1), it follows that $\mathbf{R}_S \geq 0$ and $\mathbf{R}_n > 0$. In our applications, this noise is modeled as additive so that if $x(n)$ is the "signal" and $w(n)$ the "noise," the recorded signal is

$$y(n) = x(n) + w(n).$$

Often, this additive noise is assumed to have zero mean and to be uncorrelated with the signal. In this case, the covariance of the measured data, $y(n)$, is the sum of the covariance of $x(n)$ and $w(n)$. Specifically, note that since

$$r_y(k, l) = E\{y(k)y^*(l)\} = E\{[x(k) + w(k)][x(l) + w(l)]^*\}$$
$$= E\{x(k)x^*(l)\} + E\{w(k)w^*(l)\} + E\{x(k)w^*(l)\} + E\{w(k)x^*(l)\},$$

if $x(n)$ and $w(n)$ are uncorrelated, then

$$E\{x(k)w^*(l)\} = E\{w(k)x^*(l)\} = 0$$

and it follows that

$$r_y(k, l) = r_x(k, l) + r_w(k, l). \tag{3.2}$$

Discrete-time random processes are often represented in matrix form. If

$$\mathbf{x} = [x(0), x(1), \ldots, x(p)]^T$$

is a vector of $p + 1$ values of a process $x(n)$, then the outer product

$$\mathbf{xx}^H = \begin{pmatrix} x(0)x^*(0) & x(0)x^*(1) & & x(0)x^*(p) \\ x(1)x^*(0) & x(1)x^*(1) & \cdots & \\ \vdots & & \ddots & \vdots \\ x(p)x^*(0) & x(p)x^*(1) & \cdots & x(p)x^*(p) \end{pmatrix}$$

is a $(p + 1) \times (p + 1)$ matrix. If $x(n)$ is wide-sense stationary, taking the expected value and using the Hermitian symmetry of the covariance sequence, $r_x(k) = r_x^*(-k)$, leads to the $(p + 1) \times (p + 1)$ matrix of covariance values

$$\mathbf{R}_x = E\{\mathbf{xx}^H\} = \begin{pmatrix} r_x(0) & r_x^*(1) & & r_x^*(p) \\ r_x(1) & r_x(0) & \cdots & r_x^*(p - 1) \\ \vdots & & \ddots & \vdots \\ r_x(p) & r_x(p - 1) & \cdots & r_x(0) \end{pmatrix} \tag{3.3}$$

referred to as the *covariance matrix*. The correlation matrix R_x has the following structure:

$$\mathbf{R}_x = r_x(0)\mathbf{I} + \bar{\mathbf{R}}_x, \quad \mathrm{Tr}\bar{\mathbf{R}}_x = 0, \tag{3.4}$$

where $\mathbf{I}$ is identity matrix and $\mathrm{Tr}\mathbf{A}$ represents the trace of $\mathbf{A}$. The covariance matrix has the following basic structures:

1. The covariance matrix of a WSS random process is a *Hermitian* Toeplitz matrix, $\mathbf{R}_x = \mathbf{R}_x{}^*$.
2. The covariance matrix of a WSS random process is *positive semidefinite*, $\mathbf{R}_x \geq 0$. In other words, the eigenvalues, λ_k, of this covariance matrix are real-valued and nonnegative, that is, $\lambda_k \geq 0$.

A complete list of properties is given in Table 3.1. When the mean values m_x and m_y are zero, the autocovariance and matrices are equal. We will always assume that all random processes have zero mean. Therefore, we use the two definitions interchangeably.

Example 3.1 (Covariance matrices for sinusoids and complex exponentials)

An important random process in radar and communications is the harmonic process. An example of a real-valued harmonic process is the random phase sinusoid, which is defined by

$$x(n) = A \sin (n\omega_0 + \phi),$$

where A and ω_0 are fixed constants and ϕ is a random variable that is uniformly distributed over interval $-\pi$ and π. The mean of this process can easily be shown to be zero. Thus, $x(n)$ is a zero mean process. The covariance of $x(n)$ is

$$r_x(k, l) = E\{x(k)x^*(l)\} = E\{A \sin (k\omega_0 + \phi) A \sin (l\omega_0 + \phi)\}.$$

Using the trigonometric identity

$$2 \sin A \sin B = \cos(A - B) - \cos(A + B),$$

we have

$$r_x(k, l) = \frac{1}{2}|A|^2 E\{\cos [(k - l)\omega_0)]\} - \frac{1}{2}|A|^2 E\{\cos [(k + l)\omega_0 + 2\phi)]\}.$$

Note that the first term is the expected value of a constant and the second term is equal to zero. Therefore,

$$r_x(k, l) = \frac{1}{2}|A|^2 \cos [(k - l)\omega_0)].$$

Table 3.1 Definition and properties for correlation and covariance matrices [108, p. 39]

Auto-correlation and covariance	$\mathbf{R}_x = \mathbb{E}\{xx^*\}$	$\mathbf{C}_x = \mathbb{E}\{(x - \mathbb{E}x)(x - \mathbb{E}x)^*\}$
Symmetry	$\mathbf{R}_x = \mathbf{R}_x{}^*$	$\mathbf{C}_x = \mathbf{C}_x{}^*$
Positive semidefinite	$\mathbf{R}_x \geq 0$	$\mathbf{C}_x \geq 0$
Interrelation	$\mathbf{R}_x = \mathbf{C}_x + m_x m_x{}^*$	$m_x = \mathbb{E}\{x\}, m_y = \mathbb{E}\{y\}$
Cross-correlation and cross-covariance	$\mathbf{R}_{xy} = \mathbb{E}\{xy^*\}$	$\mathbf{C}_{xy} = \mathbb{E}\{(x - m_x)(y - m_y)^*\}$
Relation to $\mathbf{R}_{yx}$ and $\mathbf{C}_{yx}$	$\mathbf{R}_{xy} = \mathbf{R}_{yx}$	$\mathbf{C}_{xy} = \mathbf{C}_{yx}$
Interrelation	$\mathbf{R}_{xy} = \mathbf{C}_{xy} + m_x m_y{}^*$	$m_x = \mathbb{E}\{x\}, m_y = \mathbb{E}\{y\}$
Orthogonal and uncorrelated	x, y orthogonal: $\mathbf{R}_{xy} = 0$	x, y uncorrelated: $\mathbf{C}_{xy} = 0$
Sum of	$\mathbf{R}_{x+y} = \mathbf{R}_x + \mathbf{R}_y$	$\mathbf{C}_{x+y} = \mathbf{C}_x + \mathbf{C}_y$
x and y	if x, y orthogonal	if x, y uncorrelated

As another example, consider the complex harmonic process

$$x(n) = Ae^{j(n\omega_0 + \phi)},$$

where, as with the random phase sinusoid, ϕ is a random variable that is uniformly distributed between $-\pi$ and π. The mean of this process is zero. The covariance is

$$r_x(k, l) = E\{x(k)x^*(l)\} = E\{Ae^{j(k\omega_0 + \phi)}A^*e^{-j(l\omega_0 + \phi)}\}$$
$$= |A|^2 E\{e^{j(k-l)\omega_0}\} = |A|^2 e^{j(k-l)\omega_0}$$

Consider a harmonic process consisting of L sinusoids

$$x(n) = \sum_{l=1}^{L} A_l \sin\left(n\omega_l + \phi_l\right).$$

Assuming the random variables ϕ_l and A_l are uncorrelated, the covariance sequence is

$$r_x(k) = \sum_{l=1}^{L} \frac{1}{2} E\{A_l^2\} \cos\left(k\omega_l\right).$$

The 2×2 covariance matrix for $L = 1$ sinusoid is

$$R_s = \frac{1}{2}|A|^2 \begin{bmatrix} 1 & \cos\omega_0 \\ \cos\omega_0 & 1 \end{bmatrix} = \frac{1}{2}|A|^2 \left(I + \sigma_1 \cos\omega_0\right).$$

The 2×2 covariance matrix for L sinusoids is

$$\mathbf{R_x} = \sum_{l=1}^{L} \frac{1}{2}E\{A_l^2\} \begin{bmatrix} 1 & \dfrac{1}{\sum_{l=1}^{L} \frac{1}{2}E\{A_l^2\}} \sum_{l=1}^{L} \frac{1}{2}E\{A_l^2\} \cos\left(\omega_l\right) \\ \dfrac{1}{\sum_{l=1}^{L} \frac{1}{2}E\{A_l^2\}} \sum_{l=1}^{L} \frac{1}{2}E\{A_l^2\} \cos\left(\omega_l\right) & 1 \end{bmatrix}$$

$$= a\left(\mathbf{I} + b\sigma_1\right),$$

where

$$a = \sum_{l=1}^{L} \frac{1}{2}E\{A_l^2\}, b = \frac{1}{\sum_{l=1}^{L} \frac{1}{2}E\{A_l^2\}} \sum_{l=1}^{L} \frac{1}{2}E\{A_l^2\} \cos\left(\omega_l\right).$$

As another example, consider the complex-valued process consisting of a sum of two complex exponentials

$$y(n) = Ae^{j(n\omega_1 + \phi_1)} + Ae^{j(n\omega_2 + \phi_2)}.$$

The covariance sequence for two uncorrelated processes is

$$r_x(k) = |A|^2 e^{jk\omega_1} + |A|^2 e^{jk\omega_2}.$$

The 2×2 covariance matrix for two complex exponentials is

$$\mathbf{R}_x = |A|^2 \begin{bmatrix} 2 & e^{-j\omega_1} + e^{-j\omega_2} \\ e^{-j\omega_1} + e^{-j\omega_2} & 2 \end{bmatrix} = \frac{1}{2}|A|^2 \left[\mathbf{I} + \left(\frac{e^{-j\omega_1} + e^{-j\omega_2}}{2} \right) \sigma_1 \right]. \quad \square$$

Example 3.2 (Covariance matrix for white noise)
The 2×2 covariance matrix for white additive noise is

$$\mathbf{R}_w = \sigma_w^2 \begin{pmatrix} 1 & 0 \\ 0 & 1 \end{pmatrix} = \sigma_w^2 \mathbf{I}.$$

In practice, we must deal with this form

$$\mathbf{R}_w = \sigma_w^2 \mathbf{I} + \sigma_w^2 \begin{pmatrix} x_{11} & x_{12} \\ x_{21} & x_{22} \end{pmatrix} = \sigma_w^2 \mathbf{I} + \sigma_w^2 \mathbf{X},$$

where the elements of $\mathbf{X}$ are approximately zero-mean random variables whose variances are 10 dB lower than the diagonal elements of $\mathbf{R}_w$. At low SNR, such as -20 dB, these random variables make $\mathbf{R}_w$ a random matrix. One realization example is

$$\mathbf{X} = \begin{pmatrix} 0.043579 & 0.10556 \\ 0.10556 & 0.14712 \end{pmatrix}. \quad \square$$

Let $\mathbf{A}$ be a Hermitian operator with $q\mathbf{I} \leq \mathbf{A} \leq Q\mathbf{I}$. The matrices $Q\mathbf{I} - \mathbf{A}$ and $\mathbf{A} - q\mathbf{I}$ are positive and commute with each other [109, p. 95]. Since $\mathbf{R}_w$ is positive (of course, Hermitian), we have

$$q\mathbf{I} \leq \mathbf{R}_w \leq Q\mathbf{I}.$$

The random matrix $\mathbf{X} = \frac{1}{\sigma_w^2} \mathbf{R}_w - \mathbf{I}$ is Hermitian, but not necessarily positive. $\mathbf{X}$ is Hermitian, since its eigenvalues must be real. The Hoffman-Wielandt is relevant in this context.

Lemma 3.1 (Hoffman-Wielandt) *[16, p. 21] Let $\mathbf{A}$ and $\mathbf{B}$ be $N \times N$ Hermitian matrices, with eigenvalues $\lambda_1^{\mathbf{A}} \leq \lambda_2^{\mathbf{A}} \leq \cdots \leq \lambda_N^{\mathbf{A}}$ and $\lambda_1^{\mathbf{B}} \leq \lambda_2^{\mathbf{B}} \leq \cdots \leq \lambda_N^{\mathbf{B}}$. Then,*

$$\sum_{i=1}^{N} \left| \lambda_i^{\mathbf{A}} - \lambda_i^{\mathbf{B}} \right| \leq \text{Tr}(\mathbf{A} - \mathbf{B})^2, \tag{3.5}$$

where $\mathbf{X}$ and $\mathbf{Y}$ are random symmetric matrices.

(3.5) can be used to bound the difference of the eigenvalues between $\mathbf{A}$ and $\mathbf{B}$.

3.3   Sample Covariance Matrix

In Section 3.2, the true covariance matrix is needed for hypothesis detection. In practice, we only have access to the sample covariance matrix which is a random matrix. We first present some basic definitions and properties related to a sample covariance matrix. The determinant of a random matrix $\mathbf{S}$, $\det \mathbf{S}$, also called generalized variance, is of special interest. It is an important measure of spread in multidimensional statistical analysis.

3.3.1   The Data Matrix

The general $(n \times N)$ data matrix will be denoted $\mathbf{X}$ or $\mathbf{X}(n \times N)$. The element in row i and column j is x_{ij}. We write the matrix $\mathbf{X} = (x_{ij})$. The rows of $\mathbf{X}$ will be written as

$$\mathbf{x}_1^T, \mathbf{x}_2^T, \ldots, \mathbf{x}_n^T.$$

Or

$$\mathbf{X} = \begin{bmatrix} \mathbf{x}_1^T \\ \mathbf{x}_2^T \\ \vdots \\ \mathbf{x}_n^T \end{bmatrix} = [\mathbf{x}_{(1)}, \mathbf{x}_{(2)}, \ldots, \mathbf{x}_{(N)}],$$

where

$$\mathbf{x}_i = \begin{bmatrix} x_{i1} \\ x_{i2} \\ \vdots \\ x_{iN} \end{bmatrix} (i = 1, 2, \ldots, n), \mathbf{x}_{(j)} = \begin{bmatrix} x_{1j} \\ x_{2j} \\ \vdots \\ x_{Nj} \end{bmatrix} (j = 1, 2, \ldots, N).$$

Example 3.3 (Random matrices)
MATLAB Code: $N=1000$; $X=randn(N,N)$; This code generates a random matrix of size 1000×1000.

r = randn(n) returns an n-by-n matrix containing pseudorandom values drawn from the standard normal distribution. randn returns a scalar. randn(size(A)) returns an array the same size as A.

(1) Generate values from a normal distribution with mean 1 and standard deviation 2.

$r = 1 + 2.*randn(100,1)$;

(2) Generate values from a bivariate normal distribution with specified mean vector and covariance matrix. $mu = [1,2]$; $Sigma = [1\ .5;\ .5\ 2]$; $R = chol(Sigma)$; $z = repmat(mu,100,1) + randn(100,2)*R$;   □

3.3.1.1   The Mean Vector and Covariance Matrix

The sample mean of the ith variable is

$$\bar{x}_i = \frac{1}{n} \sum_{l=1}^{n} x_{li}, \tag{3.6}$$

and the sample variance of the ith variable is

$$s_{ii} = \frac{1}{n} \sum_{l=1}^{n} (x_{li} - \bar{x}_i) = s_i^2, i = 1, \ldots, N. \tag{3.7}$$

The sample covariance between the ith and jth variable is

$$s_{ij} = \frac{1}{n} \sum_{l=1}^{n} (x_{li} - \bar{x}_i)(x_{lj} - \bar{x}_j). \tag{3.8}$$

The vector of means,

$$\bar{\mathbf{x}} = \begin{bmatrix} \bar{x}_1 \\ \bar{x}_2 \\ \vdots \\ \bar{x}_N \end{bmatrix},$$ (3.9)

is called the sample mean vector, or simply the "mean vector." The $N \times N$ matrix

$$\mathbf{S} = (s_{ij})$$

is called the sample covariance matrix, or simply "covariance matrix." It is more convenient to express the statistics in matrix notation. Corresponding to (3.6) and (3.9), we have

$$\bar{\mathbf{x}} = \frac{1}{n} \sum_{l=1}^{n} \mathbf{x}_l = \frac{1}{n} \mathbf{X}^T \mathbf{1},$$ (3.10)

where $\mathbf{1}$ is a column vector of n ones. On the other hand,

$$s_{ij} = \frac{1}{n} \sum_{l=1}^{n} x_{li} x_{lj} - \bar{x}_i \bar{x}_j,$$

so that

$$\mathbf{S} = \frac{1}{n} \sum_{l=1}^{n} (\mathbf{x}_l - \bar{\mathbf{x}})(\mathbf{x}_l - \bar{\mathbf{x}})^T = \frac{1}{n} \sum_{l=1}^{n} \mathbf{x}_l \mathbf{x}_l^T - \bar{\mathbf{x}} \bar{\mathbf{x}}^T.$$ (3.11)

This may be expressed as

$$\mathbf{S} = \frac{1}{n} \mathbf{X}^T \mathbf{X} - \bar{\mathbf{x}} \bar{\mathbf{x}}^T = \frac{1}{n} (\mathbf{X}^T \mathbf{X} - \frac{1}{n} \mathbf{X}^T \mathbf{1} \mathbf{1}^T \mathbf{X}),$$

using (3.10). Writing

$$\mathbf{H} = \mathbf{I} - \frac{1}{n} \mathbf{1} \mathbf{1}^T,$$

where $\mathbf{H}$ is called the centering matrix, we obtain the following standard form

$$\mathbf{S} = \frac{1}{n} \mathbf{X}^T \mathbf{H} \mathbf{X},$$ (3.12)

which is a convenient matrix representation of the sample covariance matrix. We need a total of nN points of samples to estimate the sample covariance matrix $\mathbf{S}$. Turning the table around, we can "summarize" information of nN points of samples into one single matrix $\mathbf{S}$. In spectrum sensing, we are given a long record of data about some random variables, or a random vector of large data dimensionality.

Example 3.4 (Representation of sample covariance matrix)

Given a total of 10^5 points of samples, how many sample covariance matrices are needed? Collect a data segment consisting of 1024 points to form N-dimensional vectors, where $N = 32$. These N-dimensional data vectors are used to form a sample covariance matrix $\mathbf{S}$ of $N \times N$.

By doing this, we have $K = 10^5/1025 = 97$ segments. From each segment, we estimate a sample covariance matrix. Thus we have $K = 97$ sample covariance matrices; in other words, a series of K matrices, $\mathbf{S}_1, \mathbf{S}_2, \cdots, \mathbf{S}_K$ are obtained.   □

Let us check the most important property of $\mathbf{S}$: $\mathbf{S}$ is a positive semidefinite matrix. Since $\mathbf{H}$ is a symmetric idempotent matrix: $\mathbf{H} = \mathbf{H}^T$, $\mathbf{H} = \mathbf{H}^2$, for any N-vector $\mathbf{a}$,

$$\mathbf{a}^T \mathbf{S} \mathbf{a} = \frac{1}{n} \mathbf{a}^T \mathbf{X}^T \mathbf{H}^T \mathbf{H} \mathbf{X} \mathbf{a} = \frac{1}{n} \mathbf{y}^T \mathbf{y} \geq 0,$$

where $\mathbf{y} = \mathbf{H}\mathbf{X}\mathbf{a}$. Thus, the covariance matrix $\mathbf{S}$ is positive semidefinite, writing

$$\mathbf{S} \geq 0.$$

For continuous data, we expect $\mathbf{S}$ is not only positive semidefinite, but positive definite, writing

$$\mathbf{S} > 0,$$

if $n \geq N + 1$.

It is often convenient to define the covariance matrix with a divisor of $n - 1$ instead of n. Set

$$\mathbf{S}_u = \frac{1}{n-1} \mathbf{X}^T \mathbf{H} \mathbf{X} = \frac{n}{n-1} \mathbf{S}.$$

If the data forms a random vector sample from a multivariate distribution, with finite second moments, then $\mathbf{S}_u$ is an *unbiased* estimate of the true covariance matrix. See Theorem 2.8.2 of [110, p. 50].

The sample correlation coefficient between the ith and the j variables is

$$\rho_{ij} = \frac{s_{ij}}{s_i s_j}.$$

Unlike s_{ij}, the correlation coefficient is **invariant** under both changes of scale and origin of the ith and the jth variables. This property is the foundation of detection of correlated structures among random vectors. Clearly,

$$0 \leq |\rho_{ij}| \leq 1,$$

where $|a|$ is the absolute value of a. Define the sample correlation matrix as

$$\Sigma = (\rho_{ij})$$

with $\rho_{ii} = 1$. It follows that

$$\Sigma \geq 0.$$

If $\Sigma = \mathbf{I}$, we say that the variables are uncorrelated. This is the case for white Gaussian noise. If $\mathbf{D} = \text{diag}(s_i)$, then

$$\Sigma = \mathbf{D}^{-1} \mathbf{S} \mathbf{D}^{-1}, \mathbf{S} = \mathbf{D} \Sigma \mathbf{D}.$$

3.3.1.2 Measure of Multivariate Scatter

The matrix $\mathbf{S}$ is one possible multivariate generation of the univariate notion of variance, measuring scatter above the mean. Physically, the variance is equivalent to the power of the random vector. For example, for a white Gaussian random variable, the variance is its power.

Sometimes, for example, for hypothesis testing problems, we would rather have a *single* number to measure multivariate scatter. Of course, the matrix $\mathbf{S}$ contains many more structures ("information") than this single real number. Two common such measures are

1. the *generalized variance* $\det \mathbf{S}$, or $|\mathbf{S}|$.
2. the *total variance*, $\mathrm{Tr}\mathbf{S}$.

A motivation for these measures is in principle component analysis (PCA) that will be treated later. For both measures, large values indicate a high degree of scatter about the mean vector $\bar{\mathbf{x}}$—physically larger power. Low values represent concentration about the mean vector $\bar{\mathbf{x}}$. Two different measures reflect different aspects of the variability in the data. The generalized variance plays an important role in maximum likelihood (ML) estimation while the total variance is a useful concept in principal component analysis. In the context of low SNR detection, it seems that the total variance is a more sensitive measure to decide on two alternative hypotheses.

The necessity for studying the (empirical) sample covariance matrix in statistics arose during 1950s when practitioners were searching for a scalar measure of dispersion for the multivariate data [111, Chapter 2]. This scalar measure of dispersion is relevant under the context of hypothesis testing. We need a scalar measure to set the threshold for testing.

3.3.1.3 Linear Combinations

Linear transformations can simplify the structure of the covariance matrix, making interpretation of the data more straightforward. Consider a linear combination

$$y_l = a_1 x_{l1} + a_2 x_{l2} + \cdots + a_N x_{lN}, l = 1, 2, \ldots, n,$$

where $a_1, \cdots, a_N$ are given. From (3.10), the mean is

$$\bar{y} = \frac{1}{n}\sum_{l=1}^n y_l = \frac{1}{n}\mathbf{a}^T\sum_{l=1}^n \mathbf{x}_l = \mathbf{a}^T\bar{\mathbf{x}},$$

and the variance is

$$s_y^2 = \frac{1}{n}\sum_{l=1}^n (y_l - \bar{y})^2 = \frac{1}{n}\sum_{l=1}^n \mathbf{a}^T(\mathbf{x}_l - \bar{\mathbf{x}})(\mathbf{x}_l - \bar{\mathbf{x}})^T\mathbf{a} = \mathbf{a}^T\mathbf{S}_x\mathbf{a},$$

where (3.11) is used.

For a q-dimensional linear transformation, we have

$$\mathbf{y}_l = \mathbf{A}\mathbf{x}_l + \mathbf{b}, l = 1, 2, \ldots, n,$$

which may be written as

$$\mathbf{Y} = \mathbf{X}\mathbf{A}^T + \mathbf{1}\mathbf{b}^T,$$

where $\mathbf{Y}$ is a $q \times q$ matrix and $\mathbf{b}$ is a q-vector. Usually, $q \leq N$.

The mean vector and covariance matrix of the new objects $\mathbf{y}_l$ are

$$\bar{\mathbf{y}} = \bar{\mathbf{x}} + \mathbf{b},$$

$$\mathbf{S}_y = \frac{1}{n} \sum_{l=1}^{n} (\mathbf{y}_l - \bar{\mathbf{y}})(\mathbf{y}_l - \bar{\mathbf{y}})^T = \mathbf{A}\mathbf{S}_x\mathbf{A}^T.$$

If $\mathbf{A}$ is nonsingular (so, in particular, $q = N$), then

$$\mathbf{S}_x = \mathbf{A}^{-1}\mathbf{S}_y(\mathbf{A}^T)^{-1} = \mathbf{A}^{-1}\mathbf{S}_y\mathbf{A}^{-T}.$$

Here we give three most important examples: (1) the scaling transform; (2) Mahalanobis transform; (3) principal component transformation (or analysis).

3.3.1.4   The Scaling Transform

The n vectors of dimension N are objects of interest. Define the scaling transform as

$$\mathbf{y}_l = \mathbf{D}^{-1}(\mathbf{x}_l - \bar{\mathbf{x}}), l = 1, 2, \ldots, n$$
$$\mathbf{D} = \mathrm{diag}(s_i).$$

This transformation scales each variable to have unit variance and thus eliminates the arbitrariness in the choice of scale. For example, if $\mathbf{x}_{(1)}$ measures lengths, then $\mathbf{y}_{(1)}$ will be the same. We have

$$\mathbf{S}_y = \Sigma.$$

3.3.1.5   Mahalanobis Transformation

If $\mathbf{S} > 0$, then $\mathbf{S}^{-1}$ has a unique symmetric positive definite square root $\mathbf{S}^{-1/2}$. See A.6.15 of [110]. We define the Mahalanobis transformation as

$$\mathbf{z}_l = \mathbf{S}_x^{-1/2}(\mathbf{x}_l - \bar{\mathbf{x}}), l = 1, 2, \ldots, n.$$

Then

$$\mathbf{S}_z = \mathbf{I},$$

so that this transformation eliminates the correlation between the variables and standardizes the variance of each variable.

3.3.1.6   Principle Component Analysis

In the era of high-dimensionality data processing, PCA is extremely important to reduce the dimension of the data. One is motivated to summarize the total variance using much fewer dimensions. The notion of rank of the data matrix occurs naturally in this context. For zero-mean random vector, it follows, from (3.12), that

$$\mathbf{S} = \frac{1}{n}\mathbf{X}^T\mathbf{X}.$$

This mathematical structure plays a critical role in its applications.

By spectrum decomposition theorem, the covariance matrix $\mathbf{S}$ may be written as

$$\mathbf{S} = \mathbf{U}\Lambda\mathbf{U}^T,$$

where $\mathbf{U}$ is an orthogonal matrix and Λ is a diagonal matrix of the eigenvalues of $\mathbf{S}$,

$$\Lambda = \text{diag}\left[\, \lambda_1 \;\; \lambda_2 \;\; \cdots \;\; \lambda_N \,\right].$$

The principal component transformation is defined by the unitary *rotation*

$$\mathbf{w}_l = \mathbf{U}^T(\mathbf{x}_l - \bar{\mathbf{x}}),\, l = 1, 2, \ldots, N.$$

Since

$$\mathbf{S}_w = \mathbf{U}^T \mathbf{S}_x \mathbf{U} = \Lambda,$$

the columns of $\mathbf{W}$, called principal components, represent *uncorrelated* linear combinations of the variables. In practice, one hopes to summarize most of the variability in the data using only the principal components with the highest variances, thus reducing the dimensions. This approach is the standard benchmark for dimensionality reduction.

The principal components are *uncorrelated* with variances

$$\lambda_1, \lambda_2, \cdots, \lambda_N.$$

It seems natural to define the "overall" spread of the data by some symmetric monotonically increasing function of $\lambda_1, \lambda_2, \cdots, \lambda_n$, such as the geometric mean and the arithmetic mean

$$\prod_{i=1}^{N} \lambda_i \quad \text{or} \quad \sum_{i=1}^{N} \lambda_i$$

Using the properties of linear algebra, we have

$$\det \mathbf{S}_x = \det \Lambda = \prod_{i=1}^{N} \lambda_i,$$

$$\text{Tr}\mathbf{S}_x = \text{Tr}\Lambda = \sum_{i=1}^{N} \lambda_i.$$

We have used the facts for a $N \times N$ matrix

$$\det \mathbf{A} = \prod_{i=1}^{N} \lambda_i, \quad \text{Tr}\mathbf{A} = \sum_{i=1}^{N} \lambda_i,$$

where λ_i is the eigenvalues of the matrix $\mathbf{A}$. See [110, A.6] or [112], since the geometric mean of the nonnegative sequence is always smaller than the arithmetic mean of the nonnegative sequence, or [113]

$$(a_1 a_2 \cdots a_n)^{1/n} \leq \frac{a_1 + a_2 + \cdots + a_n}{n},$$

where a_i are nonnegative real numbers. Besides, the special structure of $\mathbf{S} \geq 0$, implies that all eigenvalues are nonnegative [114, p. 160]

$$\lambda_i(\mathbf{S}) \geq 0.$$

The arithmetic mean-geometric mean inequality is thus valid for our case. Finally we obtain

$$(\det \mathbf{S}_x)^{\frac{1}{N}} \le \frac{1}{N} \mathrm{Tr} \mathbf{S}_x. \tag{3.13}$$

The rotation to principal components provides a motivation for the measures of multivariate scatter. Let us consider one application in spectrum sensing. The key idea behind covariance-based primary user's signal detection that the primary user signal received at the CR user is usually correlated because of the dispersive channels, the utility of multiple receiver antennas, or even oversampling. Such correlation can be used by the CR user to differentiate the primary signal from white noise.

Since $\mathbf{S}_x$ is a random matrix, $\det \mathbf{S}_x$ and $\mathrm{Tr} \mathbf{S}_x$ are scalar random variables. Girko studied random determinants [111]. (5.21) relates the determinant to the trace of the random matrix $\mathbf{S}_x$. In Chapter 4, the tracial functions of $\mathbf{S}_x$ are commonly encountered.

Example 3.5 (Covariance-based detection)
The received signal is

$$y(n) = \theta s(n) + w(n), 0 \le n \le N - 1$$

where $\theta = 1$ and $\theta = 0$ denote the presence and absence of the primary signal, respectively. The sample covariance matrix of the received signal is estimated as

$$\hat{\mathbf{R}}_y = \frac{1}{N} \sum_{n=1}^{N} \mathbf{y}[n] \mathbf{y}^H[n] \tag{3.14}$$
$$\mathbf{y}[n] = \left[y[n], y[n-1], \dots, y[n-L+1] \right]^T.$$

When the number of samples N approaches infinity, $\hat{\mathbf{R}}_y$ converges in probability at

$$\mathbf{R}_y = E\left\{ \mathbf{y}[n] \mathbf{y}[n]^H \right\} = \theta \mathbf{R}_s + \mathbf{R}_w,$$

where $\mathbf{R}_s$ and $\mathbf{R}_w$ are, respectively, the $L \times L$ covariance matrices of the primary signal vector and the noise vector

$$\mathbf{s}[n] = [s[n], s[n-1], \dots, s[n-L+1]]^T,$$
$$\mathbf{w}[n] = [w[n], w[n-1], \dots, w[n-L+1]]^T.$$

Our standard problem is

$$\mathcal{H}_0 : \mathbf{R}_x = \mathbf{R}_w$$
$$\mathcal{H}_1 : \mathbf{R}_x = \mathbf{R}_s + \mathbf{R}_w, \tag{3.15}$$

where $\mathbf{R}_s$ and $\mathbf{R}_w$ are, respectively, covariance matrices of signal and noise.

Based on the sample covariance matrix $\hat{\mathbf{R}}_y$, various test statistics can be used. Let $\mu_{\min}$ and $\mu_{\max}$ denote the minimum and maximum eigenvalues of $\hat{\mathbf{R}}_y$. Then

$$\mathcal{H}_0 : \sigma_n^2 \le \lambda_i \le \sigma_n^2$$
$$\mathcal{H}_1 : \alpha_{\min} + \sigma_n^2 \le \lambda_i \le \alpha_{\max} + \sigma_n^2,$$

where α_{max} and α_{min} are the maximum and minimum eigenvalues of $\mathbf{R}_s$ and $\mathbf{R}_w = \sigma_n^2 \mathbf{I}_L$ is assumed, where σ_n^2 is the noise power and $\mathbf{I}_L$ is the $L \times L$ identity matrix. Because of the correlation among the sampled signal, $\alpha_{max} > \alpha_{min}$. Thus, if there is no primary signal

$$\frac{\mu_{max}}{\mu_{min}} = 1,$$

otherwise

$$\frac{\mu_{max}}{\mu_{min}} > 1.$$

Based on the above heuristic, the max-min eigenvalue algorithm is formulated as follows:

1. Estimate the covariance matrix of the received signal according to (3.14).
2. Calculate the ratio of the the maximum and minimum eigenvalues.
3. If the ratio $\frac{\mu_{max}}{\mu_{min}} > 1$, claim $\mathcal{H}_1$ otherwise claim $\mathcal{H}_0$.   □

The max-min eigenvalue algorithm is simple and has a fairly good performance under the context of low SNR. At extremely low SNR, say $-25\,\text{dB}$, the calculated eigenvalues of the sample covariances matrix are random and look identical. This algorithm breaks down as a result of this phenomenon. Note that the eigenvalues are the variances of the principal components. The problem is that this algorithm depends on the variance of one dimension (associated with the minimum or the maximum eigenvalues).

The variances of different components are uncorrelated random variables. It is thus more natural to use the total variance or total variation.

3.4   Random Matrices with Independent Rows

We focus on a general model of random matrices, where we only assume independence of the rows rather than all entries. Such matrices are naturally generated by high-dimensional distributions. Indeed, given an arbitrary probability distribution in $\mathbb{R}^n$, one takes a sample of N independent points and arranges them as the rows of an $N \times n$ matrix $\mathbf{A}$. Recall that n is the dimension of the probability space.

Let $\mathbf{X}$ be a random vector in $\mathbb{R}^n$. For simplicity we assume that $\mathbf{X}$ is centered, or $\mathbb{E}\mathbf{X} = 0$. Here $\mathbb{E}\mathbf{X}$ denotes the expectation of $\mathbf{X}$. The covariance matrix of $\mathbf{X}$ is the $n \times n$ matrix $\Sigma = \mathbb{E}\mathbf{X}\mathbf{X}^T$. The simplest way to estimate Σ is to take some N independent samples $\mathbf{X}_i$ from the distribution and form the sample covariance matrix $\Sigma_N = \frac{1}{N}\sum_{i=1}^{N} \mathbf{X}_i\mathbf{X}_i^T$. By the law of large numbers, $\Sigma_N \to \Sigma$ almost surely as $N \to \infty$. So, taking sufficiently many samples we are guaranteed to estimate the covariance matrix as well as we want. This, however, does not address the quantitative aspect: what is the minimal *sample size* N that guarantees approximation with a given accuracy?

The relation of this question to random matrix theory becomes clear when we arrange the samples $\mathbf{X}_i =: \mathbf{A}_i$ as rows of the $N \times n$ random matrix $\mathbf{A}$. Then, the sample covariance matrix is expressed as $\Sigma_N = \frac{1}{N}\mathbf{A}^*\mathbf{A}$. Note that $\mathbf{A}$ is a matrix with independent rows but usually not independent entries. The reference of [107] has worked out the analysis of such matrices, separately for sub-Gaussian and general distributions.

We often encounter covariance estimation for sub-Gaussian distribution due to the presence of Gaussian noise. Consider a sub-Gaussian distribution in $\mathbb{R}^n$ with covariance matrix Σ, and let $\varepsilon \in (0, 1)$, $t \geq 1$. Then with probability at least $1 - 2\exp(-t^2 n)$ one has

$$\text{If } N \geq C(t/\varepsilon)^2 n, \text{ then } ||\Sigma_N - \Sigma|| \leq \varepsilon. \tag{3.16}$$

Here C depends only on the sub-Gaussian norm of a random vector taken from this distribution; the spectral norm of $\mathbf{A}$ is denoted as $||\mathbf{A}||$, which is equal to the maximum singular value of A, that is, $s_{\max} = ||\mathbf{A}||$.

Covariance estimation for arbitrary distribution is also encountered when a general noise interference model is used. Consider a sub-Gaussian distribution in $\mathbb{R}^n$ with covariance matrix Σ and supported in some centered Euclidean ball whose radius we denote $\sqrt{m}$. Let $\varepsilon \in (0, 1)$ and $t \geq 1$. Then, with probability at least $1 - n^{-t^2}$, one has

$$\text{If } N \geq C(t/\varepsilon)^2 ||\Sigma||^{-1} m \log n, \text{ then } ||\Sigma_N - \Sigma|| \leq \varepsilon ||\Sigma||. \tag{3.17}$$

Here C is an absolute constant and log denotes the natural logarithm. In (3.17), typically $m = O(||\Sigma||n)$. The required sample size is $N \geq C(t/\varepsilon)^2 n \log n$.

Low rank estimation is used, since the distribution of a signal in $\mathbb{R}^n$ lies close to a low-dimensional subspace. In this case, a much smaller sample size suffices for covariance estimation. The intrinsic dimension of the distribution can be measured with the *effective rank* of the matrix Σ, defined as

$$r(\Sigma) = \frac{\text{Tr}(\Sigma)}{||\Sigma||} \tag{3.18}$$

where $\text{Tr}(\Sigma)$ denotes the trace of Σ. One always has $r(\Sigma) \leq \text{rank}(\Sigma) \leq n$, and this bound is sharp. The effective rank $r = r(\Sigma)$ always controls the typical norm of X, as $\mathbb{E}||X||_2^2 = \text{Tr}(\Sigma) = r\Sigma$. Most of the distribution is supported in a ball of radius $\sqrt{m}$ where $m = O(r||\Sigma||)$. The conclusion of (3.17) holds with sample size $N \geq C(t/\varepsilon)^2 r \log n$.

Summarizing the above discussion, (3.16) shows that the sample size $N = O(n)$ suffices to approximate the covariance matrix of a sub-Gaussian distribution in $\mathbb{R}^n$ by the sample covariance matrix. While for arbitrary distribution, $N = O(n \log n)$ is sufficient. For distributions that are approximately low-dimensional, such as that of a signal, a much smaller sample size is sufficient. Namely, if the effective rank of Σ equals r, then a sufficient size is $N = O(r \log n)$.

Each observation of the sample covariance matrix is a random matrix. We can study the expectation of the observed random matrix. Since the expectation of a random matrix can be viewed as a convex combination, and also the positive semidefinite cone is convex [115, p. 459], expectation preserves the semidefinite order [116]:

$$\mathbf{B} \geq \mathbf{A} \geq 0 \text{ implies } \mathbb{E}\mathbf{B} \geq \mathbb{E}\mathbf{A}. \tag{3.19}$$

Noncommunicativity of two sample covariance matrices: If positive matrices $\mathbf{X}$ and $\mathbf{Y}$ commute, then the symmetrized product is: $\mathbf{X} \circ \mathbf{Y} = \frac{1}{2}(\mathbf{XY} + \mathbf{YX}) \geq 0$, which is not true,[1] if we deal with two sample covariances. A simple MATLAB simulation using two random

---

[1] This is true, if we know the *true covariance matrices*, rather than the sample covariance matrices.

matrices can verify this observation. It turns out that this observation is of an elementary nature: Quantum information is built upon this noncommunicativity of operators (matrices). If the matrices A and B commute, the problem of (3.1) is equivalent to the classical likelihood ratio test [117]. A unifying framework including classical and quantum hypothesis testing (first suggested in [117]) is developed here.

When only N samples are available, the sample covariance matrices can be used to approximate the actual ones. A random vector $\mathbf{X} \in \mathbb{R}^n$ is used to model the noise or interference. Similarly, a random vector $\mathbf{S} \in \mathbb{R}^n$ models the signal. In other words, (3.1) becomes

$$\mathcal{H}_0 : \mathbf{A} = \frac{1}{N} \sum_{i=1}^{N} \mathbf{X}_i \mathbf{X}_i^T = \hat{\mathbf{R}}_n$$

$$\mathcal{H}_1 : \mathbf{B} = \frac{1}{N} \sum_{i=1}^{N} \mathbf{S}_i \mathbf{S}_i^T + \frac{1}{N} \sum_{i=1}^{N} \mathbf{X}_i \mathbf{X}_i^T + \frac{1}{N} \sum_{i=1}^{N} \mathbf{S}_i \mathbf{X}_i^T + \frac{1}{N} \sum_{i=1}^{N} \mathbf{X}_i \mathbf{S}_i^T \qquad (3.20)$$

$$= \hat{\mathbf{R}}_S + \hat{\mathbf{R}}_n + \hat{\mathbf{R}}_{SX} + \hat{\mathbf{R}}_{XS},$$

where $\hat{\mathbf{R}}_S \geq 0; \hat{\mathbf{R}}_n > 0; \mathbf{A} > 0$.

For any $\mathbf{A} \geq 0$, all the eigenvalues of $\mathbf{A}$ are nonnegative. Since some eigenvalues of $\hat{\mathbf{R}}_{SX}$ and $\hat{\mathbf{R}}_{XS}$ are negative, they are indefinite matrices of small tracial values. Under extremely low signal-to-noise ratios, the positive term (signal) $\hat{S}$ in (3.20) is extremely small, compared with the other three terms. All these matrices are random matrices with dimension n.

Our motivation is to use the fundamental relation of (3.19). Consider (sufficiently large) K i.i.d. observations of random matrices $\mathbf{A}$ and $\mathbf{B}$:

$$\mathcal{H}_0 : \mathbb{E}\mathbf{A} \approx \frac{1}{K} \sum \mathbf{A}_k; \mathcal{H}_1 : \mathbb{E}\mathbf{B} \approx \frac{1}{K} \sum \mathbf{B}_k. \qquad (3.21)$$

The problem at hand is how the fusion center combines the information from these K observations. The justification for using the expectation is based on the basic observation of (3.20): expectation increases the effective signal-to-noise ratio. For the K observations, the signal term experiences coherent summation, while these other three random matrices go through incoherent summation.

Simulations: In (3.20), $\mathrm{Tr}\hat{\mathbf{R}}_{SX} + \mathrm{Tr}\hat{\mathbf{R}}_{XS}$ is no bigger than 0.5, so they do not have significant influence on the gap between $\mathrm{Tr}(\hat{\mathbf{R}}_S + \hat{\mathbf{R}}_n)$ and $\mathrm{Tr}\hat{\mathbf{R}}_n$. Figure 3.1 shows that this gap is very stable. A narrowband signal is used. The covariance matrix $\hat{\mathbf{R}}_S$ is 4×4. About 25 observations are sufficient to recover this matrix with acceptable accuracy. Two independent experiments are performed to obtain $\hat{\mathbf{R}}_n$ and $\hat{\mathbf{R}}_{n_0}$. In our algorithms, we need to set the threshold first for the hypothesis test of $\mathcal{H}_1$; we rely on $\hat{\mathbf{R}}_{n_0}$ for $\mathcal{H}_0$. To obtain every point in the plot, $N = 600$ is used in (3.20).

Denote the set of positive-definite matrices by $\mathbb{F}^{n \times n}$. The following theorem [115, p. 529] provides a framework: Let $\mathbf{A}, \mathbf{B} \in \mathbb{F}^{n \times n}$, assume that $\mathbf{A}$ and $\mathbf{B}$ are positive semidefinite, assume that $\mathbf{A} \leq \mathbf{B}$, assume that $f(0) = 0$, f is continuous, and f is increasing. Then,

$$\mathrm{Tr} f(\mathbf{A}) \leq \mathrm{Tr} f(\mathbf{B}). \qquad (3.22)$$

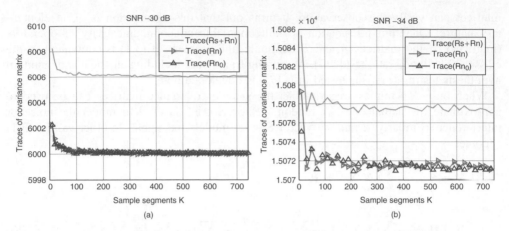

Figure 3.1 The traces of covariances at extremely low SNR as a function of K observations. (a) SNR $= -30$ dB; (b) SNR $= -34$ dB.

A trivial case is: $f(x) = x$. If $\mathbf{A}$ and $\mathbf{B}$ are random matrices, combining (3.22) with (3.19) leads to the final equation

$$\mathrm{Tr}\, f(\mathbb{E}\mathbf{A}) \leq \mathrm{Tr}\, f(\mathbb{E}\mathbf{B}). \tag{3.23}$$

Algorithm 3.1 *(1) Claim hypothesis $\mathcal{H}_1$ if matrix inequality (3.22) is satisfied; (2) otherwise, $\mathcal{H}_0$ is claimed.*

Consider a general Gaussian detection problem: $\mathcal{H}_0 : \mathbf{x} = \mathbf{w}, \mathcal{H}_1 : \mathbf{x} = \mathbf{s} + \mathbf{w}$ where $\mathbf{w} \sim \mathcal{N}(\mathbf{0}, \mathbf{C}_w)$, $\mathbf{x} \sim \mathcal{N}(\boldsymbol{\mu}_s, \mathbf{C}_s)$, and $\mathbf{s}$ and $\mathbf{w}$ are independent. The Neyman-Pearson detector decides $\mathcal{H}_1$ if $\frac{p(\mathbf{x};\mathcal{H}_1)}{p(\mathbf{x};\mathcal{H}_1)} > \gamma$. This LRT leads to a structure of a prewhitener followed by an EC [118, p. 167]. In our simulation, we assume that we know perfectly the signal and noise covariance matrices. This serves as the upper limit for the LRT detector. It is amazing to discover that Algorithm 3.1 outperforms the LRT by several dBs!

Related Work: Several sample covariance matrix based algorithms have been proposed in spectrum sensing. Maximum-minimum eigenvalue (MME)[119] and arithmetic-to-geometric mean (AGM) [120] uses the eigenvalues information, while feature template matching (FTM) [121] uses eigenvectors as prior knowledge. All these algorithms are based on covariance matrices. All the thresholds are determined by probability of false alarm.

Preliminary Results Using Algorithm 3.1: Sinusoidal signals and DTV signals captured in Washington D.C are used. For each simulation, zero-mean i.i.d. Gaussian noise is added according to different SNR. 2,000 simulations are performed on each SNR level. The threshold obtained by Monte Carlo simulations is in perfect agreement with that of the derived expression. The number of total samples contained in the segment is $N_s = 100,000$ (corresponding to about 5 ms sampling time). The smoothing factor L is chosen to be 32. Probability of false alarm is fixed with $P_{fa} = 10\%$. For simulated sinusoidal signal, the parameters are set the same.

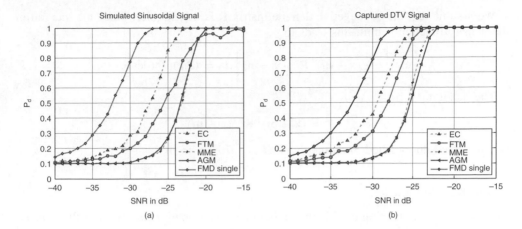

Figure 3.2 Probability of detection. (a) Simulated Narrowband Signal; (b) Measured DTV Data.

Hypothesis detection using a function of matrix detection (FMD) is based on (3.23). For more details, we see [122]. It is compared with the benchmark EC, together with AGM, FTM, MME, as shown in Figure 3.2. FMD is 3 dB better than EC. While using simulated sinusoidal signal, the gain between FMD and EC is 5 dB. The longer the data, the bigger this gain.

3.5 The Multivariate Normal Distribution

The multivariate normal (MVN) distribution is the most important distribution in science and engineering [123, p. 55]. The reasons are manifold: Central limit theorems make it the limiting distributions for some sums of random variables, its marginal distributions are normal, linear transformations of multivariate normals are also multivariate normal, and so on. Let

$$\mathbf{X} = \begin{bmatrix} X_1 & X_2 & \cdots & X_N \end{bmatrix}^T$$

denote an $N \times 1$ random vector. The mean of $\mathbf{X}$ is

$$\mathbf{m} = E\mathbf{X} = \begin{bmatrix} m_1 & m_2 & \cdots & m_N \end{bmatrix}^T.$$
$$m_i = EX_i.$$

The covariance matrix of $\mathbf{X}$ is

$$\mathbf{R} = E(\mathbf{X} - \mathbf{m})(\mathbf{X} - \mathbf{m})^T = \{r_{ij}\},$$
$$r_{ij} = E(X_i - m_i)(X_j - m_j).$$

The random vector $\mathbf{X}$ is called to be multivariate normal if its density function is

$$f(\mathbf{x}) = \frac{1}{(2\pi)^{N/2}(\det \mathbf{R})^{1/2}} \exp\left[-\frac{1}{2}(\mathbf{x} - \mathbf{m})^T \mathbf{R}^{-1}(\mathbf{x} - \mathbf{m}) \right]. \tag{3.24}$$

We assume that the nonnegetive definite matrix $\mathbf{R}$ is nonsingular. Since the integral of the density function is unit, this leads to

$$\int \exp\left[-\frac{1}{2}(\mathbf{x} - \mathbf{m})^T \mathbf{R}^{-1}(\mathbf{x} - \mathbf{m})\right] d\mathbf{x} = (2\pi)^{N/2}(\det \mathbf{R})^{1/2}.$$

The quadratic form

$$d^2 = (\mathbf{x} - \mathbf{m})^T \mathbf{R}^{-1}(\mathbf{x} - \mathbf{m})$$

is a weighted norm called *Mahalanobis* distance from $\mathbf{x}$ to $\mathbf{m}$.

Characteristic Function

The characteristic function of $\mathbf{X}$ is the multidimensional Fourier transform of the density

$$\Phi(\boldsymbol{\omega}) = Ee^{-j\boldsymbol{\omega}^T \mathbf{X}} = \int d\mathbf{x} \frac{1}{(2\pi)^{N/2}(\det \mathbf{R})^{1/2}} \exp\left[-j\boldsymbol{\omega}^T \mathbf{X} - \frac{1}{2}(\mathbf{x} - \mathbf{m})^T \mathbf{R}^{-1}(\mathbf{x} - \mathbf{m})\right].$$

By some manipulation [123], we have

$$\boldsymbol{\omega} = \exp\left\{-j\boldsymbol{\omega}^T \mathbf{m} - \frac{1}{2}\boldsymbol{\omega}^T \mathbf{R}\boldsymbol{\omega}\right\}. \tag{3.25}$$

The characteristic function itself is a multivariate normal function of the frequency variable $\boldsymbol{\omega}$.

Linear Transforms

Let $\mathbf{Y}$ be a linear transformation of a multivariate normal random variable:

$$\mathbf{Y} = \mathbf{A}^T \mathbf{X}$$
$$\mathbf{A}^T : m \times N(m \leq N).$$

The characteristic function of $\mathbf{Y}$ is

$$\Phi(\boldsymbol{\omega}) = Ee^{-j\boldsymbol{\omega}^T \mathbf{Y}} = Ee^{-j\boldsymbol{\omega}^T \mathbf{A}^T \mathbf{X}} = \exp\left\{-j\boldsymbol{\omega}^T \mathbf{A}^T \mathbf{m} - \frac{1}{2}\boldsymbol{\omega}^T \mathbf{A}^T \mathbf{R}\mathbf{A}\boldsymbol{\omega}\right\}.$$

Thus, $\mathbf{Y}$ is also a multivariate normal random variable with a new mean vector and new variance matrix

$$\mathbf{Y} = \mathbf{A}^T \mathbf{X} : \mathcal{N}[\mathbf{A}^T \mathbf{m}, \mathbf{A}^T \mathbf{R}\mathbf{A}]$$

if the matrix $\mathbf{A}^T \mathbf{R}\mathbf{A}$ is nonsingular.

Diagonalizing Transforms

The correlation matrix is symmetric and nonnegative definite. In other words, $R \geq 0$. Therefore, there exists an orthogonal matrix $\mathbf{U}$ such that

$$\mathbf{U}^T \mathbf{R}\mathbf{U} = \text{diag}\,[\lambda_1^2 \quad \cdots \quad \lambda_N^2].$$

The vector $\mathbf{Y} = \mathbf{U}^T \mathbf{X}$ is distributed as

$$\mathbf{Y} = \mathbf{U}^T \mathbf{X} : \mathcal{N}[\mathbf{U}^T \mathbf{m}, \operatorname{diag}[\lambda_1^2 \quad \cdots \quad \lambda_N^2]].$$

The random variables $Y_1, Y_2, \ldots, Y_N$ are uncorrelated since

$$E(\mathbf{Y} - \mathbf{U}^T \mathbf{m})(\mathbf{Y} - \mathbf{U}^T \mathbf{m})^T = \mathbf{U}^T \mathbf{R} \mathbf{U} = \operatorname{diag}\left[\lambda_1^2 \quad \cdots \quad \lambda_N^2\right].$$

In fact, Y_n are independent normal random variables with mean $\mathbf{U}^T \mathbf{m}$ and variances λ_n^2:

$$f(\mathbf{y}) = \prod_{n=1}^{N} (2\pi \lambda_n^2)^{-1/2} \exp\left\{-\frac{1}{2\lambda_n^2} [y_n - (\mathbf{U}^T \mathbf{m})_n]^2\right\}.$$

This transformation $\mathbf{Y} = \mathbf{U}^T \mathbf{X}$ is called a *Karhunen-Loeve* or *Hotelling* transform. It simply diagonalizes the covariance matrix

$$\mathbf{R} : \mathbf{U}^T \mathbf{R} \mathbf{U} = \Lambda^2.$$

Such a transform can be implemented in MATLAB, using a function called *eig* or *svd*.

Quadratic Forms in MVN Random Variables

Linear functions of MVN random vectors remain MVN. In GLRT, quadratic forms of MVNs are involved. The natural question is what about quadratic forms of MVNs? In some important cases, the quadratic forms have a χ^2 distributions. Let $\mathbf{X}$ denote an $\mathcal{N}[\mathbf{m}, \mathbf{R}]$ random variable. The distribution

$$Q = (\mathbf{X} - \mathbf{m})^T \mathbf{R}^{-1} (\mathbf{X} - \mathbf{m})$$

is χ_N^2 distributed. The characteristic function of Q is

$$\Phi(\omega) = E e^{-j\omega Q} = \int d\mathbf{x} \exp\left[-j\omega(\mathbf{X} - \mathbf{m})^T \mathbf{R}^{-1}(\mathbf{X} - \mathbf{m})\right]$$

$$\times \frac{1}{(2\pi)^{N/2}(\det \mathbf{R})^{1/2}} \exp\left[-\frac{1}{2}(\mathbf{x} - \mathbf{m})^T \mathbf{R}^{-1}(\mathbf{x} - \mathbf{m})\right]$$

$$= \int d\mathbf{x} \frac{1}{(1 + 2j\omega)^{N/2}} \frac{1}{(2\pi)^{N/2}(\det \mathbf{R})^{1/2}} (1 + 2j\omega)^{N/2}$$

$$\times \exp\left[-\frac{1}{2}(\mathbf{x} - \mathbf{m})^T \mathbf{R}^{-1}(\mathbf{I} + 2j\omega\mathbf{I})(\mathbf{x} - \mathbf{m})\right]$$

$$= \frac{1}{(1 + 2j\omega)^{N/2}}.$$

which is the characteristic function of a chi-squared distribution with N degrees of freedom, denoted χ_N^2. The density function for Q is the inverse Fourier transform

$$f(q) = \frac{1}{\Gamma(N/2)2^{N/2}} q^{(N/2)-1} e^{-q/2}; \quad q \geq 0.$$

The mean and variance of Q are obtained from the characteristic function

$$EQ = N$$
$$\text{Var}Q = 2N.$$

Sometimes we encounter more general quadratic forms in the symmetric matrix $\mathbf{P}$:

$$Q = (\mathbf{X} - \mathbf{m})^T \mathbf{P}(\mathbf{X} - \mathbf{m})$$
$$\mathbf{X} : N[\mathbf{m}, \mathbf{R}].$$

The mean and variance of Q are

$$EQ = \text{Tr}\mathbf{PR}$$
$$\text{Var} = 2\text{Tr}(\mathbf{PR})^2.$$

The characteristic function of Q is

$$\Phi(\omega) = \int d\mathbf{x} \frac{1}{(2\pi)^{N/2}(\det \mathbf{R})^{1/2}} \exp\left[-\frac{1}{2}(\mathbf{x} - \mathbf{m})^T (\mathbf{I} + 2j\omega \mathbf{PR})(\mathbf{x} - \mathbf{m}) \right]$$

$$= \int d\mathbf{x} \frac{1}{(2\pi)^{N/2}} \frac{1}{\{\det [\mathbf{R}(\mathbf{I} + 2j\omega \mathbf{PR})^{-1}]\}^{1/2}} \frac{1}{[\det (\mathbf{I} + 2j\omega \mathbf{PR})]^{1/2}}$$

$$\times \exp\left[-\frac{1}{2}(\mathbf{x} - \mathbf{m})^T (\mathbf{I} + 2j\omega \mathbf{PR})(\mathbf{x} - \mathbf{m}) \right]$$

$$= \frac{1}{[\det (\mathbf{I} + 2j\omega \mathbf{PR})]^{1/2}}.$$

If $\mathbf{PR}$ is symmetric: $\mathbf{PR} = \mathbf{RP}$, the characteristic function is

$$\Phi(\omega) = \frac{1}{\displaystyle\prod_{n=1}^{N} (1 + 2j\omega\lambda_n)^{1/2}},$$

where λ_n are eigenvalues of $\mathbf{PR}$. This is the characteristic function of a χ_r^2 random variable iff

$$\lambda_n = \begin{cases} 1, n = 1, 2, \ldots, r \\ 0, n = r + 1, \ldots, N \end{cases}.$$

If $\mathbf{R} = \mathbf{I}$, meaning $\mathbf{X}$ consists of indepedent components, then the quadratic form Q is χ_r^2 iff $\mathbf{P}$ is idempotent:

$$\mathbf{P}^2 = \mathbf{P}.$$

Such a matrix is called a projection matrix. We have the following result: if $\mathbf{X}$ is $\mathcal{N}[\mathbf{0}, \mathbf{I}]$ and $\mathbf{P}$ is a rank r projection, the linear transformation $\mathbf{Y} = \mathbf{PX}$ is $\mathcal{N}[\mathbf{0}, \mathbf{P}]$, and the quadratic form $\mathbf{Y}^T \mathbf{Y} = \mathbf{X}^T \mathbf{PX}$ is χ_r^2. More generally, if $\mathbf{X}$ is $\mathcal{N}[\mathbf{0}, \mathbf{R}]$, $\mathbf{R} = \mathbf{U}\Lambda^2 \mathbf{U}^T$, $\Lambda^2 = \text{diag}\left[\lambda_1^2 \ \lambda_2^2 \ \cdots \ \lambda_N^2 \right]$ and $\mathbf{P}$ is chosen to be $\mathbf{U}\Lambda_r^{-1}\mathbf{U}^T$ with $\Lambda_r^{-2} = \text{diag}\left[\lambda_1^{-2} \ \lambda_2^{-2} \ \cdots \ \lambda_N^{-2} \right]$, then $\mathbf{PRP} = \mathbf{U}I_r\mathbf{U}^T$ and the quadratic form $\mathbf{Y}^T \mathbf{Y}$ is χ_r^2.

Let $Q = (\mathbf{X} - \mathbf{m})^T \mathbf{R}^{-1} (\mathbf{X} - \mathbf{m})$ where $\mathbf{X}$ is $\mathcal{N}[\mathbf{0}, \mathbf{R}]$. Equivalently,

$$Q = \sum_{n=1}^{N} (X_n - \mu)^2 / \sigma^2$$

is a quadratic form in the i.i.d. $\mathcal{N}[\mu, \sigma^2]$ random variables $X_1, X_2, \cdots, X_N$. We have shown that Q is χ_N^2. Form the random variable

$$V = \frac{Q - N}{\sqrt{2N}}.$$

The new random variable V asymptotically has mean 0 and variance 1, that is, asymptotically $\mathcal{N}[0, 1]$.

The Matrix Normal Distribution

Let $\mathbf{X}(n \times N)$ be a matrix whose rows $\mathbf{x}_1^T, \cdots, \mathbf{x}_n^T$, are independently distributed as $\mathcal{N}(\boldsymbol{\mu}, \mathbf{R})$. Then $\mathbf{X}$ has the matrix normal distribution and represents a random matrix observation from $\mathcal{N}(\boldsymbol{\mu}, \mathbf{R})$. Using (3.24), we find that the density function of $\mathbf{X}$ is [110]

$$f(\mathbf{X}) = [\det(2\pi R)]^{-n/2} \exp\left\{ -\frac{1}{2} \sum_{i=1}^{n} (\mathbf{x}_i - \boldsymbol{\mu})^T \mathbf{R}^{-1} (\mathbf{x}_i - \boldsymbol{\mu}) \right\}$$

$$= [\det(2\pi R)]^{-n/2} \exp\left\{ -\frac{1}{2} \text{Tr}[\mathbf{R}^{-1}(\mathbf{X} - \mathbf{1}\boldsymbol{\mu}^T)^T \mathbf{R}^{-1}(\mathbf{X} - \mathbf{1}\boldsymbol{\mu}^T)] \right\},$$

where $\mathbf{1}$ is a column vector having the number 1 as its elements.

Transformation of Normal Data Matrices

We often encounter random vectors. Let

$$\mathbf{x}_1, \ldots, \mathbf{x}_n$$

be a random sample from $\mathcal{N}(\boldsymbol{\mu}, \boldsymbol{\Sigma})$ [110]. We call

$$\mathbf{X} = \begin{bmatrix} \mathbf{x}_1^T \\ \vdots \\ \mathbf{x}_n^T \end{bmatrix},$$

a data matrix from $\mathcal{N}(\boldsymbol{\mu}, \boldsymbol{\Sigma})$ or simply a "normal data matrix." This matrix is a basic building block in spectrum sensing. We must understand it thoroughly. In practice, we deal with data with high dimensionality. We use this notion as our basic information elements in data processing.

Consider linear functions

$$\mathbf{Y} = \mathbf{AXB},$$

where $A(m \times n)$ and $B(p \times q)$ are fixed matrices of real numbers. The most important linear function is the sample mean

$$\bar{\mathbf{x}} = \frac{1}{n} \sum_{i=1}^{n} \mathbf{x}_i = n^{-1} \mathbf{1}^T \mathbf{X},$$

where $\mathbf{A} = n^{-1} \mathbf{1}^T$ and $\mathbf{B} = \mathbf{I}_p$.

Theorem 3.1 (Sample mean is normal) *If* $X(n \times p)$ *is a data matrix from* $\mathcal{N}_p(\boldsymbol{\mu}, \boldsymbol{\Sigma})$, *and if* $n\bar{\mathbf{x}} = \mathbf{X}^T \mathbf{1}$, *then* $\bar{\mathbf{x}}$ *is* $\mathcal{N}_p(\boldsymbol{\mu}, n^{-1}\boldsymbol{\Sigma})$ *distribution.*

Theorem 3.2 ($\mathbf{Y} = \mathbf{AXB}$ is a normal data matrix) *If* $X(n \times p)$ *is a normal data matrix from* $\mathcal{N}_p(\boldsymbol{\mu}, \boldsymbol{\Sigma})$, *and if* $\mathbf{Y} = \mathbf{AXB}$, *then* $\mathbf{Y}$ *is a normal data matrix if and only if*

1. $\mathbf{A1} = \alpha \mathbf{1}$ *for some scalar* α, *or* $\mathbf{B}^T \boldsymbol{\mu} = \mathbf{0}$, *and*
2. $\mathbf{AA}^T = \beta \mathbf{1}$ *for some scalar* β *or* $\mathbf{B}^T \boldsymbol{\Sigma} \mathbf{B} = \mathbf{0}$.

When both conditions are satisfied, then $\mathbf{Y}$ is a normal data matrix from $\mathcal{N}_q(\alpha \mathbf{B}^T \boldsymbol{\mu}, \beta \mathbf{B}^T \boldsymbol{\Sigma} \mathbf{B})$.

Theorem 3.3 *The elements of* $\mathbf{Y} = \mathbf{AXB}$ *are independent of those of* $\mathbf{Z} = \mathbf{CXD}$.
If $X(n \times p)$ *is a normal data matrix from* $\mathcal{N}(\boldsymbol{\mu}, \boldsymbol{\Sigma})$, *and if* $\mathbf{Y} = \mathbf{AXB}$ *and* $\mathbf{Z} = \mathbf{CXD}$, *then the elements of* $\mathbf{Y}$ *are independent of* $\mathbf{Z}$ *if and only if*

1. $\mathbf{B}^T \boldsymbol{\Sigma} \mathbf{D}$ *or*
2. $\mathbf{AC}^T = \mathbf{0}$.

Under the conditions of Theorem 3.3, $\bar{\mathbf{x}} = n^{-1}\mathbf{X}^T \mathbf{1}$ is independent of $\mathbf{HX}$, and thus is independent of $\mathbf{S} = n^{-1}\mathbf{X}^T \mathbf{HX}$.

The Wishart Distribution

We often encounter the form $\mathbf{X}^T \mathbf{CX}$ where $\mathbf{C}$ is a symmetric matrix. This is a matrix-valued quadratic function. The most important special case is the sample covariance matrix obtained by putting $\mathbf{C} = n^{-1}\mathbf{H}$, where $\mathbf{H}$ is the centering matrix. These quadratic forms often lead to the Wishart distribution, which is a matrix generalization of the univariate chi-squared distribution, and has many similar properties.

If $\mathbf{M}(p \times p)$ is

$$\mathbf{M} = \mathbf{X}^T \mathbf{X},$$

where $\mathbf{X}(m \times p)$ is a data matrix from $\mathcal{N}(\mathbf{0}, \boldsymbol{\Sigma})$, then $\mathbf{M}$ is said to have a Wishart distribution with scale matrix $\boldsymbol{\Sigma}$ and degrees of freedom parameter m. We write

$$\mathbf{M} \sim \mathbf{W}_p(\boldsymbol{\Sigma}, m).$$

When $\boldsymbol{\Sigma} = \mathbf{I}_p$, the distribution is said to be in standard form.

When $p = 1$, the $W_1(\sigma^2, m)$ distribution is given by $\mathbf{x}^T \mathbf{x}$, where the elements of $\mathbf{x}(m \times 1)$ are i.i.d. $\mathcal{N}(0, \sigma^2)$ variables; that is the $W_1(\sigma^2, m)$ distribution is the same as the $\sigma^2 \chi_m^2$ distribution.

The scale matrix $\boldsymbol{\Sigma}$ plays the same role in the Wishart distribution as σ^2 does in the $\sigma^2 \chi_m^2$ distribution. We shall usually assume

$$\boldsymbol{\Sigma} > 0.$$

Theorem 3.4 (The class of Wishart matrix is closed under linear transformation) *If* $\mathbf{M} \sim \mathbf{W}_p(\mathbf{\Sigma}, m)$ *and* $\mathbf{B}$ *is a* $(p \times q)$ *matrix, then*

$$\mathbf{B}^T \mathbf{M} \mathbf{B} \sim \mathbf{W}_p(\mathbf{B}^T \mathbf{\Sigma} \mathbf{B}, m).$$

The diagonal submatrices of $\mathbf{M}$ themselves have a Wishart distribution. Also,

$$\mathbf{\Sigma}^{-1/2} \mathbf{M} \mathbf{\Sigma}^{-1/2} \sim \mathbf{W}_p(\mathbf{I}, m).$$

If $\mathbf{M} \sim \mathbf{W}_p(\mathbf{I}, m)$ and $\mathbf{B}(p \times q)$ satisfies $\mathbf{B}^T \mathbf{B} = \mathbf{I}_q$, then

$$\mathbf{B}^T \mathbf{M} \mathbf{B} \sim \mathbf{W}_q(\mathbf{I}, m).$$

Theorem 3.5 (Ratio transform) *If* $\mathbf{M} \sim \mathbf{W}_p(\mathbf{\Sigma}, m)$, *and* $\mathbf{a}$ *is any fixed p-vector such that* $\mathbf{a}^T \mathbf{\Sigma} \mathbf{a} \neq 0$, *then*

$$\frac{\mathbf{a}^T \mathbf{M} \mathbf{a}}{\mathbf{a}^T \mathbf{\Sigma} \mathbf{a}} \sim \chi_m^2.$$

Also, we have $m_{ii} \sim \sigma_i^2 \chi_m^2$.

Theorem 3.6 (The class of Wishart matrix is closed under addition) *If* $\mathbf{M}_1 \sim \mathbf{W}_p(\mathbf{\Sigma}, m_1)$ *and* $\mathbf{M}_2 \sim \mathbf{W}_q(\mathbf{\Sigma}, m_2)$, *then*

$$\mathbf{M}_1 + \mathbf{M}_2 \sim \mathbf{W}_p(\mathbf{\Sigma}, m_1 + m_2).$$

Theorem 3.7 (Cochran, 1934) *If* $\mathbf{X}(n \times p)$ *is a data matrix from* $\mathcal{N}_p(\mathbf{0}, \mathbf{\Sigma})$, *and if* $\mathbf{C}(n \times n)$ *is a symmetric matrix, then*

1. $\mathbf{X}^T \mathbf{C} \mathbf{X}$ *has the same distribution as a weighted sum of independent* $\mathbf{W}_p(\mathbf{\Sigma}, 1)$ *matrices, where the weights are eigenvalues of* $\mathbf{C}$.
2. $\mathbf{X}^T \mathbf{C} \mathbf{X}$ *has a Wishart distribution if and only if* $\mathbf{C}$ *is idempotent, in which case*

$$\mathbf{X}^T \mathbf{C} \mathbf{X} \sim \mathbf{W}_p(\mathbf{\Sigma}, r),$$

   where r is the rank $r = \mathrm{Tr}\mathbf{C} = \mathrm{rank}\mathbf{C}$;
3. $\mathbf{S} = n^{-1} \mathbf{X}^T \mathbf{H} \mathbf{X}$ *is the sample covariance matrix, then*

$$n\mathbf{S} \sim \mathbf{W}_p(\mathbf{\Sigma}, n - 1).$$

Theorem 3.8 (Craig, 1943; Lancaster, 1969, p. 23) *If the rows of* $\mathbf{X}(n \times p)$ *are i.i.d.* $\mathcal{N}_p(\boldsymbol{\mu}, \mathbf{\Sigma})$, *and if* $\mathbf{C}_1, \ldots, \mathbf{C}_k$ *are symmetric matrices, then*

$$\mathbf{X}^T \mathbf{C}_1 \mathbf{X}, \ldots, \mathbf{X}^T \mathbf{C}_k \mathbf{X}$$

are jointly independent if $\mathbf{C}_r \mathbf{C}_s = 0$ *for all* $r \neq s$.

The Hotelling T^2 Distribution

[110, p. 73] Let us study the functions such as $\mathbf{d}^T\mathbf{M}^{-1}\mathbf{d}$, where $\mathbf{d}$ is normal, $\mathbf{M}$ is Wishart, and $\mathbf{d}$ and $\mathbf{M}$ are independent. For example, $\mathbf{d}$ may be the sample mean, and $\mathbf{d}$ proportional to the sample covariance matrix. Hotelling (1931) initiated the work to derive the general distribution of quadratic forms.

If α is used to represent $m\mathbf{d}^T\mathbf{M}^{-1}\mathbf{d}$ where $\mathbf{d}$ and $\mathbf{M}$ are independently distributed as $\mathcal{N}_p(\mathbf{0}, \mathbf{I})$ and $W_p(\mathbf{I}, m)$, respectively, then we say that α has the Hotelling T^2 distribution with parameters p and m. We write $\alpha \sim T^2(p, m)$.

Theorem 3.9 (T^2 distribution) *If $\mathbf{x}$ and $\mathbf{M}$ are independently distributed as $\mathcal{N}_p(\boldsymbol{\mu}, \boldsymbol{\Sigma})$ and $\mathbf{W}(\boldsymbol{\Sigma}, m)$, respectively, then*

$$m(\mathbf{x} - \boldsymbol{\mu})^T\mathbf{M}^{-1}(\mathbf{x} - \boldsymbol{\mu}) \sim T^2(p, m).$$

If $\bar{\mathbf{x}}$ and $\mathbf{S}$ are the mean vector and covariance matrix of a sample of size n from $\mathcal{N}_p(\boldsymbol{\mu}, \boldsymbol{\Sigma})$ and $\mathbf{S}_u = (n/(n-1))\mathbf{S}$, then

$$(n-1)(\bar{\mathbf{x}} - \boldsymbol{\mu})^T\mathbf{S}^{-1}(\bar{\mathbf{x}} - \boldsymbol{\mu}) = n(\bar{\mathbf{x}} - \boldsymbol{\mu})^T\mathbf{S}_u^{-1}(\bar{\mathbf{x}} - \boldsymbol{\mu}) \sim T^2(p, n-1).$$

The T^2 statistic is invariant under any nonsingular linear transformation $\mathbf{x} \to \mathbf{Ax} + \mathbf{b}$. $\frac{\det \mathbf{M}}{\det(\mathbf{M}+\mathbf{dd}^T)} \sim B(\frac{1}{2}(m - p + 1), \frac{1}{2}p)$ where $B(\cdot, \cdot)$ is a beta variable.

Theorem 3.10 ($\mathbf{d}^T\mathbf{Md}$ is independent of $\mathbf{M} + \mathbf{d}^T\mathbf{d}$) *If $\mathbf{d}$ and $\mathbf{M}$ are independently distributed as $\mathcal{N}_p(\mathbf{0}, \mathbf{I})$ and $\mathbf{W}(\mathbf{I}, m)$, respectively. Then, $\mathbf{d}^T\mathbf{Md}$ is independent of $\mathbf{M} + \mathbf{d}^T\mathbf{d}$.*

Wilks' Lambda Distribution

Theorem 3.11 (Wilks' lambda distribution) *If $\mathbf{A} \sim \mathbf{W}(\boldsymbol{\Sigma}, m)$ and $\mathbf{B} \sim \mathbf{W}(\boldsymbol{\Sigma}, n)$ are independent and if $m \geq p$ and $n \geq p$. Then,*

$$\phi = \det(\mathbf{A}^{-1}\mathbf{B}) = \frac{\det \mathbf{B}}{\det \mathbf{A}},$$

is proporational to the product of p independent F variables, of which the i-th has degrees of freedom $n - i + 1$ and $m - i + 1$.

If $\mathbf{A} \sim \mathbf{W}(\boldsymbol{\Sigma}, m)$ and $\mathbf{B} \sim \mathbf{W}(\boldsymbol{\Sigma}, n)$ are independent and if $m \geq p$, we say that

$$\Lambda = \frac{\det \mathbf{A}}{\det(\mathbf{A} + \mathbf{B})} = \frac{1}{\det(\mathbf{I} + \mathbf{A}^{-1}\mathbf{B})} \sim \Lambda(p, m, n),$$

has a Wilks' lambda distribution with parameters p, m, n.

The Λ family of distributions occurs frequently in the context of likelihood ratio test. The parameter p is dimension. The parameter m represents the "error" degrees of freedom and n the "hypothesis" degrees of freedom. Thus $m + n$ represents the "total" degrees of freedom. Like the T^2 statistic, Wilks' lambda distribution is invariant under changes of the scale parameters of $\mathbf{A}$ and $\mathbf{B}$.

Theorem 3.12 (Independent variables Wilks' lambda distribution)

$$\Lambda(p, m, n) \sim \prod_{i=1}^{n} u_i,$$

where $u_i, \ldots, u_n$ are independent variables and

$$u_i \sim B\left(\frac{1}{2}(m + i - p), \frac{1}{2}p\right), i = 1, \ldots, n.$$

Theorem 3.13 (Total degrees of freedom) *The $\Lambda(p, m, n)$ and $\Lambda(n, m + n - p, p)$ distributions are the same.*

If $\mathbf{A} \sim \mathbf{W}(\mathbf{\Sigma}, m)$ is independent of $\mathbf{B} \sim \mathbf{W}(\mathbf{\Sigma}, n)$ where $m \geq p$. Then, the largest eigenvalue θ of $(\mathbf{A} + \mathbf{B})^{-1}\mathbf{B}$ is called the greatest root statistic and its distribution is denoted $\theta(p, m, n)$.

If λ is an eigenvalue of $\mathbf{A}^{-1}\mathbf{B}$, then $\frac{\lambda}{(1+\lambda)}$ is an eigenvalue of $(\mathbf{A} + \mathbf{B})^{-1}\mathbf{B}$. Since this is a monotone function of λ, θ is given by

$$\theta = \frac{\lambda_1}{1 + \lambda_1},$$

where λ_1 is the largest eigenvalue of $\mathbf{A}^{-1}\mathbf{B}$. Since $\lambda_1 > 0$, we see that $0 \leq \theta \leq 1$.

For multisample hypotheses, we see [110, p. 138].

Geometric Ideas

The multivariate normal distribution in N dimensions has constant density on ellipses or ellipsoids of the form

$$(\mathbf{x} - \boldsymbol{\mu})^T \mathbf{\Sigma}^{-1}(\mathbf{x} - \boldsymbol{\mu}) = c^2, \tag{3.26}$$

where c is a constant. These ellipsoids are called the contour of the distribution or the ellipsoids of equal concentration. For $\boldsymbol{\mu} = 0$, these contours are centered at $\mathbf{x} = 0$, and when $\mathbf{\Sigma} = \mathbf{I}$ the contours are circles or in higher dimensions spheres or hyperspheres.

The principal component transformation facilitates interpretation of the ellipsoids of equal concentration. Using the spectral decomposition

$$\mathbf{\Sigma} = \mathbf{\Gamma} \mathbf{\Lambda} \mathbf{\Gamma}^T,$$

where $\mathbf{\Lambda} = \text{diag}(\lambda_1, \lambda_2, \cdots, \lambda_N)$ is the matrix of eigenvalues of $\mathbf{\Sigma}$, and $\mathbf{\Gamma}$ is an orthogonal matrix whose columns are the corresponding eigenvectors. As in Section 3.3.1, define the principal component transform by

$$\mathbf{y} = \mathbf{\Gamma}^T(\mathbf{x} - \boldsymbol{\mu}).$$

In terms of $\mathbf{y}$, (3.26) becomes

$$\sum_{i=1}^{N} \frac{y_i^2}{\lambda_i} = c^2,$$

so that the components of $\mathbf{y}$ represents axes of the ellipsoid.

In the GLRT, the following difference between two ellipsoids is encountered

$$(\mathbf{x} - \boldsymbol{\mu})^T \boldsymbol{\Sigma}_1^{-1} (\mathbf{x} - \boldsymbol{\mu}) - (\mathbf{x} - \boldsymbol{\mu})^T \boldsymbol{\Sigma}_0^{-1} (\mathbf{x} - \boldsymbol{\mu}) = (\mathbf{x} - \boldsymbol{\mu})^T (\boldsymbol{\Sigma}_1^{-1} - \boldsymbol{\Sigma}_0^{-1})(\mathbf{x} - \boldsymbol{\mu}) = d^2,$$
$$(3.27)$$

where d is a constant. When the actual covariance matrices $\boldsymbol{\Sigma}_1$ and $\boldsymbol{\Sigma}_0$ are perfectly known, the problem is fine. Technical difficulty arises from the fact that sample covariance matrices $\hat{\boldsymbol{\Sigma}}_0$ and $\hat{\boldsymbol{\Sigma}}_1$ are used in replacement of $\boldsymbol{\Sigma}_0$ and $\boldsymbol{\Sigma}_1$. The fundamental problem is to guarantee that (3.27) has a geometric meaning; in other words, this implies

$$(\mathbf{x} - \boldsymbol{\mu})^T (\boldsymbol{\Sigma}_1^{-1} - \boldsymbol{\Sigma}_0^{-1})(\mathbf{x} - \boldsymbol{\mu}) \geq 0. \tag{3.28}$$

As in Section A.3, the trace function, $\mathrm{Tr}\mathbf{A} = \sum_i a_{ii}$, satisfies the following properties [110] for matrices $\mathbf{A}, \mathbf{B}, \mathbf{C}, \mathbf{D}, \mathbf{X}$ and scalar α:

$$\mathrm{Tr}\alpha = \alpha, \mathrm{Tr}(\mathbf{A} \pm \mathbf{B}) = \mathrm{Tr}\mathbf{A} \pm \mathrm{Tr}\mathbf{B}, \mathrm{Tr}\alpha\mathbf{A} = \alpha\mathrm{Tr}\mathbf{A}$$

$$\mathrm{Tr}\mathbf{C}\mathbf{D} = \mathrm{Tr}\mathbf{D}\mathbf{C} = \sum_{i,j} c_{ij} d_{ji}.$$

$$\mathrm{Tr}[(\mathbf{x} - \boldsymbol{\mu})^T (\boldsymbol{\Sigma}_1^{-1} - \boldsymbol{\Sigma}_0^{-1})(\mathbf{x} - \boldsymbol{\mu})] = \mathrm{Tr}[(\boldsymbol{\Sigma}_1^{-1} - \boldsymbol{\Sigma}_0^{-1})(\mathbf{x} - \boldsymbol{\mu})(\mathbf{x} - \boldsymbol{\mu})^T] \geq 0. \tag{3.29}$$

Using the fact that for $\mathbf{A}, \mathbf{B} \geq \mathbf{0}$, we have

$$(\mathrm{Tr}\mathbf{A})(\mathrm{Tr}\mathbf{B}) \geq \mathrm{Tr}(\mathbf{A}\mathbf{B}) \geq 0,$$

we have the necessary condition (for (3.28) to be valid)

$$\mathrm{Tr}[(\boldsymbol{\Sigma}_1^{-1} - \boldsymbol{\Sigma}_0^{-1})] \geq 0, \tag{3.30}$$

since

$$\mathrm{Tr}[(\mathbf{x} - \boldsymbol{\mu})^T (\mathbf{x} - \boldsymbol{\mu})] \geq 0. \tag{3.31}$$

The necessary condition (3.30) is easily satisfied when the actual covariance matrices $\boldsymbol{\Sigma}_1$ and $\boldsymbol{\Sigma}_0$ are known. When, in practice, the sample covariance matrices $\hat{\boldsymbol{\Sigma}}_0$ and $\hat{\boldsymbol{\Sigma}}_1$ are known, instead, the problem arises

$$\mathrm{Tr}[(\hat{\boldsymbol{\Sigma}}_1^{-1} - \hat{\boldsymbol{\Sigma}}_0^{-1})] \geq 0. \tag{3.32}$$

This leads to a natural problem of sample covariance matrix estimation and the related GLRT. The fundamental problem is that the GLRT requires the exact probability distribution functions for two alternative hypotheses. This condition is rarely met in practice. Another problem is intuition that the random vectors fit the exact probability distribution function. In fact, the empirical probability distribution function does not satisfy this condition. This problem is more obvious when we deal with high data dimensionality such as $N = 10^5, 10^6$ which are commonly feasible in real-world spectrum sensing problems. For example, $N = 100,000$ corresponding to 4.65 milliseconds sampling time.

3.6 Sample Covariance Matrix Estimation and Matrix Compressed Sensing

Fundamental Fact: The noise lies in a high-dimensional space; the signal, on the contrast, lies in a much lower-dimensional space.

If a random matrix $\mathbf{A}$ has i.i.d column A_i, then

$$\mathbf{A}^*\mathbf{A} = \sum_i A_i A_i^T,$$

where $\mathbf{A}^*$ is the adjoint matrix of $\mathbf{A}$. We often study $\mathbf{A}$ through the $n \times n$ symmetric, positive semidefinite matrix, the matrix $\mathbf{A}^*\mathbf{A}$; in other words

$$\mathbf{A}^*\mathbf{A} \geq 0.$$

The absolute matrix is defined as

$$|\mathbf{A}| = \sqrt{\mathbf{A}^*\mathbf{A}}.$$

A matrix $\mathbf{C}$ is positive semidefinite,

$$\mathbf{C} \geq 0,$$

if and only all its eigenvalues λ_i are nonnegative

$$\lambda_i \geq 0.$$

The eigenvalues of $|\mathbf{A}|$ are therefore nonnegative real numbers or

$$\mathbf{D} = |\mathbf{A}| \geq 0.$$

An immediate application of random matrices is the fundamental problem of *estimating covariance matrices* of high-dimensional distributions [107]. The analysis of the row-independent models can be interpreted as a study of sample covariance matrices.

There are deep results in random matrix theory. The main motivation of this subsection is to exploit the existing results in this field to better guide the estimate of covariance matrices, using nonasymptotic results [107].

We focus on a general model of random matrices, where we only assume independence of the rows rather than all entries. Such matrices are naturally generated by high-dimensional distributions. Indeed, given an arbitrary probability distribution in $\mathbb{R}^n$, one takes a sample of N independent points and arranges them as the rows of an $N \times n$ matrix $\mathbf{A}$. Recall that n is the dimension of the probability space.

Let $\mathbf{X}$ be a random vector in $\mathbb{R}^n$. For simplicity we assume that $\mathbf{X}$ is centered, or $\mathbb{E}\mathbf{X} = 0$. Here $\mathbb{E}\mathbf{X}$ denotes the expectation of $\mathbf{X}$. The covariance matrix of $\mathbf{X}$ is the $n \times n$ matrix

$$\mathbf{\Sigma} = \mathbb{E}\mathbf{X}\mathbf{X}^T.$$

The simplest way to estimate $\boldsymbol{\Sigma}$ is to take some N independent samples $\mathbf{X}_i$ from the distribution and form the sample covariance matrix

$$\boldsymbol{\Sigma}_N = \frac{1}{N} \sum_{i=1}^{N} \mathbf{X}_i \mathbf{X}_i^{\,T}.$$

By the law of large numbers,

$$\boldsymbol{\Sigma}_N \to \boldsymbol{\Sigma},$$

almost surely as $N \to \infty$. So, taking sufficiently many samples we are guaranteed to estimate the covariance matrix as well as we want. This, however, does not address the quantitative aspect: what is the minimal *sample size* N that guarantees approximation with a given accuracy?

The relation of this question to random matrix theory becomes clear when we arrange the samples

$$\mathbf{X}_i =: \mathbf{A}_i, i = 1, 2, \ldots, N,$$

as rows of the $N \times n$ random matrix $\mathbf{A}$. Then, the sample covariance matrix is expressed as

$$\boldsymbol{\Sigma}_N = \frac{1}{N} \mathbf{A}^* \mathbf{A}.$$

Note that $\mathbf{A}$ is a matrix with independent rows but usually not independent entries. The reference of [107] has worked out the analysis of such matrices, separately for sub-Gaussian and general distributions.

We often encounter covariance matrix estimation for sub-Gaussian distribution due to the presence of Gaussian noise. Consider a sub-Gaussian distribution in $\mathbb{R}^n$ with covariance matrix $\boldsymbol{\Sigma}$, and let $\varepsilon \in (0, 1)$, $t \geq 1$. Then with probability of at least

$$1 - 2 \exp(-t^2 n),$$

one has

$$\text{If } N \geq C(t/\varepsilon)^2 n, \text{ then } ||\boldsymbol{\Sigma}_N - \boldsymbol{\Sigma}|| \leq \varepsilon. \tag{3.33}$$

Here C depends only on the sub-Gaussian norm of a random vector taken from this distribution; the spectral norm of $\mathbf{A}$ is denoted as $||\mathbf{A}||$, which is equal to the maximum singular value of $\mathbf{A}$, that is,

$$s_{\max} = ||\mathbf{A}||.$$

Covariance matrix estimation for arbitrary distribution is also encountered when a general noise interference model is used. Consider a sub-Gaussian distribution in $\mathbb{R}^n$ with covariance matrix $\boldsymbol{\Sigma}$ and supported in some centered Euclidean ball whose radius we denote $\sqrt{m}$. Let $\varepsilon \in (0, 1)$ and $t \geq 1$. Then, with probability at least $1 - n^{-t^2}$, one has

$$\text{If } N \geq C(t/\varepsilon)^2 ||\boldsymbol{\Sigma}||^{-1} m \log n, \text{ then } ||\boldsymbol{\Sigma}_N - \boldsymbol{\Sigma}|| \leq \varepsilon ||\boldsymbol{\Sigma}||. \tag{3.34}$$

Here C is an absolute constant and log denotes the natural logarithm. In (3.34), typically

$$m = O(||\mathbf{\Sigma}||n).$$

Thus the required sample size is

$$N \geq C(t/\varepsilon)^2 n \log n.$$

Low rank estimation is often used, since the distribution of a signal in $\mathbb{R}^n$ lies close to a low-dimensional subspace. In this case, a much smaller sample size suffices for covariance estimation. The intrinsic dimension of the distribution can be measured with the *effective rank* of the matrix $\mathbf{\Sigma}$, defined as

$$r(\mathbf{\Sigma}) = \frac{\text{Tr}(\mathbf{\Sigma})}{||\mathbf{\Sigma}||}, \tag{3.35}$$

where $\text{Tr}(\mathbf{\Sigma})$ denotes the trace of $\mathbf{\Sigma}$. One always has

$$r(\mathbf{\Sigma}) \leq \text{rank}(\mathbf{\Sigma}) \leq n,$$

and this bound is sharp. The effective rank

$$r = r(\mathbf{\Sigma}),$$

always controls the typical norm of $\mathbf{X}$, as

$$\mathbb{E}||\mathbf{X}||_2^2 = \text{Tr}(\mathbf{\Sigma}) = r\mathbf{\Sigma}.$$

Most of the distribution is supported in a ball of radius $\sqrt{m}$, where

$$m = O(r||\mathbf{\Sigma}||).$$

The conclusion of (3.34) holds with sample size

$$N \geq C(t/\varepsilon)^2 r \log n.$$

Summarizing the above discussion, (3.34) shows that the sample size

$$N = O(n)$$

suffices to approximate the covariance matrix of a sub-Gaussian distribution in $\mathbb{R}^n$ by the sample covariance matrix. While for arbitrary distribution,

$$N = O(n \log n)$$

is sufficient. For distributions that are approximately low-dimensional, such as that of a signal, a much smaller sample size is sufficient. Namely, if the effective rank of $\mathbf{\Sigma}$ equals r, then a sufficient size is

$$N = O(r \log n).$$

As in Section 3.3.1, our standard problem (3.15) is

$$\mathcal{H}_0 : \mathbf{R}_x = \mathbf{R}_w,$$
$$\mathcal{H}_1 : \mathbf{R}_x = \mathbf{R}_s + \mathbf{R}_w,$$

where $\mathbf{R}_s$ and $\mathbf{R}_w$ are, respectively, covariance matrices of signal and noise. When the sample covariance matrices are used in replacement of actual covariance matrices, it follows that

$$\mathcal{H}_0 : \hat{\mathbf{R}}_x = \hat{\mathbf{R}}_w,$$
$$\mathcal{H}_1 : \hat{\mathbf{R}}_x = \hat{\mathbf{R}}_s + \hat{\mathbf{R}}_w. \qquad (3.36)$$

We must exploit the fundamental fact that $\hat{\mathbf{R}}_s$ requires only $O(r \log n)$ samples, while $\hat{\mathbf{R}}_w$ requires $O(n)$. Here the effective rank r of $\mathbf{R}_s$ is small. For one real sinusoid signal, the rank is only two, that is, $r = 2$. We can sum up the K sample covariance matrices $\bar{\mathbf{R}}_x = \sum_{k=1}^{K} \hat{\mathbf{R}}_{x,k}$, for example, $K = 200$. Let us consider the three-step algorithm:

1. break the long record of data into a total of K segments. Each segment has a length of p. In other words, a total of pK is available for signal processing.

$$\mathcal{H}_0 : \hat{\mathbf{R}}_{x,k} = \hat{\mathbf{R}}_{w,k},$$
$$\mathcal{H}_1 : \hat{\mathbf{R}}_{x,k} = \hat{\mathbf{R}}_{s,k} + \hat{\mathbf{R}}_{w,k}, k = 1, 2, \ldots, K. \qquad (3.37)$$

2. Choose the p such that $p > O(r \log n)$; so the sample covariance matrix $\hat{\mathbf{R}}_{s,k}$ accurately approximates the actual covariance matrix:

$$\mathcal{H}_0 : \hat{\mathbf{R}}_{x,k} = \hat{\mathbf{R}}_{w,k},$$
$$\mathcal{H}_1 : \hat{\mathbf{R}}_{x,k} \approx \mathbf{R}_{s,k} + \hat{\mathbf{R}}_{w,k}, k = 1, 2, \ldots K. \qquad (3.38)$$

3. We can sum up the K estimated sample covariance matrices $\hat{\mathbf{R}}_{x,k}$:

$$\mathcal{H}_0 : \bar{\mathbf{R}}_x = \sum_{k=1}^{K} \hat{\mathbf{R}}_{x,k} = \sum_{k=1}^{K} \hat{\mathbf{R}}_{w,k},$$
$$\mathcal{H}_1 : \bar{\mathbf{R}}_x = \sum_{k=1}^{K} \hat{\mathbf{R}}_{x,k} \approx K\mathbf{R}_{s,1} + \sum_{k=1}^{K} \hat{\mathbf{R}}_{w,k}, \qquad (3.39)$$

where, without loss of generality, we have assumed $\mathbf{R}_{s,k} = \mathbf{R}_{s,1}$.

In Step 3, the signal part coherently adds up and the noise part randomly adds up. This step enhances SNR, which is especially relevant at low SNR signal detection. This basic idea will be behind the chapter on quantum detection.

Another underlying idea is to develop a nonasymptotic theory for signal detection. Given a finite number of (complex) data samples collected into a random vector $\mathbf{x} \in \mathbb{C}^n$, the vector length n is very large, but not infinity, in other words, $n < \infty$. We cannot simply resort to the central limit theorem to argue that the probability distribution function of this vector $\mathbf{x}$ is Gaussian since this theorem requires $n \to \infty$.

The recent work on compressed sensing is highly relevant to this problem. Given n, we can develop a theory that is valid with an overwhelming probability. As a result, one cornerstone of our development here is compressed sensing to recover the "information" from n data points $\mathbf{x}$. Another cornerstone is the concentration of measure to study sums of random matrices. For example what are the statistics of the resultant random matrix $\bar{\mathbf{R}}_x = \sum_{k=1}^{K} \hat{\mathbf{R}}_{x,k}$?

This problem is closely related to classical multivariate analysis [110, p. 108]. Section 3.6.1 will show that the sum of the random matrices is the ML estimation of the actual covariance matrix.

3.6.1 The Maximum Likelihood Estimation

The problem of Section 3.6.1 is closely related to classical multivariate analysis [110, p. 108]. Given k independent data matrices

$$\mathbf{X}_1, \mathbf{X}_2, \cdots, \mathbf{X}_k,$$

where the rows of $\mathbf{X}_i (n \times p)$ are i.i.d.

$$\mathcal{N}_p(\boldsymbol{\mu}_i, \boldsymbol{\Sigma}_i), i = 1, 2, \ldots, k,$$

what is the ML estimation of the sample covariance matrix?

In practice, the most common constraints are

$$(\text{a}) : \boldsymbol{\Sigma}_1 = \ldots = \boldsymbol{\Sigma}_k$$

or

$$(\text{b}) : \boldsymbol{\Sigma}_1 = \ldots = \boldsymbol{\Sigma}_k \quad \text{and} \quad \boldsymbol{\mu}_1 = \ldots = \boldsymbol{\mu}_k.$$

If (b) holds, we can treat all the data matrices as constituting one matrix sample from a single population (distribution).

Suppose that $\mathbf{x}_1, \ldots, \mathbf{x}_n$ is a random vector sample from a population (distribution) with the pdf $f(\mathbf{x}, \boldsymbol{\theta})$, where $\boldsymbol{\theta}$ is a parameter vector. The likelihood function of the whole sample is

$$L(\mathbf{X}; \boldsymbol{\theta}) = \prod_{i=1}^{n} f(\mathbf{x}_i; \boldsymbol{\theta}).$$

$$l(\mathbf{X}; \boldsymbol{\theta})) = \log L(\mathbf{X}; \boldsymbol{\theta})) = \sum_{i=1}^{n} \log f(\mathbf{x}_i; \boldsymbol{\theta})).$$

Given a matrix sample $\mathbf{X}$, both $l(\mathbf{X}; \boldsymbol{\theta}))$ and $L(\mathbf{X}; \boldsymbol{\theta}))$ are considered as functions of the vector parameter $\boldsymbol{\theta}$.

Suppose $\mathbf{x}_1, \ldots, \mathbf{x}_n$ is a random sample from $\mathcal{N}_p(\boldsymbol{\mu}, \boldsymbol{\Sigma})$. We have

$$L(\mathbf{X}; \boldsymbol{\mu}; \boldsymbol{\Sigma}) = [\det (2\pi \boldsymbol{\Sigma})]^{-n/2} \exp \left[-\frac{1}{2} \sum_{i=1}^{n} (\mathbf{x}_i - \boldsymbol{\mu})^T \boldsymbol{\Sigma}^{-1} (\mathbf{x}_i - \boldsymbol{\mu}) \right],$$

and

$$l(\mathbf{X}; \boldsymbol{\mu}; \boldsymbol{\Sigma}) = \log L(\mathbf{X}; \boldsymbol{\mu}; \boldsymbol{\Sigma}) = -\frac{n}{2} \log \det (2\pi \boldsymbol{\Sigma}) - \frac{1}{2} \sum_{i=1}^{n} (\mathbf{x}_i - \boldsymbol{\mu})^T \boldsymbol{\Sigma}^{-1} (\mathbf{x}_i - \boldsymbol{\mu}).$$

$$(3.40)$$

Let us simplify these equations. When the identity

$$(\mathbf{x}_i - \boldsymbol{\mu})^T \boldsymbol{\Sigma}^{-1}(\mathbf{x}_i - \boldsymbol{\mu}) = (\mathbf{x}_i - \bar{\mathbf{x}})^T \boldsymbol{\Sigma}^{-1}(\mathbf{x}_i - \bar{\mathbf{x}}) + (\bar{\mathbf{x}} - \boldsymbol{\mu})^T \boldsymbol{\Sigma}^{-1}(\bar{\mathbf{x}} - \boldsymbol{\mu})$$
$$+ 2(\bar{\mathbf{x}} - \boldsymbol{\mu})^T \boldsymbol{\Sigma}^{-1}(\mathbf{x}_i - \bar{\mathbf{x}}),$$

is summed over the index $i = 1, \ldots, n$, the final term on the right-hand side vanishes, giving

$$\sum_{i=1}^{n}(\mathbf{x}_i - \boldsymbol{\mu})^T \boldsymbol{\Sigma}^{-1}(\mathbf{x}_i - \boldsymbol{\mu}) = \sum_{i=1}^{n}(\mathbf{x}_i - \bar{\mathbf{x}})^T \boldsymbol{\Sigma}^{-1}(\mathbf{x}_i - \bar{\mathbf{x}}) + n(\bar{\mathbf{x}} - \boldsymbol{\mu})^T \boldsymbol{\Sigma}^{-1}(\bar{\mathbf{x}} - \boldsymbol{\mu}).$$

$$(3.41)$$

Since each term $(\mathbf{x}_i - \bar{\mathbf{x}})^T \boldsymbol{\Sigma}^{-1}(\mathbf{x}_i - \bar{\mathbf{x}})$ is a scalar, it equals the trace of itself. Thus using

$$\mathrm{Tr}\mathbf{AB} = \mathrm{Tr}\mathbf{BA},$$

we have

$$(\mathbf{x}_i - \bar{\mathbf{x}})^T \boldsymbol{\Sigma}^{-1}(\mathbf{x}_i - \bar{\mathbf{x}}) = \mathrm{Tr}\boldsymbol{\Sigma}^{-1}(\mathbf{x}_i - \bar{\mathbf{x}})(\mathbf{x}_i - \bar{\mathbf{x}})^T. \qquad (3.42)$$

Summing (3.42) over index i and substituting in (3.41) gives

$$\sum_{i=1}^{n}(\mathbf{x}_i - \boldsymbol{\mu})^T \boldsymbol{\Sigma}^{-1}(\mathbf{x}_i - \boldsymbol{\mu}) = \mathrm{Tr}\boldsymbol{\Sigma}^{-1} \sum_{i=1}^{n}(\mathbf{x}_i - \bar{\mathbf{x}})(\mathbf{x}_i - \bar{\mathbf{x}})^T + n(\bar{\mathbf{x}} - \boldsymbol{\mu})^T \boldsymbol{\Sigma}^{-1}(\bar{\mathbf{x}} - \boldsymbol{\mu})^T.$$

$$(3.43)$$

Writing

$$\sum_{i=1}^{n}(\mathbf{x}_i - \bar{\mathbf{x}})(\mathbf{x}_i - \bar{\mathbf{x}})^T = n\mathbf{S}$$

and using (3.40) in (3.43), we have

$$l(\mathbf{X}; \boldsymbol{\mu}; \boldsymbol{\Sigma}) = -\frac{n}{2}\log\det(2\pi\boldsymbol{\Sigma}) - \frac{n}{2}\mathrm{Tr}\boldsymbol{\Sigma}^{-1}\mathbf{S} - \frac{n}{2}(\bar{\mathbf{x}} - \boldsymbol{\mu})^T \boldsymbol{\Sigma}^{-1}(\bar{\mathbf{x}} - \boldsymbol{\mu})^T. \qquad (3.44)$$

For the special case $\boldsymbol{\Sigma} = \mathbf{I}$ and $\boldsymbol{\mu} = \boldsymbol{\theta}$ then (3.44) becomes

$$l(\mathbf{X}; \boldsymbol{\theta}) = -\frac{np}{2}\log(2\pi) - \frac{n}{2}\mathrm{Tr}\mathbf{S} - \frac{n}{2}(\bar{\mathbf{x}} - \boldsymbol{\theta})^T \boldsymbol{\Sigma}^{-1}(\bar{\mathbf{x}} - \boldsymbol{\theta})^T. \qquad (3.45)$$

To calculate the ML estimation if (a) holds, from (3.44), we have

$$l = \sum_{l=1}^{k}[n_i \log\det(2\pi\boldsymbol{\Sigma}) + n_i \mathrm{Tr}\boldsymbol{\Sigma}^{-1}(\mathbf{S}_i + \mathbf{d}_i\mathbf{d}_i^T)], \qquad (3.46)$$

where $\mathbf{S}_i$ is the covariance matrix of the i-th matrix sample, $i = 1, .., k$, and $\mathbf{d}_i = \bar{\mathbf{x}}_i - \boldsymbol{\mu}_i$. Since there is no restriction on the population means, the ML estimation of $\boldsymbol{\mu}_i$ is $\bar{\mathbf{x}}_i$. Setting

$$n = \sum_{i=1}^{k} n_k,$$

(3.46) becomes

$$l = -\frac{1}{2}n \log \det(2\pi\mathbf{\Sigma}) - \frac{1}{2}\text{Tr}\mathbf{\Sigma}^{-1}\mathbf{W},$$
$$\mathbf{W} = \sum_{i=1}^{k} n_i \mathbf{S}_i. \tag{3.47}$$

Differentiating (3.47) with respect to $\mathbf{\Sigma}$ and equating to zero yields

$$\mathbf{\Sigma} = n^{-1}\mathbf{W} = n^{-1}\sum_{i=1}^{k} n_i \mathbf{S}_i, \tag{3.48}$$

which is the ML estimation of $\mathbf{\Sigma}$ under the conditions stated.

3.6.2 Likelihood Ratio Test (Wilks' Λ Test) for Multisample Hypotheses

Consider k independent normal matrix samples $\mathbf{X}_1, \ldots, \mathbf{X}_k$, whose likelihood is considered in Section 3.6.1.

$$\mathcal{H}_0 : \mathbf{X} : \mathcal{N}_p(\boldsymbol{\mu}, \mathbf{\Sigma}), \boldsymbol{\mu}_1 = \cdots = \boldsymbol{\mu}_k, \quad \text{given } \mathbf{\Sigma}_1 = \cdots = \mathbf{\Sigma}_k,$$
$$\mathcal{H}_1 : \boldsymbol{\mu}_1 \neq \cdots \neq \boldsymbol{\mu}_k, \quad \text{given } \mathbf{\Sigma}_1 = \cdots = \mathbf{\Sigma}_k,$$

In Section 3.6.1, the ML estimation under $\mathcal{H}_1$ is $\bar{\mathbf{x}}$ and $\mathbf{S}$, since the observation can be viewed under $\mathcal{H}_1$ as constituting a single random matrix sample. The ML estimation of $\boldsymbol{\mu}_i$ under the alternative hypothesis $\mathcal{H}_0$ is $\bar{\mathbf{x}}_i$, the i-th sample mean, and the ML estimation of the common sample covariance matrix is $n^{-1}\mathbf{W}$, where, following (3.47), we have

$$\mathbf{W} = \sum_{i=1}^{k} n_i \mathbf{S}_i,$$

is the "within-groups" sum of squares and products (SSP) matrix and $n = \sum_{i=1}^{k} n_i$. Using (3.46), the LRT is given by

$$\lambda = \left\{ \frac{\det \mathbf{W}}{\det(n\mathbf{S})} \right\}^{n/2} = [\det(\mathbf{T}^{-1}\mathbf{W})]^{n/2}. \tag{3.49}$$

Here

$$\mathbf{T} = n\mathbf{S}$$

is the "total" SSP matrix, derived by regarding all the data matrices as if they constituted a single matrix sample. In contrast, the matrix $\mathbf{W}$ is the "within-group" SSP and

$$\mathbf{B} = \mathbf{T} - \mathbf{W} = \sum_{i=1}^{k} n_i (\bar{\mathbf{x}}_i - \bar{\mathbf{x}})(\bar{\mathbf{x}}_i - \bar{\mathbf{x}})^T$$

may be regarded as the "between-groups" SSP matrix. Thus, from (3.49), we have

$$\lambda^{2/n} = \frac{\det \mathbf{W}}{\det(\mathbf{B} + \mathbf{W})} = \frac{1}{\det(\mathbf{I} + \mathbf{W}^{-1}\mathbf{B})}.$$

The matrix $\mathbf{W}^{-1}\mathbf{B}$ is an obvious generalization of the univariate variance ratio. It will tend to zero if $\mathcal{H}_0$ is true.

If $n \geq p + k$, under $\mathcal{H}_0$,

$$[\det(\mathbf{I} + \mathbf{W}^{-1}\mathbf{B})]^{-1} \sim \Lambda(p, n - k, k - 1) \tag{3.50}$$

where Wilks' Λ statistics is described in Section 3.5.

Let us derive the statistics of (3.50). Write the k matrix samples as a single data matrix

$$\mathbf{X} = \begin{bmatrix} \mathbf{X}_1 \\ \vdots \\ \mathbf{X}_k \end{bmatrix},$$

where $\mathbf{X}_i(n_i \times p)$ is the i-th matrix sample, $i = 1, \ldots, k$. Let $\mathbf{1}_i$ be the n-vector with 1 in the places corresponding to the i-th sample and 0 elsewhere, and set $\mathbf{I}_i = \text{diag}(\mathbf{1}_i)$. Then $\mathbf{I} = \sum \mathbf{I}_i$, and $\mathbf{1} = \sum \mathbf{1}_i$. Let

$$\mathbf{H}_i = \mathbf{I}_i - n_i^{-1}\mathbf{1}_i\mathbf{1}_i{}^T$$

be the centering matrix for the i-th matrix sample. The sample covariance matrix can be expressed as

$$n_i\mathbf{S}_i = \mathbf{X}_i{}^T\mathbf{H}_i\mathbf{X}.$$

Set

$$\mathbf{C}_1 = \sum \mathbf{H}_i, \mathbf{C}_2 = \sum n_i^{-1}\mathbf{1}_i\mathbf{1}_i{}^T - n^{-1}\mathbf{1}\mathbf{1}^T.$$

We can easily verify that

$$\mathbf{W} = \mathbf{X}^T\mathbf{C}_1\mathbf{X}, \mathbf{B} = \mathbf{X}^T\mathbf{C}_2\mathbf{X}.$$

Further, $\mathbf{C}_1$ and $\mathbf{C}_2$ are idempotent matrices of rank $n - k$ and $k - 1$, respectively, and $\mathbf{C}_1\mathbf{C}_2 = 0$.

Under $\mathcal{H}_0$, $\mathbf{H}$ is data matrix from $\mathcal{N}(\boldsymbol{\mu}, \boldsymbol{\Sigma})$. Thus by Theorem 3.7 and Theorem 3.8, we have

$$\mathbf{W} = \mathbf{X}^T\mathbf{C}_1\mathbf{X} \sim \mathbf{W}_p(\boldsymbol{\mu}, n - k),$$
$$\mathbf{B} = \mathbf{X}^T\mathbf{C}_2\mathbf{X} \sim \mathbf{W}_p(\boldsymbol{\mu}, k - 1),$$

and, furthermore, $\mathbf{W}$ and $\mathbf{B}$ are independent. Therefore, (3.50) is derived.

3.7 Likelihood Ratio Test

3.7.1 General Gaussian Detection and Estimator-Correlator Structure

The most general signal assumption is to allow the signal to be composed of a deterministic component and a random component. The signal then can be modeled as a random process with the deterministic part corresponding to a nonzero mean and the random part corresponding to a zero mean random processes with a given signal covarance matrix. For generality the noise covariance matrix can be assumed to be arbitrary. These assumptions lead to the general Gaussian detection problem [118, 167], which mathematically is written as

$$\mathcal{H}_0 : y[n] = w[n], n = 0, 1, \ldots, N - 1$$
$$\mathcal{H}_1 : y[n] = x[n] + w[n], n = 0, 1, \ldots, N - 1.$$

Define the column vector

$$\mathbf{y} = [y[0]y[1] \cdots y[N]]^T, \tag{3.51}$$

and similarly for $\mathbf{x}$ and $\mathbf{w}$. In vector and matrix form

$$\mathcal{H}_0 : \mathbf{y} = \mathbf{w}; \quad \mathbf{C}_y = \mathbf{C}_w = \mathbf{A}$$
$$\mathcal{H}_1 : \mathbf{y} = \mathbf{x} + \mathbf{w}; \quad \mathbf{C}_y = \mathbf{C}_x + \mathbf{C}_w = \mathbf{B},$$

where $\mathbf{w} \sim \mathcal{N}(\mathbf{0}, \mathbf{C}_w)$, and $\mathbf{x} \sim \mathcal{N}(\boldsymbol{\mu}_x, \mathbf{C}_x)$, and $\mathbf{x}$ and $\mathbf{w}$ are independent. The LRT decides $\mathcal{H}_1$ if

$$\Lambda(\mathbf{y}) = \frac{p(\mathbf{y}; \mathcal{H}_1)}{p(\mathbf{y}; \mathcal{H}_0)} > \gamma,$$

where

$$p(\mathbf{y}; \mathcal{H}_1) = \frac{1}{(2\pi)^{N/2} \det^{1/2}(\mathbf{C}_x + \mathbf{C}_w)} \exp\left[-\tfrac{1}{2}(\mathbf{y} - \boldsymbol{\mu}_x)^*(\mathbf{C}_x + \mathbf{C}_w)^{-1}(\mathbf{y} - \boldsymbol{\mu}_x)\right]$$
$$p(\mathbf{y}; \mathcal{H}_0) = \frac{1}{(2\pi)^{N/2} \det^{1/2}(\mathbf{C}_w)} \exp\left[-\tfrac{1}{2}\mathbf{y}^*\mathbf{C}_w^{-1}\mathbf{y}\right].$$

Taking the logarithm, retaining only the data-dependent terms, and scaling produces the test statistic

$$T(\mathbf{y}) = \mathbf{y}^*\mathbf{C}_w^{-1}\mathbf{y} - (\mathbf{y} - \boldsymbol{\mu}_x)^*(\mathbf{C}_x + \mathbf{C}_w)^{-1}(\mathbf{y} - \boldsymbol{\mu}_x)$$
$$= \mathbf{y}^*\mathbf{C}_w^{-1}\mathbf{y} - \mathbf{y}^*(\mathbf{C}_x + \mathbf{C}_w)^{-1}\mathbf{y} + 2\mathbf{y}^*(\mathbf{C}_x + \mathbf{C}_w)^{-1}\boldsymbol{\mu}_x - \boldsymbol{\mu}_x^*(\mathbf{C}_x + \mathbf{C}_w)^{-1}\boldsymbol{\mu}_x. \tag{3.52}$$

From the matrix inverse lemma [114, p. 43] [118, 167]

$$\mathbf{C}_w^{-1} - (\mathbf{C}_x + \mathbf{C}_w)^{-1} = \mathbf{C}_w^{-1}\mathbf{C}_x(\mathbf{C}_x + \mathbf{C}_w)^{-1}, \tag{3.53}$$

for $\boldsymbol{\mu}_x = 0$, (3.52) is rewritten as

$$T(\mathbf{y}) = \frac{1}{2}\mathbf{y}^*[\mathbf{C}_w^{-1} - (\mathbf{C}_x + \mathbf{C}_w)^{-1}]\mathbf{y} = \frac{1}{2}\mathbf{y}^*\mathbf{C}_w^{-1}\mathbf{C}_x(\mathbf{C}_x + \mathbf{C}_w)^{-1}\mathbf{y} = \frac{1}{2}\mathbf{y}^*\mathbf{C}_w^{-1}\hat{\mathbf{x}},$$

where $\hat{\mathbf{x}} = \mathbf{C}_x(\mathbf{C}_x + \mathbf{C}_w)^{-1}\mathbf{y}$ is the MMSE estimator of $\mathbf{x}$. This is a prewhitener followed by an estimator-correlator.

Using the properties of the trace operation (see Section A.3)

$$\mathrm{Tr}\alpha = \alpha, \mathrm{Tr}(\mathbf{A} \pm \mathbf{B}) = \mathrm{Tr}\mathbf{A} \pm \mathrm{Tr}\mathbf{B}, \mathrm{Tr}\alpha\mathbf{A} = \alpha\mathrm{Tr}\mathbf{A}, \mathrm{Tr}\mathbf{AB} = \mathrm{Tr}\mathbf{BA}, \tag{3.54}$$

where α is a constant, we have

$$T(\mathbf{y}) = \mathrm{Tr}[T(\mathbf{y})] = \frac{1}{2}\mathrm{Tr}\{\mathbf{y}^*[\mathbf{C}_w^{-1} - (\mathbf{C}_x + \mathbf{C}_w)^{-1}]\mathbf{y}\}$$

$$= \frac{1}{2}\mathrm{Tr}\{[\mathbf{C}_w^{-1} - (\mathbf{C}_x + \mathbf{C}_w)^{-1}]\mathbf{y}\mathbf{y}^*\} \overset{\mathcal{H}_1}{\underset{}{>}} T_0, \tag{3.55}$$

where the matrix $\mathbf{yy}^*$ is of size $N \times N$ and of rank one. We can easily choose $T_0 > 0$, since $T(\mathbf{y}) > 0$ that will be obvious later. Therefore, (3.55) can be rewritten as

$$2\sqrt{\mathrm{Tr}\{[\mathbf{C}_w^{-1} - (\mathbf{C}_x + \mathbf{C}_w)^{-1}]\mathbf{yy}^*\}} > T_1. \tag{3.56}$$

Let $\mathbf{A} \geq 0$ and $\mathbf{B} \geq 0$ be of the same size. Then [112, 329] (see Section A.6.2)

$$\begin{aligned}0 \leq \mathrm{Tr}\mathbf{AB} \leq (\mathrm{Tr}\mathbf{A})(\mathrm{Tr}\mathbf{B}), \\ 0 \leq 2\sqrt{\mathrm{Tr}\mathbf{AB}} \leq \mathrm{Tr}\mathbf{A} + \mathrm{Tr}\mathbf{B}.\end{aligned} \tag{3.57}$$

With the help of (3.57), on one hand, (3.56) becomes

$$\mathrm{Tr}[\mathbf{C}_w^{-1} - (\mathbf{C}_x + \mathbf{C}_w)^{-1}] + \mathrm{Tr}(\mathbf{yy}^*) \geq 2\sqrt{\mathrm{Tr}\{[\mathbf{C}_w^{-1} - (\mathbf{C}_x + \mathbf{C}_w)^{-1}]\mathbf{yy}^*\}} > T_1.$$

Or,

$$\mathrm{Tr}(\mathbf{yy}^*) \geq T_1 - \mathrm{Tr}[\mathbf{C}_w^{-1} - (\mathbf{C}_x + \mathbf{C}_w)^{-1}], \tag{3.58}$$

if

$$\mathrm{Tr}[\mathbf{C}_w^{-1} - (\mathbf{C}_x + \mathbf{C}_w)^{-1}] > 0. \tag{3.59}$$

Using (3.57), on the other hand, (3.55) becomes

$$\mathrm{Tr}[\mathbf{C}_w^{-1} - (\mathbf{C}_x + \mathbf{C}_w)^{-1}]\mathrm{Tr}(\mathbf{yy}^*) \geq \mathrm{Tr}\{[\mathbf{C}_w^{-1} - (\mathbf{C}_x + \mathbf{C}_w)^{-1}]\mathbf{yy}^*\} \underset{\mathcal{H}_0}{\overset{\mathcal{H}_1}{\underset{<}{>}}} T_1, \tag{3.60}$$

$$\mathrm{Tr}(\mathbf{yy}^*) \underset{\mathcal{H}_0}{\overset{\mathcal{H}_1}{\underset{<}{>}}} \frac{T_1}{\mathrm{Tr}[\mathbf{C}_w^{-1} - (\mathbf{C}_x + \mathbf{C}_w)^{-1}]} = \frac{T_1}{\mathrm{Tr}(\mathbf{A}^{-1} - \mathbf{B}^{-1})}, \tag{3.61}$$

if

$$\mathrm{Tr}[\mathbf{C}_w^{-1} - (\mathbf{C}_x + \mathbf{C}_w)^{-1}] = \mathrm{Tr}(\mathbf{A}^{-1} - \mathbf{B}^{-1}) > 0, \tag{3.62}$$

(which is identical to (3.59)) whose necessary condition to be valid is:

$$\mathbf{A} \leq \mathbf{B} \quad \text{or} \quad \mathbf{C}_w < \mathbf{C}_x + \mathbf{C}_w, \tag{3.63}$$

due to the following fact: If $\mathbf{A} \leq \mathbf{B}$ for $\mathbf{A} \geq 0$ and $\mathbf{B} \geq 0$, then (see Section A.6.2)

$$\mathbf{A}^{-1} \geq \mathbf{B}^{-1}. \tag{3.64}$$

Note that $\mathbf{A}^{-1} + \mathbf{B}^{-1} \neq (\mathbf{B} + \mathbf{A})^{-1}$.

Starting from (3.53), we have

$$\begin{aligned}\mathrm{Tr}[\mathbf{C}_w^{-1} - (\mathbf{C}_x + \mathbf{C}_w)^{-1}] &= \mathrm{Tr}[\mathbf{C}_w^{-1}\mathbf{C}_x(\mathbf{C}_x + \mathbf{C}_w)^{-1}] = \mathrm{Tr}[\mathbf{C}_x(\mathbf{C}_x + \mathbf{C}_w)^{-1}\mathbf{C}_w^{-1}] \\ &= \mathrm{Tr}\{[\mathbf{C}_w(\mathbf{C}_x + \mathbf{C}_w)]^{-1}\mathbf{C}_x\} = \mathrm{Tr}(\mathbf{A}^{-1}\mathbf{B}),\end{aligned} \tag{3.65}$$

with

$$\mathbf{A} = \mathbf{C}_w(\mathbf{C}_x + \mathbf{C}_w), \mathbf{B} = \mathbf{C}_x, \mathbf{A} > 0, \mathbf{B} > 0, \tag{3.66}$$

where (3.54) is used in the second step and, in the third step, we use the following [114, p. 43]

$$(\mathbf{AB})^{-1} = \mathbf{B}^{-1}\mathbf{A}^{-1}.$$

Let $\mathbf{A} > 0$ and $\mathbf{B} > 0$ be of the same size. Then [114, p. 181]

$$\text{Tr}\mathbf{A}^{-1}\mathbf{B} \geq \frac{\text{Tr}\mathbf{A}}{\lambda_{\max}(\mathbf{A})} \geq \frac{\text{Tr}\mathbf{B}}{\text{Tr}\mathbf{A}}, \text{Tr}\mathbf{A} > 0, \text{Tr}\mathbf{B} > 0, \qquad (3.67)$$

where $\lambda_{\max}(\mathbf{A})$ is the largest eigenvalue of $\mathbf{A}$. Due to the facts $\mathbf{A} > 0$ and $\mathbf{B} > 0$ given in (3.66), with the help of (3.65) and (3.67), (3.61) becomes

$$\text{Tr}(\mathbf{yy}^*) \underset{\mathcal{H}_0}{\overset{\mathcal{H}_1}{\underset{<}{\gtrless}}} \frac{T_1}{\text{Tr}(\mathbf{A}^{-1}\mathbf{B})} \geq T_1 \frac{\text{Tr}\mathbf{A}}{\text{Tr}\mathbf{B}} = T_1 \frac{\text{Tr}[\mathbf{C}_w(\mathbf{C}_x + \mathbf{C}_w)]}{\text{Tr}\mathbf{C}_x}. \qquad (3.68)$$

Combining (3.65) and (3.67) yields

$$\text{Tr}[\mathbf{C}_w^{-1} - (\mathbf{C}_x + \mathbf{C}_w)^{-1}] \geq \frac{\text{Tr}\mathbf{C}_x}{\text{Tr}[\mathbf{C}_w(\mathbf{C}_x + \mathbf{C}_w)]} > 0. \qquad (3.69)$$

Using (3.69), (3.58) turns out to be

$$|\text{Tr}(\mathbf{yy}^*) - T_1| > \text{Tr}[\mathbf{C}_w^{-1} - (\mathbf{C}_x + \mathbf{C}_w)^{-1}] \geq \frac{\text{Tr}\mathbf{C}_x}{\text{Tr}[\mathbf{C}_w(\mathbf{C}_x + \mathbf{C}_w)]},$$

since

$$\text{Tr}[\mathbf{C}_w^{-1} - (\mathbf{C}_x + \mathbf{C}_w)^{-1}] > 0.$$

Here $|a|$ is the absolutue value of a. The necessary condition of (3.63) is satisfied even if the covariance matrix of the signal $\mathbf{C}_x$ is extremely small, compared with that of the noise $\mathbf{C}_w$. This is critical to the practical problem at hand: sensing of an extremely weak signal in the spectrum of interest. At first sight, the necessary condition of (3.63) seems be trivial to achieve; this intuition, however, is false. At extremely low SNR, this condition is too strong to be satisfied. The lack of sufficient samples to estimate the covariance matrices $\mathbf{C}_w$ and $\mathbf{C}_x$ is the root of the technical difficulty. Fortunately, the signal covariance matrix $\mathbf{C}_x$ lies in the space of lower rank than that of the noise covariance matrix $\mathbf{C}_w$. This fundamental difference in their ranks is the departure point for most of developments in Chapter 4.

(3.68) is expressed in a form that is convenient in practice. Only the traces of these covariance matrices are needed! The traces are, of course, positive scalar real values, which, in general, are random variables. The necessary condition for (3.68) to be valid is (3.63) $\mathbf{C}_w < \mathbf{C}_x + \mathbf{C}_w$.

3.7.1.1 Divergence

Let $\mathbf{y} : \mathcal{N}[0, \mathbf{C}_y]$ denote an $N \times 1$ normal random vector with mean zero and covariance matric $\mathbf{C}_y$. The problem considered here is the test:

$$\mathcal{H}_0 : \mathbf{C}_y = \mathbf{C}_0$$
$$\mathcal{H}_1 : \mathbf{C}_y = \mathbf{C}_1 .$$

In particular, we have $\mathbf{C}_0 = \mathbf{C}_w$ and $\mathbf{C}_1 = \mathbf{C}_w + \mathbf{C}_x$. Divergence [123] is a coarse measure of how the log likelihood distinguishes between $\mathcal{H}_0$ and $\mathcal{H}_1$:

$$J = E_{\mathcal{H}_1} L(\mathbf{y}) - E_{\mathcal{H}_0} L(\mathbf{y}),$$

where

$$L(\mathbf{y}) = \mathbf{y}^* \mathbf{Q} \mathbf{y}, \mathbf{Q} = \mathbf{C}_0^{-1} - \mathbf{C}_1^{-1}.$$

The matrix $\mathbf{Q}$ can be rewritten as

$$\mathbf{Q} = \mathbf{C}_0^{-T/2} (\mathbf{I} - \mathbf{S}) \mathbf{C}_0^{-1/2}$$
$$\mathbf{C}_0 = \mathbf{C}_0^{1/2} \mathbf{C}_0^{T/2}; \mathbf{C}_0^{T/2} = (\mathbf{C}_0^{1/2})^T .$$
$$\mathbf{S} = \mathbf{C}_0^{-1/2} \mathbf{C}_1 \mathbf{C}_0^{-T/2}; \mathbf{C}_0^{-T/2} = (\mathbf{C}_0^{-1/2})^T$$

The log likelihood ratio can be rewritten as

$$L(\mathbf{y}) = \mathbf{z}^* (\mathbf{I} - \mathbf{S}^{-1}) \mathbf{z}$$
$$\mathbf{z} = \mathbf{C}_0^{-1/2} \mathbf{y}.$$

The transformed vector $\mathbf{z}$ is distributed as $\mathcal{N}[0, \mathbf{C}_z]$, with $\mathbf{C}_z = \mathbf{I}$ for $\mathcal{H}_0$ and $\mathbf{C}_z = \mathbf{S}$ for $\mathcal{H}_1$. $\mathbf{S}$ is called the signal-to-noise-ratio matrix.

3.7.1.2 Orthogonal Decomposition

The matrix $\mathbf{S}$ has an orthogonal decomposition

$$\mathbf{S} = \mathbf{C}_0^{-1/2} \mathbf{C}_1 \mathbf{C}_0^{-T/2} = \mathbf{U} \mathbf{\Lambda} \mathbf{U}^T ,$$
$$\mathbf{S} \mathbf{U} = \mathbf{U} \mathbf{\Lambda}$$

where $\mathbf{\Lambda}$ is a diagonal matrix with diagonal elements λ_i and $\mathbf{U}$ satisfies $\mathbf{U}^T \mathbf{U} = \mathbf{I}$. This implies that $(\mathbf{C}_0^{-T/2} \mathbf{U}, \mathbf{\Lambda})$ solves the generalized eigenvalue problem

$$\mathbf{C}_1 (\mathbf{C}_0^{-T/2} \mathbf{U}) - \mathbf{C}_0 (\mathbf{C}_0^{-T/2} \mathbf{U}) \mathbf{\Lambda} = 0.$$

With this representation for S, the log likelihood ratio may be expressed as

$$L(\mathbf{y}) = \mathbf{z}^* \mathbf{U} (\mathbf{I} - \mathbf{\Lambda}^{-1}) \mathbf{U}^T \mathbf{z},$$

where the random vector $\mathbf{y}$ has covariance matrix $\mathbf{U} \mathbf{\Lambda} \mathbf{U}^T$ under $\mathcal{H}_1$ and $\mathbf{I}$ under $\mathcal{H}_0$:

$$E_{\mathcal{H}_1} \mathbf{z} \mathbf{z}^* = \mathbf{S}$$
$$E_{\mathcal{H}_0} \mathbf{z} \mathbf{z}^* = \mathbf{I}.$$

3.7.1.3   Rank Reduction

A reduced-rank version of the log likelihood ratio is

$$L_r(\mathbf{y}) = \mathbf{z}^*\mathbf{U}(\mathbf{I}_r - \mathbf{\Lambda}_r^{-1})\mathbf{U}^T\mathbf{z},$$

where $\mathbf{I}_r$ and $\mathbf{\Lambda}_r^{-1}$ are reduced-rank versions of $\mathbf{I}$ and $\mathbf{\Lambda}^{-1}$ that retain r nonzero terms and $N - r$ zero terms.

This rank reduction is fine, whenever the discarded eigenvalues λ_i are unity (noise components). The problem is that nonunity eigenvalues are sometimes discarded with much penalty. We introduce a new criterion, divergence, a coarse measure of how the log likelihood distinguishes between $\mathcal{H}_0$ and $\mathcal{H}_1$

$$
\begin{aligned}
J &= E_{\mathcal{H}_1} L(\mathbf{y}) - E_{\mathcal{H}_0} L(\mathbf{y}) = \text{Tr}\mathbf{U}(\mathbf{I} - \mathbf{\Lambda}^{-1})\mathbf{U}^T\mathbf{U}\mathbf{\Lambda}\mathbf{U}^T - \text{Tr}\mathbf{U}(\mathbf{I} - \mathbf{\Lambda}^{-1})\mathbf{U}^T\mathbf{U}\mathbf{U}^T \\
&= \text{Tr}(\mathbf{\Lambda} + \mathbf{\Lambda}^{-1} - 2\mathbf{I}) \\
&= \sum_{n=1}^{N}(\mathbf{\Lambda}_n + \mathbf{\Lambda}_n^{-1} - 2) = \text{Tr}(\mathbf{S} + \mathbf{S}^{-1} - 2\mathbf{I}).
\end{aligned}
$$

We emphasize that it is the sum of $\mathbf{\Lambda}_n + \mathbf{\Lambda}_n^{-1}$, not $\mathbf{\Lambda}_n$, that determines the contribution of an eigenvalue. It will lead to penalty when we discard the small eigenvalues. At extremely low SNR, the eigenvalues are almost uniformly distributed, it is difficult to do reduced-rank processing. The trace sum of $\mathbf{S}$ and $\mathbf{S}^{-1}$ must be considered as a whole. Obviously, we require that

$$\frac{1}{2}(\mathbf{S} + \mathbf{S}^{-1}) \geq \mathbf{I},$$

since

$$\mathbf{A} \geq \mathbf{B} \Rightarrow \text{Tr}\mathbf{A} \geq \text{Tr}\mathbf{B},$$

where $\mathbf{A}, \mathbf{B}$ are Hermitian matrices. The matrix inequality condition $\frac{1}{2}(\mathbf{S} + \mathbf{S}^{-1}) \geq \mathbf{I}$ implies

$$J = \text{Tr}(\mathbf{S} + \mathbf{S}^{-1} - 2\mathbf{I}) \geq 0,$$

or,

$$J = \text{Tr}[(\mathbf{C}_0^{-1} - \mathbf{C}_1^{-1})\mathbf{C}_1] - \text{Tr}[(\mathbf{C}_0^{-1} - \mathbf{C}_1^{-1})\mathbf{C}_0] = \text{Tr}[(\mathbf{C}_0^{-1} - \mathbf{C}_1^{-1})(\mathbf{C}_0 - \mathbf{C}_1)] \geq 0.$$

The rank r divergence is identical to the full-rank divergence when $N - r$ of the eigenvalues in the original diagonal matrix $\mathbf{\Lambda}$ are unity. To illustrate, consider the case

$$\mathcal{H}_0 : \mathbf{C}_0 = \mathbf{Q} + \sum_{i=1}^{p} \sigma_i^2 u_i u_i^*$$

$$\mathcal{H}_1 : \mathbf{C}_1 = \mathbf{Q} + \sum_{i=1}^{p} \sigma_i^2 v_i v_i^*.$$

The difference between $\mathbf{C}_1$ and $\mathbf{C}_0$ is

$$
\begin{aligned}
\mathbf{C}_1 - \mathbf{C}_0 &= \sum_{i=1}^{p} \sigma_i^2 v_i v_i^* - \sum_{i=1}^{p} \sigma_i^2 u_i u_i^* = \mathbf{L}\mathbf{D}\mathbf{L}^* \\
\mathbf{L} &= [v_{p+1}, v_{p+2}, \cdots v_{p+q}, u_1, \cdots, u_p] \\
\mathbf{D} &= \text{diag}[\sigma_{p+1}^2, \sigma_{p+2}^2, \cdots, \sigma_{p+q}^2, -\sigma_1^2, -\sigma_2^2 \cdots, \sigma_p^2].
\end{aligned}
$$

Assuming $\sigma_i^2 > 0$, then $\text{rank}(\mathbf{R}_1 - \mathbf{R}_0) = \text{rank}(\mathbf{L})$ and a $\text{rank}(\mathbf{L})$ detector has the same divergence as a full-rank detector.

Example 3.6 (An optimal rank-one detector)

Consider building a low-rank detector for $\mathcal{H}_0$ versus $\mathcal{H}_1$ when the observed data is distributed as $\mathcal{N}(\mathbf{0}, \mathbf{R}_0)$ under $\mathcal{H}_0$ and $\mathcal{N}(\mathbf{0}, \mathbf{R}_1)$ under $\mathcal{H}_1$:

$$\mathcal{H}_0 : \mathbf{C}_0 = \sigma^2 \mathbf{I} + \beta^2 \mathbf{w}\mathbf{w}^*$$
$$\mathcal{H}_1 : \mathbf{C}_1 = \sigma^2 \mathbf{I} + \beta^2 \mathbf{w}\mathbf{w}^* + \mathbf{v}\mathbf{v}^*.$$

After some manipulation, we get the following signal-to-noise-ratio matrix:

$$\mathbf{S} = \mathbf{I} + \mathbf{U}\boldsymbol{\Sigma}\mathbf{v}^*$$
$$\boldsymbol{\Sigma} = \text{diag}\left[\mathbf{v}^*\mathbf{C}_0\mathbf{v}, 0, 0, \cdots, 0\right].$$

The eigenvalues of $\mathbf{S}$ are

$$\lambda_1 = 1 + \mathbf{v}^*\mathbf{C}_0\mathbf{v}, \lambda_2 = 1, \lambda_3 = 1, \cdots, \lambda_N = 1.$$

A rank-one detector is optimum for this problem. It is constructed using the eigenvector corresponding to λ_1. At extremely low SNR, the eigenvector is a more reliable feature to detect than the eigenvalue. In N-dimension geometric terms, the eigenvector is the direction of the data point, while the eigenvalue the length of this data point. $\square$

Example 3.7 (Gaussian signal plus Gaussian noise)

We can extend the previous example to a general Gaussian signal problem

$$\mathcal{H}_0 : \mathbf{C}_0$$
$$\mathcal{H}_1 : \mathbf{C}_1 = \mathbf{C}_0 + \mathbf{C}_S,$$

yielding

$$\mathbf{S} = \mathbf{I} + \mathbf{C}_0^{-1/2}\mathbf{C}_S\mathbf{C}_0^{-T/2}.$$

Clearly, all eigenvalues are greater than one. This does not mean, however, that eigenvalues close to one cannot be discarded in order to approximate log likelihood with a low-rank detector. At extremely low SNR, this approximation will most often not work since maybe all eigenvalues are close to one. $\square$

3.7.2 Tests with Repeated Observations

Consider a binary hypothesis problem [124] with a sequence of independent distributed random vectors $\mathbf{y}_k \in \mathcal{C}^N$. If

$$\Lambda(\mathbf{y}) = \frac{p(\mathbf{y}; \mathcal{H}_1)}{p(\mathbf{y}; \mathcal{H}_0)}$$

denotes the likelihood radio function of a single observation, an LRT for this problem takes the form

$$\prod_{k=1}^{K} \Lambda(\mathbf{y}_k) \underset{\mathcal{H}_0}{\overset{\mathcal{H}_1}{\underset{<}{\gtrless}}} \tau(K), \tag{3.70}$$

where $\tau(K)$ denotes a threshold that may depend on the number of observations, K. Taking the logarithms on both sides of (3.70), and denoting $Z_k = \ln(\Lambda(\mathbf{y}_k))$, we find that

$$S_K = \frac{1}{K}\sum_{k=0}^{K-1} Z_k \underset{\mathcal{H}_0}{\overset{\mathcal{H}_1}{\underset{<}{>}}} \gamma(K) = \frac{\ln(\tau(K))}{K}.$$

When $\tau(K)$ tends to a constant as $K \to \infty$, $\gamma(K) \overset{\Delta}{=} \lim_{K\to\infty}\gamma(K) = 0$.

Taking the logarithm, retaining only the data-dependent terms, and scaling produces the test statistic (setting μ_x for brevity)

$$T(\mathbf{y}) = \frac{1}{2}\sum_{k=0}^{K-1} \mathbf{y}_k^*[\mathbf{C}_w^{-1} - (\mathbf{C}_x + \mathbf{C}_w)^{-1}]\mathbf{y}_k \underset{\mathcal{H}_0}{\overset{\mathcal{H}_1}{\underset{<}{>}}} T_0. \qquad (3.71)$$

Using the following fact

$$\text{Tr}\sum_{k=0}^{K-1} \mathbf{x}_k^*\mathbf{A}\mathbf{x}_k = \text{Tr}(\mathbf{A}\mathbf{X}), \text{ where } \mathbf{X} = \sum_{k=0}^{K-1}\mathbf{x}_k\mathbf{x}_k^*, \qquad (3.72)$$

(3.71) becomes

$$\text{Tr}\left\{[\mathbf{C}_w^{-1} - (\mathbf{C}_x + \mathbf{C}_w)^{-1}]\frac{1}{K}\sum_{k=0}^{K-1}\mathbf{y}_k^*\mathbf{y}_k\right\} \underset{\mathcal{H}_0}{\overset{\mathcal{H}_1}{\underset{<}{>}}} \frac{T_1}{K}. \qquad (3.73)$$

If the following sufficient condition for the LRT for repeated observations is satisfied

$$\mathbf{C}_w^{-1} - (\mathbf{C}_x + \mathbf{C}_w)^{-1} > 0, \qquad (3.74)$$

implying (see Section A.6.2)

$$\text{Tr}[\mathbf{C}_w^{-1} - (\mathbf{C}_x + \mathbf{C}_w)^{-1}] > 0,$$

using (3.57), we have

$$\text{Tr}\{[\mathbf{C}_w^{-1} - (\mathbf{C}_x + \mathbf{C}_w)^{-1}]\}\text{Tr}\left(\frac{1}{K}\sum_{k=0}^{K-1}\mathbf{y}_k^*\mathbf{y}_k\right) \overset{\mathcal{H}_1}{>} \frac{1}{K}T_0. \qquad (3.75)$$

Or,

$$\text{Tr}\left(\frac{1}{K}\sum_{k=0}^{K-1}\mathbf{y}_k^*\mathbf{y}_k\right) \overset{\mathcal{H}_1}{>} \frac{1}{K}\frac{1}{\text{Tr}[\mathbf{C}_w^{-1} - (\mathbf{C}_x + \mathbf{C}_w)^{-1}]}T_0. \qquad (3.76)$$

The covariance matrix of $\mathbf{y}$ is defined as

$$\mathbf{C}_y = \mathbb{E}(\mathbf{y}\mathbf{y}^*).$$

The asymptotic case of $K \to \infty$ is

$$\text{Tr}\mathbf{C}_y \overset{\mathcal{H}_1}{>} \frac{1}{K}\frac{1}{\text{Tr}[\mathbf{C}_w^{-1} - (\mathbf{C}_x + \mathbf{C}_w)^{-1}]}T_0 \to 0, \qquad (3.77)$$

since the sample covariance matrix converges to the true covariance matrix, that is

$$\mathbf{C}_y = \lim_{K \to \infty} \frac{1}{K} \sum_{k=0}^{K-1} \mathbf{y}_k^* \mathbf{y}_k.$$

To guarantee (3.74), the following stronger condition is enough

$$\mathbf{C}_w \le \mathbf{C}_x + \mathbf{C}_w, \tag{3.78}$$

due to (3.64).

3.7.2.1   Case 1. Diagonal Covariance Matrix on $\mathcal{H}_0$: Equal Variances

When $x[n]$ is a complex Gaussian random process with zero mean and covariance matrix $\mathbf{C}_x$ and $w[n]$ is complex white Gaussian noise (CWGN) with variance matrix σ_n^2. The probability distribution functions (PDFs) are given by

$$\begin{aligned} p(\mathbf{y}; \mathcal{H}_1) &= \frac{1}{\pi^N \det(\mathbf{C}_x + \sigma_n^2 \mathbf{I})} \exp[-\mathbf{y}^* (\mathbf{C}_x + \sigma_n^2 \mathbf{I})^{-1} \mathbf{y}] \\ p(\mathbf{y}; \mathcal{H}_0) &= \frac{1}{\pi^N \sigma_n^{2N}} \exp[-\frac{1}{\sigma_n^2} \mathbf{y}^* \mathbf{y}] \end{aligned},$$

where $\mathbf{I}$ is the identity matrix of $N \times N$, and $\mathrm{Tr}\mathbf{I} = N$.

The log-likelihood ratio is

$$\ln L(\mathbf{y}) = -\mathbf{y}^* [(\mathbf{C}_x + \sigma_n^2 \mathbf{I})^{-1} - \frac{1}{\sigma_n^2} \mathbf{I}] \mathbf{y} - \ln \det (\mathbf{C}_x + \sigma_n^2 \mathbf{I}) + \ln \det \sigma_n^{2N}.$$

Consider the following special case:

$$\mathbf{C}_w^{-1} = \frac{1}{\sigma_n^2} \mathbf{I}; \ \mathbf{C}_x = \sigma_s^2 \mathbf{I}; \ (\mathbf{C}_x + \mathbf{C}_w)^{-1} = (\sigma_s^2 \mathbf{I} + \sigma_n^2 \mathbf{I})^{-1} = \frac{1}{\sigma_s^2 + \sigma_n^2} \mathbf{I},$$

and

$$\mathrm{Tr}\mathbf{C}_w^{-1} \mathrm{Tr}\mathbf{C}_x \mathrm{Tr}(\mathbf{C}_x + \mathbf{C}_w)^{-1} = \frac{N \sigma_s^2}{\sigma_n^2 (\sigma_s^2 + \sigma_n^2)}.$$

The LRT of (3.79) becomes

$$\mathrm{Tr}\left(\frac{1}{K} \sum_{k=0}^{K-1} \mathbf{y}_k^* \mathbf{y}_k \right) \overset{\mathcal{H}_1}{\underset{}{>}} \frac{1}{K} \frac{1}{\mathrm{Tr}\{[\mathbf{C}_w^{-1} - (\mathbf{C}_x + \mathbf{C}_w)^{-1}]\}} T_0. \tag{3.79}$$

Using the matrix inverse lemma

$$\mathbf{C}_w^{-1} - (\mathbf{C}_x + \mathbf{C}_w)^{-1} = \mathbf{C}_w^{-1} \mathbf{C}_x (\mathbf{C}_x + \mathbf{C}_w)^{-1},$$

we have

$$\mathrm{Tr}\left(\frac{1}{K} \sum_{k=0}^{K-1} \mathbf{y}_k^* \mathbf{y}_k \right) \overset{\mathcal{H}_1}{\underset{}{>}} \frac{1}{K} \frac{T_0}{\mathrm{Tr}\mathbf{C}_w^{-1} \mathrm{Tr}\mathbf{C}_x \mathrm{Tr}(\mathbf{C}_x + \mathbf{C}_w)^{-1}} = \frac{1}{K} T_0 \frac{\sigma_n^2 (\sigma_s^2 + \sigma_n^2)}{N \sigma_s^2} \approx T_0 \frac{\sigma_n^4}{K N \sigma_s^2}, \tag{3.80}$$

when the signal is very weak, or $\sigma_s^2 \ll \sigma_n^2$. $\mathrm{Tr}\mathbf{C}_y$ is the total power (variation) of the received signal plus noise. Note $\mathbf{y}_k$ are vectors of length N.

Consider the single observation case, $K = 1$, we have

$$\mathrm{Tr}(\mathbf{yy}^*) = \sum_{i=0}^{N-1} |y[i]|^2,$$

which is the energy detector. Intuitively, if the signal is present, the energy of the received data increases. In fact, the equivalent test statistic $T = \frac{1}{N} \sum_{i=0}^{N-1} |y[i]|^2$ can be regarded as an estimator of the variance. Comparing this to a threshold recognizes that the variance under $\mathcal{H}_0$ is σ_n^2 but under $\mathcal{H}_1$ is $\sigma_s^2 + \sigma_n^2$.

3.7.2.2 Case 2. Correlated Signal

Now assume $N = 2$ and

$$\mathbf{C}_x = \sigma_s^2 \begin{pmatrix} 1 & \rho \\ \rho & 1 \end{pmatrix},$$

where ρ is the correlation coefficient between $x[0]$ and $x[1]$.

$$\mathbf{C}_w^{-1} = \frac{1}{\sigma_n^2}\mathbf{I}; \ (\mathbf{C}_x + \mathbf{C}_w)^{-1} = \begin{pmatrix} \sigma_s^2 + \sigma_n^2 & \rho\sigma_s^2 \\ \rho\sigma_s^2 & \sigma_s^2 + \sigma_n^2 \end{pmatrix}^{-1} = [(\sigma_s^2 + \sigma_n^2)\mathbf{I} + \rho\sigma_s^2\mathbf{Q}]^{-1},$$

where

$$\mathbf{Q} = \begin{pmatrix} 0 & 1 \\ 1 & 0 \end{pmatrix},$$

$$\mathbf{A} = \begin{pmatrix} a & b \\ b & a \end{pmatrix}, \mathbf{A}^{-1} = \frac{1}{a^2 - b^2} \begin{pmatrix} a & -b \\ -b & a \end{pmatrix}, \mathrm{Tr}\mathbf{A}^{-1} = \mathrm{Tr}(\mathbf{C}_x + \mathbf{C}_w)^{-1} = \frac{2a}{a^2 - b^2},$$

where

$$\mathbf{A} = \mathbf{C}_x + \mathbf{C}_w, a = \sigma_s^2 + \sigma_n^2, b = \rho\sigma_s^2.$$

If $\mathbf{A} > 0$, then

$$\left(\sum_{i=1}^{n} a_{ii}\right)^{-1} \leq \mathrm{Tr}\mathbf{A}^{-1}.$$

Obviously, $\mathbf{C}_w + \mathbf{C}_x > 0$. From (3.77), we have

$$\mathrm{Tr}\left(\frac{1}{K}\sum_{k=0}^{K-1} \mathbf{y}_k^*\mathbf{y}_k\right) \underset{\mathcal{H}_1}{\overset{\mathcal{H}_1}{>}} \frac{1}{K}\frac{1}{\mathrm{Tr}[\mathbf{C}_w^{-1} - (\mathbf{C}_x + \mathbf{C}_w)^{-1}]}T_0. \tag{3.81}$$

3.7.3 Detection Using Sample Covariance Matrices

If we know the covariance matrices perfectly, we have

$$\mathcal{H}_0 : \mathbf{R}_y = \mathbf{A}, \mathbf{A} \geq 0$$
$$\mathcal{H}_1 : \mathbf{R}_y = \mathbf{A} + \mathbf{B}, \mathbf{B} > 0 \ .$$

In practice, estimated covariance matrices such as sample covariance matrices must be used:

$$\mathcal{H}_0 : \mathbf{R}_y = \mathbf{A}_0, \mathbf{A}_0 \geq 0$$
$$\mathcal{H}_1 : \mathbf{R}_y = \mathbf{A}_1 + \mathbf{B}, \mathbf{A}_1, \mathbf{B} > 0,$$

where $\mathbf{A}_0$ and $\mathbf{A}_1$ are sample covariance matrices for the noise while $\mathbf{B}$ is the sample covariance matrix for the signal. The eigenvalues λ_i of

$$(\mathbf{A} + \mathbf{B})^{-1}\mathbf{A},$$

where $\mathbf{A}$ is positive semidefinite and $\mathbf{B}$ positive definite, satisfy

$$0 \leq \lambda_i \leq 1.$$

Let $\mathbf{A}$ be positive definite and $\mathbf{B}$ symmetric such that $\det(\mathbf{A} + \mathbf{B}) \neq 0$, then

$$(\mathbf{A} + \mathbf{B})^{-1}\mathbf{B}(\mathbf{A} + \mathbf{B})^{-1} \leq \mathbf{A}^{-1} - (\mathbf{A} + \mathbf{B})^{-1}.$$

The inequality, known as Olkin's inequality, is strict if and only if $\mathbf{B}$ is nonsingular.

Wilks' lambda test [110, p. 335]

$$\Lambda = \frac{\det \mathbf{A}}{\det(\mathbf{A} + \mathbf{B})} = \prod_{j=1}^{p}(1 + \lambda_j)^{-1},$$

where λ_j are the eigenvalues of $\mathbf{A}^{-1}\mathbf{B}$.

The equicorrelation matrix is [112, p. 241]

$$\mathbf{E} = \begin{pmatrix} 1 & \rho \cdots & \rho \\ \rho & 1 \cdots & \rho \\ \vdots & \vdots & \vdots \\ \rho & \rho \cdots & 1 \end{pmatrix},$$

or,

$$\mathbf{E} = (1 - \rho)\mathbf{I} + \rho\mathbf{J},$$

where ρ is any real number and $\mathbf{J}$ is a unit matrix $\mathbf{J}_p = \mathbf{1}\mathbf{1}^T$, $\mathbf{1} = (1, \ldots, 1)^T$. Then $e_{ii} = 1$, $e_{ij} = \rho$, for $i \neq j$. For statistical purposes this matrix is most useful for $-(p-1)^{-1} < \rho < 1$. Direct verification shows that, if $\rho \neq 1, -(p-1)^{-1}$, then $\mathbf{E}^{-1}$ exists and is given by

$$\mathbf{E}^{-1} = (1 - \rho)^{-1}\{\mathbf{I} - \rho[1 + (p-1)\rho]^{-1}\mathbf{J}\}.$$

Its determinant is given by

$$\det \mathbf{E} = (1 - \rho)^{p-1}[1 + (p-1)\rho].$$

Since $\mathbf{J}$ has rank 1 with eigenvalues p and corresponding eigenvector $\mathbf{1}$, we see that the equicorrelation matrix $\mathbf{E} = (1 - \rho)\mathbf{I} + \rho\mathbf{J}$ has eigenvalues:

$$\lambda_1 = 1 + (p-1)\rho, \lambda_2 = \cdots = \lambda_p = 1 + (p-1)\rho,$$

and the same eigenvectors as $\mathbf{J}$. $\log \det \mathbf{A}$ is bounded:

$$\frac{\det \mathbf{A}}{\det (\mathbf{A} + \mathbf{B})} \leq \exp(\mathrm{Tr}(\mathbf{A}^{-1}\mathbf{B})),$$

where $\mathbf{A}$ and $\mathbf{A} + \mathbf{B}$ are positive definite, with equality if and only if $\mathbf{B} = 0$. If $\mathbf{A} \geq 0$ and $\mathbf{B} \geq 0$, then [112, 329]

$$\begin{aligned} 0 \leq \mathrm{Tr}\mathbf{AB} &\leq (\mathrm{Tr}\mathbf{A})(\mathrm{Tr}\mathbf{B}) \\ \sqrt{\mathrm{Tr}\mathbf{AB}} &\leq (\mathrm{Tr}\mathbf{A} + \mathrm{Tr}\mathbf{B})/2. \end{aligned}$$

3.7.4 GLRT for Multiple Random Vectors

The data is modeled as a complex Gaussian random vector $\mathbf{x} : \Omega \to \mathbb{C}^N$ with probability density function

$$p(\mathbf{x}) = \frac{1}{\pi^N \det \mathbf{R}_{xx}} \exp[-(\mathbf{x} - \boldsymbol{\mu}_x)^H \mathbf{R}_{xx}^{-1}(\mathbf{x} - \boldsymbol{\mu}_x)],$$

mean $\boldsymbol{\mu}_x$, and covariance matrix $\mathbf{R}_{xx}$. Consider M independent, identically distributed (i.i.d.) random vectors

$$\mathbf{x} = [\mathbf{x}_1, \mathbf{x}_2, \cdots, \mathbf{x}_M],$$

drawn from this distribution. The joint probability density function of these vectors is written as

$$\begin{aligned} p(\mathbf{x}) &= \frac{1}{\pi^{MN}(\det \mathbf{R}_{xx})^M} \exp\left[-\sum_{m=1}^{M}(\mathbf{x}_m - \boldsymbol{\mu}_x)^H \mathbf{R}_{xx}^{-1}(\mathbf{x}_m - \boldsymbol{\mu}_x)\right] \\ &= \pi^{-MN}(\det \mathbf{R}_{xx})^{-M} \exp[-M\mathrm{Tr}(\mathbf{R}_{xx}^{-1}\mathbf{S}_{xx})], \end{aligned}$$

where $\mathbf{S}_{xx}$ is the sample covariance matrix

$$\mathbf{S}_{xx} = \frac{1}{M}\sum_{m=1}^{M}(\mathbf{x}_m - \boldsymbol{\mu}_x)(\mathbf{x}_m - \boldsymbol{\mu}_x)^H = \frac{1}{M}\mathbf{x}\mathbf{x}^H - \mathbf{m}_x,$$

and $\mathbf{m}_x$ is the sample mean vector

$$\mathbf{m}_x = \frac{1}{M}\sum_{m=1}^{M}\mathbf{x}_m.$$

Our task is to test whether $\mathbf{R}_{xx}$ has structure 0 or the alternative structure 1:

$$\mathcal{H}_0 : \mathbf{R}_{xx} \in \mathbb{R}_0,$$
$$\mathcal{H}_1 : \mathbf{R}_{xx} \in \mathbb{R}_1.$$

The GLRT statistic is

$$L = \frac{\max\limits_{\mathbf{R}_{xx} \in \mathbb{R}_0} p(\mathbf{x})}{\max\limits_{\mathbf{R}_{xx} \in \mathbb{R}_1} p(\mathbf{x})}.$$

The actual covariance matrices are not known, they are replaced with their ML estimates computed using the M random vectors. If we denote by $\hat{\mathbf{R}}_0$ the ML estimate of $\mathbf{R}_{xx}$ under $\mathcal{H}_0$ and by $\hat{\mathbf{R}}_1$ the ML estimate of $\mathbf{R}_{xx}$ under $\mathcal{H}_1$, we have

$$L = \det{}^M (\hat{\mathbf{R}}_0^{-1} \hat{\mathbf{R}}_1) \exp[-M\mathrm{Tr}(\hat{\mathbf{R}}_0^{-1} \mathbf{S}_{xx} - \hat{\mathbf{R}}_1^{-1} \mathbf{S}_{xx})].$$

If we assume further that $\mathbb{R}_1$ is the set of positive definite matrices (no special restrictions are imposed), then $\hat{\mathbf{R}}_1 = \mathbf{S}_x x$, and

$$L = \det{}^M (\hat{\mathbf{R}}_0^{-1} \mathbf{S}_{xx}) \exp[MN - \mathrm{Tr}(\hat{\mathbf{R}}_0^{-1} \mathbf{S}_{xx})].$$

The generalized likelihood ratio for testing whether $\mathbf{R}_{xx}$ has structure $\mathbb{R}_0$ is

$$l = L^{1/(MN)} = g \exp(1 - a),$$

where a and g are the arithmetic and geometric means of the eigenvalues of $\hat{\mathbf{R}}_0^{-1} \mathbf{S}_{xx}$:

$$a = \tfrac{1}{N}\mathrm{Tr}(\hat{\mathbf{R}}_0^{-1} \mathbf{S}_{xx}),$$
$$g = [\det (\hat{\mathbf{R}}_0^{-1} \mathbf{S}_{xx})]^{1/N}.$$

Based on the desirable probability of false alarm or probability of detection, we can choose a threshold l_0. So if $l > l_0$, we accept hypothesis $\mathcal{H}_0$, and if $l < l_0$, we reject it. Consider a special case

$$\mathbb{R}_0 = \{\mathbf{R}_{xx} = \sigma_x^2 \mathbf{I}\},$$

where σ_x^2 is the variance of each component of $\mathbf{x}$. The ML estimate of $\mathbf{R}_{xx}$ under $\mathcal{H}_0$ is $\hat{\mathbf{R}}_0 = \hat{\sigma}_x^2 \mathbf{I}$, where the variance is estimated as $\hat{\sigma}_x^2 = \tfrac{1}{N}\mathrm{Tr}\mathbf{S}_{xx}$. Therefore, the GLRT is

$$l = \frac{(\det \mathbf{S}_{xx})^{1/N}}{\tfrac{1}{N}\mathrm{Tr}\mathbf{S}_{xx}}.$$

This test is invariant under scale and unitary transformation.

3.7.5 Linear Discrimination Functions

Two random vectors can be stochastically ordered by using their likelihood ratio. Here, we take our liberty in freely borrowing materials from [123]. Linear discrimination may be used to approximate a quadratic likelihood ratio when testing $\mathbf{y} : \mathcal{N}(\mathbf{0}, \mathbf{C}_i)$ under hypothesis $\mathcal{H}_i$. The Neyman-Pearson test of $\mathcal{H}_0$ versus $\mathcal{H}_1$ will have us compare the log likelihood ratio to a threshold:

$$L(\mathbf{y}) = \ln \frac{f_{\theta_1}(\mathbf{y})}{f_{\theta_0}(\mathbf{y})} \underset{\mathcal{H}_0}{\overset{\mathcal{H}_1}{\underset{<}{\geq}}} \eta.$$

We hope, on the average, $L(\mathbf{y})$ will be larger than η under $\mathcal{H}_1$ and smaller than η under $\mathcal{H}_0$. An incomplete measure of how the test of $\mathcal{H}_0$ versus $\mathcal{H}_1$ will perform is the difference in terms of $L(y)$ under two hypotheses:

$$J = E_{\theta_1} L(\mathbf{y}) - E_{\theta_0} L(\mathbf{y}) = E_{\theta_1} \ln \frac{f_{\theta_1}(\mathbf{y})}{f_{\theta_0}(\mathbf{y})} - E_{\theta_0} \ln \frac{f_{\theta_1}(\mathbf{y})}{f_{\theta_0}(\mathbf{y})}.$$

This function is the J-divergence between $\mathcal{H}_0$ versus $\mathcal{H}_1$ introduced previously in Section 3.7.1. It is related to information that a random sample can bring about the hypothesis $\mathcal{H}_i$. The J-divergence for the multivariate normal problem $\mathcal{H}_i : \mathbf{y} : N(0, C_i)$ is computed by carrying out the expectations:

$$\begin{aligned} J &= \mathrm{Tr}[(\mathbf{C}_0^{-1} - \mathbf{C}_1^{-1})\mathbf{C}_1] - \mathrm{Tr}[(\mathbf{C}_0^{-1} - \mathbf{C}_1^{-1})\mathbf{C}_0] \\ &= \mathrm{Tr}[(\mathbf{C}_0^{-1} - \mathbf{C}_1^{-1})(\mathbf{C}_0 - \mathbf{C}_1)] \\ &= \mathrm{Tr}(\mathbf{C}_1\mathbf{C}_0^{-1} + \mathbf{C}_0\mathbf{C}_1^{-1} - 2I) \geq 0. \end{aligned}$$

This expression does not completely characterize the performance of a likelihood ratio statistic, but it does bring useful information about the "distance" between $\mathcal{H}_0$ and $\mathcal{H}_1$.

3.7.5.1 Linear Discrimination

Assume that the data $\mathbf{y}$ is used to form the linear discriminant function (or statistic)

$$z = \mathbf{w}^*\mathbf{y}.$$

This statistic is distributed as $\mathcal{N}[0, \mathbf{w}^*\mathbf{R}_i\mathbf{w}]$ under $\mathcal{H}_i$. If a log likelihood ratio is formed using the new variable z, then the divergence between $\mathcal{H}_0$ versus $\mathcal{H}_1$ is

$$J = \frac{1}{2}\left(\frac{\mathbf{w}^*\mathbf{R}_1\mathbf{w}}{\mathbf{w}^*\mathbf{R}_0\mathbf{w}} + \frac{\mathbf{w}^*\mathbf{R}_0\mathbf{w}}{\mathbf{w}^*\mathbf{R}_1\mathbf{w}} - 2\right).$$

Let us define the following ratio of quadratic forms

$$\lambda[\mathbf{Q}] = \frac{\mathbf{w}^*\mathbf{Q}\mathbf{w}}{\mathbf{w}^*\mathbf{R}_0\mathbf{w}}.$$

Then we may rewrite the divergence as

$$J = \frac{1}{2}\left[\lambda[\mathbf{R}_1] + \frac{1}{\lambda[\mathbf{R}_1]} - 2\right].$$

It is remarkable to note that the choice of the discriminant w that maximizes divergence is also the choice of w that maximizes a function of a quadratic form. The maximization of quadratic forms is formulated as a generalized eigenvalue problem [123, p. 163]. Let us rewrite divergence as

$$J = \frac{1}{2}\left[\lambda[\mathbf{R}_1] + \frac{1}{\lambda[\mathbf{R}_1]} - 2\right] = \frac{1}{2}\left[\lambda^{1/2}[\mathbf{R}_1] - \frac{1}{\lambda^{1/2}[\mathbf{R}_1]}\right]^2.$$

The function of J is convex in λ. It achieves its maximum either at λ_{max} or at λ_{min}, where λ_{max} and λ_{min}, respectively, are maximum and minimum values of

$$\mathbf{C}_1\mathbf{C}_0^{-1}.$$

The divergence is maximized as follows:

$$\mathbf{w} = \begin{cases} \mathbf{w}_{max}, & \text{if } \lambda_{max} > \frac{1}{\lambda_{min}} \\ \mathbf{w}_{min}, & \text{if } \lambda_{max} < \frac{1}{\lambda_{min}}. \end{cases}$$

The linear discriminant function is either the maximum eigenvector of $\mathbf{C}_1\mathbf{C}_0^{-1}$ or the minimum eigenvector of $\mathbf{C}_1\mathbf{C}_0^{-1}$, depending on the nature of the maximum and minimum values. It is not always the maximum eigenvector. The choice of $\mathbf{w} = \mathbf{w}_{max}$ or $\mathbf{w} = \mathbf{w}_{min}$ is also called a principal component analysis (PCA). Without loss of generality, $\mathbf{w}$ may be normalized as $\mathbf{w}^*\mathbf{w} = 1$. Then, if $\mathbf{R}_0 = \mathbf{I}$, the linear discriminant function is distributed as follows

$$z = \mathbf{w}^*\mathbf{y} : \begin{cases} \mathcal{N}[0, 1] \text{ under } \mathcal{H}_0 \\ \mathcal{N}[0, \lambda_{max}] \text{ or } \mathcal{N}[0, \lambda_{min}] \text{ under } \mathcal{H}_1. \end{cases}$$

3.7.6 Detection of Correlated Structure for Complex Random Vectors

For the assessment of multivariate association between two complex random vectors $\mathbf{x}$ and $\mathbf{y}$, our treatment here draws materials from [125, 126]. Consider two real zero-mean vectors $\mathbf{x} \in \mathbb{R}^m$ and $\mathbf{y} \in \mathbb{R}^n$ with two correlation matrices

$$\mathbf{R}_{xx} = E\mathbf{x}\mathbf{x}^T, \mathbf{R}_{yy} = E\mathbf{y}\mathbf{y}^T.$$

We assume both correlation matrices are invertible. The cross-correlation properties between $\mathbf{x}$ and $\mathbf{y}$ are described by the cross-correlation matrix

$$\mathbf{R}_{xy} = E\mathbf{x}\mathbf{y}^T,$$

but this matrix is generally difficult to explain. In order to illuminate the underlying structure, many correlation analysis techniques transform $\mathbf{x}$ and $\mathbf{y}$ into p-dimensional internal representation

$$\boldsymbol{\xi} = \mathbf{A}\mathbf{x}, \boldsymbol{\omega} = \mathbf{B}\mathbf{Y},$$

with $p = \min\{m, n\}$. The full rank matrices $\mathbf{A}, \mathbf{B}$ are chosen such that all partial sums over the absolute values of the correlations

$$k_i = E\xi_i\omega_i,$$

are maximized

$$\max_{\mathbf{A},\mathbf{B}} \sum_{i=1}^{r} |k_i|, \quad r = 1, \dots, p. \tag{3.82}$$

The solution to the maximization problem (3.82) leads to a diagonal cross-correlation matrix between $\boldsymbol{\xi}$ and $\boldsymbol{\omega}$

$$\mathbf{K} = E\boldsymbol{\xi}\boldsymbol{\omega}^T = \mathrm{diag}(k_1, k_2, \cdots, k_p), \tag{3.83}$$

with

$$k_1 \geq k_2 \geq \cdots > k_p \geq 0.$$

In order to summarize the correlation between $\mathbf{x}$ and $\mathbf{y}$, an overall correlation coefficient ρ is defined as a function of the diagonal correlations $\{k_i\}$. This correlation coefficient shares the invariance of the $\{k_i\}$. Because of the maximization (3.82), the assessment of correlation is allowed in a lower-dimensional subspace of rank

$$r \leq p = \min(m, n).$$

There is a variety of possible correlation coefficients that can be defined based on the first r canonical correlations $\{k_{C,i}{}_{i=1}^{r}\}$ for a given rank r.

$$\rho_{C_1} = \frac{1}{p} \sum_{i-1}^{r} k_{C,i}^2,$$

$$\rho_{C_2} = 1 - \prod_{i=1}^{r} (1 - k_{C,i}^2),$$

$$\rho_{C_3} = \frac{\displaystyle\sum_{i=1}^{r} \frac{k_{C,i}^2}{1-k_{C,i}^2}}{\displaystyle\sum_{i=1}^{r} \frac{1}{1-k_{C,i}^2} + (p-r)}.$$

For $r = p$, these coefficients can be expressed in terms of the original correlation matrices

$$\rho_{C_1} = \tfrac{1}{p}\mathrm{Tr}(\mathbf{R}_{xx}^{-1}\mathbf{R}_{xy}\mathbf{R}_{yy}^{-1}\mathbf{R}_{xy}^T) = \tfrac{1}{p}\mathrm{Tr}(\mathbf{CC}^T),$$

$$\rho_{C_2} = 1 - \det(\mathbf{I} - \mathbf{R}_{xx}^{-1}\mathbf{R}_{xy}\mathbf{R}_{yy}^{-1}\mathbf{R}_{xy}^T) = 1 - \det(\mathbf{I} - \mathbf{CC}^T),$$

$$\rho_{C_3} = \frac{\mathrm{Tr}[\mathbf{R}_{xy}\mathbf{R}_{yy}^{-1}\mathbf{R}_{xy}^T(\mathbf{R}_{xx}-\mathbf{R}_{xy}\mathbf{R}_{yy}^{-1}\mathbf{R}_{xy}^T)^{-1}]}{\mathrm{Tr}[\mathbf{R}_{xx}(\mathbf{R}_{xx}-\mathbf{R}_{xy}\mathbf{R}_{yy}^{-1}\mathbf{R}_{xy}^T)^{-1}]} = \frac{\mathrm{Tr}[\mathbf{CC}^T(\mathbf{I}-\mathbf{CC}^T)^{-1}]}{\mathrm{Tr}(\mathbf{I}-\mathbf{CC}^T)^{-1}}$$

where

$$\mathbf{C} = \mathbf{R}_{xx}^{-1/2}\mathbf{R}_{xy}\mathbf{R}_{yy}^{-T/2}.$$

These coefficients share the invariance of the canonical correlations, that is, they are invariant under a nonsingular linear transformation of $\mathbf{x}$ and $\mathbf{y}$. For jointly Gaussian $\mathbf{x}$ and

$\mathbf{y}$, ρ_{C_2} determines the mutual information between $\mathbf{x}$ and $\mathbf{y}$

$$I(\mathbf{x}; \mathbf{y}) = -\frac{1}{2} \sum_{i=1}^{r} \log(1 - k_{C,i}^2) = -\frac{1}{2} \log(1 - \rho_{C_2}).$$

The complex version of correlation analysis is discussed in [125, 126].

4

Hypothesis Detection of Noncommutative Random Matrices

4.1 Why Noncommutative Random Matrices?

The most basic building block for quantum information is the covariance matrix. We are dealing with the matrix space whose elements are covariance matrices. The sufficient and necessary conditions for a matrix to be a covariance matrix are semidefinite positive. As a result, the basic elements for us to manipulate are the SDP matrices. Naturally, convex optimization (SDP matrices are of course convex) is the new calculus under this context.

For any two elements (matrices) $\mathbf{A}$ and $\mathbf{B}$, we need to define the basic metric to order them. If they are random matrices, we call this order the stochastic order, for example,

$$\mathbf{B} \overset{st}{\geqslant} \mathbf{A},$$

if $\mathbf{B}$ is stochastically greater than $\mathbf{A}$.

More generally, $\mathbf{A}$ and $\mathbf{B}$ are two matrix-valued random variables, in contrast with the scalar random variables. Recall that every entry of $\mathbf{A}$ and $\mathbf{B}$ is a scalar random variable. The focus of the current engineering curriculum is on the scalar random variable. When we deal with "Big Data" [1] in a high-dimensional vector space, the most natural objects of mathematical operations are such (SDP) matrix-valued random variables.

The matrix operation is fundamentally different from its scalar counterpart in that the matrix multiplication is not communicative. The quantum mechanics is built upon this mathematical fact.

When we process the data, we argue in this chapter that the so-called quantum information [127] must be preserved and extracted. Data mining is about quantum information processing [128, 129]. For more details, we refer to the standard text [128].

Now, random matrices are our new objects of interest. We will dedicate an entire chapter to study this connection. The fundamental reason for us to study random matrices

Cognitive Radio Communications and Networking: Principles and Practice, First Edition.
Robert C. Qiu, Zhen Hu, Husheng Li and Michael C. Wicks.
© 2012 John Wiley & Sons, Ltd. Published 2012 by John Wiley & Sons, Ltd.

is that a sample covariance matrix (in practice, we do not know the exact covariance matrix) is a large-dimensional random matrix. Random matrices are a special case of noncommunicative (matrix-valued[1]) random variables.

See Appendix A.5 for details on noncommunicative matrix-valued random variables: random matrices are their special cases.

4.2   Partial Orders of Covariance Matrices: A < B

Example 4.1 (Positivity of covariance matrices)
Consider the 2×2 covariance matrix of form

$$\mathbf{R}_s = \begin{pmatrix} 1 & \xi \\ \xi & 1 \end{pmatrix}.$$

What is the condition that guarantees the positivity of $\mathbf{R}_s$? A Hermitian matrix $\mathbf{A}$ is positive if and only if all eigenvalues of $\mathbf{A}$ are positive. The eigenvalues of $\mathbf{R}_s$ are

$$\lambda_1 = 1 + \xi$$
$$\lambda_2 = 1 - \xi.$$

The condition $|\xi| \leq 1$ is sufficient to make two eigenvalues nonnegative, thus $\mathbf{R}_s$ positive. The covariance matrices illustrated in Example 3.1 are special cases of this example.   □

For a general 2×2 matrix, it is easy to check the positivity:

$$\mathbf{R}_s = \begin{pmatrix} a & b \\ \bar{b} & c \end{pmatrix} \geq 0 \text{ if } a \geq 0 \text{ and } b\bar{b} \leq ac$$

since

$$\lambda_1 = a/2 + c/2 + \tfrac{1}{2}(a^2 - 2ac + c^2 + 4b\bar{b})^{1/2}$$
$$\lambda_2 = a/2 + c/2 - \tfrac{1}{2}(a^2 - 2ac + c^2 + 4b\bar{b})^{1/2}.$$

If the entries are $n \times n$ matrices, then the condition for positivity is similar but it is more complicated. Matrices with matrix entries are called block-matrices.

Theorem 4.1 (Positivity of block matrices) *The self-adjoint block-matrix*

$$\begin{pmatrix} \mathbf{A} & \mathbf{B} \\ \mathbf{B}^* & \mathbf{C} \end{pmatrix}$$

is positive if and only if $\mathbf{A}, \mathbf{C} \geq 0$ *and there exists an operator* $\mathbf{X}$ *such that* $\|\mathbf{X}\| \leq 1$ *and* $\mathbf{B} = \mathbf{C}^{\frac{1}{2}}\mathbf{X}^{\frac{1}{2}}$. *When* $\mathbf{A}$ *is invertible, then this condition is equivalent to*

$$\mathbf{B}\mathbf{A}^{-1}\mathbf{B}^* \leq \mathbf{C}.$$

Theorem 4.2 (Schur) *Let* $\mathbf{A}$ *and* $\mathbf{B}$ *be positive* $n \times n$ *matrices. Then*

$$\mathbf{C}_{ij} = \mathbf{A}_{ij}\mathbf{B}_{ij} \quad (1 \leq i, j \leq n)$$

determines a positive matrix.

---

[1] After we get used to this notion, we can drop the words of "matrix-valued."

The matrix $\mathbf{C}$ of the previous theorem is called the Hadamard (or Schur) product of the matrices $\mathbf{A}$ and $\mathbf{B}$. In notation, $\mathbf{C} = \mathbf{A} \circ \mathbf{B}$.

Let $\mathbf{A}$ and $\mathbf{B}$ be self-adjoint operators. $\mathbf{A} \leq \mathbf{B}$ if $\mathbf{B} - \mathbf{A}$ is positive. The inequality $\mathbf{A} \leq \mathbf{B}$ implies $\mathbf{X}\mathbf{A}\mathbf{X}^* \leq \mathbf{X}\mathbf{B}\mathbf{X}^*$ for every operator $\mathbf{X}$. The partial order between $\mathbf{A}$ and $\mathbf{B}$ can be defined. It is called Loewner's order [109, 114, 130–133]. Generally, stochastic order [132] can be defined for two random operators $\mathbf{A}$ and $\mathbf{B}$.

Example 4.2 (Hypothesis testing in terms of covariance matrices)
From Example 3.2 in Chapter 3, we have

$$\begin{aligned} \mathcal{H}_0 &: \mathbf{A} = \sigma_w^2 \mathbf{I} + \sigma_w^2 \mathbf{X} \\ \mathcal{H}_1 &: \mathbf{B} = \mathbf{R_x} + \sigma_w^2 \mathbf{I} + \sigma_w^2 \mathbf{X}, \end{aligned} \tag{4.1}$$

where $\mathbf{R_x}$ is the covariance matrix of the signal. For the complex exponentials, $\mathbf{R_x}$ is given in Example 3.1. Thus, we have

$$\begin{aligned} \mathcal{H}_0 &: \mathbf{A} = \sigma_w^2 \mathbf{I} + \sigma_w^2 \mathbf{X} \\ \mathcal{H}_1 &: \mathbf{B} = \mathbf{R_x} + \sigma_w^2 \mathbf{I} + \sigma_w^2 \mathbf{X} = \tfrac{1}{2}|\mathbf{A}|^2 \left(\mathbf{I} + a\sigma_1\right) + \sigma_w^2 \mathbf{I} + \sigma_w^2 \mathbf{X}. \end{aligned}$$

Without loss of generality, we set $\sigma_w^2 = 1$. It follows that

$$\begin{aligned} \mathcal{H}_0 &: \mathbf{A} = \mathbf{I} + \mathbf{X} \\ \mathcal{H}_1 &: \mathbf{B} = SNR\left(\mathbf{I} + a\sigma_1\right) + \mathbf{I} + \mathbf{Y}. \end{aligned} \tag{4.2}$$

$\square$

The hypothesis testing problem, see Example 4.2 for an illustraion, can be viewed as a problem of partial ordering of two covariance matrices $\mathcal{H}_0 : \mathbf{A}$ and $\mathcal{H}_1 : \mathbf{B}$ for two hypotheses. Matrix inequalities are the basis of the proposed formalism. Often, Hermitian matrices (or finite-dimensional self-adjoint operators) are objects of study. The positivity of these matrices is required for many recent results developed in quantum information theory. The fundamental role of positivity of covariance matrices is emphasized here.

For positive operators $\mathbf{A}$ and $\mathbf{B}$,

$$\|\mathbf{A} - \mathbf{B}\|_1^2 + 4(\mathrm{Tr}(\mathbf{A}^{1/2}\mathbf{B}^{1/2}))^2 \leq (\mathrm{Tr}(\mathbf{A} + \mathbf{B}))^2. \tag{4.3}$$

Let $\mathbf{A}$ and $\mathbf{B}$ be positive operators, then for $0 \leq s \leq 1$,

$$\mathrm{Tr}(\mathbf{A}^{1/2}\mathbf{B}^{1/2}) \geq \mathrm{Tr}(\mathbf{A} + \mathbf{B} - |\mathbf{A} - \mathbf{B}|)/2 \tag{4.4}$$

or

$$2\mathrm{Tr}(\mathbf{A}^{1/2}\mathbf{B}^{1/2}) + \mathrm{Tr}|\mathbf{A} - \mathbf{B}| = 2\mathrm{Tr}(\mathbf{A}^{1/2}\mathbf{B}^{1/2}) + ||\mathbf{A} - \mathbf{B}||_1 \geq \mathrm{Tr}(\mathbf{A} + \mathbf{B}). \tag{4.5}$$

If f is convex then

$$f(x) - f(y) - (x - y)f'(y) \geq 0$$

and

$$\mathrm{Tr}f(\mathbf{B}) \geq \mathrm{Tr}f(\mathbf{A}) + \mathrm{Tr}(\mathbf{B} - \mathbf{A})\mathbf{f}'(\mathbf{B}). \tag{4.6}$$

In particular, for $f(t) = t \log t$, the relative entropy of two states is positive:

$$S(\mathbf{A}||\mathbf{B}) = \text{Tr}\mathbf{A} \log \mathbf{A} - \text{Tr}\mathbf{B} \log \mathbf{B} \geq \text{Tr}(\mathbf{B} - \mathbf{A}). \tag{4.7}$$

This is the original Klein inequality. A stronger estimate is obtained [34, p. 174]:

$$S(\mathbf{A}||\mathbf{B}) \geq \tfrac{1}{2}\text{Tr}(\mathbf{B} - \mathbf{A})^2. \tag{4.8}$$

From (4.3) to (4.8), the only requirement is that $\mathbf{A}$ and $\mathbf{B}$ are positive operators (matrices). Of course, they are valid for $\mathbf{A} < \mathbf{B}$.

Let $\mathbf{A}, \mathbf{B} \in M_n$ be positive semidefinite. Then for any complex number z, and any unitarily invariant norm [133],

$$||\mathbf{A} - |z|\mathbf{B}|| \leq ||\mathbf{A} + z\mathbf{B}|| \leq ||\mathbf{A} + |z|\mathbf{B}||.$$

4.3 Partial Ordering of Completely Positive Mappings: $\Phi(\mathbf{A}) < \Phi(\mathbf{B})$

It has long been realized that trace-preserving, completely positive maps seem to be the appropriate mathematical structure needed to model noise in quantum communication channels and quantum computers [134].

We define a quantum operation Φ as a map from the set of density operators of the input space Q_1 to the set of density operators for the output space Q_2, with the following *three axiomatic properties* [128]:

- **A1**: First, $\text{Tr}[\Phi(\rho)]$ is the probability that the transformation $\rho \rightarrow \Phi\rho$ takes place; $0 \leq \text{Tr}[\Phi(\rho)] \leq 1$ for any state ρ.
- **A2**: Second, Φ is a *convex-linear map* on the set of density operators, that is, for probabilities $\{p_i\}$ of states ρ_i,

$$\Phi\left(\sum_i p_i \rho_i\right) = \sum_i p_i \Phi(\rho_i). \tag{4.9}$$

- **A3**: Third, Φ is a *completely positive* map. That is, if Φ maps density operators of system Q_1 to density operators of system Q_2, then $\Phi(\mathbf{A})$ must be positive for any positive operator $\mathbf{A}$. Furthermore, if we introduce an extra system R of arbitrary dimensionality, it must be true that $(\mathcal{I} \otimes \Phi)(\mathbf{A})$ is positive for any positive operator $\mathbf{A}$ on the combined system RQ_1, where $\mathcal{I}$ denotes the identity map on system R.

The following theorem is fundamental to the adopted formalism: The map Φ satisfies axioms **A1**, **A2**, **A3** *if and only if*

$$\Phi(\rho) = \sum_i E_i \rho E_i^*, \tag{4.10}$$

for some set of operators E_i which map the input Hilbert space to the output Hilbert space, and $\sum_i E_i E_i^* \leq I$ where I is the identity operator and $*$ denotes the conjugate and transpose. Φ is obviously linear. The map Φ sends a density matrix into another one, thus ΦA and ΦB are density matrices that satisfy the conditions for (3.22). The hypothesis test (3.22) is, thus, generalized by replacing the expectation with the map Φ:

$$\text{Tr}\Phi A \leq \text{Tr}\Phi B. \tag{4.11}$$

Algorithm 4.1 *(1) Claim hypothesis* $\mathcal{H}_1$ *if matrix inequality (4.11) is satisfied; (2) otherwise,* $\mathcal{H}_0$ *is claimed.*

The map Φ in (4.11) is very general. The whole body of knowledge of quantum information theory [127] can be borrowed. Two maps are of the most important significance: (1) positive linear maps; (2) completely positive maps. The mathematical foundation is treated in textbooks [109, 130]. A positive linear map (also unital) Φ may be thought as a noncommutative analogue of an expectation map.

Since positivity is a useful and interesting property, it is natural to ask what linear transformations preserve it [109, Chapter 2]. It is instructive to think of positive maps as noncommutative (matrix) averaging operations [109, 115, 130, 133].

In this section we use the symbol Φ for a linear map from $\mathcal{M}_n$ to $\mathcal{M}_k$. When $k = 1$, such a map is called a linear functional, and we use the lower-case symbol φ for it. A linear map $\Phi: \mathcal{M}_n \to \mathcal{M}_n$ is called *positive* if $\Phi(A) \geq 0$ where $A \geq 0$ and $\mathcal{M}_n$ is the space of $n \times n$ matrices. It is said to be *unital* if $\Phi(I) = I$. We say Φ is strictly positive if $\Phi(A) > 0$ where $A > 0$. It is easy to see that a positive linear map is strictly positive if and only if $\Phi(I) > 0$.

Any positive linear combination of positive maps is positive. Any convex combination of positive, unital maps is positive and unital. There are ten basic examples in [109, Chapter 2]. The combination of these basic maps allows us to form many combined maps that are suitable for specific needs across the layers of the cognitive radio network. This subtask needs further investigation.

From (4.11), it is required that: (1) the map Φ is *positive*: positive matrices are mapped to positive matrices, that is, $\Phi A \geq 0$ for any $A \geq 0$; (2) the map is *trace-preserving*, that is, $\text{Tr}\Phi A = \text{Tr}A$. This special class of positive maps, called completely positive, trace-preserving (CPTP) linear maps [109, Chapter 3], is central to the proposed research. The map in (4.10) is such a map. A CPTP linear operation takes statistical operators to statistical operators. Such maps in (4.10) are also called *quantum channels* in quantum information theory.

4.4   Partial Ordering of Matrices Using Majorization: $\mathbf{A} \prec \mathbf{B}$

$\mathbf{B} > \mathbf{A}$ is very strong condition at extremely low SNR such as -20 dB. The weak majorization $\mathbf{A} \prec_w \mathbf{B}$ is equivalent to $\sigma_k(\mathbf{A}) \leq \sigma_k(\mathbf{B})$ for all k. This is hardly satisfied at extremely low SNR, due to the presence of two random matrices, for example, $\mathbf{X}$ and $\mathbf{Y}$ in (4.2). The majorization $\mathbf{A} \prec \mathbf{B}$ holds if and only if

$$\mathbf{A} + a\mathbf{I} \prec \mathbf{B} + a\mathbf{I} \tag{4.12}$$

for some $a \in \mathcal{R}$. By shifting a self-adjoint matrix, we can make it to be positive always. When discussing the properties of majorization, we can restrict ourselves to positive (definite) matrices.

Theorem 4.3 (Majorization) *Let ρ_1 and ρ_2 be states. The following statements are equivalent.*

1. $\rho_1 \prec \rho_2$.
2. ρ_1 is more mixed than ρ_2.
3. $\rho_1 = \sum_{i=1}^{n} \lambda_i U_i \rho_2 U_i^*$ for some convex combination λ_i and for some unitaries U_i.
4. $\operatorname{Tr} f(\rho_1) \leq \operatorname{Tr} f(\rho_2)$ for any convex function $f : \mathcal{R} \to \mathcal{R}$.

Theorem 4.4 (Wehrl) *Let ρ be a density matrix of finite quantum system $\mathbf{B}(\mathbb{H})$ and $f : \mathbb{R}^+ \to \mathbb{R}^+$ a convex function with $f(0) = 0$. The ρ is majorized by the density*

$$\rho_f = \frac{f(\rho)}{\operatorname{Tr} f(\rho)}. \tag{4.13}$$

Theorem 4.5 (Majorization for nonnegative increasing convex function [135]) *If f is a nonnegative increasing convex function on $[0, \infty]$ with $f(0) = 0$, then*

$$\lambda(f(\mathbf{A}) + f(\mathbf{B})) \prec_w \lambda(f(\mathbf{A} + \mathbf{B})) \tag{4.14}$$

for all $\mathbf{A}, \mathbf{B} \geq 0$, or equivalently

$$|||(f(\mathbf{A}) + f(\mathbf{B}))||| \prec_w |||f(\mathbf{A} + \mathbf{B})|||. \tag{4.15}$$

Here, $||| \cdot |||$ stands for the symmetric, unitarily invariant norm. Given two covariances $\bar{\mathbf{A}}$ and $\bar{\mathbf{B}}$, these covariance matrices are affected by random signals experiencing fading and network control. It is difficult to guarantee that the covariance matrix of the noise or interference, $\bar{\mathbf{B}} = \mathbf{R}_w$, is known (due to noise power uncertainty). We can work on the "blind" version of the algorithms. The covariance matrices can be normalized by their traces. The impact of this normalization process is described by (4.13) in Wehrl's theorem.

Example 4.3 (Positive operator valued hypothesis testing)
This example is continued from Examples 3.1 and 4.2. For sinusoidal signals, we have

$$\mathcal{H}_0 : \mathbf{A} = \sigma_w^2 \mathbf{I} + \sigma_w^2 \mathbf{X}$$
$$\mathcal{H}_1 : \mathbf{B} = \frac{1}{2}|\mathbf{A}|^2 \begin{bmatrix} 1 & \cos \omega_0 \\ \cos \omega_0 & 1 \end{bmatrix} + \sigma_w^2 \mathbf{I} + \sigma_w^2 \mathbf{Y} = \frac{1}{2}|\mathbf{A}|^2 \left(\mathbf{I} + \bar{\mathbf{R}}_x \right) + \sigma_w^2 \mathbf{I} + \sigma_w^2 \mathbf{Y},$$

where $\bar{\mathbf{R}}_x = \sigma_1 \cos \omega_0$. Obviously, $\operatorname{Tr} \bar{\mathbf{R}}_x = 0$ since $\operatorname{Tr} \sigma_1 = 0$. If we set $\sigma_w^2 = 1$, then we can define SNR as $SNR = \frac{|\mathbf{A}|^2}{2\sigma_w^2}$. $\square$

Using the structure of (3.4) and considering the unit power of additive noise (without loss of generality), $\sigma_w^2 = 1$, we have

$$\mathcal{H}_0 : \mathbf{A} = \mathbf{R}_w = \mathbf{I} + \mathbf{X}, \mathbf{A} > 0, \operatorname{Tr} \mathbf{X} = 0 \tag{4.16}$$
$$\mathcal{H}_1 : \mathbf{B} = \mathbf{R}_s + \mathbf{R}_w = SNR \left(\mathbf{I} + \tilde{\mathbf{R}}_x \right) + \mathbf{I} + \mathbf{Y}, \mathbf{B} > 0, \operatorname{Tr} \tilde{\mathbf{R}}_x = 0, \operatorname{Tr} \mathbf{Y} = 0$$

$$\mathbf{B} + \mathbf{A} = (2 + SNR)\mathbf{I} + \mathbf{X} + \mathbf{Y}$$
$$\mathbf{B} - \mathbf{A} = SNR\,\mathbf{I} + SNR\,\tilde{R}_x + \mathbf{Y} - \mathbf{X}, \mathrm{Tr}(\mathbf{B} - \mathbf{A}) = SNR \qquad (4.17)$$

With the aid of (4.17) and $\mathrm{Tr}(\mathbf{A} + \mathbf{B}) = \mathrm{Tr}\mathbf{A} + \mathrm{Tr}\mathbf{B}$, one detection algorithm using the preset threshold η_0 can be stated as following:

Algorithm 4.2 (Threshold detection algorithm using the traces of two hypotheses)

1. Claim $\mathcal{H}_1$, if $\mathrm{Tr}(\mathbf{B}) > \mathrm{Tr}(\mathbf{A}) + \eta_0$, with $\eta_0 = SNR$.
2. Otherwise, claim $\mathcal{H}_0$.

The beauty of Algorithm 4.2 is that $\mathrm{Tr}(\mathbf{A})$ is independent of the measured signals. We can use the statistics of the additive noise (interference), $\mathrm{Tr}\mathbf{A}$, a random variable, to set the threshold for the measured signals plus noise, $\mathrm{Tr}\mathbf{B}$, also a random variable.

If we have the prior knowledge of $\mathbf{R}_s$, we can consider

$$\mathcal{H}_0 : \mathbf{R}_s^* \mathbf{R}_w, \mathbf{R}_s > 0, \mathbf{R}_w > 0$$
$$\mathcal{H}_1 : \mathbf{R}_s^*(\mathbf{R}_s + \mathbf{R}_w) = \mathbf{R}_s^*\mathbf{R}_s + \mathbf{R}_s^*\mathbf{R}_w = |\mathbf{R}_s|^2 + \mathbf{R}_s^*\mathbf{R}_w, \mathbf{R}_s > 0, \mathbf{R}_w > 0, \qquad (4.18)$$

where $\mathbf{R}_s^*$ is used to match the signal covariance matrix $\mathbf{R}_s$ to get the absolute value $|\mathbf{R}_s|^2$. Recall that $|\mathbf{A}| = (\mathbf{A}^*\mathbf{A})^{\frac{1}{2}}$.

Consider K independent copies $\mathbf{A}_k, k = 1, 2, \ldots, K$

$$\mathcal{H}_0 : \mathbf{A}_k = \mathbf{R}_{w,k}, \mathbf{A}_k > 0,$$
$$\mathcal{H}_1 : \mathbf{B}_k = \mathbf{R}_{s,k} + \mathbf{R}_{w,k}, \mathbf{B}_k > 0. \qquad (4.19)$$

Let $\mathbf{C}_k \geq 0$ and $\mathbf{D}_k \geq 0$ be of the same size. Then [114, p. 166]

$$\mathbf{C}_k + \mathbf{D}_k \geq \mathbf{D}_k, k = 1, 2, \ldots, K. \qquad (4.20)$$

For $\mathcal{H}_1$, with the aid of (4.20), both sides of these K inequalities in (4.19) are summed up to yield

$$\sum_{k=1}^{K} \mathbf{B}_k = (\mathbf{R}_{s,1} + \mathbf{R}_{s,2} + \cdots + \mathbf{R}_{s,K}) + (\mathbf{R}_{w,1} + \mathbf{R}_{w,2} + \cdots + \mathbf{R}_{w,K}) \geq \mathbf{R}_{w,1}$$
$$+ \mathbf{R}_{w,2} + \cdots + \mathbf{R}_{w,K}. \qquad (4.21)$$

Algorithm 4.3 (Threshold detection algorithm using the traces of two hypotheses (many copies))

1. Claim $\mathcal{H}_1$, if $\mathrm{Tr}(\mathbf{B}_1 + \mathbf{B}_2 + \cdots + \mathbf{B}_K) > \mathrm{Tr}(\mathbf{A}_1 + \mathbf{A}_2 + \cdot + \mathbf{A}_K) + \eta$, where $\eta = \sum_{k=1}^{K} \mathbf{R}_{s,k} > 0$.
2. Otherwise, claim $\mathcal{H}_0$.

$$\mathrm{Tr}|\mathbf{B}_1 + \mathbf{B}_2 + \cdot + \mathbf{B}_K| > \mathrm{Tr}|\mathbf{A}_1 + \mathbf{A}_2 + \cdot + \mathbf{A}_K| + \eta \qquad (4.22)$$
$$\mathrm{Tr}|\mathbf{A}_1 + \mathbf{A}_2 + \cdots + \mathbf{A}_K| \leq \mathrm{Tr}|\mathbf{A}_1| + \mathrm{Tr}|\mathbf{A}_2| + \cdots + \mathrm{Tr}|\mathbf{A}_K|. \qquad (4.23)$$

Two covariances $\bar{\mathbf{A}}$ and $\bar{\mathbf{B}}$ are normalized first using (4.13) in Wehrl's theorem:

$$\mathcal{H}_0 : \mathbf{A} = \tfrac{\bar{\mathbf{A}}}{\mathrm{Tr}\mathbf{A}} = \tfrac{1}{N}\mathbf{I} + \mathbf{X}, \mathrm{Tr}\mathbf{A} = 1, \mathbf{A} > 0$$
$$\mathcal{H}_1 : \mathbf{B} = \tfrac{\bar{\mathbf{B}}}{\mathrm{Tr}\bar{\mathbf{B}}} = \tfrac{1}{N}\mathbf{I} + \tilde{\mathbf{R}}_x + \mathbf{Y}, \mathrm{Tr}\mathbf{B} = 1, \mathbf{B} > 0,$$

(4.24)

where $\mathbf{X}$, $\mathbf{Y}$ and $\tilde{\mathbf{R}}_x$ are self-adjoint random matrices with $\mathrm{Tr}\mathbf{X} = 0$, $\mathrm{Tr}\mathbf{Y} = 0$ and $\mathrm{Tr}\tilde{\mathbf{R}}_x = 0$, and $N = \mathrm{Tr}\mathbf{I}$ is the dimensionality of identity matrix $\mathbf{I}$. $\mathbf{X}$ and $\mathbf{Y}$ are two independent, identical distributed copies whose rows are independent (see Section 3.4). It follows that

$$\mathbf{A} + \mathbf{B} = \tfrac{2}{N}\mathbf{I} + \tilde{\mathbf{R}}_x + \mathbf{Y} + \mathbf{X}$$
$$\mathbf{B} - \mathbf{A} = \tilde{\mathbf{R}}_x + \mathbf{Y} - \mathbf{X}.$$

(4.25)

Note that $\mathrm{Tr}(\mathbf{B} - \mathbf{A}) = 0$ which implies that $\mathrm{Tr}\mathbf{U}^*(\mathbf{B} - \mathbf{A})\mathbf{U} = 0$, where U is an arbitrary unitary matrix. Consider

$$|\mathbf{B} - \mathbf{A}| = |\tilde{\mathbf{R}}_x + \mathbf{Y} - \mathbf{X}|.$$

Using (A.6) [114, p. 239]: $\mathrm{Tr}|\mathbf{A} + \mathbf{B}| \le \mathrm{Tr}|\mathbf{A}| + \mathrm{Tr}|\mathbf{B}|$, it follows that

$$\mathrm{Tr}|\mathbf{B} - \mathbf{A}| = \sum_i |\lambda_i(\mathbf{B}) - \lambda_i(\mathbf{A})| \le \mathrm{Tr}|\tilde{\mathbf{R}}_x| + \mathrm{Tr}|\mathbf{Y} - \mathbf{X}|,$$

(4.26)

where $||\mathbf{X} - \mathbf{Y}||_1 = \mathrm{Tr}|\mathbf{Y} - \mathbf{X}|$ is the distance between two random matrices, also called trace norm. λ_i is the i-th eigenvalue. If $\mathrm{Tr}|\tilde{\mathbf{R}}_x| = 0$ and $\mathrm{Tr}|\mathbf{Y} - \mathbf{X}|$, then $\mathrm{Tr}|\mathbf{B} - \mathbf{A}| = 0$, which implies that $\mathbf{A}$ and $\mathbf{B}$ cannot be distinguished from each other.

In (4.25), generally we can not claim that $\mathbf{B} - \mathbf{A}$ is positive, although $\mathbf{B} - \mathbf{A}$ is still Hermitian. Let $\mathbf{A}$ and $\mathbf{B}$ be positive operators, then for $0 \le s \le 1$,

$$\mathrm{Tr}(\mathbf{B}^s\mathbf{A}^{1-s}) \ge \mathrm{Tr}(\mathbf{A} + \mathbf{B} - |\mathbf{B} - \mathbf{A}|)/2.$$

(4.27)

In general, if $\mathbf{A}, \mathbf{B} \ge 0$, we have

$$\mathrm{Tr}\mathbf{A}\mathbf{B} \ge 0.$$

(4.28)

However, the product of $\mathbf{A}\mathbf{B}$ is not a Hermitian matrix. Note that although $\mathbf{A}\mathbf{B} + \mathbf{B}\mathbf{A}$ is Hermitian, it is generally not positive semidefinite. In (4.27), we are interested in the absolute value of $\mathbf{B} - \mathbf{A}$ only, in terms of $||\mathbf{A} - \mathbf{B}||_1 = \mathrm{Tr}|\mathbf{B} - \mathbf{A}|$. This trace norm $||\mathbf{A} - \mathbf{B}||_1$ is a natural distance between complex $n \times n$ matrices $\mathbf{A}$ and $\mathbf{B}$, $\mathbf{A}, \mathbf{B} \in M_n(\mathbb{C})$. Similarly,

$$||\mathbf{A} - \mathbf{B}||_2 = \left(\sum_{i,j} |\mathbf{A}_{ij} - \mathbf{B}_{ij}|^2 \right)^{1/2}$$

is also a natural distance. We can define the $p-$ norm as

$$||\mathbf{X}||_p = (\mathrm{Tr}(\mathbf{X}^*\mathbf{X})^{2/p})^{1/p}, 1 \le p, \mathbf{X} \in M_n(\mathbb{C}).$$

It was Von Neumann who showed first that the Hoelder inequality remains true in the matrix setting

$$||\mathbf{AB}||_1 \leq ||\mathbf{A}||_p ||\mathbf{B}||_q, \frac{1}{p} + \frac{1}{q} = 1.$$

For $\mathbf{A} \in M_n(\mathcal{C})$, the absolute value $|\mathbf{A}|$ is defined as $\sqrt{\mathbf{A}^*\mathbf{A}}$ and it is a positive matrix. If $\mathbf{A}$ is a self-adjoint and written as

$$\mathbf{A} = \sum_i \lambda_i \mathbf{e}_i \mathbf{e}_i^*,$$

where the vector $\mathbf{e}_i$ forms an orthonormal basis, then it is defined as

$$\{\mathbf{A} \geq 0\} = \mathbf{A}_+ = \sum_{i:\lambda_i \geq 0} \lambda_i \mathbf{e}_i \mathbf{e}_i^*; \{\mathbf{A} < 0\} = \mathbf{A}_- = \sum_{i:\lambda_i < 0} \lambda_i \mathbf{e}_i \mathbf{e}_i^*.$$

Then $\mathbf{A} = \{\mathbf{A} \geq 0\} + \{\mathbf{A} < 0\} = \mathbf{A}_+ + \mathbf{A}_-$ and $|\mathbf{A}| = \{\mathbf{A} \geq 0\} - \{\mathbf{A} < 0\} = \mathbf{A}_+ - \mathbf{A}_-$. The decomposition is called the Jordan decomposition of $\mathbf{A}$.

4.5 Partial Ordering of Unitarily Invariant Norms: $|||\mathbf{A}||| < |||\mathbf{B}|||$

Theorem 4.6 (A matrix subadditivity inequality for a nonnegative function of matrix [136]) *Let* $\mathbf{A}, \mathbf{B} \geq 0$ *and let* $f : [0, \infty] \rightarrow [0, \infty]$ *be a convex function with* $f(0) = 0$. *Then, for all symmetric (or unitarily invariant) norms*

$$|||f(\mathbf{A} + \mathbf{B})||| \geq |||f(\mathbf{A}) + f(\mathbf{B})|||. \tag{4.29}$$

Let $\mathbf{A}, \mathbf{B} \geq 0$ *and let* $g : [0, \infty] \rightarrow [0, \infty]$ *be a concave function with* $g(0) = 0$. *Then, for all symmetric norms*

$$|||g(\mathbf{A} + \mathbf{B})||| \leq |||g(\mathbf{A}) + g(\mathbf{B})|||. \tag{4.30}$$

For the trace norm, Theorem 4.6 is a classical inequality. Recall that $||\mathbf{A}||_1 = \text{Tr}(\mathbf{A}^*\mathbf{A})^{\frac{1}{2}} = \sum_i \sigma_i$, where σ_i is the singular value. Special cases: (1) $f(t) = t^m, m = 1, 2, \ldots$; (2) $g(t) = \sqrt{t}$.

4.6 Partial Ordering of Positive Definite Matrices of Many Copies: $\sum_{k=1}^{K} \mathbf{A}_k \leq \sum_{k=1}^{K} \mathbf{B}_k$

Theorem 4.7 (Unitarily invariant norms with nonnegative convex/concave function [135]) *Let* $\mathbf{A}_1, \mathbf{A}_2, \ldots, \mathbf{A}_K \geq 0$. *Then for every nonnegative convex function* f *on* $[0, \infty]$ *with* $f(0) = 0$ *and for every unitarily invariant norm* $||| \cdot |||$

$$|||f(\mathbf{A}_1) + f(\mathbf{A}_2) + \cdots f(\mathbf{A}_K)||| \leq |||f(\mathbf{A}_1 + \mathbf{A}_2 + \cdots \mathbf{A}_K)|||. \tag{4.31}$$

If g is a nonnegative concave function, the inequality of (4.31) is reversed:

$$|||g(\mathbf{A}_1) + g(\mathbf{A}_2) + \cdots g(\mathbf{A}_K)||| \geq |||g(\mathbf{A}_1 + \mathbf{A}_2 + \cdots \mathbf{A}_K)|||. \tag{4.32}$$

The function $f : [0, \infty] \to \mathbb{R}$, defined by $f(x) = \frac{1}{2}((x-1) + |x-1|)$ satifies the inequality of (4.32). We interpret Theorem 4.7 as a norm-matrix generation of the scalar inequality $f(a) + f(b) \leq f(a+b)$, where $a, b \geq 0$ and $f : [0, \infty] \to [0, \infty]$ is a convex function with $f(0) = 0$.

4.7 Partial Ordering of Positive Operator Valued Random Variables: Prob($A \leq X \leq B$)

Consider K matrix-valued observations:

$$
\begin{aligned}
\mathcal{H}_0 : \mathbf{A}_k &= \mathbf{R}_{n,k} = \sigma_{n,k}^2(\mathbf{I} + \mathbf{X}_k), \mathrm{Tr}\mathbf{X}_k = 0, \\
\mathcal{H}_1 : \mathbf{B}_k &= \mathbf{R}_{s,k} + \mathbf{R}_{n,k} = \sigma_{s,k}^2(\mathbf{I} + \mathbf{S}_k) + \sigma_{n,k}^2(\mathbf{I} + \mathbf{Y}_k), \mathrm{Tr}\mathbf{S}_k = 0, \mathrm{Tr}\mathbf{Y}_k = 0,
\end{aligned}
\tag{4.33}
$$

where $\mathbf{X}_k$, $\mathbf{Y}_k$, and $\mathbf{S}_k$ are of zero trace and denote the nondiagonal elements of the covariance matrices.

$$
\begin{aligned}
\mathcal{H}_0 : \sum_{k=1}^{K} \mathbf{A}_k &= \left(\sum_{k=1}^{K} \sigma_{n,k}^2\right)\mathbf{I} + \sum_{k=1}^{K} \sigma_{n,k}^2\mathbf{X}_k = \left(\mathrm{Tr}\sum_{k=1}^{K} \mathbf{A}_k\right)\mathbf{I} + \sum_{k=1}^{K} \sigma_{n,k}^2\mathbf{X}_k \\
\mathcal{H}_1 : \sum_{k=1}^{K} \mathbf{B}_k &= \left[\sum_{k=1}^{K} (\sigma_{s,k}^2 + \sigma_{n,k}^2)\right]\mathbf{I} + \sum_{k=1}^{K} \sigma_{s,k}^2\mathbf{S}_k + \sum_{k=1}^{K} \sigma_{n,k}^2\mathbf{Y}_k \\
&= \left(\mathrm{Tr}\sum_{k=1}^{K} \mathbf{B}_k\right)\mathbf{I} + \sum_{k=1}^{K} \sigma_{s,k}^2\mathbf{S}_k + \sum_{k=1}^{K} \sigma_{n,k}^2\mathbf{Y}_k,
\end{aligned}
\tag{4.34}
$$

where the diagonal terms are associated with $\mathbf{I}$ with

$$\mathrm{Tr}\sum_{k=1}^{K} \mathbf{A}_k = \sum_{k=1}^{K} \mathrm{Tr}\mathbf{A}_k = \sum_{k=1}^{K} \sigma_{n,k}^2, \mathrm{Tr}\sum_{k=1}^{K} \mathbf{B}_k = \sum_{k=1}^{K} \mathrm{Tr}\mathbf{B}_k = \sum_{k=1}^{K} (\sigma_{s,k}^2 + \sigma_{n,k}^2).$$

Using the central limit theorem, the total trace (or total power) can be reduced to (scalar) Gaussian random variables.

Algorithm 4.4 (Detection using traces of sums of covariance matrices)

1. Claim $\mathcal{H}_1$ if

$$\mathrm{Tr}\sum_{k=1}^{K} \mathbf{A}_k = \xi \leq \mathrm{Tr}\sum_{k=1}^{K} \mathbf{B}_k,$$

2. Otherwise, claim $\mathcal{H}_0$.

Only diagonal elements are used in Algorithm 4.3; in (4.34), however, nondiagonal elements $\sum_{k=1}^{K} \sigma_{s,k}^2\mathbf{S}_k$ contain information of use to detection. The exponential of a matrix provides one tool. See Example 4.4. In particular, we have

$$\mathrm{Tr} e^{\mathbf{A}+\mathbf{B}} \leq \mathrm{Tr} e^{\mathbf{A}} e^{\mathbf{B}}.$$

The following matrix inequality

$$\mathrm{Tr} e^{\mathbf{A}+\mathbf{B}+\mathbf{C}} \leq \mathrm{Tr} e^{\mathbf{A}} e^{\mathbf{B}} e^{\mathbf{C}}$$

is known to be false.

Let $\mathbf{A}$ and $\mathbf{B}$ be two Hermitian matrices of the same size. If $\mathbf{A} - \mathbf{B}$ is positive semidefinite, we write [114]

$$\mathbf{A} \geq \mathbf{B} \text{ or } \mathbf{B} \leq \mathbf{A}. \tag{4.35}$$

$\geq$ is a partial ordering, referred to as Löwner partial ordering, on the set of Hermitian matrices, that is,

1. $\mathbf{A} \geq \mathbf{A}$ for every Hermitian matrix $\mathbf{A}$,
2. if $\mathbf{A} \geq \mathbf{B}$ and $\mathbf{B} \geq \mathbf{A}$, then $\mathbf{A} = \mathbf{B}$, and
3. if $\mathbf{A} \geq \mathbf{B}$ and $\mathbf{B} \geq \mathbf{C}$, then $\mathbf{A} \geq \mathbf{C}$.

The statement $\mathbf{A} \geq 0 \Leftrightarrow \mathbf{X}^*\mathbf{A}\mathbf{X} \geq 0$ is generalized as follows:

$$\mathbf{A} \geq \mathbf{B} \Leftrightarrow \mathbf{X}^*\mathbf{A}\mathbf{X} \geq \mathbf{X}^*\mathbf{B}\mathbf{X}, \tag{4.36}$$

for every complex matrix $\mathbf{X}$.

A hypothesis detection problem can be viewed as a problem of partially ordering the measured matrices for individual hypotheses. If many (K) copies of the measured matrices $\mathbf{A}_k$ and $\mathbf{B}_k$ are at our disposal, it is nature to ask this fundamental question:

Is $\mathbf{B}_1 + \mathbf{B}_2 + \cdots + \mathbf{B}_K$ (statistically) larger than $\mathbf{A}_1 + \mathbf{A}_2 + \cdots + \mathbf{A}_K$?

To answer this question motivates this whole section. It turns out that a new theory is needed. We freely use [137] that contains a relatively complete appendix for this topic.

The theory of real random variables provides the framework of much of modern probability theory, such as laws of large numbers, limit theorems, and probability estimates for large deviations, when sums of independent random variables are involved. Researchers develop analogous theories for the case that the algebraic structure of the reals is substituted by more general structures such as groups, vector spaces, etc.

At the hands of our current problem of hypothesis detection, we focus on a structure that has vital interest in quantum probability theory and names the algebra of operators[2] on a (complex) Hilbert space. In particular, the real vector space of self-adjoint operators (Hermitian matrices) can be regarded as a partially ordered generalization of the reals, as reals are embedded in the complex numbers.

A matrix-valued random variable $X : \Omega \to \mathbf{A}_s$, where

$$\mathbf{A}_s = \{\mathbf{A} \in \mathcal{A} : \mathbf{A} = \mathbf{A}^*\} \tag{4.37}$$

is the self-adjoint part of the $\mathbf{C}^*-$ algebra $\mathcal{A}$ [138], which is a real vector space. For more details, we refer to Appendix A.4. Let $\mathcal{L}(\mathcal{H})$ be the full operator algebra of the

---

[2] The finite-dimensional operators and matrices are used interchangeably.

complex Hilbert space $\mathcal{H}$. We denote $d = \dim(\mathcal{H})$, which is assumed to be finite. Here dim means the dimensionality of the vector space. In the general case, $d = \mathrm{Tr}\mathbf{I}$, and $\mathcal{A}$ can be embedded into $\mathcal{L}(\mathcal{C}^d)$ as an algebra, preserving the trace.

The real cone

$$\mathcal{A}_+ = \{\mathbf{A} \in \mathcal{A} : \mathbf{A} = \mathbf{A}^* \geq 0\} \qquad (4.38)$$

induces a partial order $\leq$ in $\mathcal{A}_s$. We can introduce some convenient notation: for $\mathbf{A}, \mathbf{B} \in \mathcal{A}_s$ the closed interval $[\mathbf{A}, \mathbf{B}]$ is defined as

$$[\mathbf{A}, \mathbf{B}] = \{\mathbf{X} \in \mathbf{A}_s : \mathbf{A} \leq \mathbf{X} \leq \mathbf{B}\}. \qquad (4.39)$$

Similarly, open and half-open intervals $(\mathbf{A}, \mathbf{B})$, $[\mathbf{A}, \mathbf{B})$, etc.

For simplicity, the space Ω on which the random variable lives is discrete. Some remarks on the operator order is as follows.

1. $\leq$ is not a total order unless $\mathcal{A} = \mathcal{C}$, in which case $\mathcal{A}_s = \mathcal{R}$. Thus in this case (classical case), the theory developed below reduces to the study of the real random variables.
2. $\mathbf{A} \geq 0$ is equivalent to saying that all eigenvalues of $\mathbf{A}$ are nonnegative. These are d nonlinear inequalities:

$$\begin{aligned}\mathbf{A} \geq 0 &\Leftrightarrow \forall \rho \text{ density operator } \mathrm{Tr}(\rho\mathbf{A}) \geq 0 \\ &\Leftrightarrow \forall \pi \text{ one } - \dim.\text{projector } \mathrm{Tr}(\pi\mathbf{A}) \geq 0.\end{aligned} \qquad (4.40)$$

3. The operator mapping $\mathbf{A} \mapsto \mathbf{A}^s$, for $s \in [0, 1]$ and $\mathbf{A} \mapsto \log\mathbf{A}$ are defined on $\mathcal{A}_+$, and both are operator monotone and operator concave. In contrast, $\mathbf{A} \mapsto \mathbf{A}^s$, for $s > 2$ and $\mathbf{A} \mapsto \exp\mathbf{A}$ are neither operator monotone nor operator convex. Remarkably, $\mathbf{A} \mapsto \mathbf{A}^s$, for $s \in [1, 2]$ is operator convex (though not operator monotone).
4. The mapping $\mathbf{A} \mapsto \mathrm{Tr}\exp\mathbf{A}$ is monotone and convex.
5. Golden-Thompson-inequality: for $\mathbf{A}, \mathbf{B} \in \mathcal{A}_s$

$$\mathrm{Tr}\exp(\mathbf{A} + \mathbf{B}) \leq \mathrm{Tr}((\exp\mathbf{A})(\exp\mathbf{B})). \qquad (4.41)$$

Note that a rarely few of mappings (functions) are operator convex (concave) or operator monotone. Fortunately, we are interested in the trace functions that have much bigger sets. Take a look at (4.42) for example. In (4.33), since $\mathcal{H}_0 : \mathbf{A} = \mathbf{I} + \mathbf{X}$, and $\mathbf{A} \in \mathcal{A}_s$ (even stronger $\mathbf{A} \in \mathcal{A}_+$), it follows from (4.42) that

$$\mathcal{H}_0 : \mathrm{Tr}\exp(\mathbf{A}) = \mathrm{Tr}\exp(\mathbf{I} + \mathbf{X}) \leq \mathrm{Tr}((\exp\mathbf{I})(\exp\mathbf{X})). \qquad (4.42)$$

The use of (4.42) allows us to separately study the diagonal part and the nondiagonal part of the covariance matrix of the noise, since all the diagonal elements are equal for a WSS random process (see (3.4)). At low SNR, the goal is to find some ratio or threshold that is statistically stable over a large number of Monte Carlo trials.

Algorithm 4.5 (Ratio detection algorithm using the trace exponentials)

1. *Claim* $\mathcal{H}_1$, *if* $\xi = \frac{\mathrm{Tr}\exp\mathbf{A}}{\mathrm{Tr}((\exp\mathbf{I})(\exp\mathbf{X}))} \geq 1$, *where* $\mathbf{A}$ *is the measured covariance matrix with or without signals and* $\mathbf{X} = \frac{\mathbf{R}_w}{\sigma_w^2} - \mathbf{I}$.
2. *Otherwise, claim* $\mathcal{H}_0$.

Example 4.4 (Exponential of the 2 × 2 matrix)
The 2×2 covariance matrix for L sinusoidal signals in Example 3.1 has symmetric structure with identical diagonal elements

$$\mathbf{R}_s = \mathrm{Tr}\mathbf{R}_s(\mathbf{I} + b\sigma_1),$$

where

$$\sigma_1 = \begin{pmatrix} 0 & 1 \\ 1 & 0 \end{pmatrix}$$

and b is a positive number. Obviously, $\mathrm{Tr}\sigma_1 = 0$. We can study the diagonal elements and nondiagonal elements separately. The two eigenvalues of the 2×2 matrix [126]

$$A = \begin{pmatrix} a & b \\ c & d \end{pmatrix}$$

are

$$\lambda_{1,2} = \tfrac{1}{2}\mathrm{Tr}\mathbf{A} \pm \tfrac{1}{2}\sqrt{\mathrm{Tr}^2\mathbf{A} - 4\det\mathbf{A}}$$

and the corresponding eigenvectors are, respectively,

$$u_1 = \frac{1}{\|u_1\|} \begin{pmatrix} b \\ \lambda_1 - a \end{pmatrix}; u_2 = \frac{1}{\|u_2\|} \begin{pmatrix} b \\ \lambda_2 - a \end{pmatrix}.$$

To study how the zero-trace 2×2 matrix σ_1 affects the exponential, consider

$$\mathbf{X} = \begin{pmatrix} 0 & b \\ a^{-1} & 0 \end{pmatrix}.$$

The exponential of the matrix $\mathbf{X}$, $e^{\mathbf{X}}$, has positive entries, and in fact [139]

$$e^{\mathbf{X}} = \begin{pmatrix} \cosh\sqrt{\frac{b}{a}} & \sqrt{ab}\sinh\sqrt{\frac{b}{a}} \\ \frac{1}{\sqrt{ab}}\sinh\sqrt{\frac{b}{a}} & \cosh\sqrt{\frac{b}{a}} \end{pmatrix}. \qquad \square$$

Theorem 4.8 (Markov inequality) *Let* $\mathbf{X}$ *a random variable with values in* $\mathcal{A}_+$ *and expectation*

$$M = \mathbb{E}X = \sum_x \mathrm{Pr}\{X = x\}x, \qquad (4.43)$$

and $\mathbf{A} \geq 0$. *Then*

$$\mathrm{Pr}\{X \nleq \mathbf{A}\} \leq \mathrm{Tr}(\mathbf{M}\mathbf{A}^{-1}). \qquad (4.44)$$

Theorem 4.9 (Chebyshev inequality) *Let* $\mathbf{X}$ *a random variable with values in* $\mathcal{A}_s$, *expectation* $\mathbf{M} = \mathbb{E}\mathbf{X}$, *and variance*

$$\mathrm{Var}X = S^2 = \mathbb{E}((X - M)^2) = \mathbb{E}(X^2) - M^2. \qquad (4.45)$$

For $\mathbf{\Delta} \geq 0$,

$$\mathrm{Pr}\{|X - M| \nleq \mathbf{\Delta}\} \leq \mathrm{Tr}(S^2\mathbf{\Delta}^{-2}). \qquad (4.46)$$

Recall that

$$|\mathbf{X} - \mathbf{M}| \leq \boldsymbol{\Delta} \Leftarrow (\mathbf{X} - \mathbf{M})^2 \leq \boldsymbol{\Delta}^2$$

since $\sqrt{(\cdot)}$ is operator monotone.

If $\mathbf{X}, \mathbf{Y}$ are independent, then $\mathrm{Var}(\mathbf{X} + \mathbf{Y}) = \mathrm{Var}\mathbf{X} + \mathrm{Var}\mathbf{Y}$. This is the same as in the classical case but one has to pay attention to the noncommunicativity that causes technical difficulty.

Corollary 4.1 (Weak law of large numbers) *Let* $\mathbf{X}, \mathbf{X}_1, \mathbf{X}_2, \ldots, \mathbf{X}_n$ *be identically, independently, distributed (i.i.d.) random variables with values in* $\mathcal{A}_s$, *expectation* $\mathbf{M} = \mathbb{E}\mathbf{X}$, *and variance* $\mathrm{Var}\mathbf{X} = \mathbf{S}^2$. *For* $\boldsymbol{\Delta} \geq 0$, *then*

$$\Pr\left\{\frac{1}{n}\sum_{n=1}^{n} X_i \notin [M - \boldsymbol{\Delta}, M + \boldsymbol{\Delta}]\right\} \leq \frac{1}{n}\mathrm{Tr}\left(S^2\boldsymbol{\Delta}^{-2}\right),$$

$$\Pr\left\{\sum_{n=1}^{n} X_i \notin \left[nM - \sqrt{n}\boldsymbol{\Delta}, nM - \sqrt{n}\boldsymbol{\Delta}\right]\right\} \leq \frac{1}{n}\mathrm{Tr}\left(S^2\boldsymbol{\Delta}^{-2}\right).$$
(4.47)

Lemma 4.1 (Large deviations and Bernstein trick) *For a random variable* $\mathbf{Y}, \mathbf{B} \in \mathcal{A}_s$, *and* $\mathbf{T} \in \mathcal{A}$ *such that* $\mathbf{T}^*\mathbf{T} > 0$

$$\Pr\left\{Y \not\leq \mathbf{B}\right\} \leq \mathrm{Tr}\left(\mathbb{E}\exp\left(TYT^* - TBT^*\right)\right).$$
(4.48)

Theorem 4.10 (i.i.d random variables) *Let* $\mathbf{X}, \mathbf{X}_1, \ldots, \mathbf{X}_n$ *be i.i.d. random variables with values in* $\mathcal{A}_s$, $\mathbf{A} \in \mathcal{A}_s$. *Then for* $\mathbf{T} \in \mathcal{A}$, $\mathbf{T}^*\mathbf{T} > 0$

$$\Pr\left\{\sum_{n=1}^{n} X_i \not\leq n\mathbf{A}\right\} \leq d \cdot ||\mathrm{Tr}\left(\mathcal{E}\exp(TXT^* - TAT^*)\right)||^n.$$
(4.49)

Define the binary I-divergence as

$$D(u||v) = u(\log u - \log v) + (1 - u)(\log(1 - u) - \log(1 - v)).$$
(4.50)

Theorem 4.11 (Chernoff) *Let* $\mathbf{X}, \mathbf{X}_1, \ldots, \mathbf{X}_n$ *be i.i.d. random variables with values in* $[\mathbf{0}, \mathbf{I}] \in \mathcal{A}_s$, $\mathbb{E}\mathbf{X} \leq m\mathbf{I}$, $\mathbf{A} \geq a\mathbf{I}$, $1 \geq a \geq m \geq 0$. *Then*

$$\Pr\left\{\sum_{n=1}^{n} X_i \not\leq n\mathbf{A}\right\} \leq d \cdot \exp\left(-nD\left(a||m\right)\right),$$
(4.51)

Similarly, $\mathbb{E}\mathbf{X} \geq m\mathbf{I}$, $\mathbf{A} \leq a\mathbf{I}$, $0 \leq a \leq m \leq 1$. *Then*

$$\Pr\left\{\sum_{n=1}^{n} X_i \not\geq n\mathbf{A}\right\} \leq d \cdot \exp\left(-nD\left(a||m\right)\right),$$
(4.52)

As a consequence, we get, for $\mathbb{E}\mathbf{X} = \mathbf{M} \geq \mu\mathbf{I}$ *and* $0 \leq \varepsilon \leq \frac{1}{2}$, *then*

$$\Pr\left\{\frac{1}{n}\sum_{n=1}^{n} X_i \notin [(1 - \varepsilon)M, (1 + \varepsilon)M]\right\} \leq 2d \cdot \exp\left(-n \cdot \frac{\varepsilon^2\mu}{2\ln 2}\right).$$
(4.53)

4.8 Partial Ordering Using Stochastic Order: $\mathbf{A} \leq_{st} \mathbf{B}$

If $\mathbf{x} \leq_{st} \mathbf{y}$, then $\mathbb{E}\mathbf{x} \leq \mathbb{E}\mathbf{y}$.

Let x have a multivariate normal density with mean vector zero and variance matrix Σ_1. Let y have a multivariate normal density with mean vector zero and variance matrix $\Sigma_1 + \Sigma_2$, where Σ_2 is a nonnegative definite matrix. Then [132, p. 14]

$$||\mathbf{x}||_2^2 \leq_{st} ||\mathbf{y}||_2^2, \tag{4.54}$$

where $|| \cdot ||$ is the Euclidean norm defined as $||\mathbf{x}||_2 = \left(\sum_{i=1}^n |\mathbf{x}(i)|^2 \right)^{\frac{1}{2}} = \sqrt{\mathbf{x}^*\mathbf{x}}$, for $\mathbf{x} \in \mathbb{R}^n$.

4.9 Quantum Hypothesis Detection

We consider the two hypotheses $\mathcal{H}_0$ (null):ρ and $\mathcal{H}_1$ (alternative):σ. We identify a state with a density operator, that is, a linear positive operator with trace one on finite-dimensional Hilbert space $\mathbb{H}$. Physically discriminating between the two hypotheses corresponds to performing a generalized (POVM) measurement on the quantum system. In analogy to the classical proceeding, one accepts $\mathcal{H}_0$ or $\mathcal{H}_1$ according to a decision rule based on the outcome of the measurement. There is no loss of generality assuming the POVM consists of only two elements, which denotes by $\{I - \Pi, \Pi\}$, where Π may be any linear operator on $\mathcal{H}$ with $0 \leq \Pi \leq I$ and I is identity operator. Neyman and Pearson introduces the idea of similarly making a distinction between type I and type II errors: (1) The type I error or false positive, denoted by α, is the error of accepting the alternative hypothesis when in reality the null hypothesis holds; (2) The type II error or false negative, denoted by β, is the error of accepting the null hypothesis when the alternative hypothesis is the true state of nature. The type-I and type-II error probabilities α and β are the probabilities of mistaking σ for ρ, and vice-versa, and are given by

$$\alpha = \text{Tr}(\Pi\rho)$$
$$\beta = \text{Tr}[(I - \Pi)\sigma]$$

The average error probability P_e is given by

$$P_e = \pi_0\alpha + \pi_1\beta = \pi_0\text{Tr}(\Pi\rho) + \pi_1\text{Tr}[(I - \Pi)\sigma] \tag{4.55}$$

The Bayesian distinguishably problem consists of finding the Π that minimizes P_e. A special case is the symmetric one where the prior probabilities π_0 and π_1 are equal.

Let us first introduce some basic notations. Abusing terminology, we will use the term 'positive' for 'positive semidefinite'(denoted $\mathbf{A} \geq 0$). We use the positive semidefinite ordering on the linear operators on $\mathcal{H}$ throughout, that is, $\mathbf{A} \geq \mathbf{B}$ if and only if $\mathbf{A} - \mathbf{B} \geq 0$. For each linear operator $\mathbf{A} \in \mathcal{B}(\mathcal{H})$ the absolute value $|\mathbf{A}|$ is defined as $|\mathbf{A}| = (\mathbf{A}^*\mathbf{A})^{\frac{1}{2}}$ where $\mathbf{A}^*$ is the transpose and conjugate (Hermitian) of $\mathbf{A}$. The Jordan decomposition of a self-adjoint operator $\mathbf{A}$ is given by $\mathbf{A} = \mathbf{A}_+ - \mathbf{A}_-$, where

$$\mathbf{A}_+ = (|\mathbf{A}| + \mathbf{A})/2, \mathbf{A}_- = (|\mathbf{A}| - \mathbf{A})/2 \tag{4.56}$$

are the positive part and negative part of $\mathbf{A}$, respectively. Both parts are positive by definition, and $\mathbf{A}_+\mathbf{A}_- = 0$. There is a very useful variational characterization of the trace of the postitive part of a self-adjoint operator $\mathbf{A}$:

$$\mathrm{Tr}(\mathbf{A}_+) = \max_{\mathbf{X}}\{\mathrm{Tr}(\mathbf{AX}) : 0 \leq \mathbf{X} \leq \mathbf{I}\}. \tag{4.57}$$

In other words, the maximum is taken over all positive contractive operators. Since the extremal values of the set of positive contractive operators are exactly the orthogonal projector, we also have

$$\mathrm{Tr}(\mathbf{A}_+) = \max_{P}\{\mathrm{Tr}(\mathbf{A}P) : P \geq 0, P = P^2\}. \tag{4.58}$$

The maximizer on the right-hand side is the orthogonal projector onto the range of $\mathbf{A}_+$.

Lemma 4.2 (Quantum Neyman-Pearson Lemma) *Let ρ and σ be the density operators associated to hypotheses $\mathcal{H}_0$ and $\mathcal{H}_1$, respectively. Let c be a fixed positive number. Consider the POVM with elements $\{\mathbf{I} - \Pi^*, \Pi^*\}$ where Π^* is the projector onto the range of $(c\sigma - \rho)_+$, and let $\alpha^* = \mathrm{Tr}(\Pi^*\rho)$ and $\beta^* = \mathrm{Tr}(\mathbf{I} - \Pi^*)\sigma$ be the associated errors. For any other POVM $\{\mathbf{I} - \Pi, \Pi\}$, with associated errors $\alpha = \mathrm{Tr}(\Pi\rho)$ and $\beta = \mathrm{Tr}[(\mathbf{I} - \Pi)\sigma]$, we have*

$$\alpha + c\beta \geq \alpha^* + c\beta^* = c - \mathrm{Tr}[(c\sigma - \rho)_+]. \tag{4.59}$$

Thus if $\alpha \leq \alpha^*$, then $\beta \geq \beta^*$.

Proof 4.1 *By formulae (4.57) and (4.58), for all $0 \leq \Pi \leq \mathbf{I}$ we have*

$$\mathrm{Tr}[\Pi (c\sigma - \rho) \leq \mathrm{Tr}[\Pi(c\sigma - \rho)_+ = \mathrm{Tr}[\Pi^* (c\sigma - \rho). \tag{4.60}$$

In terms of α, β, α^*, β^*, this reads

$$c(1 - \beta) - \alpha \leq c(1 - \beta^*) - \alpha^*,$$

which is equivalent to the statement of the Lemma. ◇

The Lemmas say that the POVM $\{\mathbf{I} - \Pi^*, \Pi^*\}$ is the optimal one when the goal is to minimize the quantity $\alpha + c\beta$. In symmetric hypothesis testing the positive number c is taken to be the ratio π_1/π_0 of the prior probabilities. The goal of the Bayesian distinguishability problem is to minimize the average error probabilities P_e defined in (4.55) and can be rewritten as

$$P_e = \pi_1 - \mathrm{Tr}[\Pi(\pi_1\sigma - \pi_0\rho)].$$

By the Neyman-Pearson Lemma, the optimal test is given by the projector Π^* onto the range of $(\pi_1\sigma - \pi_0\rho)_+$, and the obtained minimal error probability is given by

$$\begin{aligned} P_e^* &= \pi_1 - \mathrm{Tr}[(\pi_1\sigma - \pi_0\rho)_+] = \pi_1 - \mathrm{Tr}(\pi_1\sigma - \pi_0\rho) - \mathrm{Tr}[|\pi_1\sigma - \pi_0\rho|/2] \\ &= \tfrac{1}{2}(1 - \|\pi_1\sigma - \pi_0\rho\|_1), \end{aligned} \tag{4.61}$$

where $\|\mathbf{A}\|_1 = \mathrm{Tr}|\mathbf{A}|$ is the trace norm. We call Π^* the Holevo-Helstrom projector. Note $\mathrm{Tr}\rho = \mathrm{Tr}\sigma = 1$ since ρ and σ are arbitrary density operators. Our goal in this task is to establish the connection of the heuristic hypothesis testing defined by (3.23) with quantum hypothesis testing. Consider a quantum system $\mathcal{H}$ whose state is represented by the density matrix ρ and σ; more precisely, $\mathcal{H}_0 : \rho$ and $\mathcal{H}_1 : \sigma$. This procedure may be expressed as a Hermitian matrix.

Let us define the projection $\{\mathbf{X} \geq 0\}$ with respect to a Hermitian matrix $\mathbf{X}$ with a spectral decomposition $\mathbf{X} = \sum_i x_i E_{\mathbf{X},i}$:

$$\{\mathbf{X} \geq 0\} = \sum_{x_i \geq 0} E_{\mathbf{X},i}.$$

When the state is ρ, the probability of the set $\{x_i \geq 0\}$ is $\sum_{x_i \geq 0} \mathrm{Tr}\rho E_{\mathbf{X},i} = \mathrm{Tr}\rho\{\mathbf{X} \geq 0\}$. This notation generalizes the concept of the subset to the noncommunicative case. It is known that two noncommunicative Hermitian matrices $\mathbf{X}$ and Y cannot be diagonalized simultaneously by a common orthonormal basis. This fact causes many technical difficulties.

The two-valued POVM $\{\mathbf{T}, \mathbf{I} - \mathbf{T}\}$ for a Hermitian matrix $\mathbf{T}$ satisfying $\mathbf{I} \geq \mathbf{T} \geq 0$ allows us to perform the discrimination. Thus, $\mathbf{T}$ will be called a *test*. The following theorem [140, 141] holds for an arbitrary real number $c > 0$: The average probability of error is

$$\min_{\mathbf{I} \geq \mathbf{T} \geq 0} (\mathrm{Tr}\rho(\mathbf{I} - \mathbf{T}) + c\mathrm{Tr}\sigma\mathbf{T}) = \mathrm{Tr}\rho\{\rho - c\sigma \leq 0\} + c\mathrm{Tr}\sigma\{\rho - c\sigma > 0\} \tag{4.62}$$

The minimum value is achieved when $\mathbf{T} = \{\rho - \sigma \geq 0\}$. In particular, if $c = 1$[3], it follows that

$$\min_{\mathbf{I} \geq \mathbf{T} \geq 0} (\mathrm{Tr}\rho(\mathbf{I} - \mathbf{T}) + c\mathrm{Tr}\sigma\mathbf{T}) = 1 - \tfrac{1}{2}\|\rho - \sigma\|_1. \tag{4.63}$$

The optimal average probability of correct discrimination is

$$\tfrac{1}{2}\min_{\mathbf{I} \geq \mathbf{T} \geq 0} (\mathrm{Tr}\rho(\mathbf{I} - \mathbf{T}) + c\mathrm{Tr}\sigma\mathbf{T}) = \mathrm{Tr}\rho\{\rho - \sigma \leq 0\} + \mathrm{Tr}\sigma\{\rho - \sigma > 0\} = \tfrac{1}{2} + \tfrac{1}{4}\|\rho - \sigma\|_1. \tag{4.64}$$

Therefore, the trace norm gives a measure for the discrimination of two states. Here $\|\mathbf{A}\|_1 = \mathrm{Tr}|\mathbf{A}|$ and the absolute value $|\mathbf{A}|$ is defined as $|\mathbf{A}| = \sqrt{\mathbf{A}\mathbf{A}^*}$. From (4.63), the necessary condition for quantum detection is: $\|\rho - \sigma\|_1 = \mathrm{Tr}|\rho - \sigma| > 0$. Since only the absolute value is involved, the trace norm distance is symmetric. Without loss of generality, considering $\sigma \geq \rho \geq 0$ the necessary condition reduces to

$$\mathrm{Tr}\sigma \geq \mathrm{Tr}\rho \quad \text{or} \quad \mathrm{Tr}f(\mathbb{E}\mathbf{A}) \geq \mathrm{Tr}f(\mathbb{E}\mathbf{B}), \tag{4.65}$$

if $\rho = f(\mathbb{E}\mathbf{A})$ and $\sigma = f(\mathbb{E}\mathbf{B})$. Condition (4.65) is exactly identical to (3.22) used in Algorithm 3.1. Therefore, it is shown that Algorithm 3.1 is equivalent to the Holevo-Helstrom tests [142, 143], which are noncommunicative generalizations of the classical LRT. The above "proof" paves the way for systematically exploiting the deep work done for quantum hypothesis testing [142, 144–217]. This subtask may lead to algorithms for spectrum sensing with unprecedented performance.

---

[3] Two hypotheses have two equal prior probabilities in this Bayesian test.

4.10   Quantum Hypothesis Testing for Many Copies

A single copy of the quantum system is not enough for a good decision. One should make independent measurement on several identical copies, or joint measurements. The basic problem is to identify how the error probability P_e behaves in the asymptotic limit, that is, when one has to discriminate between the hypotheses $\mathcal{H}_0$ and $\mathcal{H}_1$ corresponding to either n copies of ρ or n copies of σ. To do so, we need to study the quantity

$$P_{e,n}^* = (1 - \|\pi_1 \sigma^{\otimes n} - \pi_0 \rho^{\otimes n}\|_1)/2. \tag{4.66}$$

where $\rho^{\otimes n} = \underbrace{\rho \otimes \rho \cdots \otimes \rho}_{n}$ is the nth-tensor powers of ρ. Such states can be regarded as the quantum version of independent, identical distributions (i.i.d). It turns out that $P_{e,n}^*$ exponentially decreases in n: $P_{e,n}^* \sim \exp(-n\xi_{QCB})$. This exponential decrease is very desirable for cooperative sensing of RF spectrum, where a large number n of copies are feasible.

Theorem [34, 142, 143]: For any two states ρ and σ on a finite-dimensional Hilbert space, occurring with prior probabilities π_1 and π_2, respectively, the rate limit of $P_{e,n}^*$, as defined by (4.66), exists and is equal to the quantum Chernoff distance ξ_{QCB}

$$\lim_{n \to \infty} \left(-\frac{1}{n} \log P_{e,n}^* \right) = \xi_{QCB} = -\log \left(\inf_{0 \le s \le 1} \mathrm{Tr}(\rho^{1-s} \sigma^s) \right). \tag{4.67}$$

This recent result provides a convenient tool for quantifying the asymptotic limit of the cooperative sensing of RF spectrum. For a general test with n different states $\mathcal{H}_0 : \bar{\rho} = \rho_1 \otimes \cdots \otimes \rho_n$ and $\mathcal{H}_1 : \bar{\sigma} = \sigma_1 \otimes \cdots \otimes \sigma_n$, the necessary condition for (4.66) to be valid takes a new look:

$$0 < \|\rho_1 \otimes \cdots \otimes \rho_n - \sigma_1 \otimes \cdots \otimes \sigma_n\|_1 \le \sum_{i=1}^{n} \|\rho_i - \sigma_i\|_1 = \sum_{i=1}^{n} \mathrm{Tr}|\rho_i - \sigma_i|$$

which, if $\sigma_i > \rho_i$, reduces to

$$\mathrm{Tr} \sum_{i=1}^{n} \rho_i < \mathrm{Tr} \sum_{i=1}^{n} \sigma_i \, \mathrm{or} \, \sum_{i=1}^{n} \mathrm{Tr}\rho_i < \sum_{i=1}^{n} \mathrm{Tr}\sigma_i.$$

This is equivalent to a special form of (3.23): by replacing the expectation with the average of n copies and letting $f(x) = x$ in (3.23).

This subtask can borrow from the use of many copies for coding, basic to quantum information [34, 117, 127, 129, 140–143, 218–250].

5

Large Random Matrices

5.1 Large Dimensional Random Matrices: Moment Approach, Stieltjes Transform and Free Probability

The necessity of studying the spectra of large dimensional random matrices, in particular, the Wigner matrices, arose in nuclear physics in the 1950s. In quantum mechanics, the energy levels of quantum are not directly observable (very similar to many problems in today's wireless communications and the Smart Grid), but can be characterized by the eigenvalues of a matrix of observations [10].

Let X_{ij} be i.i.d. standard normal variables of $n \times p$ matrix $\mathbf{X}$

$$
\mathbf{X} = \begin{bmatrix} X_{11} & X_{12} & \cdots & X_{1n} \\ X_{21} & X_{22} & \cdots & X_{2n} \\ \vdots & \vdots & \vdots & \vdots \\ X_{p1} & X_{p2} & \cdots & X_{pn} \end{bmatrix}_{p \times n}.
$$

The sample covariance matrix is defined as

$$
\mathbf{S}_n = \left(\frac{1}{n} \sum_{k=1}^{n} X_{ki} X_{kj} \right)_{i,j=1}^{p},
$$

where n vector samples of a p-dimensional zero-mean random vector with population matrix I.

The classical limit theorem are no longer suitable for dealing with large dimensional data analysis. In the early 1980s, major contributions on the existence of the limiting spectral distribution (LSD) were made. In recent years, research on random matrix theory has turned toward second-order limiting theorems, such as the central limit theorem for linear spectral statistics, the limiting distributions of spectral spacings, and extreme eigenvalues.

Cognitive Radio Communications and Networking: Principles and Practice, First Edition.
Robert C. Qiu, Zhen Hu, Husheng Li and Michael C. Wicks.
© 2012 John Wiley & Sons, Ltd. Published 2012 by John Wiley & Sons, Ltd.

Many applied problems require an estimate of a covariance matrix and/or of its inverse, where the matrix dimension is large compared to the sample size [20]. In such situations, the usual estimator, the sample covariance matrix, is known to perform poorly. When the matrix dimension p is larger than the number n of observations available, the sample covariance matrix is not even invertible. When the ratio p/n is less than one but not negligible, the sample covariance matrix is invertible but numerically ill-conditioned, which means that inverting it amplifies estimation error dramatically. For a large value of p, it is difficult to find enough observations to make p/n negligible, and therefore it is important to develop a well-conditioned estimator for large-dimensional covariance matrices such as in [20].

Suppose $\mathbf{A}_N$ is an $N \times N$ matrix with eigenvalues $\lambda_1(\mathbf{A}_N), \ldots, \lambda_N(\mathbf{A}_N)$. If all these eigenvalues are real (e.g., if $\mathbf{A}_N$ is Hermitian), we can define a one-dimensional distribution function. The empirical cumulative distribution of the eigenvalues, also called the empirical spectrum distribution (ESD), of an $N \times N$ Hermitian matrix $\mathbf{A}$ is denoted by $F_{\mathbf{A_N}}$

$$F_{\mathbf{A}_N}(x) = \frac{\text{Number of eigenvalues of } \mathbf{A}_N \leq x}{N} = \frac{1}{N} \sum_{i=1}^{N} 1\{\lambda_i(\mathbf{A}_N) \leq x\}, \qquad (5.1)$$

where $1\{\}$ is the indicator function.

Following [10], we divide available techniques into three categories: (1) Moment approach; (2) Stieltjes transform; (3) Free probability. Applications for these basic techniques will be covered.

The significance of ESD is due to the fact that many important statistics in multivariate analysis can be expressed as functionals of the ESD of some random matrices. For example, the determinant and the rank functions are the most common examples. The most significant theorem relevant to our applications is the convergence of the sample covariance matrix: the Marchenko-Pastur law.

Theorem 5.1 (Marchenko-Pastur law [251]) *Consider a $p \times N$ matrix $\mathbf{X}$, whose entries are independent, zero-mean complex (or real) random variables, with variance $\frac{\sigma^2}{N}$ and fourth moments of order $O\left(\frac{1}{N^2}\right)$. As*

$$p, N \to \infty \quad \text{with} \quad \frac{p}{N} \to \alpha, \qquad (5.2)$$

the empirical distribution of $\mathbf{X}\mathbf{X}^H$ converges almost surely to a nonrandom limiting distribution with density

$$f(x) = (1 - \alpha^{-1})^+ \delta(x) + \frac{\sqrt{(x-a)^+ (b-x)^+}}{2\pi\alpha x}, \qquad (5.3)$$
$$a = \sigma^2(1 - \sqrt{\alpha})^2, b = \sigma^2(1 + \sqrt{\alpha})^2.$$

Example 5.1 (Determinant of a positive definite matrix)
Let $\mathbf{A}_N$ be a positive definite matrix of $N \times N$. Then

$$\det(\mathbf{A}_N) = \prod_{j=1}^{N} \lambda_j = \exp\left(N \int_0^{\infty} \log x \, F_{\mathbf{A}_N}(dx)\right).$$

When $N \to \infty$, the determinant of $\mathbf{A}_N$, $\det(\mathbf{A}_N)$, is approaching a nonrandom limit value. □

Example 5.2 (Hypothesis testing)
Let the covariance matrix of the received signal have the form [14, p. 5]

$$\mathbf{\Sigma}_N = \mathbf{\Sigma}_q + \sigma^2 \mathbf{I},$$

where the dimension of $\mathbf{\Sigma}_N$ is p and the rank of $\mathbf{\Sigma}_q$ is $q(< p)$. Note that N and p are different. Suppose $\mathbf{S}_N$ is the sample covariance matrix based on N i.i.d. vector samples drawn from the signal. The eigenvalues of $\mathbf{S}_N$ are

$$\lambda_1 \geq \lambda_2 \cdots \geq \lambda_p.$$

The test statistic for the hypothesis problem

$$\begin{aligned}
\mathcal{H}_0 &: \text{rank } (\mathbf{\Sigma}_q) = q, \\
\mathcal{H}_1 &: \text{rank } (\mathbf{\Sigma}_q) > q,
\end{aligned} \tag{5.4}$$

is given by

$$\begin{aligned}
T &= \frac{1}{p-q} \sum_{j=q+1}^{p} \lambda_j^2 - \left(\frac{1}{p-q} \sum_{j=q+1}^{p} \lambda_j \right)^2 \\
&= \frac{1}{p-q} \int_0^{\lambda_q} x^2 F_{\mathbf{S}_N}(dx) - \left[\frac{1}{p-q} \int_0^{\lambda_q} x F_{\mathbf{S}_N}(dx) \right]^2.
\end{aligned} \tag{5.5}$$

where T is the variance of the sequence of eigenvalues. □

The ultimate goal of hypothesis testing is to search for some metrics that are "robust" for decision making by setting a threshold. For example, the trace functions are commonly used. To represent the trace functions, we suggest four methodologies: moment method, Stieltjes transform, orthogonal polynomial decomposition and free probability. We only give the basic definitions and their relevance to our problems of spectral sensing. We refer to [14] for details.

The goal of random matrix theory is to present several aspects of the asymptotics of random matrix "macroscopic" quantities [252] such as

$$L_N = \frac{1}{n} \text{Tr}(\mathbf{A}_{i_1}^n \cdots \mathbf{A}_{i_p}^n),$$

where $i_k \in \{1, \ldots, m\}$, $1 \leq k \leq p$ and $(\mathbf{A}_p^n)_{1 \leq p \leq m}$ are some $n \times n$ random matrices whose size n goes to infinity. $(\mathbf{A}_p^n)_{1 \leq p \leq m}$ are most often Wigner matrices, that is Hermitian matrices with independent entries, and Wishart matrices.

5.2 Spectrum Sensing Using Large Random Matrices

5.2.1 System Model

The most remarkable intuition of random matrices is that in many cases the eigenvalues of matrices with random entries turn out to converge to some *fixed distribution*, when both the dimensions of the signal matrix tend to infinity with the same order [253]. For Wishart matrices, the limiting joint distribution called Marchenko-Pastur Law has been known since 1967 [251]. Then, most recently, the marginal distribution of single ordered eigenvalues have been found. By exploiting these results, we are able to express the largest and the smallest eigenvalues of sample covariance matrices using their asymptotic values in closed form. The closed-form, exact expression for the standard condition number (defined as the ratio of the largest to the smallest eigenvalue) is available.

We often treat the asymptotic limiting results for large matrices to the finite-size matrices. The power of large random matrices is such that the approximate technique is often stunningly precise. If the matrices under consideration are larger than 8×8, the asymptotic results are accurate enough to approximate the simulated results.

The received signal contains L vectors $\mathbf{y}_l, l = 1, \ldots, L$

$$
\begin{aligned}
&\mathcal{H}_0 : y_l[i] = w_l[i], i = 1, \ldots N \\
&\mathcal{H}_1 : y_l[i] = h_l[i]s_l[i] + w_l[i], i = 1, \ldots N
\end{aligned}
\tag{5.6}
$$

where $h_l[i]$ is the channel gain (often having a Rayleigh fading distribution) for the i-th sample time of the l-th sensor. This is similar for signal vector $\mathbf{s}_l$ and noise vector $\mathbf{w}_l$. Let $\mathbf{y}$ be a $n \times 1$ vector modeled as

$$
\mathbf{y} = \mathbf{Hs} + \mathbf{w},
$$

where $\mathbf{H}$ is an $n \times L$ matrix, $\mathbf{s}$ is an $L \times 1$ "signal" vector and $\mathbf{w}$ is an $n \times 1$ "noise" vector. This model appears frequently in many signal processing and communications applications. If $\mathbf{s}$ and $\mathbf{w}$ are modeled as independent Gaussian vectors with independent elements with zero mean and unit variance matrix (identity covariance matrix), then $\mathbf{y}$ is a multivariate Gaussian with zero mean and covariance matrix written as

$$
\boldsymbol{\Sigma} = \mathbf{R} = E\{\mathbf{yy}^H\} = \mathbf{HH}^H + \mathbf{I}.
\tag{5.7}
$$

In most practical applications, the true covariance matrix is unknown. Instead, it is estimated from N independent observations ("snapshots") $\mathbf{y}_1, \mathbf{y}_2, \ldots, \mathbf{y}_N$ as:

$$
\mathbf{S}_Y = \frac{1}{N} \sum_{i=1}^{N} \mathbf{y}_i \mathbf{y}_i^H = \frac{1}{N} \mathbf{Y}_n \mathbf{Y}_n^H,
$$

where $\mathbf{Y}_n = [\mathbf{y}_1, \mathbf{y}_2, \ldots, \mathbf{y}_N]$ is referred to as the "data matrix" and $\mathbf{S}_Y$ is the sample covariance matrix.

When n is fixed and $N \to \infty$, it is well-known that the sample covariance matrix converges to the true covariance matrix. However, when both $n, N \to \infty$ with

$$
n/N = \alpha > 0,
$$

this is no longer true. Such a scenario is very relevant in practice where stationarity constraints limit the amount of data (N) that can be used to form the sample covariance matrix. Free probability is an invaluable tool in such situations when attempting to understand the structure of the resulting sample covariance matrices [254].

In matrix form, we have the following $L \times N$ matrix:

$$\mathbf{Y} = \begin{pmatrix} y_1[1] & y_1[2] & \cdots & y_1[N] \\ y_2[1] & y_2[2] & \cdots & y_2[N] \\ \vdots & \vdots & \cdots & \vdots \\ y_L[1] & y_L[2] & \cdots & y_L[N] \end{pmatrix}_{L \times N} . \tag{5.8}$$

Similarly, we do this for $\mathbf{H}, \mathbf{S}, \mathbf{W}$. (5.8) can be rewritten in matrix form as

$$\mathbf{Y} = \mathbf{HS} + \mathbf{W} = \mathbf{X} + \mathbf{W}. \tag{5.9}$$

where $\mathbf{X} = \mathbf{HS}$. Using (5.9), (5.6) becomes our standard form:

$$\begin{aligned} \mathcal{H}_0 &: \mathbf{YY}^H = \mathbf{WW}^H, \\ \mathcal{H}_1 &: \mathbf{YY}^H = \mathbf{XX}^H + \mathbf{WW}^H, \end{aligned} \tag{5.10}$$

where we have made the assumption that

$$(\mathbf{X} + \mathbf{W})(\mathbf{X} + \mathbf{W})^H = \mathbf{XX}^H + \mathbf{WW}^H. \tag{5.11}$$

(5.11) can be justified rigorously using random matrix theory.

In general, knowing the eigenvalues of two matrices, say $\mathbf{A}, \mathbf{B}$, is not enough to find the eigenvalues of the sum or product of the two matrices, unless they commute. Free probability gives us a certain sufficient condition, called asymptotic freeness, under which the asymptotic spectrum of the sum $\mathbf{A} + \mathbf{B}$ or product $\mathbf{AB}$ can be obtained from the individual asymptotic spectra, without involving the structure of the eigenvectors of the matrices [255]. [13, p. 9] [256]

Theorem 5.2 (Wishart matrices) *If* $\mathbf{W}$ *has a Wishart distribution with* m *degrees of freedom and true covariance matrix* $\mathbf{\Sigma}$, *write* $W_p(\mathbf{\Sigma}, m)$, *and* $\mathbf{C}$ *is a* $q \times p$ *matrix of rank* q, *then*

$$\mathbf{CWC}^H \sim W_q(\mathbf{CWC}^H, m).$$

The sample covariance matrix $\mathbf{S}_Y$ based on $\mathbf{Y}$, which contains N samples and L column vectors, is

$$\mathbf{S}_Y = \frac{1}{N}\mathbf{YY}^H.$$

The sample covariance matrix $\mathbf{S}_Y$ is related to the true covariance matrix $\mathbf{\Sigma}_Y$ by the property of Wishart distribution (see Theorem 5.2)

$$\mathbf{S}_Y = \mathbf{\Sigma}_Y^{1/2}\mathbf{ZZ}^H\mathbf{\Sigma}_Y^{1/2}, \tag{5.12}$$

where $\mathbf{Z}$ is a $L \times N$ i.i.d. Gaussian zero mean matrix. In fact,

$$\mathbf{W}(\alpha) = \frac{1}{N}\mathbf{ZZ}^H \tag{5.13}$$

is the Wishart matrix.

For a standard signal plus noise model, the true covariance matrix $\mathbf{\Sigma}_Y$ has the form

$$\mathbf{\Sigma}_Y = \mathbf{\Sigma}_X + \mathbf{\Sigma}_W. \tag{5.14}$$

where $\mathbf{\Sigma}_X$ and $\mathbf{\Sigma}_W$ are, respectively, the true covariance matrix of the signal and the noise; also $\mathbf{\Sigma}_W = \sigma^2 \mathbf{I}$ if the white noise is assumed.

Comparing the true covariance matrix (5.14) with its sample counterpart (5.11) reveals the fundamental role of a rigorous random matrix theory. We really cannot say much about the relation between the two versions of equations, generally for small sample size N. Luckily, when the sample size N is very large, the two versions can be proven equivalent (which will be justified later). This is the reason why random matrix theory is so relevant to wireless communications since a majority of wireless systems can be expressed in the form of (5.9). For example, CDMA, MIMO and OFDM systems can be expressed in such a form.

5.2.2 Marchenko-Pastur Law

The Marchenko Pastur law stated in Theorem 5.1 serves as a theoretical prediction under the assumption that the matrix is "all noise" [255]. Deviations from this theoretical limit in the eigenvalue distribution should indicate nonnoisy components, in other words, they should suggest information about the matrix.

Example 5.3 (Spectrum sensing using the ratio $\lambda_{max}/\lambda_{min}$ [119, 255, 257, 258])
We mainly follow [255] in this example. (5.8) is repeated here for convenience. In matrix form, the received signal model is expressed as the following $L \times N$ matrix

$$\mathbf{Y} = \begin{pmatrix} y_1[1] & y_1[2] & \cdots & y_1[N] \\ y_2[1] & y_2[2] & \cdots & y_2[N] \\ \vdots & \vdots & \cdots & \vdots \\ y_L[1] & y_L[2] & \cdots & y_L[N] \end{pmatrix}_{L \times N}, \tag{5.15}$$

where N samples are recorded at L sensors.

For a fixed L and $N \to \infty$, the sample covariance matrix $\frac{1}{N}\mathbf{Y}\mathbf{Y}^H$ converges to $\sigma^2 \mathbf{I}$. This is the consequence of using the Central Limit Theorem. However, in practice, N can be of the same order of magnitude as L; this scenario is what the random matrix theory offers.

In the case where the entries of $\mathbf{Y}$ are independent (irrespective of the specific probability distribution, which corresponds to the case where no signal is present$-\mathcal{H}_0$), results from asymptotic random matrix theory can be used. Theorem 5.1 proposed by Marchenko and Pastur (1967) is valid for this case as $L, N \to \infty$ with $\frac{L}{N} \to \alpha$.

Interestingly, the support of the eigenvalues is finite, even if there is no signal. The theoretical prediction offered by the Marchenko-Pastur law can be used to set the threshold for detection.

To illustrate, let us consider the case when only one signal is present for $\mathcal{H}_1$:

$$
\mathbf{Y} = \begin{pmatrix} h_1 & \sigma & & 0 \\ \vdots & & \ddots & \\ h_L & 0 & & \sigma \end{pmatrix} \begin{pmatrix} s[1] & \cdots & s[N] \\ z_1[1] & \cdots & z_1[N] \\ \vdots & \ddots & \vdots \\ z_L[1] & \cdots & z_L[N] \end{pmatrix},
$$

where $s[i]$ and $z[i] = \sigma n_l[i]$ are, respectively, the independent signal and noise with unit variance at instant i and sensor l. Let us denote by $\mathbf{T}$ the matrix:

$$
\mathbf{T} = \begin{pmatrix} h_1 & \sigma & & 0 \\ \vdots & & \ddots & \\ h_L & 0 & & \sigma \end{pmatrix}.
$$

Clearly, $\mathbf{TT}^H$ has only one "significant" eigenvalue

$$
\lambda_1 = \sum_{j=1}^{L} |h_j|^2 + \sigma^2, \lambda_i = \sigma^2, i = 1, 2, \ldots, \min(L, K).
$$

The behavior of $\frac{1}{N}\mathbf{TT}^H$ is related to the study of the eigenvalues of large sample covariance matrices of spiked population models [26]. Let us define the signal to noise ratio γ as

$$
\gamma = \frac{\sum_{j=1}^{L} |h_j|^2}{\sigma^2}.
$$

Baik and colleagues [26, 259] show recently that, when

$$
\frac{L}{N} < 1 \text{ and } \gamma > \sqrt{\frac{L}{N}},
$$

the maximum eigenvalue of $\frac{1}{N}\mathbf{TT}^H$ converges almost surely to

$$
\mathcal{H}_1 : b_1 = \left(\sum_{j=1}^{L} |h_j|^2 + \sigma^2 \right) \left(1 + \tfrac{\alpha}{\gamma} \right),
$$

$$
\mathcal{H}_0 : b = \sigma^2 (1 + \sqrt{\alpha})^2,
$$

where b_1 is superior to b_0 that is also defined in Theorem 5.1. The difference between b_1 and b_0 can be used to sense the spectrum. Whenever the distribution of the eigenvalues of the sample covariance matrix $\frac{1}{N}\mathbf{YY}^H$ —all entries are observable and the size of the matrix is finite—departs from the predicted distribution obtained using the Marchenko-Pastur law, the detector knows that the signal is present. This approach of sensing non-null hypothesis is standard. But the metric and the mathematical tools are novel.   □

5.2.2.1 Noise Distribution Unknown, Variance Known

The criteria is

$$\text{decison} = \begin{cases} \mathcal{H}_0 : \lambda_i \in [a, b], \\ \mathcal{H}_1 : \text{otherwise}, \end{cases}$$

where a and b are defined in Theorem 5.1. The results are based on the asymptotic eigenvalue distribution.

5.2.2.2 Both Noise Distribution and Variance are Unknown

The ratio of the maximum and the minimum eigenvalues in the $\mathcal{H}_0$ case does not depend on the variance of the noise. This allows us to circumvent the need for the knowledge of the noise:

$$\text{decision} = \begin{cases} \mathcal{H}_0 : \frac{\lambda_{\max}}{\lambda_{\min}} \leq \frac{(1+\sqrt{\alpha})^2}{(1-\sqrt{\alpha})^2}, \\ \mathcal{H}_1 : \text{otherwise}. \end{cases}$$

The test $\mathcal{H}_1$ provides a good estimator of the SNR γ. The ratio of the largest eigenvalue b_1 and the smallest a of $\frac{1}{N}\mathbf{Y}\mathbf{Y}^H$ is related solely to γ and α

$$\frac{b_1}{a} = \frac{(1+\gamma)\left(1+\frac{\alpha}{\gamma}\right)}{(1-\sqrt{\alpha})^2}.$$

For extremely low SNR such as $\gamma = 0.01$, that is, $-20\,\text{dB}$, the above relation becomes

$$\frac{b_1}{a} = \frac{(1+\gamma)\left(1+\frac{\alpha}{\gamma}\right)}{(1-\sqrt{\alpha})^2} \cong \frac{(1+100\alpha)}{(1-\sqrt{\alpha})^2}.$$

Typically, we have $\alpha = 1/2$ and $\alpha = 1/10$.

Example 5.4 (Spectrum sensing using the ratio $\lambda_{max}/\lambda_{min}$ [260])
The example is continued from Example 5.3. We define the normalized covariance matrix as

$$\tilde{\mathbf{R}} = \frac{1}{\sigma^2}\mathbf{Y}\mathbf{Y}^H,$$

whose largest eigenvalue and smallest one are, respectively, l_{max} and l_{min}. In contrast, λ_{max} and λ_{min} are the corresponding ones of the sample covariance matrix $\frac{1}{N}\mathbf{Y}\mathbf{Y}^H$. Under $\mathcal{H}_0$, $\tilde{\mathbf{R}}$ turns out to be complex white Wishart matrix and by the Machenko-Pastur law, the eigenvalue support is finite [10]. Under $\mathcal{H}_1$, the covariance matrix belongs to the class of "spiked population models" and its largest eigenvalue increases outside the Marchenko-Pastur support [26]. This property suggests using

$$T = \frac{l_{\max}}{l_{\min}} = \frac{\lambda_{\max}}{\lambda_{\min}},$$

as test statistic for signal detection. Denoting T_0 the decision threshold, the detector claims $\mathcal{H}_1$ if $T > T_0$; otherwise, it claims $\mathcal{H}_0$.

Example 5.4 uses the asymptotic properties of Wishart matrices. The smallest and largest eigenvalues of $\tilde{\mathbf{R}}$ under $\mathcal{H}_0$ almost surely to

$$
\begin{aligned}
l_{\max} &\to a_{\max} = (\sqrt{N} + \sqrt{L})^2, \\
l_{\min} &\to a_{\min} = (\sqrt{N} - \sqrt{L})^2,
\end{aligned}
\tag{5.16}
$$

in the limit

$$
N, L \to \infty \text{ with } \frac{L}{N} \to \alpha
\tag{5.17}
$$

where $\alpha \in (0, 1)$ is a constant.

A semi-asymptotic approach [257] can be used. It is shown in [22] that under the same assumption of (5.17), the random variable

$$
L_{\max} = \frac{l_{\max} - a_{\max}}{\nu},
$$

with

$$
\nu = (\sqrt{N} + \sqrt{L})\left(\frac{1}{\sqrt{N}} + \frac{1}{\sqrt{L}}\right)^{1/3},
$$

converges in distribution to the Trace-Widow law of order 2 defined in (5.50). The decision threshold [257] can be linked to the probability of false alarm defined as

$$
P_{fa} = P(T > T_0 | \mathcal{H}_0),
$$

by using the asymptotic limit for the smallest eigenvalue (5.16) and the Tracy-Widom culmination distribution function for the largest one. The threshold can be expressed as

$$
T_0 = \frac{a_{\max}}{a_{\min}} \cdot \left(1 + \frac{(\sqrt{N} + \sqrt{L})^{-2/3}}{(NL)^{1/6}} F_{TW2}^{-1}(1 - P_{fa})\right)
$$

where F_{TW2}^{-1} is the inverse Tracy-Widom culmination distribution function of order 2.

Recently, it is established [261] that the smallest eigenvalue also converges to the Tracy-Widom culmination distribution as $K, L \to \infty$, up to a proper rescaling factor. In particular, the random variable

$$
L_{\min} = \frac{l_{\min} - a_{\min}}{\mu},
$$

with

$$
\mu = (\sqrt{L} - \sqrt{N})\left(\frac{1}{\sqrt{L}} - \frac{1}{\sqrt{N}}\right)^{1/3},
$$

converges to the Tracy-Widom culmination distribution function of order 2.

As a consequence of (5.17), μ is always negative in the considered range of α. The test statistic may be expressed as

$$
T = \frac{l_{\max}}{l_{\min}} = \frac{\nu L_{\max} + a_{\max}}{\mu L_{\min} + a_{\min}}.
$$

The test statistic can be linked to the probability of false alarm. For details, we refer
to [260].   □

5.2.2.3 On the Empirical Distribution of Eigenvalues of Large Dimensional Information-Plus-Noise Type Matrices

Sample covariance matrices for systems with noise are the starting point in many problems, for example, spectrum sensing. Multiplicative free deconvolution has been shown in
[262] to be a method. This method can assist in expressing limit eigenvalues distributions
for sample covariance matrices, and to simplify estimators for eigenvalue distributions of
covariance matrices.

We adopt a problem formulation from [263]. Let $\mathbf{X}_n$ be $n \times N$ containing i.i.d. complex
entries and unit variance (sum of variances of real and imaginary parts equals 1), $\sigma > 0$
constant, and $\mathbf{R}_n$ an $n \times N$ random matrix independent of $\mathbf{X}_n$. Assume, almost surely,
as $n \to \infty$, the empirical distribution function (e.d.f.) of the eigenvalues of $\frac{1}{N}\mathbf{R}_n\mathbf{R}_n^H$
converges in distribution to a nonrandom probability distribution function (p.d.f.), and the
ratio $\frac{n}{N}$ tends to a positive number. Then it is shown that, almost surely, the e.d.f. of the
eigenvalues of

$$\mathbf{C}_N = \frac{1}{N}(\mathbf{R}_n + \sigma\mathbf{X}_n)(\mathbf{R}_n + \sigma\mathbf{X}_n)^H \tag{5.18}$$

converges in distribution. The limit is nonrandom and is characterized in terms of its
Stieltjes transform, which satisfies a certain equation. n and N both converge to infinity
but their ratio $\frac{n}{N}$ converges to a positive quantity c. The aim of [263] is to show that,
almost surely, $F_{\mathbf{C}_N}$ converges in distribution to a nonrandom p.d.f. F. (5.18) can be
thought of as the sample covariance matrices of random vectors $\mathbf{r}_n + \sigma\mathbf{x}_n$, where $\mathbf{r}_n$ can
be a vector containing the system information and $\mathbf{x}_n$ is additive noise, with σ a measure
of the strength of the noise.

The matrix $\mathbf{C}_N$ can be viewed as the sample correlation matrix of the columns of
$\mathbf{R}_n + \sigma\mathbf{X}_n$, which models situations where relevant information is contained in the $\mathbf{R}_{.i}$'s
and can be extracted from $\frac{1}{N}\mathbf{R}_n\mathbf{R}_n^H$. Since $\mathbf{R}_{.i}$ is corrupted by $\mathbf{X}_{.i}$, the creation of this
matrix $\mathbf{C}_N$ is hindered. If the number of samples N is sufficiently large and if the noise
is centered, then $\mathbf{C}_N$ would be a reasonable estimate of $\frac{1}{N}\mathbf{R}_n\mathbf{R}_n^H + \sigma^2\mathbf{I}$ ($\mathbf{I}$ denoting the
$n \times n$ identity matrix), which could also yield significant (if not all) information. Under
the assumption

$$\frac{n}{N} \to c > 0, \tag{5.19}$$

$\mathbf{C}_N$ models situations where, due to the size of n, the number of samples N needed
to adequately approximate $\frac{1}{N}\mathbf{R}_n\mathbf{R}_n^H + \sigma^2\mathbf{I}$ is unattainable, but is of the same order of
magnitude as n. (5.19) is typical of many situations arising in signal processing where one
can gather only a limited number of observations during which the characteristics of the
signal do not change. This is the case for spectrum sensing when fading changes rapidly.

One application of the matrix $\mathbf{C}_N$ defined in (5.18) is the problem of spectrum sensing,
for example, in Example 5.3. The beauty of the above model is that σ is arbitrary. Of
course, this model applies to the low SNR detection problem for spectrum sensing.

Assume that N observations for n sensors. These sensors form a random vector $\mathbf{r}_n +
\sigma\mathbf{x}_n$, and the observed values form a realization of the sample covariance matrix $\mathbf{C}_n$.

Based on the fact that $\mathbf{C}_n$ is known, one is interested in inferring as much as possible about the random vector $\mathbf{r}_n$, and hence on the system (5.18). Within this setting, one would like to connect the following quantities:

1. the eigenvalue distribution of $\mathbf{C}_n$;
2. the eigenvalue distribution of $\frac{1}{N}\mathbf{R}_n\mathbf{R}_n^H$.

5.2.2.4 Statistical Eigen-Inference from Large Wishart Matrices

The measurements are of the form

$$\mathbf{x}_i = \mathbf{A}\mathbf{s}_i + \mathbf{z}_i, i = 1, \ldots, n,$$

where $\mathbf{z}_i \sim \mathcal{N}_p(0, \mathbf{\Sigma}_z)$ denotes a p-dimensional (real or complex) Gaussian noise vector with covariance matrix $\mathbf{\Sigma}_z$, $\mathbf{s}_i \sim \mathcal{N}_k(0, \mathbf{\Sigma}_s)$ denotes a k-dimensional zero-mean (real or complex) Gaussian signal vector with covariance matrix $\mathbf{\Sigma}_s$, and $\mathbf{A}$ is a $p \times k$ unknown nonrandom matrix.

5.3 Moment Approach

Most of the material in this section can be found in [10]. Throughout this section, only Hermitian matrices are considered. Real symmetric matrices are treated as special cases.

Let $\mathbf{A}$ be an $n \times n$ Hermitian matrix, and its eigenvalues be denoted by

$$\lambda_1 \geq \lambda_2 \cdots \geq \lambda_n.$$

Then, from the Definition 5.1, the k-th moment of $F_\mathbf{A}$ can be expressed as

$$\beta_{n.k}(\mathbf{A}) = \int_{-\infty}^{\infty} x^k F_\mathbf{A}(dx) = \frac{1}{n}\text{Tr}(\mathbf{A}^k). \tag{5.20}$$

(5.20) plays a fundamental role in random matrix theory. Most results in finding limiting spectral distribution were obtained by estimating the mean, variance or higher moments of $\frac{1}{n}\text{Tr}(\mathbf{A}^k)$.

To motivate our development, let us see an example first.

Example 5.5 (Moments-based hypothesis testing)
Continued from Example 5.5.

The hypothesis problem (5.4) is reformulated as

$$\mathcal{H}_0 : \text{Tr}(\mathbf{\Sigma}_q) = q\|\mathbf{\Sigma}_q\|,$$
$$\mathcal{H}_1 : \text{Tr}(\mathbf{\Sigma}_q) > q\|\mathbf{\Sigma}_q\|,$$

by using the effective rank r

$$r = \frac{\text{Tr}(\mathbf{A})}{\|\mathbf{A}\|},$$

where $\|\mathbf{A}\|$ is the maximum singular value of $\mathbf{A}$, and the matrix inequality (this bound is sharp) [107]

$$r(\mathbf{A}) \leq \text{rank}(\mathbf{A}) \leq n.$$

Claim $\mathcal{H}_1$, if the test statistic (5.5) is replaced with the new statistic k-th moment

$$T = \frac{1}{n}\sum_{k=1}^{M} \text{Tr}(\mathbf{A}^k) > T_0.$$

For the case of the moment $M = 1$, it has been found that the algorithm performs very well. $\square$

When sample covariance matrices $\mathbf{S}$ that are random matrices are used instead of Σ, the moments of $\mathbf{S}$ are scalar random variables. Girko studied the random determinants $\det \mathbf{S}$ for decades [111]. Repeat (5.21) here for convenience:

$$(\det \mathbf{S})^{\frac{1}{N}} \leq \frac{1}{N}\text{Tr}\mathbf{S}. \tag{5.21}$$

5.3.1 Limiting Spectral Distribution

To show that $F_{\mathbf{A}}$ converges to a limit, say F, we often employ the Moment Convergence Theorem,

$$\beta_k(\mathbf{A}) \to \beta_k = \int x^k F(dx),$$

in some sense, for example, almost surely (a.s.) or in probability and the Carleman's condition

$$\sum_{k=1}^{\infty} \beta_k^{-1/(2k)} \leq \infty.$$

Thus the Moment Convergence Theorem can be used to show the existence of the limiting spectral distribution.

5.3.1.1 Wigner Matrix

The celebrated semicircle law (distribution) is related to a Wigner matrix. A Wigner matrix $\mathbf{W}$ of order n is defined as an $n \times n$ Hermitian matrix whose entries above the diagonal are i.i.d. complex random variables with variance σ^2, and whose diagonal elements are i.i.d. real random variables (without any moment requirement). We have the following theorem.

Theorem 5.3 (Semicircle law) *under the conditions described above, as $n \to \infty$, with probability 1, the empirical spectral distribution tends to the semicircle law with scale parameter σ, whose density is given by*

$$p_\sigma(x) = \begin{cases} \frac{1}{2\pi\sigma^2}\sqrt{4\sigma^2 - x^2}, & \text{if } |x| \leq 2\sigma, \\ 0, & \text{otherwise.} \end{cases} \tag{5.22}$$

For each n, the entries above the diagonal of $\mathbf{W}$ are independent complex random variables with mean zero and variance σ^2, but they may not be identically distributed and depend on n. We have the following theorem.

Theorem 5.4 *If* $\mathbb{E}(w_{jk}^{(n)}) = 0$, $\mathbb{E}|w_{jk}^{(n)}|^2 = \sigma^2$, *and for any* $\delta > 0$

$$\lim_{n \to \infty} \frac{1}{\delta^2 n^2} \sum_{jk} \mathbb{E}|w_{jk}^{(n)}|^2 I_{(|w_{jk}^{(n)}| > \delta\sqrt{n})} = 0, \tag{5.23}$$

where $I(\cdot)$ *is the indication function, then the conclusion of Theorem 5.3 holds.*

In Girko's book (1990) [111], (5.23) is stated as a necessary and sufficient condition for the conclusion of Theorem 5.4.

5.3.1.2 Sample Covariance Matrix

Suppose that x_{jn}, $j, n = 1, 2, \ldots$ is a double array of i.i.d. complex random variables with mean zero and variance σ^2. Write

$$\mathbf{x}_n = [x_{1n}, \ldots, x_{pn}]^T, \mathbf{X} = [\mathbf{x}_1, \ldots, \mathbf{x}_N].$$

The sample covariance matrix is defined as

$$\mathbf{S} \triangleq \frac{1}{N} \sum_{n=1}^{N} \mathbf{x}_n \mathbf{x}_n^H = \frac{1}{N} \mathbf{X} \mathbf{X}^H.$$

Marchenko and Pasture (1967) [251] had the first success in finding the limit spectral distribution of $\mathbf{S}$. The work also provided the tool of Stieltjes transform. Afterwards, Bai and Yin (1988) [264], Grenander and Silverstein (1977) [265], Jonsson (1982) [266], Wachter (1978) [267] and Yin (1986) [268] did further research on the sample covariance matrix.

Theorem 5.5 ([268]) *Suppose that* $\frac{p}{N} \to c \in (0, \infty)$. *Under the assumptions stated at the beginning of this subsection, the empirical spectral distribution of* $\mathbf{S}$ *tends to a limiting distribution with density*

$$f(x) = \begin{cases} \frac{1}{2\pi c \sigma^2 x} \sqrt{(b-x)(x-a)}, & \text{if } a \le x \le b \\ 0, & \text{otherwise} \end{cases}$$

and a point mass $1 - c^{-1}$ *at the origin if* $c > 1$, *where*

$$a = \sigma^2(1 - \sqrt{c})^2, b = \sigma^2(1 + \sqrt{c})^2.$$

The limit distribution of Theorem 5.5 is called the Marchenko-Pastur law (distribution) with ratio index c and scale index σ^2. The existence of the second moment of the entries is necessary and sufficient for the Marchenko-Pastur law since the limit spectral distribution involves the parameter σ^2. The condition of zero mean can be relaxed to have a common mean.

Sometimes, in practice, the entries of $\mathbf{X}$ depend on N and for each N, they are independent but not identically distributed. We have the following theorem.

Theorem 5.6 *Suppose that for each N, the entries of $\mathbf{X}_N$ are independent complex variables, with a common mean and variance σ^2. Assume that $\frac{p}{N} \to c \in (0, \infty)$, and that for any $\delta > 0$,*

$$\frac{1}{\delta^2 N p} \sum_{jk} \mathbb{E}|x_{jk}^{(N)}|^2 I_{(|x_{jk}^{(N)}| > \delta\sqrt{N})} \to 0. \tag{5.24}$$

Then, $F_{\mathbf{S}}$ tends almost surely to the Marchenko-Pastur distribution with ratio index c and scale index σ^2.

Now consider the case $p \to \infty$, but $\frac{p}{N} \to 0$. Almost all eigenvalues tend to 1 and thus the empirical spectrum distribution of $\mathbf{S}$ tend to a degenerate one. For convenience, we consider instead the matrix

$$\mathbf{W} = \sqrt{\frac{p}{N}}(\mathbf{S} - \sigma^2\mathbf{I}) = \frac{1}{\sqrt{pN}}(\mathbf{XX}^H - N\sigma^2\mathbf{I}).$$

When $\mathbf{X}$ is real, under the existence of the fourth moment, Bai and Yin (1988) [264] showed that its empirical spectrum distribution tends to the semicircle law almost surely as $p \to \infty$. Bai (1988) [10] gives a generalization of this result.

Theorem 5.7 ([10]) *Suppose that, for each N, the entries of the matrix $\mathbf{X}_N$ are independent complex random variables with a common mean and variance σ^2. Assume that, for any constant $\delta > 0$, as $p \to \infty$ with $p/N \to 0$,*

$$\frac{1}{p\delta^2\sqrt{Np}} \sum_{jk} \mathbb{E}|x_{jk}^{(N)}|^2 I_{(|x_{jk}^{(N)}| > \delta\sqrt[4]{Np})} = o(1), \tag{5.25}$$

and

$$\frac{1}{Np^2} \sum_{jk} \mathbb{E}|x_{jk}^{(N)}|^4 I_{(|x_{jk}^{(N)}| > \delta\sqrt[4]{Np})} = o(1). \tag{5.26}$$

Then, with probability 1, the empirical spectral distribution of $\mathbf{W}$ tends to the semicircular law with scale index σ^2.

Conditions (5.25) and (5.26) hold if the entries of $\mathbf{X}$ have bounded fourth moments.

Theorem 5.8 (Theorem 4.10 of [14]) *Let $\mathbf{F} = \mathbf{S}_{N1}\mathbf{S}_{N2}^{-1}$, where $\mathbf{S}_{N1}$ and $\mathbf{S}_{N2}$ are sample covariance matrices with dimension p and sample size N_1 and N_2 with an underlying distribution of mean 0 and variance 1. If $\mathbf{S}_{N1}$ and $\mathbf{S}_{N2}$ are independent,*

$$p/N_1 \to y_1 \in (0, \infty), p/N_2 \to y_2 \in (0, 1).$$

Then, the limit spectral density F_{y_1, y_2} of $\mathbf{F}$ exists and has a density

$$F'_{y_1, y_2}(x) = \begin{cases} \frac{(1-y_2)\sqrt{(b-x)(x-a)}}{2\pi x(y_1 + xy_2)}, & a < x < b, \\ 0, & otherwise, \end{cases}$$

Further, if $y_1 > 0$, then F_{st} has a point mass $1 - 1/y_1$ at the origin.

Example 5.6

Consider an example to apply Theorem 5.8. Consider

$$\mathcal{H}_0 : \mathbf{S}_N = \mathbf{W}_N$$
$$\mathcal{H}_1 : \mathbf{S}_N = \mathbf{B}_N + \mathbf{W}_N,$$

where $\mathbf{W}_N$ is an underlying distribution of mean 0 and variance 1 and

$$\mathcal{H}_0 : \mathbf{S}_{N1} \mathbf{S}_{N2}^{-1} : \mathbf{W}_N$$
$$\mathcal{H}_1 : \mathbf{S}_{N1} \mathbf{S}_{N2}^{-1} : \mathbf{B}_N + \mathbf{W}_N.$$

Under $\mathcal{H}_0$, we can apply Theorem 5.8 to get the density function. Under $\mathcal{H}_1$, the density is different from that of $\mathcal{H}_0$. □

5.3.1.3 Product of Two Random Matrices

The motivation of studying products of two random matrices arises from the fact that the true covariance matrix $\mathbf{\Sigma}$ is not a multiple of an identity matrix $\mathbf{I}$, and that of multivariate $\mathbf{F} = \mathbf{S}_1 \mathbf{S}_2^{-1}$. When $\mathbf{S}_1$ and $\mathbf{S}_2$ are independent Wishart, the limit spectral distribution of $\mathbf{F}$ follows from Wachter (1980) [267].

Theorem 5.9 ([10]) *Suppose that the entries of $\mathbf{X}$ are independent complex random variables satisfying (5.24), and assume that $\mathbf{T}(= \mathbf{T_N})$ is a sequence of $p \times p$ Hermitian matrices independent of $\mathbf{X}$, such that its empirical spectral distribution tends to a nonrandom and nondegenerate distribution H in probability (or almost surely). Further, assume that*

$$\frac{p}{N} \to c \in (0, \infty).$$

Then, the empirical spectral distribution of the matrix product $\mathbf{ST}$ tends to a nonrandom limit in probability (or almost surely).

5.3.2 Limits of Extreme Eigenvalues

5.3.2.1 Limits of Extreme Eigenvalues of the Wigner Matrix

The real case of the following theorem is obtained in [269] and the complex case is in [10].

Theorem 5.10 ([10, 269]) *Suppose that the diagonal entries of the Wigner matrix $\mathbf{W}$ are i.i.d. real random variables, the entries above the diagonal are i.i.d. complex random variables, and all these variables are independent. Then, the largest eigenvalue λ_1 of $N^{-\frac{1}{2}} \mathbf{W}$ tends to $2\sigma > 0$ with probability 1 if and only if the following four conditions are true:*

1. $\mathbb{E}((w_{11}^+)^2) < \infty$;
2. $\mathbb{E}(w_{12})$ is real and ≤ 0;
3. $\mathbb{E}\left(|w_{12} - \mathbb{E}(w_{12})|^2\right) = \sigma^2$;
4. $\mathbb{E}(|w_{12}^4|) < \infty$;

where $x^+ = \max(x, 0)$.

For the Wigner matrix, symmetry between the largest and smallest eigenvalues exists. Thus, Theorem 5.10 actually proves the following: the necessary and sufficient conditions (for both the largest and the smallest eigenvalues) to have finite limits almost surely are (1) the diagonal entries have finite second moments; (2) the off-diagonal entries have zero mean and finite fourth moments.

5.3.2.2 Limits of Extreme Eigenvalues of Sample Covariance Matrix

Geman (1980) [270] proved that, as $\frac{p}{N} \to c$, the largest eigenvalue of a sample covariance matrix tends to $b(c)$ almost surely, assuming a certain growth condition on the moments of the underlying distribution, where $b(c) = \sigma^2(1 + \sqrt{c})^2$ defined in Theorem 5.5. The real case of the following theorem is in [271], and their result is extended to the complex case in [10].

Theorem 5.11 ([10, 271]) *In addition to the assumptions of Theorem 5.5, we assume that the entries of* $\mathbf{X}$ *have finite fourth moment. Then,*

$$-2c\sigma^2 \leq \lim_{N \to \infty} \inf \lambda_{\min}(\mathbf{S} - \sigma^2(1 + c)\mathbf{I}) \leq \lim_{N \to \infty} \inf \lambda_{\max}(\mathbf{S} - \sigma^2(1 + c)\mathbf{I}) \leq 2c\sigma^2, a.s.$$

If we define the smallest eigenvalue as the $(p - N + 1)$-st smallest eigenvalue of $\mathbf{S}$ when $p > N$, then from Theorem 5.11, we immediately reach the following conclusion:

Theorem 5.12 ([10]) *Under the assumptions of Theorem 5.11, we have*

$$\lim_{N \to \infty} \lambda_{\min}(\mathbf{S}) = \sigma^2(1 - \sqrt{c})^2, a.s.$$
$$\lim_{N \to \infty} \lambda_{\max}(\mathbf{S}) = \sigma^2(1 + \sqrt{c})^2, a.s.$$

The first work to exploit Theorem 5.12 for spectrum sensing is [258] with their conference version published in 2007. Denote the eigenvalues of $\mathbf{S}_N$ by $\lambda_1 \leq \lambda_2 \leq \cdots \leq \lambda_N$. Write $\lambda_{max} = \lambda_N$ and

$$\lambda_{\min} = \begin{cases} \lambda_1, & p \leq N, \\ \lambda_{p-N+1}, & p > N. \end{cases}$$

Using the convention above, Theorem 5.12 is true [14] for all $c \in (0, \infty)$.

Theorem 5.13 (Theorem 5.9 of [14]) *Suppose that the entries of the matrix* $\mathbf{X}_N = \{x_{jkN}, j \leq p, k \leq N\}$ *are independent (not necessarily identically distributed) and satisfy*

1. $\mathbb{E}(x_{jkN}) = 0$;
2. $|x_{jkN}| \le \sqrt{N}\delta_N$;
3. $\max_{j,k} |\mathbb{E}|x_{jkN}|^2 - \sigma^2| \to 0$, *as $N \to \infty$; and*
4. $\mathbb{E}|x_{jkN}|^l \le b(\sqrt{N}\delta_N)^{l-3}$ *for all $l \ge 3$;*

where $\delta_N \to 0$ and $b > 0$. Let $\mathbf{S}_N = \frac{1}{N}\mathbf{X}_N\mathbf{X}_N^H$. Then, for any $x > \varepsilon > 0$ and integers $j, k \ge 2$, we have

$$P[\lambda_{\max}(\mathbf{S}_N) \ge \sigma^2(1 + \sqrt{c})^2 + x] \le CN^{-k}[\sigma^2(1 + \sqrt{c})^2 + x - \varepsilon]^{-k}$$

for some constant $C > 0$.

5.3.2.3 Limiting Behavior of Eigenvectors

Relatively less work has been done on the limiting behavior of eigenvectors than eigenvalues. See [272] for the latest additions to the literature.

There is a good deal of evidence that the behavior of large random matrices is asymptotically distribution-free. In other words, it is asymptotically equivalent to the case where the basic entries are i.i.d. mean 0 normal, provided that some moment requirements are met.

5.3.2.4 Miscellanea

The norm $(N^{-1/2}\mathbf{X})^k$ is sometimes important.

Theorem 5.14 ([269]) *If $\mathbb{E}(|w_{11}^4|) < \infty$, then*

$$\lim_{N\to\infty} \sup \|(N^{-1/2}\mathbf{X})^k\| \le (1 + k)\sigma^k, a.s., \quad \text{for all } k.$$

The following theorem is proven independently by [273] and [269].

Theorem 5.15 ([269, 273]) *If $\mathbb{E}(|w_{11}^4|) < \infty$, then*

$$\lim_{N\to\infty} \sup \max_{j \le N} |\lambda_j(N^{-1/2}\mathbf{X})| \le \sigma, a.s.$$

5.3.2.5 Circular Law–Non-Hermitian Matrices

We consider the non-Hermitian matrix. Let

$$\mathbf{Q} = \frac{1}{\sqrt{N}}(x_{jn})$$

be an $N \times N$ complex matrix with i.i.d. entries x_{jn} of mean zero and variance 1. The eigenvalues of $\mathbf{Q}$ are complex and thus the empirical spectral distribution of $\mathbf{Q}$, denoted by $F_N(x, y)$, is defined in the complex plane. Since the 1950s, it has been conjectured that $F_N(x, y)$ tends to the uniform distribution over the unit disc in the complex plane, called the circular law. The problem was open until Bai (1997) [274].

Theorem 5.16 (Circular Law [274]) *Suppose that the entries have finite $(4 + \varepsilon)$-th moments, and that the joint distribution of the real and imaginary parts of the entries, or the conditional distribution of the real part given the imaginary part, has a uniformly bounded density. Then, the circular law holds.*

5.3.3 Convergence Rates of Spectral Distributions

5.3.3.1 Wigner Matrix

Consider the model of Theorem 5.4, and assume that the entries of **W** above or on the diagonal are independent and satisfy

$$\mathbb{E}(w_{jk}) = 0, \text{ for all } 1 \leq k \leq j \leq N;$$
$$\mathbb{E}(|w_{jk}^2|) = 1, \text{ for all } 1 \leq k \leq j \leq N;$$
$$\mathbb{E}(|w_{jj}^2|) = 1, \text{ for all } 1 \leq j \leq N; \tag{5.27}$$
$$\sup_N \max_{1 \leq k \leq j \leq N} \mathbb{E}(|w_{jk}^4|) \leq M < \infty.$$

Theorem 5.17 ([275]) *Under the conditions in (5.27), we have*

$$\|\mathbb{E}F_{(N^{-1/2}\mathbf{W})} - F\| = \mathcal{O}(N^{-1/4}),$$

where F is the semicircular law with scalar parameter 1.

Theorem 5.18 ([276]) *Under the four conditions in (5.27), we have*

$$\|F_{(N^{-1/2}\mathbf{W})} - F\| = \mathcal{O}_p(N^{-1/4}),$$

where "p" stands for probability.

5.3.3.2 Sample Covariance Matrix

Assume the following conditions are true.

$$\mathbb{E}(x_{jk}) = 0, \mathbb{E}(|x_{jk}^2|) = 1, \text{ for all } j, k, n$$
$$\sup_N \sup_{j,k} \mathbb{E}(|x_{jk}^4|) I_{(|x_{jk}| \geq M)} \to 0, \text{ as } M \to \infty. \tag{5.28}$$

Theorem 5.19 ([275]) *Under the assumptions in (5.28), for $0 < \theta < \Theta < 1$ or $1 < \theta < \Theta < \infty$,*

$$\sup_{c_p \in (\theta, \Theta)} \|\mathbb{E}F_{\mathbf{S}} - F_{c_p}\| = \mathcal{O}(N^{-1/4}),$$

where $c_p = p/N$ and F_c is the Marchenko-Pastur distribution with dimension-ratio c and parameter $\sigma^2 = 1$.

Theorem 5.20 ([275]) *Under the assumptions in (5.28), for any $0 < \varepsilon < 1$,*

$$\sup_{c_p \in (1-\varepsilon, 1+\varepsilon)} \|\mathbb{E}F_\mathbf{S} - F_{c_p}\| = \mathcal{O}(N^{-5/48}).$$

Theorem 5.21 ([275]) *Under the assumptions in (5.28), the conclusions in Theorems 5.19 and 5.20 can be improved to*

$$\sup_{c_p \in (\theta, \Theta)} \|F_\mathbf{S} - F_{c_p}\| = \mathcal{O}_p(N^{-1/4}),$$

and

$$\sup_{c_p \in (1-\varepsilon, 1+\varepsilon)} \|F_\mathbf{S} - F_{c_p}\| = \mathcal{O}_p(N^{-5/48}).$$

Consider $\mathbf{S}_N = \frac{1}{N}\mathbf{T}_N^{1/2}\mathbf{X}_N\mathbf{X}_N^H\mathbf{T}_N^{1/2}$, where $\mathbf{X}_N = (x_{ij})$ is a $p \times p$ matrix consisting of independent complex entries with mean zero and variance one, $\mathbf{T}_N$ is a $p \times p$ nonrandom positive definite Hermitian matrix with spectral norm uniformly bounded in p. If

$$\sup_N \sup_{i,j} \mathbb{E}|x_{ij}|^8 < \infty,$$

and $c_N = p/N < 1$ uniformly as $N \to \infty$, we obtain [277] that the rate of the expected empirical spectral distribution of $\mathbf{S}_N$ converging to its limit spectral distribution is $\mathcal{O}(N^{-\frac{1}{2}})$. Under the same assumption, it can be proved that for any $\eta > 0$, the rates of the convergence of the empirical spectral distribution of $\mathbf{S}_N$ in probability and the almost sure convergence are $\mathcal{O}(N^{-\frac{2}{5}})$ and $\mathcal{O}(N^{-\frac{2}{5}+\eta})$.

5.3.4 Standard Vector-In, Vector-Out Model

Random vectors are our basic building blocks in our signal processing. We define the standard vector-in, vector-out model (VIVO)[1] as

$$\mathbf{y}_n = \mathbf{H}\mathbf{x}_n + \mathbf{w}_n, n = 1, \ldots, N$$

where $\mathbf{y}_n$ is an $M \times 1$ complex vector of observations collected from M sensors, $\mathbf{x}_n$ is $K \times 1$ complex vector of transmitted waveform, $\mathbf{H}$ is an $M \times K$ matrix, and $\mathbf{w}_n$ is an $M \times 1$ complex vector of additive Gaussian noise with mean zero and variance σ_w^2.
  Defining

$$\mathbf{Y} = [\mathbf{y}_1, \ldots, \mathbf{y}_N], \mathbf{X} = [\mathbf{x}_1, \ldots, \mathbf{x}_N], \mathbf{W} = [\mathbf{w}_1, \ldots, \mathbf{w}_N],$$

[1] Multiple-input, multiple-output (MIMO) has a special meaning in wireless communications.

we have
$$\mathbf{Y} = \mathbf{HX} + \mathbf{W}.$$

The sample covariance matrix is defined as
$$\mathbf{S} = \frac{1}{N}\mathbf{YY}^H = \frac{1}{N}(\mathbf{HX} + \mathbf{W})(\mathbf{HX} + \mathbf{W})^H.$$

For the noise-free case, that is, $\sigma_w^2 = 0$, we have
$$\mathbf{S} = \frac{1}{N}(\mathbf{HX})(\mathbf{HX})^H = \frac{1}{N}\mathbf{HXX}^H\mathbf{H}^H.$$

We can formulate the problem as a hypothesis testing problem
$$\mathcal{H}_0 : \mathbf{S} = \frac{1}{N}\mathbf{WW}^H,$$

$$\mathcal{H}_1 : \mathbf{S} = \frac{1}{N}(\mathbf{HX} + \mathbf{W})(\mathbf{HX} + \mathbf{W})^H.$$

5.3.5 Generalized Densities

In the generalized densities, the moments of the matrix play a critical role. Assume that the matrix $\mathbf{A}$ has a density
$$p_N(\mathbf{A}) = H(\lambda_1, \ldots, \lambda_n).$$

The joint density function of its eigenvalues is of the form
$$p_N(\lambda_1, \ldots, \lambda_n) = cJ(\lambda_1, \ldots, \lambda_n)H(\lambda_1, \ldots, \lambda_n),$$

$$H(\lambda_1, \ldots, \lambda_n) = \prod_{k=1}^{n} g(\lambda_k),$$

$$J = \prod_{i<j} (\lambda_i - \lambda_j)^\beta \prod_{k=1}^{n} h_n(\lambda_k).$$

For example for a real Gaussian matrix, $\beta = 1$ and $h_n = 1$, for a complex Gaussian matrix, $\beta = 2$ and $h_n = 1$, for a quaternion Gaussian matrix, $\beta = 4$ and $h_n = 1$, and for a real Wishart matrix with $n \geq p$, $\beta = 1$ and $h_n = x^{n-p}$. The following examples illustrate this.

1. Real Gaussian matrix, that is, symmetric, $\mathbf{A}^T = \mathbf{A}$:
$$p_N(\mathbf{A}) = c \exp\left(-\frac{1}{4\sigma^2}\mathrm{Tr}(\mathbf{A}^2)\right).$$

   The diagonal entries of $\mathbf{A}$ are i.i.d. real $\mathcal{N}(0, 2\sigma^2)$ and entries above diagonal are i.i.d. real $\mathcal{N}(0, \sigma^2)$.
2. Complex Gaussian matrix, that is, Hermitian, $\mathbf{A}^* = \mathbf{A}$:
$$p_N(\mathbf{A}) = c \exp\left(-\frac{1}{2\sigma^2}\mathrm{Tr}(\mathbf{A}^2)\right).$$

   The diagonal entries of $\mathbf{A}$ are i.i.d. real $\mathcal{N}(0, \sigma^2)$ and entries above diagonal are i.i.d. complex $\mathcal{N}(0, \sigma^2)$ (whose real and imaginary parts are i.i.d. $\mathcal{N}(0, \sigma^2/2)$).

3. Real Wishart matrix, of order $p \times n$:

$$p_N(\mathbf{A}) = c \exp\left(-\frac{1}{2\sigma^2}\mathrm{Tr}(\mathbf{A}^*\mathbf{A})\right).$$

The entries of $\mathbf{A}$ are i.i.d. real $\mathcal{N}(0, \sigma^2)$.

4. Complex Wishart matrix, of order $p \times n$:

$$p_N(\mathbf{A}) = c \exp\left(-\frac{1}{\sigma^2}\mathrm{Tr}(\mathbf{A}^*\mathbf{A})\right).$$

The entries of $\mathbf{A}$ are i.i.d. complex $\mathcal{N}(0, \sigma^2)$.

For generalized densities, we have

1. Symmetric matrix:

$$p_N(\mathbf{A}) = c \exp(-\mathrm{Tr}G(\mathbf{A})),$$

where $G(t^2)$ is a polynomial of even orders with a positive leading coefficient, such as $G(t^2) = 4t^4 + 2t^2 + 3$.

2. Hermitian matrix:

$$p_N(\mathbf{A}) = c \exp(-\mathrm{Tr}G(\mathbf{A})),$$

where $G(t^2)$ is a polynomial of even orders with a positive leading coefficient.

3. Real covariance matrix, of dimension p and degrees of freedom n:

$$p_N(\mathbf{A}) = c \exp(-\mathrm{Tr}G(\mathbf{A}^T\mathbf{A})),$$

where $G(t)$ is a polynomial with a positive leading coefficient, such as $G(t) = 4t^3 + 2t^2 + 3t + 5$.

4. Complex covariance matrix, of dimension p and degrees of freedom n:

$$p_N(\mathbf{A}) = c \exp(-\mathrm{Tr}G(\mathbf{A}^*\mathbf{A})),$$

where $G(t)$ is a polynomial with a positive leading coefficient.

The book of [14] mainly concentrates on results without assuming density conditions.

5.4 Stieltjes Transform

We follow [10] closely for the definition of the Stieltjes transform. Let G be a function of bounded variation defined on the real line. Then, its Stieltjes transform is defined by

$$m(z) \triangleq \int_{-\infty}^{\infty} \frac{1}{x - z} G(dx), \tag{5.29}$$

where $z = u + iv$ with $v > 0$. The integrand in (5.29) is bounded by $1/v$, the integral always exists, and

$$\frac{1}{\pi}\text{Im}(m(z)) = \int_{-\infty}^{\infty} \frac{v}{\pi[(x-u)^2 + v^2]} G(dx).$$

This is the convolution of G with a Cauchy density with a scale parameter v. If G is a distribution function, then its Stieltjes transform always has a positive imaginary part. Thus, we can easily verify that, for any continuity points $x_1 < x_2$ of G,

$$\lim_{v \to 0} \int_{x_1}^{x_2} \frac{1}{\pi}\text{Im}(m(z)) du = G(x_2) - G(x_1). \tag{5.30}$$

(5.30) provides a continuity theorem between the family of distribution functions and the family of their Stietjes transforms.

Also, if $\text{Im}(m(z))$ is continuous at $x_0 + i0$, then $G(x)$ is differentiable at $x = x_0$ and its derivative equals $\frac{1}{\pi}\text{Im}(m(x_0 + i0))$. (5.30) gives an easy way to find the density of a distribution if its Stieltjes transform is known.

Let G be the empirical spectral distribution of a Hermitian matrix $\mathbf{A}_N$ of $N \times N$. It is seen that

$$m_G(z) = \frac{1}{N}\text{Tr}(\mathbf{A} - z\mathbf{I})^{-1} = \frac{1}{N}\sum_{i=1}^{N} \frac{1}{A_{ii} - z - \boldsymbol{\alpha}_i^H(\mathbf{A}_i - z\mathbf{I}_{N-1})^{-1}\boldsymbol{\alpha}_i} \tag{5.31}$$

where $\boldsymbol{\alpha}_i$ is the i-th column vector of $\mathbf{A}$ with the i-th entry removed and $\mathbf{A}_i$ is the matrix obtained from $\mathbf{A}$ with the i-th row and column deleted. (5.31) is a powerful tool in analyzing the spectrum of large random matrix. As mentioned above, the mapping from distribution functions to their Stieltjes transforms is continuous.

Example 5.7 (Limiting spectral distributions of the wigner matrix)
As an illustration of how to use (5.31), let us consider the Wigner matrix to find its limiting spectral distribution.

Let $m_N(z)$ be the Stieltjes transform of the empirical spectral distribution of $N^{-1/2}\mathbf{W}$. By (5.31), and noticing $w_{ii} = 0$, we have

$$m_N(z) = \frac{1}{N}\sum_{i=1}^{N} \frac{1}{-z - \frac{1}{N}\boldsymbol{\alpha}_i^H(N^{-1/2}\mathbf{W}_i - z\mathbf{I}_{N-1})^{-1}\boldsymbol{\alpha}_i}$$

$$= \frac{1}{N}\sum_{i=1}^{N} \frac{1}{-z - \sigma^2 m_N(z) + \varepsilon_i} = -\frac{1}{-z + \sigma^2 m_N(z)} + \delta_N,$$

where

$$\varepsilon_i = \sigma^2 m_N(z) - \frac{1}{N}\boldsymbol{\alpha}_i^H(N^{-1/2}\mathbf{W}_i - z\mathbf{I}_{N-1})^{-1}\boldsymbol{\alpha}_i,$$

$$\delta_N = \frac{1}{N}\sum_{i=1}^{N} \frac{-\varepsilon_i}{(-z - \sigma^2 m_N(z) + \varepsilon_i)(-z - \sigma^2 m_N(z))}.$$

For any fixed $v_0 > 0$ and $B > 0$, with $z = u + iv$, we have (omitting the proof)

$$\sup_{|u| \le B, v_0 \le v \le B} |\delta_N(z)| = o(1), a.s. \tag{5.32}$$

Omitting the middle steps, we have

$$m_N(z) = -\frac{1}{2\sigma^2} \left[z + \delta_N \sigma^2 - \sqrt{(z - \delta_N \sigma^2)^2 - 4\sigma^2} \right]. \tag{5.33}$$

From (5.33) and (5.32), it follows that, with probability 1, for every fixed z with $v > 0$,

$$m_N(z) \to m(z) = -\frac{1}{2\sigma^2} \left[z - \sqrt{z^2 - 4\sigma^2} \right].$$

Letting $v \to 0$, we find the density of the semicircle law as given in (5.22). □

Let $\mathbf{A}_N$ be an $N \times N$ Hermitian matrix and $F_{\mathbf{A}_N}$ be its empirical spectral distribution. If the measure μ admits a density $f(x)$ with support Ω:

$$d\mu(x) = f(x) \, dx \text{ on } \Omega,$$

Then, the Stieltjest transform of $F_{\mathbf{A}_N}$ is given for complex arguments by

$$S_{\mathbf{A}_N}(z) = \Psi_\mu(z) = \int \frac{1}{x-z} dF_{\mathbf{A}_N}(x) = \frac{1}{N} \mathrm{Tr}(\mathbf{A}_N - z\mathbf{I})^{-1}$$
$$= -\sum_{k=0}^{\infty} z^{-(k+1)} \left(\int_\Omega x^k f(x) \, dx \right) = -\sum_{k=0}^{\infty} z^{-(k+1)} M_k, \tag{5.34}$$

where $M_k = \int_\Omega x^k f(x) \, dx$ is the k-th moment of F. This provides a link between the Stieltjes transform and the moments of $\mathbf{A}_N$. The moments of random Hermitian matrices become practical if direct use of the Stieltjes transform is too difficult.

Let $\mathbf{A} \in \mathbb{C}^{N \times M}, \mathbf{B} \in \mathbb{C}^{M \times N}$, such that $\mathbf{AB}$ is Hermitian. Then, for $z \in \mathbb{C}\backslash\mathbb{R}$, we have [12, p. 37]

$$\frac{M}{N} m_{F_{\mathbf{BA}}}(z) = m_{F_{\mathbf{AB}}}(z) + \frac{N - M}{N} \frac{1}{z}.$$

In particular, we can apply $\mathbf{AB} = \mathbf{XX}^H$.

Let $\mathbf{X} \in \mathbb{C}^{N \times N}$ be Hermitian and a be a nonzero real. Then, for $z \in \mathbb{C}\backslash\mathbb{R}$

$$m_{F_{a\mathbf{X}}}(z) = \frac{1}{a} m_{F_{\mathbf{X}}}(z).$$

There are only a few kinds of random matrices for which the corresponding asymptotic eigenvalue distributions are known explicitly [278]. For a wider class of random matrices, however, explicit calculation of the moments turns out to be unfeasible. The task of finding an unknown probability distribution given its moments is known as the problem of moments. It was addressed by Stieltjes in 1894 using the integral transform defined in (5.34). A simple Taylor series expansion of the kernel of the Stieltjes transform

$$-\lim_{s \to \infty} \frac{d^m}{dx^m} \frac{G(s^{-1})}{s} = m! \int x^m dF(x)$$

shows how the moments can be found given the Stieltjes transform, without the need for integration. The probability density function can be obtained from the Stieltjes transform, simply taking the limit

$$p(x) = \lim_{y \to 0+} \frac{1}{\pi} \mathrm{Im} G(x + jy),$$

which is called the Stieltjes inverse formula [11].

We follow [279] for the following properties:

1. Identical sign for imaginary part

$$\mathrm{Im} \Psi_{\mu}(z) = \mathrm{Im}(z) \int_{\Omega} \frac{f(\lambda)}{(\lambda - x)^2} d\lambda,$$

where $\Im$ is the imaginary part of $z \in \mathbb{C}$.

2. Monotonicity. If $z = x \in \mathbb{R} \backslash \Omega$, then $\Psi_{\mu}(z)$ is well defined and

$$\Psi'_{\mu}(z) = \int_{\Omega} \frac{f(\lambda)}{(\lambda - x)^2} d\lambda > 0 \Rightarrow \Psi'_{\mu}(z) \nearrow \ \text{on} \ \ \backslash \Omega.$$

3. Inverse formula

$$f(x) = \frac{1}{\pi} \lim_{y \to 0^+} \mathrm{Im} \Psi(x + jy). \tag{5.35}$$

Note that if $x \in \mathbb{R} \backslash \Omega$, then $\Psi_{\mu}(x) \in \mathbb{R} \Rightarrow f(x) = 0$.

4. Dirac measure. Let δ_x be the Dirac measure at x

$$\delta_x(A) = \begin{cases} 1 & \text{if } x \in A, \\ 0 & \text{else.} \end{cases}$$

Then,

$$\Psi_{\delta_x}(z) = \frac{1}{x - z}; \ \Psi_{\delta_0}(z) = -\frac{1}{z}.$$

An important example is

$$L_M = \frac{1}{M} \sum_{k=1}^{M} \delta_{\lambda_k} \Rightarrow \Psi_{L_M}(z) = \frac{1}{M} \sum_{k=1}^{M} \frac{1}{\lambda_k - z}.$$

5. Link with the resolvent. Let $\mathbf{X}$ be a $M \times M$ Hermitian matrix

$$\mathbf{X} = \mathbf{U} \begin{pmatrix} \lambda_1 & & 0 \\ & \ddots & \\ 0 & & \lambda_M \end{pmatrix} \mathbf{U}^H$$

and consider its resolvent $\mathbf{Q}(z)$ and spectral measure L_M

$$\mathbf{Q}(z) = (\mathbf{X} - z\mathbf{I})^{-1}, \ L_M = \frac{1}{M} \sum_{k=1}^{M} \delta_{\lambda_k}.$$

The Stieltjes transform of the spectral measure is the normalized trace of the resolvent

$$\Psi_{L_M}(z) = \frac{1}{M}\text{Tr}\mathbf{Q}(z) = \frac{1}{M}\text{Tr}(\mathbf{X} - z\mathbf{I})^{-1}.$$

Gaussian tools [280] are useful. Let the Z_i's be independent complex Gaussian random variables denoted by $\mathbf{z} = (Z_1, \cdots, Z_n)$.

1. Integration by part formula

$$\mathbb{E}(Z_k \Phi(\mathbf{z}, \bar{\mathbf{z}})) = \mathbb{E}|Z_k|^2 \mathbb{E}\left(\frac{\partial \Phi}{\partial \bar{Z}_k}\right).$$

2. Poincaré-Nash inequality

$$\text{var}(\Phi(\mathbf{z}, \bar{\mathbf{z}})) \le \sum_{k=1}^{n} |Z_k|^2 \left(\left|\frac{\partial \Phi}{\partial Z_k}\right|^2 + \left|\frac{\partial \Phi}{\partial \bar{Z}_k}\right|^2\right).$$

5.4.1 Basic Theorems

Theorem 5.22 ([281]) *Let $m_F(z)$ be the Stieltjes transform of a distribution function F, then*

1. *m_F is analytic over $\mathbb{C}^+$;*
2. *if $z \in \mathbb{C}^+$, then $m_F(z) \in \mathbb{C}^+$;*
3. *if $z \in \mathbb{C}^+$, $|m_F(z)| \le \frac{1}{\text{Im}(z)}$ and $\text{Im}(\frac{1}{m_F(z)}) \le -\text{Im}(z)$;*
4. *if $F(0^-) = 0$, then m_F is analytic over $\mathbb{C}\backslash\mathbb{R}^+$. Moreover, $z \in \mathbb{C}^+$ implies $zm_F(z) \in \mathbb{C}^+$ and we have the inequalities*

$$|m_F(z)| \le \begin{cases} \frac{1}{|\text{Im}(z)|}, z \in \mathbb{C}\backslash\mathbb{R} \\ \frac{1}{|z|}, z < 0 \\ \frac{1}{\text{dist}(z,\mathbb{R}^+)}, z \in \mathbb{C}\backslash\mathbb{R}^+ \end{cases}$$

with dist *being the Euclidean distance.*

Conversely, if $m_F(z)$ is a function analytical on $\mathbb{C}^+$ such that $m_F(z) \in \mathbb{C}^+$ if $z \in \mathbb{C}^+$ and

$$\lim_{y\to\infty} -iym_F(iy) = 1,$$

then $m_F(z)$ is the Stieltjes transform of a distribution function F given by

$$F(b) - F(a) = \lim_{y\to 0} \frac{1}{\pi} \int_a^b \text{Im}\,(m_F(x + jy))dx.$$

If, moreover, $zm_F(z) \in \mathbb{C}^+$ for $z \in \mathbb{C}^+$, then $F(0^-) = 0$, in which case $m_F(z)$ has an analytic continuation on $\mathbb{C}\backslash\mathbb{R}^+$.

Our version of the above theorem is close to [12] with slightly different notation.

Let $t > 0$ and $m_F(z)$ be the Stieltjes transform of a distribution function F. Then, for $z \in \mathbb{C}^+$ we have [12]

$$\left| \frac{1}{1 + t m_F(z)} \right| \leq \frac{|z|}{\text{Im}(z)}.$$

Let $\mathbf{x} \in \mathbb{C}^N$, $t > 0$ and $\mathbf{A} \in \mathbb{C}^{N \times N}$ be Hermitian, nonnegative definite. Then, for $z \in \mathbb{C}^+$ we have [12]

$$\left| \frac{1}{1 + t \mathbf{x}^H (\mathbf{A} - z\mathbf{I})^{-1} \mathbf{x}} \right| \leq \frac{|z|}{\text{Im}(z)}.$$

The fundamental result in the following theorem [282] states the equivalence between pointwise convergence of Stieltjes transform and weak onvergence of probability measures.

Theorem 5.23 (Equivalence) *Let (μ_n) be probability measures on $\mathbb{R}$ and (Ψ_{μ_n}), Ψ_{μ_n} the associated Stieltjes transform. Then the following two statements are equivalent:*

1. $\Psi_{\mu_n}(z) \underset{n \to \infty}{\to} \Psi_\mu(z)$ for all $z \in \mathbb{C}^+$;

2. $\mu_n \underset{n \to \infty}{\overset{w}{\to}} \mu$.

Let the random matrix $\mathbf{W}$ be square $N \times N$ with i.i.d. entries with zero mean and variance $\frac{1}{N}$. Let Ω be the set containing eigenvalues of $\mathbf{W}$. The empirical distribution of the eigenvalues

$$P_{\mathbf{H}}(z) \overset{\Delta}{=} \frac{1}{N} |\{\lambda \in \Omega : \text{Re}\lambda < \text{Re}z \text{ and } \text{Im } \lambda < \text{ Im } z\}|$$

converges a nonrandom distribution functions as $N \to \infty$. Table 5.2 lists commonly used random marices and their density functions.

Table 5.1 compiles some moments for commonly encountered matrices from [278]. Calculating eigenvalues λ_k of a matrix $\mathbf{X}$ is not a linear operation. Calculation of the moments of the eigenvalue distribution is, however, conveniently done using a normalized trace since

$$\frac{1}{N} \sum_{k=1}^{N} \lambda_k^m = \frac{1}{N} \text{Tr}(\mathbf{X}^m).$$

Thus, in the large matrix limit, we define $\text{tr}(\mathbf{X})$ as

$$\text{tr}(\mathbf{X}) \overset{\Delta}{=} \lim_{N \to \infty} \frac{1}{N} \text{Tr}(\mathbf{X}).$$

Table 5.2 is made self-contained and only some remarks are made here. For Haar distribution, all eigenvalues lie on the complex unit circle since the matrix $\mathbf{T}$ is unitary. The essential nature is that the eigenvalues are uniformly distributed. Haar distribution demands for Gaussian distributed entries in the random matrix $\mathbf{W}$. This condition does

Table 5.1 Common random matrices and their moments (The entries of $\mathbf{W}$ are i.i.d. with zero mean and variance $\frac{1}{N}$; $\mathbf{W}$ is square $N \times N$, unless otherwise specified. $\mathrm{tr}(\mathbf{H}) \triangleq \lim_{N \to \infty} \frac{1}{N} \mathrm{Tr}(\mathbf{H})$)

Convergence Laws	Definitions	Moments
Full-Circle Law	$\mathbf{W}$ square $N \times N$	
Semicircle Law	$\mathbf{K} = \frac{\mathbf{W} + \mathbf{W}^H}{\sqrt{2}}$	$\mathrm{tr}(\mathbf{K}^{2m}) = \frac{1}{m+1} \begin{pmatrix} 2m \\ m \end{pmatrix}$
Quarter Circle Law	$\mathbf{Q} = \sqrt{\mathbf{W}\mathbf{W}^H}$	$\mathrm{tr}(\mathbf{Q}^m) = \frac{2^{2m}}{\pi m} \frac{1}{\left(\frac{m}{2}+1\right)} \begin{pmatrix} m-1 \\ \frac{m-1}{2} \end{pmatrix} \forall\ m\ odd$
Deformed Quarter Circle Law	$\mathbf{Q}^2$ $\mathbf{R} = \sqrt{\mathbf{W}^H\mathbf{W}}$, $\mathbf{W} \in \mathbb{C}^{N \times \beta N}$	
	$\mathbf{R}^2$	$\mathrm{tr}(\mathbf{R}^{2m}) = \frac{1}{m} \sum_{i=1}^{m} \begin{pmatrix} m \\ i \end{pmatrix} \begin{pmatrix} m \\ i-1 \end{pmatrix} \beta^i$
Haar Distribution	$\mathbf{T} = \mathbf{W}\left(\mathbf{W}^H\mathbf{W}\right)^{-\frac{1}{2}}$	
Inverse Semicircle Law	$\mathbf{Y} = \mathbf{T} + \mathbf{T}^H$	

Table 5.2 Definition of commonly encountered random matrices for convergence laws (the entries of $\mathbf{W}$ are i.i.d. with zero mean and variance $\frac{1}{N}$; $\mathbf{W}$ is square $N \times N$, unless otherwise specified)

Convergence Laws	Definitions	Density Functions
Full-Circle Law	$\mathbf{W}$ square $N \times N$	$p_{\mathbf{W}}(z) = \begin{cases} \frac{1}{\pi} & |z| < 1 \\ 0 & \text{elsewhere} \end{cases}$
Semicircle Law	$\mathbf{K} = \frac{\mathbf{W}+\mathbf{W}^H}{\sqrt{2}}$	$p_{\mathbf{K}}(z) = \begin{cases} \frac{1}{2\pi}\sqrt{4-x^2} & |x| < 2 \\ 0 & \text{elsewhere} \end{cases}$
Quarter Circle Law	$\mathbf{Q} = \sqrt{\mathbf{W}\mathbf{W}^H}$	$p_{\mathbf{Q}}(z) = \begin{cases} \frac{1}{\pi}\sqrt{4-x^2} & 0 \le x \le 2 \\ 0 & \text{elsewhere} \end{cases}$
	$\mathbf{Q}^2$	$p_{\mathbf{Q}^2}(z) = \begin{cases} \frac{1}{2\pi}\sqrt{\frac{4-x}{x}} & 0 \le x \le 4 \\ 0 & \text{elsewhere} \end{cases}$
Deformed Quarter Circle Law	$\mathbf{R} = \sqrt{\mathbf{W}^H\mathbf{W}}$, $\mathbf{W} \in \mathbb{C}^{N \times \beta N}$	$p_{\mathbf{R}}(z) = \begin{cases} \frac{\sqrt{4\beta-(x^2-1-\beta)^2}}{\pi x} & a \le x \le b \\ \left(1-\sqrt{\beta}\right)^+\delta(x) & \text{elsewhere} \end{cases}$ $a = \left|1-\sqrt{\beta}\right|, b = 1+\sqrt{\beta}$
	$\mathbf{R}^2$	$p_{\mathbf{R}^2}(z) = \begin{cases} \frac{\sqrt{4\beta-(x-1-\beta)^2}}{2\pi x} & a^2 \le x \le b^2 \\ \left(1-\sqrt{\beta}\right)^+\delta(x) & \text{elsewhere} \end{cases}$
Haar Distribution	$\mathbf{T} = \mathbf{W}\left(\mathbf{W}^H\mathbf{W}\right)^{-\frac{1}{2}}$	$p_{\mathbf{T}}(z) = \frac{1}{2\pi}\delta\left(|z|-1\right)$
Inverse Semicircle Law	$\mathbf{Y} = \mathbf{T} + \mathbf{T}^H$	$p_{\mathbf{Y}}(z) = \begin{cases} \frac{1}{\pi}\frac{1}{\sqrt{4-x^2}} & |x| < 2 \\ 0 & \text{elsewhere} \end{cases}$

Table 5.3 Table of Stieltjes, R- and S-transforms

Stieltjes Transform	R-Transform	S-Transform
$G(z) \triangleq \int \frac{1}{x-z} dP(x),$ $\mathrm{Im}\, z > 0,\, \mathrm{Im}\, G(z) \geq 0$	$R(z) \triangleq G^{-1}(-z) - z^{-1}$	$S(z) \triangleq \frac{1+z}{z} \Upsilon^{-1}(z),$ $\Upsilon(z) \triangleq -z^{-1} G^{-1}\left(z^{-1}\right) - 1$
$G_{\alpha\mathbf{I}}(z) = \frac{1}{\alpha-z}$	$R_{\alpha\mathbf{I}}(z) = \alpha$	$S_{\alpha\mathbf{I}}(z) = \frac{1}{\alpha}$,
$G_{\mathbf{K}}(z) =$ $\frac{z}{2}\sqrt{1-\frac{4}{z^2}} - \frac{z}{2}$	$R_{\mathbf{K}}(z) = z$	$S_{\mathbf{K}}(z) = $ undefined
$G_{\mathbf{Q}}(z) =$ $\sqrt{1-\frac{4}{z^2}}\left(\frac{z}{2} - \arcsin\frac{2}{z}\right) - \frac{z}{2} - \frac{1}{2\pi}$	$R_{\mathbf{Q}^2}(z) = \frac{1}{1-z}$	$S_{\mathbf{Q}^2}(z) = \frac{1}{1+z}$
$G_{\mathbf{Q}^2}(z) = \frac{1}{2}\sqrt{1-\frac{4}{z}} - \frac{1}{2}$	$R_{\mathbf{R}^2}(z) = \frac{\beta}{1-z}$	$S_{\mathbf{R}^2}(z) = \frac{1}{\beta+z}$
$G_{\mathbf{R}^2}(z) =$ $\sqrt{\frac{(1-\beta)^2}{4z^2} - \frac{1+\beta}{2z} + \frac{1}{4}} - \frac{1}{2} - \frac{(1-\beta)}{2z}$	$R_{\mathbf{Y}}(z) = \frac{-1+\sqrt{1+4z^2}}{z}$	$S_{\mathbf{Y}}(z) = $ undefined
$G_{\mathbf{Y}}(z) = \frac{-\mathrm{sign}(\mathrm{Re} z)}{\sqrt{z^2-4}}$	$R_{\alpha\mathbf{X}}(z) = \alpha R_{\mathbf{X}}(\alpha z)$	$S_{\mathbf{AB}}(z) = S_{\mathbf{A}}(z) S_{\mathbf{B}}(z)$
$G_{\lambda^2}(z) = \frac{G_\lambda(\sqrt{z}) - G_\lambda(-\sqrt{z})}{2\sqrt{z}}$	$\lim_{z\to\infty} R(z) = \int x\, dP(x)$	
$G_{\mathbf{XX}^H}(z) = \beta G_{\mathbf{X}^H\mathbf{X}}(z) + \frac{\beta-1}{z},$ $\mathbf{X} \in \mathbb{C}^{N\times\beta N}$	$R_{\mathbf{A+B}}(z) = R_{\mathbf{A}}(z) R_{\mathbf{B}}(z)$	
	$G_{\mathbf{A+B}}\left(R_{\mathbf{A+B}}(-z) - z^{-1}\right) = z$	

$G_{\mathbf{X+WYW}^H}(z) =$
$G_{\mathbf{X}}\left(z - \beta \int \frac{y\, dP_{\mathbf{Y}}(x)}{1+y G_{\mathbf{X+WYW}^H}(z)}\right)$
$\mathrm{Im}\, z > 0, \mathbf{X}, \mathbf{Y}, \mathbf{W}$ jointly independent.

$G_{\mathbf{WW}^H}(z) = \int_0^1 u(x,z)dx,$
$u(x,z) =$
$\left[-z + \int_0^\beta \frac{w(x,y)dy}{1+\int_0^1 u(x',z)w(x',y)dx'}\right]^{-1},$
$x \in [0,1]$

not seem to be necessary, but allowing for any complex distribution with zero mean and finite variance is not sufficient.

Table 5.3[2] lists some transforms (Stieltjes, R-, S- transforms) and their properties. The Stieltjes transform is more fundamental since both R-transform and S-transform can be expressed in terms of the Stieltjes transform.

---

[2] This table is primarily compiled from [278].

5.4.1.1 Products of Random Matrices

Almost surely, the eigenvalue distribution of the matrix product

$$\mathbf{P} = \mathbf{W}^H \mathbf{W} \mathbf{X}$$

converges in distribution, as $K, N \to \infty$ but $\beta = K/N$.

5.4.1.2 Sums of Random Matrices

Consider the limiting distribution of random Hermitian matrices of the form [251, 283]

$$\mathbf{A} + \mathbf{W} \mathbf{D} \mathbf{W}^H,$$

where $\mathbf{W}(N \times K)$, $\mathbf{D}(K \times K)$, $\mathbf{A}(N \times N)$ are independent, with $\mathbf{W}$ containing i.i.d. entries having second moments, $\mathbf{D}$ is diagonal with real entries, and $\mathbf{A}$ is Hermitian. The asymptotic regime is

$$K/N \to \alpha \text{ as } N \to \infty.$$

The behavior is expressed using the limiting distribution function $F_{\mathbf{A}+\mathbf{W}\mathbf{D}\mathbf{W}^H}(x)$. The remarkable result is that the convergence of

$$F_{\mathbf{A}+\mathbf{W}\mathbf{D}\mathbf{W}^H}(x)$$

to a nonrandom F.

Theorem 5.24 ([251, 283]) *Let $\mathbf{A}$ be an $N \times N$ Hermitian matrix, nonrandom, for which $F_{\mathbf{A}}(x)$ converge weakly as $N \to \infty$ to a distribution function $\mathbb{A}$. Let $F_{\mathbf{D}}(x)$ converges weakly as $N \to \infty$ to a distribution function denoted $\mathbb{D}$. Suppose the entries of $\sqrt{N}\mathbf{W}$ i.i.d. for fixed N with unit variance (sum of the variances of the real and imaginary parts in the complex case). Then, the eigenvalue distribution of $\mathbf{A} + \mathbf{W}\mathbf{D}\mathbf{W}^H$ converges weakly to a deterministic F. Its Stieltjes transform $G(z)$ satisfies the equation:*

$$G(z) = G_{\mathbb{A}}\left(z - \alpha \int \frac{\tau}{1 + \tau G(z)} d\mathbb{T}(\tau)\right).$$

Theorem 5.25 ([284]) *Assume*

1. *$\mathbf{X}_n = \frac{1}{\sqrt{n}}(X_{ij}^{(n)})$, where $1 \le i \le n$, $1 \le j \le p$, and $X_{i,j,N}$ are independent real random variables with a common mean and variance σ^2, satisfying*

$$\frac{1}{n^2 \varepsilon_n^2} \sum_{i,j} X_{ij}^2 I(|X_{ij}| \ge \varepsilon_n \sqrt{n}) \underset{n \to \infty}{\to} 0,$$

where $I(x)$ is an indication function and ε_n^2 is a positive sequence tending to zero;
2. *$\frac{p}{n} \to y > 0$ as $n \to \infty$;*
3. *$\mathbf{T}_n$ is an $p \times p$ random symmetric matrix with $F_{\mathbf{T}_n}$ converging almost surely to a distribution $H(t)$ as $n \to \infty$;*

4. $\mathbf{B}_n = \mathbf{A}_n + \mathbf{X}_n \mathbf{T}_n \mathbf{X}_n^H$, where $\mathbf{A}_n$ is a random $p \times p$ symmetric matrix with $F_{\mathbf{A}_n}$ almost surely to $F_{\mathbf{A}}$, a (possibly defective) nonrandom distribution;
5. $\mathbf{X}_N, \mathbf{T}_N, \mathbf{A}_N$ are independent.

Then, as $n \to \infty$, $F_{\mathbf{B}_n}$ converges almost surely to a nonrandom distribution F, whose Stieltjes transform $m(z)$ satisfies

$$m(z) = m_{\mathbf{A}}(z) \left(z - y \int \frac{x}{1 + xm(z)} dH(x) \right).$$

Theorem 5.26 ([285]) Let $\mathbf{S}_n$ denote the sample covariance matrix of n pure noise vectors distributed $\mathcal{N}(0, \sigma^2 \mathbf{I}_p)$. Let l_1 be the largest eigenvalue of $\mathbf{S}_n$. In the joint limit $p, n \to \infty$, with $p/n \to c \geq 0$, the distribution of the largest eigenvalues of $\mathbf{S}_n$ converges to a Tracy-Widom distribution

$$\Pr \left\{ \frac{l_1/\sigma^2 - \mu_{n,p}}{\xi_{n,p}} \right\} \to F_\beta(s),$$

with $\beta = 1$ for real valued noise and $\beta = 2$ for complex valued noise. The centering and scaling parameters, $\mu_{n,p}$ and $\xi_{n,p}$ are functions of n and p only.

Theorem 5.27 ([285]) Let l_1 be the largest eigenvalue as in Theorem 5.26. Then,

$$\Pr \left\{ l_1/\sigma^2 > \left(1 + \sqrt{\frac{p}{n}} \right)^2 + \varepsilon \right\} \leq \exp(-n J_{LAG}(\varepsilon))$$

where

$$J_{LAG}(\varepsilon) = \int_1^x (x - y) \frac{(1+c)y + 2\sqrt{c}}{(y+B)^2} \frac{dy}{\sqrt{y^2 - 1}},$$

$$c = p/n, x = 1 + \frac{\varepsilon}{2\sqrt{c}}, B = \frac{1+c}{2\sqrt{c}}.$$

Consider the standard model for signals with p sensors. Let $\{\mathbf{x}_i = \mathbf{x}(t_i)\}_{i=1}^n$ denote p-dimensional i.i.d. observations of the form

$$\mathbf{x}(t) = \mathbf{A}\mathbf{s}(t) + \sigma \mathbf{n}(t), \tag{5.36}$$

sampled at n distinct times t_i, where $\mathbf{A} = [\mathbf{a}_1, \ldots, \mathbf{a}_K]^T$ is the $p \times K$ matrix of K linearly independent p-dimensional vectors. The $K \times 1$ vector $\mathbf{s}(t) = [s_1(t), \ldots, s_K(t)]^T$ represents the random signals, assumed zero-mean and stationary with full rank covariance matrix. σ is the unknown noise level, and $bfn(t)$ is a $p \times 1$ additive Gaussian noise vector, distributed $\mathcal{N}(0, \mathbf{I}_p)$ and independent of $\mathbf{s}(t)$.

Theorem 5.28 ([285]) Let $\mathbf{S}_n$ denote the sample covariance matrix of n observations from (5.36) with a single signal of strength λ. Then, in the joint limit $p, n \to \infty$, with $p/n \to c \geq 0$, the largest eigenvalue of $\mathbf{S}_n$ converges almost surely to

$$\lambda_{\max}(\mathbf{S}_n) \overset{a.s.}{\to} \begin{cases} \sigma^2(1 + \sqrt{p/n})^2 & \lambda \leq \sigma^2 \sqrt{p/n} \\ (\lambda + \sigma^2)(1 + \frac{p}{n} \frac{\sigma^2}{\lambda}) & \lambda > \sigma^2 \sqrt{p/n} \end{cases}.$$

Theorem 5.29 ([286]) *Let* $\mathbf{C} \in {}^{p \times p}$ *be positive semidefinite. Fix an integer* $l \leq p$ *and assume the tail*

$$\{\lambda_i(\mathbf{C})\}_{i > l}$$

of the spectrum of $\mathbf{C}$ *decays sufficiently fast that*

$$\sum_{i > l} \lambda_i(\mathbf{C}) = \mathcal{O}(\lambda_1(\mathbf{C})).$$

Let $\{\mathbf{x}_i\}_{i=1}^n \in \mathbb{R}^p$ *be i.i.d. samples drawn from a* $\mathcal{N}(\mathbf{0}, \mathbf{C})$ *distribution. Define the sample covariance matrix*

$$\hat{\mathbf{C}} = \frac{1}{n} \sum_{i=1}^n \mathbf{x}_i \mathbf{x}_i^H.$$

Let κ_l *be the condition number associated with a dominant l-dimensional invariance subspace of* $\mathbf{C}$,

$$\kappa_l = \frac{\lambda_1(\mathbf{C})}{\lambda_l(\mathbf{C})}.$$

If

$$n = \Omega(\varepsilon^{-2} \kappa_l^2 l \log p),$$

then with high probability

$$|\lambda_k(\hat{\mathbf{C}}_n) - \lambda_k(\mathbf{C}_n)| \leq \varepsilon \lambda_k(\mathbf{C}_n), \text{for} k = 1, \ldots, l.$$

Theorem 5.29 says, assuming sufficiently fast decay of the residual eigenvalues, $n = \Omega(\varepsilon^{-2} \kappa_l^2 l \log p)$ samples ensure that the top l eigenvalues are captured with relative precision.

5.4.2 Large Random Hankel, Markov and Toepltiz Matrices

Two most significant matrices, whose limiting spectral distributions have been extensively studied, are the Wigner and the sample covariance matrices. Here, we study other structured matrices. The important papers include Bryc, Dembo, and Jiang [287], Bose *et al.* [288–294], and Miller *et al.* [295, 296]. We mainly follow Bryc, Dembo, and Jiang (2006) [287] for this development. For a symmetric $n \times n$ matrix $\mathbf{A}$, let $\lambda_j(\mathbf{A}), 1 \leq \mathbf{j} \leq \mathbf{n}$, denote the eigenvalues of the matrix $\mathbf{A}$, written in a nonincreasing order. The spectral measure of $\mathbf{A}$, denoted $\hat{\mu}(\mathbf{A})$, is the empirical spectral distribution (ESD) of its eigenvalues, namely

$$\hat{\mu}(\mathbf{A}) = \frac{1}{n} \sum_{j=1}^n \delta_{\lambda_j(\mathbf{A})},$$

where δ_x is the Dirac delta measure at x. So when $\mathbf{A}$ is a random matrix, $\hat{\mu}(\mathbf{A})$ is a random measure on $(\mathbb{R}, \mathbb{B})$.

The ensembles of random matrices are studied here. Let $X_k : k = 0, 1, 2, \ldots$ be a sequence of i.i.d. real-valued random variables. We can visualize the Wigner matrix as

$$
\mathbf{W}_n = \begin{pmatrix}
X_{11} & X_{12} & X_{13} & \cdots & X_{1(n-1)} & X_{1n} \\
X_{21} & X_{22} & X_{23} & \cdots & X_{2(n-1)} & X_{2n} \\
& & & \vdots & & \\
X_{p1} & X_{p2} & X_{p3} & \cdots & X_{p(n-1)} & X_{pn}
\end{pmatrix}.
$$

It is well known that almost surely, the limiting spectral distribution of $n^{-1/2}(\mathbf{W}_p)$ is the semicircle law.

The sample covariance matrix $\mathbf{S}$ is defined as

$$
\mathbf{S}_p = \frac{1}{n} \mathbf{W}_p \mathbf{W}_p^T,
$$

where $\mathbf{W}_p = ((X_{ij}))_{1 \le i \le p, 1 \le j \le n}$.

1. If $p \to \infty$ and $p/n \to 0$, then almost surely, the limiting spectral distribution of $\sqrt{\frac{n}{p}}(\mathbf{S}_p - \mathbf{I}_p)$ is the semicircle law.
2. If $p \to \infty$ and $p/n \to c \in (0, \infty)$, then almost surely, $\mathbf{S}_p$ is the Marchenko-Pastur law.

In view of the above discussion, it is thus natural to study the limiting spectral distribution of matrices of the form $\mathbf{S}_p = \frac{1}{n} \mathbf{X}_p \mathbf{X}_p^T$ where $\mathbf{X}_p$ is a $p \times n$ suitably patterned (asymmetric) random matrix. Asymmetry is used very loosely. It just means that $\mathbf{X}_n$ is not necessarily symmetric. One may ask the following questions [290]:

1. Suppose that $p/n \to c, 0 < c < \infty$. When does the limiting spectral distribution of $\mathbf{S}_p = \frac{1}{n} \mathbf{X}_p \mathbf{X}_p^T$ exist?
2. Suppose that $p/n \to 0$. When does the imiting spectral distribution of $\sqrt{\frac{n}{p}} \left(\frac{1}{n} \mathbf{X}_p \mathbf{X}_p^T - \mathbf{I}_p \right)$ exist?

For $n \in \mathbb{N}$, define a random $n \times n$ Hankel matrix $\mathbf{H}_n = [X_{i+j-1}]_{1 \le i, j \le n}$,

$$
\mathbf{H}_n = \begin{pmatrix}
X_1 & X_2 & \cdots & \cdots & X_{n-1} & X_n \\
X_2 & X_3 & \ddots & \ddots & X_n & X_{n+1} \\
\vdots & \vdots & \ddots & \ddots & X_{n+1} & X_{n+2} \\
X_{n-2} & X_{n-1} & \ddots & \ddots & \vdots & \vdots \\
X_{n-1} & X_n & \ddots & & X_{2n-3} & X_{2n-2} \\
X_n & X_{n+1} & \cdots & \cdots & X_{2n-2} & X_{2n-1}
\end{pmatrix}
$$

and a random $n \times n$ Toeplitz matrix $\mathbf{T}_n = [X_{|i-j|}]_{1 \le i, j \le n}$,

$$
\mathbf{T}_n = \begin{pmatrix}
X_0 & X_1 & X_2 & \cdots & X_{n-2} & X_{n-1} \\
X_1 & X_0 & X_1 & \ddots & \ddots & X_{n-2} \\
X_2 & X_1 & X_0 & \ddots & \ddots & \vdots \\
\vdots & \ddots & \ddots & \ddots & \ddots & X_2 \\
X_{n-2} & \ddots & \ddots & \ddots & X_0 & X_1 \\
X_{n-1} & X_{n-2} & \cdots & X_2 & X_1 & X_0
\end{pmatrix}.
$$

Theorem 5.30 (Toeplitz matrices by Bryc, Dembo, and Jiang (2006) [287]) *Let $X_k : k = 0, 1, 2, \ldots$ be a sequence of i.i.d. real-valued random variables with variance one $\mathrm{Var}(X_1) = 1$. Then, with probability 1, the empirical spectral distribution of $\frac{1}{\sqrt{n}}\mathbf{T}_n$, or $\hat{\mu}(\mathbf{T}_n/\sqrt{n})$, converges weakly, as $n \to \infty$, to a nonrandom symmetric probability measure, γ_T, which does not depend on the distribution of the entries of X_1 and has unbounded support.*

Theorem 5.31 (Hankel matrices by Bryc, Dembo, and Jiang (2006) [287]) *Let $X_k : k = 0, 1, 2, \ldots$ be a sequence of i.i.d. real-valued random variables with variance one $\mathrm{Var}(X_1) = 1$. Then, with probability 1, the empirical spectral distribution of $\frac{1}{\sqrt{n}}\mathbf{H}_n$, or $\hat{\mu}(\mathbf{H}_n/\sqrt{n})$, converges weakly, as $n \to \infty$, to a nonrandom symmetric probability measure, γ_H, which does not depend on the distribution of the entries of X_1 and has unbounded support and is not unimodal.*

A symmetric distribution ν is said to be unimodal, if the function $x \mapsto \nu((-\infty, x])$ is convex for $x < 0$.

To state the theorem on the Markov matrices, define the free convolution of two probability measures μ and ν as the probability measure whose nth cumulant is the sum of the nth cumulants of μ and ν.

Let us define the Markov matrices $\mathbf{M}_n$. Let $X_{ij} : j \geq i \geq 1$ be an infinite upper triangular array of i.i.d. random variables and define $X_{ij} = X_{ji}$ for $j \geq i \geq 1$. Let $\mathbf{M}_n$ be a random $n \times n$ symmetric matrix given by

$$\mathbf{M}_n = \mathbf{X}_n - \mathbf{D}_n,$$

where $\mathbf{X}_n = [X_{ij}]_{1 \leq i,j \leq n}$ and $\mathbf{D}_n = \mathrm{diag}\left(\sum_{j=1}^{n} X_{ij}\right)_{1 \leq i \leq n}$ is a diagonal matrix, so each of the rows of $\mathbf{M}_n$ has a zero sum. The values of X_{ij} are irrelevant for $\mathbf{M}_n$.

Wigner's classical result says that $\hat{\mu}(\mathbf{X}_n/\sqrt{n})$ converges weakly as $n \to \infty$ to the (standard) semicircle law with the density $\sqrt{4 - x^2}/(2\pi)$ on $(-2, 2)$. For normal $\mathbf{X}_n$ and normal i.i.d. diagonal $\tilde{\mathbf{D}}_n$ independent of $\mathbf{X}_n$, the weak limit of $\hat{\mu}(\mathbf{X}_n - \tilde{\mathbf{D}}_n/\sqrt{n})$ is the free convolution of the semicircle and standard normal measures; see [297] and the references therein. The predicted result holds for the Markov matrix $\mathbf{M}_n$, but the problem is nontrivial since $\mathbf{D}_n$ strongly depends on $\mathbf{X}_n$.

$$\mathbf{M}_n = \begin{pmatrix} -\sum_{j=2}^{n} X_{1j} & X_{12} & X_{13} & \cdots & X_{1n} \\ X_{21} & -\sum_{j\neq 2}^{n} X_{2j} & X_{23} & \cdots & X_{2n} \\ \vdots & \ddots & \ddots & & \vdots \\ X_{k1} & X_{k2} & \cdots & -\sum_{j\neq k}^{n} X_{kj} & \cdots & X_{kn} \\ \vdots & \vdots & & \ddots & \vdots \\ X_{n1} & X_{n2} & \cdots & & -\sum_{j=1}^{n-1} X_{nj} \end{pmatrix}.$$

Theorem 5.32 (Markov matrices by Bryc, Dembo, and Jiang (2006) [287]) *Let the entries of a Markov matrix* $\mathbf{M}_n$ *be i.i.d. real-valued random variables with mean zero and variance one. Then, with probability one, the empirical spectral distribution of* $\frac{1}{\sqrt{n}}\mathbf{M}_n$ *converges weakly, as* $n \to \infty$, *to the free convolution of the semicircle and standard normal measures. This measure is a nonrandom symmetric probability measure with smooth bounded density, which does not depend on the distribution of the entries of the underlying random variables and has unbounded support.*

5.4.3 Information Plus Noise Model of Random Matrices

We follow [282] for this subsection. We consider $M, N \in \mathbb{N}$ such that $N = M(N)$, $M < N$ and $c_N = M/N \to c \in (0, 1)$ as $N \to \infty$. A Gaussian information plus noise model matrix is a $M \times N$ random matrix defined by

$$\mathbf{\Sigma}_N = \mathbf{B}_N + \mathbf{W}_N, \tag{5.37}$$

where matrix $\mathbf{B}_N$ is deterministic such that

$$\sup \|\mathbf{B}_N\| \le B_{\max} < \infty,$$

and the entries $W_{i,j,N}$ of $\mathbf{W}_N$ are i.i.d. and satisfy

$$W_{i,j,N} \sim \mathbb{E}\mathcal{N}(0, \sigma^2).$$

Most results can be also extended to the non-Gaussian case.

The convergence of the empirical spectral measure of $\mathbf{\Sigma}_N\mathbf{\Sigma}_N{}^H$ defined by

$$\hat{\mu}_N \triangleq \frac{1}{M}\sum_{i=1}^{M}\delta_{\hat{\lambda}_{i,N}},$$

with δ_x is the Dirac measure at point x.

We define the resolvent of matrix $\mathbf{\Sigma}_N\mathbf{\Sigma}_N{}^H$ by

$$\mathbf{Q}_N(z) = (\mathbf{\Sigma}_N\mathbf{\Sigma}_N{}^H - z\mathbf{I}_M)^{-1},$$

$z \in \mathbb{C}\backslash\mathbb{R}^+$. Its normalized trace $\frac{1}{M}\mathrm{Tr}\mathbf{Q}_N(z)$ can be written as the Stieltjes transform of $\hat{m}_N(z)_N(z)$ defined as

$$\hat{\mu}_N(z) = \frac{1}{M}\mathrm{Tr}\mathbf{Q}_N(z) = \int_{\mathbb{R}^+}\frac{1}{\lambda - z}d\hat{\mu}_N(\lambda).$$

The weak convergence of $\hat{\mu}_N(z)$ can be studied by characterizing the convergence of $\frac{1}{M}\mathrm{Tr}\mathbf{Q}_N(z)$ as $N \to \infty$, with the aid of (5.4.3). The main result is summarized in this theorem.

Theorem 5.33 *There exists a deterministic probability measure* μ_N, *satisfying* $\mathrm{supp}(\mu_N) \in \mathbb{R}^+$, *and such that* $\hat{\mu}_N - \mu_N \to 0$ *as* $N \to \infty$ *with probability one. Equivalently, the*

Stieltjes transform $m_N(z)$ *of* μ_N *satisfies* $\hat{m}_N(z) - m_N(z) \to 0$ *almost surely* $\forall z \in \mathbb{C}\backslash\mathbb{R}^+$. *Moreover,* $\forall z \in \mathbb{C}\backslash\mathbb{R}^+$, $m_N(z)$ *is the unique solution of the equation*

$$m_N(z) = \frac{1}{M}\text{Tr}\mathbf{T}_N(z)$$

$$= \frac{1}{M}\text{Tr}\left(-z(1 + \sigma^2 c_N m_N(z))\mathbf{I}_M + \sigma^2(1 - c_N)\mathbf{I}_M + \frac{\mathbf{B}_N\mathbf{B}_N^H}{1 + \sigma^2 c_N m_N(z)}\right)^{-1} \quad (5.38)$$

satisfying $\text{Im}(m_N(z)) > 0$ *for* $z \in \mathbb{C}^+$.

This result was first proven by Girko [298] and later Dozier-Silverstein [263]. This result is also valid for the non-Gaussian case.

If the spectral distribution

$$F_N(x) \triangleq \frac{1}{M}\text{card}\{k : \lambda_{k,N} \leq x\}$$

of matrix $\mathbf{B}_N\mathbf{B}_N^H$ converges to the distribution $F(x)$ as $N \to \infty$, then

$$\mu_N \overset{w}{\to} \mu$$

with μ probability measure, whose Stieltjes transform

$$m(z) \triangleq \int_{\mathbb{R}} \frac{1}{\lambda - z} d\mu(\lambda)$$

satisfies

$$m(z) = \int_{\mathbb{R}} \frac{1}{\frac{\lambda}{1 + \sigma^2 cm(z)} - z(1 + c\sigma^2 m(z)) + \sigma^2(1 - c)} dF(\lambda).$$

The convergence of $\hat{m}_N(z)$ can be guaranteed by the following theorem.

Theorem 5.34 (The convergence of $\hat{m}_N(z)$) *For all* $z \in \mathbb{C}\backslash\mathbb{R}$,

$$|(\hat{m}_N(z)) - m_N(z)| \leq \frac{1}{N^2} P_1(|z|) P_2\left(\frac{1}{|\text{Im}(z)|}\right),$$

for all large N, *with* P_1, P_2 *two polynomials with positive coefficients independent of* N, z.

According to Theorem 5.33, $\frac{1}{M}\text{Tr}\mathbf{Q}_N(z)$ is a good approximation of $\frac{1}{M}\text{Tr}\mathbf{T}_N(z)$. The following theorem shows that the entries of $\mathbf{Q}_N(z)$ also approximate the entries of $\mathbf{T}_N(z)$.

Theorem 5.35 (The entries of $\mathbf{Q}_N(z)$ approximate the entries of $\mathbf{T}_N(z)$) *Let* $\mathbf{T}_N(z)$ *be defined in (5.38). Let* $(\mathbf{d}_{1,N})$ *and* $(\mathbf{d}_{2,N})$ *be two sequences of deterministic vectors such that*

$$\sup_N\|\mathbf{d}_{1,N}\|, \sup_N\|\mathbf{d}_{2,N}\| < \infty.$$

Then,

$$\mathbf{d}_{1,N}^H(\mathbf{Q}_N(z) - \mathbf{T}_N(z))\mathbf{d}_{2,N} \underset{N\to\infty}{\to} 0$$

almost surely for all $z \in \mathbb{C}\backslash\mathbb{R}$. Moreover,

$$\left|\left[\mathbf{d}_{1,N}^H (\mathbf{Q}_N(z) - \mathbf{T}_N(z)) \mathbf{d}_{2,N}^H\right]\right| \leq \frac{1}{N^{3/2}} P_1(|z|) P_2\left(\frac{1}{|\text{Im}(z)|}\right),$$

for all large N, with P_1, P_2 two polynomials with positive coefficients independent of N, z.

Theorem 5.35 is valid for the non-Gaussian case that is proven in [299].

Definition 5.1 (Assumption 5.1) *Matrix $\mathbf{B}_N$ has rank $K = K(N) < M$, and the eigenvalues of $\mathbf{B}_N \mathbf{B}_N^H$ has multiplicity one for all N.*

Definition 5.2 (Assumption 5.2) *The rank $K > 0$ of $\mathbf{B}_N \mathbf{B}_N^H$ does not depend on N and for all $k = 1, \ldots, K$, the positive sequence $\{\lambda_{M-K+k,N}\}$ is expressed as*

$$\lambda_{M-K+k,N} = \gamma_k + \varepsilon_{k,N},$$

with

$$\lim_{N \to \infty} \varepsilon_{k,N} = 0$$

and increasing values

$$\gamma_1 < \ldots < \gamma_K.$$

The support of μ_N is studied in [300]. Under further assumption such as Assumption 5.1, this is studied in [299]. Assumption 5.2 is stronger than Assumption 5.1. Assumption 5.2 says the rank of $\mathbf{B}_N \mathbf{B}_N^H$ is independent of N.

Theorem 5.36 (Exact separation of the eigenvalues for the spiked model [299]) *Under Assumption 5.2, define*

$$K_s \triangleq \frac{1}{M} \text{card}\{k : \lambda_k > \sigma^2 \sqrt{c}\}$$

and assume that

$$\sigma^2 \sqrt{c} \notin \{\gamma_1, \ldots, \gamma_K\},$$

that is,

$$\gamma_1 < \ldots < \gamma_{K-K_s} < \sigma^2 \sqrt{c} < \gamma_{K-K_s+1} < \cdots < \gamma_K.$$

Thus, for N large enough, the support Ω_N has $Q = K_s + 1$ clusters, that is,

$$\Omega_N = \cup_{q=1}^{K_s+1} [x_{q,N}^-, x_{q,N}^+].$$

The first cluster is associated with $\lambda_{1,N}, \ldots, \lambda_{M-K_s,N}$ and is given by

$$x_{1,N}^- = \sigma^2 (1 - \sqrt{c_N})^2 + \mathcal{O}^+\left(\frac{1}{N}\right) \text{ and } x_{1,N}^+ = \sigma^2 (1 + \sqrt{c_N})^2 + \mathcal{O}^+\left(\frac{1}{N}\right).$$

For $q = 2, 3, \ldots, K_s + 1$ *and* $k = q - 1$, *the cluster* $[x_{q,N}^-, x_{q,N}^+]$ *is associated with* $\lambda_{M-K+k,N}$ *and*

$$x_{q,N}^- = g(\lambda_{M-K+k,N}, c_N) - \mathcal{O}^+\left(\frac{1}{\sqrt{N}}\right),$$
$$x_{q,N}^+ = g(\lambda_{M-K+k,N}, c_N) + \mathcal{O}^+\left(\frac{1}{\sqrt{N}}\right),$$
$$g(\lambda, c) = \frac{(\lambda + \sigma^2 c)(\lambda + c)}{\lambda},$$

and $\mathcal{O}^+\left(\frac{1}{\sqrt{N}}\right)$ *is a positive* $\mathcal{O}\left(\frac{1}{\sqrt{N}}\right)$ *term.*

Under the spiked model assumption, measure μ_N is intuitively expected to be very close to the Marchenko-Pastur distribution μ, and particularly Ω_N should be close to

$$\text{supp}(\mu) = [\sigma^2(1 - \sqrt{c_N})^2, \sigma^2(1 + \sqrt{c_N})^2].$$

Theorem 5.36 shows that the first cluster $[x_{1,N}^-, x_{1,N}^+]$ is very close to the support of the Marchenko-Pastur distribution; we have the presence of additional clusters, if the eigenvalues of $\mathbf{B}_N\mathbf{B}_N{}^H$ are large enough. Indeed, if K_s eigenvalues of $\mathbf{B}_N\mathbf{B}_N{}^H$ converge to different limits, above the threshold $\sigma^2\sqrt{c}$, then there will be K_s additional clusters in the support of Ω_N for all large N.

Theorem 5.36 also states that the smallest $M - K_s$ eigenvalues of $\mathbf{B}_N\mathbf{B}_N{}^H$ are associated with the first cluster, or equivalently that

$$\mu_N[x_{1,N}^-, x_{1,N}^+] = \frac{M - K_s}{M},$$

and that

$$\mu_N[x_{k,N}^-, x_{k,N}^+] = \frac{1}{M}, \quad k = 2, \ldots, K_s.$$

The conditions for the support Ω_N to split into several clusters depend in a nontrivial way on σ, the eigenvalues of $\mathbf{B}_N\mathbf{B}_N{}^H$, the distance between them. However, under stronger Assumption 5.2 (K independent of N and convergence of the eigenvalues to different limits), explicit conditions for the separation of the eigenvalues are obtained: an eigenvalue of $\mathbf{B}_N\mathbf{B}_N{}^H$ is separated from the others if its limit is greater than $\sigma^2\sqrt{c}$. The nonseparated eigenvalues are those associated with $\mu_N[x_{1,N}^-, x_{1,N}^+]$. Therefore, in the spiked model case, the behaviors of the clusters of Ω_N are completely characterized.

The spectral decomposition of $\mathbf{B}_N\mathbf{B}_N^H$ and $\mathbf{\Sigma}_N\mathbf{\Sigma}_N^H$ are expressed as

$$\mathbf{B}_N\mathbf{B}_N^H = \mathbf{U}_N\mathbf{\Lambda}\mathbf{U}_N^H \text{ and } \mathbf{\Sigma}_N\mathbf{\Sigma}_N^H = \tilde{\mathbf{U}}_N\tilde{\mathbf{\Lambda}}\tilde{\mathbf{U}}_N^H$$

with $\mathbf{U}_N, \tilde{\mathbf{U}}_N$ unitary matrices and $\mathbf{\Lambda} = \text{diag}(\lambda_{1,N}, \ldots, \lambda_{M,N})$, $\tilde{\mathbf{\Lambda}} = \text{diag}(\tilde{\lambda}_{1,N}, \ldots, \tilde{\lambda}_{M,N})$. The eigenvalues of $\mathbf{B}_N\mathbf{B}_N^H$ and $\mathbf{\Sigma}_N\mathbf{\Sigma}_N^H$ are decreasingly ordered such that $0 \leq \lambda_{1,N} \leq \ldots \leq \lambda_{M,N}$ and $0 \leq \tilde{\lambda}_{1,N} \leq \ldots \leq \tilde{\lambda}_{M,N}$, respectively.

Theorem 5.37 *Under Assumption 5.2,*

$$\tilde{\lambda}_{M-K_s+k,N} \xrightarrow[N\to\infty]{a.s.} \begin{cases} \sigma^2(1 + \sqrt{c}), & k = 0 \\ g(\gamma_k, c) & k = 1, \ldots, K. \end{cases}$$

Let us consider the eigenvectors of $\mathbf{B}_N\mathbf{B}_N^H$ and $\mathbf{\Sigma}_N\mathbf{\Sigma}_N^H$. Let us first start with a problem of DOA estimation, and then convert the problem into the standard information plus noise model defined in (5.37).

The observed M-dimensional time series $\mathbf{y}_n$ for the n-vector sample are expressed as

$$\mathbf{y}_n = \sum_{k=1}^K \mathbf{a}_k s_{k,n} + \mathbf{v}_n = \mathbf{A}\mathbf{s}_n + \mathbf{v}_n, n = 1, \ldots, N$$

with

$$\mathbf{s}_n = (s_{1,n}, \ldots, s_{K,n})^T, \mathbf{A} = (\mathbf{a}_1, \ldots, \mathbf{a}_K),$$

where $\mathbf{s}_n$ collects $K < M$ nonobservable "source signals," the matrix $\mathbf{A}$ of $M \times K$ is deterministic with an unknown rank $K < M$, and $(\mathbf{v}_n)_{n\in\mathbb{Z}}$ is additive white Gaussian noise such that $\mathbb{E}(\mathbf{v}_n\mathbf{v}_n^H) = \sigma^2\mathbf{I}_M$. Here $\mathbb{Z}$ denotes the set of all integers.

In matrix form, we have $\mathbf{Y}_N = (\mathbf{y}_1, \ldots, \mathbf{y}_N)$, observation matrix of $M \times N$. Similarly, we do this for $\mathbf{S}_N$ and $\mathbf{V}_N$. Then,

$$\mathbf{Y}_N = \mathbf{A}\mathbf{S}_N + \mathbf{V}_N.$$

Using the normalized matrices

$$\mathbf{\Sigma}_N = \frac{1}{\sqrt{N}}\mathbf{Y}_N, \mathbf{B}_N = \frac{1}{\sqrt{N}}\mathbf{A}\mathbf{S}_N, \mathbf{W}_N = \frac{1}{\sqrt{N}}\mathbf{V}_N,$$

we obtain the standard model

$$\mathbf{\Sigma}_N = \mathbf{B}_N + \mathbf{W}_N. \tag{5.39}$$

which is identical to (5.37). Recall that

- $\mathbf{B}_N$ is a rank K deterministic matrix;
- $\mathbf{W}_N$ is a complex Gaussian matrix with i.i.d. entries having zero mean and variance σ^2/N.

The "noise subspace" is defined as

$$\{\mathbf{u}_{1,N}, \ldots, \mathbf{u}_{M-K,N}\},$$

that is, the eigenstate associated with 0 of $\mathbf{B}_N\mathbf{B}_N^H$, and the "signal space" the orthogonal complement, that is, the eigenspace associated with the non-null eigenvalues of $\mathbf{B}_N\mathbf{B}_N^H$. The goal of subspace estimation is to find the projection matrix onto the noise subspace, that is,

$$\mathbf{\Pi}_N = \sum_{k=1}^{M-K} \mathbf{u}_{1,N}\mathbf{u}_{k,N}^H.$$

The subspace estimation problem we consider here is to find a consistent estimation of

$$\eta_N = \mathbf{d}_N\mathbf{\Pi}\mathbf{d}_N^H, \text{ when } N \to \infty,$$

where $(\mathbf{d}_N)$ is a sequence of deterministic vectors such that $\sup_N \|\mathbf{d}_N\| < \infty$.

Traditionally, η_N is estimated by

$$\eta_N = \mathbf{d}_N \tilde{\boldsymbol{\Pi}}_N \mathbf{d}_N^H = \sum_{k=1}^{M-K} \mathbf{d}_N \tilde{\mathbf{u}}_{k,N}^H \tilde{\mathbf{u}}_{k,N} \mathbf{d}_N^H,$$

in other words, by replacing the eigenvectors of true signal covariance $\mathbf{B}_N \mathbf{B}_N^H$ (information only) with those of their empirical estimates $\boldsymbol{\Sigma}_N \boldsymbol{\Sigma}_N^H$ (information plus noise). This estimator only makes sense in the regime where M does not depend on N (thus $c_N \to 0$), because from the classical law of large numbers, we have

$$\|\boldsymbol{\Sigma}_N \boldsymbol{\Sigma}_N^H - (\mathbf{B}_N \mathbf{B}_N^H + \sigma^2 \mathbf{I}_M)\| \overset{a.s.}{\underset{N \to \infty}{\to}} 0.$$

whose convergence is not true in general, if $c_N \to c > 0$. It can be shown that $\eta_N - \tilde{\eta}_N$ does not converge to zero.

Fortunately, we can derive a consistent estimate of η_N by using the results concerning the convergence of bilinear forms of the resolvent of $\boldsymbol{\Sigma}_N \boldsymbol{\Sigma}_N^H$.

Theorem 5.38 (Consistent estimate for the spiked model [282]) *Let*

$$\tilde{\eta}_{spike,N} = \mathbf{d}_N^H \tilde{\boldsymbol{\Sigma}}_N \mathbf{d}_N + \sum_{k=M-K+1}^{M} \mathbf{d}_N^H \tilde{\mathbf{u}}_{k,N} \tilde{\mathbf{u}}_{k,N}^H \mathbf{d}_N \left(1 - \frac{\Gamma'(\tilde{\lambda}_{K,N})}{\Gamma(\tilde{\lambda}_{K,N}) m(\tilde{\lambda}_{K,N})} \right),$$

where $\Gamma(x) = x m(x) \tilde{m}(x)$, and $m(x)$ is the Stieltjes transform of the Marchenko-Pastur law, expressed as

$$m(z) = \frac{1}{-z(1 + c\sigma^2 m(z) + \sigma^2(1-c))},$$

and

$$\tilde{m}(x) = c m(x) - \frac{1-c}{x}.$$

Then, under Assumption 5.2, if

$$\lim_{N \to \infty} \lambda_{M-K+1,N} = \gamma_1 > \sigma^2 \sqrt{c},$$

it holds that

$$\tilde{\eta}_{spike,N} - \eta_N \overset{a.s.}{\underset{N \to \infty}{\to}} 0.$$

This theorem is derived using a different method [301].

5.4.4 Generalized Likelihood Ratio Test Using Large Random Matrices

The material in this subsection can be found in [302]. Denote by N the number of observed samples

$$\mathcal{H}_0 : \mathbf{y}[n] = \mathbf{w}[n], \quad n = 0, 1, \ldots, N-1,$$
$$\mathcal{H}_1 : \mathbf{y}[n] = \mathbf{h}s[n] + \mathbf{w}[n], \quad n = 0, 1, \ldots, N-1,$$

where

- $(\mathbf{w}[n]), n = 0, \ldots, N - 1$ represents an indepdent and identically distributed (i.i.d.) process of $K \times 1$ vectors with circular complex Gaussian entries with mean zero and covariance matrix $\sigma^2 \mathbf{I}_K$;
- vector $\mathbf{h} \in \mathbb{C}^{k \times 1}$ is deterministic, signal $s[n], n = 0, \ldots, N - 1$ denotes a scalar i.i.d. circular complex Gaussian process with zero mean and unit variance;
- $(\mathbf{w}[n]), n = 0, \ldots, N - 1$ and $s[n], n = 0, \ldots, N - 1$ are assumed to be independent processes.

We stack the observed data into a $K \times N$ matrix

$$\mathbf{Y} = [\mathbf{y}[0], \mathbf{y}[1], \mathbf{y}[N - 1]].$$

Denote by $\hat{\mathbf{R}}$ the sample covariance matrix defined as

$$\hat{\mathbf{R}} = \frac{1}{N} \mathbf{Y}\mathbf{Y}^H. \tag{5.40}$$

We denote by $p_0(\mathbf{Y}; \sigma^2)$ and $p_1(\mathbf{Y}; \sigma^2)$ the likelihood functions of the observation matrix $\mathbf{Y}$ indexed by the unknown parameters $\mathbf{h}$ and σ^2 under hypotheses $\mathcal{H}_0$ and $\mathcal{H}_1$.

As $\mathbf{Y}$ is a $K \times N$ matrix whose columns are i.i.d Gaussian vectors with covariance matrix $\boldsymbol{\Sigma}$

$$\mathcal{H}_0 : \boldsymbol{\Sigma} = \sigma^2 \mathbf{I}_K,$$
$$\mathcal{H}_1 : \boldsymbol{\Sigma} = \mathbf{h}\mathbf{h}^H + \sigma^2 \mathbf{I}_K.$$

When parameters $\mathbf{h}$ and σ^2 are known, the Neyman-Pearson procedure gives a uniformly most power test, defined by the likelihood function

$$L = \frac{p_1(\mathbf{Y}; \sigma^2)}{p_0(\mathbf{Y}; \sigma^2)}.$$

In practice, this is not the case: parameters $\mathbf{h}$ and σ^2 are not known. We will deal with this case in the following. No simple procedure guarantees a uniformly most powerful test, and a classical approach called GLRT considers

$$L_N = \frac{\sup_{\mathbf{h},\sigma^2} p_1(\mathbf{Y}; \sigma^2)}{\sup_{\mathbf{h},\sigma^2} p_0(\mathbf{Y}; \sigma^2)}.$$

The GLRT rejects hypothesis $\mathcal{H}_0$ when L_N is above some threshold ξ_N

$$L_N \overset{\mathcal{H}_1}{\underset{\mathcal{H}_0}{\gtrless}} \xi_N, \tag{5.41}$$

where ξ_N is selected in order that the probability of false alarm $\mathbb{P}_0(L_N > \xi_N)$ does not exceed a given level α.

With the aid of [303, 304], the closed form expression of the GLRT L_N is derived in [302]. Denote by

$$\lambda_1 > \lambda_2 > \cdots > \lambda_K \geq 0$$

the ordered eigenvalues of $\hat{\mathbf{R}}$ (all distinct with probability one).

Proposition 5.1 *Let T_N be defined by*

$$T_N = \frac{\lambda_1}{\frac{1}{K} Tr \hat{\mathbf{R}}}. \tag{5.42}$$

Then, the GLRT writes

$$L_N = \frac{C}{(T_N)^N (1 - \frac{1}{K} T_N)^{(K-1)N}} = \phi_{N,K}(T_N), \tag{5.43}$$

where

$$C = \left(1 - \frac{1}{K} \right)^{(1-K)N}.$$

Since $T_N \in (1, K)$ and $\phi(\cdot)$ is an increasing function in this interval, (5.43) is equivalent to

$$T_N = \phi_{N,K}^{-1}(L_N). \tag{5.44}$$

Using (5.44), (5.41) is rewritten as

$$T_N \overset{H_1}{\underset{H_0}{\gtrless}} \gamma_N, \tag{5.45}$$

with

$$\gamma_N = \phi_{N,K}^{-1}(\xi_N).$$

The GLRT (5.45) requires setting the threshold γ_N which is a function of N. Let $p_N(t)$ be the complementary c.d.f. of the statistics T_N under the null hypothesis $\mathcal{H}_0$

$$p_N(t) = \mathbb{P}_0(T_N > t).$$

The threshold γ_N is thus defined as

$$\gamma_N = \phi_{N,K}^{-1}(\alpha),$$

which guarantees that the probability of false alarm $\mathbb{P}_0(T_N > t)$ is kept under a desired level $\alpha \in (0, 1)$.

Since $p_N(t)$ is continuous and decreasing from 1 to 0 in the interval $t \in [0, \infty)$, the threshold $p_N^{-1}(\alpha)$ is well defined. It is more convenient to rewrite the GLRT (5.45) as the final form

$$p_N(T_N) \underset{H_0}{\overset{H_1}{\gtrless}} \alpha. \tag{5.46}$$

The exact expression required in (5.46) has been derived in [302]. The fundamental point is that T_N is only a function of the eigenvalues of $\lambda_1, \ldots, \lambda_K$ of the sample covariance matrix $\hat{\mathbf{R}}$ defined in (5.40). The adopted approach is to study the asymptotic behavior of the complex c.d.f. p_N as the number of observations N goes to infinity. The asymptotic regime is defined as the joint limit where both the number K of sensors and the number N of snapshots go to infinity at the same speed

$$\text{Asymptotic regime}: \quad N \to \infty, K \to \infty, c_N \triangleq \frac{K}{N} \to c, \text{ with } 0 < c < 1. \tag{5.47}$$

This asymptotic regime (5.47) is relevant in cases where the sensing system must be able to perform source detection in a moderate amount of time, that is, both the number K of sensors and the number N of snapshots are of the same order. Very often, the number of sensors is lower than the number of snapshots; hence, the ratio c is lower than 1.

(5.47) is particularly the case for "cognitive radio network as sensors" presented in Chapter 12. The basic idea behind this concept is that spectrum sensing is required in the cognitive radio systems. The availability of so much information that is used for spectrum sensing can also be exploited for sensing the radio environment (as "sensors"); in this manner, a cognitive radio network is used as sensors. Note that the cognitive radio network has much more information at its disposal than the traditional sensors such as ZigBee and Wi-Fi. The programmability of software defined radios must be exploited. Waveforms are programmable in these systems. Waveform diversity for remote sensing is thus enabled.

Under hypothesis $\mathcal{H}_0$, $\Sigma = \mathbf{I}_K$. Sample covariance matrix $\hat{\mathbf{R}}$ is a complex Wishart matrix. Its mathematical properties are well studied.

Under hypothesis $\mathcal{H}_1$, $\Sigma = \mathbf{I}_K + \mathbf{h}\mathbf{h}^H$. Sample covariance matrix $\hat{\mathbf{R}}$ follows a single spiked model, in which all the population eigenvalues are one except for a few fixed eigenvalues [26].

The sample covariance matrix is not only central to the GLRT, but also to multivariate statistics. In many examples, indeed, a few eigenvalues of the sample covariance matrix are separated from the rest of the eigenvalues. Many practical examples show that the samples have non-null covariance. It is natural to ask whether it is possible to determine which non-null population model can possibly lead to the few sample eigenvalues separated from the Marchenko-Pastur density.

The simplest non-null case would be when the population covariance is finite rank perturbation of a multiple of the identity matrix. In other words, all but finitely many eigenvalues of the population covariance matrix are the same, say equal to 1. Such a population model has been called "spiked population model": a null or purely noise model "spiked" with a few significant eigenvalues. The spiked population model was first proposed by [22]. The question is how the eigenvalues of the sample covariance matrix would depend on the nonunit population eigenvalues as $N, K \rightarrow \infty$, as for example, a few large population eigenvalues would possibly pull up a few sample eigenvalues.

Since the behavior of T_N is not affected if the entries of $\mathbf{Y}$ are multiplied by a given constant, we find it convenient to consider the model

$$\boldsymbol{\Sigma} = \mathbf{I}_K + \mathbf{h}\mathbf{h}^H.$$

Define the signal-to-noise ratio (SNR) as

$$\rho_K = \frac{\|\mathbf{h}\|^2}{\sigma^2}.$$

The matrix

$$\boldsymbol{\Sigma} = \mathbf{U}\mathbf{D}\mathbf{U}^H,$$

where $\mathbf{U}$ is a unitary matrix and

$$\mathbf{D} = \mathrm{diag}\,(\rho_K, 1, \ldots, 1).$$

The limiting behavior of the largest eigenvalue λ_1 can change, if the signal-to-noise ratio ρ_K is large enough, above a threshold.

The support of the Marchenko-Pastur distribution is defined as $[\lambda^-, \lambda^+]$, with λ^- the left edge and λ^+ the right edge, where

$$\begin{aligned}\lambda^- &= (1 - \sqrt{c})^2, \\ \lambda^+ &= (1 + \sqrt{c})^2.\end{aligned} \tag{5.48}$$

A further result due to Johnstone [22] and Nadler [305] gives its speed of convergence $\mathbb{O}(N^{-2/3})$. Let Λ_1 be defined as

$$\Lambda_1 = N^{2/3}\left(\frac{T_N - (1 + \sqrt{c_N})^2}{b_N}\right),$$

$$\text{with} \quad b_N = (1 + \sqrt{c_N})\left(\frac{1}{\sqrt{c_N}} + 1\right)^{1/3}, \tag{5.49}$$

then Λ_1 converges in distribution toward a standard Tracy-Widom random variable with c.d.f. F_{TW} defined in (5.50). The Tracy-Widom distribution was first introduced in [24, 25], as the asymptotic distribution of the centered and rescaled large eigenvalue of a matrix from the Gaussian Unitary Ensemble.

Definition 5.3 (Trace-Widom Law [24])

$$F_{TW2}(s) = \exp\left(-\int_s^{+\infty}(x - s)q^2(x)\,dx\right), \forall x \in \mathbb{R}, \tag{5.50}$$

where $q(s)$ is the solution of the Painleve II differential equation

$$\frac{d^2q(s)}{ds^2} = sq(s) + 2q^3(s),$$

satisfying the condition $q(s) \sim -Ai(s)$ (the Airy function) as $s \rightarrow +\infty$.

Tables of the Tracy-Widom law are available, for example, in [306], and a practical algorithm [307] is used to efficiently evaluate (5.50). Refer to [19] for an excellent survey.

Definition 5.4 (Assumption 5.1) *The following constant $\rho \in \mathbb{R}$ exists*

$$\rho = \lim_{K \to \infty} \frac{\|\mathbf{h}\|^2}{\sigma^2} (= \lim_{K \to \infty} \rho_K).$$

We call ρ the limiting SNR. We also define

$$\lambda_{\text{spk}}^{\infty} = (1 + \rho)\left(1 + \frac{c}{\rho}\right).$$

Under hypothesis $\mathcal{H}_1$, the largest eigenvalue has the following asymptotic behavior [26] as $N, K \to \infty$

$$\lambda_1 \overset{a.s.}{\underset{\mathcal{H}_1}{\to}} \begin{cases} \lambda_{\text{spk}}^{\infty}, & \rho > \sqrt{c} \\ \lambda^+, \end{cases}$$

where $\lambda_{\text{spk}}^{\infty}$ is strictly larger than the right edge λ^+. In other words, if the perturbation is large enough, the largest eigenvalue converges outside the support of Marchenko-Pastur distribution $[\lambda^-, \lambda^+]$. The condition for the detectability of the rank one perturbation is

$$\rho > \sqrt{c}. \tag{5.51}$$

Proposition 5.2 (Limiting behavior of T_N under $\mathcal{H}_0$ and $\mathcal{H}_1$) *Let Assumption 5.1 hold true and further assume (5.51) is true, that is, $\rho > \sqrt{c}$. Then,*

$$T_N \overset{a.s.}{\underset{\mathcal{H}_0}{\to}} (1 + \sqrt{c})^2, \text{ and}$$

$$T_N \overset{a.s.}{\underset{\mathcal{H}_1}{\to}} (1 + \rho)\left(1 + \frac{c}{\rho}\right), \text{ as } N, K \to \infty.$$

In Theorem 5.39, we take advantage of the fundamental fact: the largest eigenvalues of the sample covariance matrix $\hat{\mathbf{R}}$, defined in (5.40), converge in the asymptotic regime, defined in (5.47). The threshold and the p-value of interest can be expressed in terms of Tracy-Widom quantiles. Related work includes [24, 25, 308–311], Johnstone [9, 19, 22, 312–318], and Nadler [305].

Theorem 5.39 (Limiting behavior of GLRT [319]) *Consider a fixed level $\alpha \in (0, 1)$ and let γ_N be the threshold for which the power of (5.45) is maximum, that is,*

$$T_N \underset{H_0}{\overset{H_1}{\underset{<}{\gtrless}}} \gamma_N, \tag{5.52}$$

with

$$\gamma_N = \phi_{N,K}^{-1}(\xi_N).$$

Then,

1. *The following convergence is true*

$$\zeta_N \triangleq \frac{N^{2/3}}{b_N}(\gamma_N - (1 - \sqrt{c_N})^2) \underset{N \to \infty, K \to \infty}{\longrightarrow} F_{TW}^{-1}(\alpha).$$

2. *The probability of false alarm of the following test*

$$T_N \overset{H_1}{\underset{H_0}{\gtrless}} (1 + \sqrt{c_N})^2 + \frac{N^{2/3}}{b_N} F_{TW}^{-1}(\alpha)$$

   converges to α.

3. *The p-value* $p_N(T_N)$ *associated with the GLRT can be approximated by*

$$\tilde{p}_N(T_N) = \bar{F}_{TW}^{-1}\left(\frac{N^{2/3}(T_N - (1 + \sqrt{c_N})^2)}{b_N}\right)$$

   in the sense that

$$p_N(T_N) - \tilde{p}_N(T_N) \to 0.$$

Definition 5.5 (Hypothesis test of the condition number) *Define the random variable of the condition number* χ_N *as*

$$\chi_N \triangleq \frac{\lambda_1}{\lambda_K},$$

where λ_1 *and* λ_K *are the largest and the lowest eigenvalue of the sample covariance matrix* $\hat{\mathbf{R}}$ *defined as (5.40).*

A related test [257] uses the ratio of the maximum to the minimum of the eigenvalues of the sample covariance matrix. As for T_N, χ_N is independent of the unknown noise power σ^2. This test χ_N is based on an observation based on (5.48).

Under hypothesis $\mathcal{H}_0$, the spectral measure of $\hat{\mathbf{R}}$ weakly converges to the Marchenko-Pastur distribution with support (λ^-, λ^+) with λ^- and λ^+ defined in (5.48). The largest eigenvalue of $\hat{\mathbf{R}}$, λ_1, converges toward λ^+ under $\mathcal{H}_0$, and $\lambda_{\text{spk}}^\infty$ under $\mathcal{H}_1$.

The lowest eigenvalue of $\hat{\mathbf{R}}$, λ_K, converges to [26, 271, 320]

$$\lambda_K \overset{a.s.}{\to} \lambda^- = \sigma^2(1 - \sqrt{c})^2,$$

under both $\mathcal{H}_0$ and $\mathcal{H}_1$. Therefore, the statistic χ_N admits the following limit

$$\chi = \frac{\lambda_1}{\lambda_K} \overset{a.s.}{\underset{\mathcal{H}_0}{\to}} \frac{\lambda^+}{\lambda^-} = \frac{(1 + \sqrt{c})^2}{(1 - \sqrt{c})^2},$$

$$\chi = \frac{\lambda_1}{\lambda_K} \overset{a.s.}{\underset{\mathcal{H}_1}{\to}} \frac{\lambda_{\text{spk}}^\infty}{\lambda^-} = \frac{(1 + \rho)(1 + \frac{c}{\rho})}{(1 - \sqrt{c})^2}, \quad \text{for } \rho > \sqrt{c}$$

$$\text{with } \lambda_{\text{spk}}^\infty = (1 + \rho)\left(1 + \frac{c}{\rho}\right).$$

The test is based on the observation that the limit of χ_N under the alternative $\mathcal{H}_1$ is strictly larger than the ratio $\frac{\lambda^+}{\lambda^-}$, at least when the SNR ρ is large enough.

The threshold must be determined before using the condition number test. It is proven in [302] that T_N outperforms χ_N. Λ_1 is defined in (5.49) (repeated below) and Λ_K is defined as

$$\Lambda_1 = N^{2/3} \left(\frac{T_N - (1 + \sqrt{c_N})^2}{b_N} \right),$$

$$\Lambda_K = N^{2/3} \left(\frac{\lambda_K - (1 + \sqrt{c_N})^2}{(\sqrt{c_N} - 1)\left(\frac{1}{\sqrt{c_N}} - 1\right)^{1/3}} \right).$$

Then, both Λ_1 and Λ_K converge toward Tracy-Widom random variables

$$(\Lambda_1, \Lambda_K) \underset{N \to \infty, K \to \infty}{\to} (X, Y),$$

where X and Y are independent random variables, both distributed according to $F_{TW}(x)$. A direct use of the Delta method [321, Chapter 3] gives the following convergence in distribution

$$N^{2/3} \left(\frac{\lambda_1}{\lambda_K} - \frac{(1 + \sqrt{c_N})^2}{(1 - \sqrt{c_N})^2} \right) \to (aX + bY)$$

where

$$a = \frac{(1 + \sqrt{c})^2}{(1 - \sqrt{c})^2} \left(\frac{1}{\sqrt{c}} + 1 \right)^{1/3},$$

$$b = \frac{(1 + \sqrt{c})^2}{(\sqrt{c} - 1)^2} \left(\frac{1}{\sqrt{c}} - 1 \right)^{1/3}.$$

The optimal threshold is found to be

$$\xi_N \triangleq N^{2/3} \left(\gamma_N - \frac{(1 + \sqrt{c_N})^2}{(1 - \sqrt{c_N})^2} \right) \underset{N \to \infty, K \to \infty}{\to} \bar{F}^{-1}_{aX+bY}(\alpha).$$

with $\alpha = \mathbb{P}_0(\chi_N > \gamma_N), \alpha \in (0, 1)$.

In particular, ξ_N is bounded as $N, K \to \infty$.

5.4.5   Detection of High-Dimensional Signals in White Noise

We mainly follow [322] for this development. We observe M samples ("snapshots") of possibly signal bearing N-dimensional snapshot vectors $\mathbf{x}_1, \ldots, \mathbf{x}_M$. For each i,

$$\mathbf{x}_i \sim \mathcal{N}_N(0, \sigma^2 \mathbf{I}),$$

where $\mathbf{x}_i$ are mutually independent. The snapshot vectors are modelled as

$$\mathcal{H}_0 : \mathbf{x}_i = \mathbf{z}_i, \text{No signal}$$
$$\mathcal{H}_1 : \mathbf{x}_i = \mathbf{H}\mathbf{s}_i + \mathbf{z}_i, \text{Signal present, i} = 1, \ldots, M,$$

where

- $\mathbf{z}_i \sim \mathcal{N}_N(0, \sigma^2 \mathbf{I})$, denotes an N-dimensional (real or circularly symmetric complex) Gaussian noise vector whose σ^2 is assumed to be unknown;
- $\mathbf{s}_i \sim \mathcal{N}_K(0, \mathbf{R}_s)$ denotes a K-dimensional (real or circularly symmetric complex) Gaussian signal vector with $\mathbf{R}_s$;
- and $\mathbf{H}$ is a $N \times K$ unknown nonrandom matrix.
- $\mathbf{H}$ encodes the parameter vector associated with the j-th signal whose magnitude is described by the j-th element of $\mathbf{s}_i$.

Since the signal and noise vectors are independent of each other, the covariance matrix of $\mathbf{x}_i$ can be decomposed as

$$\mathbf{R} = \tilde{\mathbf{R}}_s + \sigma^2 \mathbf{I}$$

where

$$\tilde{\mathbf{R}}_s = \mathbf{H R}_s \mathbf{H}^H.$$

The sample covariance matrix is defined as

$$\hat{\mathbf{R}} = \frac{1}{M} \sum_{i=1}^{M} \mathbf{x}_i \mathbf{x}_i^H = \frac{1}{M} \mathbf{X X}^H,$$

where

$$\mathbf{X} = [\mathbf{x}_1 | \dots | \mathbf{x}_M]$$

is the matrix of observations (samples).

It is assumed that the rank of $\tilde{\mathbf{R}}_s$ is K. Equivalently, the $N - K$ smallest eigenvalues of $\tilde{\mathbf{R}}_s$ are equal to zero. Denote the eigenvalues of $\mathbf{R}$ by

$$\lambda_1 \geq \lambda_2 \geq \cdots \geq \lambda_N,$$

then the smallest $N - K$ eigenvalues of $\mathbf{R}$ are all equal to σ^2 so that

$$\lambda_{K+1} = \lambda_{K+2} = \cdots = \lambda_N = \lambda = \sigma^2.$$

In practice, we have to estimate the value of K, so called rank estimation.

We assume $M > N$ and $\mathbf{x}_i \in \mathbb{C}^N$. Similarly to the case of the true covariance matrix, the eigenvalues of $\hat{\mathbf{R}}$ are ordered

$$l_1 \geq l_2 \geq \cdots \geq l_N.$$

Our estimator developed here is robust to high-dimensionality and sample size constraints.

A central object in the study of large random matrices is the empirical distribution function (e.d.f.) of the eigenvalues. Under $\mathcal{H}_0$, the e.d.f. of $\hat{\mathbf{R}}$ converges to the Marchenko-Pastur density $F_W(x)$. The almost sure convergence of the e.d.f. of the signal-free sample covariance matrix (SCM) implies that the moments of the eigenvalues converge almost surely, so that

$$\frac{1}{N} \sum_{i=1}^{N} l_i^k \overset{a.s.}{\to} \int x^k dF_W(x) = M_k^W,$$

where [266]

$$M_k^W = \lambda^k \sum_{j=0}^{k-1} c^j \frac{1}{j+1} \binom{k}{j} \binom{k-1}{j}.$$

For finite N and M, the sample moments, that is, $\frac{1}{N} \sum_{i=1}^{N} l_i^k$, will fluctuate about these limiting values.

Proposition 5.3 (Convergence of moments in distribution [322]) *Let $\hat{\mathbf{R}}$ denote a signal-free sample covariance matrix found from a $N \times M$ matrix of observations with i.i.d. Gaussian samples of mean zero and variance $\lambda = \sigma^2$. For the asymptotic regime*

$$N, M \to \infty, \quad and \quad c_M = \frac{N}{M} \to c \in (0, \infty),$$

we have

$$N \left(\begin{bmatrix} \frac{1}{N} \sum_{i=1}^{N} l_i \\ \frac{1}{N} \sum_{i=1}^{N} l_i^2 \end{bmatrix} - \begin{bmatrix} \lambda \\ \lambda^2(1+c) \end{bmatrix} \right)$$

$$\overset{\mathcal{D}}{\to} \mathcal{N} \left(\underbrace{\begin{bmatrix} 0 \\ (\frac{2}{\beta} - 1)\lambda^2 c \end{bmatrix}}_{\boldsymbol{\mu}_Q} - \frac{2}{\beta} \underbrace{\begin{bmatrix} \lambda^2 c & 2\lambda^3 c(1+c) \\ 2\lambda^3 c(c+1) & 2\lambda^4 c(2c^2 + 5c + 2) \end{bmatrix}}_{\mathbf{Q}} \right),$$

where the convergence is in distribution.

Proposition 5.4 (Convergence of the statistic q_N) *Assume $\hat{\mathbf{R}}$ satisfies the hypothesis of Proposition 5.3 for some λ. Consider the statistic*

$$q_N = \frac{\frac{1}{N} \sum_{i=1}^{N} l_i^2}{\left(\frac{1}{N} \sum_{i=1}^{N} l_i \right)^2}.$$

Then, as

$$N, M \to \infty, \quad and \quad c_M = \frac{N}{M} \to c \in (0, \infty),$$

we have

$$N[q_N - (1+c)] \xrightarrow{\mathcal{D}} \mathcal{N}\left(\left(\frac{2}{\beta} - 1\right)c, \frac{4}{\beta}c^2\right),$$

where the convergence is in distribution.

The two Propositions 5.3 and 5.4 deal with $\mathcal{H}_0$. Now we introduce the two propositions in the signal-bearing case $\mathcal{H}_1$. In the signal-bearing case, a so-called phase transition phenomenon is observed, in that the largest eigenvalue will converge to a limit different from that in the signal-free case only if the "signal" eigenvalues are above a certain threshold.

Proposition 5.5 (Convergence of the eigenvalues of $\hat{\mathbf{R}}$) *Let $\hat{\mathbf{R}}$ denote a sample covariance matrix formed from a $N \times M$ matrix of observations with i.i.d. Gaussian observations whose columns are independent of each other and identically distributed with mean zero and variance $\mathbf{R}$. Denote the eigenvalues of $\mathbf{R}$ by*

$$\lambda_1 \geq \lambda_2 \geq \cdots \geq \lambda_K > \lambda_{K+1} = \cdots \lambda_N = \lambda.$$

Let l_j be the j-th largest eigenvalue of $\hat{\mathbf{R}}$. Then,

$$N, M \to \infty, \text{ with } c_M = \frac{N}{M} \to c \in (0, \infty),$$

we have

$$l_j = \begin{cases} \lambda_j(1 + \frac{\lambda c}{\lambda_j - \lambda}) & \text{if } \lambda_j > \lambda(1 + \sqrt{c})^2 \\ \lambda(1 + \sqrt{c})^2 & \text{if } \lambda_j \leq \lambda(1 + \sqrt{c})^2 \end{cases} \text{ for } j = 1, \ldots, K,$$

where convergence is almost certain.

This result appears in [26] for a very general setting. A matrix theoretic proof for the real-valued SCM case may be found in [27] while a determinental proof for the complex case may be found in [259]. A heuristic derivation appears in [323].

The "signal" eigenvalues strictly below the threshold described in Proposition 5.5 exhibit, on rescaling, fluctuations described by the Tracy-Widom distribution [24, 25]. An excellent survey is given in [19].

Proposition 5.6 (Convergence of the eigenvalues of $\hat{\mathbf{R}}$) *Assume $\mathbf{R}$ and $\hat{\mathbf{R}}$ satisfies the hypotheses of Proposition 5.5. If $\lambda_j > \lambda(1 + \sqrt{c})$ has multiplicity 1 and if $\sqrt{M}|c - N/M| \to 0$, then*

$$\sqrt{N}\left[l_j - \lambda_j\left(1 + \frac{\lambda c}{\lambda_j - \lambda}\right)\right] \xrightarrow[a.s.]{\mathcal{D}} \mathcal{N}\left(0, \frac{2}{\beta}\lambda_j^2\left(1 - \frac{c}{(\lambda_j - \lambda)^2}\right)\right)$$

where the convergence in distribution is almost sure.

A matrix theoretic proof for the real-valued SCM case may be found in [27] while a determinental proof for the complex case may be found in [259]. The result has been strengthened for non-Gaussian situations by Baik and Silverstein for general $c \in (0, \infty)$.

Theorem 5.40 (The eigenvalues of R and $\hat{\mathbf{R}}$ converge to the same limit) *Let* $\mathbf{R}$ *and* $\hat{\mathbf{R}}$ *be two* $N \times N$ *sized covariance matrices whose eigenvalues are related as*

$$\boldsymbol{\Lambda} = diag\ (\lambda_1, \ldots, \lambda_p, \lambda_{p+1}, \ldots, \lambda_K, \lambda, \ldots, \lambda)$$
$$\tilde{\boldsymbol{\Lambda}} = diag\ (\lambda_1, \ldots, \lambda_p, \lambda, \ldots, \lambda),$$

where for some $c \in (0, \infty)$, *and*

$$\lambda < \lambda_i \le \lambda(1 + \sqrt{c}), \quad all\ i = p + 1, \ldots, K.$$

Let $\mathbf{R}$ *and* $\hat{\mathbf{R}}$ *be the associated sample covariance matrices formed from* M *snapshots. Then, for*

$$every\ N, M(N) \to \infty,\ and\ c_M = \frac{N}{M} \to c \in (0, \infty)$$
$$Prob\ (\hat{K} = j | \mathbf{R}) \to Prob\ (\hat{K} = j | \tilde{\mathbf{R}})\ for\ j = 1, \ldots, p$$

and

$$Prob\ (\hat{K} > p | \mathbf{R}) \to Prob\ (\hat{K} > p | \tilde{\mathbf{R}})\ for\ j = 1, \ldots, p,$$

where the convergence is almost surely and $\hat{K}$ *is the estimate of the number of signals obtained using the algorithm in [322].*

By Proposition 5.3, we heuristically define the effective number of (identifiable) signals as

$$K_{eff}(\mathbf{R}) = \text{Number of eigenvalues of } \mathbf{R} > \sigma^2 \left(1 + \sqrt{\frac{N}{M}} \right).$$

Consider an example of

$$\mathbf{R} = \sigma_{S1}^2 \mathbf{v}_1 \mathbf{v}_1^H + \sigma_{S2}^2 \mathbf{v}_2 \mathbf{v}_2^H + \sigma^2 \mathbf{I},$$

which has the $N - 2$ smallest eigenvalues $\lambda_3 = \cdots = \lambda_N = \sigma^2$ and the two largest eigenvalues

$$\lambda_1 = \sigma^2 + \frac{(\sigma_{S1}^2 \|\mathbf{v}_1\|^2 + \sigma_{S2}^2 \|\mathbf{v}_2\|^2)}{2} + \frac{\sqrt{(\sigma_{S1}^2 \|\mathbf{v}_1\|^2 - \sigma_{S2}^2 \|\mathbf{v}_2\|^2) + 4\sigma_{S1}^2 \sigma_{S2}^2 |\langle \mathbf{v}_1, \mathbf{v}_2 \rangle|^2}}{2},$$

$$\lambda_2 = \sigma^2 + \frac{(\sigma_{S1}^2 \|\mathbf{v}_1\|^2 + \sigma_{S2}^2 \|\mathbf{v}_2\|^2)}{2} - \frac{\sqrt{(\sigma_{S1}^2 \|\mathbf{v}_1\|^2 - \sigma_{S2}^2 \|\mathbf{v}_2\|^2) + 4\sigma_{S1}^2 \sigma_{S2}^2 |\langle \mathbf{v}_1, \mathbf{v}_2 \rangle|^2}}{2}$$

respectively. Applying the result in Proposition 5.3, we can express the effective number of signals as

$$K_{eff} = \begin{cases} 2, & \text{if } \sigma^2 \left(1 + \sqrt{\frac{N}{M}} \right) < \lambda_2 \\ 1, & \text{if } \lambda_2 \le \sigma^2 \left(1 + \sqrt{\frac{N}{M}} \right) < \lambda_1 \\ 0, & \text{if } \lambda_1 \le \sigma^2 \left(1 + \sqrt{\frac{N}{M}} \right). \end{cases}$$

In the special situation when

$$\|\mathbf{v}_1\| = \|\mathbf{v}_2\| = \|\mathbf{v}\| \quad \text{and} \quad \sigma_{S1}^2 = \sigma_{S2}^2 = \sigma_S^2,$$

we can (in an asymptotic sense) reliably detect the presence of both signals from the sample eigenvalues alone, whenever we have the following condition

$$\text{Asymptotic identifiability condition: } \sigma_S^2 \|\mathbf{v}\|^2 \left(1 - \frac{|\langle \mathbf{v}_1, \mathbf{v}_2 \rangle|}{\|\mathbf{v}\|}\right) > \sigma^2 \sqrt{\frac{N}{M}}.$$

We define Z_j^{Sep} as

$$Z_j^{\text{Sep}} = \frac{\lambda_j \left(1 + \frac{\sigma^2 N}{M(\lambda_j - \sigma^2)}\right) - \sigma^2 \left(1 + \sqrt{\frac{N}{M}}\right)^2}{\sqrt{\frac{2}{\beta N} \lambda_j^2 \left(1 - \frac{N}{M(\lambda_j - \sigma^2)}\right)}},$$

which measures the (theoretical) separation of the j-th "signal" eigenvalue from the largest "noise" eigenvalue in standard deviations of the j-th signal eigenvalue's fluctuations. Simulations suggest that reliable detection (with the empirical probability greater than 90%) of the effective number of signals is possible if Z_j^{Sep} lies between 5 and 15.

5.4.6 Eigenvalues of $(\mathbf{A} + \mathbf{B})^{-1}\mathbf{B}$ and Applications

Roy's largest root test [324] is relevant under this context. We follow [325] for this development. Let $\mathbf{X}$ be an $m \times p$ normal data matrix: each row is an independent observation from $\mathcal{N}_p(\mathbf{0}, \boldsymbol{\Sigma})$. A $p \times p$ matrix $\mathbf{A} = \mathbf{X}^{\text{H}}\mathbf{X}$ is then said to have a Wishart distribution

$$\mathbf{A} \sim \mathcal{W}_p(\boldsymbol{\Sigma}, m).$$

Let

$$\mathbf{B} \sim \mathcal{W}_p(\boldsymbol{\Sigma}, n).$$

Assume that $m \geq p$; then $\mathbf{A}^{-1}$ exists and the nonzero eigenvalues of $\mathbf{A}^{-1}\mathbf{B}$ generalize the univariate F ratio. The scale matrix $\boldsymbol{\Sigma}$ has no effect on the distribution of these eigenvalues; without loss of generality we assume that $\boldsymbol{\Sigma} = \mathbf{I}$. We follow the definition of [110, p. 84] for the greatest root statistic.

Definition 5.6 (Greatest root statistic) *Let $\mathbf{A} \sim \mathcal{W}_p(\boldsymbol{\Sigma}, m)$ is independent of $\mathbf{B} \sim \mathcal{W}_p(\boldsymbol{\Sigma}, n)$, where $m \geq p$. Then the largest eigenvalue θ of $(\mathbf{A} + \mathbf{B})^{-1}\mathbf{B}$ is called the greatest root statistic and a random variate having this distribution is denoted $\lambda_1(p, m, n)$ or $\lambda_{1,p}$ for short.*

Since $\mathbf{A}$ is positive definite, $0 < \lambda_i < 1$, for the i-th eigenvalue. Equivalently, $\lambda_1(p, m, n)$ is the largest root of the determinantal equation

$$\det\left[\mathbf{B} - \lambda(\mathbf{A} + \mathbf{B})\right] = 0.$$

The parameter p stands for dimension, m the "error" degrees of freedom and n the "hypothesis" degrees of freedom. Thus $m + n$ represents the "total" degrees of freedom.

The greatest root distribution has the property

$$\lambda_1(p, m, n) = \lambda_1(n, m + n - p, p), \tag{5.53}$$

useful in the case when $n < p$. [110, p. 84]

Assume p is even and that $p, m = m(p), n = n(p)$ all go to infinity together such that

$$\lim_{p \to \infty} \frac{\min(p, n)}{m + n} > 0, \ \lim_{p \to \infty} \frac{p}{m} < 1. \tag{5.54}$$

Define the logit transform W_p as

$$W_p = \text{logit}\lambda_{1,p} = \log\left(\frac{\lambda_{1,p}}{1 - \lambda_{1,p}}\right).$$

Johnstone (2008) [325] shows W_p, is, with appropriate centering and scaling, approximately Tracy-Widom distributed:

$$\frac{W_p - \mu_p}{\sigma_p} \overset{D}{\Rightarrow} \mathcal{Z}_1 \sim \mathcal{F}_1.$$

The distribution function $\mathcal{F}_1$ was found by Tracy and Widom to be the limiting law of the largest eigenvalue of a $p \times p$ Gaussian symmetric matrix [25].

The centering and scaling parameters are

$$\mu_p = 2\log\tan\left(\frac{\varphi + \gamma}{2}\right), \sigma_p^3 = \frac{16}{(m + n - 1)^2} \frac{1}{\sin^2(\varphi + \gamma)\sin\varphi\sin\gamma}, \tag{5.55}$$

where the angle parameters γ, φ are defined by

$$\sin^2\left(\frac{\gamma}{2}\right) = \frac{\min(p, n) - 1/2}{m + n - 1}, \sin^2\left(\frac{\varphi}{2}\right) = \frac{\max(p, n) - 1/2}{m + n - 1}. \tag{5.56}$$

Theorem 5.41 (Johnstone (2008) [325]) *Assume that $m(p), n(p) \to \infty$ as $p \to \infty$ through even values of p according to (5.54). For each $t_0 \in \mathbb{R}$, there exists $C > 0$ such that for $t > t_0$,*

$$|P\{W_p \leq \mu_p + \sigma_p t\} - \mathcal{F}_1(t)| \leq Cp^{-2/3}e^{-t/2}.$$

Here C depends on (γ, φ) and also on t_0 if $t_0 < 0$.

Data matrices $\mathbf{X}$ based on complex-valued data rises frequently in signal processing and communications. If the rows of $\mathbf{X}$ are drawn independently from a complex normal distribution $\mathbb{C}\mathcal{N}(\boldsymbol{\mu}, \boldsymbol{\Sigma})$, then we say

$$\mathbf{A} = \mathbf{X}^H\mathbf{X} \sim \mathbb{C}\mathcal{W}_p(\boldsymbol{\Sigma}, n).$$

In parallel with the real case definition, if

$$\mathbf{A} \sim \mathbb{C}\mathcal{W}_p(\mathbf{I}, m) \quad \text{and} \quad \mathbf{B} \sim \mathbb{C}\mathcal{W}_p(\mathbf{I}, n)$$

are independent, then the joint density of the eigenvalues

$$1 \geq \lambda_1 \geq \lambda_2 \geq \cdots \geq \lambda_p \geq 0$$

of $(\mathbf{A} + \mathbf{B})^{-1}\mathbf{B}$, or equivalently

$$\det [\mathbf{B} - \lambda(\mathbf{A} + \mathbf{B})] = 0,$$

is given by [326]

$$f(\lambda) = c \prod_{i=1}^{p} (1 - \lambda_i)^{m-p} \lambda_i^{n-p} \prod_{i<j} (\lambda_i - \lambda_j).$$

The largest eigenvalue $\lambda^C(p, m, n)$ of $(\mathbf{A} + \mathbf{B})^{-1}\mathbf{B}$ is called the greatest root statistic, with distribution $\lambda^C(p, m, n)$. The property (5.53) carries over to the complex case.

Again, we define

$$W_p^C = \mathrm{logit}\lambda_{1,p}^C = \log \left(\frac{\lambda_{1,p}^C}{1 - \lambda_{1,p}^C} \right).$$

Theorem 5.42 (Johnstone (2008) [325]) *Assume that* $m(p), n(p) \to \infty$ *as* $p \to \infty$ *according to (5.54). For each* $t_0 \in \mathbb{R}$, *there exists* $C > 0$ *such that for* $t > t_0$,

$$|P\{W_p^C \leq \mu_p^C + \sigma_p^C t\} - \mathcal{F}_2(t)| \leq Cp^{-2/3}e^{-t/2}.$$

Here C *depends on* (γ, φ) *and also on* t_0 *if* $t_0 < 0$.

The centering μ_p^C and scaling σ_p^C are given in [325]. Software implementation is also available. See [325] for details.

We are now in a position to consider several settings in multivariate statistics using double Wishart models.

5.4.7 Canonical Correlation Analysis

Suppose that there are N observations on each of $L + M$ variables. For definiteness, assume that $L \leq M$. The first L variables are grouped into an $N \times L$ data matrix

$$\mathbf{X} = [\mathbf{x}_1 \mathbf{x}_2 \cdots \mathbf{x}_L]$$

and the last M into $N \times L$ data matrix

$$\mathbf{Y} = [\mathbf{y}_1 \mathbf{y}_2 \cdots \mathbf{y}_M].$$

Write

$$\mathbf{S}_{XX} = \mathbf{X}^T \mathbf{X}, \mathbf{S}_{XY} = \mathbf{X}^T \mathbf{Y}, \mathbf{S}_{YY} = \mathbf{Y}^T \mathbf{Y},$$

for cross-product matrices. Canonical correlation analysis (CCA), or more precisely, the zero-mean version of CCA, seeks the *linear combinations* $\mathbf{a}^T \mathbf{x}$ and $\mathbf{b}^T \mathbf{y}$ that are most highly correlated, that is, to maximize

$$\rho = \text{Corr}(\mathbf{a}^T \mathbf{x}, \mathbf{b}^T \mathbf{y}) = \frac{\mathbf{a}^T \mathbf{S}_{XY} \mathbf{b}}{\sqrt{\mathbf{a}^T \mathbf{S}_{XX} \mathbf{a}} \sqrt{\mathbf{b}^T \mathbf{S}_{YY} \mathbf{b}}}. \tag{5.57}$$

This leads to a maximal correlation ρ_1 and associated canonical vectors $\mathbf{a}_1$ and $\mathbf{b}_1$, usually each taken to have unit length. The procedure may be iterated. We restrict the search to vectors that are orthogonal to those already found:

$$\rho_k = \max \left\{ \begin{array}{l} \mathbf{a}^T \mathbf{S}_{XY} \mathbf{b} : \mathbf{a}^T \mathbf{S}_{XX} \mathbf{a} = \mathbf{b}^T \mathbf{S}_{YY} \mathbf{b} = 1, \text{ and} \\ \mathbf{a}^T \mathbf{S}_{XX} \mathbf{a}_j = \mathbf{b}^T \mathbf{S}_{YY} \mathbf{b}_j = 1, \text{ for } 1 \le j \le k \end{array} \right\}.$$

The successive canonical correlations $\rho_1 \ge \rho_2 \ge \cdots \ge \rho_L \ge 0$ may be found as the roots of the determinantal equation

$$\det (\mathbf{S}_{XY} \mathbf{S}_{YY}^{-1} \mathbf{S}_{YX} - \rho^2 \mathbf{S}_{XX}) = 0. \tag{5.58}$$

See, for example, [110, p. 284]. A typical question in applications is how many of the ρ_k are significantly different from zero.

After some manipulations, (5.58) becomes

$$\det (\mathbf{B} - \rho^2 (\mathbf{A} + \mathbf{B})) = 0. \tag{5.59}$$

Now assume that $\mathbf{Z} = [\mathbf{XY}]$ is an $N \times (L + M)$ Gaussian data matrix with mean zero. The covariance matrix is partitioned into

$$\mathbf{\Sigma} = \left(\begin{array}{cc} \mathbf{\Sigma}_{XX} & \mathbf{\Sigma}_{XY} \\ \mathbf{\Sigma}_{YX} & \mathbf{\Sigma}_{YY} \end{array} \right).$$

Under these Gaussian assumptions, $\mathbf{X}$ and $\mathbf{Y}$ variable sets will be independent if and only if

$$\mathbf{\Sigma}_{XY} = 0.$$

This is equivalent to asserting that

$$\mathcal{H}_0 : \rho_1 = \rho_2 = \cdots = \rho_L = 0.$$

The canonical correlations $(\rho_1, \ldots, \rho_L)$ are invariant under block diagonal transformations

$$(\mathbf{x}_i, \mathbf{y}_i) \rightarrow (\mathbf{B}\mathbf{x}_i, \mathbf{C}\mathbf{y}_i)$$

of the data (for $\mathbf{B}$ and $\mathbf{C}$ nonsingular $L \times L$ and $M \times M$ matrices, respectively). It follows that under hypothesis

$$\mathcal{H}_0 : \mathbf{\Sigma}_{XY} = 0,$$

the distribution of the canonical correlations can be found (without loss of generality) by assuming that

$$\mathcal{H}_0 : \mathbf{\Sigma}_{XX} = \mathbf{I}_L, \mathbf{\Sigma}_{YY} = \mathbf{I}_M.$$

In this case, the matrices $\mathbf{A}$ and $\mathbf{B}$ of (5.59) are

$$\mathbf{A} \sim \mathbb{C}\mathcal{W}_L(\mathbf{I}, M), \mathbf{B} \sim \mathbb{C}\mathcal{W}_L(\mathbf{I}, N - M).$$

From the definition, the largest squared canonical correlation $\lambda_1 = \rho_1^2$ has the $\Lambda(L, N - M, M)$ distribution under the null hypothesis $\mathbf{\Sigma}_{XY} = 0$.

In practice, it is more common to allow each variable to have a separate, unknown mean. One can correct the mean using the approach, for example, in [325].

5.4.8 Angles and Distances between Subspaces

The cosine of the angle between two vectors $\mathbf{u}, \mathbf{v} \in \mathbb{R}^N$ is given by

$$\cos \vartheta = \sigma(\mathbf{u}, \mathbf{v}) = \frac{|\mathbf{u}^T \mathbf{v}|}{\|\mathbf{u}\|_2 \|\mathbf{v}\|_2}.$$

(5.57) becomes

$$\rho = \sigma(\mathbf{Xa}, \mathbf{Yb}).$$

5.4.9 Multivariate Linear Model

In the multivariate model,

$$\mathbf{Y} = \mathbf{HX} + \mathbf{W}$$

where

1. $\mathbf{Y}$ of $N \times M$ is an observed matrix of M response variables on each of N individuals (sensors);
2. $\mathbf{H}$ of $N \times K$ is a known design matrix (channel response);
3. $\mathbf{X}$ of $K \times M$ is a matrix of unknown regression parameters;
4. $\mathbf{W}$ of $N \times M$ is a matrix of unobserved random distributions (additive white Gaussian noise). It is assumed that $\mathbf{W}$ is a normal matrix of N vector samples from $\mathcal{N}_M(\mathbf{0}, \mathbf{\Sigma})$, so that the rows are independent Gaussian, each with mean 0 and common covariance matrix $\mathbf{\Sigma}$.

The model matrix $\mathbf{H}$ remains the same for each response; however, there are separate vectors of unknown coefficients and errors for each response; these are organized into $\mathbf{X}$ of regression coefficients and $N \times M$ matrix $\mathbf{E}$ of errors [327]. Assuming for now that the model matrix $\mathbf{H}$ has full rank, the least squares estimator is

$$\hat{\mathbf{X}} = (\mathbf{H}^T \mathbf{H})^{-1} \mathbf{H}^T \mathbf{Y}.$$

Consider the linear hypothesis

$$\mathcal{H}_0 : \mathbf{C}_1 \mathbf{X} = \mathbf{0},$$

where $\mathbf{C}_1$ is a $r \times K$ matrix of rank r. For more details about $\mathbf{C}_1$, we refer to [327].

The hypothesis sums and errors sums of squares and product matrices become

$$\mathbf{E} = \mathbf{Y}^T \mathbf{PX} = \mathbf{Y}^T (\mathbf{I} - \mathbf{H}(\mathbf{H}^T \mathbf{H})^{-1} \mathbf{H}^T) \mathbf{Y},$$
$$\mathbf{D} = \mathbf{Y}^T \mathbf{P}_2 \mathbf{Y} = (\mathbf{C}_1 \hat{\mathbf{X}})^T (\mathbf{C}_1 (\mathbf{H}^T \mathbf{H})^{-1} \mathbf{C}_1^T) \mathbf{C}_1 \hat{\mathbf{X}}.$$

It follows [327] that

$$\mathbf{E} \sim \mathcal{W}_M(\mathbf{I}, N - K),$$

and under hypothesis $\mathcal{H}_0$,

$$\mathbf{D} \sim \mathcal{W}_M(\mathbf{I}, r);$$

in addition, $\mathbf{D}$ and $\mathbf{E}$ are independent. Generalization of the $\mathcal{F}$-test is obtained from the eigenvalues of the matrix $\mathbf{E}^{-1}\mathbf{D}$, or equivalently, the eigenvalues of $(\mathbf{D} + \mathbf{E})^{-1}\mathbf{D}$.

Thus, under the null hypothesis $\mathbf{C}_1\mathbf{X} = \mathbf{0}$, Roy's maximum root statistic λ_1 has null distribution

$$\lambda_1 \sim \Lambda(M, N - K, r) \text{ where}$$
$$M = \text{ dimension}, r = \text{rank}(\mathbf{C}_1), K = \text{rank}(\mathbf{H}), N = \text{samples}.$$

5.4.10 Equality of Covariance Matrices

Suppose that independent samples from two normal distributions $\mathcal{N}_M(\boldsymbol{\mu}_1, \boldsymbol{\Sigma}_1)$ and $\mathcal{N}_M(\boldsymbol{\mu}_2, \boldsymbol{\Sigma}_2)$ lead to covariance estimates $\mathbf{S}_1$ and $\mathbf{S}_2$ which are independent and Wishart distributed on N_1 and N_2 degrees of freedom:

$$\mathbf{A}_i = N_i\mathbf{S}_i \sim \mathcal{W}_M(\boldsymbol{\Sigma}_i, N_i), i = 1, 2.$$

Then the largest root test of the null hypothesis

$$\mathcal{H}_0 : \boldsymbol{\Sigma}_1 = \boldsymbol{\Sigma}_2$$

is based on the largest eigenvalue λ of $(\mathbf{A}_1 + \mathbf{A}_2)^{-1}\mathbf{A}_2$, which under $\mathcal{H}_0$ has the $\Lambda(M, N_1, N_2)$ distribution [328].

5.4.11 Multiple Discriminant Analysis

Suppose that there are K populations, the i-th population being assumed to follow an M-variate normal distribution $\mathcal{N}_M(\boldsymbol{\mu}_i, \boldsymbol{\Sigma}_i)$, with the covariance matrix assumed to be unknown, but common to all populations. A sample of size N_i (vector) observations is available from the i-th population, leading to a total of $N = \sum N_i$ observations. Multiple discriminant analysis uses the "within groups" and "between groups" sums of squares and products matrices $\mathbf{W}$ and $\mathbf{B}$ to construct linear discriminant functions based on eigenvectors of $\mathbf{W}^{-1}\mathbf{B}$. A test of the null hypothesis that discrimination is not worthwhile

$$\mu_1 = \cdots = \mu_K$$

can be based, for example, on the largest root of $\mathbf{W}^{-1}\mathbf{B}$, which leads to use of the $\Lambda(M, N - K, K - 1)$ distribution [110, pages 318 and 138].

5.5   Case Studies and Applications

5.5.1   Fundamental Example of Using Large Random Matrix

We follow [279] for this development. Define an $M \times N$ complex matrix as

$$\mathbf{X} = \begin{bmatrix} X_{11} & X_{12} & \cdots & X_{1N} \\ X_{21} & X_{22} & \cdots & X_{2N} \\ \vdots & \vdots & \vdots & \vdots \\ X_{M1} & X_{M2} & \cdots & X_{MN} \end{bmatrix}$$

where $(X_{ij})_{1 \le i \le M, 1 \le j \le N}$ are (a number of MN) i.i.d. complex Gaussian variables $\mathcal{CN}(0, \sigma^2)$. $\mathbf{x}_1, \mathbf{x}_2, \ldots, \mathbf{x}_N$ are columns of $\mathbf{X}$. The covariance matrix $\mathbf{R}$ is

$$\mathbf{R} = \mathbb{E}\mathbf{x}\mathbf{x}^H = \sigma^2 \mathbf{I}_M.$$

The empirical covariance matrix is defined as

$$\hat{\mathbf{R}} = \frac{1}{N} \sum_{n=1}^{N} \mathbf{x}_n \mathbf{x}_n^H.$$

In practice, we are interested in the behavior of the empirical distribution of the eigenvalues of $\hat{\mathbf{R}}$ for large M and N. For example, how do the histograms of the eigenvalues $(\lambda_i)_{i=1,\ldots,M}$ of $\hat{\mathbf{R}}$ behave when M and N increase? It is well known that when M is fixed, but N increases, that is, $\frac{M}{N}$ is small, the large law of large numbers requires

$$\lim_{\substack{N \to \infty \\ \text{fixed } M}} \frac{1}{N} \sum_{n=1}^{N} \mathbf{x}_n \mathbf{x}_n^H \approx \mathbb{E}\mathbf{x}\mathbf{x}^H = \sigma^2 \mathbf{I}_M.$$

In other words, if $N \gg M$, the eigenvalues of $\frac{1}{N}\mathbf{X}\mathbf{X}^H$ are concentrated around σ^2.

On the other hand, let us consider the practical case when M and N are of the same order of magnitude. As

$$M, N \to +\infty \text{ such that } \frac{M}{N} = c \in [a, b], a > 0, b < +\infty, \tag{5.60}$$

it follows that

$$\hat{\mathbf{R}}_{ij} = \sigma^2 \delta_{i-j}$$

but

$$\|\hat{\mathbf{R}}_{ij} - \sigma^2 \mathbf{I}_M\|$$

does not converge toward zero. Here $\| \cdot \|$ denotes the norm of a matrix. It is remarkable to find (by Marchenko and Pastur [251]) that the histograms of the eigenvalues of $\hat{\mathbf{R}}$ tend to concentrate around the probability density of the so-called Marchenko-Pastur distribution

$$p_c(x) = \begin{cases} \frac{1}{2\pi cx}\sqrt{(a-x)(x-b)}, & x \in [a,b] \\ 0, & \text{otherwise} \end{cases} \quad (5.61)$$

with

$$a = \sigma^2(1-\sqrt{c})^2, b = \sigma^2(1+\sqrt{c})^2.$$

(5.61) is still true in the non Gaussian case. One application of (5.61) is to evaluate the asymptotic behavior of linear statistics

$$\frac{1}{M}\sum_{k=1}^{M} f(\lambda_k) = \frac{1}{M}\text{Tr}(f(\hat{\mathbf{R}})) \approx \int f(x)p_c(x)\,dx, \quad (5.62)$$

where $f(x)$ is an arbitrary continuous function. The use of (5.61) allows many problems to be treated in closed forms. To illustrate, let us consider several examples:

1. $f(x) = \frac{1}{\rho^2+x}$. Using (5.62), it follows that

$$\frac{1}{M}\text{Tr}(\hat{\mathbf{R}} + \sigma^2\mathbf{I}_M)^{-1} \approx \int \frac{1}{\rho^2+x}p_c(x)\,dx = m_N(-\rho^2),$$

where $m_N(-\rho^2)$ is a unique positive solution of the equation

$$m_N(-\rho^2) = \frac{1}{\rho^2 + \frac{\sigma^2}{1+c\sigma^2 m_N(-\rho^2)}}.$$

2. $f(x) = \log(1 + \frac{x}{\rho^2})$. Using (5.62), it is found that the expression

$$\frac{1}{M}\log\det\left(\mathbf{I}_M + \frac{1}{\rho^2}\hat{\mathbf{R}}\right)$$

is nearly equal to

$$\frac{1}{c}\log(1 + c\sigma^2 m_N(-\rho^2)) + \log\left(1 + c\sigma^2 m_N(-\rho^2) + (1-c)\frac{\sigma^2}{\rho^2}\right)$$
$$-\rho^2\sigma^2 m_N(-\rho^2)\left(cm_N(-\rho^2) + \frac{1-c}{\rho^2}\right). \quad (5.63)$$

The fluctuations of the linear statistics (5.62) can be cast into closed forms. The bias of the linear estimator is

$$\mathbb{E}\left[\frac{1}{M}\text{Tr}(f(\hat{\mathbf{R}}))\right] = \int f(x)p_c(x)\,dx + \mathcal{O}\left(\frac{1}{M^2}\right).$$

The variance of the linear estimator is

$$M\left[\frac{1}{M}\text{Tr}(f(\hat{\mathbf{R}})) - \int f(x)p_c(x)\,dx\right] \to \mathcal{N}(0, \Delta^2),$$

where Δ^2 is the variance and $\mathcal{N}$ denotes the normal Gaussian distribution. In other words,

$$\frac{1}{M}\text{Tr}(f(\hat{\mathbf{R}})) - \int f(x)p_c(x)\,dx \approx \mathcal{N}\left(0, \frac{\Delta^2}{M^2}\right).$$

5.5.2 Stieltjes Transform

We here follow [329]. Let us consider

$$\mathbf{W}\mathbf{W}^H + \sigma^2 \mathbf{I},$$

where $\sqrt{N}\mathbf{W}$ is an $N \times K$ matrix with i.i.d. entries with zero mean and variance one, for

$$K/N \to \alpha \text{ as } N \to \infty.$$

Denote $\mathbf{A} = \sigma^2 \mathbf{I}$ and $\mathbf{D} = \mathbf{I}_{K,K}$. For this case, we have

$$d\mathbb{A}(x) = \delta(x - \sigma^2)$$
$$d\mathbb{D}(x) = \delta(x - 1).$$

Applying Theorem 5.25, it follows that

$$
\begin{aligned}
G(z) &= G_{\sigma^2 \mathbf{I}}\left(z - \alpha \int \frac{\tau \delta(\tau-1)}{1+\tau G(z)} d\tau\right) \\
&= G_{\sigma^2 \mathbf{I}}(z - \frac{\alpha}{1+G(z)}) \\
&= \int \frac{\delta(\sigma^2-x)}{x-z+\frac{\alpha}{1+G(z)}} dx \\
&= \frac{1}{\sigma^2-z+\frac{\alpha}{1+G(z)}}.
\end{aligned}
\tag{5.64}
$$

$G_{\sigma^2 \mathbf{I}}(z)$ is the Cauchy transform of the eigenvalue distribution of matrix $\sigma^2 \mathbf{I}$. The solution of (5.64) gives

$$G(z) = \frac{1-\alpha}{2(\sigma^2 - z)} - \frac{1}{2} \pm \frac{1}{2(\sigma^2 - z)}\sqrt{(\sigma^2 - z + \alpha - 1)^2 + 4(\sigma^2 - z)}.$$

The asymptotic eigenvalue distribution is given by

$$
f(x) = \begin{cases}
(1-\alpha)^+ \delta(x) & \text{if } \sigma^2 + (\sqrt{\alpha} - 1)^2 \\
\quad + \frac{\alpha}{\pi(x-\sigma^2)}\sqrt{x - \sigma^2 - \frac{1}{4}(x - \sigma^2 + 1 - \alpha)} & \quad \leq x \leq \sigma^2 + (\sqrt{\alpha} + 1)^2 \\
0 & \text{otherwise}
\end{cases}
$$

where $\delta(x)$ is a unit mass at 0 and $[z]^+ = \max(0, z)$.

Another example is the standard vector-input, vector-output (VIVO) model[3]

$$\mathbf{y} = \mathbf{H}\mathbf{x} + \mathbf{n},\tag{5.65}$$

where $\mathbf{x}$ and $\mathbf{y}$ are, respectively, input and output vectors, and $\mathbf{H}$ and $\mathbf{n}$ are channel transfer function and additive white Gaussian noise with zero mean and variance σ^2. Here $\mathbf{H}$ is a random matrix. (5.65) covers a number of systems including CDMA, OFDM, MIMO, cooperative spectrum sensing and sensor network. The mutual information between the

---

[3] MIMO has a special meaning in the context of wireless communications. This informal name VIVO captures our perception of the problem. Vector nature is fundamental. Vector space is the fundamental mathematical space for us to optimize the system.

input vector $\mathbf{x}$ and the output vector $\mathbf{y}$ is a standard result in information theory

$$C = \frac{1}{N} I(\mathbf{x}; \mathbf{y}) = \frac{1}{N} \log \det (\mathbf{I} + \mathbf{HH}^H)$$

$$= \frac{1}{N} \sum_{i=1}^{N} \log \left(1 + \frac{1}{\sigma^2} \lambda_i (\mathbf{HH}^H)\right)$$

$$= \int \log \left(1 + \frac{1}{\sigma^2} \lambda\right) \frac{1}{N} \sum_{i=1}^{N} \delta(\lambda - \lambda_i(\mathbf{HH}^H)) d\lambda$$

$$= \int \log \left(1 + \frac{1}{\sigma^2} \lambda\right) F_{\mathbf{HH}^H}(\lambda) \, d\lambda.$$

Differentiating C with respect to σ^2 yields

$$\frac{1}{N} \frac{\partial C}{\partial \sigma^2} = \int \log \left(\frac{-\frac{1}{\sigma^4}\lambda}{1+\frac{1}{\sigma^2}\lambda}\right) F_{\mathbf{HH}^H}(\lambda) \, d\lambda$$

$$= -\frac{1}{\sigma^2} \int \log \left(\frac{1+\frac{1}{\sigma^2}\lambda - 1}{1+\frac{1}{\sigma^2}\lambda}\right) F_{\mathbf{HH}^H}(\lambda) \, d\lambda$$

$$= -\frac{1}{\sigma^2} + \int \log \left(1 + \frac{1}{\sigma^2}\lambda\right) F_{\mathbf{HH}^H}(\lambda) \, d\lambda$$

$$= -\frac{1}{\sigma^2} + m_{\mathbf{HH}^H}(-\sigma^2).$$

It is interesting to note that we get the closed form in terms of the Stieltjes transform.

5.5.3 Free Deconvolution

We follow the definitions and notations of the example shown in Section 5.5.1. For more details, we see [12, 329, 330]. For a number of N vector observations of $\mathbf{x}_i$, $i = 1, \ldots, N$, the sample covariance matrix is defined as

$$\hat{\mathbf{R}} = \frac{1}{N} \sum_{n=1}^{N} \mathbf{x}_n \mathbf{x}_n^H$$
$$= \mathbf{R}^{1/2} \mathbf{WW}^H \mathbf{R}^{1/2}. \tag{5.66}$$

Here, $\mathbf{W}$ is an $M \times N$ matrix consisting of i.i.d. zero-mean, Gaussian vectors of variance $1/N$. The main advantage of free deconvolution techniques is that asymptotic "kick-in" at a much earlier stage than other techniques available up to now [329]. Often, we know the values of $\mathbf{R}$ which are the theoretical values. We would like to find $\hat{\mathbf{R}}$. If we know the behavior of the matrix $\mathbf{WW}^H$, with the aid of (5.66), $\hat{\mathbf{R}}$ can be obtained. Thus, our problem of finding $\hat{\mathbf{R}}$ is reduced to understand $\mathbf{WW}^H$. Fortunately, the limiting distribution of the eigenvalues of $\mathbf{WW}^H$ is well-known to be the Marchenko-Pastur law.

Due to our invariance assumption on one of the matrices (here $\mathbf{WW}^H$), the eigenvector structure does not matter. The result enables us to compute the eigenvalues of $\mathbf{R}$, by knowing only the eigenvalues of $\hat{\mathbf{R}}$. The invariance assumption "frees," in some sense, one matrix from the other by "disconnecting" their eigenspaces.

5.5.4 Optimal Precoding of MIMO Systems

Given M receive antennas and N transmit antennas, the standard vector channel model is

$$\mathbf{y} = \mathbf{Hx} + \mathbf{n}, \tag{5.67}$$

where the complex entries of $\mathbf{H}$ of $M \times N$ are the MIMO channel gains and $\mathbf{H}$ is a nonobservable Gaussian random matrix with known (or well estimated) second order statistics. Here, $\mathbf{x}$ is the transmitted signal vector and $\mathbf{n}$ is the additive Gaussian noise vector at the receiver with $\mathbb{E}\mathbf{n}\mathbf{n}^H = \rho^2 \mathbf{I}_M$.

The optimum precoding problem is to find the covariance matrix $\mathbf{Q}$ of $\mathbf{x}$ in order to maximize some figure of merit of the system. For example, the optimization problem can be expressed as

$$\text{Maximize } I(\mathbf{Q}) = E\left[\log \det \left(\mathbf{I}_M + \tfrac{1}{\rho^2}\mathbf{H}\mathbf{Q}\mathbf{H}^H\right)\right]$$
$$\text{Subject to } \mathbf{Q} \geq 0; \tfrac{1}{M}\text{Tr}(\mathbf{Q}) \leq 1.$$

A possible alternative is to maximize a large system approximation of $I(\mathbf{Q})$. Closed-form expressions (5.63) can be used [331]. For more details, see [279].

5.5.5 Marchenko and Pastur's Probability Distribution

We follow [279] for this development. The Stieltjes transform is one of the numerous transforms associated with a measure. It is well suited to study large random matrices and was first introduced in this context by Marchenko and Pastur [251]. The Stieltjes transform is defined in (5.34).

Consider

$$\mathbf{W}_N = \frac{1}{N}\mathbf{V}_N \tag{5.68}$$

where $\mathbf{V}_N$ is a $M \times N$ matrix with i.i.d. complex Gaussian random variables $\mathcal{CN}(0, \sigma_2)$. Our aim is the limiting spectral distribution of $\mathbf{X} = \mathbf{W}_N \mathbf{W}_N^{\ H}$. Consider the associated resolvent and its Stieltjes transform

$$\mathbf{Q}(z) = (\mathbf{X} - z\mathbf{I})^{-1}, \hat{m}_N(z) = \frac{1}{M}\text{Tr}\mathbf{Q}(z) = \frac{1}{M}\text{Tr}(\mathbf{X} - z\mathbf{I})^{-1}. \tag{5.69}$$

The main assumption is: The ratio $c_N = \frac{M}{N}$ is bounded away from zero and upper bounded, as $M, N \to \infty$.

The approach is sketched here. First one derives the equation that is satisfied by the Stieltjes transform of the limiting spectral distribution $\hat{m}_N(z)$ defined in (5.69). Afterwards, one relies on the inverse formula (see (5.35)) of Stieltjes transform, to obtain the so-called Marchenko-Pastur distribution.

There are three main steps:

1. To prove that $\text{var}(\hat{m}_N(z)) = \mathcal{O}(N^{-2})$. This enables us to replace $\hat{m}_N(z)$ by its expectation $\mathbb{E}\hat{m}_N(z)$ in the derivation.
2. To establish the limiting equation satisfied by $\mathbb{E}\hat{m}_N(z)$.
3. To recover the probability distribution, with the help of the inverse formula of Stieltjes transform (5.35).

The Stieltjes transform in this work of large random matrices plays a role analogous to the Fourier transform in a linear, time-invariant (LTI) system.

5.5.6 Convergence and Fluctuations Extreme Eigenvalues

We here follow [279] for presentation. Consider the $\mathbf{WW}^H$ defined in (5.68). Denote by

$$\hat{\lambda}_{1,N} \geq \hat{\lambda}_{2,N} \geq \cdots \geq \hat{\lambda}_{N,N}$$

the ordered eigenvalues of $\mathbf{WW}^H$. The support of Marchenko-Pastur distribution is

$$(\sigma^2(1 - \sqrt{c_N})^2, \sigma^2(1 + \sqrt{c_N})^2).$$

One theorem is: If $c_N \to c_*$, we have

$$\hat{\lambda}_{1,N} \xrightarrow[N,M\to\infty]{a.s.} \sigma^2(1 + \sqrt{c_N})^2,$$

$$\hat{\lambda}_{N,N} \xrightarrow[N,M\to\infty]{a.s.} \sigma^2(1 - \sqrt{c_N})^2,$$

where "a.s." denotes "almost surely." The ratio of two limit expressions is used for spectrum sensing in Example 5.4.

A central limit theorem holds for the largest eigenvalue of matrix $\mathbf{WW}^H$, as $M, N \to \infty$. The limiting distribution is known as Tracy-Widom Law's distribution (see (5.50)) for fluctuations of $\hat{\lambda}_{1,N}$.

The function $F_{TW2}(s)$ stands for Tracy-Widom culmination distribution function. MAT-LAB codes are available to compute [332].

Let us $c_N \to c_*$. When corrected centered and rescaled, $\hat{\lambda}_{1,N}$ converges to a Tracy-Widom distribution:

$$\frac{N^{2/3}}{\sigma^2} \times \frac{\hat{\lambda}_{1,N} - \sigma^2(1 + \sqrt{c_N})^2}{(1 + \sqrt{c_N})\left(\frac{1}{\sqrt{c_N}} + 1\right)^{1/3}} \xrightarrow[N,M\to\infty]{L} F_{TW2}.$$

5.5.7 Information plus Noise Model and Spiked Models

We refer to [263, 279, 299, 300, 302, 333–335] for more details. The observed M-dimensional time series $\mathbf{y}_n$ for the n-vector sample are expressed as

$$\mathbf{y}_n = \sum_{k=1}^{K} \mathbf{a}_k s_{k,n} + \mathbf{v}_n = \mathbf{A}\mathbf{s}_n + \mathbf{v}_n, n = 1, \ldots, N$$

with

$$\mathbf{s}_n = (s_{1,n}, \ldots, s_{K,n})^T, \mathbf{A} = (\mathbf{a}_1, \ldots, \mathbf{a}_N),$$

where $\mathbf{s}_n$ collects $K < M$ nonobservable "source signals," the matrix $\mathbf{A}$ is deterministic with an unknown rank $K < M$, and $(\mathbf{v}_n)_{n \in \mathbb{Z}}$ is additive white Gaussian noise such that $\mathbb{E}(\mathbf{v}_n \mathbf{v}_n^H) = \sigma^2 \mathbf{I}_M$. Here $\mathbb{Z}$ denotes the set of all integers.

In matrix form, we have $\mathbf{Y}_N = (\mathbf{y}_1, \ldots, \mathbf{y}_N)^T$, observation matrix of $M \times N$. We do this for $\mathbf{S}_N$ and $\mathbf{V}_N$. Then,

$$\mathbf{Y}_N = \mathbf{A}\mathbf{S}_N + \mathbf{V}_N.$$

Using the normalized matrices

$$\boldsymbol{\Sigma}_N = \frac{1}{\sqrt{N}}\mathbf{Y}_N, \mathbf{B}_N = \frac{1}{\sqrt{N}}\mathbf{AS}_N, \mathbf{W}_N = \frac{1}{\sqrt{N}}\mathbf{V}_N,$$

we obtain

$$\boldsymbol{\Sigma}_N = \mathbf{B}_N + \mathbf{W}_N. \tag{5.70}$$

Detection of the presence of signal(s) from matrix $\boldsymbol{\Sigma}_N$ is to tell whether $K = 1$ versus $K = 0$ (noise only) to simplify. Since K does not scale with M, that is, $K \ll M$, a spiked model is reached.

We assume that the number of sources K is $K \gg N$. (5.39) is a model of

$$\boldsymbol{\Sigma}_N = \text{Matrix with Gaussian iid elements} + \text{fixed rank perturbation.}$$

The asymptotic regime is defined as

$$N \rightarrow \infty, M/N \rightarrow c_*, \text{ and } K \text{ is fixed.}$$

Let us further assume that $\mathbf{S}_N$ is a random matrix with independent $\mathcal{CN}(0, 1)$ elements (Gaussian iid source signals), and $\mathbf{A}_N$ is deterministic. It follows that

$$\boldsymbol{\Sigma}_N = (\mathbf{A}_N\mathbf{A}_N^H + \sigma^2\mathbf{I}_M)^{1/2}\mathbf{X}_N$$

where $\mathbf{B}_N$ is $M \times N$ with independent $\mathcal{CN}(0, 1)$ elements.

Consider a spectral factorization of $\mathbf{A}_N\mathbf{A}_N^H$

$$\mathbf{A}_N\mathbf{A}_N^H = \mathbf{U}_N \begin{pmatrix} \lambda_1 & & & & 0 \\ & \ddots & & & \\ & & \lambda_K & & \\ 0 & & & & 0 \end{pmatrix} \mathbf{U}_N^H.$$

Let $\mathbf{P}_N$ be the $M \times M$ matrix

$$\mathbf{P}_N = \text{diag}\left(\sqrt{\frac{\lambda_1 + \sigma^2}{\sigma^2}}, \sqrt{\frac{\lambda_2 + \sigma^2}{\sigma^2}}, \cdots, \sqrt{\frac{\lambda_K + \sigma^2}{\sigma^2}}, 1, \cdots, 1 \right).$$

Then

$$\mathbf{U}_N^H\boldsymbol{\Sigma}_N = \sigma\mathbf{P}_N\mathbf{U}_N^H\mathbf{X}_N \overset{\mathcal{D}}{=} \mathbf{P}_N\mathbf{W}_N$$

where $\mathbf{W}_N$ is $M \times N$ with independent $\mathcal{CN}(0, 1)$ elements and $\mathcal{D}$ denotes weak convergence. Since $\mathbf{P}_N$ is a fixed rank perturbation of identity, we reach the so-called multiplicative spike model

$$\text{eigenvalues of } \boldsymbol{\Sigma}_N\boldsymbol{\Sigma}_N^H \equiv \text{eigenvalues of } \mathbf{P}_N\mathbf{W}_N\mathbf{W}_N^H\mathbf{P}_N^H.$$

Similarly, we can define the additive spike model. Let us assume that $\mathbf{S}_N$ is a deterministic matrix and

$$\mathbf{B}_N = N^{-1/2}\mathbf{A}_N\mathbf{S}_N$$

is such that

$$\text{rank } (\mathbf{B}_N) = K \text{ (fixed)} .$$

The additive spike model is defined as

$$\boldsymbol{\Sigma}_N = \mathbf{B}_N + \mathbf{W}_N.$$

A natural question arises: What is the impact of $\tilde{\mathbf{B}}_N$ on the spectrum of $\boldsymbol{\Sigma}_N \boldsymbol{\Sigma}_N{}^H$ in the asymptotic regime?

Let $\tilde{F}_N$ and F_N be the distribution functions of the spectral measures of $\boldsymbol{\Sigma}_N \boldsymbol{\Sigma}_N{}^H$ and $\mathbf{W}_N \mathbf{W}_N^H$, respectively. Then

$$\sup_x |\tilde{F}_N - F_N| \le \frac{1}{M} \text{ rank } (\boldsymbol{\Sigma}_N \boldsymbol{\Sigma}_N{}^H - \mathbf{W}_N \mathbf{W}_N^H) \underset{N \to \infty}{\to} 0.$$

Thus $\boldsymbol{\Sigma}_N \boldsymbol{\Sigma}_N{}^H$ and $\mathbf{W}_N \mathbf{W}_N^H$ have identical (Marchenko-Pastur) limit spectral measure, either for the multiplicative or the additive spike model.

We use our measured data to verify the Marchenko-Pastur law. There are five USRP platforms serving as sensor nodes. The data acquired from one USRP platform are segmented into twenty data blocks. All these data blocks are used to build large random matrices. In this way, we emulate the network with 100 sensor nodes. If there is no signal, the spectral distribution of noise sample covariance matrix is shown in Figure 5.1(a) which follows the Marchenko-Pastur law in (5.3). When signal exists, the spectral distribution of sample covariance matrix of signal plus noise is show in Figure 5.1(b). The experimental results well agree with the theory. The support of the eigenvalues is finite. The theoretical prediction offered by the Marchenko-Pastur law can be used to set the threshold for detection.

Main results on the eigenvalues can be summarized into the theorem [279].

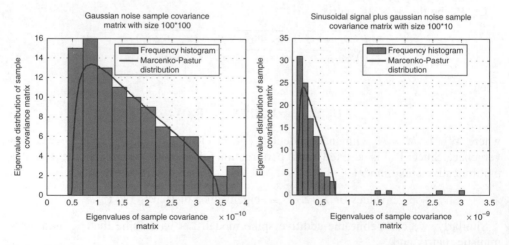

Figure 5.1 Spectral distribution. (a) Spectral distribution of noise sample covariance matrix; (b) Spectral distribution of sample covariance matrix of signal plus noise.

Theorem 5.43 (Main result on the eigenvalues) *The additive spike model is*

$$\mathbf{\Sigma}_N = \mathbf{B}_N + \mathbf{W}_N,$$

where $\mathbf{B}_N$ is a deterministic rank-K matrix such that

$$\lambda_{k,N} \to \rho_k$$

for $k = 1, \ldots, K$, and $\mathbf{W}_N$ is a $M \times N$ random matrix with independent $\mathcal{CN}(0, \sigma^2/N)$ elements. Let $i \le K$ be the maximum index for which $\rho_i > \sigma^2 \sqrt{c_*}$. Then, for $k = 1, \ldots, i$,

$$\lambda_{k,N} \underset{N \to \infty}{\overset{a.s.}{\to}} \gamma_k = \frac{(\sigma^2 c_* + \rho_k)(\sigma^2 + \rho_k)}{\rho_k} > \sigma^2 (1 + \sqrt{c_*})^2, \mathcal{H}_1$$
$$\lambda_{i+1,N} \underset{N \to \infty}{\overset{a.s.}{\to}} \sigma^2 (1 + \sqrt{c_*})^2, \mathcal{H}_0.$$

where $\mathcal{H}_1$ denotes the presence of signal(s) while $\mathcal{H}_0$ the absence of signal(s).

5.5.8   Hypothesis Testing and Spectrum Sensing

This example is continued from the example shown in Section 5.5.7. For more details, we see [263, 279, 299, 300, 302, 333–335]. One motivation is to exploit the asymptotic limiting distribution for spectrum sensing.

The hypothesis test is formulated as

$$\begin{aligned}\mathcal{H}_1 &: \mathbf{\Sigma}_N = \mathbf{B}_N + \mathbf{W}_N \quad \text{(noise)}\\ \mathcal{H}_0 &: \mathbf{\Sigma}_N = \mathbf{W}_N \quad \text{(Information + noise)}.\end{aligned}$$

Assume further $K = 1$ source for convenience.

$$\mathbf{B}_N = N^{-1/2} \mathbf{a}_{1,N} \mathbf{s}_{1,N},$$

is a rank one matrix such that

$$\|\mathbf{B}_N\|^2 \underset{N \to \infty}{\to} \rho > 0.$$

The GLRT is

$$T_N = \frac{\lambda_{1,N}}{M^{-1} \text{Tr}(\mathbf{\Sigma}_N \mathbf{\Sigma}_N{}^H)}. \tag{5.71}$$

The natural question is: what is the asymptotic performance of T_N under the assumption of large random matrices?

Under $\mathcal{H}_0$ and $\mathcal{H}_1$, we have

$$M^{-1} \mathbf{\Sigma}_N \mathbf{\Sigma}_N{}^H \underset{N \to \infty}{\overset{a.s.}{\to}} \sigma^2.$$

As a consequence of Theorem 5.43, under $\mathcal{H}_1$, if $\rho > \sigma^2 \sqrt{c_*}$, then

$$\lambda_{1,N} \underset{N \to \infty}{\overset{a.s.}{\to}} \gamma_1 = \frac{(\sigma^2 c_* + \rho)(\sigma^2 + \rho)}{\rho} > \sigma^2 (1 + \sqrt{c_*})^2,$$
$$\lambda_{2,N} \underset{N \to \infty}{\overset{a.s.}{\to}} \sigma^2 (1 + \sqrt{c_*})^2.$$

If $\rho \le \sigma^2 \sqrt{c_*}$, then

$$\lambda_{1,N} \xrightarrow[N\to\infty]{a.s.} \sigma^2(1 + \sqrt{c_*})^2.$$

Using the above result in (5.71), under $\mathcal{H}_0$, we have

$$T_N \xrightarrow[N\to\infty]{a.s.} (1 + \sqrt{c_*})^2.$$

Under $\mathcal{H}_1$, if $\rho > \sigma^2 \sqrt{c_*}$, we have

$$T_N \xrightarrow[N\to\infty]{a.s.} \frac{(\sigma^2 c_* + \rho)(\sigma^2 + \rho)}{\sigma^2 \rho} > (1 + \sqrt{c_*})^2,$$

If $\rho \le \sigma^2 \sqrt{c_*}$, we have

$$T_N \xrightarrow[N\to\infty]{a.s.} (1 + \sqrt{c_*})^2.$$

Recall that

$$c_N = \frac{M}{N} \to c_*.$$

The limit of detectability by the GLRT is given by

$$\rho > \sigma^2 \sqrt{c_*}.$$

Defining $SNR = \frac{\rho}{\sigma^2}$, we have

$$SNR > \sqrt{c_*}.$$

For extremely low SNR, it follows that c_* must be very small, implying

$$N \gg M.$$

With the help of the Tracy-Widom law, false alarm probability can be evaluated and linked with the decision threshold T_N.

For finite, low rank perturbation of large random matrices, the eigenvalues and eigenvectors are studied in [335].

Example 5.8 (Dozier and Silverstein [263, 279, 300])
According to Dozeir and Silverstein [263, 279, 300] it exists a deterministic probability measure μ_N by $\mathbb{R}^+$ such that

$$\frac{1}{M} \sum_{k=1}^{M} \delta(\lambda - \lambda_{k,N}) - \mu_N \to 0 \text{ weakly almost surely.}$$

Consider the additive spike model (5.70) repeated here for convenience

$$\Sigma_N = \mathbf{B}_N + \mathbf{W}_N. \tag{5.72}$$

The approach to characterize μ_N is sketched here: The Stieltjes transform of μ_N is defined on $\mathbb{C} - \mathbb{R}^+$ as

$$m_N(z) = \int_+ \frac{1}{\lambda - z} \mu_N(d\lambda),$$

$$m_N(z) = \frac{1}{M} \mathrm{Tr} \mathbf{T}_N(z)$$

with

$$\mathbf{T}_N(z) = \left(\frac{\mathbf{B}_N \mathbf{B}_N^H}{1 + \sigma^2 c_N m_N(z)} - z(1 + \sigma^2 c_N m_N(z)) \mathbf{I}_M + \sigma^2 (1 - c_N) \mathbf{I}_M \right)^{-1}. \qquad \square$$

5.5.9 Energy Estimation in a Wireless Network

Consider a wireless (primary) network [330] in which K entities are transmitted data simultaneously on the same frequency resource. Transmitter $k \in (1, \ldots, K)$ has transmitted power P_k and is equipped with n_k antennas. We denote

$$n = \sum_{k=1}^{K} n_k$$

the total number of transmit antennas of the primary network.

Consider a secondary network composed of a total of N, $N \geq n$, sensing devices: they may be N single antennas devices or multiple devices embedded with multiple antennas whose sum is equal to N. The N sensors are collectively called the receiver. To ensure that every sensor in the second network roughly captures the same amount of energy from a given transmitter, it is assumed that the respective transmitter-sensor distances are alike. This is realistic assumption for anb in-house femtocell network.

Denote $\mathbf{H}_k \in \mathbb{C}^{N \times n_k}$ the multiple antenna channel matrix between transmitter k and the receiver. We assume that the entries of $\sqrt{N} \mathbf{H}_k$ are independent and identically distributed (i.i.d.), with zero mean, unit variance, and finite fourth-order moment.

At time instant m, transmitter k emits the multi-antenna signal vector $\mathbf{x}_k^{(m)} \in \mathbb{C}^{n_k}$, whose entries are assumed to be i.i.d., with zero mean, unit variance, and finite fourth-order moment.

Further, we assume that at time instant m, the received signal vector is impaired by an additive white Gaussian noise (AWGN) vector, denoted $\sigma \mathbf{w}^{(m)} \in \mathbb{C}^N$, whose entries are assumed to be i.i.d., with zero mean, variance σ_2, and finite fourth-order moment on every sensor. The entries of $\sigma \mathbf{w}_k^{(m)}$ have unit variance.

At time m, the receiver senses the signal $\mathbf{y}^{(m)} \in \mathbb{C}^N$ defined as

$$\mathbf{y}^{(m)} = \sum_{k=1}^{K} \sqrt{P_k} \mathbf{H}_k \mathbf{x}_k^{(m)} + \sigma \mathbf{w}_k^{(m)}.$$

It is assumed that at least M consecutive sampling periods, the channel fading coefficients are constant. We concatenate M successive signal realizations into

$$\mathbf{Y} = [\mathbf{y}^{(1)}, \ldots, \mathbf{y}^{(M)}] \in \mathbb{C}^{N \times M},$$

we have

$$\text{with } \mathbf{X}_k = [\mathbf{x}^{(1)}, \ldots, \mathbf{x}^{(M)}] \in \mathbb{C}^{n_k \times M}, \mathbf{W}_k = [\mathbf{w}^{(1)}, \ldots, \mathbf{w}^{(M)}] \in \mathbb{C}^{N \times M},$$

for every k. This can be further recast into the final form

$$\mathbf{Y} = \mathbf{H}\mathbf{P}^{\frac{1}{2}}\mathbf{X} + \sigma\mathbf{W} \tag{5.73}$$

where $\mathbf{P} \in \mathbb{R}^{n \times n}$ is diagonal with first n_1 entries P_1, subsequent n_2 entries P_2, ..., last n_K entries P_K,

$$\mathbf{H} = [\mathbf{H}_1, \ldots, \mathbf{H}_K], \text{ and } \mathbf{X} = [\mathbf{X}_1^T, \ldots, \mathbf{X}_K^T] \in \mathbb{C}^{n \times M}.$$

By convention, it is assumed that

$$P_1 \leq \cdots \leq P_K.$$

$\mathbf{H}, \mathbf{W}$ and $\mathbf{X}$ have independent entries of finite fourth-order moment. The entries of $\mathbf{X}$ need not be identically distributed, but may originate from a maximum of K distinct distributions.

Our objective is to infer the values of $P_1, \cdots, P_K$ from the realization of the random matrix $\mathbf{Y}$. The problem at hand is to exploit the eigenvalue distribution of $\frac{1}{M}\mathbf{Y}\mathbf{Y}^H$ as N, n and M grow large at the same rate.

Theorem 5.44 (Stieltjes transform of $\frac{1}{M}\mathbf{Y}\mathbf{Y}^H$) *Let*

$$\mathbf{B}_N = \frac{1}{M}\mathbf{Y}\mathbf{Y}^H,$$

where $\mathbf{Y}$ is defined in (5.73). Then, for M, N, n growing large with limit ratios

$$M, N, n \to \infty, \frac{M}{N} \to c, \frac{N}{n_k} \to c_k, 0 < c, c_1, \ldots, c_K < \infty,$$

the eigenvalue distribution function $F_{\mathbf{B}_N}$ of $\mathbf{B}_N$, referred to as the empirical spectral function (e.s.d.) of $\mathbf{B}_N$, converges almost surely to the deterministic distribution function F, referred to as the limit spectral function (l.s.d.) of $\mathbf{B}_N$, whose Stieltjes transform $m_F(z)$ satisfies, for $z \in \mathbb{C}^+$

$$m_F(z) = cm_{\underline{F}}(z) + (c - 1)\frac{1}{z}$$

where $m_{\underline{F}}(z)$ is the unique solution with positive imaginary part of the implicit equation in $m_{\underline{F}}$

$$\frac{1}{m_{\underline{F}}} = -\sigma^2 + \frac{1}{f} - \sum_{k=1}^{K} \frac{1}{c_k}\frac{P_k}{1 + P_K f}$$

in which we denote f the value

$$f = (1 - c)m_{\underline{F}} - czm_{\underline{F}}^2.$$

For Assumption 5.3 and Assumption 5.4—too long to be covered in this context—that are used in the following theorem, we refer to [330].

5.5.10 Multisource Power Inference

Let $\mathbf{B}_N \in \mathbb{C}^{N \times N}$ be defined as in Theorem 5.45, and

$$\boldsymbol{\lambda} = (\lambda_1, \ldots, \lambda_N), \lambda_1 \leq \cdots \leq \lambda_N,$$

be the vector of the ordered eigenvalues of $\mathbf{B}_N$. Further assume that the limiting ratios $c, c_1, \ldots, c_K$ and $\mathbf{P}$ are such that Assumptions 5.3 and 5.4 are fulfilled for some $k \in \{1, \ldots, K\}$. Then, as N, n, M grow large, we have $\hat{P}_k - P_k \xrightarrow{a.s.} 0$ where the estimates $\hat{P}_k$ is given by

- if $M \neq N$,

$$\hat{P}_k = \frac{NM}{n_k(M - N)} \sum_{i \in \mathcal{N}_k} (\eta_i - \mu_i)$$

- if $M = N$,

$$\hat{P}_k = \frac{N}{n_k(M - N)} \sum_{i \in \mathcal{N}_k} \left(\sum_{j=1}^{N} \frac{\eta_i}{(\lambda_j - \eta_i)^2} \right)^{-1}$$

in which

$$\mathcal{N}_k = \left\{ \sum_{i=1}^{k-1} n_i + 1, \ldots, \sum_{i=1}^{k} n_i \right\},$$

$(\eta_1, \ldots, \eta_N)$ are ordered eigenvalues of the matrix $\mathrm{diag}(\boldsymbol{\lambda}) - \frac{1}{N}\sqrt{\boldsymbol{\lambda}}\sqrt{\boldsymbol{\lambda}}$ and $(\mu_1, \ldots, \mu_N)$ are the ordered eigenvalues of the matrix $\mathrm{diag}(\boldsymbol{\lambda}) - \frac{1}{M}\sqrt{\boldsymbol{\lambda}}\sqrt{\boldsymbol{\lambda}}$.

A blind multisource power estimation has been derived in [330]. Under the assumptions that the ratio between the number of sensors and the number of signals are not too small, and the source transmit powers are sufficiently distinct from one another, they derive a method to infer the individual source powers if the number of sources are known. This novel method outperforms the alternative estimation techniques in the medium to high SNR regime. This method is robust to small system dimensions. As such, it is particularly suited to the blind detection of primary mobile users in future cognitive radio networks.

5.5.11 Target Detection, Localization, and Reconstruction

We follow [336] for this development. A point reflector can model a small dielectric anomaly in electromagnetism; a small density anomaly in acoustics, or more generally, a local variation of the index of refraction in the scalar wave equation. The contrast of the anomaly can be of order one but its volume is small compared to the wavelength. In such a situation, it is possible to expand the solution of the wave equation around the background solution.

Consider the scalar wave equation in a d-dimensional homogeneous medium with the index of refraction n_0. The reference speed of propagation is denoted by c. It is assumed that the target is a small reflector of inclusion D with the index of refraction $n_{ref} \neq n_0$. The support of the inclusion is of the form $D = \mathbf{x}_{ref} + B$, where B is a domain with small volume. Thus the scalar wave equation with the source $S(t, x)$ takes the form

$$\frac{n^2(\mathbf{x})}{c^2}\partial_t^2 E - \Delta_x E = S(t, \mathbf{x}),$$

where the index of refraction is given by

$$n(\mathbf{x}) = n_0 + (n_{ref} - n_0)\mathbf{1}_D(\mathbf{x}).$$

For any $\mathbf{y}_n, \mathbf{z}_m$ far from $\mathbf{x}_{ref}$ the field $\mathrm{Re}[(\mathbf{y}_n, \mathbf{z}_m)e^{-j\omega t}]$, observed at $\mathbf{y}_n$, when a point source emits a time-harmonic signal with frequency ω at $\mathbf{z}_m$, can be expanded as powers of the volume as

$$\hat{E}(\mathbf{y}_n, \mathbf{z}_m) = \hat{G}(\mathbf{y}_n, \mathbf{z}_m) + k_0^2 \rho_{ref}\hat{G}(\mathbf{y}_n, \mathbf{x}_{ref})\hat{G}(\mathbf{x}_{ref}, \mathbf{y}_n) + \mathcal{O}\left(|B|^{\frac{d+1}{d}}\right),$$

where $k_0 = n_0\omega/c$ is the homogeneous wavenumber, ρ_{ref} is the scattering amplitude

$$\rho_{ref} = \left(\frac{n_{ref}^2}{n_0^2} - 1\right)|B|,$$

and $\hat{G}(\mathbf{y}, \mathbf{z})$ is the Green's function or fundamental solution of the Helmhotz equation with a point source at $\mathbf{z}$:

$$\Delta_x \hat{G}(\mathbf{x}, \mathbf{z}) + k_0^2 \hat{G}(\mathbf{x}, \mathbf{z}) = -\delta(\mathbf{x} - \mathbf{z}).$$

More explicitly, we have

$$\hat{G}(\mathbf{x}, \mathbf{z}) = \begin{cases} \frac{i}{4}H_0^{(1)}(k_0|\mathbf{x} - \mathbf{z}|) & d = 2, \\ \frac{e^{jk_0|\mathbf{x}-\mathbf{z}|}}{4\pi|\mathbf{x}-\mathbf{z}|} & d = 3, \end{cases}$$

where $H_0^{(1)}$ is the Hankel function of the first kind of order zero.

When there are M sources $(\mathbf{z}_m)_{m=1,...,M}$ and N receivers $(\mathbf{y}_n)_{n=1,...,N}$, the *response matrix* is the $N \times M$ matrix

$$\mathbf{H}_0 = (H_{0nm})_{n=1,...,N, m=1,...,M}$$

defined by

$$H_{0nm} = \hat{E}(\mathbf{y}_n, \mathbf{z}_m) - \hat{G}(\mathbf{y}_n, \mathbf{z}_m).$$

This matrix has rank one:

$$\mathbf{H}_0 = \sigma_{ref}\mathbf{u}_{ref}\mathbf{v}_{ref}^H.$$

The nonzero singular value is

$$\sigma_{ref} = k_0^2 \rho_{ref}\left(\sum_{l=1}^{N}|\hat{G}(\mathbf{y}_l, \mathbf{x})|^2\right)^{1/2}\left(\sum_{l=1}^{N}|\hat{G}(\mathbf{z}_l, \mathbf{x})|^2\right)^{1/2}. \tag{5.74}$$

The associated left and right singular vectors $\mathbf{u}_{ref}$ and $\mathbf{v}_{ref}$ are given by

$$\mathbf{u}_{ref} = \mathbf{u}(\mathbf{x}_{ref}),\ \mathbf{u}_{ref} = \mathbf{v}(\mathbf{x}_{ref}),$$

where the normalized vectors of Green's functions are defined as

$$\mathbf{u}(\mathbf{x}) = \left(\frac{\hat{G}(\mathbf{y}_n, \mathbf{x})}{\left(\sum_{l=1}^{N} |\hat{G}(\mathbf{y}_l, \mathbf{x})|^2 \right)^{1/2}} \right)_{n=1,\dots,N}, \quad \mathbf{v}(\mathbf{x}) = \left(\frac{\hat{G}^*(\mathbf{z}_m, \mathbf{x})}{\left(\sum_{l=1}^{M} |\hat{G}(\mathbf{z}_l, \mathbf{x})|^2 \right)^{1/2}} \right)_{m=1,\dots,M},$$

where $*$ denotes the conjugation of the function.

The matrix $\mathbf{H}_0$ is the complete data set that can be collected. In practice, the measured matrix is corrupted by electronic or measurement noise that has the form of an additive noise. The standard acquisition gives

$$\mathbf{H} = \mathbf{H}_0 + \mathbf{W}$$

where the entries of $\mathbf{W}$ are independent complex Gaussian random variables with zero mean and variance σ_n^2 / M. We assume that $N \geq M$.

The detection of a target can be formulated as a standard hypothesis testing problem

$$\mathcal{H}_0 : \mathbf{H} = \mathbf{W}$$
$$\mathcal{H}_1 : \mathbf{H} = \mathbf{H}_0 + \mathbf{W}.$$

Without target $\mathcal{H}_0$, the behavior of $\mathbf{W}$ is has been extensively studied. With target $\mathcal{H}_1$, the singular values of the perturbed random response matrix are of interest. This model is also called the information plus noise model or the spiked population model. The critical regime of practical interest is that the singular values of an unperturbed matrix are of the same order, as the singular values of the noise, that is, σ_{ref} is of the same order of magnitude as σ. Related work is in [24, 25, 308–311], Johnstone [9, 19, 22, 312–318], and Nadler [305].

Proposition 5.7 (The singular values of the perturbed random response matrix [336]) *In the regime $M \to \infty$,*

1. The normalized l^2-norm of the singular values satisfies

$$M \left[\frac{1}{M} \sum_{j=1}^{M} (\sigma_j^{(M)})^2 - \gamma \sigma^2 \right] \overset{M \to \infty}{\underset{D}{\to}} \sigma_{ref}^2 + \sqrt{2} \sigma^2 Z_0,$$

where Z_0 follows a Gaussian distribution with zero mean and variance one and "D" denotes convergence in distribution.

2. If $\sigma_{ref} < \gamma^{1/4} \sigma$, then the maximum singular value satisfies

$$\sigma_1^{(M)} \cong \sigma \left[\gamma^{1/2} + 1 + \frac{1}{2M^{2/3}} (1 + \gamma^{-1/2})^{1/3} Z_2 + o \left(\frac{1}{M^{2/3}} \right) \right] \text{ in distribution,}$$

where Z_2 follows a type-2 Tracy-Widom distribution.

3. If $\sigma_{ref} = \gamma^{1/4}\sigma$, $\sigma_{ref} < \gamma^{1/4}\sigma$, then the maximum singular value satisfies

$$\sigma_1^{(M)} \cong \sigma \left[\gamma^{1/2} + 1 + \frac{1}{2M^{2/3}} (1 + \gamma^{-1/2})^{1/3} Z_3 + o\left(\frac{1}{M^{2/3}}\right) \right] \text{ in distribution,}$$

where Z_2 follows a type-3 Tracy-Widom distribution.

4. If $\sigma_{ref} > \gamma^{1/4}\sigma$, then the maximal singular value has Gaussian distribution with the mean and variance given by

$$\mathbb{E}\,[\sigma_1^{(M)}] = \sigma_{ref} \left[1 + (1 + \gamma)\frac{\sigma^2}{\sigma_{ref}^2} + \gamma\frac{\sigma^4}{\sigma_{ref}^4} + o\left(\frac{1}{M^{1/2}}\right) \right],$$

$$\mathrm{Var}\,[\sigma_1^{(M)}] = \frac{\sigma^2}{2M} \left[\frac{1 - \gamma\frac{\sigma^4}{\sigma_{ref}^4}}{1 + (1 + \gamma)\frac{\sigma^2}{\sigma_{ref}^2} + \gamma\frac{\sigma^4}{\sigma_{ref}^4}} + o(1) \right].$$

The type-3 Tracy-Widom distribution has the cdf $\Phi_{TW3}(z)$ given by

$$\Phi_{TW3}(z) = \exp\left(-\int_z^\infty [\varphi(x) + (x - z)\varphi^2(x)]dx \right).$$

The expectation of Z_3 is $\mathbb{E}[Z_3] = -0.49$ and its variance is $\mathrm{Var}[Z_3] = 1.22$.

The singular eigenvectors of the perturbed response matrix are described in the following proposition. Define the scalar product as

$$\langle \mathbf{u}, \mathbf{v} \rangle = \mathbf{u}^H \mathbf{v}.$$

Proposition 5.8 (The singular vectors of the perturbed random response matrix [336]) *In the regime* $M \to \infty$,

1. If $\sigma_{ref} < \gamma^{1/4}\sigma$, then the angles satisfy

$$|\langle \mathbf{u}_{ref}, \mathbf{u}_1^{(M)} \rangle|^2 = 0 + o(1) \quad \text{in probability,}$$

$$|\langle \mathbf{v}_{ref}, \mathbf{v}_1^{(M)} \rangle|^2 = 0 + o(1) \quad \text{in probability.}$$

2. If $\sigma_{ref} > \gamma^{1/4}\sigma$, then the angles satisfy

$$|\langle \mathbf{u}_{ref}, \mathbf{u}_1^{(M)} \rangle|^2 = \frac{1 - \gamma\frac{\sigma^4}{\sigma_{ref}^4}}{1 + \gamma\frac{\sigma^4}{\sigma_{ref}^4}} + o(1) \quad \text{in probability,}$$

$$|\langle \mathbf{v}_{ref}, \mathbf{v}_1^{(M)} \rangle|^2 = \frac{1 - \gamma\frac{\sigma^4}{\sigma_{ref}^4}}{1 + \gamma\frac{\sigma^4}{\sigma_{ref}^4}} + o(1) \quad \text{in probability.}$$

A standard imaging function for target localiztion is the MUSIC function defined by

$$I_{MUSIC}(\mathbf{x}) = \left\| \mathbf{u}(\mathbf{x}) - ((\mathbf{u}_1^{(M)})^H \mathbf{u}(\mathbf{x})) \mathbf{u}_1^{(M)} \right\|^{-1/2} = \left(1 - |\mathbf{u}^H(\mathbf{x}) \mathbf{u}_1^{(M)}|^2 \right)^{-1/2},$$

where $\mathbf{u}(\mathbf{x})$ is the normalized vector of Green's function. It is a nonlinear function of a weighted subspace migration functional

$$I_{SM}(\mathbf{x}) = 1 - I_{MUSIC}(\mathbf{x})^{-2} |\mathbf{u}^H(\mathbf{x}) \mathbf{u}_1^{(M)}|^2.$$

The reconstruction can be formulated in this context. Using Proposition 5.7, we can see that the quantity

$$\hat{\sigma}_{ref} = \frac{\hat{\sigma}}{\sqrt{2}} \left\{ \left(\frac{\sigma_1^{(M)}}{\hat{\sigma}} \right)^2 - 1 - \gamma + \left(\left[\left(\frac{\sigma_1^{(M)}}{\hat{\sigma}} \right)^2 - 1 - \gamma \right]^2 - 4\gamma \right)^{1/2} \right\}^{1/2} \qquad (5.75)$$

is an estimator of σ_{ref}, provided that $\sigma_{ref} > \gamma^{1/4} \hat{\sigma}$. From (5.74), we can estimate the scattering amplitude ρ_{ref} of the inclusion by

$$\hat{\rho}_{ref} = \frac{c_0^2}{\omega^2} \left(\sum_{n=1}^{N} |\hat{G}(\omega, \hat{\mathbf{X}}_{ref}, \mathbf{y}_n)|^2 \right)^{-1/2} \left(\sum_{m=1}^{M} |\hat{G}(\omega, \hat{\mathbf{X}}_{ref}, \mathbf{z}_m)|^2 \right)^{-1/2} \hat{\sigma}_{ref},$$

with $\hat{\sigma}_{ref}$ the estimator of (5.75) of σ_{ref} and $\hat{\mathbf{x}}_{ref}$ is an estimator of the position of the inclusion. This estimator is not biased asymptotically since it compensates for the level repulsion of the first singular value due to the noise.

5.5.12 State Estimation and Malignant Attacker in the Smart Grid

A natural situation to use the large random matrices is in the Smart Grid where the big network is met. We use one example to illustrate this potential. We follow the model of [337] for our setting. State estimation and a malignant attack on it are considered in the context of large random matrices.

Power network state estimators are broadly used to obtain an optimal estimate from redundant noisy measurements, and to estimate the state of a network branch which, for economical or computational reasons, is not directly monitored.

The state of a power network at a certain instant of time is composed of the voltage angles and magnitudes at all the system buses. Explicitly, let $\mathbf{x} \in \mathbb{R}^n$ and $\mathbf{z} \in \mathbb{R}^p$ be, respectively, the state and measurements vector. Then, we have

$$\mathbf{z} = h(\mathbf{x}) + \boldsymbol{\eta}, \qquad (5.76)$$

where $h(\mathbf{x})$ is a nonlinear measurement function, and $\boldsymbol{\eta}$ is a zero mean random vector satisfying

$$\mathbb{E}[\boldsymbol{\eta}\boldsymbol{\eta}^T] = \boldsymbol{\Sigma}_{\eta} = \boldsymbol{\Sigma}_{\eta}^T > 0.$$

The network state could be obtained by measuring directly the voltage phasors by means of phasor measurement devices. We adopt the approximated estimation model that follows from the linearization around the origin of (5.76)

$$\mathbf{z} = \mathbf{H}\mathbf{x} + \mathbf{v},$$

where

$$\mathbf{H} \in \mathbb{R}^{p \times n}, \mathbb{E}[\mathbf{v}] = 0, \mathbb{E}[\mathbf{v}\mathbf{v}^T] = \mathbf{\Sigma} = \mathbf{\Sigma}^T > 0.$$

Because of the interconnection structure of the power network, the measurement matrix $\mathbf{H}$ is sparse.

We assume that $\mathbf{z}_i$ is available from $i = 1$ to $i = N$. We denote by $\mathbf{Z}_N$ the $p \times N$ observation matrix. (5.76) can be rewritten as

$$\mathbf{Z}_N = \mathbf{H}\mathbf{X}_N + \mathbf{V}_N \tag{5.77}$$

where

$$\mathbf{Z}_N = [\mathbf{z}_1, \ldots, \mathbf{z}_N], \mathbf{X}_N = [\mathbf{x}_1, \ldots, \mathbf{x}_N], \mathbf{V}_N = [\mathbf{v}_1, \ldots, \mathbf{v}_N].$$

From this matrix $\mathbf{Z}_N$, we can define the sample covariance matrix of the observation as

$$\hat{\mathbf{R}}_N = \frac{1}{N}\mathbf{Z}_N\mathbf{Z}_N^H,$$

while the empirical spatial correlation matrix associated with the noiseless observation will take the form

$$\frac{1}{N}\mathbf{H}\mathbf{X}_N\mathbf{X}_N^H\mathbf{H}^H.$$

To simplify the notation in the future, we define the matrices

$$\mathbf{\Sigma}_N = \frac{\mathbf{Z}_N}{\sqrt{N}}, \mathbf{B}_N = \frac{\mathbf{H}\mathbf{X}_N}{\sqrt{N}}, \mathbf{W}_N = \frac{\mathbf{V}_N}{\sqrt{N}},$$

so that (5.77) can be equivalently formulated as

$$\mathbf{\Sigma}_N = \mathbf{B}_N + \mathbf{W}_N, \tag{5.78}$$

where $\mathbf{\Sigma}_N$ is the (normalized) matrix of observations, $\mathbf{B}_N$ is a deterministic matrix containing the signals contribution, and $\mathbf{W}_N$ is a complex Gaussian white noise matrix with i.i.d. entries that have zero mean and variance σ^2/N.

If $N \to \infty$ while M is fixed, the sample covariance matrix of the observations

$$\hat{\mathbf{R}}_N = \mathbf{\Sigma}_N\mathbf{\Sigma}_N^H$$

of $\mathbf{Z}_N$ converges toward the matrix

$$\mathbf{R}_N = \mathbf{B}_N\mathbf{B}_N^H + \sigma^2\mathbf{I}_p,$$

in the sense that

$$\|\mathbf{R}_N - \mathbf{B}_N\mathbf{B}_N^H - \sigma^2\mathbf{I}_p\| \to 0 \quad \text{almost surely (a.s.).} \tag{5.79}$$

However, in the joint limits

$$\text{asymptotic region } N \to \infty, \, p \to \infty, \text{ but } \frac{p}{N} \to c,$$

which is the practical case, (5.79) is no longer true. The random matrix theory must be used to derive the consequences. (5.78) is a standard form in [282, 333, 338, 339].

Given the distributed nature of a power system and the increasing reliance on local area networks to transmit data to a control center, it is possible for an attacker to attack the network functionality by corrupting the measurements vector $\mathbf{z}$. When a malignant agent corrupts some of the measurements, the new state to measurements relation becomes

$$\begin{aligned} \mathcal{H}_0 &: \mathbf{z} = \mathbf{Hx} + \mathbf{v}, \\ \mathcal{H}_1 &: \mathbf{z} = \mathbf{Hx} + \mathbf{v} + \mathbf{a}, \end{aligned} \tag{5.80}$$

where $\mathbf{a} \in \mathbb{R}^p$ is chosen by the attacker, and thus, it is unknown and unmeasurable by any of the monitoring stations.

(5.80) is a standard hypothesis testing problem. The GLRT can thus be used, together with the random matrix theory. Following the same standard procedure as above, we have

$$\begin{aligned} \mathcal{H}_0 &: \mathbf{Z}_N = \mathbf{HX}_N + \mathbf{V}_N, \\ \mathcal{H}_1 &: \mathbf{Z}_N = \mathbf{HX}_N + \mathbf{V}_N + \mathbf{A}_N, \end{aligned}$$

where

$$\mathbf{A}_N = [\mathbf{a}_1, \dots, \mathbf{a}_N].$$

By studying the sample covariance matrix

$$\hat{\mathbf{R}}_N = \frac{1}{N} \mathbf{Z}_N \mathbf{Z}_N^H,$$

we are able to infer different behavior under hypothesis $\mathcal{H}_0$ or $\mathcal{H}_1$. It seems that this result for this example is reported for the first time.

5.5.13 Covariance Matrix Estimation

We see [340] for more details. Consider a discrete-time complex-valued K-user N-dimensional vector channel with M channel uses. We define $\alpha \triangleq \frac{M}{N}$ and $\beta \triangleq \frac{M}{N}$. We assume the system load $\beta < 1 (K < N)$; otherwise the signal subspace is simply the entire N-vector space. In the m-th channel use, the signal at the receiver can be represented by an N-vector defined by

$$\mathbf{y}(m) = \sum_{k=1}^{K} h_{km} \mathbf{x}_k + \mathbf{w}(m) \tag{5.81}$$

where h_{km} is the channel symbol of user k, having unit power, $\mathbf{x}_k$ is the signature waveform of user k (note that $\mathbf{s}_k$ is independent of the sample index m), and $\mathbf{w}(m)$ is additive noise. By defining

$$\mathbf{X}_{N \times K} = [\mathbf{x}_1, \dots, \mathbf{x}_K], \, \mathbf{h}_{K \times 1}(m) = [h_{1m}^*, \dots, h_{Km}^*]^H,$$

(5.81) is rewritten as

$$\mathbf{y}(m) = \mathbf{Xh}(m) + \mathbf{w}(m). \tag{5.82}$$

We do not assume specific distribution laws of the entries in $\mathbf{H}$, $\mathbf{x}$, $\mathbf{w}$, thereby making the channel model more general [340]:

- The entries of $\mathbf{X}$ are mutually independent random variables, each having zero expectation and variance $\frac{1}{\sqrt{N}}$. Therefore, $\forall k$, $\|\mathbf{x}_k\| \to 1$ almost surely, as $N \to \infty$.
- The entries of $\mathbf{h}(m)$ are mutually independent random variables. The random vectors $\{\mathbf{h}(m)\}_{m=1,\dots,M}$ are mutually independent for different values of m and satisfying

$$\mathbb{E}\{\mathbf{h}(m)\mathbf{h}^H(m)\} = \mathbf{I}_{K \times K}, \mathbb{E}\{\mathbf{h}(m)\mathbf{h}^T(m)\} = \mathbf{0}_{K \times K}.$$

- The entries of $\mathbf{w}(m)$ are mutually independent random variables. The random vectors $\mathbf{w}(m)_{m=1,\dots,M}$ are mutually independent for different values of m and satisfy

$$\mathbb{E}\{\mathbf{w}(m)\mathbf{w}^H(m)\} = \sigma_w^2 \mathbf{I}_{N \times N}, \mathbb{E}\{\mathbf{x}(m)\mathbf{x}^T(m)\} = \mathbf{0}_{N \times N}.$$

- $\mathbf{X}$, $\mathbf{h}(m)$, $\mathbf{w}(m)$ are jointly independent.

Such a model is useful for CDMA and MIMO systems.

The covariance matrix of the received signal (5.81) is given by

$$\mathbf{R} \triangleq \mathbb{E}\{\mathbf{y}(m)\mathbf{y}^H(m)\} = \mathbf{X}\mathbf{X}^H + \sigma_w^2 \mathbf{I}_{N \times N}. \tag{5.83}$$

Based on (5.82) and

$$\mathbb{E}\{\mathbf{w}(m)\mathbf{w}^H(m)\} = \sigma_w^2 \mathbf{I}_{N \times N},$$

the unbiased sample covariance matrix estimate is defined as

$$\hat{\mathbf{R}} = \frac{1}{M} \sum_{m=1}^M \mathbf{y}(m)\mathbf{y}^H(m) = \frac{1}{M}(\mathbf{X}\mathbf{H} + \mathbf{W})(\mathbf{X}\mathbf{H} + \mathbf{W})^H, \tag{5.84}$$

where

$$\mathbf{H} \triangleq [\mathbf{h}_1, \dots, \mathbf{h}_M], \mathbf{W} = [\mathbf{w}_1, \dots, \mathbf{w}_M].$$

By applying the theory of noncrossing partitions, one can obtain explicit expressions for the asymptotic eigenvalue moments of the covariance matrix estimate [340]. Here we only give some key results.

5.5.13.1 Noise-Free Case

When $\sigma_w^2 = 0$, the sample covariance matrix is given by

$$\hat{\mathbf{R}} = \frac{1}{M}(\mathbf{X}\mathbf{H})(\mathbf{X}\mathbf{H})^H = \frac{1}{M}\mathbf{X}\mathbf{H}\mathbf{H}^H\mathbf{X}.$$

The generic eigenvalue of $\hat{\mathbf{R}}$ is denoted by $\hat{\lambda}$ and one defines the eigenvalue moments as

$$\hat{\lambda}_p = \lim_{K,N,M \to \infty} \mathbb{E}\{\lambda^p\}.$$

The explicit expressions are derived in [340].

Corollary 5.1 ([340]) *The eigenvalue moments of the matrix $\frac{1}{M}\mathbf{Z}\mathbf{X}\mathbf{X}^H\mathbf{Z}^H$, where $\mathbf{Z}$ is an $M \times N$ matrix with mutually independent entries having unit variance, are the same as those of the matrix $\frac{1}{M}\mathbf{H}\mathbf{X}\mathbf{X}^H\mathbf{H}^H$.*

The Stieltjes transform of $\hat{\lambda}$ is denoted by $m_{\hat{\lambda}}(z)$.

Corollary 5.2 ([340]) *When σ_w^2, the Stieltjes transform of $\hat{\lambda}$ satisfies*

$$z^2 m_{\hat{\lambda}}^3(z) + (2 - \alpha - \beta)z m_{\hat{\lambda}}^2(z) - (\alpha z - (1 - \beta)(1 - \alpha))m_{\hat{\lambda}}(z) - \alpha = 0. \qquad (5.85)$$

(5.85) can be used to derive the cumulation distribution function (CDF) and the probability distribution function (PDF) of $\hat{\lambda}$, through the inverse formula for the Stieltjes transform.

Lemma 5.1 ([340]) *There exist a constant $C > 0$ and $p_0 \in \mathbb{N}$ such that*

$$\hat{\lambda}_p < C^p, \forall p > p_0.$$

Theorem 5.45 ([340]) *The distribution of $\hat{\lambda}$ converges weakly to a unique distribution determined by the eigenvalue moments as $K, N, M \to \infty$.*

Theorem 5.46 ([340]) *When σ_w^2, $\forall x > 0$, the PDF $\hat{f}(x)$ of the random variable $\hat{\lambda}$ is given by*

$$\hat{f}(x) = \frac{1}{\pi}\mathrm{Im}\,(m_{\hat{\lambda}}(x)).$$

The closed-form PDF of $\hat{\lambda}$ within its support has been derived in [340] and is too long to be included here.

Theorem 5.47 ([340]) *The PDF $\hat{f}(x)$ has the following properties:*

1. the support of $\hat{f}(x)$ is given by $(\hat{\lambda}_{\max}, \hat{\lambda}_{\min})$, where

$$\hat{\lambda}_{\min} \triangleq \inf_{\hat{\lambda} > 0}(\hat{\lambda}), \hat{\lambda}_{\max} \triangleq \sup_{\hat{\lambda} > 0}(\hat{\lambda}).$$

2. $\lambda_{\max} \leq \hat{\lambda}_{\max} \leq \lambda_{\max}\left(1 + \min\left(\sqrt{\frac{\beta}{\alpha}}, \sqrt{\frac{\alpha}{\beta}}\right)\right)^2$;

3. for sufficiently large α, $(\lambda_{\min}, \lambda_{\max}) \subset (\hat{\lambda}_{\min}, \hat{\lambda}_{\max})$; and
4. for sufficiently small $\alpha < \beta$, $\hat{\lambda}_{\min} \leq \lambda_{\max}$.

5.5.13.2 Noisy Case

We extend the anaysis to the general case of $\sigma_w^2 \geq 0$. When $\sigma_w^2 > 0$, the exact covariance matrix is of full rank and there is a mass point at $\lambda = \sigma_w^2$ with probability $1 - \beta$.

Theorem 5.48 ([340]) *The distribution of eigenvalues of the matrix $\frac{1}{M}(\mathbf{X}\mathbf{H} + \mathbf{W})(\mathbf{X}\mathbf{H} + \mathbf{W})^H$ is the same as that of the matrix $\frac{1}{M}\mathbf{Z}(\mathbf{X}\mathbf{X}^H + \sigma_w^2\mathbf{I}_{N\times N})\mathbf{Z}^H$, as $K, M, N \to \infty$, where $\mathbf{Z}$ is an $M \times N$ matrix, whose entries are mutually independent random variables with unit variance.*

Similar to the noise-free case, the eigenvalue moments of $\mathbf{XX}^H + \sigma_w^2 \mathbf{I}_{N \times N}$ are derived in a closed form in [340]. Let us give the first four moments

$$\mathbb{E}\{\hat{\lambda}\} = \sigma_w^2 + \beta$$

$$\mathbb{E}\{\hat{\lambda}^2\} = \left(\frac{1}{\alpha} + 1\right)(\sigma_w^2 + \beta)^2 + \beta$$

$$\mathbb{E}\{\hat{\lambda}^3\} = \left(\frac{1}{\alpha^2} + \frac{3}{\alpha} + 1\right)(\sigma_w^2 + \beta)^3 + 3\left(\frac{1}{\alpha} + 1\right)\beta(\sigma_w^2 + \beta) + \beta$$

$$\mathbb{E}\{\hat{\lambda}^4\} = \left(\frac{1}{\alpha^3} + \frac{6}{\alpha^2} + \frac{6}{\alpha} + 1\right)(\sigma_w^2 + \beta)^4 + \left(\frac{6}{\alpha^2} + \frac{16}{\alpha} + 6\right)\beta(\sigma_w^2 + \beta)^2$$

$$+ \frac{1}{\alpha}(4\beta\sigma_w^2 + 6\beta^2) + 6\beta^2 + 4\beta\sigma_w^2 + \beta.$$

The asymptotic eigenvalue moments of the estimated covariance matrix are larger than those of the exact covariance matrix (except for the expectation). This is true for both noisy and noise-free cases.

The Stieltjes transform of the eigenvalue $\hat{\lambda}$, denoted by $m_{\hat{\lambda}}$, is given by

$$\sigma_w^2 z^2 m_{\hat{\lambda}}^4(z) + (\alpha z^2 + 2(1-\alpha)\sigma_w^2 z)m_{\hat{\lambda}}^3(z) + ((1-\alpha)^2\sigma_w^2 + \alpha(2 - \alpha - \beta - \sigma_w^2)z)m_{\hat{\lambda}}^2(z)$$

$$- \alpha(\alpha z - (1-\alpha)(1-\beta-\sigma_w^2))m_{\hat{\lambda}}(z) - \alpha^2 = 0.$$

We define

$$\hat{\lambda}_{\min} \triangleq \inf_{f(\hat{\lambda}) > 0, \hat{\lambda} > 0} (\hat{\lambda}), \quad \hat{\lambda}_{\max} \triangleq \sup_{f(\hat{\lambda}) > 0, \hat{\lambda} > \sigma_w^2} (\hat{\lambda}).$$

Their counterparts for the exact covariance matrix, denoted by λ_{min}, λ_{max}, and λ'_{min}, are given by $(1 + \sqrt{\beta})^2 + \sigma_w^2$, σ_w^2, and $(1 - \sqrt{\beta})^2 + \sigma_w^2$, respectively.

Theorem 5.49 ([340]) *There is no mass point for any positive eigenvalue $\hat{\lambda}$. The support of $\hat{f}$ satisfies the following properties:*

1. for sufficiently large α, the support of $\hat{\lambda}$ is not continuous interval when $\sigma_w^2 > 0$;

2. $\lambda_{\max} \leq \hat{\lambda}_{\max} \leq \lambda_{\max}\left(1 + \min\left(\sqrt{\frac{1}{\alpha}}, \sqrt{\alpha}\right)\right)^2$;

3. for sufficiently large α, $(\lambda_{\min}, \lambda_{\max}) \subset (\hat{\lambda}_{\min}, \hat{\lambda}_{\max})$; and
4. for sufficiently small $\alpha < \beta$, $\hat{\lambda}_{\min} \leq \lambda_{\max}$.

The properties 3 and 4 in Theorem 5.47 are the same as in Theorem 5.49. Property 1 is completely different. The essential reason is the existence of a mass point at σ_w^2. When $\sigma_w^2 = 0$, the mass point at 0 always exists with probability $1 - \beta$ and the support on positive eigenvalues is continuous. When $\sigma_w^2 > 0$, and $1 < \alpha < \infty$, the estimated covariance matrix is of full rank and there is no mass point. When $\alpha \to \infty$, the support of positive eigenvalues has to be separated into at least two disjoint intervals such that the support around σ_w^2 shrinks to a point.

5.5.14 Deterministic Equivalents

Deterministic equivalents for certain functions of large random matrices are of interest. The most important references are [281, 341–344]. Let us follow [281] for this presentation. Consider an $N \times n$ random matrix $\mathbf{Y}_n = (Y_{ij}^n)$, where the entries are given by

$$Y_{ij}^n = \frac{\sigma_{ij}(n)}{\sqrt{n}} X_{ij}^n,$$

n. Here $(\sigma_{ij}(n), 1 \leq i \leq N, 1 \leq j \leq n)$ is a bounded sequence of real numbers called a variance profile; the X_{ij}^n are centered with unit variance, independent and identically distributed (i.i.d.) with finite $4 + \varepsilon$ moment. Consider now a deterministic $N \times n$ matrix $\mathbf{A}_n$ whose columns and rows are uniformly bounded in the Euclidean norm.

Let

$$\Sigma_n = \mathbf{Y}_n + \mathbf{A}_n.$$

This model has two interesting features: the random variables are independent but not i.i.d. since the variance may vary and $\mathbf{A}_n$, the centering perturbation of $\mathbf{Y}_n$, can have a very general form. The purpose of our problem is to study the behavior of

$$\frac{1}{N} \mathrm{Tr}(\Sigma_n \Sigma_n^T - z \mathbf{I}_N)^{-1}, z \in \mathbb{C} - \mathbb{R},$$

that is, the Stieltjes transform of the empirical eigenvalue distribution of $\Sigma_n \Sigma_n^T$ when $n \to \infty$, and $N \to \infty$ in such a way that $\frac{N}{n} \to c, 0 < c < \infty$.

There exists a deterministic $N \times N$ matrix-valued function $\mathbf{T}_n(z)$ analytic in $\mathbb{C} - \mathbb{R}$ such that, almost surely,

$$\lim_{n \to +\infty, N/c \to c} \left(\frac{1}{N} \mathrm{Tr}(\Sigma_n \Sigma_n^T - z \mathbf{I}_N)^{-1} - \frac{1}{N} \mathrm{Tr} \mathbf{T}_n(z) \right) = 0.$$

In other words, there exists a deterministic equivalent to the empirical Stieltjes transform of the distribution of the eigenvalues of $\Sigma_n \Sigma_n^T$. It is also proved that $\frac{1}{N} \mathrm{Tr} \mathbf{T}_n(z)$ is the Stieltjes transform of a probability measure $\pi_n(d\lambda)$, and that for every bounded continuous function f, the following convergence holds almost surely

$$\frac{1}{N} \sum_{k=1}^{N} f(\lambda_k) - \int_0^\infty f(\lambda) \pi_n(d\lambda) \underset{n \to \infty}{\to} 0,$$

where the $(\lambda_k)_{1 \leq k \leq N}$ are the eigenvalues of $\Sigma_n \Sigma_n^T$. The advantage of considering $\frac{1}{N} \mathrm{Tr} \mathbf{T}_n(z)$ as a deterministic approximation instead of $\mathbb{E} \frac{1}{N} \mathrm{Tr}(\Sigma_n \Sigma_n^T - z \mathbf{I}_N)^{-1}$ (which is deterministic as well) lies in the fact that $\mathbf{T}_n(z)$ is in general far easier to compute than $\mathbb{E} \frac{1}{N} \mathrm{Tr}(\Sigma_n \Sigma_n^T - z \mathbf{I}_N)^{-1}$ whose computation relies on Monte Carlo simulations. These Monte Carlo simulations become increasingly heavy as the size of the matrix Σ_n increases.

This work is motivated by the MIMO wireless channels. The performance of these systems depends on the so-called channel matrix $\mathbf{H}_n$ whose entries $(H_{ij}^n, 1 \leq i \leq N, 1 \leq j \leq n)$ represent the gains between transmit antenna j and receive antenna i. Matrix $\mathbf{H}_n$ is often modeled as a realization of a random matrix. In certain context, the Gram matrix $\mathbf{H}_n\mathbf{H}_n^*$ is unitarily equivalent to a matrix $(\mathbf{Y}_n + \mathbf{A}_n)(\mathbf{Y}_n + \mathbf{A}_n)^*$ where $\mathbf{A}_n$ is a possibly full rank deterministic matrix. As an application, we derive a deterministic equivalent to the mutual information:

$$C_n(\sigma^2) = \frac{1}{N}\mathbb{E}\log\det\left(\mathbf{I}_N + \frac{\Sigma_n\Sigma_n^T}{\sigma^2}\right),$$

where σ^2 is a known parameter.

Let us consider the extension of the above work. Consider

$$Y_{ij}^n - \frac{\sigma_{ij}(n)}{\sqrt{n}}X_{ij}^n,$$

where $(\sigma_{ij}(n), 1 \le i \le N, 1 \le j \le n)$ is uniformly bounded sequence of real numbers, and the random variables X_{ij}^n are complex, centered, i.i.d. with unit variance and finite 8th moment.

We are interested in the fluctuations of the random variable

$$I_n(\rho) = \frac{1}{N}\log\det\left(\mathbf{Y}_n\mathbf{Y}_n^* + \rho\mathbf{I}_N\right)$$

where $\mathbf{Y}_n^*$ is the Hermitian adjoint of $\mathbf{Y}_n$ and $\rho > 0$ is an additional parameter. It is proved [342] that when centered and properly scaled, this random variable satisfies a Center Limit Theorem (CLT) and has a Gaussian limit whose parameters are identified. Understanding its fluctuations and in particular being able to approximate its standard deviation is of major interest for various applications such as for instance the computation of the so-called outage probability.

Consider the following linear statistics of the eigenvalues

$$I_n(\rho) = \frac{1}{N}\log\det\left(\mathbf{Y}_n\mathbf{Y}_n^* + \rho\mathbf{I}_N\right) = \frac{1}{N}\sum_{i=1}^{N}\log(\lambda_i + \rho),$$

where λ_i is the eigenvalue of matrix $\mathbf{Y}_n\mathbf{Y}_n^*$. This functional is of course the mutual information for the MIMO channel. The purpose of [342] is to establish a CLT for $I_n(\rho)$ whenever $n \to \infty$, $\frac{N}{n} \to c$, $0 < c < \infty$.

There exists a sequence of deterministic probability measure π_n such that the mathematical expectation $\mathbb{E}I_n$ satisfies

$$\mathbb{E}I_n(\rho) - \int\log(\lambda + \rho)\pi_n(d\lambda) \underset{n\to\infty}{\to} 0.$$

We study the fluctuations of

$$\frac{1}{N}\log\det\left(\mathbf{Y}_n\mathbf{Y}_n^* + \rho\mathbf{I}_N\right) - \int\log(t + \rho)\pi_n(dt),$$

and prove that this quantity properly rescaled converges toward a Gaussian random variable. In order to prove the CLT, we study the quantity

$$N(I_n(\rho) - \mathbb{E}I_n(\rho)).$$

from which the fluctuations arise and the quantity

$$N(\mathbb{E}I_n(\rho) - \int \log(\lambda + \rho)\pi_n(d\lambda)),$$

which yields a bias.

The variance of Θ^2 of $N(I_n(\rho) - \mathbb{E}I_n(\rho))$ takes a remarkably simple closed-form expression. In fact, there exists a $n \times n$ deterministic matrix $\mathbf{A}_n$ whose entries depend on the variance profile σ_{ij} such that the variance takes the form:

$$\Theta_n^2 = \log \det (\mathbf{I}_n - \mathbf{A}_n) + \kappa \operatorname{Tr}\mathbf{A}_n,$$

where $\kappa = \mathbb{E}|X_{11}|^4 - 2$ in the fourth cumulant of the complex variable X_{11} and the CLT is expressed as:

$$\frac{N}{\Theta_n^2}(I_n(\rho) - \mathbb{E}I_n(\rho)) \xrightarrow[n\to\infty]{\mathcal{L}} \mathcal{N}(0, 1).$$

The bias can be also modeled. There exists a deterministic quantity $\mathbf{B}_n$ such that:

$$N\left(\mathbb{E}I_n(\rho) - \int \log(\lambda + \rho)\pi_n(d\lambda)\right) - \mathbf{B}_n(\rho) \xrightarrow[n\to\infty]{} 0.$$

In [343], they study the fluctuations of the random variable:

$$I_n(\rho) = \frac{1}{N}\log \det (\Sigma_n \Sigma_n^T + \rho\mathbf{I}_N) = \frac{1}{N}\sum_{i=1}^{N}\log(\lambda_i + \rho), \rho > 0,$$

where

$$\Sigma_n = n^{-1/2}\mathbf{D}_n^{1/2}\mathbf{X}_n\tilde{\mathbf{D}}_n^{1/2} + \mathbf{A}_n,$$

as the dimensions of the matrices go to infinity at the same pace. Matrices $\mathbf{X}_n$ and $\mathbf{A}_n$ are respectively random and deterministic $N \times n$ matrices; matrices $\mathbf{D}_n$ and $\tilde{\mathbf{D}}_n$ are deterministic and diagonal. Matrix $\mathbf{X}_n$ has centered, i.i.d., entries with unit variance, either real and complex. They study the fluctuations associated to noncentered large random matrices. Their contribution is to establish the CLT regardless of specific assumptions on the real or complex nature of the underlying random variables. It is in particular not assumed that the random variables are Gaussian, neither that whenever the random variables X_{ij} are complex, their second moment $\mathbb{E}X_{ij}^2$ is zero nor is it assumed that the random variables are circular.

The mutual information I_n has a strong relationship with the Stieltjes transform

$$f_n(z) = \frac{1}{N}\operatorname{Tr}(\Sigma_n\Sigma_n^T - z\mathbf{I}_N)^{-1}$$

of the spectral measure of $\Sigma_n\Sigma_n^T$:

$$I_n(\rho) = \log \rho + \int_\rho^\infty \left(\frac{1}{w} - f_n(-w)\mathbf{d}w\right).$$

Accordingly, the study of the fluctuations of I_n is also an important step toward the study
of general linear statistics of the eigenvalues of $\Sigma_n \Sigma_n^T$ which can be expressed via the
Stieltjes transform:

$$\frac{1}{N}\mathrm{Tr}h(\Sigma_n\Sigma_n^T) = \frac{1}{N}\sum_{i=1}^{N}h(\lambda_i) = -\frac{1}{2i\pi}\oint_C h(z)f_n(z)\mathbf{d}z.$$

5.5.15   Local Failure Detection and Diagnosis

The joint fluctuations of the extreme eigenvalues and eigenvectors are studied for a large
dimensional sample covariance matrix [345], when the associated population covariance
matrix is a finite-rank perturbation of the identity matrix, corresponding to the so-called
spiked model in random matrix theory. The asymptotic fluctuations, as the matrix size
grows large, are shown to be intimately linked with matrices from the Gaussian unitary ensemble (GUE). When the spiked population eigenvalues have unit multiplicity,
the fluctuations follow a central limit theorem. This result is used to develop an original framework for the detection and diagnosis of local failure in large sensor networks,
from known or unknown failure magnitude. This approach is relevant to the Cognitive
Radio Network and the Smart Grid. This approach is to perform fast and computationally
reasonable detection and localization of multiple failure in large sensor networks through
this general hypothesis testing framework. Practical simulations suggest that the proposed
algorithms allow for high failure detection and localization performance even for networks
of small sizes, although for those much more observations than theoretically predicted
are in general demanded.

5.6   Regularized Estimation of Large Covariance Matrices

Estimation of population covariance matrices from samples of multivariate data has always
been important for a number of reasons [344, 346, 347]. Principals among these are:

1. estimation of principal components and eigenvalues in order to get an interpretable
   low-dimensional data representation (principal component analysis, or PCA);
2. construction of linear discriminant functions for classification of Gaussian data (linear
   discriminant analysis, or LDA);
3. establishing independence and conditional independence relations between components
   using exploratory data analysis and testing;
4. setting confidence intervals on linear functions of the means of the components.

(1) requires estimation of the eigenstructure of the covariance matrix while (2) and (3)
require estimation of the inverse. In signal processing and wireless communication, the
covariance matrix is always the starting point.

Exact expressions were cumbersome, and multivariate data were rarely Gaussian. The
remedy was asymptotic theory for large sample and fixed relatively small dimensions.
Recently, due to the rising vision of "big data" [1], datasets that do not fit into this
framework have been very common—the data are very high-dimensional and sample
sizes can be very small relative to dimension.

It is well known by now that the empirical covariance matrix for samples of size n from a p-variate Gaussian distribution, $\mathcal{C}(\boldsymbol{\mu}, \boldsymbol{\Sigma}_p)$, is not a good estimator of the population covariance if p is large. Johnstone and his students [9, 19, 22, 312–318, 325, 327] are relevant here.

The empirical covariance matrix for samples of size n from a p-variate Gaussian distribution has unexpected features if both p and n are large. If $p/n \to c \in (0, 1)$, and the covariance matrix $\boldsymbol{\Sigma}_p = \mathbf{I}$ (the identity), then the empirical distribution of the eigenvalues of the sample covariance matrix $\boldsymbol{\Sigma}_p$ follows the Marchenko-Pastur law [348], which is supported on

$$[(1 - \sqrt{c})^2, (1 + \sqrt{c})^2].$$

Thus, the larger p/n (thus c), the more spread out the eigenvalues.

Two broad classes of covariance estimators [347] have emerged: (1) those that rely on a natural ordering among variables, and assume that variables far apart in the ordering are only weakly correlated, and (2) those invariant to variable permutations. However, there are many applications for which there is no notion of distance between variables at all.

Implicitly, some approaches, for example, [312], postulate different notions of sparsity. Thresholding of the sample covariance matrix has been proposed in [347] as a simple and permutation-invariant method of covariance regulation. A class of regularized estimators of (large) empirical covariance matrices corresponding to stationary (but not necessarily Gaussian) sequences is obtained by *banding* [344].

We follow [346] for notation, motivation, and background.

We observe $\mathbf{X}_1, \ldots, \mathbf{X}_n$, i.i.d. p-variate random variables with mean $\mathbf{0}$ and covariance matrix $\hat{\boldsymbol{\Sigma}}_p$, and write

$$\mathbf{X}_i = (X_{i1}, \ldots, X_{ip})^T.$$

For now, we assume that $\mathbf{X}_i$ are multivariate normal. We want to study the behavior of estimates of $\boldsymbol{\Sigma}_p$ as both p and $n \to \infty$. It is well known that the ML estimation of $\boldsymbol{\Sigma}_p$, the sample covariance matrix,

$$\hat{\boldsymbol{\Sigma}}_p = \frac{1}{n} \sum_{i=1}^{n} (\mathbf{X}_i - \bar{\mathbf{X}})(\mathbf{X}_i - \bar{\mathbf{X}})^T$$

behaves optimally if p is fixed, converging to $\boldsymbol{\Sigma}_p$ at rate $n^{-1/2}$. If $p \to \infty$, $\hat{\boldsymbol{\Sigma}}_p$ can behave very badly, unless it is "regularized" in some fashion.

5.6.1 Regularized Covariance Estimates

5.6.1.1 Banding the Sample Covariance Matrix

For any matrix $\mathbf{A} = [a_{ij}]_{p \times p}$, and any $0 \le k \le p$, define

$$\mathcal{B}_k(\mathbf{A}) = \left[a_{ij} \mathbf{1}(|i - j| \le k) \right]$$

and estimate the covariance $\hat{\boldsymbol{\Sigma}}_{k,p} \equiv \hat{\boldsymbol{\Sigma}}_k = \mathcal{B}(\hat{\boldsymbol{\Sigma}}_p)$. This kind of regularization is ideal in the situation where the indexes have been arranged in a such a way that in $\boldsymbol{\Sigma}_p = [\sigma_{ij}]$ we have

$$|i - j| > k \Rightarrow \sigma_{ij} = 0.$$

This assumption holds, for example, if Σ_p is the covariance matrix of $Y_1, \ldots, Y_p$, where $Y_1, \ldots, Y_p$ is a finite inhomogeneous moving average (MA) process,

$$Y_t = \sum_{j=1}^{k} \alpha_{t,t-1} x_j,$$

and x_j are i.i.d. mean 0. Banding an arbitrary covariance matrix does not guarantee positive definitiveness.

All our sets will be the subsets of the so-called *well-conditioned covariance matrices*, Σ_p, such that, for all p,

$$0 < \varepsilon \leq \lambda_{\min}(\Sigma) \leq \lambda_{\max}(\Sigma) \leq 1/\varepsilon < \infty.$$

Here, $\lambda_{\max}(\Sigma)$ and $\lambda_{\min}(\Sigma)$ are the maximum and minimum eigenvalue of Σ_p, and ε is independent of p.

Examples of such matrices [349] include

$$Y_i = X_i + W_i, i = 1, 2, \ldots$$

where X_i is a stationary ergodic process, and W_i is a noise process independent of $\{X_i\}$. This model also includes the "spiked model" of Paul [27], since a matrix of bounded rank is Hilbert-Schmidt. We discuss this model in detail elsewhere.

We define the first class of positive definite symmetric well conditioned matrices $\Sigma = [\sigma_{ij}]$ as follows:

$$\mathcal{U}(\varepsilon_0, \alpha, C) = \left\{ \begin{array}{l} \Sigma : \max_j \sum_i \{|\sigma_{ij}| : |i - j| > k\} \leq Ck^{-\alpha} \text{ for all } k > 0, \\ \text{and } 0 < \varepsilon_0 \leq \lambda_{\min}(\Sigma) \leq \lambda_{\max}(\Sigma) \leq 1/\varepsilon_0 < \infty \end{array} \right\}. \quad (5.86)$$

The class $\mathcal{U}$ in (5.86) contains the Toeplitz class $\mathcal{T}$ defined by

$$\mathcal{T}(\varepsilon_0, m, C) = \left\{ \begin{array}{l} \Sigma : \sigma_{ij} = \sigma(i - j) \text{(Toeplitz) with spectral density } f_{\Sigma} \\ \text{and } 0 < \varepsilon_0 \leq \|f_{\Sigma}\|_{\infty} \leq \varepsilon_0^{-1} < \infty, \|f_{\Sigma}^{(m)}\|_{\infty} \leq C \end{array} \right\},$$

where $f_{\Sigma}^{(m)}$ denotes the mth derivative of f. By [350], Σ is symmetric, Toeplitz, $\Sigma = [\sigma(i - j)]$, with $\sigma(-k) = \sigma(k)$, and Σ has an absolutely continuous spectral distribution with Radon-Nikodym derivative $f_{\Sigma}(t)$, which is continuous on $(-1, 1)$, then

$$\|\Sigma\| = \sup_t |f_{\Sigma}(t)|, \|\Sigma^{-1}\| = [\inf_t |f_{\Sigma}(t)|]^{-1}.$$

A second uniformity class of nonstationary covariance matrices is defined by

$$\mathcal{K}(m, C) = \{\Sigma : \sigma_{ii} \leq Ci^{-m}, \text{ all } i\}.$$

The bound independent of dimension identifies any limit as being of "trace class" as operator for $m > 1$.

The main work is summarized in the following theorem.

Theorem 5.50 (Bickel and Levina (2008) [346]) *Suppose that* $\mathbf{X}$ *is Gaussian and* $\mathcal{U}(\varepsilon_0, \alpha, C)$ *is the class of covariance matrices defined in (5.86). Then, if* $k_n \simeq (n^{-1} \log p)^{-1/(2(\alpha+1))}$,

$$\|\mathbf{\Sigma}_{k_n,p} - \mathbf{\Sigma}_p\| = \mathcal{O}_P\left(\frac{\log p}{n}\right)^{\alpha/(2(\alpha+1))} = \|\mathbf{\Sigma}_{k_n,p}^{-1} - \mathbf{\Sigma}_p^{-1}\| \tag{5.87}$$

uniformly on $\mathbf{\Sigma} \in \mathcal{U}$.

5.6.2 Banding the Inverse

Suppose we have

$$\mathbf{X} = (X_1, \ldots, X_p)^T$$

defined on a probability space, with probability measure $\mathcal{P}$, which is $\mathcal{N}_p(\mathbf{0}, \mathbf{\Sigma}_p)$, $\mathbf{\Sigma}_p = [\sigma_{ij}]$. Let

$$\hat{X}_j = \sum_{t=1}^{j-1} a_{jt} X_t = \mathbf{Z}_j^T \mathbf{a}_j \tag{5.88}$$

be the $\mathcal{L}_2(\mathcal{P})$ projection of $\hat{X}_j$ on the linear span of $X_1, \ldots X_{j-1}$, with $\mathbf{Z}_j = (X_1, \ldots, X_{j-1})^T$ the vector of coordinates up to $j-1$, and $\mathbf{a}_j = (a_{j1}, \ldots, a_{j,j-1})^T$ the vector of coefficients. If $j = 1$, let $\hat{\mathbf{X}}_1 = 0$. Each vector $\mathbf{a}_j$ can be computed as

$$\mathbf{a}_j = (\text{var}(\mathbf{Z}_j))^{-1} \text{Cov}(\mathbf{X}_j, \mathbf{Z}_j). \tag{5.89}$$

Let the lower triangular matrix $\mathbf{A}$ with zeros on the diagonal contain the coefficients $\mathbf{a}_j$ arranged in rows. Let $\varepsilon_j = X_j - \hat{X}_j$, $d_j^2 = \text{Var}(\varepsilon_j)$ and let $\mathbf{D} = \text{diag}(d_1^2, \ldots, d_p^2)$ be a diagonal matrix. The geometry of $\mathcal{L}_2(\mathcal{P})$ or standard regression theory implies independence of the residuals. After applying the covariance operator to the identity

$$\boldsymbol{\varepsilon} = (\mathbf{I} - \mathbf{A})\mathbf{X},$$

we obtain the modified Cholesky decomposition of $\mathbf{\Sigma}_p$ and $\mathbf{\Sigma}_p^{-1}$:

$$\begin{aligned} \mathbf{\Sigma}_p &= (\mathbf{I} - \mathbf{A})^{-1} \mathbf{D}[(\mathbf{I} - \mathbf{A})^{-1}]^T, \\ \mathbf{\Sigma}_p^{-1} &= (\mathbf{I} - \mathbf{A})^T \mathbf{D}^{-1} (\mathbf{I} - \mathbf{A}). \end{aligned} \tag{5.90}$$

Suppose now that $k < p$. It is natural to define an approximation to $\mathbf{\Sigma}_p$ by restricting the variables in regression (5.88) to

$$\mathbf{Z}_j^{(k)} = (X_{\max\{j-k,1\}}, \ldots, X_{j-1})^T.$$

In other words, in (5.88), *we regress each* X_j *on its closest* k *predecessors only.* Let $\mathbf{A}_k$ be the k-banded lower triangular matrix containing the new vectors of coefficients $\mathbf{a}_j^{(k)}$, and let $\mathbf{D}_k = \text{diag}(d_{j,k}^2)$ be the diagonal matrix containing the corresponding residual variance. Population k-banded approximations $\mathbf{\Sigma}_{k,p}$ and $\mathbf{\Sigma}_{k,p}^{-1}$ are obtained by plugging in $\mathbf{A}_k$ and $\mathbf{D}_k$ in (5.90) for $\mathbf{A}$ and $\mathbf{D}$.

If

$$\mathbf{\Sigma}^{-1} = \mathbf{T}^T(\mathbf{\Sigma})\mathbf{D}^{-1}(\mathbf{\Sigma})\mathbf{T}(\mathbf{\Sigma})$$

with $\mathbf{T}(\mathbf{\Sigma})$ lower triangular, $\mathbf{T}(\mathbf{\Sigma}) \equiv [t_{ij}(\mathbf{\Sigma})]$, let

$$\mathcal{U}^{-1}(\varepsilon_0, \alpha, C) = \left\{ \begin{array}{l} \mathbf{\Sigma} : \max_i \sum_{j < i-k} |t_{ij}(\mathbf{\Sigma})| \le Ck^{-\alpha} \text{ for all } k \le p-1, \\ \text{and } 0 < \varepsilon_0 \le \lambda_{\min}(\mathbf{\Sigma}) \le \lambda_{\max}(\mathbf{\Sigma}) \le \varepsilon_0^{-1} < \infty \end{array} \right\}. \qquad (5.91)$$

Theorem 5.51 (Bickel and Levina (2008) [346]) *Uniformly for* $\mathbf{\Sigma} \in \mathcal{U}^{-1}(\varepsilon_0, \alpha, C)$*, if* $k_n \asymp (n^{-1} \log p)^{-1/(2(\alpha+1))}$*, and* $n^{-1} \log p = \mathcal{O}(1)$*,*

$$\|\tilde{\mathbf{\Sigma}}_{k_n,p}^{-1} - \mathbf{\Sigma}_p^{-1}\| = \mathcal{O}_P \left(\frac{\log p}{n} \right)^{\alpha/(2(\alpha+1))} = \|\tilde{\mathbf{\Sigma}}_{k_n,p} - \mathbf{\Sigma}_p\|. \qquad (5.92)$$

Corollary 5.3 (Bickel and Levina (2008) [346]) *For* $m \ge 2$*, uniformly on* $\mathcal{T}(\varepsilon_0, m, C)$*, if* $k_n \asymp (n^{-1} \log p)^{-1/2m}$*,*

$$\|\tilde{\mathbf{\Sigma}}_{k_n,p}^{-1} - \mathbf{\Sigma}_p^{-1}\| = \mathcal{O}_P \left(\frac{\log p}{n} \right)^{\frac{(m-1)}{2m}} = \|\tilde{\mathbf{\Sigma}}_{k_n,p} - \mathbf{\Sigma}_p\|. \qquad (5.93)$$

5.6.3 Covariance Regularization by Thresholding

Bickel and Levina (2008) [347] considers regularizing a covariance matrix of p variables estimated from n (vector) observations, by hard thresholding. They show that the thresholded estimate is consistent in the operator norm as long as the true covariance matrix is sparse in a suitable sense, the variables are Gaussian or sub-Gaussian, and $(\log p)/n \to 0$, and obtain explicit rates.

The approach of thresholding of the sample covariance matrix is a simple and permutation-invariant method of covariance regularization. We define the thresholding operator by

$$\mathbf{T}_s(\mathbf{A}) = \left[a_{ij} \mathbf{1}(|a_{ij}| \ge s) \right],$$

which we refer to as $\mathbf{A}$ thresholded at s. $\mathbf{T}_s$ preserves symmetry and is invariant under permutations of variables labels, but does not necessarily preserve positive definiteness. However, if

$$\|\mathbf{T}_s - \mathbf{T}_0\| \le \varepsilon \text{ and } \lambda_{\min}(\mathbf{A}) > \varepsilon,$$

then $\mathbf{T}_s(\mathbf{A})$ is necessarily positive definite, since for all vectors $\mathbf{v}$ with $||\mathbf{v}||_2 = 1$, we have

$$\mathbf{v}^T \mathbf{T}_s \mathbf{A} \mathbf{v} \ge \mathbf{v}^T \mathbf{A} \mathbf{v} - \varepsilon \ge \lambda_{\min}(\mathbf{A}) - \varepsilon > 0.$$

Here, $\lambda_{\min}(\mathbf{A})$ stands for the minimum eigenvalue of $\mathbf{A}$.

$\mathcal{U}(\varepsilon_0, \alpha, C)$ in (5.86) defines the uniformity class of "approximately bandable" covariance matrices. Here, we define the uniformity class of covariance matrices invariant under permutations by

$$\mathcal{U}_\tau(q, c_0(p), \mathbf{A}) = \left\{ \Sigma : \sigma_{ii} \leq \mathbf{A}, \sum_{j=1}^{p} |\sigma_{ij}|^q \leq c_0(p), \text{ for all } i, 0 \leq q < 1 \right\}.$$

If $q = 0$, we have

$$\mathcal{U}_\tau(0, c_0(p), \mathbf{A}) = \left\{ \Sigma : \sigma_{ii} \leq \mathbf{A}, \sum_{j=1}^{p} \mathbf{1}(\sigma_{ij} \neq 0) \leq c_0(p) \right\},$$

is a class of sparse matrices. Naturally, there is a class of covariance matrices $\mathcal{V}(\varepsilon_0, \alpha, C)$ that satisfy both banding and thresholding conditions. Define a subset of $\mathcal{U}(\varepsilon_0, \alpha, C)$ by

$$\mathcal{V}(\varepsilon_0, \alpha, C) = \left\{ \begin{array}{l} \Sigma : |\sigma_{ii}| \leq C |i - j|^{-(\alpha+1)}, \text{ for all } i, j : |i - j| \geq 1, \\ \text{and } 0 < \varepsilon_0 \leq \lambda_{\min}(\Sigma) \leq \lambda_{\max}(\Sigma) \leq 1/\varepsilon_0 \end{array} \right\},$$

for $\alpha > 0$.

We consider n i.i.d. p-dimensional observations $\mathbf{X}_1, \ldots, \mathbf{X}_n$ distributed according to a distribution $\mathcal{F}$, with $\mathbb{E}\mathbf{X} = 0$ (without loss of generality), and $\mathbb{E}(\mathbf{X}\mathbf{X}^T) = \Sigma$. We define the empirical (sample) covariance matrix by

$$\hat{\Sigma} = \frac{1}{n} \sum_{k=1}^{n} (\mathbf{X}_k - \hat{\mathbf{X}})(\mathbf{X}_k - \hat{\mathbf{X}})^T,$$

where $\hat{\mathbf{X}} = n^{-1} \sum_{k=1}^{n} \mathbf{X}_k$, and write $\hat{\Sigma} = [\hat{\sigma}_{ij}]$.

Theorem 5.52 (Bickel and Levina (2008) [347]) *Suppose $\mathcal{F}$ is Gaussian. Then, uniformly on $\mathcal{U}_\tau(0, c_0(p), \mathbf{A})$, for sufficiently large M', if*

$$t_n = M' \sqrt{\frac{\log p}{n}}$$

and $\frac{\log p}{n} = o(1)$, then

$$\|\mathbf{T}_{t_n}(\hat{\Sigma}) - \Sigma\| = \mathcal{O}_P\left(c_0(p) \left(\frac{\log p}{n} \right)^{(1-q)/2} \right)$$

and uniformly on $\mathcal{U}_\tau(q, c_0(p), \mathbf{A})$,

$$\left\| (\mathbf{T}_{t_n}(\hat{\Sigma}))^{-1} - \Sigma^{-1} \right\| = \mathcal{O}_P\left(c_0(p) \left(\frac{\log p}{n} \right)^{(1-q)/2} \right).$$

This theorem is in parallel with the banding result of Theorem 5.50.

5.6.4 Regularized Sample Covariance Matrices

Let us follow [344] to state a certain central limit theorem for regularized sample covariance matrices. We just treated how to band the covariance matrix $\mathbf{\Sigma}$; here we consider how to band the sample covariance matrix $\hat{\mathbf{\Sigma}} = \mathbf{X}^T \mathbf{X}$. We consider regularization by banding, that is, by replacing those entries of $\mathbf{X}^T \mathbf{X}$ that are, at distance, exceeding $b = b(p)$ away from the diagonal by 0. Let $\mathbf{Y} = \mathbf{Y}^{(p)}$ denote the thus regularized empirical matrix.

Let $X_1, \ldots, X_k$ be real random variables on a common probability space with moments of all orders, in which the characteristic function

$$\mathbb{E} \exp \left(\sum_{i=1}^{k} j t_i X_i \right)$$

is an infinitely differentiable function of the real variables $t_1, \ldots, t_k$. One defines the joint cumulant $\mathbf{C}(X_1, \ldots, X_k)$ by the formula

$$\mathbf{C}(X_1, \ldots, X_k) = \mathbf{C}\{X_i\}_{i=1}^{k} = j^{-k} \frac{\partial^k}{\partial t_1 \cdots \partial t_k} \log \mathbb{E} \exp \left(\sum_{i=1}^{k} j t_i X_i \right) \Bigg|_{t_1 = \cdots t_k = 0} . \tag{5.94}$$

(The middle expression is a convenient abbreviated notation.) The quantity $\mathbf{C}(X_1, \ldots, X_k)$ depends symmetrically and $\mathbb{R}$-multilinearly on $X_1, \ldots, X_k$. Moreover, dependence is continuous with respect to the $\mathcal{L}^k$-norm. One has in particular

$$\mathbf{C}(X) = \mathbb{E}X, \mathbf{C}(X, X) = \mathrm{var}X, \mathbf{C}(X, Y) = \mathrm{cov}(X, Y).$$

Lemma 5.2 *If there exists $0 < l < k$ such that the σ-fields $\sigma\{X_i\}_{i=1}^{l}$ and $\sigma\{X_i\}_{i=l+1}^{l}$ are independent, then $\mathbf{C}(X_1, \ldots, X_k) = 0$.*

Lemma 5.3 *The random vector $X_1, \ldots, X_k$ has a Gaussian joint distribution if and only if $\mathbf{C}(X_{i_1}, \ldots, X_{i_r}) = 0$ for every integer $r \geq 3$ and sequence $i_1, \ldots, i_r \in 1, \ldots, k$.*

Let

$$\{Z_i\}_{i=-\infty}^{\infty}$$

be a stationary sequence of real random variables, satisfying the following conditions:

1. **Assumption 5.5.** As $p \to \infty$, we have $b \to \infty, n \to \infty$ and $b/n \to 0$, with $b \leq p$.
2. **Assumption 5.6.**

$$\mathbb{E}(|Z_0|^k) < \infty \text{ for all } k \geq 1 \tag{5.95}$$

$$\mathbb{E}Z_0 = 0, \tag{5.96}$$

$$\sum_{i_1} \cdots \sum_{i_r} |\mathbf{C}(Z_0, Z_{i_1}, \ldots, Z_{i_r})| \text{ for all } r \geq 1 \tag{5.97}$$

Let us turn to random matrices. Let

$$\{\{Z_j^{(i)}\}_{j=-\infty}^{\infty}\}_{i=1}^{\infty}$$

be an i.i.d. family of copies of $\{Z_j\}_{j=-\infty}^{\infty}$. Let $\mathbf{X} = \mathbf{X}^{(p)}$ be the $n \times p$ random matrices with entries

$$\mathbf{X}(i, j) = \mathbf{X}_{ij} = \frac{1}{\sqrt{n}} Z_j^{(i)}.$$

Let $\mathbf{B} = \mathbf{B}^{(p)}$ be the $p \times p$ deterministic matrix with entries

$$\mathbf{B}(i, j) = \mathbf{B}_{ij} = \begin{cases} 1, & \text{if } |i - j| \le b, \\ 0, & \text{if } |i - j| > b. \end{cases}$$

Let $\mathbf{Y} = \mathbf{Y}^{(p)}$ be the $p \times p$ random symmetric matrix with entries

$$\mathbf{Y}(i, j) = \mathbf{Y}_{ij} = \mathbf{B}_{ij}(\mathbf{X}^T\mathbf{X})_{ij} \qquad (5.98)$$

and eigenvalues $\{\lambda_i^{(p)}\}_{i=1}^{p}$.

For integers j, let

$$R(j) = \mathrm{Cov}(Z_0, Z_j) = \mathbf{C}(Z_0, Z_j).$$

For integers $m > 0$ and all integers i and j, we write

$$Q_{ij} = \sum_{l \in \mathbb{Z}} \mathbf{C}(Z_i, Z_0, Z_{j+l}, Z_l),$$

$$R_i^{(m)} = \underbrace{R \star R \star \cdots \star R(i)}_{m}, \ R_i^{(0)} = \delta_{i0}. \qquad (5.99)$$

Here, the convolution $\star$ is defined for any two summable functions $F, G : \mathbb{Z} \to \mathbb{R}$:

$$(F \star G)(j) = \sum_{k \in \mathbb{Z}} F(j - k)G(k).$$

Now we are in a position to state a central limit theorem.

Theorem 5.53 (Anderson and Zeitouni (2008) [344]) *Let Assumption 5.5 and Assumption 5.6 hold. Let $\mathbf{Y} = \mathbf{Y}^{(p)}$ be as in (5.98). Let Q_{ij} and $R_i^{(m)}$ be as in (5.99). Then the process*

$$\left\{ \sqrt{\frac{n}{p}} (Tr\mathbf{Y}^k - \mathbb{E}Tr\mathbf{Y}^k) \right\}_{k=1}^{\infty}$$

converge in distribution, as $p \to \infty$, to a zero mean Gaussian process $\{G_k\}_{k=1}^{\infty}$, with covariance specified by

$$\frac{1}{kl} \mathbb{E}G_k G_l = 2R_0^{(k+l)} + \sum_{i,j \in \mathbb{Z}} R_i^{(k-1)} Q_{ij} R_j^{(l-1)}.$$

Example 5.9 (Some stationary sequences satisfying Assumption 5.6 [344])
Fix a summable function $h : \mathbb{Z} \to \mathbb{R}$ and an i.i.d. sequence $\{W_l\}_{l=-\infty}^{\infty}$ of mean zero random variables with moments of all orders. Now convolve: put

$$Z_j = \sum_l h(j + l)W_l, \text{ for every } j.$$

It is obvious that (5.95) and (5.96) hold. To see the summability condition (5.97) on joint cumulants, assume at first that h has finite support. Then, by standard properties of joint cumulants (Lemma 5.2), we get the formula

$$\mathbf{C}(Z_{j_0}, \ldots, Z_{j_r}) = \sum_l h(j_0 + l) \cdots h(j_r + l) \mathbf{C} \underbrace{(W_0, \ldots, W_0)}_{r+1}.$$

By a straightforward calculation, this leads the analogous formula without the assumption of finite support of h, whence in turn verification of (5.97). □

5.6.5   Optimal Rates of Convergence for Covariance Matrix Estimation

Despite recent progress on covariance matrix estimation, there has been remarkably little fundamental theoretical study on optimal estimation. Cai, Zhang and Zhou (2010) [351] establish the optimal rates for estimating the covariance matrix under both the operator norm and Frobenius norm. Optimal procedures under two norms are different, and consequently matrix estimation under the operator norm is fundamentally different from vector estimation. The minimax upper bound is reached by constructing a special class of tapering estimators and by studying their risk properties. The banding estimator treated previously in Section 5.6.1 is suboptimal and the performance can be significantly improved using the technique to be covered now.

We write $a_n \asymp b_n$ if there are positive constants c and C independent of n such that $c \le a_n/b_n \le C$. For matrix $\mathbf{A}$, its operator norm is defined as $||\mathbf{A}|| = \sup_{||\mathbf{x}||_2=1} ||\mathbf{Ax}||_2$. We assume that $p \le \exp(\gamma n)$ for some constant $\gamma > 0$.

$$\mathcal{F}_\alpha = \mathcal{F}_\alpha(M_0, M) = \left\{ \begin{array}{l} \boldsymbol{\Sigma} : \max_j \sum_i \left\{ |\sigma_{ij}| : |i - j| > k \right\} \le Mk^{-\alpha} \text{ for all } k, \\ \text{and } \lambda_{\max}(\boldsymbol{\Sigma}) \le M_0 \end{array} \right\}. \quad (5.100)$$

where $\lambda_{max}(\boldsymbol{\Sigma})$ is the maximum eigenvalue of the matrix $\boldsymbol{\Sigma}$, and $\alpha > 0$, $M > 0$ and $M_0 > 0$.

Theorem 5.54 (Minimax risk by Cai, Zhang and Zhou (2010) [351]) *The minimax risk of estimating the covariance matrix $\boldsymbol{\Sigma}$ over the class $\mathcal{P}_\alpha$ satisfies*

$$\inf_{\hat{\boldsymbol{\Sigma}}} \sup_{\mathcal{P}_\alpha} \mathbb{E} \| \hat{\boldsymbol{\Sigma}} - \boldsymbol{\Sigma} \|^2 \asymp \min \left\{ n^{-2\alpha/(2\alpha+1)} + \frac{\log p}{n}, \frac{p}{n} \right\}. \quad (5.101)$$

The proposed procedure does not attempt each row/column optimally as a vector. This procedure does not optimally trade bias and variance for each row/column. This proposed estimator has good numerical performance; it nearly uniformly outperforms the banding estimator.

Example 5.10 (Tapering estimator [351])
For a given even integer with $1 \le k \le p$, we define a tapering estimator

$$\hat{\boldsymbol{\Sigma}} = \hat{\boldsymbol{\Sigma}}_k = (w_{ij}\sigma_{ij}^*)_{p \times p}, \quad (5.102)$$

where σ_{ij}^* are the entries in the ML estimator $\hat{\boldsymbol{\Sigma}}^*$ and the weights

$$w_{ij} = k_h^{-1}\left\{(k - |i - j|)_+ - (k_h - |i - j|)_+\right\},$$

where $k_h = k/2$. Without loss of generality, we assume that k is even. The weights w_{ij} can be rewritten as

$$w_{ij} = \begin{cases} 1, & \text{when } |i - j| \le k_h, \\ 2 - \frac{|i-j|}{k_h}, & \text{when } k_h < |i - j| \le k, \\ 0, & \text{otherwise.} \end{cases} \tag{5.103}$$

The tapering estimators are different from the banding estimators used in [346]. See also Section 5.6.1. $\square$

Lemma 5.4 *The tapering estimator* $\hat{\boldsymbol{\Sigma}}$ *given in (5.102) can be expressed as*

$$\hat{\boldsymbol{\Sigma}} = k_h^{-1}(\mathbf{S}^{*(k)} - \mathbf{S}^{*(k_h)}). \tag{5.104}$$

Assume that the distribution of the $\mathbf{X}_1$'s are sub-Gaussian in the sense that there is $\rho > 0$ such that

$$\mathbb{P}\left\{|\mathbf{v}^T(\mathbf{X}_1 - \mathbb{E}\mathbf{X})\mathbf{v}| > t\right\} \le e^{-t^2\rho/2} \text{ for all } t > 0 \text{ and } \|\mathbf{v}\|_2 = 1. \tag{5.105}$$

Let $\mathcal{P}_\alpha = \mathcal{P}_\alpha(M_0, M, \rho)$ denote the set of distributions of $\mathbf{X}_1$ that satisfy (5.100) and (5.105).

Theorem 5.55 (Upper bound by Cai, Zhang and Zhou (2010) [351]) *The tapering estimator* $\hat{\boldsymbol{\Sigma}}$, *defined in (5.104), of the covariance matrix* $\boldsymbol{\Sigma}_{p\times p}$ *with* $p > n^{1/(2\alpha+1)}$ *satisfies*

$$\sup_{\mathcal{P}_\alpha} \mathbb{E}\|\hat{\boldsymbol{\Sigma}} - \boldsymbol{\Sigma}\|^2 \le C\frac{k + \log p}{n} + Ck^{-2\alpha} \tag{5.106}$$

for $k = o(n)$, $\log p = o(n)$ *and some constant* $C > 0$. *In particular, the estimator* $\hat{\boldsymbol{\Sigma}} - \hat{\boldsymbol{\Sigma}}_k$ *with* $k = n^{1/(2\alpha+1)}$ *satisfies*

$$\sup_{\mathcal{P}_\alpha} \mathbb{E}\|\hat{\boldsymbol{\Sigma}} - \boldsymbol{\Sigma}\|^2 \le Cn^{-2\alpha/(2\alpha+1)} + C\frac{\log p}{n}. \tag{5.107}$$

From (5.106), it is clear that the optimal choice of k is of order $n^{-2\alpha/(2\alpha+1)}$. The upper bound given in (5.107) is thus rated optimal, among the class of the tapering estimators defined in (5.104). The minimax lower bound derived in Theorem 5.56 shows that the estimator $\hat{\boldsymbol{\Sigma}}_k$ with $k = n^{-2\alpha/(2\alpha+1)}$ is in fact rated optimal among all estimators.

Theorem 5.56 (Lower bound by Cai, Zhang and Zhou (2010) [351]) *Suppose* $p \le exp(\gamma n)$ *for some constant* $\gamma > 0$. *The minimax risk for estimating the covariance matrix* $\boldsymbol{\Sigma}$ *over* $\mathcal{P}_\alpha$ *under the operator norm satisfies*

$$\inf_{\hat{\boldsymbol{\Sigma}}} \sup_{\mathcal{P}_\alpha} \mathbb{E}\|\hat{\boldsymbol{\Sigma}} - \boldsymbol{\Sigma}\|^2 \ge cn^{-2\alpha/(2\alpha+1)} + c\frac{\log p}{n}.$$

Theorem 5.55 and Theorem 5.56 together show that the minimax risk for estimating the covariance matrices over the distribution space $\mathcal{P}_\alpha$ satisfies, for $p > n^{1/(2\alpha+1)}$,

$$\inf_{\hat{\Sigma}} \sup_{\mathcal{P}_\alpha} \mathbb{E} \| \hat{\Sigma} - \Sigma \|^2 \asymp n^{-2\alpha/(2\alpha+1)} + \frac{\log p}{n}. \tag{5.108}$$

The results also show that the tapering estimator $\hat{\Sigma}_k$ with tapering parameter $k = n^{1/(2\alpha+1)}$ attains the optimal rate of convergence $n^{-2\alpha/(2\alpha+1)} + \frac{\log p}{n}$.

It is interesting to compare the tapering estimator with the banding estimator of [346]. A banding estimator with bandwidth $k = \left(\frac{\log p}{n} \right)^{1/(2\alpha+1)}$ was proposed and the rate of convergence of $\left(\frac{\log p}{n} \right)^{\alpha/(\alpha+1)}$ was proven.

Both the tapering estimator and the banding estimator are not necessarily positive semidefinite. A practical proposal to avoid this would projecting the estimator $\hat{\Sigma}$ to the space of positive semidefinite matrices under the operator norm. One may first diagonalize $\hat{\Sigma}$ and then replace negative eigenvalues by zeros. The resulting estimator will be then positive semidefinite.

In addition to the operator norm, the Frobenius norm is another commonly used matrix norm. The Frobenius norm of a matrix $\mathbf{A}$ is defined as the l_2 vector norm of all entries in the matrix

$$\| \mathbf{A} \|_F = \sqrt{\sum_{i,j} a_{ij}^2}.$$

This is equivalent to treating the matrix $\mathbf{A}$ as a vector of length p^2. It is easy to see that the operator norm is bounded by the Frobenius norm, that is, $\| \mathbf{A} \| \leq \| \mathbf{A} \|_F$.

Consider estimating the covariance matrix Σ from the sample $\{ \mathbf{X}_1, \ldots, \mathbf{X}_n \}$. We have considered the parameter space $\mathcal{F}_\alpha$ defined in (5.100). Other similar parameter spaces can be also considered. For example, in time series analysis it is often assumed the covariance $|\sigma_{ij}|$ decays at the rate $|i - j|^{-(\alpha-1)}$ for some $\alpha > 0$. Consider the collection of positive-definite symmetric matrices satisfying the following conditions

$$\mathcal{G}_\alpha = \mathcal{G}_\alpha(M_0, M) = \left\{ \Sigma : |\sigma_{ij}| \leq M_1 |i - j|^{-(\alpha+1)} \text{ for } i \neq j \text{ and } \lambda_{\max}(\Sigma) \leq M_0 \right\}, \tag{5.109}$$

where $\lambda_{\max}(\Sigma)$ is the maximum eigenvalue of the matrix Σ. $\mathcal{G}_\alpha$ is a subset of $\mathcal{F}_\alpha(M_0, M)$ as long as $M_1 \leq \alpha M$.

Let $\mathcal{P}_\alpha = \mathcal{P}'_\alpha(M_0, M)$ denote the set of distribution of $\mathbf{X}_1$ that satisfies (5.105) and (5.109).

Theorem 5.57 (Minimax risk under Frobenius norm by Cai, Zhang and Zhou (2010) [351]) *The minimax risk under Frobenius norm satisfies*

$$\inf_{\hat{\Sigma}} \sup_{\mathcal{P}_\alpha} \mathbb{E} \frac{1}{p} \| \hat{\Sigma} - \Sigma \|_F^2 \asymp \inf_{\hat{\Sigma}} \sup_{\mathcal{P}'_\alpha} \mathbb{E} \frac{1}{p} \| \hat{\Sigma} - \Sigma \|_F^2 \asymp \min\{ n^{-(2\alpha+1)/(2(\alpha+1))}, \frac{p}{n} \}. \tag{5.110}$$

The inverse of the covariance matrix $\boldsymbol{\Sigma}$ is of significant interest. For this purpose, we require the minimum eigenvalue of $\boldsymbol{\Sigma}$ to be bounded away from zero. For $\delta > 0$, we define

$$\mathbf{L}_\delta = \{\boldsymbol{\Sigma} : \lambda_{\min}(\boldsymbol{\Sigma}) \geq \delta\}. \qquad (5.111)$$

Let $\tilde{\mathcal{P}}_\alpha = \tilde{\mathcal{P}}_\alpha(M_0, M, \rho, \delta)$ denote the set of distributions of $\mathbf{X}_1$ that satisfy (5.100), (5.105), and (5.111), and similarly, distribution in $\tilde{\mathcal{P}}'_\alpha = \tilde{\mathcal{P}}'_\alpha(M_0, M, \rho, \delta)$ that satisfy (5.105), (5.109), and (5.111).

Theorem 5.58 (Minimax risk of estimating the inverse covariance matrix by Cai, Zhang and Zhou (2010) [351]) *The minimax risk under Frobenius norm satisfies*

$$\inf_{\hat{\Sigma}} \sup_{\tilde{P}} \mathbb{E}\|\hat{\boldsymbol{\Sigma}}^{-1} - \boldsymbol{\Sigma}^{-1}\|^2 \asymp \min\{n^{-2\alpha/(2(\alpha+1))} + \frac{\log p}{n}, \frac{p}{n}\}, \qquad (5.112)$$

where $\tilde{P}$ denotes either $\tilde{\mathcal{P}}_\alpha$ or $\tilde{\mathcal{P}}'_\alpha$.

5.6.6 Banding Sample Autocovariance Matrices of Stationary Processes

Nonstationary covariance estimators by banding a sample covariance matrix or its Cholesky factor were considered in [352] and [346] in the context of longitudinal and multivariate data. Estimation of covariance matrices of stationary processes was considered in [353]. Under a short-range dependent condition for a wide class of nonlinear processes, it is shown that the banded covariance matrix estimates converge, in operator norm, to the true covariance matrix with explicit rates of convergence. Their consistency was established under some regularity conditions when

$$n, p \to \infty \quad and \quad n^{-1} \log p \to 0,$$

where n and p are the number of subjects and variables, respectively. Many good references are included in [353].

Given a realization of $X_1, \ldots, X_n$ of a mean-zero stationary process $\{X_t\}$, its autocovariance function $\sigma_k = cov(X_0, X_k)$ can be estimated by

$$\hat{\sigma}_k = \frac{1}{n} \sum_{i=1}^{n-|k|} X_i X_{i+k}, \quad k = 0, \pm 1, \ldots, \pm(n-1). \qquad (5.113)$$

It is known that for fixed $k \in \mathbb{Z}$, under ergodicity condition, $\hat{\sigma}_k \to \sigma_k$ in probability. Entry-wise convergence, however, does not automatically imply that $\hat{\boldsymbol{\Sigma}}_n = (\hat{\sigma}_{i-j})_{1 \leq i, j \leq n}$ is a good estimator of $\boldsymbol{\Sigma}_n = (\sigma_{i-j})_{1 \leq i, j \leq n}$. Indeed, although positive definite, $\hat{\boldsymbol{\Sigma}}_n$ is not uniformly close to the population (true) covariance matrix $\boldsymbol{\Sigma}_n$, in the sense that the largest eigenvalue or the operator norm of $\hat{\boldsymbol{\Sigma}}_n - \boldsymbol{\Sigma}_n$ does not converge to zero. Such uniform convergence is important when studying the rate of convergence of the finite predictor coefficients and performance of various classification methods in time series.

Not necessarily positive definite, the covariance matrix estimator is of the form

$$\hat{\boldsymbol{\Sigma}}_n = (\sigma_{i-j} \mathbf{1}_{|i-j| \leq l})_{1 \leq i, j \leq n}, \qquad (5.114)$$

where $l \geq 0$ is an integer. It is a truncated version of $\hat{\Sigma}_n$ preserving the diagonal and the $2l$ main subdiagonals; if $l \geq n - 1$, then $\hat{\Sigma}_{n,l} = \hat{\Sigma}_n$. By following [346], $\hat{\Sigma}_{n,l}$ is called the banded covariance matrix estimate and l its band parameter.

Hannan and Deistler (1988) [354] have considered certain linear ARMA processes and obtained the uniform bound

$$\|\hat{\Sigma}_{n,l} - \hat{\Sigma}_n\|_\infty = \mathcal{O}(\sqrt{\log \log n}/\sqrt{n}), l \leq (\log n)^\alpha, \alpha < \infty.$$

Here, we consider the comparable results for nonlinear processes, mainly following the notation and results of [353].

Let $\varepsilon_i, i \in \mathbb{Z}$, be independent and identically distributed (i.i.d.) random variables. Assume that $\{X_i\}$ is a causal process of the form

$$X_i = g(\cdots, \varepsilon_{i-1}, \varepsilon_i), \tag{5.115}$$

where g is a measurable function such that X_i is well-defined and $\mathbb{E}(X_i^2) < \infty$. Many stationary processes fall within the framework of (5.115).

To introduce the dependent structure, let $(\varepsilon_i')_{i \in \mathbb{Z}}$ be an independent copy of $(\varepsilon_i)_{i \in \mathbb{Z}}$ and $\xi_i = (\cdots, \varepsilon_{i-1}, \varepsilon_i)$. Following [355], for $i \geq 0$, let

$$\xi_i' = (\cdots, \varepsilon_{-1}, \varepsilon_0', \varepsilon_1, \ldots, \varepsilon_{i-1}, \varepsilon_i), X_i' = g(\xi_i').$$

For $\alpha > 0$, define the physical dependence of measure

$$\delta_\alpha(i) = \|X_i - X_i'\|_\alpha. \tag{5.116}$$

Here, for a random variable Z, we write $Z \in \mathcal{L}^\alpha$, if

$$\|Z\|_\alpha \equiv [\mathbb{E}(|Z|^\alpha)]^{1/\alpha} < \infty,$$

and write $\| \cdot \| = \| \cdot \|_2$. Observe that $X_i' = g(\xi_i')$ is a coupled version of $X_i = g(\xi_i)$ with ε_0 in the latter replaced by an i.i.d. copy ε_0'. The quantity $\delta_p(i)$ measures the dependence of X_i on ε_0. We say that $\{X_i\}$ is short-range dependent with moment α if

$$\Delta_\alpha \equiv \sum_{i=0}^{\infty} \delta_\alpha(i) < \infty. \tag{5.117}$$

That is, the cumulative impact of ε_0 on future values of the process or $\{X_i\}_{i \geq 0}$ is finite, thus implying a short-range dependence.

Example 5.11 ([353])
Let

$$X_j = g\left(\sum_{i=0}^{\infty} a_i \varepsilon_{j-i}\right),$$

where a_i are real coefficients with $\sum_{i=0}^{\infty} |a_i| < \infty$, ε_i are i.i.d. with $\varepsilon_i \in \mathcal{L}^\alpha, \alpha > 1$, and g is a Lipschitz continuous function. Then, $\sum_{i=0}^{\infty} a_i \varepsilon_{j-i}$ is a well-defined random variable and $\delta_\alpha(i) = \mathcal{O}(|a_i|)$. Hence we have (5.117). □

Example 5.12 ([353])
Let ε_i be i.i.d. random variables and set

$$X_i = g(X_{i-1}, \varepsilon_i),$$

where g is a bivariate function. Many nonlinear time series models follow this framework.   □

Let $\rho^2(\mathbf{A})$ is the largest eigenvalue of $\mathbf{A}^T\mathbf{A}$. The $n \times n$ matrix $\mathbf{A}$ has the operator norm $\rho(\mathbf{A})$.

We define the project operator $\mathcal{P}_k$ as

$$\mathcal{P}_k \cdot = \mathbb{E}(\cdot | \xi_k) - \mathbb{E}(\cdot | \xi_{k-1}), k \in \mathbb{Z}.$$

Theorem 5.59 (No convergence in probability [353]) *Assume that the process X_i in (5.115) satisfies*

$$\sum_{i=0}^{\infty} \|\mathcal{P}_k X_i\| < \infty.$$

If $\sum_{i=0}^{\infty} \|\mathcal{P}_k X_i\| > 0$, then, $\rho(\hat{\mathbf{\Sigma}}_n - \mathbf{\Sigma}_n)$ does not converge to zero in probability.

Theorem 5.60 (Convergence in probability [353]) *Let $2 < \alpha \leq 4$ and $q = \alpha/2$. Assume (5.117) and $0 \leq l < n - 1$. Then,*

$$\|\rho(\hat{\mathbf{\Sigma}}_{n,l} - \mathbf{\Sigma}_n)\|_q \leq c_\alpha (l+1) n^{1/q-1} \|X_1\|_\alpha \Delta_\alpha + \frac{2}{n} \sum_{j=1}^{l} j |\sigma_j| + 2 \sum_{j=l+1}^{n} |\sigma_j|, \quad (5.118)$$

where $c_\alpha > 0$ is a constant depending only on α.

5.7 Free Probability

In quantum detection, tensor products are needed. For a large number of random matrices, tensor products are too computationally expensive for our problem at hand. Free probability is a highly noncommunicative probability theory with independence based on *free products* instead of tensor products [356]. Basic examples include the asymptotic behavior of large Gaussian random matrices. The freeness (its beauty and fruitfulness) is the central concept [357].

Independent symmetric Gaussian matrices which are random matrices (also noncommunitative matrix-valued random variables) are asymptotic free. See Appendix A.5 for details on noncommunicative matrix-valued random variables: random matrices are their special cases.

In this subsection, we take the liberty of drawing material from [12, 13]. Here we are motivated for spectrum sensing and (possible) other applications in cognitive radio network. Free probability is a mathematical theory that studies noncommunitative random variables. The "freeness" is the analogue of the classical notation of independence, and it is connected with free products. This theory was initiated by Dan Voiculescu around 1986, who made the statement [16]:

free probability theory = noncommunitative probability theory + free independence.

His first motivation was to study the von Neumann algebras of free groups. One of Voiculescu's central observations was that such groups can be equipped with tracial states (also called states), which resemble expectations in classical probability.

What is the spectrum of the sum $\mathbf{A} + \mathbf{B}$ [358]? For deterministic matrices $\mathbf{A}$ and $\mathbf{B}$ one cannot in general determine the eigenvalues of $\mathbf{A} + \mathbf{B}$ from those of $\mathbf{A}$ and $\mathbf{B}$ alone, as they depend on the eigenvectors of $\mathbf{A}$ and $\mathbf{B}$ as well. However, it turns out that for large random matrices $\mathbf{A}$ and $\mathbf{B}$ satisfying a property called *freeness*, the limiting spectrum of the sum $\mathbf{A} + \mathbf{B}$ can indeed be determined from the individual spectra of $\mathbf{A}$ and $\mathbf{B}$. This is a central result in *free probability theory*.

Define the functional φ as

$$\varphi(\mathbf{A}_n^k) = \frac{1}{n}\mathrm{Tr}(E\mathbf{A}_n^k).$$

ϕ stands for the normalized expected trace of a random matrix.

The matrices $\mathbf{A}_1, \ldots, \mathbf{A}_m$ are called free if

$$\varphi\left([p_1(\mathbf{A}_{i_1}) \cdots p_k(\mathbf{A}_{i_k})]\right) = 0$$

whenever

- $p_1, \ldots, p_k$ are polynomials in one variable;
- $i_1 \neq i_2 \neq i_3 \cdots \neq i_k$ (only neighboring elements are required to be distinct);
- $\varphi(p_j(\mathbf{A}_{i_j})) = 0$ for all $j = 1, \ldots, k$.

For independent random variables, the joint distribution can be specified completely by the marginal distributions [359]. For free random variables, the same result can be proven, directly from definition. In particular, if $\mathbf{X}$ and $\mathbf{Y}$ are free, then the moments $\varphi[(\mathbf{X} + \mathbf{Y})^n]$ of $\mathbf{X}$ and $\mathbf{Y}$ can be completely specified by the moments of $\mathbf{X}$ and the moments of $\mathbf{Y}$. The distribution is naturally called free convolution of the two marginal distributions. Classical convolution can be computed via transforms: the log moment generating function of the distribution of $\mathbf{X} + \mathbf{Y}$ is the sum of the log moment generating function of the individual distributions of $\mathbf{X}$ and $\mathbf{Y}$. In contrast, for free convolution, the appropriate transform is called the R-transform. This is defined via the Steltjes transform given by (5.34).

Asymptotic Freeness

To apply the theory of free probability to random matrix theory, we need to extend the definition of free to asymptotic freeness, by replacing the state functional φ by ϕ

$$\phi(\mathbf{A}) = \lim_{n \to \infty} \frac{1}{n}\mathbb{E}\mathrm{Tr}(\mathbf{A}_n).$$

The expected asymptotic pth moment is $\phi(\mathbf{A}^p)$ and $\phi(\mathbf{I}) = 1$. The definition of asymptotic freeness is analogous to the concept of independent random variables. However, statistical independence does not imply asymptotic freeness.

The Hermitian random matrices $\mathbf{A}$ and $\mathbf{B}$ are asymptotic free if for all l and for all polynomials $p_i(\cdot)$ and $q_i(\cdot)$ with $1 \le i \le l$ such that

$$\phi(p_i(\mathbf{A})) = \phi(q_i(\mathbf{B})) = 0,$$
$$\phi(p_1(\mathbf{A})q_1(\mathbf{A}) \cdots p_l(\mathbf{A})q_l(\mathbf{A})) = 0.$$

We state the following useful relationships for asymptotically free $\mathbf{A}$ and $\mathbf{B}$

$$\phi(\mathbf{A}^k\mathbf{B}^l) = \phi(\mathbf{A}^k)\phi(\mathbf{B}^l),$$
$$\phi(\mathbf{ABAB}) = \phi^2(\mathbf{B})\phi(\mathbf{A}^2) + \phi^2(\mathbf{A})\phi(\mathbf{B}^2) - \phi^2(\mathbf{A})\phi^2(\mathbf{B}).$$

One approach to characterize the asymptotic spectrum of a random matrix is to obtain its moments of all orders. The moments of a noncommunicative polynomial $p(\mathbf{A}, \mathbf{B})$ of two asymptotically free random matrices can be computed from the individual moments of $\mathbf{A}$ and $\mathbf{B}$. Thus, if $p(\mathbf{A}, \mathbf{B})$, $\mathbf{A}$ and $\mathbf{B}$ are Hermitian, the asymptotic spectrum of $p(\mathbf{A}, \mathbf{B})$ depends on only those of $\mathbf{A}$ and $\mathbf{B}$, even if they do not have the same eigenvectors!

Example 5.13 (Moments of polynomial matrix function $p(\mathbf{A}, \mathbf{B}) = \mathbf{A} + \mathbf{B}$)
Let us consider the important special case of $p(\mathbf{A}, \mathbf{B}) = \mathbf{A} + \mathbf{B}$. Under $\mathcal{H}_1$, the sample covariance matrix has the form

$$\phi(\mathbf{A} + \mathbf{B}) = \phi(\mathbf{A}) + \phi(\mathbf{B}),$$

$$\phi[(\mathbf{A} + \mathbf{B})^2] = \phi(\mathbf{A}^2) + \phi(\mathbf{B}^2) + 2\phi(\mathbf{A})\phi(\mathbf{B}),$$

$$\phi[(\mathbf{A} + \mathbf{B})^3] = \phi^3(\mathbf{A}) + \phi(\mathbf{B}^3) + 3\phi(\mathbf{A})\phi(\mathbf{B}^2) + 3\phi(\mathbf{B})\phi(\mathbf{A}^2),$$

$$\phi[(\mathbf{A} + \mathbf{B})^4] = \phi^4(\mathbf{A}) + \phi(\mathbf{B}^4) + 4\phi(\mathbf{A})\phi(\mathbf{B}^3)$$
$$+ 4\phi(\mathbf{B})\phi(\mathbf{A}^3) + 2\phi^2(\mathbf{B})\phi(\mathbf{A}^2)$$
$$+ 2\phi^2(\mathbf{A})\phi(\mathbf{B}^2) + 2\phi(\mathbf{B}^2)\phi(\mathbf{A}^2).$$

All higher moments can be computed analogously.   □

[13] compiles a list of some of the most useful instances of asymptotic freeness that have been shown so far. Let us list some here:

1. Any random matrix and the identity are asymptotically free.
2. Independent Gaussian standard Wigner matrices are asymptotically free.
3. Let $\mathbf{X}$ and $\mathbf{Y}$ be independent standard Gaussian matrices. Then $\{\mathbf{X}, \mathbf{X}^H\}$ and $\{\mathbf{Y}, \mathbf{Y}^H\}$ are asymptotically free.
4. Independent standard Wigner matrices are asymptotically free.

Sum of Asymptotic Free Random Matrices

Free probability is useful mainly due to the following theorem.

Theorem 5.61 (Sum of two asymptotic free random matrices) *If* $\mathbf{A}$ *and* $\mathbf{B}$ *are asymptotically free random matrices, then the R-transform of their sum satisfies*

$$R_{\mathbf{A}+\mathbf{B}}(z) = R_{\mathbf{A}}(z) + R_{\mathbf{B}}(z).$$

In particular, the following translation property is valid

$$R_{\mathbf{A}+\gamma\mathbf{I}}(z) = R_{\mathbf{A}}(z) + R_{\gamma\mathbf{I}}(z) = R_{\mathbf{A}}(z) + \gamma.$$

Theorem 5.62 (Free probability central limit theorem) *If* $\mathbf{A}_1, \mathbf{A}_2, \ldots$ *are a sequence of* $N \times N$ *asymptotically free random matrices. Assume that* $\phi(\mathbf{A}_i) = 0$ *and* $\phi(\mathbf{A}_i^2) = 1$. *Further assume that* $\sup_i |\phi(\mathbf{A}_i^k)| < \infty$ *for all k. Then, as* $m, N \to \infty$, *the asymptotic spectrum of*

$$\frac{1}{\sqrt{m}}(\mathbf{A}_1 + \mathbf{A}_2 + \cdots \mathbf{A}_m)$$

converges in distribution to the semicircle law, that is, for every k,

$$\phi\left(\frac{1}{\sqrt{m}}(\mathbf{A}_1 + \mathbf{A}_2 + \cdots \mathbf{A}_m)^k\right) \to \begin{cases} 0, k \text{ odd} \\ \frac{1}{1+\frac{k}{2}}\binom{k}{\frac{k}{2}}, k \text{ even.} \end{cases}$$

Let us revisit the problem of sum of K random matrices in Section 3.6. The K sample covariance matrices are asymptotic free.

Example 5.14 ($\mathbf{HH}^H$ [13])
Let $\mathbf{H}$ be an $N \times m$ random matrix whose entries are zero-mean i.i.d. Gaussian random variables with variance $\frac{1}{\sqrt{mN}}$ and denote

$$\frac{1}{N}\sqrt{m} = \varsigma.$$

We can represent

$$\mathbf{HH}^H = \frac{1}{\sqrt{m}}\sum_{i=1}^{m} \mathbf{s}_i\mathbf{s}_i^H \tag{5.119}$$

with $\mathbf{s}_i$ an N-dimensional vector whose entries are zero-mean i.i.d. with variance $\frac{1}{\sqrt{N}}$, it can be shown that as $N, m \to \infty$ with $\frac{N}{m} \to 0$, the asymptotic spectrum of the matrix

$$\mathbf{HH}^H - \varsigma\sqrt{N}\mathbf{I}$$

is the semicircle law.   □

Example 5.15 (Sum of K (random) sample covariance matrices in Section 3.6)
The sample covariance matrices have the form

$$\mathbf{S}_k = \frac{1}{N}\mathbf{Y}_k\mathbf{Y}_k^H, k = 1, 2, \ldots, K,$$

where $\mathbf{Y}_k$ have m row vectors and N column vectors. A long data record is divided into K segments; each segment can be used to estimate the sample covariance matrix. The sum of K sample covariance matrices is

$$\mathbf{S_Y} = \sum_{k=1}^{K} \mathbf{S}_k = \frac{1}{N} \sum_{k=1}^{K} \mathbf{Y}_k \mathbf{Y}_k^H .$$

Under $\mathcal{H}_0$: only Gaussian noise is present, each $\mathbf{S}_k$ is of the form of (5.119). Thus the sum has the form

$$\mathbf{S_Y} = \frac{1}{\sqrt{mK}} \sum_{i=1}^{mK} \mathbf{s}_i \mathbf{s}_i^H .$$

The sum of K sample covariance matrices will make the asymptotic spectrum more like the semicircle law since in practice $\frac{N}{mK} \to \infty$ with faster rate. $\square$

Products of Asymptotic Free Random Matrices

The S-transform plays an analogous role to the R-transform for products (instead of sum) of asymptotically free matrices.

Theorem 5.63 *Let* $\mathbf{A}$ *and* $\mathbf{B}$ *be nonnegative asymptotically free random matrices. The S-transform of their products satisfies*

$$\Sigma_{\mathbf{A+B}}(x) = \Sigma_{\mathbf{A}}(x)\Sigma_{\mathbf{B}}(x).$$

The S-transform is the free analog of the Mellin transform in classical probability theory, whereas the R-transform is the free analog of the log-moment generating function in classical probability theory.

There are useful theorems [11] to calculate $\phi[(\mathbf{A} + \mathbf{B})^n]$ and $\phi[(\mathbf{AB})^n]$.

Moments of the Sums and Products

Theorem 5.64 ([13]) *Consider matrices* $\mathbf{A}_1, \ldots, \mathbf{A}_l$ *whose size is such that the product* $\mathbf{A}_1, \ldots, \mathbf{A}_l$ *is defined. Some of these matrices are allowed to be identical. Omitting repetitions, assume that the matrices are asymptotically free. Let* ρ *be the partition of* $\{1, \ldots, l\}$ *determined by the equivalence relation* $j \equiv k$ *if* $i_j = i_k$. *For each partition* ϖ *of* $\{1, \ldots, l\}$, *let*

$$\phi_{\varpi} = \prod_{\substack{\{j_1, \ldots, j_r\} \in \varpi \\ j_1 < \cdots < j_r}} \phi(\mathbf{A}_{j_1} \cdots \mathbf{A}_{j_r}).$$

There exist universal coefficients $c(\varpi, \rho)$ *such that*

$$\phi(\mathbf{A}_1 \cdots \mathbf{A}_l) = \sum_{\varpi \leq \rho} c(\varpi, \rho)\phi_{\varpi}$$

where $\varpi \leq \rho$ *indicates that* ϖ *is finer than* ρ.

Finding an explicit formula for the coefficients $c(\varpi, \rho)$ is a nontrivial combinatorial problem that has been solved by Speicher [360]. From Theorem 5.64, $\phi(\mathbf{A}_1 \cdots \mathbf{A}_l)$ is completely determined by the moments of the individual matrices.

Theorem 5.65 ([11]) *Let* $\mathbf{A}$ *and* $\mathbf{B}$ *be nonnegative asymptotically free random matrices. Then, the moments of their sum* $\mathbf{A} + \mathbf{B}$ *are expressed by the free cumulants of* $\mathbf{A}$ *and* $\mathbf{B}$ *as*

$$\phi[(\mathbf{A} + \mathbf{B})^n] = \sum_{\varpi} \prod_{V \in \varpi} (c_{|V|}(\mathbf{A}) + c_{|V|}(\mathbf{B}))$$

where the summation is over all noncrossing partitions of $1, \ldots, n$, $c_l(\mathbf{A})$ *denotes the lth free cumulant of* $\mathbf{A}$, *and* $|V|$ *denotes the cardinality of V.*

Theorem 5.65 is based on the fact that, if $\mathbf{A}$ and $\mathbf{B}$ be nonnegative asymptotically free random matrices, the free cumulants of the sum satisfy

$$c_l(\mathbf{A} + \mathbf{B}) = c_l(\mathbf{A}) + c_l(\mathbf{B}).$$

Theorem 5.66 ([11]) *Let* $\mathbf{A}$ *and* $\mathbf{B}$ *be nonnegative asymptotically free random matrices. Then, the moments of their sum* $\mathbf{A} + \mathbf{B}$ *are expressed as*

$$\phi[(\mathbf{AB})^n] = \sum_{\varpi_1, \varpi_2} \prod_{V_1 \in \varpi_1} c_{|V_1|}(\mathbf{A}) \prod_{V_2 \in \varpi_2} c_{|V_2|}(\mathbf{B})$$

where the summation is over all noncrossing partitions of $1, \ldots, n$.

5.7.1 Large Random Matrices and Free Convolution

5.7.1.1 Random Matrices and Free Random Variables

In free probability, large random matrices is an example of "free" random variables. Let $\mathbf{A}_N$ be an $N \times N$ symmetric (or Hermitian) random matrix with real eigenvalues. So the two-dimensional complex problem is converted into a one-dimensional real-value problem. The probability measure on the set of its eigenvalues

$$\lambda_1, \lambda_2, \ldots, \lambda_N$$

(counted with multiplicities) is given by

$$\mu_{\mathbf{A}_N} = \frac{1}{N} \sum_{i=1}^{N} \delta_{\lambda_i}.$$

We are interested in the limiting spectral measure $\mu_{\mathbf{A}}$ as $N \to \infty$. This limiting spectral measure is uniquely characterized by its moments, when compactly supported. We refer to $\mathbf{A}$ as an element of the "algebra" with probability measure $\mu_{\mathbf{A}}$ and moments above.

For two random matrices $\mathbf{A}_N$ and $\mathbf{B}_N$ with limiting probability distribution $\mu_{\mathbf{A}}$ and $\mu_{\mathbf{B}}$, we would like to compute the limiting probability distribution for $\mathbf{A}_N + \mathbf{B}_N$ and $\mathbf{A}_N \mathbf{B}_N$ in terms of the moments of $\mu_{\mathbf{A}}$ and $\mu_{\mathbf{B}}$. As treated above, the appropriate structure of "freeness," analogous to independence for "classical" random variables, is what we need to impose on $\mathbf{A}_N$ and $\mathbf{B}_N$, in order to compute these distributions. Since $\mathbf{A}$ and $\mathbf{B}$ do not commute we are dealing with noncommutative algebra. Since all possible products of $\mathbf{A}$ and $\mathbf{B}$ are allowed, we have the "free" products, that is, all words in $\mathbf{A}$ and $\mathbf{B}$ are allowed. We have already dealt with how to compute the moments of these products. The connection with random matrices comes in, because a pair of random matrices $\mathbf{A}_N$ and $\mathbf{B}_N$ are asymptotically free, that is, in the limit of $N \to \infty$, so long as at least one of $\mathbf{A}_N$ or $\mathbf{B}_N$ has what amounts to eigenvectors that are uniformly distributed with Haar measure. This result is stated precisely in [356].

Table 5.3 lists definitions of R-transform and S-transform and their properties.

5.7.1.2 Free Additive Convolution

When $\mathbf{A}_N$ and $\mathbf{B}_N$ are asymptotically free, the (limiting) spectral measure $\mu_{\mathbf{AB}}$ for random matrices of the form

$$\mathbf{A}_N + \mathbf{B}_N$$

is given by the *free additive convolution* of the probability measures $\mu_{\mathbf{A}}$ and $\mu_{\mathbf{B}}$ and written as [356]

$$\mu_{\mathbf{A+B}} = \mu_{\mathbf{A}} \boxplus \mu_{\mathbf{B}}, \tag{5.120}$$

An algorithm in terms of the so-called R-transform exists for computing $\mu_{\mathbf{A+B}}$ from $\mu_{\mathbf{A}}$ and $\mu_{\mathbf{B}}$. See [356] for details and [361] for computational issues.

5.7.1.3 Free Multiplicative Convolution

When $\mathbf{A}_N$ and $\mathbf{B}_N$ are asymptotically free, the (limiting) spectral measure $\mu_{\mathbf{AB}}$ for random matrices of the form

$$\mathbf{A}_N \mathbf{B}_N$$

is given by the *free multiplicative convolution* of the probability measures $\mu_{\mathbf{A}}$ and $\mu_{\mathbf{B}}$ and written as [356]

$$\mu_{\mathbf{AB}} = \mu_{\mathbf{A}} \boxtimes \mu_{\mathbf{B}}, \tag{5.121}$$

The algorithm for computing $\mu_{\mathbf{AB}}$ is given in [254, 361–364].

The convolution operators on the noncommunicative algebra of large random matrices exist, and can be computed efficiently (e.g., in MATLAB codes). Symbolic computational tools are now available to perform these nontrial computations efficiently [361, 362]. These tools enable us to analyze the structure of sample covariance matrices and design algorithms that take advantage of this structure [254].

5.7.1.4 Applications to Rank Estimation and Spectrum Sensing

Since the Wishart matrix so formed in (5.13) has eigenvectors that are uniformly distributed with Haar measure, the matrices $\mathbf{R}$ and $\mathbf{W}(\alpha)$ are asymptotically free! Thus the limiting probability measure $\mu_{\hat{\mathbf{R}}}$ can be obtained using *free multiplication convolution* as

$$\mu_{\hat{\mathbf{R}}} = \mu_{\mathbf{R}} \boxtimes \mu_{\mathbf{W}}, \tag{5.122}$$

where $\mu_{\hat{R}}$ is the limiting probability measure on the true covariance matrix $\mathbf{R}$ and $\mu_{\mathbf{W}}$ is the Marchenko-Pastur density [251], which is defined in (5.3). As given in (5.7), the limiting spectral measure of $\mathbf{R}$ is simply

$$\mu_{\mathbf{R}} = p\delta(x - \rho - 1) + (1 - p)\delta(x - 1).$$

The free probability results are exact when $N \to \infty$, but the predictions are very accurate for $N \approx 8$, for rank estimation [254].

Example 5.16 (Rank estimation)
Let $\mathbf{HH}^H$ in (5.7) have np of its eigenvalues of magnitude ρ and $n(1 - p)$ of its eigenvalues of manitude 0 where $p < 1$. This corresponds to $\mathbf{H}$ being an $n \times L$ matrix with $L < n$ with $p = \frac{L}{n}$ so that L of its singular values are of magnitude $\sqrt{\rho}$ while the eigenvectors of $\mathbf{H}$ are unknown or random. Since free multiplicative convolution predicts the spectrum of the sample covariance matrix $\hat{\mathbf{R}}$ so accurately such that we can use free multiplicative *deconvolution*, to infer the parameters of the underlying covariance matrix model from just one realization of the sample covariance matrix!

The first three moments of $\hat{\mathbf{R}}$ can be analytically parameterized in terms of the unknown parameters β, ρ and the known parameter $c = n/N$ as

$$\varphi(\hat{\mathbf{R}}) = 1 + p\rho,$$

$$\varphi(\hat{\mathbf{R}}^2) = 1 + p\rho^2 + c + 1 + 2p\rho c + 2p\rho + cp^2\rho^2,$$

$$\varphi(\hat{\mathbf{R}}^3) = 1 + 3c + c^2 + 3\rho^2 p + 3\rho^3 cp^2 + 3p\rho + 9p\rho c + 6p^2\rho^2 c$$
$$+ 3c\rho^2 p + 3\rho p c^2 + 3\rho^2 c^2 p^2 + \rho^3 p^3 c^2 + \rho^3 p,$$

Given an $n \times N$ observation matrix $\mathbf{Y}_n$, we can compute estimates of the first three moments as

$$\hat{\varphi}(\hat{\mathbf{R}}^k) = \frac{1}{n}\text{Tr}\left[\left(\frac{1}{N}\mathbf{Y}_n\mathbf{Y}_n^H\right)^k\right], k = 1, 2, 3.$$

Since we know $c = n/N$, we can estimate ρ, p by simply solving the nonlinear system of equations (minimizing the least squares)

$$(\hat{\rho}, \hat{p}) = \underset{\rho > 0, p > 0}{\arg\min} \|\varphi(\hat{\mathbf{R}}^k) - \hat{\varphi}(\hat{\mathbf{R}}^k)\|^2.$$

For the example of $n = 200$ and $p = 0.5$, the estimated rank is within 1 dimension of the true rank of the system which is $np = 100$.   □

Example 5.17 (Spectrum sensing)
Consider the standard form of (5.10), which is repeated here for convenience,

$$\begin{aligned} \mathcal{H}_0 &: \hat{\mathbf{R}} = \mathbf{YY}^H = \mathbf{WW}^H, \\ \mathcal{H}_1 &: \hat{\mathbf{R}} = \mathbf{YY}^H = \mathbf{XX}^H + \mathbf{WW}^H, \end{aligned} \qquad (5.123)$$

The true covariance matrices are

$$
\begin{aligned}
\mathcal{H}_0 &: \mathbf{R} = \sigma^2 \mathbf{I}, \\
\mathcal{H}_1 &: \mathbf{R} = \mathbf{H}\mathbf{H}^H + \sigma^2 \mathbf{I},
\end{aligned}
\tag{5.124}
$$

The conventional approach to find the power of the received signal plus noise is to use (5.124). In practice, the usual approaches are to use large sample covariance matrices through (5.123). Indeed, the sample covariance matrix is connected with the true covariance matrix by the property of Wishart distribution through (5.12).

Using (5.122), we can convert the problem of calculating the sample covariance matrix $\hat{\mathbf{R}}$ into the problem of calculating the true covariance matrix $\mathbf{R}$, with the help of the Wishart matrix $\mathbf{W}(c)$! Recall that $\mathbf{W}(c) = \frac{1}{N}\mathbf{Z}\mathbf{Z}$ is formed from an $n \times N$ Gaussian random matrix. Once again, c is defined as the limit $n/N \to c > 0$ as $n, N \to \infty$. Under $\mathcal{H}_1$, we have the form of (5.7). We can thus calculate the limiting probability measure $\mu_{\hat{\mathbf{R}}}$ using (5.12). □

5.7.2 Vandermonde Matrices

For notation and some key theorems, we follow [365] closely. Vandermonde matrices have a central role in signal processing such as the fast Fourier transform or Hadamard transforms. A Vandermonde matrix with complex entries on the unit circle has the following form

$$
\mathbf{V} = \frac{1}{\sqrt{N}}
\begin{bmatrix}
1 & \cdots & 1 \\
e^{-j\omega_1} & \cdots & e^{-j\omega_L} \\
\vdots & \ddots & \vdots \\
e^{-j(N-1)\omega_1} & \cdots & e^{-j(N-1)\omega_L}
\end{bmatrix},
\tag{5.125}
$$

where the factor $\frac{1}{\sqrt{N}}$ and the assumption of $e^{-j\omega_i}$ are included to ensure that the analysis will give limiting asymptotic behavior defined in the asymptotic regime of

$$
\text{Asymptotic regime:} \quad N \to \infty, L \to \infty, \quad \text{but } \frac{L}{N} \to c.
\tag{5.126}
$$

We are interested in the case where $\omega_1, \ldots, \omega_L$ are independent and identically distributed (i.i.d.), taking values in $[0, 2\pi]$. The ω_i is called phase distributions. $\mathbf{V}$ will be only to denote Vandermonde matrices in this section with a given phase distribution, and the dimensions of the Vandermonde matrices will always be $N \times L$.

[111] has some related results. The overwhelming majority of the known results are concerned about Gaussian matrices or matrices with independent entries. Very few results are available in the literature on matrices whose structure is strongly related to the Vandermonde case.

Often, we are interested in only the moments. It will be shown that asymptotically, the moments of the Vandermonde matrices $\mathbf{V}$ depend only on the ratio c and the phase distributions, and have explicit expressions. Moments are useful for performing deconvolution. The normalized trace is defined as

$$
\text{tr}(\mathbf{A}) = \frac{1}{L}\text{Tr}(\mathbf{A}).
$$

The matrices $\mathbf{D}_r(N)$, $1 \le r \le n$ will denote nonrandom diagonal $L \times L$ matrices, where we implicitly assume that $\frac{L}{N} \to c$.

We say the $\{\mathbf{D}_r(N)\}_{1 \le r \le n}$ have the joint limit distribution as $N \to \infty$ if the limit

$$D_{i_1,\ldots,i_n} = \lim_{N \to \infty} \mathrm{tr}(\mathbf{D}_{i_1}(N) \cdots \mathbf{D}_{i_n}(N))$$

exists for all choices of $i_1, \ldots, i_s \in \{1, \ldots, n\}$.

The concepts from partition theory are needed. We denote by $\mathcal{P}(n)$ the set of all partitions of $\{1, \ldots, n\}$, and use ρ as notation for a partition in $\mathcal{P}(n)$. We write $\rho = \{W_1, \ldots, W_k\}$, where W_j will be used to denote the blocks of ρ. $|\rho| = k$ denotes the number of blocks in ρ and $|W_j|$ will represent the number of entries in a given block.

For $\rho = \{W_1, \ldots, W_k\}$, with $W_i = \{\omega_{i1}, \ldots, \omega_{i|W_i|}\}$, we define

$$D_{W_i} = D_{i_{\omega i1}, \ldots, i_{\omega i|W_i|}}$$

$$D_\rho = \prod_{i=1}^{k} D_{W_i}.$$

For $\rho \in \mathcal{P}(n)$, define

$$K_{\rho,\omega,N} = \frac{1}{N^{n+1-|\rho|}} \int_{(0,2\pi)^{|\rho|}} \prod_{i=1}^{k} \frac{1 - e^{jN(\omega_{b(k-1)} - \omega_{b(k)})}}{1 - e^{j(\omega_{b(k-1)} - \omega_{b(k)})}} d\omega_1 \cdots d\omega_{|\rho|}$$

where

$$\omega_{W_1}, \ldots, \omega_{W_{|\rho|}} \tag{5.127}$$

are i.i.d. (indexed by the blocks of ρ), all with the same distribution as ω, and where $b(k)$ is the block of ρ which contains k (notation is cyclic, that is, $b(0) = b(n)$). If the limit

$$K_{\rho,\omega} = \lim_{N \to \infty} K_{\rho,\omega,N}$$

exists, then we call it a Vandermonde mixed moment expansion coefficient.

Theorem 5.67 ([365]) *Assume that the $\{\mathbf{D}_r(N)\}_{1 \le r \le n}$ have a joint limit distribution as $N \to \infty$. Assume also that all Vandermonde mixed moment expansion coefficients $K_{\rho,\omega}$ exist. Then, the limit*

$$M_n = \lim_{N \to \infty} \mathbb{E}[tr(\mathbf{D}_1(N)\mathbf{V}^H\mathbf{V}\mathbf{D}_2(N)\mathbf{V}^H\mathbf{V} \times \cdots \mathbf{D}_n(N)\mathbf{V}^H\mathbf{V})]$$

also exists when $\frac{L}{N} \to c$, and equals

$$\sum_{\rho \in \mathcal{P}(n)} K_{\rho,\omega} c^{|\rho|-1} D_\rho.$$

For the case of Vandermonde matrices with uniform phase distribution, the noncrossing partitions play a central role. Let u denote the uniform distribution on $[0, 2\pi]$.

Theorem 5.68 ([365]) *Assume* $\mathbf{D}_1(N) = \mathbf{D}_2(N) = \cdots = \mathbf{D}_n(N)$, *set* $c = \frac{L}{N}$, *and define*

$$m_n^{(N,L)} = c\mathbb{E}[tr(\mathbf{D}_2(N)\mathbf{V}^H\mathbf{V})^n]$$
$$d_n^{(N,L)} = ctr(\mathbf{D}(N))^n.$$

When $\omega = \mu$, *we have that*

$$m_1^{(N,L)} = d_1^{(N,L)}$$

$$m_2^{(N,L)} = (1 - N^{-1})d_2^{(N,L)} + (d_1^{(N,L)})^2$$

$$m_3^{(N,L)} = (1 - 3N^{-1} + 2N^{-2})d_3^{(N,L)} + 3(1 - N^{-1})d_1^{(N,L)}d_2^{(N,L)} + (d_1^{(N,L)})^3$$

$$m_4^{(N,L)} = \left(1 - \frac{20}{3}N^{-1} + 12N^{-2} - \frac{19}{3}N^{-3}\right)d_4^{(N,L)} + (4 - 12N^{-1} + 8N^{-2})d_3^{(N,L)}d_1^{(N,L)}$$

$$+ \left(\frac{8}{3} - 6N^{-1} + \frac{10}{3}N^{-2}\right)(d_2^{(N,L)})^2 + 6(1 - N^{-1})d_2^{(N,L)}(d_1^{(N,L)})^2 + (d_1^{(N,L)})^4.$$

Let us consider generalized Vandermonde matrices defined as

$$\mathbf{V} = \frac{1}{\sqrt{N}}\begin{bmatrix} e^{-j[Nf(0)]\omega_1} & \cdots & e^{-j[Nf(0)]\omega_L} \\ e^{-j[Nf(\frac{1}{N})]\omega_1} & \cdots & e^{-j[Nf(\frac{1}{N})]\omega_L} \\ \vdots & \ddots & \vdots \\ e^{-j\left[Nf\left(\frac{N-1}{N}\right)\right]\omega_1} & \cdots & e^{-j[Nf(\frac{N-1}{N})]\omega_L} \end{bmatrix}, \tag{5.128}$$

where f is called the power distribution, and is a function from $[0, 1]$ to $[0, 1]$. We also consider the more general case when f is replaced with a random variable λ,

$$\mathbf{V} = \frac{1}{\sqrt{N}}\begin{bmatrix} e^{-jN\lambda_1\omega_1} & \cdots & e^{-jN\lambda_L\omega_L} \\ e^{-jN\lambda_2\omega_1} & \cdots & e^{-jN\lambda_2\omega_L} \\ \vdots & \ddots & \vdots \\ e^{-jN\lambda_N\omega_1} & \cdots & e^{-j\lambda_L\omega_L} \end{bmatrix}, \tag{5.129}$$

with the λ_i i.i.d. and distributed as λ, defined and taking values in $[0, 1]$, and also independent from the ω_j.

For (5.128) and (5.129), define

$$K_{\rho,\omega,f,N} = \frac{1}{N^{n+1-|\rho|}} \int\limits_{(0,2\pi)^{|\rho|}} \prod_{k=1}^{n}\left(\sum_{r=0}^{N-1} p_{f_N}(r)e^{jr(\omega_{b(k-1)} - \omega_{b(k)})}\right)d\omega_1 \cdots d\omega_{|\rho|}$$

$$K_{\rho,\omega,\lambda,N} = \frac{1}{N^{n+1-|\rho|}} \int\limits_{(0,2\pi)^{|\rho|}} \prod_{k=1}^{n}\left(\int_0^1 Ne^{jN\lambda(\omega_{b(k-1)} - \omega_{b(k)})}d\lambda\right)d\omega_1 \cdots d\omega_{|\rho|},$$

where $\omega_{W_1,\ldots,W_{|\rho|}}$ are defined as in (5.127). If the limits

$$K_{\rho,\omega,f} = \lim_{N\to\infty} K_{\rho,\omega,f,N}$$
$$K_{\rho,\omega,\lambda} = \lim_{N\to\infty} K_{\rho,\omega,\lambda,N},$$

exist, then they are called Vandermonde mixed moment expansion coefficients.

Theorem 5.69 ([365]) *Theorem 5.67 holds also when Vandermonde matrices (5.125) are replaced with generalized Vandermonde matrices on either form (5.128) or (5.129), and with $K_{\rho,\omega}$ replaced with either $K_{\rho,\omega,f}$ or $K_{\rho,\omega,\lambda}$.*

Theorem 5.70 ([365]) *Assume that the $\{\mathbf{D}_r(N)\}_{1 \le r \le n}$ have a joint limit distribution as $N \to \infty$. Assume also that $\mathbf{V}_1, \mathbf{V}_2, \ldots$ are independent Vandermonde matrices with the same phase distribution ω_i, and that the density of ω is continuous. Then, the limit*

$$\lim_{N \to \infty} \mathbb{E}[\mathbf{D}_1(N)\mathbf{V}_{i_1}^H \mathbf{V}_{i_2} \mathbf{D}_2(N)\mathbf{V}_{i_2}^H \mathbf{V}_{i_3} \times \cdots \times \mathbf{D}_n(N)\mathbf{V}_{i_n}^H \mathbf{V}_{i_1}]$$

exists when $\frac{L}{N} \to c$. The limit is 0 when n is odd, and equals

$$\sum_{\rho \le \sigma \in \mathcal{P}(n)} K_{\rho,\omega} c^{|\rho|-1} D_\rho \tag{5.130}$$

where

$$\sigma = \{\sigma_1, \sigma_2\} = \{\{1, 3, 5, \ldots,\}, \{2, 4, 6, \ldots,\}\}$$

is the partition where two blocks are the even numbers, and the odd numbers.

Corollary 5.4 ([365]) *The first three mixed moments*

$$V_n^{(2)} = \lim_{N \to \infty} \mathbb{E}[(\mathbf{V}_1^H \mathbf{V}_2 \mathbf{V}_2^H \mathbf{V}_1)^n]$$

of independent Vandermonde matrices $\mathbf{V}_1, \mathbf{V}_2$ are given by

$$V_1^{(2)} = I_2$$
$$V_2^{(2)} = \frac{2}{3}I_2 + I_3 + I_4$$
$$V_3^{(2)} = \frac{11}{20}I_2 + 4I_3 + 9I_4 + 6I_5 + I_6,$$

where

$$I_k = (2\pi)^{k-1} \left(\int_0^{2\pi} p_\omega(x)^k dx \right).$$

In particular, when the phase distribution is uniform, the first three moments are given by

$$V_1^{(2)} = 1, \ V_2^{(2)} = \frac{11}{3}, \ V_3^{(2)} = \frac{411}{20}.$$

Theorem 5.71 ([365]) *Assume that $\{\mathbf{V}_i\}_{1 \le i \le s}$ are independent Vandermonde matrices, where $\mathbf{V}_i$ has continuous phase distribution ω_i. Denoted by p_{ω_i}, the density of ω_i. Then, (5.130) still holds, with $K_{\rho,\omega}$ replaced by*

$$K_{\rho,u}(2\pi)^{|\rho|-1} \int_0^{2\pi} \prod_{i=1}^s p_{\omega_i}(x)^{|\rho_i|} dx,$$

where ρ_i consists of all numbers k such that $i_k = i$.

Example 5.18 (Detection of the number of sources [365])

In this example, d is the distance between the antennas whereas λ is the wavelength. The ratio $\frac{d}{\lambda}$ is a figure of the resolution with which the system will be able to separate users in space. Let us consider a central node equipped with N receiving antennas, and with L mobiles (each with a single antenna). The received signal at the central node is given by

$$\mathbf{y}_i = \mathbf{V}\mathbf{P}^{1/2}\mathbf{x}_i + \mathbf{w}_i, \tag{5.131}$$

where

- $\mathbf{y}_i$ is the $N \times 1$ received vector,
- $\mathbf{x}_i$ is the $L \times 1$ transmit vector by the L users; we assume $\mathbb{E}[\mathbf{x}_i\mathbf{x}_i^H] = \mathbf{I}_L$,
- $\mathbf{w}_i$ is $N \times 1$ additive, white, Gaussian noise of variance $\frac{\sigma}{\sqrt{N}}$, and
- all components in $\mathbf{x}_i$ and $\mathbf{w}_i$ are assumed independent.

In the case of line of sight between the users and the central node, for a uniform linear array (ULA), the matrix $\mathbf{V}$ has the following form

$$\mathbf{V} = \frac{1}{\sqrt{N}} \begin{bmatrix} 1 & \cdots & 1 \\ e^{-j2\pi \frac{d}{\lambda} \sin\theta_1} & \cdots & e^{-j2\pi \frac{d}{\lambda} \sin\theta_L} \\ \vdots & \ddots & \vdots \\ e^{-j(N-1)\frac{d}{\lambda} \sin\theta_1} & \cdots & e^{-j(N-1)\frac{d}{\lambda} \sin\theta_L} \end{bmatrix}. \tag{5.132}$$

Here, θ_i is the angle of the user and is supposed to be uniformly distributed over $[-\alpha, \alpha]$. $\mathbf{P}^{1/2}$ is an $L \times L$ diagonal power matrix due to the different distances from which the users emit. The phase distribution has been assumed to have the form $2\pi \frac{d}{\lambda} \sin\theta$ with θ uniformly distributed on $[-\alpha, \alpha]$.

By taking inverse function, the density is, for $\frac{2\pi \sin\alpha}{\lambda} < 1$,

$$p_\omega(x) = \frac{1}{2\alpha\sqrt{\frac{4\pi^2 d^2}{\lambda^2} - x^2}}$$

on $[-\frac{2\pi \sin\alpha}{\lambda}, \frac{2\pi \sin\alpha}{\lambda}]$, and 0 elsewhere.

By defining

$$\mathbf{Y} = [\mathbf{y}_1, \ldots, \mathbf{y}_N], \mathbf{X} = [\mathbf{x}_1, \ldots, \mathbf{x}_K], \mathbf{W} = [\mathbf{w}_1, \ldots, \mathbf{w}_N], \tag{5.133}$$

(5.131) is rewritten as

$$\mathbf{Y} = [\mathbf{y}_1, \ldots, \mathbf{y}_N] = \mathbf{V}\mathbf{P}^{1/2}[\mathbf{x}_1, \ldots, \mathbf{x}_K] + [\mathbf{w}_1, \ldots, \mathbf{w}_N] = \mathbf{V}\mathbf{P}^{1/2}\mathbf{X} + \mathbf{W}.$$

The sample covariance matrix can be written as

$$\mathbf{S} = \frac{1}{N}\mathbf{Y}\mathbf{Y}^H = \frac{1}{N}(\mathbf{V}\mathbf{P}^{1/2}\mathbf{X} + \mathbf{W})(\mathbf{V}\mathbf{P}^{1/2}\mathbf{X} + \mathbf{W})^H.$$

If we have only the sample covariance matrix $\mathbf{S}$, in order to get an estimate of $\mathbf{P}$, we have three independent parts to deal with: $\mathbf{X}, \mathbf{W}, \mathbf{V}$. We can achieve this by combining Gaussian decomposition [366] and Vandermonde deconvolution by the following steps:

1. Estimate the moments of $\frac{1}{N}\mathbf{V}\mathbf{P}^{1/2}\mathbf{X}\mathbf{X}^H\mathbf{P}^{1/2}\mathbf{V}^H$ using multiplicative free convolution [262]. This is the denoising part.
2. Estimate the moments of $\mathbf{P}\mathbf{V}\mathbf{V}^H$, using multiplicative free convolution.
3. Estimate the moments of $\mathbf{P}$ using Vandermonde deconvolution in the paper of [365]. □

Proposition 5.9 ([365]) *Define*

$$I_n = (2\pi)^{n-1} \int_0^{2\pi} p_\omega(x)^n dx$$

and denote the moments of $\mathbf{P}$ and $\mathbf{S}$ by

$$P_i = tr(\mathbf{P}^i), \ S_i = tr(\mathbf{S}^i).$$

Then, the equations

$$S_1 = c_2 P_1 + \sigma^2$$

$$S_2 = c_2 P_2 + (c_2^2 I_2 + c_2 c_3)(P_1)^2 + 2\sigma^2(c_2 + c_3)P_1 + \sigma^4(1 + c_1)$$

$$S_3 = c_3 P_3 + (3c_2^2 I_2 + c_2 c_3)P_1 P_2 + (c_2^3 I_3 + 3c_2^2 c_3 I_2 + c_2 c_3^2)(P_1)^3 + 3\sigma^2(1 + c_1)c_2 P_2$$

$$\qquad + 3\sigma^2((1 + c_1)c_2^2 I_2 + c_3(c_3 + 2c_2))(P_1)^2 + 3\sigma^4(c_1^2 + 3c_1 + 1)c_2 P_1$$

$$\qquad + \sigma^6(c_1^2 + 3c_1 + 1)$$

provide an asymptotically unbiased estimator for the moments P_i from the moments of S_i (or vice versa) when

$$\lim_{N \to \infty} \frac{N}{K} \to c_1, \ \lim_{N \to \infty} \frac{L}{N} \to c_2, \ \lim_{N \to \infty} \frac{L}{K} \to c_3.$$

Example 5.19 (Estimation of the number of paths [365])
Consider a multipath channel

$$h(\tau) = \sum_{i=1}^L x_i \delta(\tau - \tau_i).$$

Here x_i are i.d. Gaussian random variables with power P_i and τ_i are uniformly distributed delay over $[0, T]$. The x_i represent the attenuation factors due to different physical mechanisms such as reflections, refractions, or diffractions. L is the total number of paths. In the frequency domain, the channel is

$$H(f) = \sum_{i=1}^L x_i G(f) e^{-j2\pi f \tau_i}.$$

□

A generalized multipath model that has taken into account the per-path pulse distortion [367–373] is relevant to the context. The so-called scatter centers that are used for the radar community are mathematically modeled by the multiple maths that are used in wireless communications. As a result, this work bridges the gap between two communities. Deeper research can be pursued using this mathematical analogy between two different systems. Physically, the two systems are equivalent.

By sampling the continuous frequency signal at sampling rate $f_i = i\frac{B}{N}$ where B is the bandwidth (in Hertz), we have (for a given channel realization)

$$\mathbf{H} = \mathbf{V}\mathbf{P}^{1/2}\mathbf{x} \tag{5.134}$$

where

$$\mathbf{V} = \frac{1}{\sqrt{N}} \begin{bmatrix} 1 & \cdots & 1 \\ e^{-j2\pi\frac{B}{N}\tau_1} & \cdots & e^{-j2\pi\frac{B}{N}\tau_L} \\ \vdots & \ddots & \vdots \\ e^{-j2\pi(N-1)\frac{B}{N}\tau_1} & \cdots & e^{-j2\pi(N-1)\frac{B}{N}\tau_L} \end{bmatrix}.$$

We set here $B = T = 1$, which implies that the ω_i of (5.125) are uniformly distributed over $[0, 2\pi)$. When additive noise $\mathbf{w}$ is taken into account, our model again becomes that of (5.131): The only difference is that the phase distribution of the Vandermonde matrix now is uniform. L now is the number of paths, N the number of frequency samples, and $\mathbf{P}$ is the unknown $L \times L$ diagonal power matrix. Taking K observations, we reach the same form as in (5.133). We can do even better than Proposition 5.9. Our estimators for the moments are unbiased for any number of observations K and frequency samples N.

Proposition 5.10 ([365]) *Assume that* $\mathbf{V}$ *has uniform phase distribution, and let* P_i *be the moments of* $\mathbf{P}$*, and* $S_i = \mathrm{tr}(\mathbf{S}^i)$ *the moments of the sample covariance matrix. Define also*

$$\frac{N}{K} = c_1, \quad \frac{L}{N} = c_2, \quad \frac{L}{K} = c_3.$$

Then,

$$\mathbb{E}[S_1] = c_2 P_1 + \sigma^2$$

$$\mathbb{E}[S_2] = c_2 \left(1 - \frac{1}{N}\right) P_2 + c_2(c_2 + c_3)(P_1)^2 + 2\sigma^2(c_2 + c_3)P_1 + \sigma^4(1 + c_1)$$

$$\mathbb{E}[S_3] = c_2 \left(1 + \frac{1}{K^2}\right)(1 - \frac{3}{N} + \frac{2}{N^2})P_3 + \left(1 - \frac{1}{N}\right)\left(3c_2^2\left(1 + \frac{1}{K^2}\right) + 3c_2 c_3\right) P_1 P_2$$

$$+ \left(c_2^3\left(1 + \frac{1}{K^2}\right) + 3c_2^2 c_3 + c_2 c_3^2\right)(P_1)^3 + 3\sigma^2 \left((1 + c_1)c_2 + \frac{c_1 c_2^2}{KL}\right)\left(1 - \frac{1}{N}\right) P_2$$

$$+ 3\sigma^2 \left(\frac{c_1 c_2^3}{KL} + c_2^2 + c_3^2 + 3c_2 c_3\right)(P_1)^2 + 3\sigma^4 \left(c_1^2 + 3c_1 + 1 + \frac{1}{K^2}\right) c_2 P_1$$

$$+ \sigma^6 \left(c_1^2 + 3c_1 + 1 + \frac{1}{K^2}\right).$$

Wavelength in (5.132) can be also estimated. See [365] for details.

Example 5.20 (Signal reconstruction and estimation of the sampling distribution [365])
Consider the signal $y(t)$ as a superposition of its N frequency components

$$y(t) = \frac{1}{\sqrt{N}} \sum_{k=0}^{N-1} x_k e^{-j\frac{2\pi}{N}kt}. \tag{5.135}$$

We sample the continuous signal $y(t)$ at time instants $t = [t_1, \ldots, t_L]$ with $t_i \in [1]$. (5.135) can be written equivalently as

$$y(\omega) = \frac{1}{\sqrt{N}} \sum_{k=0}^{N-1} x_k e^{-jk\omega} \text{ory} = \mathbf{V}^T \mathbf{x}.$$

In the presence of noise, one has

$$\mathbf{y} = \mathbf{V}^T \mathbf{x} + \mathbf{w}$$

with

$$\mathbf{y} = [y(\omega_1), \ldots, y(\omega_L)],$$

and $\mathbf{x}$ and $\mathbf{w}$ are defined in (5.131). $\mathbf{V}$ is defined as our standard model (5.125). [374] has a similar analysis for such cases.
    We define

$$\mathbf{Y} = [\mathbf{y}_1, \ldots, \mathbf{y}_K] = \mathbf{V}^T [\mathbf{x}_1, \ldots, \mathbf{x}_K] + [\mathbf{w}_1, \ldots, \mathbf{w}_K] = \mathbf{V}^T \mathbf{X} + \mathbf{W}$$
$$\mathbf{S} = \frac{1}{K} \mathbf{Y} \mathbf{Y}^H = \frac{1}{K} (\mathbf{V}^T \mathbf{X} + \mathbf{W})(\mathbf{V}^T \mathbf{X} + \mathbf{W})^H.$$

Consider the asymptotic regime

$$\lim_{N \to \infty} \frac{N}{K} \to c_1, \; \lim_{N \to \infty} \frac{L}{N} \to c_2, \; \lim_{N \to \infty} \frac{L}{K} \to c_3.$$

□

Proposition 5.11 ([365])

$$\mathbb{E}[tr(\mathbf{S})] = c_2 P_1 + \sigma^2$$

$$\mathbb{E}[tr(\mathbf{S}^2)] = c_2 I_2 + (1 + c_3)(1 + \sigma^2)^2 \tag{5.136}$$

$$\mathbb{E}[tr(\mathbf{S}^3)] = 1 + 3c_2(1 + c_3)I_2 + 3c_3 + c_3^2 + c_3^2 I_3 + 3\sigma^2(1 + 3c_3 + c_3^2 + c_2(1 + c_3)I_2)$$
$$+ 3\sigma^4 c_2(c_3^2 + 3c_3 + 1) + \sigma^6(c_1^2 + 3c_1 + 1),$$

where I_n is defined in Proposition 5.19.

    Consider a phase distribution ω which is uniform on $[0, \alpha]$, and 0 elsewhere. The density is thus $\frac{2\pi}{\alpha}$ on $[0, \alpha]$, and 0 elsewhere. In this case we have

$$I_2 = \frac{2\pi}{\alpha}, I_3 = \left(\frac{2\pi}{\alpha}\right)^2.$$

The first of these equations, combined with (5.136), enable us to estimate α.

Certain matrices similar to Vandermonde matrices have analytical expressions for the moments. In [375], the matrices with entries of the form $A_{i,j} = F(\omega_i, \omega_j)$ are considered. This is relevant to the Vandermonde matrices since

$$\frac{1}{N}(\mathbf{V}^H\mathbf{V})_{i,j} = \frac{\sin\left(\frac{N}{2}(\omega_i - \omega_j)\right)}{N\sin\left(\frac{1}{2}(\omega_i - \omega_j)\right)}.$$

Example 5.21 (Vandermonde matrices with unit complex entries [376])
Consider the network with M mobile users talking to a base station with N antenna elements, arranged in a uniform linear array. The antenna array response is a Vandermonde matrix. We refer to [376] for this example. $\qquad\square$

5.7.3 Convolution and Deconvolution with Vandermonde Matrices

In the large dimensional limit, certain random matrices a deterministic behavior of the eigenvalue distribution [377]. In particular, one can obtain the eigenvalue distribution of $\mathbf{AB}$ and $\mathbf{A}+\mathbf{B}$, based on only the individual eigenvalue distributions of $\mathbf{A}$ and $\mathbf{B}$, when the matrices are independent and large. This operation is called convolution, and the inverse operation is called deconvolution.

Gaussian-like matrices fit into this setting, since the concept of freeness [11] can be used. [364] used large Wishart matrices. Random matrix theory was used in [9]; other deterministic equivalents [17, 281, 298, 378] are used; Although used successfully [366], all these techniques can only treat very simple models, that is, one of these considered matrices are unitarily invariant.

The method of moments, which is the focus in this section, is very appealing and powerful when freeness does not apply, for which we still do not have a general framework. It requires the combinatorial skills and can be used for a large class of random matrices. Compared with the Stieltjes transform, this approach has the main drawback that it rarely provides the exact eigenvalue distribution. In many applications, however, we only need a subset of the moments. We mainly follow Ryan and Debbah (2011) [377] for our development.

A $N \times N$ Vandermonde matrix $\mathbf{V}$ is defined in (5.125). We repeat it here for convenience:

$$\mathbf{V} = \frac{1}{\sqrt{N}}\begin{bmatrix} 1 & \cdots & 1 \\ e^{-j\omega_1} & \cdots & e^{-j\omega_L} \\ \vdots & \ddots & \vdots \\ e^{-j(N-1)\omega_1} & \cdots & e^{-j(N-1)\omega_L} \end{bmatrix}. \tag{5.137}$$

The $\omega_1, \ldots, \omega_L$, also called phase distributions, will be assumed i.i.d., taking values in $[0, 2\pi]$. Similarly, we consider the asymptotic regime defined in (5.126): N and L go to infinity at the same rate, and write $c = \lim_{N\to\infty} \frac{L}{N}$.

In Section 5.7.2, the limit eigenvalue distributions of combinations of $\mathbf{V}^H\mathbf{V}$ and diagonal matrices $\mathbf{D}(N)$ were shown to be dependent on the limit eigenvalue distributions of the two matrices.

Define

$$\lim_{N \to \infty} \mathrm{tr}(\mathbf{D}_1(N)\mathbf{V}_{i_1}^H \mathbf{V}_{i_2} \times \cdots \times \mathbf{D}_n(N)\mathbf{V}_{i_{2n-1}}^H \mathbf{V}_{i_{2n}}), \tag{5.138}$$

where $\mathbf{V}_1, \mathbf{V}_2, \ldots$ are assumed independent, with phase distributions $\omega_1, \ldots, \omega_L$. Consider the following four expressions:

1. $\lim_{N \to \infty} \mathbf{D}(N)\mathbf{V}^H \mathbf{V}$ and $\lim_{N \to \infty} \mathbf{D}(N) + \mathbf{V}^H \mathbf{V}$
2. $\lim_{N \to \infty} \mathbf{D}(N)\mathbf{V}\mathbf{V}^H$ and $\lim_{N \to \infty} (\mathbf{D}(N) + \mathbf{V}\mathbf{V}^H)$
3. $\lim_{N \to \infty} \mathbf{V}_1^H \mathbf{V}_1 \mathbf{V}_2^H \mathbf{V}_2$ and $\lim_{N \to \infty} (\mathbf{V}_1^H \mathbf{V}_1 + \mathbf{V}_2^H \mathbf{V}_2)$
4. $\lim_{N \to \infty} \mathbf{V}_1 \mathbf{V}_1^H \mathbf{V}_2 \mathbf{V}_2^H$ and $\lim_{N \to \infty} (\mathbf{V}_1 \mathbf{V}_1^H + \mathbf{V}_2 \mathbf{V}_2^H)$.

Theorem 5.72 ([377]) *Let $\mathbf{V}_i$ be independent $N_i \times L$ Vandermonde matrices with aspect ratios $c_i = \lim_{N \to \infty} \frac{L}{N_i}$ and phase distributions ω_i with continuous densities in $[0, 2\pi]$. The limit*

$$\lim_{N \to \infty} \mathbb{E}[\mathrm{tr}(\mathbf{D}_1(N)\mathbf{V}_{i_1}^H \mathbf{V}_{i_2} \mathbf{D}_2(N)\mathbf{V}_{i_2}^H \mathbf{V}_{i_3} \times \cdots \times \mathbf{D}_n(N)\mathbf{V}_{i_{2n-1}}^H \mathbf{V}_{i_{2n}})] \tag{5.139}$$

always exists, when $\mathbf{D}_i(N)$ have a joint limit distribution, whenever the matrix product is well-defined and square. Moreover, (5.138) converges almost surely in distribution to the limit in (5.139). When $\sigma \geq [0, 1]_n$ (that is, there are no terms in the form of $\mathbf{V}_r^H \mathbf{V}_s$ with $\mathbf{V}_r$ and $\mathbf{V}_s$ independent and with different phase distributions), (5.139) can be expressed as a formula in the aspect ratio c_i, σ, and the individual moments

$$\begin{aligned} \mathbf{V}_n^{(r)} &= \lim_{N \to \infty} \mathbb{E}[\mathrm{tr}(\mathbf{V}_r^H \mathbf{V}_s)^n] \\ D_{i_1, \ldots, i_s} &= \mathrm{tr}(\mathbf{D}_{i_1}(N) \cdots \mathbf{D}_{i_s}(N)). \end{aligned} \tag{5.140}$$

A special case of Theorem 5.72 is considered here. This theorem states in particular that

$$\mathrm{tr}((\mathbf{V}_1 + \mathbf{V}_2 + \cdots)^H (\mathbf{V}_1 + \mathbf{V}_2 + \cdots))^p$$

depends only on the moments. This expression characterizes the singular law of a sum of independent Vandermonde matrices. Also, expressions 1 and 3 are found to only rely on the spectra of the component matrices. For convolution expression 1, we have the following corollary.

Corollary 5.5 ([377]) *Assume that $\mathbf{V}$ has a phase distribution with continuous density, and define*

$$\begin{aligned} V_n &= \lim_{N \to \infty} \mathrm{tr}((\mathbf{V}^H \mathbf{V})^n) \\ D_n &= c \lim_{N \to \infty} \mathrm{tr}(\mathbf{D}(N)^n) \\ M_n &= c \lim_{N \to \infty} \mathrm{tr}((\mathbf{D}(N)\mathbf{V}^H \mathbf{V})^n) \\ N_n &= c \lim_{N \to \infty} \mathrm{tr}((\mathbf{D}(N) + \mathbf{V}^H \mathbf{V})^n), \end{aligned}$$

where $c = \lim\limits_{N\to\infty} \frac{L}{N}$. *Whenever either* $\{M_n\}_{1\le n\le k}$ *or* $\{N_n\}_{1\le n\le k}$ *are known, and* $\{V_n\}_{1\le n\le k}$ *(or* $\{D_n\}_{1\le n\le k}$*) are known, then* $\{D_n\}_{1\le n\le k}$ *(or* $\{V_n\}_{1\le n\le k}$*) are uniquely determined.*

For expression 3, we have the following corollary.

Corollary 5.6 ([377]) *Assume that* $\mathbf{V}_1$ *and* $\mathbf{V}_2$ *are independent Vandermonde matrices where the phase distributions have continuous densities, and set*

$$V_1^{(n)} = \lim_{N\to\infty} tr((\mathbf{V}_1^H\mathbf{V}_1)^n)$$
$$V_2^{(n)} = \lim_{N\to\infty} tr((\mathbf{V}_2^H\mathbf{V}_2)^n)$$
$$M_n = c \lim_{N\to\infty} tr((\mathbf{V}_1^H\mathbf{V}_1\mathbf{V}_2^H\mathbf{V}_2)^n)$$
$$N_n = c \lim_{N\to\infty} tr((\mathbf{V}_1^H\mathbf{V}_1 + \mathbf{V}_2^H\mathbf{V}_2)^n).$$

M_n *and* N_n *are completely determined by* $V_2^{(i)}$, $V_3^{(i)}$, ... *and the aspect ratios*

$$c_1 = \lim_{N\to\infty} \frac{L}{N_1}, c_2 = \lim_{N\to\infty} \frac{L}{N_2}.$$

Also, whenever either $\{M_n\}_{1\le n\le k}$ *or* $\{N_n\}_{1\le n\le k}$ *are known, and* $\{V_1^{(n)}\}_{1\le n\le k}$ *are known, then* $\{V_2^{(n)}\}_{1\le n\le k}$ *are uniquely determined.*

For expression 4, we have the following corollary.

Corollary 5.7 ([377]) *Assume that* $\mathbf{V}_1$ *and* $\mathbf{V}_2$ *are independent Vandermonde matrices with the same continuous density, and set*

$$V_n^{(i)} = \lim_{N\to\infty} tr((\mathbf{V}_i^H\mathbf{V}_i)^n)$$
$$M_n = \lim_{N\to\infty} tr((\mathbf{V}_1^H\mathbf{V}_2\mathbf{V}_2^H\mathbf{V}_1)^n).$$

Then, $\{M_n\}_{1\le n\le N}$ *are uniquely determined from* $\{V_n\}_{1\le n\le 2N}$.

The spectral separability seems to be a phenomenon for large N-limit. We are only aware of Gaussian and deterministic matrices where spectral separability occur in finite case [379]. The moments of Hankel, Markov, and Toeplitz matrices [287] are relevant to this context.

A practical example is studied in [377]:

1. From observations of the form $\mathbf{D}(N)\mathbf{V}^H\mathbf{V}$ or $\mathbf{D}(N) + \mathbf{V}^H\mathbf{V}$, one can infer on either the spectrum of $\mathbf{D}(N)$, or the spectrum or phase distribution of $\mathbf{V}$, when exactly one of these is unknown.
2. From observations of the form $\mathbf{V}_1^H\mathbf{V}_2\mathbf{V}_2^H\mathbf{V}_1$ or $\mathbf{V}_1^H\mathbf{V}_1 + \mathbf{V}_2^H\mathbf{V}_2$, one can infer on the spectrum or phase distribution of one of the Vandermonde matrices, when one of the Vandermonde matrices is known.

The example only makes an estimate of the first moments of the component matrix $\mathbf{D}(N)$. These moments can give valuable information: in cases where it is known that there are few distinct eigenvalues, and the multiplicities are known, only some lower order moments are needed, in order to get an estimate of these eigenvalues.

5.7.4   Finite Dimensional Statistical Inference

We follow [379] for the development here, converting to our notation. Given $\mathbf{X}$ and $\mathbf{Y}$ are two $N \times N$ independent square Hermitian (or symmetric) random matrices:

1. Can one derive the eigenvalues distribution of $\mathbf{X}$ from the ones of $\mathbf{X}+\mathbf{Y}$ and $\mathbf{Y}$? If feasible in the large N-limit, this operation is named additive free deconvolution.
2. Can one derive the eigenvalues distribution of $\mathbf{X}$ from the ones of $\mathbf{XY}$ and $\mathbf{Y}$? If feasible in the large N-limit, this operation is named multiplicative free deconvolution.

The method of moments [380] and the Stieltjes transform method [381] can be used. The expression is simple, if some kind of asymptotic freeness [11] of the matrices is assumed. Freeness, however, is not valid for finite matrices. Remarkably, the method of moments can still be used for this purpose. The general finite-dimensional statistical inference framework was proposed [379], and the codes for MATLAB implementation are available [382]. The calculations are tedious. Only Gaussian matrices are addressed. But other matrices such as Vandermonde matrices can also be implemented in the same vein. The general case is more difficult.

Consider the doubly correlated Wishart matrix [383]. Let M, N be positive integers, $\mathbf{W}$ be $M \times N$ standard, complex, Gaussian, and $\mathbf{D}$ (deterministic) $M \times M$ and $\mathbf{E}$ $N \times N$. Given any positive integer $p_,$, the following moments

$$\mathbb{E}\left[\text{tr}\left(\frac{1}{N}(\mathbf{DWEW}^H)^p\right)\right]$$

$$\mathbb{E}\left[\text{tr}\left(\frac{1}{N}((\mathbf{D}+\mathbf{W})(\mathbf{E}+\mathbf{W})^H)^p\right)\right]$$

exist and can be calculated [379].

The framework of [379] enables us to compute the moments of many types of combinations of independent, Gaussian- and Wishart random matrices, without any assumptions on the matrix dimensions. Since the method of moments only encode information about the lower order moments, it lacks much information which is encoded naturally into the Stieltjes transform; spectrum estimation based on the Stieltjes transform is more accurate than the case when a few moments are used. One interesting question is to ask how many moments are typically required, in order to reach the performance close to that of the Stieltjes transform.

Example 5.22 (MIMO rate estimation [379])
One has K noisy symbol-observations of the channel

$$\mathbf{Y}_i = \mathbf{D} + \sigma\mathbf{W}_i, \quad i = 1, 2, \ldots, K,$$

where $\mathbf{D}$ is an $M \times N$ deterministic channel matrix, $\mathbf{W}_i$ is an $M \times N$ standard, complex, Gaussian representing the noise, and σ is the noise variance. The channel $\mathbf{D}$ is assumed to stay constant during K symbols measurements. The rate estimator is given by

$$C = \frac{1}{M}\log_2\det\left(\mathbf{I}_M + \frac{\rho}{N}\mathbf{DD}^H\right) = \frac{1}{M}\log_2\det\left(\prod_{i=1}^{M}(1+\rho\lambda_i)\right)$$

where $\rho = \frac{1}{\sigma^2}$ is the SNR, and λ_i are the eigenvalues of $\frac{1}{N}\mathbf{DD}^H$.

The problem falls within the framework suggested before. The extra parameter did not appear in any of the main theorems of [379]. An unbiased estimator for the expression of

$$\prod_{i=1}^{M} (1 + \rho \lambda_i)$$

has been derived in [379].   □

Example 5.23 (Understanding network in a finite time [379])
In cognitive MIMO network, one must learn and control the "black box" (wireless channel) with vector inputs and vector outputs. Let $\mathbf{y}$ be the output vector, and $\mathbf{x}$ and $\mathbf{w}$, respectively, the input signal and the noise vector,

$$\mathbf{y} = \mathbf{x} + \sigma \mathbf{w}. \tag{5.141}$$

By defining

$$\mathbf{Y} = [\mathbf{y}_1, \ldots, \mathbf{y}_K], \mathbf{X} = [\mathbf{x}_1, \ldots, \mathbf{x}_K], \mathbf{W} = [\mathbf{w}_1, \ldots, \mathbf{w}_K],$$

we have

$$\mathbf{Y} = \mathbf{X} + \sigma \mathbf{W}.$$

In the Gaussian case, the rate is given by

$$C = H(\mathbf{y}) - H(\mathbf{y}|\mathbf{x}) = \log_2 \det (\pi e \mathbf{R}_Y) - \log_2 \det (\pi e \mathbf{R}_W) = \log_2 \left(\frac{\det \mathbf{R}_Y}{\det \mathbf{R}_W} \right), \tag{5.142}$$

where $\mathbf{R}_Y$ is the covariance of the output signal vector, and $\mathbf{R}_W$ is the covariance of the noise vector. According to (5.142), one can fully find the information transfer of the system, by knowing only the eigenvalues of $\mathbf{R}_Y$ and $\mathbf{R}_W$. Unfortunately, the receiver has only access to a limited number (samples) of N observations of the output vector $\mathbf{y}$, not the covariance matrix $\mathbf{R}_Y$. In other words, the system has access to only the sample covariance matrix $\hat{\mathbf{R}}_Y$, not the true covariance matrix $\mathbf{R}_Y$. Here, we define

$$\hat{\mathbf{R}}_Y = \frac{1}{K} \sum_{i=1}^{K} \mathbf{y}_i \mathbf{y}_i^H = \frac{1}{K} \mathbf{Y} \mathbf{Y}^H = \frac{1}{K} (\mathbf{X} + \mathbf{W})(\mathbf{X} + \mathbf{W})^H.$$

When $\mathbf{x}$ and $\mathbf{w}$ in (5.141) are both Gaussian vectors, we can write $\mathbf{y}$ as

$$\mathbf{y} = \mathbf{R}_Y^{1/2} \mathbf{z} \tag{5.143}$$

where $\mathbf{z}$ is an i.i.d. standard Gaussian vector. The problem falls, therefore, in the realm of inference with a correlated Wishart model defined by

$$\hat{\mathbf{R}}_Y = \frac{1}{K} \sum_{i=1}^{K} \mathbf{y}_i \mathbf{y}_i^H = \mathbf{R}_Y^{1/2} \left(\frac{1}{L} \sum_{i=1}^{L} \mathbf{z}_i \mathbf{z}_i^H \right) \mathbf{R}_Y^{1/2} = \mathbf{R}_Y^{1/2} \hat{\mathbf{R}}_Z \mathbf{R}_Y^{1/2},$$

where

$$\hat{\mathbf{R}}_Z = \frac{1}{L}\sum_{i=1}^{L} \mathbf{z}_i\mathbf{z}_i^H = \frac{1}{L}\mathbf{Z}\mathbf{Z}^H, \mathbf{Z} = [\mathbf{z}_1,\ldots,\mathbf{z}_L],$$

□

Example 5.24 (Power estimation [379])
Under the assumption of a large number of observations, the finite-dimensional inference framework was not strictly required in the above two examples. The observations can, instead, be stacked into a large matrix, where asymptotic results are more suitable. This example illustrates a model, where it is unclear how to apply such a stacking strategy, thus, making the finite-dimensional results more useful. In many multiuser MIMO applications, one needs to determine the power of each user. Consider the system given by

$$\mathbf{y}_i = \mathbf{H}\mathbf{P}^{1/2}\mathbf{x}_i + \sigma\mathbf{w}_i, i = 1, 2, \ldots, K,$$

where $\mathbf{H}, \mathbf{P}, \mathbf{s}_i, \mathbf{w}_i$ are, respectively, the $N \times M$ channel gain matrix, the $M \times M$ diagonal power matrix due to the different distances from which the users transmit, the $M \times 1$ vector of signals and the $N \times 1$ vector representing the noise with variance σ. In particular, $\mathbf{P}, \mathbf{s}_i, \mathbf{w}_i$ are independent standard, complex, Gaussian matrices and vectors. We suppose that we have K observations of the received signal vector $\mathbf{y}_i$, during which the channel gain matrix $\mathbf{H}$ stays constant.

Consider the 2×2 matrix

$$\mathbf{P}^{1/2} = \begin{pmatrix} 1 & 0 \\ 0 & 0.5 \end{pmatrix}.$$

We can estimate the moments of the matrix $\mathbf{P}$ from the moments of the matrix $\mathbf{Y}\mathbf{Y}^H$, where $\mathbf{Y} = [\mathbf{y}_1,\ldots,\mathbf{y}_K]$ is the component observation matrix.

We assume that we have an increasing number of observations K of the matrix $\mathbf{Y}$, and perform an averaging over the estimated moments—we average across a number of block fading channels. From the estimated moments of $\mathbf{P}$, we can then estimate its eigenvalues. When K increases, the prediction is close to the true eigenvalues of $\mathbf{P}$. $K = 1,200$ was considered in [379]. □

6

Convex Optimization

Optimization refers to minimizing or maximizing the objective function by systematically choosing the values of optimization variables from or within an allowed set defined by the constraint functions. Many engineering problems can be effectively characterized in the form of optimization. Thus, optimization theory is a powerful tool to solve engineering problems. In order to map from engineering problem to optimization issue, objectives, constraints, and variables should be extracted from the engineering problem and expressed in a mathematical fashion. Objective can be the key performance metric we care about. In wireless communication, objective can be capacity or throughput. For the radar system, detection rate can be the design goal. For smart grid, the total cost for purchasing power should be minimized. Constraints are the physical limits of the system or the performance requirements. Variables can be the adjustable or controllable parameters in the system, for example, weights, gains, power, and so on. Besides, optimality, feasibility, and sensitivity should also be taken into account. Reasonable constraints should be set for the optimization problem, and active constraints should be given more attention.

There are many categories of optimization formats:

- Linear optimization and nonlinear optimization.
- Discrete optimization and continuous optimization.
- Deterministic optimization and stochastic programming.
- Constrained optimization and unconstrained optimization.
- Convex optimization and nonconvex optimization.

Convex optimization is a subfield of optimization theory, which studies the problem of minimizing convex objective function based on a compact convex set. The strength of convex optimization is if a local minimum exists, then it is a global minimum. Thus, if the engineering problem can be formulated as convex optimization, then global optimal solution can be obtained. That is one reason why convex optimization has recently become popular.

The other reason for the popularity of convex optimization is that convex optimization can be solved by cutting plane method, ellipsoid method, subgradient method, or interior point method. Thereinto, interior point method is widely used. This method consists of a self-concordant barrier function used to encode the convex set and reaches an optimal solution by traversing the interior of the feasible region. The interior point method can guarantee that the number of iterations is bounded by a polynomial in the dimension and accuracy of the solution.

Convex optimization can be used in any engineering field. The popular topic in sensing and image processing is compressive sensing (CS) which finds the sparse solution to the underdetermined linear equations using the prior knowledge that the solution is sparse or compressible. CS is formulated as minimizing the l_1 norm which is convex optimization. Though the core of CS is optimization theory, CS can be still treated as a dedicated theory because of its particularity and significance. Though the sparse signal reconstruction has existed for at least four decades, this field has recently exploded, partially due to several important results by David Donoho, Emmanuel Candes, Justin Romberg, and Terence Tao. Besides, Lawrence Carin and his colleagues have built a new Bayesian framework for solving the inverse problem of CS [384, 385] and estimating a distribution for the unknown parameters. CS has been used for radar imaging in [386]. In cognitive radar network, though the data are huge, a sparse representation of data is still preferred. Thus, we should explore the method to learn the optimal dictionary for data representation [387]. Meanwhile, CS shows that physically sparse signal can be recovered from far fewer samples than the signal dimension [387]. Hence we should also find the optimal sensing matrix to project signals to the small amount of data with improving performance of reconstruction accuracy. In this way, the amount of data or information needed for radar signal processing can be greatly reduced. Furthermore, the overhead of cognitive radar network can be reduced.

It is also worth noting that the contribution of convex optimization to machine learning is significant. Learning the kernel matrix with SDP has been discussed in [388]. Learning multiresolution models with in-scale conditional covariance is formulated as convex optimization in [389]. E. J. Candes and his colleagues discuss robust PCA and try to decompose a data matrix into a low rank component and a sparse component by solving a convex program called principal component pursuit (PCP). Thus, it is safely foreseeable that convex optimization will play an important role in the function of cognition in the near future.

The standard format of convex optimization problem is [8],

$$\begin{aligned} &\text{minimize}\\ &f_0(x)\\ &\text{subject to}\\ &f_m(x) \le c_m, m = 1, 2, \ldots, M \end{aligned} \tag{6.1}$$

where $f_0(x), f_1(x), \ldots, f_M(x)$ are all convex functions, which means,

$$f_m(\theta x_1 + (1 - \theta)x_2) \le \theta f_m(x_1) + (1 - \theta)f_m(x_2), m = 0, 1, 2, \ldots, M \tag{6.2}$$

for any θ with $0 \le \theta \le 1$ and all x_1 as well as x_2 which lies in a convex set [8].

In the convex optimization problem (6.1), x is the optimization variable. x can be a scaler, a vector or even a matrix. $f_0(x)$ is the objective function. $f_m(x), m = 1, 2, \ldots, M$ are called constraint functions.

In mathematics, a concave function is the negative of a convex function. A function $f(x)$ is concave over a convex set if and only if $-f(x)$ is a convex function over the same set. If we would like to maximize one concave function, we can minimize its corresponding convex function.

The well-known convex functions or concave functions are listed as follows [8, 390]. The readers can refer to [8] for the definitions of notations.

- $f(x) = e^{ax}$ is convex on $\mathbf{R}$ for any $a \in \mathbf{R}$.
- $f(x) = \log x$ is concave on $\mathbf{R}_{++}$.
- $f(x) = x^a$ is convex on $\mathbf{R}_{++}$ if $a \geq 1$ or $a \leq 0$ and concave if $0 \leq a \leq 1$.
- Every norm on $\mathbf{R}^N$ is convex.
- $f(x) = x \log x$ is convex on $\mathbf{R}_{++}$.
- $f(\mathbf{x}) = \max\{x_1, x_2, \ldots, x_N\}$ is convex on $\mathbf{R}^N$.
- $f(\mathbf{x}) = \log(e^{x_1} + e^{x_2} + \cdots + e^{x_N})$ is convex on $\mathbf{R}^N$.
- The geometric mean $f(\mathbf{x}) = (\prod_{n=1}^{N} x_n)^{\frac{1}{N}}$ is concave on $\mathbf{R}_{++}^N$.
- $f(x) = \frac{x^2}{y}$ is convex on $\mathbf{R} \times \mathbf{R}_{++}$.
- $f(\mathbf{X}) = \log \det \mathbf{X}$ is concave on $\mathbf{S}_{++}^N$.
- $f(\mathbf{X}) = \lambda_{\max}(\mathbf{X})$ is convex on $\mathbf{X} \in \mathbf{S}^N$ where $\lambda_{\max}(\mathbf{X})$ means maximum eigenvalue of a matrix.
- $f(\mathbf{X}) = \text{trace}(\mathbf{X}^{-1})$ is convex on $\mathbf{X} \in \mathbf{S}_{++}^N$.
- $f(\mathbf{X}) = (\det \mathbf{X})^{\frac{1}{N}}$ is concave on $\mathbf{S}_{++}^N$.
- If $\mathbf{A}$ is positive definite matrix $\mathbf{A} \in \mathbf{C}^{N \times N}$, $f(\mathbf{X}) = \text{trace}(\mathbf{X}\mathbf{A}\mathbf{X}^H)$ is strictly convex.

6.1 Linear Programming

If the objective and constraint functions are all linear, the optimization problem is called a linear programming. Linear programming is one kind of convex optimization problems. A general linear programming has the form [8],

$$
\begin{aligned}
&\text{minimize} \\
&\mathbf{a}^T \mathbf{x} + b \\
&\text{subject to} \\
&\mathbf{C}\mathbf{x} = \mathbf{d} \\
&\mathbf{G}\mathbf{x} \preceq \mathbf{h},
\end{aligned}
\tag{6.3}
$$

where $\mathbf{x} \in \mathbf{R}^N$, $\mathbf{a} \in \mathbf{R}^N$, $\mathbf{C} \in \mathbf{R}^{M \times N}$, $\mathbf{d} \in \mathbf{R}^M$, $\mathbf{G} \in \mathbf{R}^{L \times N}$, and $\mathbf{h} \in \mathbf{R}^L$.

A standard form linear programming is expressed as [8],

$$
\begin{aligned}
&\text{minimize} \\
&\mathbf{a}^T \mathbf{x} + b \\
&\text{subject to} \\
&\mathbf{C}\mathbf{x} = \mathbf{d} \\
&\mathbf{x} \succeq 0,
\end{aligned}
\tag{6.4}
$$

where the only inequality constraints are component-wise nonnegativity constraints $\mathbf{x} \succeq 0$. An inequality form linear programming is written as [8],

$$
\begin{aligned}
& \text{minimize} \\
& \mathbf{a}^T \mathbf{x} + b \\
& \text{subject to} \\
& \mathbf{Gx} \preceq \mathbf{h},
\end{aligned}
\tag{6.5}
$$

where no equality constraints exist.

6.2 Quadratic Programming

If the linear objective function in linear programming are replaced by the convex quadratic objective function, the corresponding optimization problem is called quadratic programming which can be expressed as [8],

$$
\begin{aligned}
& \text{minimize} \\
& \tfrac{1}{2}\mathbf{x}^T \mathbf{P}\mathbf{x} + \mathbf{q}^T \mathbf{x} + r \\
& \text{subject to} \\
& \mathbf{Cx} = \mathbf{d} \\
& \mathbf{Gx} \preceq \mathbf{h},
\end{aligned}
\tag{6.6}
$$

where $\mathbf{P} \in \mathbf{S}_{+}^{N}$ and $\mathbf{q} \in \mathbf{R}^{N}$.

Furthermore, if the inequality constraints $\mathbf{Gx} \preceq \mathbf{h}$ in the quadratic programming (6.6) is replaced by the convex quadratic constraints, the corresponding optimization problem is called quadratically constrained quadratic programming (QCQP) which can be expressed as [8],

$$
\begin{aligned}
& \text{minimize} \\
& \tfrac{1}{2}\mathbf{x}^T \mathbf{P}_0 \mathbf{x} + \mathbf{q}_0^T \mathbf{x} + r_0 \\
& \text{subject to} \\
& \mathbf{Cx} = \mathbf{d} \\
& \tfrac{1}{2}\mathbf{x}^T \mathbf{P}_m \mathbf{x} + \mathbf{q}_m^T \mathbf{x} + r_m \leq 0, m = 1, 2, \ldots, M,
\end{aligned}
\tag{6.7}
$$

where $\mathbf{P}_m \in \mathbf{S}_{+}^{N}$ and $\mathbf{q}_m \in \mathbf{R}^{N}, m = 0, 1, 2, \ldots, M$.

The norm cone related to the norm $\| \cdot \|$ is the convex set which can be expressed as [8],

$$
C = \{(\mathbf{x}, t) | \|\mathbf{x}\| \leq t\} \subseteq \mathbf{R}^{N+1}.
\tag{6.8}
$$

If l_2 norm is considered, the corresponding cone is called second-order cone, quadratic cone, or ice-cream cone.

If the convex quadratic constraints in QCQP are replaced by the convex second-cone constraints, the corresponding optimization problem is called SOCP [8] ,

$$
\begin{aligned}
& \text{minimize} \\
& \mathbf{a}^T \mathbf{x} \\
& \text{subject to} \\
& \mathbf{Cx} = \mathbf{d} \\
& \|\mathbf{F}_m \mathbf{x} + \mathbf{e}_m\|_2 \leq \mathbf{q}_m^T \mathbf{x} + r_m, m = 1, 2, \ldots, M,
\end{aligned}
\tag{6.9}
$$

where $\mathbf{F}_m \in \mathbf{R}^{L_m \times N}$ and $\mathbf{e}_m \in \mathbf{R}^{L_m}, m = 1, 2, \ldots, M$.

6.3   Semidefinite Programming

If $\mathbf{X} \in \mathbf{S}_+^N$, $\mathbf{X}$ is a positive-semidefinite matrix or nonnegative-definite matrix which means

$$\mathbf{u}^T \mathbf{X} \mathbf{u} \geq 0 \tag{6.10}$$

for all $\mathbf{u} \in \mathbf{R}^N$. If $\mathbf{X}$ is a positive-semidefinite matrix, then all eigenvalues of $\mathbf{X}$ are nonnegative and all diagonal entries in $\mathbf{X}$ are nonnegative.

SDP is a subfield of convex optimization. SDP tries to optimize a linear objective function over the intersection of the cone of positive semidefinite matrices with an affine space. SDP based signal processing is becoming more and more popular recently. It can be applied to control theory, machine learning, statistics, circuit design, graph theory, quantum mechnics [164], and so on. The reasons for this are

- More and more practical problems can be formulated as SDP.
- Many combinatorial and nonconvex optimization problems can be relaxed to SDP.
- Most of the interior-point methods for linear programming have been generalized to SDP [391].
- The computational capability is increased greatly and SDP can be solved efficiently.

Hence, SDP serves as a core convex optimization format.

SDP has the form [8],

$$
\begin{aligned}
&\text{minimize} \\
&\mathbf{a}^T \mathbf{x} \\
&\text{subject to} \\
&\mathbf{C}\mathbf{x} = \mathbf{d} \\
&\left(\sum_{n=1}^{N} x_n \mathbf{F}_n \right) + \mathbf{E} \preceq 0,
\end{aligned}
\tag{6.11}
$$

where $\mathbf{F}_1, \mathbf{F}_2, \dots \mathbf{F}_N, \mathbf{E} \in \mathbf{S}^K$.

Similar to linear programming, a standard form SDP is expressed as [8],

$$
\begin{aligned}
&\text{minimize} \\
&\text{trace} (\mathbf{A}\mathbf{X}) \\
&\text{subject to} \\
&\text{trace} (\mathbf{F}_m \mathbf{X}) = e_m, m = 1, 2, \dots, M \\
&\mathbf{X} \succeq 0,
\end{aligned}
\tag{6.12}
$$

where $\mathbf{A}, \mathbf{F}_1, \mathbf{F}_2, \dots, \mathbf{F}_M \in \mathbf{S}^N$ and a matrix nonnegativity constraint is imposed on the variable $\mathbf{X} \in \mathbf{S}^N$.

6.4   Geometric Programming

Geometric programming is a class of optimization problems. The standard form of geometric programming itself is nonlinear and nonconvex. However, geometric programming can be easily transformed to convex optimization problem [8, 392]. In this way, a global optimum can be obtained.

If a function is defined as,

$$h(\mathbf{x}) = c x_1^{a_1} x_2^{a_2} \cdots x_N^{a_N}, \tag{6.13}$$

where $c, x_1, x_2, \ldots, x_N \in \mathbf{R}_{++}$ and $a_1, a_2, \ldots, a_N \in \mathbf{R}$, this function is called a monomial function, or simply, a monomial [8].

A sum of monomials is a posynomial function, or simply, a posynomial [8],

$$f(\mathbf{x}) = \sum_{k=1}^{K} c_k x_1^{a_{1k}} x_2^{a_{2k}} \cdots x_N^{a_{Nk}}, \tag{6.14}$$

where $c_k \in \mathbf{R}_{++}$ and $a_{1k}, a_{2k}, \ldots, a_{Nk} \in \mathbf{R}, k = 1, 2, \ldots, K$.

A standard form of geometric programming has the form [8],

$$
\begin{aligned}
&\text{minimize} \\
&f_0(\mathbf{x}) \\
&\text{subject to} \\
&f_m(\mathbf{x}) \le 1, m = 1, 2, \ldots, M \\
&h_l(\mathbf{x}) = 1, l = 1, 2, \ldots, L \\
&\mathbf{x} \succ 0,
\end{aligned} \tag{6.15}
$$

where $f_0, f_1, f_2, \ldots, f_M$ are posynomials and $h_1, h_2, \ldots, h_L$ are monomials.

Define,

$$y_n = \log x_n, n = 1, 2, \ldots, N \tag{6.16}$$

then,

$$x_n = e^{y_n}, n = 1, 2, \ldots, N. \tag{6.17}$$

A monomial can be transformed to [8]

$$
\begin{aligned}
h(\mathbf{x}) &= c x_1^{a_1} x_2^{a_2} \cdots x_N^{a_N} \\
&= c(e^{y_1})^{a_1} (e^{y_2})^{a_2} \cdots (e^{y_N})^{a_N} \\
&= e^{y_1 a_1 + y_2 a_2 \cdots y_N a_N + b} \\
&= e^{\mathbf{a}^T \mathbf{y} + b},
\end{aligned} \tag{6.18}
$$

where $b = \log c$. The change of variables turns a monomial function into the exponential of an affine function [8].

Similarly, a posynomial can be transformed to [8]

$$f(x) = \sum_{k=1}^{K} e^{\mathbf{a}_k^T \mathbf{y} + b_k}, \tag{6.19}$$

where $\mathbf{a}_k = (a_{1k}, a_{2k}, \ldots, a_{Nk})^T$ and $b_k = \log c_k, k = 1, 2, \ldots, K$.

The geometric programming (6.15) can be expressed in terms of $\mathbf{y} \in \mathbf{R}^N$ as [8],

$$
\begin{aligned}
&\text{minimize} \\
&\sum_{k=1}^{K_0} e^{\mathbf{a}_{0k}^T \mathbf{y} + b_{0k}} \\
&\text{subject to} \\
&\sum_{k=1}^{K_m} e^{\mathbf{a}_{mk}^T \mathbf{y} + b_{mk}} \le 1, m = 1, 2, \dots, M \\
&e^{\mathbf{g}_l^T \mathbf{y} + p_l} = 1, l = 1, 2, \dots, L,
\end{aligned}
\tag{6.20}
$$

where $\mathbf{a}_{mk} \in \mathbf{R}^N$, $m = 0, 1, 2, \dots, M$ and $\mathbf{g}_l \in \mathbf{R}^N$, $l = 1, 2, \dots, L$.

Finally, we perform logarithm operation to the objective function and constraint functions in the geometric programming (6.20) to get the convex form of geometric programming [8],

$$
\begin{aligned}
&\text{minimize} \\
&\tilde{f}_0(\mathbf{y}) = \log\left(\sum_{k=1}^{K_0} e^{\mathbf{a}_{0k}^T \mathbf{y} + b_{0k}}\right) \\
&\text{subject to} \\
&\tilde{f}_m(\mathbf{y}) = \log\left(\sum_{k=1}^{K_m} e^{\mathbf{a}_{mk}^T \mathbf{y} + b_{mk}}\right) \le 0, m = 1, 2, \dots, M \\
&\tilde{h}_m(\mathbf{y}) = \mathbf{g}_l^T \mathbf{y} + p_l = 0, l = 1, 2, \dots, L.
\end{aligned}
\tag{6.21}
$$

If the objective function and constraint functions in the geometric programming (6.21) are all monomials, then the geometric programming (6.21) reduces to a linear programming. Hence, geometric programming can be treated as an extension of linear programming [8].

Extensions of geometric programming are documented in [392] together with applications in communication systems. These applications include channel capacity, coding, network resource allocation, network congestion control, and so on [392].

6.5 Lagrange Duality

In optimization theory, the duality theory states that optimization problems may be viewed from either of two perspectives, the primal problem or the dual problem. No matter whether the primal problem is convex or not, the dual problem is certainly concave. Thus, the dual problem is easy to solve. The solution of the dual problem provides a lower bound to the solution of the primal problem.

Mathematically speaking, if the primal problem, which is not necessarily convex, is expressed as

$$
\begin{aligned}
&\text{minimize} \\
&f_0(\mathbf{x}) \\
&\text{subject to} \\
&f_m(\mathbf{x}) \le 0, m = 1, \dots, M \\
&h_l(\mathbf{x}) = 0, l = 1, \dots, L
\end{aligned}
\tag{6.22}
$$

and its optimal value is

$$
p^* = f_0(\mathbf{x}^*),
\tag{6.23}
$$

then the corresponding dual problem is

$$\begin{array}{l} \text{maximize} \\ g(\lambda, \upsilon) \\ \text{subject to} \\ \lambda \geq \mathbf{0}, \end{array} \tag{6.24}$$

where $g(\lambda, \upsilon)$ is Lagrange dual function defined as [8]

$$g(\lambda, \upsilon) = \inf L(\mathbf{x}, \lambda, \upsilon)$$

$$= \inf \left(f_0(\mathbf{x}) + \sum_{m=1}^{M} \lambda_m f_m(\mathbf{x}) + \sum_{l=1}^{L} \upsilon_l h_l(\mathbf{x}) \right) \tag{6.25}$$

and $\mathbf{x}$ satisfies constraints in the primal problem (6.22).

Denote the optimal value of the dual problem by d^*. Weak duality always holds for convex and nonconvex problems,

$$d^* \leq p^* \tag{6.26}$$

and strong duality usually holds for convex problems,

$$d^* = p^*. \tag{6.27}$$

6.6   Optimization Algorithm

Two categories of algorithms, that is, deterministic algorithms and stochastic algorithms, are widely used to solve optimization problems. For deterministic algorithms, interior point methods are very popular recently.

6.6.1   Interior Point Methods

Interior point methods are a class of algorithms to solve linear and nonlinear convex optimization problems. Ideally, any convex optimization problems can be solved by interior point methods. The key element of these methods is to use a self-concordant barrier function to encode the convex set [393]. A barrier function is a continuous function whose value on a point increases to infinity as the point approaches the boundary of the feasible region. Thus, interior point methods reach an optimal solution by going through the interior of the feasible region.

The ideal barrier function should be [8],

$$I(u) = \begin{cases} 0, & u \leq 0 \\ \infty, & u > 0. \end{cases} \tag{6.28}$$

In reality, logarithmic barrier is used as an approximation,

$$I(u) = -\frac{1}{t} \log(-u), \tag{6.29}$$

where $t > 0$ and approximation improves as t goes to infinity [8]. Meanwhile, logarithmic barrier function is convex and twice continuously differentiable.

6.6.2 Stochastic Methods

Stochastic methods or random search methods generate and use random variables to get the solution to the optimization problem. Stochastic methods do not need to explore the structures of objective functions and constraints, that is, derivative or gradient information. Stochastic methods will be suitable for the nonconvex optimization problems or the relatively large-scale high-dimension optimization problems. Stochastic methods cannot guarantee the global optimum, but there are often no other choices.

Stochastic methods can include but are not limited to

- simulated annealing;
- stochastic hill climbing;
- genetic algorithm;
- ant colony optimization;
- particle swarm optimization (PSO).

Therein, genetic algorithm, which is one kind of evolutionary algorithm techniques, has been widely used for multiobjective optimization or multiobjective decision making. Take PSO as an example [394]. Power allocation problem for time reversal with array gain in MIMO UWB system is formulated as a nonconvex optimization issue. Even though the first and second derivatives of the objective functions and constraints can be easily derived [394], because the objective function is a nonlinear and nonconvex function, it is hard to use deterministic algorithm to solve this optimization problem. PSO is applied [394]. PSO is a swarm intelligence based algorithm to find a solution to an optimization problem [395]. There are many particles with a position and a velocity in the swarm. Particles in a swarm communicate good positions to each other and adjust their own position and velocity based on these good positions.

Suppose there are N particles. After K iterations, the algorithm is stopped. When the k-th iteration begins, the position of particle i is $\mathbf{L}_i^{k-1}$. The velocity of particle i is $\mathbf{V}_i^{k-1}$. The local best position of particle i is

$$\mathbf{L}_{i\text{best}}^k = \arg \max_{\{\mathbf{L}_{i\text{best}}^{k-1}, \mathbf{L}_i^{k-1}\}} C(\mathbf{L}), \tag{6.30}$$

where $C(\mathbf{L})$ is the utility function. The global best position is

$$\mathbf{L}_{\text{gbest}}^k = \arg \max_{\{\mathbf{L}_{i\text{best}}^k, \, i=1,2,...,N\}} C(\mathbf{L}). \tag{6.31}$$

Then the velocity of particle i in the k-th iteration is

$$\mathbf{V}_i^k = w \times \mathbf{V}_i^{k-1} + c_1 \times \text{rand} \times (\mathbf{L}_{i\text{best}}^k - \mathbf{L}_i^{k-1}) + c_2 \times \text{rand} \times (\mathbf{L}_{\text{gbest}}^k - \mathbf{L}_i^{k-1}) \tag{6.32}$$

and the new position of particle i is $\mathbf{L}_i^k = \mathbf{L}_i^{k-1} + \mathbf{V}_i^k$. In Equation (6.32) rand means random value drawn from a uniform distribution on the unit interval; w is the inertia weight; c_1 and c_2 are two positive constants, called the cognitive and social parameter respectively.

6.7 Robust Optimization

The optimization issue with uncertainty is becoming hotter and hotter in many research fields, such as operation research, finance, industrial management, transportation scheduling, wireless communication, smart grid, and so on, because most of the optimization issues are from the dynamic complex system, and most of the variables in the optimization issue cannot be deterministic or known for sure. There are two approaches to deal with the optimization issue with uncertainty. One is robust optimization, and the other is stochastic optimization. In robust optimization, the uncertainty model is deterministic and set based [396]. However, in stochastic optimization, the uncertainty model is assumed to be random [396]. Robust optimization, which is a conservative approach [397], can guarantee the performance for all the cases within the set based uncertainty. In other words, robustness means the performance is stable with the bounded errors. However, stochastic optimization can only guarantee the performance on average for the uncertainty with known or partially known probability distribution [397] information. Thus, there is a tradeoff between robustness and performance. Robust optimization will materialize by waveform diversity.

Waveform diversity is a key research issue in the current wireless communication system, the radar system, and the sensing or image system. Waveform should be designed or optimized according to the different requirements or objectives of system performance and should be adapted or diversified dynamically to the operating environment in order to achieve a performance gain [398]. For example, the waveform should be designed to carry more information to the receiver in terms of capacity. If the energy detector is employed at the receiver, the waveform should be optimized such that the energy of the signal in the integration window at the receiver should be maximized [399–401]. For navigation and geolocation, the ultra short waveform should be used to increase the resolution. For multi-target identification, the waveform should be designed so that the returns of radar signals can bring more information back. In clutter dominant environment, maximizing the target energy and minimizing the clutter energy should be considered simultaneously.

Multiple input single output (MISO) system is one kind of multi-antenna systems in which there are multiple antennas at the transmitter and one antenna at the receiver. MISO system can explore the spatial diversity and execute the transmitter beamforming to focus energy on the desired direction or point and avoid interference to other radio systems. It is well known that waveform and spatially diverse capabilities are made possible today due to the advent of lightweight digital programming waveform generator [402] or AWG. Waveform diversity can also be applied to the wideband system. Waveform design or optimization for wideband multi-antenna system is documented in [403]. From a theoretical point of view, the contribution of [403] can be summarized as follows. The equivalent baseband waveforms are designed for the passband system. Different waveforms for different transmitter antennas are jointly optimized to obtain the global optimality. At the receiver, the received signals from different transmitter antennas will be combined together over the air such that the receiver antenna will see only one copy of the signal from the transmitter. In order to achieve this kind of over the air coherency for the passband signals, all the individual oscillators should be tied together at the transmitter [402] to make the carrier phase consistent.

In the context of cognitive radio, waveform design gives us flexibility to design radio, which can coexist with other cognitive radios and primary radios. From cognitive radio's point of view, spectral mask constraint at the transmitter and the interference cancellation at the receiver should be seriously considered for waveform design or optimization, in addition to the traditional communication objectives and constraints. Spectral mask constraint is imposed on the transmitted waveform such that cognitive radio has limited or no interference to primary radio. At the same time, the interference cancellation scheme is implemented at the receiver to cancel the interference from primary radio to cognitive radio.

Though the thought of waveform diversity for the radar system can be traced back to World War II, due to the computational capability and hardware limitation, a lot of waveform design algorithms cannot be implemented into the radar system [398] for many years. Nowadays, these bottlenecks are broken, and waveform diversity becomes a hotspot afresh in the radar society. Time reversal or phase conjugating waveform, colored waveform, sparse and regular nonuniform Doppler waveform, noncircular waveform, and so on are dealt with based on advanced mathematics tools in [404]. New trends in coded waveform design for radar applications are presented in [405]. The modern SDP and the novel algorithm on Hermitian matrix rank one decomposition are exploited to perform code selection which can maximize the detection performance and control the Doppler estimation accuracy and the similarity with a prefixed radar code [405]. Meanwhile, another force to propel the research on waveform diversity is the introduction of cognition to the radar system, that is, cognitive radar which means the radar can actively learn about the environment, and the whole radar system forms a dynamic closed feedback loop including the transmitter, environment, and receiver [406]. Waveform diversity will play an important role in cognitive radar. The radar transmitter can adjust its illumination of the environment in an intelligent, effective, adaptive, and robust manner, taking into account the results of learning and perception [406]. Thus, the philosophy of sequential testing [407] can be embraced under the umbrella of cognitive radar smoothly. Several rounds of illuminations will be used until the belief that the decision is correct is made. The waveform and the transceiver scheme for each round can be adjustable according to the results of the previous illuminations. For example, adaptive CS [384] gives us the hint to this research field. Though the thought of waveform diversity for the radar system can be traced back to World War II, due to the computational capability and hardware limitation, a lot of waveform design algorithms cannot be implemented into the radar system [398] for many years. Nowadays, these bottlenecks are broken, and waveform diversity becomes a hotspot afresh in the radar society. Time reversal or phase conjugating waveform, colored waveform, sparse and regular nonuniform Doppler waveform, noncircular waveform, and so on are dealt with based on advanced mathematics tools in [404]. New trends in coded waveform design for radar applications are presented in [405]. The modern SDP and the novel algorithm on Hermitian matrix rank one decomposition are exploited to perform code selection which can maximize the detection performance and control the Doppler estimation accuracy and the similarity with a prefixed radar code [405]. Meanwhile, another force to propel the research on waveform diversity is the introduction of cognition to the radar system, that is, cognitive radar which means the radar can actively learn about the environment, and the whole radar system forms a dynamic closed feedback loop including the transmitter, environment, and receiver [406]. Waveform diversity will play an important role in cognitive radar. The radar transmitter can adjust its illumination of the environment in an intelligent, effective, adaptive, and robust manner, taking into account the results of learning and perception [406]. Thus, the philosophy of sequential testing [407] can be embraced under the umbrella of cognitive radar smoothly. Several rounds of illuminations will be used until the belief that the decision is correct is made. The waveform and the transceiver scheme for each round can be adjustable according to the results of the previous illuminations. For example, adaptive CS [384] gives us the hint to this research field.

The previous theoretical researches on waveform diversity do not take the robustness into account. There are several reasons for this:

- The theory of robust optimization was not that mature in the old days.
- Robustness makes waveform diversity complex.
- The research on waveform diversity was only limited to computer simulation.

Nowadays, as the theory of robust optimization becomes mature and bottlenecks of computation and implementation are broken, robust optimization for waveform diversity, that is, robust waveform diversity, will bring more attention. Meanwhile, robustness is the bridge between the theoretical work and the practical situation.

Robust optimization is the key mathematical tool for robust waveform diversity, which gives the optimal waveform with robustness. Robust optimization is systematically introduced in [396]. The most frequently used optimization formats within robust optimization are robust linear programming [396, 408, 409], robust least squares [8, 410], robust mean square error (MSE) [411–417], and robust SDP [418, 409]. If the optimization issues can be formulated as robust linear programming, robust least squares, or robust MSE with some specific uncertainty models, these optimization issues will be solvable and tractable. For example, if the uncertainty model is the ellipsoidal uncertainty set, robust linear programming becomes an SOCP and a robust SOCP becomes an SDP [419], which can be efficiently solved via interior point methods. However, a robust SDP with the ellipsoidal uncertainty set is NP-hard to solve [419]. Because SDP is harder than SOCP, and SOCP is harder than linear programming taking the complexity of solving method into account, robustness increases the difficulty of the optimization issue [419]. Robust least squares with the finite uncertainty set, norm bound error, uncertainty ellipsoids, and norm bounded error with linear structure are discussed in [8]. Meanwhile, the solution of robust least squares where the coefficient matrices are unknown but bounded is also given in [410]. The worst case residual is minimized, and the corresponding optimization problem can be formulated as SOCP. The work on robust MSE is done by [411–417] from classical estimation's point of view. Robust MSE can also be called minimax MSE. The core idea of the competitive minimax approach [412] is to seek the linear estimator that minimizes the worst case regret with the assumption that the covariance of parameter vector is subject to uncertainties. The minimax MSE estimator, the linear estimator that minimizes the worst case MSE among all parameter vectors with bounded norm [414], can be found by solving an SDP. Similarly, robust MSE with noise covariance uncertainty is dealt with in [415]. Robust MSE is extended to a multisignal estimation issue in [416] where both model and noise uncertainties are considered.

Transmitter power control can be treated as one kind of waveform diversity schemes. Traditionally, power control or power allocation was implemented above the physical layer as one kind of radio resource management issues. Power control can be implemented in the physical layer. In this way, the period of power control loop will be greatly reduced. Different power control patterns will synthesize different transmitted waveforms to meet different requirements. Robust transmitter power control is documented in [397]. OFDM modulation scheme is adopted in the physical layer. Every cognitive radio should dynamically control the transmission powers for its own subcarriers in order to maximize the total benefits. Thus, the objective of this optimization issue is the maximization of total capacities of all cognitive radios, and the constraints consist of the individual power constraints and interference constraints. Because there is no central node in cognitive radio network, the feasible algorithm to solve this optimization issue should be implemented in a decentralized manner. For the nonrobust version of optimization issue, the classical iterative waterfilling can be used, and the convergence of the solution can be guaranteed. However, the cognitive radio network has a dynamic nature [397] due to the random mobility of cognitive radios and primary radios. Thus the noise plus interference term includes two components: a nominal term and a perturbation term to form the robust version of optimization issue. The price of the robustness is that a convex optimization problem becomes a nonconvex optimization problem. Most of the algorithms for convex optimization cannot be used. A new numerical technique to solve the nonconvex robust optimization issue is proposed in [420]. Neighborhood searches and robust local moves are applied iteratively to achieve the robust solutions [397]. Similar to transmitter power control, transmitter beamforming can be thought of as another kind of waveform diversity schemes. Robust transmitter beamforming in multiuser MISO cognitive radio networks is considered in [421]. Channel state information in [421] is assumed to be imperfectly known, and the imperfectness of channel state information is modeled using an Euclidean ball shaped uncertainty set [421]. Specifically speaking, the objective is to design the optimal beamforming weights for different cognitive radios with the least total transmitted power at the central node while simultaneously the least possible received signal to interference plus noise ratio (SINR) for each cognitive radio should be equal to or greater than a threshold defined by the quality of service (QoS) requirement. The interference for each primary radio should be equal to or less than a threshold to make primary radio work properly. Robust transmitter beamforming with partial channel state information for cognitive radio can also be seen in [422]. Because of the limitation of sounding system and feedback system, robustness to partial channel state information, channel state information error, or the limited feedback is very important to dealing with transmitter power control and beamforming. Meanwhile, due to the perturbation of the radio environment and the fading of the radio channel, how to deal with outdated channel state information is still worth studying. Sometimes, without channel state information, the directional beam for far field can still be formed at the transmitter using array manifold, steering vector, or spatial signature.

Robust waveform diversity is applied to MIMO radar system in [423]. The design criteria are mutual information and MMSE estimation for target identification and classification. Target PSD is assumed to lie in an uncertainty class of spectra bounded by known upper and lower bounds [423]. With this kind of prior information, the designed waveform can well match the target and bring back more information. The minimax robust waveforms can bound the worst case performance at an acceptable level [423]. Optimal and robust waveform design for MIMO radars with the consideration of the signal dependent noise, that is, clutter, is studied in [424]. Robust waveforms to minimize the estimation error of the worst case target realization are obtained [424].

It has been widely accepted that waveform diversity is implemented at the transmitter. But waveform diversity should have broader meaning and significance. First of all, any type of signal processing in the waveform level at the receiver should also be included into the waveform diversity framework. The most common signal processing is the receiver beamforming including the narrowband beamforming and wideband beamforming. Robust receiver beamforming is dealt with in [425, 426]. The uncertainty comes from the mismatch of steering vector and the estimation error of the sampled covariance matrix for interference plus noise. The worse case performance of the minimum variance beamformer or Capon beamformer is taken care of. SDP or SOCP can be exploited to solve the corresponding robust optimization issues. Robust minimum variance beamformer with probabilistic constraint is mentioned in [427], and the relationship between probability constrained and worst case optimization is discussed. Robust least squares are applied to antenna design in [409]. The optimal solution obtained from the nominal least squares is completely unstable w.r.t. small implementation errors [409]. However, robust least squares will bring stable results to combat the uncertainty. Robust wideband beamforming is addressed and presented by [428]. Similarly, error from steering vector brings instability to the system and inevitably degrades the beamformer's performance [428]. Hybrid steepest descent method is proposed to find the unique minimizer of the cost function over the feasible convex set [428].

6.8 Multiobjective Optimization

Practical optimization problems, especially the engineering design optimization problems, seem to have a multi-objective nature much more frequently than a single objective one [429]. For example, to form wideband beampattern with arbitrary shape, we need to consider at least four objectives: main beam, sidelobe, nulling, and frequency invariant property.

In terms of solution, the difference between the multiobjective optimization and the single-objective optimization is the former has a set of Pareto-optimal solutions while the latter has a single global optimum if such a solution exists. The term "Pareto-optimal solution" refers to a solution around which there is no way of improving any objective without worsening at least one other objective [429]. The set of Pareto-optimal solutions can be characterized by Pareto front—a hypersurface in the objective function space in which the Pareto-optimal points are located [429].

How to get the solution for the multiobjective optimization is based on how the individual objective should be weighted in relation to all others. Thus, four kinds of methods can be applied in terms of preference [429, 430]:

- **A Priori Preference**. We can specify the preferences before running the optimization algorithm. Most likely, a single utility function is developed to combine all the objectives.
- **Progressive Preference**. We can interact with the optimization algorithm and change the preferences during the optimization process.
- **A Posteriori Preference**. No preferences is given before or during the optimization process. We can choose the solution from a set of candidates provided by the optimization algorithm.

- **No Preference**. No preferences are needed in the whole process of the multiobjective optimization.

If preferences are given beforehand, the weighted sum method is the simplest approach and probably the most widely used classical method. This method transforms the multi-objective optimization problem into a single objective one by multiplying each objective with a predetermined weight and adding all the weighted objectives together. The solution to the single objective problem is Pareto-optimal if the weights are positive for all objectives. However, the weighted sum method cannot guarantee that any Pareto-optimal solution can be obtained using a positive weight vector. Meanwhile, if preferences are not given beforehand, we have to find a set of candidate solutions as completely as possible.

For deterministic strategy, ε-constraint method, weighted metric methods, rotated weighted metric method, value function method, and so on can be exploited. Besides, stochastic algorithms, especially evolutionary algorithms, seem to be more popular than deterministic algorithms to solve the multiobjective optimization problem [430–433]. Convergence and diversity are two important issues for multiobjective evolutionary algorithms [434]. An efficient evolutionary method to approximate the Pareto optimal set in multiobjective optimization has been proposed in [435]. A relevant example is to use strength Pareto evolutionary algorithm 2 to design simultaneous multimission waveforms [436]. A genetic algorithm is also applied in [437] to obtain OFDM radar waveform for target detection with consideration of error bound and Mahalanobis-Distance. Similarly, in order to make algorithms scalable, parallel genetic algorithms [438] are worth using.

The performance optimization of cognitive radio or cognitive radio network itself is a multi-objective optimization problem. First of all, multiple objectives exist from physical layer to application layer in cognitive radio network [439]. Different layers may have different performance metrics. Different applications may have different QoS requirements. Different users may have different subjective performance needs. Hence, multiple objectives should be taken into account simultaneously. Meanwhile, the external radio environment and internal network state determine the validity, feasibility, and sensitivity of objectives. Specifically speaking, bit error rate (BER) minimization, out-of-band interference minimization, power consumption minimization, and overall throughput maximization have been achieved using a multiobjective fitness function in the framework of distributed optimization [440, 441]. Genetic algorithm and its variants are widely exploited [441–449]. Besides, PSO can also be used for spectrum allocation in cognitive radio network with consideration of sum bandwidth reward and access fairness of secondary users [450]. From the perspective of artificial intelligence, a case-based reasoning method using the divide-and-conquer concept has been explored to generate solutions for problems with multiple objectives in cognitive radio [451].

6.9 Optimization for Radio Resource Management

Radio resource management is the system-level control for the wireless communication system [452–457]. Generally, radio resource management tries to optimize the utilization of various radio resources such that the performance of radio system can be improved. Mathematical optimization, especially convex optimization, is the main tool supporting radio resource management [458]. Meanwhile, radio resource management will be the basic function in cognitive radio [459]. Spectrum related management for spectrum sensing spectrum access, spectrum sharing, and so on, will be the feature for cognitive radio [460, 461].

Radio resource management includes but is not limited to

- power control [462–467];
- frequency band allocation;
- time slot allocation;
- adaptive modulation and coding [468–470];
- rate control [471];
- antenna selection [472–475];
- scheduling [471, 476–479];
- handover [480–482];
- admission control [483–489];
- congestion control [484, 490–494];
- load control [495];
- routing plan [496–498];
- base station deployment.

The work about radio resource management can also be found in [499–507]. Capacity, communication rate, spectrum efficiency, or capacity region is used frequently as performance metric for radio resource management. Besides, MIMO related radio resource management and OFDM related radio resource management will also be mentioned in the following chapters.

6.10 Examples and Applications

The examples and applications will show the beauty and benefit of mathematical optimization.

6.10.1 Spectral Efficiency for Multiple Input Multiple Output Ultra-Wideband Communication System

It is assumed that there are N_t transmitter antennas and N_r receiver antennas in the system.

The channel transfer function is $\mathbf{H}(f)$ with bandwidth $W = f_1 - f_0$ where $f_0(>0)$ is the starting frequency and $f_1(>0)$ is the end frequency

$$\mathbf{H}(f) = \begin{bmatrix} H_{11}(f) & H_{12}(f) & \cdots & H_{1N_t}(f) \\ H_{21}(f) & H_{22}(f) & \cdots & H_{2N_t}(f) \\ \vdots & \vdots & \vdots & \vdots \\ H_{N_r1}(f) & H_{N_r2}(f) & \cdots & H_{N_rN_t}(f) \end{bmatrix}, \tag{6.33}$$

where $\mathbf{H}_{mn}(f)$ is channel transfer function from the transmitter antenna n to the receiver antenna m. Its corresponding channel impulse response is

$$\mathbf{H}(t) = \begin{bmatrix} H_{11}(t) & H_{12}(t) & \cdots & H_{1N_t}(t) \\ H_{21}(t) & H_{22}(t) & \cdots & H_{2N_t}(t) \\ \vdots & \vdots & \vdots & \vdots \\ H_{N_r1}(t) & H_{N_r2}(t) & \cdots & H_{N_rN_t}(t) \end{bmatrix}. \tag{6.34}$$

The spectrum shaping filter at the transmitter side is

$$\mathbf{X}(t) = \begin{bmatrix} X_{11}(t) & X_{12}(t) & \cdots & X_{1N_s}(t) \\ X_{21}(t) & X_{22}(t) & \cdots & X_{2N_s}(t) \\ \vdots & \vdots & \vdots & \vdots \\ X_{N_t1}(t) & X_{N_t2}(t) & \cdots & X_{N_tN_s}(t) \end{bmatrix} \tag{6.35}$$

and its corresponding transfer function is

$$\mathbf{X}(f) = \begin{bmatrix} X_{11}(f) & X_{12}(f) & \cdots & X_{1N_s}(f) \\ X_{21}(f) & X_{22}(f) & \cdots & X_{2N_s}(f) \\ \vdots & \vdots & \vdots & \vdots \\ X_{N_t1}(f) & X_{N_t2}(f) & \cdots & X_{N_tN_s}(f) \end{bmatrix}. \tag{6.36}$$

The input of the spectrum shaping filter is the transmitted signal vector $\mathbf{a}(t)$. The entries of $\mathbf{a}(t)$ are $a_1(t), a_2(t), \ldots$, and $a_{N_s}(t)$,

$$\mathbf{a}(t) = \begin{bmatrix} a_1(t) \\ a_2(t) \\ \vdots \\ a_{N_s}(t) \end{bmatrix}, \tag{6.37}$$

all of which are independent white Gaussian random processes with zero mean and unit PSD.

The transmitted signal at the transmitter array is

$$\mathbf{S}(t) = \mathbf{X}(t) \otimes \mathbf{a}(t), \tag{6.38}$$

where "$\otimes$" denotes convolution operation and each entry of $\mathbf{S}(t)$ is

$$S_i(t) = \sum_{j=1}^{N_s} (X_{ij}(t) \otimes a_j(t)), \quad i = 1, 2, \ldots, N_t. \tag{6.39}$$

Hence, the PSD of the transmitted signal at the transmitter array is

$$\mathbf{R_S}(f) = \mathbf{X}(f)\mathbf{X}^H(f). \tag{6.40}$$

The received signal at the receiver array is

$$\mathbf{R}(t) = \mathbf{H}(t) \otimes \mathbf{S}(t) + \mathbf{N}(t) \tag{6.41}$$

where $\mathbf{N}(t)$ is AWGN the entries of which are independent random processes with zero mean and one-sided PSD N_0.

If a one-sided situation is considered, then the transmitted power is

$$P = \int_{f_0}^{f_1} \text{trace} \left[\mathbf{R_S}(f) \right] df. \tag{6.42}$$

The equivalent ratio of the transmitted signal power to the received noise power (TX SNR) is defined as

$$\rho = \frac{P}{N_0 W}. \tag{6.43}$$

The spectral efficiency is [508]

$$\frac{C}{W} = \frac{1}{W} \int_{f_0}^{f_1} \log_2 \left| \mathbf{I}_{N_r}(f) + \frac{\mathbf{H}(f)\mathbf{R_S}(f)\mathbf{H}^H(f)}{N_0} \right| df, \tag{6.44}$$

where $| \bullet |$ represents the determinant of the matrix.

The methods for the design of spectrum shaping filter are

- water filling;
- constant power water filling;
- time reversal;
- channel inverse;
- constant power spectral density;
- MMSE.

6.10.1.1   Water Filling

It is well known that the spectral efficiency of water filling is greater than that of any other spectrum shaping scheme. Let $\lambda_i(f)$, $i = 1, 2, \ldots, N_t$ denote the set of eigenvalues of $N_0 \mathbf{H}^{-1}(f)[\mathbf{H}^{-1}(f)]^H$. So SVD of $N_0 \mathbf{H}^{-1}(f)[\mathbf{H}^{-1}(f)]^H$ can be written as

$$N_0 \mathbf{H}^{-1}(f)[\mathbf{H}^{-1}(f)]^H = \mathbf{U}(f)\text{diag}\{\lambda_i(f)\}\mathbf{U}^H(f), \tag{6.45}$$

where $\text{diag}(\mathbf{a})$, if $\mathbf{a}$ is a vector with n components, returns an n-by-n diagonal matrix having $\mathbf{a}$ as its main diagonal. Because of the property of unitary matrix, $\frac{\mathbf{H}^H(f)\mathbf{H}(f)}{N_0}$ can be expressed as

$$\frac{\mathbf{H}^H(f)\mathbf{H}(f)}{N_0} = \mathbf{U}(f)\text{diag}\{\lambda_i^{-1}(f)\}\mathbf{U}^H(f). \tag{6.46}$$

Then, $\mathbf{R_S}(f)$ can be given by

$$\mathbf{R_S}(f) = \mathbf{U}(f)\text{diag}\{\Lambda_i(f)\}\mathbf{U}^H(f), \tag{6.47}$$

where $\Lambda_i(f) = (\mu - \lambda_i(f))^+$, $i = 1, 2, \ldots, N_t$ and $(x)^+ = \max[0, x]$. Here, the constant μ is the water level chosen to satisfy the power constraint with equality

$$\sum_{i=1}^{N_t} \int_{f_0}^{f_1} \Lambda_i(f)\, df = P. \tag{6.48}$$

So, the spectral efficiency $\frac{C}{W}$ in this case is [509]

$$\frac{1}{W} \sum_{i=1}^{N_t} \int_{f_0}^{f_1} \left(\log_2 \left(\frac{\mu}{\lambda_i(f)} \right) \right)^+ df. \tag{6.49}$$

6.10.1.2   Constant Power Water Filling

Constant power water filling is well studied in [510]. For water filling, the power allocation scheme is $\Lambda_i(f) = (\mu - \lambda_i(f))^+$, $i = 1, 2, \ldots, N_t$. While for constant power water filling, the power allocation scheme is

$$\Lambda_i(f) = \begin{cases} p_0, & \text{if } \lambda_i(f) \le \lambda_0, \\ 0, & \text{if } \lambda_i(f) > \lambda_0. \end{cases} \tag{6.50}$$

How to get the optimal p_0 and λ_0 is the key point of constant power water filling. Similarly, the frequency band sets Ω_i, $i = 1, 2, \ldots, N_t$ are defined as

$$\Omega_i = \{f : \lambda_i(f) \le \lambda_0; \ f_0 \le f \le f_1\}. \tag{6.51}$$

The measure of Ω_i is θ_i, and

$$\theta = \sum_{i=1}^{N_t} \theta_i. \tag{6.52}$$

λ_0 should be selected to meet the condition that $\min\{\lambda_i(f), f \in \Omega_i, i = 1, 2, \ldots, N_t\} + \frac{P}{\theta}$ is equal to

$$\max \{\lambda_i(f), f \in \Omega_i, i = 1, 2, \ldots, N_t\}. \tag{6.53}$$

Meanwhile, $p_0 = \frac{P}{\theta}$.

6.10.1.3   Time Reversal

For time reversal, it follows that

$$\mathbf{X}(f) = \alpha \mathbf{H}^H(f), \tag{6.54}$$

where the constant α is the scale factor chosen to satisfy the power constraint with equality,

$$P = \int_{f_0}^{f_1} \text{trace}\,[\mathbf{R_S}(f)]\, df$$

$$= \int_{f_0}^{f_1} \text{trace}\,[\mathbf{X}(f)\mathbf{X}^H(f)]\, df$$

$$= \alpha^2 \int_{f_0}^{f_1} \text{trace}\,[\mathbf{H}^H(f)\mathbf{H}(f)]\,df$$

$$= \alpha^2 \int_{f_0}^{f_1} \sum_{i=1}^{N_r} \sum_{j=1}^{N_t} |H_{ij}(f)|^2\,df. \tag{6.55}$$

Hence

$$\alpha = \sqrt{\frac{P}{\int_{f_0}^{f_1} \sum_{i=1}^{N_r} \sum_{j=1}^{N_t} |H_{ij}(f)|^2\,df}} \tag{6.56}$$

and

$$\mathbf{X}(f) = \sqrt{\frac{P}{\int_{f_0}^{f_1} \sum_{i=1}^{N_r} \sum_{j=1}^{N_t} |H_{ij}(f)|^2\,df}}\,\mathbf{H}^H(f). \tag{6.57}$$

The spectral efficiency $\frac{C}{W}$ in this case is [509]

$$\frac{1}{W} \int_{f_0}^{f_1} \log_2 \left| \mathbf{I} + \frac{\rho W \mathbf{H}(f)\mathbf{H}^H(f)\mathbf{H}(f)\mathbf{H}^H(f)}{\int_{f_0}^{f_1} \sum_{i=1}^{N_r} \sum_{j=1}^{N_t} |H_{ij}(f)|^2\,df} \right| df. \tag{6.58}$$

6.10.1.4 Channel Inverse

For channel inverse, it follows that

$$\mathbf{X}(f) = \alpha \mathbf{H}^H(f)[\mathbf{H}(f)\mathbf{H}^H(f)]^{-1}, \tag{6.59}$$

where the constant α is the scale factor chosen to satisfy the power constraint with equality,

$$P = \int_{f_0}^{f_1} \text{trace}\,[\mathbf{R}_S(f)]\,df$$

$$= \int_{f_0}^{f_1} \text{trace}\,[\mathbf{X}(f)\mathbf{X}^H(f)]\,df$$

$$= \alpha^2 \int_{f_0}^{f_1} \text{trace}\,[[\mathbf{H}(f)\mathbf{H}^H(f)]^{-1}]\,df. \tag{6.60}$$

Hence

$$\alpha = \sqrt{\frac{P}{\int_{f_0}^{f_1} \text{trace}\,[[\mathbf{H}(f)\mathbf{H}^H(f)]^{-1}]\,df}} \tag{6.61}$$

and

$$\mathbf{X}(f) = \sqrt{\frac{P}{\int_{f_0}^{f_1} \text{trace}\left[[\mathbf{H}(f)\mathbf{H}^H(f)]^{-1}\right] df}} \mathbf{H}^H(f)[\mathbf{H}(f)\mathbf{H}^H(f)]^{-1}. \qquad (6.62)$$

The spectral efficiency $\frac{C}{W}$ in this case is [509]

$$\frac{C}{W} = N_r \log_2\left(1 + \frac{\rho W}{\int_{f_0}^{f_1} \text{trace}\left[[\mathbf{H}(f)\mathbf{H}^H(f)]^{-1}\right] df}\right). \qquad (6.63)$$

6.10.1.5 Constant Power Spectral Density

If power is equally allocated to each transmitter antenna, then

$$\mathbf{R}_S(f) = \frac{P}{W N_t}\mathbf{I}(f). \qquad (6.64)$$

The spectral efficiency $\frac{C}{W}$ in this case is [509]

$$\frac{C}{W} = \frac{1}{W}\int_{f_0}^{f_1} \log_2\left|\mathbf{I} + \frac{\rho \mathbf{H}(f)\mathbf{H}^H(f)}{N_t}\right| df. \qquad (6.65)$$

6.10.1.6 Minimum Mean Square Error

For MMSE, it follows that

$$\mathbf{X}(f) = \alpha \mathbf{H}^H(f)[\mathbf{H}(f)\mathbf{H}^H(f) + \frac{N_r}{\rho}\mathbf{I}]^{-1}, \qquad (6.66)$$

where the constant α is the scale factor chosen to satisfy the power constraint with equality,

$$P = \int_{f_0}^{f_1} \text{trace}\left[\mathbf{R}_S(f)\right] df$$

$$= \int_{f_0}^{f_1} \text{trace}\left[\mathbf{X}(f)\mathbf{X}^H(f)\right] df. \qquad (6.67)$$

So α is equal to

$$\sqrt{\frac{P}{\int_{f_0}^{f_1} \text{tr}[\mathbf{H}^H(f)[\mathbf{H}(f)\mathbf{H}^H(f) + \frac{N_r}{\rho}\mathbf{I}]^{-2}\mathbf{H}(f)] df}}. \qquad (6.68)$$

Similarly, the spectral efficiency in this case can be calculated by Equation (6.40) and Equation (6.44).

6.10.2 Wideband Waveform Design for Single Input Single Output Communication System with Noncoherent Receiver

OOK modulation is considered and the transmitted signal is

$$s(t) = \sum_{j=-\infty}^{\infty} d_j p(t - jT_b), \qquad (6.69)$$

where T_b is the bit duration; $p(t)$ is the transmitted bit waveform defined over $[0, T_p]$; and $d_j \in \{0, 1\}$ is j-th transmitted bit. Without loss of generality, assume the minimal propagation delay is equal to zero. The energy of $p(t)$ is

$$\int_0^{T_p} p^2(t)\, dt = E_p. \qquad (6.70)$$

The received noisy signal at the output of the receiver front-end is

$$r(t) = h(t) \otimes s(t) + n(t)$$

$$= \sum_{j=-\infty}^{\infty} d_j x(t - jT_b) + n(t), \qquad (6.71)$$

where $h(t), t \in [0, T_h]$ is the multipath impulse response that takes into account the effects of channel impulse response and the RF front-ends of the transceivers including antennas. $h(t)$ is available at the transmitter [511, 512]. $n(t)$ is a low-pass additive zero mean Gaussian noise with one-sided bandwidth W and one-sided PSD N_0. $x(t)$ is the received noiseless bit-"1" waveform defined as

$$x(t) = h(t) \otimes p(t). \qquad (6.72)$$

We further assume that $T_b \geq T_h + T_p = T_x$, that is, no existence of intersymbol interference (ISI).

An energy detector performs nonlinear square operation to $r(t)$ without any explicit analog filter at the receiver. Then the integrator does the integration over a given integration window T_I. Corresponding to the time index k, the k-th decision statistic at the output of the integrator is given by

$$z_k = \int_{kT_b+T_{I0}}^{kT_b+T_{I0}+T_I} r^2(t)\, dt$$

$$= \int_{kT_b+T_{I0}}^{kT_b+T_{I0}+T_I} (d_k x(t - kT_b) + n(t))^2 dt, \qquad (6.73)$$

where T_{I0} is the starting time of integration for each bit, and $0 \leq T_{I0} < T_{I0} + T_I \leq T_x \leq T_b$.

An approximately equivalent SNR for the energy detector receiver, which provides the same detection performance when applied to a coherent receiver, is given as [400]

$$\text{SNR}_{\text{eq}} = \frac{2\left(\int_{T_{I0}}^{T_{I0}+T_I} x^2(t)\, dt\right)^2}{2.3 T_I W N_0^2 + N_0 \int_{T_{I0}}^{T_{I0}+T_I} x^2(t)\, dt}. \qquad (6.74)$$

For the best performance, the equivalent SNR SNR$_{eq}$ should be maximized. Define

$$E_I = \int_{T_{I0}}^{T_{I0}+T_I} x^2(t)\,dt. \tag{6.75}$$

For given T_I, N_0, and W, SNR$_{eq}$ is an increasing function of E_I. So the maximization of SNR$_{eq}$ in Equation (6.74) is equivalent to the maximization of E_I in Equation (6.75). The optimization problem to get the optimal waveform is shown as

$$
\begin{aligned}
&\text{maximize} \\
&\int_{T_{I0}}^{T_{I0}+T_I} x^2(t)\,dt \\
&\text{subject to} \\
&\int_0^{T_p} p^2(t)\,dt = E_p.
\end{aligned}
\tag{6.76}
$$

In order to solve the optimization problem (6.76), numerical approach is employed. In other words, $p(t)$, $h(t)$, and $x(t)$ are uniformly sampled, and the optimization problem (6.76) will be converted to its corresponding discrete time form. Assume the sampling period is T_s. $T_p/T_s = N_p$. $T_h/T_s = N_h$. $T_x/T_s = N_x$. So $N_x = N_p + N_h$.

$p(t)$, $h(t)$, and $x(t)$ are represented by $p_i, i = 0, 1, \ldots, N_p$, $h_i, i = 0, 1, \ldots, N_h$, and $x_i, i = 0, 1, \ldots, N_x$, respectively [400].

Define

$$\mathbf{p} = [p_0 \ p_1 \ \cdots \ p_{N_p}]^T \tag{6.77}$$

and

$$\mathbf{x} = [x_0 \ x_1 \ \cdots \ x_{N_x}]^T. \tag{6.78}$$

Construct channel matrix $\mathbf{H}_{(N_x+1)\times(N_p+1)}$

$$(\mathbf{H})_{i,j} = \begin{cases} h_{i-j}, & 0 \leq i - j \leq N_h \\ 0, & \text{else} \end{cases}, \tag{6.79}$$

where $(\bullet)_{i,j}$ denotes the entry in the i-th row and j-th column of the matrix or vector. Meanwhile, for vector, taking $\mathbf{p}$ as an example, $(\mathbf{p})_{i,1}$ is equivalent to p_{i-1}.

The matrix expression of Equation (6.72) is

$$\mathbf{x} = \mathbf{H}\mathbf{p} \tag{6.80}$$

and the constraint in the optimization problem (6.76) can be expressed as

$$\|\mathbf{p}\|_2^2 T_s = E_p, \tag{6.81}$$

where "$\|\bullet\|_2$" denotes the Euclidean norm of the vector. In order to make the whole document consistent, we further assume

$$\|\mathbf{p}\|_2^2 = 1. \tag{6.82}$$

Let $T_I/T_s = N_I$ and $T_{I0}/T_s = N_{I0}$. The entries in $\mathbf{x}$ within integration window constitute $\mathbf{x}_I$ as,

$$\mathbf{x}_I = [x_{N_{I0}} \, x_{N_{I0}+1} \, \cdots \, x_{N_{I0}+N_I}]^T \tag{6.83}$$

and E_I in Equation (6.75) can be equivalently shown as

$$E_I = \|\mathbf{x}_I\|_2^2 T_s. \tag{6.84}$$

Simply dropping T_s in E_I will not affect the optimization objective, so E_I is redefined as

$$E_I = \|\mathbf{x}_I\|_2^2. \tag{6.85}$$

Similar to Equation (6.80), $\mathbf{x}_I$ can be obtained by

$$\mathbf{x}_I = \mathbf{H}_I \mathbf{p}. \tag{6.86}$$

where $(\mathbf{H}_I)_{i,j} = (\mathbf{H})_{N_{I0}+i,j}$, and $i = 1, 2, \ldots, N_I + 1$ as well as $j = 1, 2, \ldots, N_p + 1$. The optimization problem (6.76) can be represented by its discrete time form as,

$$\begin{array}{l} \text{maximize} \\ E_I \\ \text{subject to} \\ \|\mathbf{p}\|_2^2 = 1. \end{array} \tag{6.87}$$

The optimal solution $\mathbf{p}^*$ for the optimization problem (6.87) is the dominant eigenvector in the following eigendecomposition [400]

$$\mathbf{H}_I^T \mathbf{H}_I \mathbf{p} = \lambda \mathbf{p}. \tag{6.88}$$

Furthermore, E_I^* will be obtained by Equation (6.85) and Equation (6.86).

6.10.2.1 Tradeoff between Energies Within and Outside Integration Window

In order to reduce ISI, the energies within and outside of integration window should be considered simultaneously, which means there is a tradeoff between energies within and outside integration window [401].

The entries in $\mathbf{x}$ outside of integration window constitute $\mathbf{x}_{\bar{I}}$ as

$$\mathbf{x}_{\bar{I}} = [x_0 \, \cdots \, x_{N_{I0}-1} \, x_{N_{I0}+N_I+1} \, \cdots \, x_{N_x}]^T \tag{6.89}$$

and the energy outside of integration window $E_{\bar{I}}$ can be expressed as

$$E_{\bar{I}} = \|\mathbf{x}_{\bar{I}}\|_2^2. \tag{6.90}$$

Similar to Equation (6.86), $\mathbf{x}_{\bar{I}}$ can be obtained by

$$\mathbf{x}_{\bar{I}} = \mathbf{H}_{\bar{I}} \mathbf{p}, \tag{6.91}$$

where $(\mathbf{H}_{\bar{I}})_{i,j} = (\mathbf{H})_{i,j}$ when $i = 1, \ldots, N_{I0}$ and $(\mathbf{H}_{\bar{I}})_{i-(N_I+1),j} = (\mathbf{H})_{i,j}$ when $i = N_{I0} + N_I + 2, \ldots, N_x + 1$ as well as $j = 1, 2, \ldots, N_p + 1$.

In order to balance energies within and outside of integration window, the tradeoff factor α is introduced. The range of α is from 0 to 1. How to choose α depends on the performance requirement. Given α, the optimization problems is formulated as

$$
\begin{aligned}
&\text{maximize}\\
&\alpha E_I - (1-\alpha)E_{\bar{I}}\\
&\text{subject to}\\
&\|\mathbf{p}\|_2^2 = 1.
\end{aligned}
\tag{6.92}
$$

The optimal solution $\mathbf{p}^*$ for the optimization problem (6.92) is the dominant eigenvector in the following eigendecomposition [401]

$$
[\alpha \mathbf{H}_I^T \mathbf{H}_I - (1-\alpha)\mathbf{H}_{\bar{I}}^T \mathbf{H}_{\bar{I}}]\mathbf{p} = \lambda \mathbf{p}.
\tag{6.93}
$$

6.10.2.2 Binary Waveform

If the transmitted waveform is constrained to the binary waveform because of the hardware limitation or implementation simplicity, which means $p_i, i = 0, 1, \ldots, N_p$ is equal to $-\dfrac{1}{\sqrt{1+N_p}}$ or $\dfrac{1}{\sqrt{1+N_p}}$, then the optimization problem is

$$
\begin{aligned}
&\text{maximize}\\
&E_I\\
&\text{subject to}\\
&[(\mathbf{p})_{i,1}]^2 = \tfrac{1}{1+N_p}, i = 0, 1, \ldots, N_p.
\end{aligned}
\tag{6.94}
$$

One suboptimal solution $\mathbf{p}_{b1}^*$ to the optimization problem (6.94) is derived from the optimal solution $\mathbf{p}^*$ of the optimization problem (6.87). When $\mathbf{p}^*$ is obtained, then

$$
(\mathbf{p}_{b1}^*)_{i,1} =
\begin{cases}
\dfrac{1}{\sqrt{1+N_p}}, & (\mathbf{p}^*)_{i,1} \geq 0\\[2ex]
-\dfrac{1}{\sqrt{1+N_p}}, & (\mathbf{p}^*)_{i,1} < 0.
\end{cases}
\tag{6.95}
$$

This simple method can lead to the optimal solution to the optimization problem (6.94) when $T_I \to 0$, which can be proved by Cauchy Schwarz inequality, but if T_I is greater than zero, there is still an improvement potential to this suboptimal solution obtained from Equation (6.95).

Define

$$
\mathbf{P} = \mathbf{p}\mathbf{p}^T
\tag{6.96}
$$

$\mathbf{P}$ should be a symmetric positive semidefinite matrix, that is, $\mathbf{P} \succeq 0$, and rank of $\mathbf{P}$ should be equal to 1. Reformulate E_I as

$$
\begin{aligned}
E_I &= \mathbf{p}^T \mathbf{H}_I^T \mathbf{H}_I \mathbf{p}\\
&= \text{trace}\,(\mathbf{H}_I^T \mathbf{H}_I \mathbf{p}\mathbf{p}^T)\\
&= \text{trace}\,(\mathbf{H}_I^T \mathbf{H}_I \mathbf{P}).
\end{aligned}
\tag{6.97}
$$

Rank constraint is nonconvex constraint, so after dropping it, the optimization problem (6.94) is relaxed to

$$
\begin{aligned}
&\text{maximize} \\
&\text{trace } (\mathbf{H}_I^T \mathbf{H}_I \mathbf{P}) \\
&\text{subject to} \\
&(\mathbf{P})_{i,i} = \frac{1}{1+N_p}, i = 0, 1, \ldots, N_p \\
&\mathbf{P} \succeq 0.
\end{aligned}
\tag{6.98}
$$

The optimal solution $\mathbf{P}^*$ of the optimization problem (6.98) can be obtained by using CVX tool [513], and the value of the objective function in the optimization problem (6.98) gives the upper bound of the optimal value in the optimization problem (6.94). Projecting the dominant eigenvector of $\mathbf{P}^*$ on $-\frac{1}{\sqrt{1+N_p}}$ and $\frac{1}{\sqrt{1+N_p}}$ based on Equation (6.95), the suboptimal solution $\mathbf{p}_{b2}^*$ is achieved [514].

Finally, the designed binary waveform is [401]

$$
\mathbf{p}_b^* = \arg \max_{\mathbf{p} \in \{\mathbf{p}_{b1}^*, \mathbf{p}_{b2}^*\}} \mathbf{p}^T \mathbf{H}_I^T \mathbf{H}_I \mathbf{p}.
\tag{6.99}
$$

6.10.2.3 Ternary Waveform

If the transmitted waveform is constrained to the ternary waveform, which means p_i, $i = 0, 1, \ldots, N_p$ is equal to three levels, that is, $-c$, 0 or c, then the optimization problem is expressed as

$$
\begin{aligned}
&\text{maximize} \\
&E_I \\
&\text{subject to} \\
&[(\mathbf{p})_{i,1}]^2 = c^2 \text{ or } 0, i = 0, 1, \ldots, N_p \\
&\|\mathbf{p}\|_2^2 = 1.
\end{aligned}
\tag{6.100}
$$

The optimization problem (6.100) is still NP-hard and can be approximately reformulated as

$$
\begin{aligned}
&\text{maximize} \\
&E_I \\
&\text{subject to} \\
&\text{Cardinality}(\mathbf{p}) \leq k \\
&1 \leq k \leq N_p + 1 \\
&\|\mathbf{p}\|_2^2 = 1
\end{aligned}
\tag{6.101}
$$

where Cardinality$(\mathbf{p})$ denotes the number of nonzero entries of $\mathbf{p}$, and cardinality constraint is also a nonconvex constraint.

Because k is the integer number between 1 and $N_p + 1$, the optimization problem (6.101) can be decomposed into $N_p + 1$ independent, and parallel subproblems and each subproblem is shown as

$$
\begin{aligned}
&\text{maximize} \\
&E_I \\
&\text{subject to} \\
&\text{Cardinality}(\mathbf{p}) \le k \\
&\|\mathbf{p}\|_2^2 = 1
\end{aligned}
\tag{6.102}
$$

where k is equal to $1, 2, \cdots$, or $N_p + 1$.

Problem (6.102) can be solved in parallel, and then the solutions are combined to get the solution of the original optimization problem (6.100). The definition in Equation (6.96) is reused, and problem (6.102) can be converted to the following SDP by semidefinite relaxation combined with l_1 heuristic [514].

$$
\begin{aligned}
&\text{maximize} \\
&\text{trace } (\mathbf{H}_I^T \mathbf{H}_I \mathbf{P}) \\
&\text{subject to} \\
&\text{trace } (\mathbf{P}) = 1 \\
&\mathbf{a}^T |\mathbf{P}|\mathbf{a} \le k \\
&\mathbf{P} \succ= 0,
\end{aligned}
\tag{6.103}
$$

where $\mathbf{a}$ is the column vector with all ones and

$$
\begin{aligned}
\|\mathbf{p}\|_2^2 &= \mathbf{p}^T \mathbf{p} \\
&= \text{trace } (\mathbf{p}\mathbf{p}^T) \\
&= \text{trace } (\mathbf{P}).
\end{aligned}
\tag{6.104}
$$

The CVX tool [513] is also operated to get the optimal solution $\mathbf{P}_k^*$ of SDP (6.103). From the dominant eigenvector $\mathbf{p}_k^*$ of $\mathbf{P}_k^*$ and the threshold $p_{\text{th}k}$, the solution for the subproblem (6.102) can be achieved as

$$
(\mathbf{p}_{tk}^*)_{i,1} =
\begin{cases}
c_k, & (\mathbf{p}_k^*)_{i,1} > p_{\text{th}k} \\
0, & |(\mathbf{p}_k^*)_{i,1}| \le p_{\text{th}k} \\
-c_k, & (\mathbf{p}_k^*)_{i,1} < p_{\text{th}k}
\end{cases},
\tag{6.105}
$$

where

$$
\begin{aligned}
p_{\text{th}k} &= \arg \max_{\{p_{\text{th}}\}} (\mathbf{p}_{tk}^*)^T \mathbf{H}_I^T \mathbf{H}_I \mathbf{p}_{tk}^* \\
&\text{subject to} \\
&\text{Cardinality}(\mathbf{p}_{tk}^*) \le k
\end{aligned}
\tag{6.106}
$$

and

$$
c_k = \frac{1}{\sqrt{\text{Cardinality}(\mathbf{p}_{tk}^*)}}.
\tag{6.107}
$$

Finally, the designed ternary waveform is [401]

$$
\mathbf{p}_t^* = \arg \max_{\mathbf{p} \in \{\mathbf{p}_{tk}^*, k=1,2,\ldots,N_p+1\}} \mathbf{p}^T \mathbf{H}_I^T \mathbf{H}_I \mathbf{p}.
\tag{6.108}
$$

6.10.3   Wideband Waveform Design for Multiple Input Single Output Cognitive Radio

MISO system is one kind of multiantenna systems in which there are multiple antennas at the transmitter and one antenna at the receiver. MISO system can explore the spatial diversity and execute the beamforming to focus energy on the desired direction or point and avoid interference to other radio systems. It is well known that waveform and spatially diverse capabilities are made possible today due to the advent of lightweight digital programming waveform generator [402] or AWG.

6.10.3.1   Cauchy–Schwarz Inequality-Based Iterative Algorithm

There are N antennas at the transmitter, and one antenna at the receiver. OOK modulation is used for transmission. The transmitted signal at the transmitter antenna n is

$$s_n(t) = \sum_{j=-\infty}^{\infty} d_j p_n(t - jT_b), \tag{6.109}$$

where T_b is the bit duration; $p_n(t)$ is the transmitted bit waveform defined over $[0, T_p]$ at the transmitter antenna n; and $d_j \in \{0, 1\}$ is j-th transmitted bit. The energy of transmitted waveforms is

$$\sum_{n=1}^{N} \int_0^{T_p} p_n^2(t) \, df = E_p. \tag{6.110}$$

The received noisy signal at the output of LNA is

$$r(t) = \sum_{n=1}^{N} h_n(t) \otimes s_n(t) + n(t)$$

$$= \sum_{j=-\infty}^{\infty} d_j \sum_{n=1}^{N} x_n(t - jT_b) + n(t), \tag{6.111}$$

where $h_n(t), t \in [0, T_h]$ is the multipath impulse response. $h_n(t)$ is available at the transmitter [511, 512]. $\int_0^{T_h} h_n^2(t) \, dt = E_{nh}$. $n(t)$ is AWGN. $x_n(t)$ is the received noiseless bit-"1" waveform defined as

$$x_n(t) = h_n(t) \otimes p_n(t). \tag{6.112}$$

We further assume that $T_b \geq T_h + T_p = T_x$, that is, no existence of ISI.

If the waveforms at different transmitter antennas are assumed to be synchronized, the k-th decision statistic is

$$r(kT_b + t_0) = \sum_{j=-\infty}^{\infty} d_j \sum_{n=1}^{N} x_n(kT_b + t_0 - jT_b) + n(t)$$

$$= d_k \sum_{n=1}^{N} x_n(t_0) + n(t). \tag{6.113}$$

In order to maximize the system performance, $\sum_{n=1}^{N} x_n(t_0)$ should be maximized. The optimization problem can be formulated as follows to get the optimal waveforms $p_n(t)$.

$$\text{maximize} \quad \sum_{n=1}^{N} x_n(t_0)$$

$$\text{subject to} \qquad\qquad\qquad\qquad (6.114)$$

$$\sum_{n=1}^{N} \int_0^{T_p} p_n^2(t)\, dt \leq E_p$$

$$0 \leq t_0 \leq T_b.$$

An iterative method is proposed here to give the optimal solution to the optimization problem (6.114). This method is a computationally efficient algorithm. For simplicity in the following presentation, t_0 is assumed to be zero, which will not degrade the optimum of the solution if such solution exists.

$$x(t) = \sum_{n=1}^{N} x_n(t). \qquad\qquad (6.115)$$

From inverse Fourier transform,

$$x_{nf}(f) = h_{nf}(f) p_{nf}(f) \qquad\qquad (6.116)$$

and

$$x_f(f) = \sum_{n=1}^{N} h_{nf}(f) p_{nf}(f), \qquad\qquad (6.117)$$

where $x_{nf}(f)$, $h_{nf}(f)$, and $p_{nf}(f)$ are the frequency domain representations of $x_n(t)$, $h_n(t)$, and $p_n(t)$, respectively. $x_f(f)$ is frequency domain representation of $x(t)$. Thus, $x(0) = \sum_{n=1}^{N} x_n(0)$ and $x_n(0) = \int_{-\infty}^{\infty} x_{nf}(f)\, df$.

If there is no spectral mask constraint, then according to the Cauchy–Schwarz inequality,

$$x(0) = \sum_{n=1}^{N} \int_{-\infty}^{\infty} h_{nf}(f) p_{nf}(f)\, df$$

$$\leq \sum_{n=1}^{N} \sqrt{\int_{-\infty}^{\infty} |h_{nf}(f)|^2\, df \int_{-\infty}^{\infty} |p_{nf}(f)|^2\, df}$$

$$\leq \sqrt{\sum_{n=1}^{N} \int_{-\infty}^{\infty} |h_{nf}(f)|^2\, df \sum_{n=1}^{N} \int_{-\infty}^{\infty} |p_{nf}(f)|^2\, df}$$

$$= \sqrt{E_p \sum_{n=1}^{N} E_{nh}}, \qquad\qquad (6.118)$$

where when $p_{nf}(f) = \alpha h_{nf}(f)$ for all f and n, two equalities are obtained. Hence,

$$\alpha = \sqrt{\frac{E_p}{\sum_{n=1}^{N} \int_{-\infty}^{\infty} |h_{nf}(f)|^2 \, df}}. \tag{6.119}$$

In this case, $p_n(t) = \alpha h_n(-t)$, which means the optimal waveform $p_n(t)$ is the corresponding time reversed multipath impulse response $h_n(t)$.

If there is spectral mask constraint, then the following optimization problem will become complicated:

$$\begin{aligned} &\text{maximize} \\ &x(0) \\ &\text{subject to} \\ &\sum_{n=1}^{N} \int_{0}^{T_p} p_n^2(t) \, dt \leq E_p \\ &|p_{nf}(f)|^2 \leq c_{nf}(f), \end{aligned} \tag{6.120}$$

where $c_{nf}(f)$ represents the arbitrary spectral mask constraint at the transmitter antenna n.

Because $p_{nf}(f)$ is the complex value, the phase and the modulus of $p_{nf}(f)$ should be determined.

Meanwhile

$$x(0) = \int_{-\infty}^{\infty} x_f(f) \, df \tag{6.121}$$

and

$$x_f(f) = \sum_{n=1}^{N} |h_{nf}(f)||p_{nf}(f)|e^{j2\pi(\arg(h_{nf}(f))+\arg(p_{nf}(f)))} \tag{6.122}$$

where the angular component of the complex value is $\arg(\bullet)$.

For the real value signal $x(t)$, $x_f(f)$ is equal to the conjugate of $x_f(-f)$. Hence,

$$x_f(-f) = \sum_{n=1}^{N} |h_{nf}(f)||p_{nf}(f)|e^{-j2\pi(\arg(h_{nf}(f))+\arg(p_{nf}(f)))} \tag{6.123}$$

and $x_f(f) + x_f(-f)$ is equal to

$$\sum_{n=1}^{N} |h_{nf}(f)||p_{nf}(f)| \cos(2\pi(\arg(h_{nf}(f)) + \arg(p_{nf}(f)))). \tag{6.124}$$

If $h_{nf}(f)$ and $|p_{nf}(f)|$ are given for all f and n, maximization of $x(0)$ is equivalent to

$$\arg(h_{nf}(f)) + \arg(p_{nf}(f)) = 0, \tag{6.125}$$

which means the angular component of $p_{nf}(f)$ is the negative angular component of $h_{nf}(f)$.

The optimization problem (6.120) can be simplified as

maximize

$$\sum_{n=1}^{N} \int_{-\infty}^{\infty} |h_{nf}(f)||p_{nf}(f)|\,df$$

subject to   (6.126)

$$\sum_{n=1}^{N} \int_{-\infty}^{\infty} |p_{nf}(f)|^2\,df \le E_p$$

$$|p_{nf}(f)|^2 \le c_{nf}(f).$$

$$|h_{nf}(f)| = |h_{nf}(-f)|$$ (6.127)

$$|p_{nf}(f)| = |p_{nf}(-f)|$$ (6.128)

$$|c_{nf}(f)| = |c_{nf}(-f)|$$ (6.129)

for all f and n. Thus uniformly discrete frequency points $f_0, \ldots, f_M$ are considered in the optimization problem (6.126). Meanwhile, f_0 corresponds to the DC component, and $f_1, \ldots, f_M$ correspond to the positive frequency components.

Define column vectors $\mathbf{h}_f, \mathbf{h}_{1f}, \ldots, \mathbf{h}_{Nf}$

$$\mathbf{h}_f = [\mathbf{h}_{1f}^T\ \mathbf{h}_{2f}^T\ \cdots\ \mathbf{h}_{Nf}^T]^T$$ (6.130)

$$(\mathbf{h}_{nf})_i = \begin{cases} |h_{nf}(f_{i-1})|, & i = 1 \\ \sqrt{2}|h_{nf}(f_{i-1})|, & i = 2, \ldots, M+1. \end{cases}$$ (6.131)

Define column vectors $\mathbf{p}_f, \mathbf{p}_{1f}, \ldots, \mathbf{p}_{Nf}$

$$\mathbf{p}_f = [\mathbf{p}_{1f}^T\ \mathbf{p}_{2f}^T\ \cdots\ \mathbf{p}_{Nf}^T]^T$$ (6.132)

$$(\mathbf{p}_{nf})_i = \begin{cases} |p_{nf}(f_{i-1})|, & i = 1 \\ \sqrt{2}|p_{nf}(f_{i-1})|, & i = 2, \ldots, M+1. \end{cases}$$ (6.133)

Define column vectors $\mathbf{c}_f, \mathbf{c}_{1f}, \ldots, \mathbf{c}_{Nf}$

$$\mathbf{c}_f = [\mathbf{c}_{1f}^T\ \mathbf{c}_{2f}^T\ \cdots\ \mathbf{c}_{Nf}^T]^T$$ (6.134)

$$(\mathbf{c}_{nf})_i = \begin{cases} \sqrt{|c_{nf}(f_{i-1})|}, & i = 1 \\ \sqrt{2|c_{nf}(f_{i-1})|}, & i = 2, \ldots, M+1. \end{cases}$$ (6.135)

The discrete version of the optimization problem (6.126) is shown as

maximize
$$\mathbf{h}_f^T \mathbf{p}_f$$
subject to   (6.136)
$$\|\mathbf{p}_f\|_2^2 \le E_p$$
$$0 \le \mathbf{p}_f \le \mathbf{c}_f.$$

An iterative algorithm is shown as follows to give the optimal solution $\mathbf{p}_f^*$ to the optimization problem (6.136) [2],

1. Initialization: $P = E_p$ and $\mathbf{p}_f^*$ is set to be the column vector with all zeros.
2. Solve the following optimization problem to get the optimal $\mathbf{q}_f^*$ using Cauchy-Schwarz inequality:

$$\begin{aligned}
&\text{maximize} \\
&\mathbf{h}_f^T \mathbf{q}_f \\
&\text{subject to} \\
&\|\mathbf{q}_f\|_2^2 \leq P.
\end{aligned} \tag{6.137}$$

3. Find i, such that $(\mathbf{q}_f^*)_i$ is the maximal value in the set $\{(\mathbf{q}_f^*)_j | (\mathbf{q}_f^*)_j > (\mathbf{c}_f)_j\}$. If $\{i\} = \emptyset$, then the method is terminated and $\mathbf{p}_f^* := \mathbf{p}_f^* + \mathbf{q}_f^*$. Otherwise go to step 4.
4. Set $(\mathbf{p}_f^*)_i = (\mathbf{c}_f)_i$.
5. $P := P - (\mathbf{c}_f)_i^2$ and set $(\mathbf{h}_f)_i$ to zero. If $\|\mathbf{h}_f\|_2^2$ is equal to zero, then the algorithm is terminated; otherwise go to step 2.

When $\mathbf{p}_f^*$ is obtained for the optimization problem (6.136), from Equation (6.125), Equation (6.132), and Equation (6.133), the optimal $p_{nf}(f)$ and the corresponding $p_n(t)$ can be smoothly achieved.

6.10.3.2 SDP-Based Iterative Algorithm

The $p_n(t)$ and the $h_n(t)$ are uniformly sampled at Nyquist rate. Assume the sampling period is T_s. $T_p/T_s = N_p$ and N_p is assumed to be even, $T_h/T_s = N_h$. $p_n(t)$ and $h_n(t)$ are represented by $p_{ni}, i = 0, 1, \ldots, N_p$ and $h_{ni}, i = 0, 1, \ldots, N_h$, respectively.
Define

$$\mathbf{p}_n = [p_{n0}\ p_{n1}\ \cdots\ p_{nN_p}]^T \tag{6.138}$$

and

$$\mathbf{h}_n = [h_{nN_h}\ h_{n(N_h-1)}\ \cdots\ h_{n0}]^T. \tag{6.139}$$

If $N_p = N_h$, then $\sum_{n=1}^{N} x_n(t_0)$ can be equivalent to $\sum_{n=1}^{N} \mathbf{h}_n^T \mathbf{p}_n$. Define

$$\mathbf{p} = [\mathbf{p}_1^T\ \mathbf{p}_2^T\ \cdots\ \mathbf{p}_N^T]^T \tag{6.140}$$

and

$$\mathbf{h} = [\mathbf{h}_1^T\ \mathbf{h}_2^T\ \cdots\ \mathbf{h}_N^T]^T. \tag{6.141}$$

Thus,

$$\sum_{n=1}^{N} \mathbf{h}_n^T \mathbf{p}_n = \mathbf{h}^T \mathbf{p}. \tag{6.142}$$

Maximization of $\mathbf{h}^T\mathbf{p}$ is the same as maximization of $(\mathbf{h}^T\mathbf{p})^2$ as long as $\mathbf{h}^T\mathbf{p}$ is equal to or greater than zero.

$$
\begin{aligned}
(\mathbf{h}^T\mathbf{p})^2 &= (\mathbf{h}^T\mathbf{p})^T(\mathbf{h}^T\mathbf{p}) \\
&= \mathbf{p}^T\mathbf{h}\mathbf{h}^T\mathbf{p} \\
&= \text{trace}\,(\mathbf{h}\mathbf{h}^T\mathbf{p}\mathbf{p}^T) \\
&= \text{trace}\,(\mathbf{H}\mathbf{P}),
\end{aligned}
\tag{6.143}
$$

where $\mathbf{H} = \mathbf{h}\mathbf{h}^T$ and $\mathbf{P} = \mathbf{p}\mathbf{p}^T$. $\mathbf{P}$ should be rank one positive semidefinite matrix. However, rank constraint is nonconvex constraint, which will be omitted in the following optimization problems. The optimization objective in the optimization problem (6.120) can be reformulated as

$$
\text{maximize trace}\,(\mathbf{H}\mathbf{P}).
\tag{6.144}
$$

Meanwhile

$$
\begin{aligned}
\|\mathbf{p}\|_2^2 &= \mathbf{p}^T\mathbf{p} \\
&= \text{trace}\,(\mathbf{p}\mathbf{p}^T) \\
&= \text{trace}\,(\mathbf{P}).
\end{aligned}
\tag{6.145}
$$

The energy constraint in the optimization problem (6.120) can be reformulated as

$$
\text{trace}\,(\mathbf{P}) \le E_p.
\tag{6.146}
$$

For cognitive radio, there is a spectral mask constraint for the transmitted waveform. Based on the previous discussion, $\mathbf{p}_n$ is assumed to be the transmitted waveform, and $\mathbf{F}$ is the discrete time Fourier transform operator. The frequency domain representation of $\mathbf{p}_n$ is

$$
\mathbf{p}_{fn} = \mathbf{F}\mathbf{p}_n,
\tag{6.147}
$$

where $\mathbf{p}_{fn}$ is a complex value vector. If the i-th row of $\mathbf{F}$ is $\mathbf{f}_i$, then each complex value in $\mathbf{p}_{fn}$ can be represented by

$$
(\mathbf{p}_{fn})_{i,1} = \mathbf{f}_i\mathbf{p}_n, \quad i = 1, 2, \ldots, \frac{N_p}{2} + 1.
\tag{6.148}
$$

Define

$$
\mathbf{F}_i = \mathbf{f}_i^H\mathbf{f}_i, \quad i = 1, 2, \ldots, \frac{N_p}{2} + 1.
\tag{6.149}
$$

Given the spectral mask constraint in terms of power spectral density $\mathbf{c}_n = \left[c_{n1} c_{n2} \cdots c_{n\frac{N_p}{2}+1} \right]^T$,

$$
\begin{aligned}
|(\mathbf{p}_{fn})_{i,1}|^2 &= |\mathbf{f}_i\mathbf{p}_n|^2 \\
&= \mathbf{p}_n^T\mathbf{f}_i^H\mathbf{f}_i\mathbf{p}_n \\
&= \mathbf{p}_n^T\mathbf{F}_i\mathbf{p}_n \\
&\le c_{ni}, \quad i = 1, 2, \ldots, \frac{N_p}{2} + 1.
\end{aligned}
\tag{6.150}
$$

Define selection matrix $\mathbf{S}_n \in R^{(N_P+1)\times(N_P+1)N}$

$$(\mathbf{S}_n)_{i,j} = \begin{cases} 1, & j = i + (N_p + 1)(n - 1) \\ 0, & \text{else} \end{cases} \tag{6.151}$$

$$\mathbf{p}_n = \mathbf{S}_n \mathbf{p} \tag{6.152}$$

and

$$\begin{aligned}
|(\mathbf{p}_{fn})_{i,1}|^2 &= \mathbf{p}_n^T \mathbf{F}_i \mathbf{p}_n \\
&= \mathbf{p}^T \mathbf{S}_n^T \mathbf{F}_i \mathbf{S}_n \mathbf{p} \\
&= \text{trace}\,(\mathbf{S}_n^T \mathbf{F}_i \mathbf{S}_n \mathbf{p}\mathbf{p}^T) \\
&= \text{trace}\,(\mathbf{S}_n^T \mathbf{F}_i \mathbf{S}_n \mathbf{P}).
\end{aligned} \tag{6.153}$$

The optimization problem (6.120) can be reformulated as SDP [515]:

$$\begin{aligned}
&\text{maximize} \\
&\text{trace}\,(\mathbf{H}\mathbf{P}) \\
&\text{subject to} \\
&\text{trace}\,(\mathbf{P}) \leq E_p \\
&\text{trace}\,(\mathbf{S}_n^T \mathbf{F}_i \mathbf{S}_n \mathbf{P}) \leq c_{ni} \\
&i = 1, 2, \ldots, \tfrac{N_P+1}{2} \\
&n = 1, 2, \ldots N.
\end{aligned} \tag{6.154}$$

If the optimal solution $\mathbf{P}^*$ to the optimization problem (6.154) is the rank one matrix, then the optimal waveforms can be obtained from the dominant eigenvector of $\mathbf{P}^*$. Otherwise, E_p in the optimization problem (6.154) should be decreased to get the rank one optimal solution $\mathbf{P}^*$ to satisfy all the other constraints.

An SDP based iterative algorithm is presented to get the rank one optimal solution $\mathbf{P}^*$ [515]:

1. Initialization of E_p.
2. Solve the optimization problem (6.154) and get the optimal solution $\mathbf{P}^*$.
3. If the ratio of dominant eigenvalue of $\mathbf{P}^*$ to trace $(\mathbf{P}^*)$ is less than 0.99, then set E_p to be trace $(\mathbf{P}^*)$ and go to step 2; otherwise, the algorithm is terminated.

The optimal waveforms can be obtained from the dominant eigenvector of $\mathbf{P}^*$ and Equation (6.140).

6.10.4 Wideband Beamforming Design

Wideband beamforming is a hot research topic in both communication and radar society, partly due to the advent of powerful real-time FPGA processing. The array working with wide frequency band can operate in both spatial domain and frequency domain simultaneously.

The architecture of wideband beamforming consists of LUT, high performance computing engine, and a two-dimensional filter bank. Look-up table (LUT) is explored to remove the presteering delay component in the traditional wideband beamforming architecture. This component is hard to implement and manipulate either in the analog domain or in the digital domain. If the presteering delay component is designed in the analog domain, the unfixed delay line with the delay from subnanosecond to nanosecond should be implemented. If the presteering delay component is designed in the digital domain, fractional delay filter bank should be implemented [516]. In this novel architecture, the data sampled by ADC from the impulse response of each RF chain with the consideration of assumed angle of arrival will be stored in LUT. The impact of channel imbalances and fractional delay will be taken care of in the general optimization issue. The coefficients of the filter bank will be calculated in the high performance computing engine. Thus, this architecture reduces the implementation burden at the cost of computational complexity. However, the computational capability has grown much faster over the last few years and the price of computation is lower than the implementation cost.

There are M antennas in the linear array. The distance between antennas is d. The mutual coupling among antennas is not considered here. The system works with the central frequency of f_c and the bandwidth of B. The equivalent baseband complex response of RF chain related to each antenna is given by $h_m(t), t \in [0, T], m = 0, 1, \ldots, M - 1$. Because of the limitation of ADC, it is hard for us to obtain continuous time $h_m(t)$. If the sampling rate of ADC is $1/T_s$ and $1/T_s \geq 2B$, the discrete time counterpart of $h_m(t)$ is $h_m[k]$ which is measured for each RF chain

$$h_m[k] = h_m(kT_s). \tag{6.155}$$

In the calibration phase, LUT should be set up. First the interpolation is performed on $h_m[k]$ to get high sampling rate data to emulate $h_m(t)$. Assume $\delta(t)$ is the signal in the far field of the system and impinges on it from the angle θ. The equivalent baseband complex response of each RF chain after ADC is defined as $h_{m,\theta}[k]$.

If the signal from far field reaches the first antenna at time $T_0 = \frac{(M-1)d}{c}$, then $h_m(t)$ will be extended to

$$h_{m,\theta}(t) = 0, t \in \left[0, T_0 + \frac{md \cos \theta}{c}\right) \tag{6.156}$$

$$h_{m,\theta}(t) = a_{m,\theta} h_m(t), t \in \left[T_0 + \frac{md \cos \theta}{c}, T + T_0 + \frac{md \cos \theta}{c}\right] \tag{6.157}$$

$$h_{m,\theta}(t) = 0, t \in \left[T + T_0 + \frac{md \cos \theta}{c}, T + 2T_0\right), \tag{6.158}$$

where c is the speed of light and $a_{m,\theta}$ is the response of antenna m to the angle θ. Without loss of generality, $a_{m,\theta}$ is assumed to be 1 here. Hence,

$$h_{m,\theta}[k] = h_{m,\theta}(kT_s) \exp\left\{-\sqrt{-1}2\pi f_c \frac{md \cos \theta}{c}\right\}. \tag{6.159}$$

Finally, $h_{m,\theta}[k]$ are saved in LUT for the following wideband beamforming.

If angles of arrival of interest are in the set $\Omega_\theta = \{\theta_1, \theta_2, \ldots, \theta_{L+1}\}$, the output of LUT will be $h_{m,\theta}[k], \theta \in \Omega_\theta$. The vector representation of $h_{m,\theta}[k]$ is $\mathbf{h}_{m,\theta}$. $\mathbf{F}$ is the discrete Fourier transform operator. Thus, the baseband response of each RF chain after ADC in the frequency domain is $\mathbf{h}_{m,\theta}^{f} = \mathbf{F}\mathbf{h}_{m,\theta}$. If the frequency points of interest $\Omega_f = \{f_1, f_2, \ldots, f_{J+1}\}$ correspond to the entries from index to index $+ J$ in $\mathbf{h}_{m,\theta}^{f}$, where index can be any reasonable integer value such that $f_{J+1} - f_1 \approx B$,

$$(\tilde{\mathbf{h}}_{m,\theta}^{f})_{1:J+1,1} = (\mathbf{h}_{m,\theta}^{f})_{\text{index:index}+J,1} \tag{6.160}$$

where $(\bullet)_{a:b,c:d}$ means the entries in the matrix from the a-th row to the b-th row and from the c-th column to the d-th column.

After a two-dimensional filter bank, the array response is defined as $B(f_j, \theta_l)$, which can be expressed as

$$B(f_j, \theta_l) = \sum_{m=0}^{M-1} \sum_{n=0}^{N-1} w_{m,n} (\tilde{\mathbf{h}}_{m,\theta_l}^{f})_{j,1} \exp\{-\sqrt{-1}n2\pi f_j T_s\}, \tag{6.161}$$

where $w_{m,n}$ is the coefficient at the $(n+1)$-th tap of the $(m+1)$-th filter.

The array response can be reformulated as the vector representation as

$$B(f_j, \theta_l) = \mathbf{s}(f_j, \theta_l)\mathbf{w}, \tag{6.162}$$

where $\mathbf{w}$ is the coefficient vector defined as

$$\mathbf{w} = [\mathbf{w}_0^H \ \mathbf{w}_1^H \ \cdots \ \mathbf{w}_{M-1}^H]^H \tag{6.163}$$

and

$$\mathbf{w}_m = [w_{m,0} \ w_{m,1} \ \cdots \ w_{m,N-1}]^H. \tag{6.164}$$

$\mathbf{s}(f_j, \theta_l)$ is the $M \times N$ steering vector. Define $1 \le i \le M \times N$

$$m = \lfloor \frac{i-1}{N} \rfloor \tag{6.165}$$

and

$$n = i - m \times N - 1. \tag{6.166}$$

Each entry in $\mathbf{s}(f_j, \theta_l)$ is

$$(\mathbf{s}(f_j, \theta_l))_{1,i} = (\tilde{\mathbf{h}}_{m,\theta_l}^{f})_{j,1} \exp\{-\sqrt{-1}n2\pi f_j T_s\}. \tag{6.167}$$

The core task of wideband beamforming is to design coefficients $\mathbf{w}$ of a two-dimensional filter bank such that the array response $B(f_j, \theta_l)$ aims at:

- desired main lobe shape with consideration of magnitude and phase;
- overall constrained side lobes;
- nulling at given angles and frequency points;
- frequency invariant property for the given angle rang and frequency range.

The aforementioned approaches are only suitable for the simple shapes of wideband beam patterns with the small number of optimization objectives and constraints. If the shape of wideband beam pattern is complex or the size of optimization issue for wideband beamforming is large, we need to resort to advanced signal processing scheme to perform the general tasks of wideband beamforming. SDP based approach can be competent for these general tasks. SDP is widely used in narrowband beamforming not only for the radar system [517, 518] but also for the communication system [421, 519]. Several papers [520, 521] formulates the design of the two-dimensional filter bank for wideband beamforming as SDP or second order cone programming (SOCP), which can be efficiently solved by SeDuMi [522] or CVX [8, 513].

Based on the architecture, we will present the general formulation of optimization issue for wideband beamforming with the consideration of 4 mentioned tasks [523].

If the look direction is at the angle θ_{l_0}, the desired main beam pattern at this angle is $P(f_j, \theta_{l_0})$, the optimization objective is to minimize the Euclidean distance between $P(f_j, \theta_{l_0})$ and $B(f_j, \theta_{l_0})$ [523]

$$\text{minimize} \sum_{f_j \in \Omega_f} |P(f_j, \theta_{l_0}) - B(f_j, \theta_{l_0})|^2. \tag{6.168}$$

For each frequency point, we would like to constrain the total energy of array response except the energy for the look direction [523]

$$\sum_{\theta_l \in \Omega_\theta - \theta_{l_0}} |B(f_j, \theta_l)|^2 \le \varepsilon(f_j)$$
$$f_j \in \Omega_f, \tag{6.169}$$

where $\varepsilon(f_j)$ is the energy threshold for each frequency point.

If there are nullings at frequency points in the set $\Omega_{f_{\text{nulling}}}$ and angles in the set $\Omega_{\theta_{\text{nulling}}}$, then [523]

$$|B(f_j, \theta_l)|^2 \le \varepsilon_{\text{nulling}}(f_j, \theta_l)$$
$$f_j \in \Omega_{f_{\text{nulling}}}$$
$$\theta_l \in \Omega_{\theta_{\text{nulling}}}, \tag{6.170}$$

where $\varepsilon_{\text{nulling}}(f_j, \theta_l)$ is the nulling threshold for the frequency f_j and the angle θ_l.

Assume the frequency invariant property is imposed on the frequency range $\Omega_{f_{\text{FIB}}}$ and the angle range $\Omega_{\theta_{\text{FIB}}}$. Similar to the concept of spatial variation [524], $f_{\text{re}} \in \Omega_{f_{\text{FIB}}}$ is chosen as the reference frequency point and the spatial variation should be bounded [523]

$$\sum_{f_j \in \Omega_{f_{\text{FIB}}} - f_{\text{re}}} \sum_{\theta_l \in \Omega_{\theta_{\text{FIB}}}} |B(f_j, \theta_l) - B(f_{\text{re}}, \theta_l)|^2 \le \varepsilon_{\text{sv}}, \tag{6.171}$$

where ε_{sv} is the threshold for spatial variation to keep the frequency invariant property.

The general optimization issue for wideband beamforming by combining (6.168) (6.169) (6.170) (6.171) can be presented as [523]

$$
\begin{aligned}
&\text{minimize} \\
&\sum_{f_j \in \Omega_f} |P(f_j, \theta_{l_0}) - B(f_j, \theta_{l_0})|^2 \\
&\text{subject to} \\
&\sum_{\theta_l \in \Omega_\theta - \theta_{l_0}} |B(f_j, \theta_l)|^2 \leq \varepsilon(f_j) \\
&f_j \in \Omega_f \\
&|B(f_j, \theta_l)|^2 \leq \varepsilon_{\text{nulling}}(f_j, \theta_l) \\
&f_j \in \Omega_{f_{\text{nulling}}} \\
&\theta_l \in \Omega_{\theta_{\text{nulling}}} \\
&\sum_{f_j \in \Omega_{f_{\text{FIB}}} - f_{\text{re}}} \sum_{\theta_l \in \Omega_{\theta_{\text{FIB}}}} |B(f_j, \theta_l) - B(f_{\text{re}}, \theta_l)|^2 \leq \varepsilon_{\text{sv}}.
\end{aligned}
\tag{6.172}
$$

The optimization problem can be efficiently solved by CVX [8, 513]. Because CVX can only give the real value solution, in order to use CVX, $B(f_j, \theta_l)$ in Equation (6.161) should be reformulated as

$$
B(f_j, \theta_l) = [\mathbf{s}(f_j, \theta_l) \ \sqrt{-1}\mathbf{s}(f_j, \theta_l)] \begin{bmatrix} \text{re}(\mathbf{w}) \\ \text{im}(\mathbf{w}) \end{bmatrix},
\tag{6.173}
$$

where re($\bullet$) gets the real part of complex value and im($\bullet$) gets the image part of complex value. CVX will return the optimization solution as

$$
\begin{bmatrix} \text{re}(\mathbf{w}^*) \\ \text{im}(\mathbf{w}^*) \end{bmatrix},
\tag{6.174}
$$

if such a solution exists. Then the optimal coefficients for the two-dimensional filter bank is

$$
\mathbf{w}^* = \text{re}(\mathbf{w}^*) + \sqrt{-1}\,\text{im}(\mathbf{w}^*).
\tag{6.175}
$$

6.10.5 Layering as Optimization Decomposition for Cognitive Radio Network

6.10.5.1 Background

We would like to design and assess innovative solutions to create cognitive cross-layer wireless networking architectures and protocols to achieve automatic network resiliency in contested RF spectrum.

Although, the highly advanced technologies, for example, MIMO, multiuser detection, interference cancellation, noncontinuous OFDM (NC-OFDM), low-density parity-check (LDPC) code, together with sophisticated radio resource management methods are exploited in the modern wireless communication systems, for example, LTE, WiMAX, and so on, to push the data rate to beat the fundamental limits, spectrum is still a scarce radio resource. There are at least two reasons for this conclusion. One is most of the spectra that can be reasonably used for wireless communication are rigidly allocated and licensed [525]. However, these licensed spectra are underutilized to make spectral efficiency and utilization very low. The other reason is the spectrum is becoming increasingly crowded by the ever-increasing number of users with their competing and conflicting data rate requirements in some military and commercial wireless applications, for example, Electronic Warfare, Central Business District in a big city, and so on. Hence, the concept of cognitive radio was proposed and widely studied to address the radio resource shortage issue. Basically, cognitive radio can be treated as one approach of implementing DSA on software defined radio (SDR) platforms [525]. However, cognitive radio is more than DSA. Traditionally, a cognitive radio user is the unlicensed user or the secondary user without licensed spectrum. Cognitive radio users can only access the spectrum when primary users do not use it. That means cognitive radio users cannot interfere with primary users. Meanwhile, primary users have no obligation to cooperate with cognitive radio users. All the burden is imposed on cognitive radio users. Thus, cognitive radio should have the capability of self-awareness, observation, learning, decision making, as well as DSA.

Cognitive radio only solves the point to point wireless communication issues to improve spectrum efficiency and utilization. From an application's point of view, cognitive radio network from physical layer to application layer should be set up to perform different application tasks. This is the complex and dynamic system. How to make this kind of system work involves a lot of challenging issues. Because of the introduction of cognition to wireless network, the design of architectures and protocols confronts unprecedented difficulties. Cognition can undoubtedly bring benefits to the system. For example, overall spectrum efficiency and utilization can be increased. However, cognition is a two-edged sword. First, more functions are needed to support cognition. In the current stage, spectrum sensing, spectrum decision, spectrum sharing, and spectrum mobility [525] are at least required. More functions will make the system more complex. Second, cognition can lead to uncertainty. No matter how sophisticated cognition is, the capability of cognition is finite. The output of cognition depends on several factors, for example, the scheme of decision making, the method of machine learning, the input data as well as their modeling. Any deviation of these factors or incomplete information will cause the wrong decision which will make the performance of the system even worse than without cognition capability. Thus, cognition should be carefully exploited. Third, cognition demands more information to support the stable network operation, which means the overhead of system will be inevitably increased. It can be foreseen that the protocols for cognitive wireless network will be more substantial than those in any traditional wireless network. Meanwhile, the acquisition and delivery of such information may lead to significant and uncontrollable delay, which will be very harmful for network operations and some real-time applications.

The basic network model is the OSI model [526] shown in Figure 6.1. The OSI model divides a communication system into smaller parts called layers. Each layer performs a different set of similar functions to provide services to the upper layer and receive services from the lower layer. The basic functions of each layer are also shown in Figure 6.1. The idea of the OSI model is simple but it works very well. The design of each layer can be independent from all the others, which breaks the complex problem into small manageable pieces. Meanwhile, the functions in each layer can be modified and upgraded in a decoupled fashion as long as the service interface is maintained. Thus, information

	7. Application	Applications for different services
Host Layers	6. Presentation	Data representation, encryption, decryption
	5. Session	Interhost communication
	4. Transport	End-to-end connection, flow control, congestion control, TCP/UDP protocol
Media Layers	3. Network	Routing, IP protocol
	2. Data Link (LLC/MAC)	Power control, scheduling, addressing
	1. Physical	Coding, modulation, array signal processing, binary transmission

Figure 6.1   OSI model [526].

hiding, decoupling change, implementation and specification separation can be achieved in the OSI model.

The virtually strict boundaries between layers in an OSI model make the design of networks not globally optimal. Toward the goal of global optimum in the context of network design, cross-layer optimization was proposed and has recently become one of the popular approaches to design and optimize the network architecture [527]. Based on an OSI model, cross-layer optimization treats the system as a whole and designs the functions in different layers jointly. More information will be exchanged between layers, and more dependencies among layers will be taken into account. In order to implement cross-layer optimization, a cross-layer optimization engine should be added into the OSI model to perform design and optimization centrally. The inputs of the cross-layer optimization can be internal or external parameters of the network, for example, channel state information, traffic information, internal buffer information, and so on. The engine is responsible for determining a set of internal operating parameters and functionalities for different layers based on the inputs and design objectives. The overall objectives of cross-layer optimization are to improve application performance, to increase user satisfaction, and to enhance efficiency of network utilization. Some simple cross-layer optimization techniques have already been deployed in the current advanced wireless networking system. Take the 3G network as an example. Power control is used to increase the throughput and minimize the interference. Hybrid automatic repeat request (HARQ) is exploited to make link condition stable. Orthogonal frequency-division multiple access (OFDMA) is a promising multiple access technique to allocate different subcarriers to different users.

Cross-layer optimization opens a wide space for the network design and optimization, but the full cross-layer optimization from physical layer through application layer is still unfeasible for implementation at the current stage. It is impossible to build a network with fully central control and design for global optimality. Thus, there is a contradiction between network design and network implementation. How can this dilemma be bypassed? Some research pioneers Mung Chiang, Steven H. Low, A. Robert Calderbank, and John C. Doyle gave a mathematical theory of network architectures, that is, layering as optimization decomposition [528]. This theoretical framework will be the analytic foundation of the work for the design of architectures and protocols for cognitive cross-layer wireless networking system. Network Utility Maximization (NUM) is exploited as the design objective in a globally optimal fashion. While for network implementation, layering as optimization decomposition is explored to decompose the master problem into several subproblems. Different subproblems correspond to different layers. Different decomposition schemes determine different layering architectures. The basic difference between the traditional OSI model and layering as optimization decomposition is that the separation of the whole network system in the OSI model is based on experiences and human intuition while the decomposition for the latter has the solid background of mathematical theory. Meanwhile, optimization decomposition will also lead to the distributed and modularized algorithm which can be implemented in disparate network nodes. The distributed algorithms rely on the local information to perform the tasks. In this way, the overhead of system can be greatly reduced.

6.10.5.2 Design Philosophy

Currently, there is no general approach to cross-layer design for wireless network. From a theoretical point of view, layering as optimization decomposition [528] is one of the general and analytic methodologies for network design. It uses common mathematical language for thinking, deriving, and comparing. Two key concepts behind it are network as an optimizer and layering as decomposition [528]. In this mathematical framework, network architecture relates to the decomposition scheme of the global optimization problem and answers the questions of how to or how not to determine different layers [528]. There are two main decompositions, that is, vertical decomposition and horizontal decomposition [528].

Vertical decomposition maps an optimization problem into several subproblems which correspond to different layers. Different functionalities are allocated to different layers to solve these subproblems. Functions of primal or dual variables coordinating subproblems will be treated as the interfaces among layers [528]. For example, cross-layer congestion control, routing, and scheduling design in ad hoc wireless networks have been studied in [529]. Jointly optimal congestion control and power control are explored to balance transport layer and physical layer in wireless multihop networks [491].

Horizontal decomposition is executed within one functionality and decomposes central computation into distributed computation over geographically different network nodes [528]. For example, congestion control protocols can be modeled as distributed algorithms for NUM [490, 530, 531]. The contention resolution algorithm in backoff based random access wireless media access control (MAC) protocols is implicitly participating in a noncooperative game [532], which is a distributed and selfish action.

Vertical decomposition across the layers and horizontal decomposition across the network nodes can be conducted together to decompose the optimization problem systematically [528]. Meanwhile, decomposition structures are not limited to aforementioned vertical decomposition and horizontal decomposition. Partial decomposition, multilevel

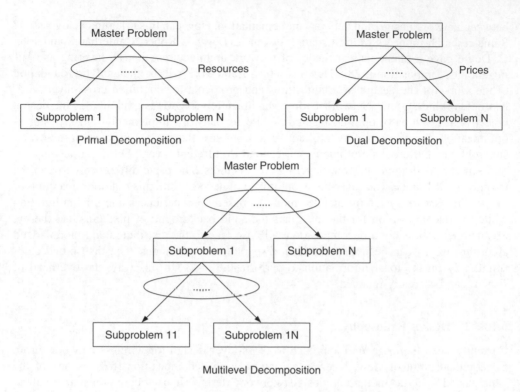

Figure 6.2 Basic decomposition schemes [533].

decomposition, and their versatile combinations can lead to many alternative decompositions [533]. These alternative decompositions can be exploited as a way to obtain different
novel network architectures [533]. Figure 6.2 shows the basic decomposition schemes
[533]. The original master problem is decomposed into several solvable subproblems
which are coordinated through some kind of signaling [533]. For primal decomposition, the master problem properly allocates the available resources to each subproblem.
Resource is the signaling between master problem and subproblems [533]. In dual decomposition, the master problem uses the price set for resource as the control signaling and
subproblems should determine the amount of resources they would like to use based on
price [533]. In multilevel decomposition, primal decomposition or dual decomposition will
be used repeatedly to divide the master problem into smaller and smaller subproblems.
These subproblems can be solved in different layers or in different network nodes.

6.10.5.3 Cognitive Capability

Cognition is the key capability and foundation of cognitive cross-layer wireless networking system, which differentiates cognitive network from the traditional wireless networks.
According to the Oxford English Dictionary, cognition is knowing, perceiving, or conceiv-
ing as an act [406]. Cognitive network is far more than cognitive radio which only covers
layer one and layer two. In a cognitive network, all layers in all network nodes should

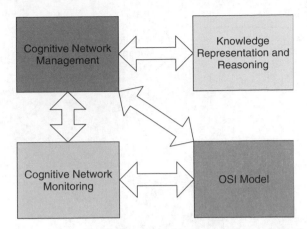

Figure 6.3   Abstract architecture of cognitive capability.

have the capability of cognition. However, different layers or different nodes may have different levels of cognition. The upper layer should be more intelligent than the lower layer. Take spectrum usage as an example: spectrum sensing, spectrum decision, spectrum mobility, and spectrum sharing are the basic functions corresponding to cognition. Spectrum sensing is implemented in physical layer. Spectrum sensing obtains information of radio environment and provides it to the upper layer. The upper layer will make the decision which spectrum can be used for transmission. Spectrum mobility means cognitive radio users can move away from the licensed spectrum once the primary user occupies this spectrum again. If multiple cognitive radio users compete for the limited available spectrum, a spectrum sharing scheme should be set up to coordinate different users and different requirements.

Cognitive cross-layer wireless networking system is a highly dynamic system. Network topology, user behavior, and radio environment are rapidly changing. Cognition is an imperative capability for the networking system to work adaptively and intelligently. For example, if link stability is not maintained in harsh and dynamic RF environments or links are determined unsuitable for the following communication requirement, routing selection should be performed with consideration of spectrum occupancy, network topology, and user demand.

The abstract architecture of cognitive capability of a cognitive cross-layer wireless networking system is shown in Figure 6.3. There are three main modules to support cognition: cognitive network management, cognitive network monitoring, as well as knowledge representation and reasoning. Cognitive network management is the brain of cognitive cross-layer wireless networking system to determine network behavior intelligently. Network management refers to the activities, methods, procedures, and tools that relate to the operation, administration, maintenance, and provisioning of networking systems. Thus, the basic functions of cognitive network management are shown in Figure 6.4.

Cognitive network monitoring is to monitor the internal and external network data under the control of cognitive network management. These data can be spectrum sensing results, traffic information, buffer state information, channel state information, quality of connectivity, and so on. Recently, network tomography [534, 535] has been proposed

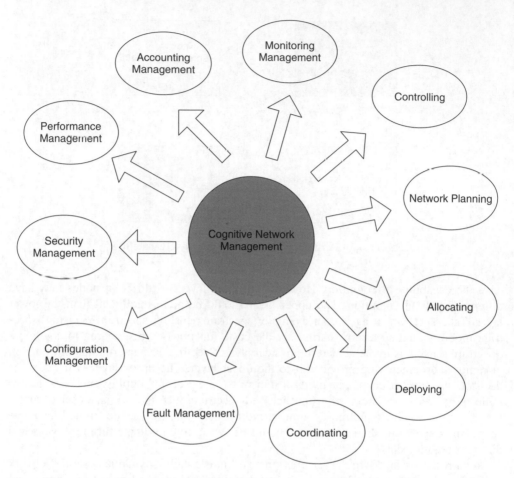

Figure 6.4 The basic functions of cognitive network management.

to extract a network's internal characteristics using information derived from end point data. Originally, tomography is imaging by sections or sectioning, through the use of waves of energy, which is widely used in medical imaging, for example, Computerized Tomography. Network monitoring and inference have a strong resemblance to tomography [535] because the internal characteristics of an objective cannot be observed directly but can be inferred from external observations. In the current literature, two issues of network tomography have been addressed. One is link level parameter estimation from end-to-end path level traffic measurements [535]. The other is path level traffic intensity estimation based on link level traffic measurements [535]. The measurements of network tomography may be passive or active. Passive measurement will monitor the existing traffic flows. However, the temporal and spatial structure of the traffic process may make the measurement sample biased [535]. Active measurement will generate probe traffic into the network. If so, the probe traffic should not distort the network state for the existing traffic [535].

Knowledge representation and reasoning is to represent knowledge in a manner that facilitates inferencing from knowledge. Cognitive network can be treated as a wireless communication network augmented by this kind of knowledge plane that can span vertically over layers and horizontally across nodes [536]. There are at least two categories of functionalities in knowledge representation and reasoning. One is a representation of relevant knowledge. The other is a cognition loop using artificial intelligence, for example, machine learning technique. Besides, prediction is also the main function. Prediction results are very important information for cognitive network management to make the decision beforehand and to tackle the possible situations in the future. In this way, the operation of the networking system will be smooth and stable.

6.10.5.4 Potential Architectures

The key design of layering as optimization decomposition is that versatile network architectures can be rigorously obtained from the decomposition of an underlying cross-layer optimization problem [533].

Cross-layer routing and dynamic spectrum allocation in cognitive radio ad hoc networks have been studied in [537]. The main contribution in [537] is that a distributed and localized algorithm was derived for joint dynamic routing and spectrum allocation called ROSA for multihop cognitive radio networks. The cross-layer ROSA algorithm aims to maximize throughput through opportunistic routing, dynamic spectrum allocation, scheduling, and power control in a distributed fashion from transport layer to MAC layer. It is a good example to explore optimization decomposition for the network design.

Based on design philosophy and cognitive requirement, several network architectures will be presented using multilevel decomposition. As mentioned before, NUM is widely used as a design objective. QoS will be measured as NUM which implicitly covers many network performance metrics, for example, capacity, latency, security, stability, and so on. The cross-layer optimization issue is to maximize the sum of QoSs for different applications in the cognitive wireless networking system. Different applications or different services may have different weights in the design objective. This is the case in the context of multiobjective optimization. In the cognitive wireless networking system, there are many restrictions and limitations for the network operation, which will be formulated as the constraints in the optimization issue. These constraints at least include:

- network carrying capacity;
- limited power and limited computing capability in each network node;
- different spectral availabilities in different locations and at different times;
- interference tolerance;
- no interference to primary user;
- queue and buffer limitation.

Cognitive capability, for example, monitoring and inference, will be integrated into the cross-layer optimization issue. Because of the uncertainty introduced by cognition, the idea from robust optimization should be explored. Meanwhile, the overhead used for cognition will be formulated as the constraints for the cross-layer optimization issue. There are several potential architectures:

- **Layered and distributed architecture**. Vertical decomposition is performed to the cross-layer optimization issue first. Different functionalities are allocated to different layers to solve the optimization issue jointly. In transport layer, traffic control including congestion control and flow control are executed. Multipath routing and dynamic routing selection are exploited in network layer. Here, multipath routing can improve the robustness of data delivery in the dense deployment of network nodes [538]. Sophisticated scheduling, power control, and DSA are implemented in MAC layer for heterogeneous traffic. After vertical decomposition, horizontal decomposition will be carried out for each layer, respectively. Then, the same functionality will be distributed to the different network nodes.

- **Distributed and layered architecture**. The cross-layer optimization issue is divided into several subproblems by horizontal decomposition. Different subproblems will be solved by different network nodes. And then the task for each network node can be partitioned by vertical decomposition. There is an essential difference between the first architecture and the second architecture, because the first level decomposition plays a more important role for network architecture than the second level decomposition.

- **Hybrid architecture**. The cross-layer optimization issue will be decomposed completely by multilevel decomposition. Several different indecomposable subproblems will be assigned to one network node. The rule of assignment is that different nodes can share less information and use less coordination to solve these subproblems. This architecture breaks the standard layered architecture. Each node should have the capability of recomposing its functionalities flexibly and dynamically. Hybrid architecture is fully adaptive in the function level. Thus, in some situations, some nodes may have the light burden, and others may have heavy duty. Meanwhile, some nodes can even hibernate without any tasks for power saving. For dynamic routing selection in wireless sensor network, battery is the key issue for the sensor's life time. The sensor with less energy cannot be chosen as the next hop in the routing path even if the radio environment around this sensor is very suitable for wireless communication.

- **Cluster based architecture**. The cross-layer optimization issue is divided into several subproblems by horizontal decomposition. Different subproblems will be solved by different clusters in the wireless networking system. The cluster consists of a cluster head and several nodes around the cluster head. The cluster head is more powerful than any other node in the cluster. The cluster head is responsible for exchanging control information among different clusters and supervising other nodes in the cluster. Thus, the first level subproblem obtained by horizontal decomposition can be further partitioned. More functionalities will be allocated to the cluster head. The rest will be distributed to other nodes based on node capability and radio environment. The cluster based architecture is a good scheme to balance the central control and the distributed implementation.

- **Mobility based architecture**. The key point of this architecture is that the node, for example, unmanned aerial vehicle (UAV), has the mobile capability. The node can at least search for the available spectrum in different locations intelligently. The movement of nodes can change the existing network topology. However, this change is still under some level of control. Mobility based architecture can undoubtedly achieve autonomous network resiliency in the contested RF spectra. If the relay node is out of the communication range or there is no available spectrum for the relay node to use, this node can intelligently change its location to maintain the connectivity of wireless communication.

6.10.5.5 Physical Layer Consideration

In order to support the potential cognitive cross-layer wireless networking architectures and protocols, NC-OFDM will be exploited as the basic physical layer transmission technique. NC-OFDM is a noncontiguous version of OFDM with some unused subcarriers. OFDM is a highly recognized signal waveform for the current advanced wireless communication system, for example, 3G network, WiFi, WiMAX, and so on. DSA as well as OFDMA can be implemented based on NC-OFDM. Cognitive radio users can easily turn off some subcarriers which a primary user occupies and use other available subcarriers to transmit data.

How to efficiently implement NC-OFDM transceiver will be studied. At the transmitter, an FFT pruning algorithm and spectral shaping technique should be used to generate arbitrary NC-OFDM signaling. Because channel state information, primary user occupancy, and throughput requirement for cognitive radio users vary over time, an FFT pruning algorithm should be able to design an efficient FFT implementation every time conditions change [539]. Besides, PAPR issue should be taken into account in synthesizing NC-OFDM signaling from implementation consideration.

The other challenge for NC-OFDM transceiver is the synchronization at the receiver, especially for blind synchronization [540]. It is hard for cognitive radio to set up a dedicated control channel between the transmitter and the receiver. If the transmitter changes subcarriers to be used for data transmission, the receiver should have a way to detect or track this change, and jump to the correct subcarriers for receiving data without any control information aided from the transmitter. Meanwhile, due to the presence of the primary user, time domain correlation fails [540], even if the predetermined preamble is used. Thus, for blind synchronization, spectrum detection should be performed to find a new transmission first [540]. And then, the preamble is learned from those subcarriers for the new transmission. The regenerated preamble will be exploited to correlate with the following incoming signal [540]. Cognitive radio has no licensed spectrum. The reliable transmission between transceiver should be built as quickly as possible if some parts of spectra are available. Thus, it is worthwhile to implement a fast and effective synchronization scheme even at the cost of computational and implementation complexity.

MIMO will be also exploited in physical layer. MIMO technique or array signal processing can bring array gain, spatial diversity gain, and spatial multiplexing gain. Interference alignment [541] has been performed based on MIMO to explore degree of freedom in the spatial domain. Meanwhile, widely studied beamforming technique can be used together with routing selection and scheduling to improve spatial reuse [542]. Directional beam patterns can increase the communication range and reduce the interference to other directions.

6.11   Summary

In this chapter, optimization theory, especially convex optimization has been presented. Convex optimization is a powerful signal processing tool which can be exploited anywhere, for example, system control, machine learning, operation research, management, and so on. Linear programming, quadratic programming, geometric programming, Lagrange duality, optimization algorithm, robust optimization, and multiobjective optimization have been covered. This chapter can give readers the whole picture of optimization theory. Some examples have been shown in this chapter to help readers to understand how to use convex optimization to solve engineering problems or improve the system performances. If the engineering problems can be formulated as convex optimization problems, these problems will be solved without doubt. In cognitive radio network, optimization theory can be widely used for spectrum sensing [543, 544], cross-layer design, resource allocation, sensing disruption from adversary [545], and so on.

7

Machine Learning

Artificial intelligence [546–554] aims at making intelligent machines where an intelligent machine or agent is a system that perceives its environment and takes actions to maximize its own utility. The central problems in artificial intelligence include deduction, reasoning [555], problem solving, knowledge representation, learning, and so on.

In order to understand how the brain learns and how the computer or system achieves intelligent behavior, the interdisciplinary study of neuroscience, computer science, cognitive psychology, mathematics, and statistics gives a new research direction of artificial intelligence, called computational neuroscience research. Computational neuroscience tries to build artificial systems and mathematical models to explore the computational principles for perception, cognition, memory, and motion. More related information can be found in Computational Neuroscience Research at Carnegie Mellon University. Leonid Perlovsky, who won the John McLucas Award in 2007, the highest US Air Force Award for science, uses knowledge instinct and dynamic logic to express and model the brain mechanisms of perception and cognition [556]. Especially, dynamic logic is a mathematical description of the knowledge instinct which describes mathematically a fundamental mind mechanism of interactions between bottom-up signals and top-down signals as a process of adaptation from vague to crisp concepts [557]. Besides, bionics also motivates the study of artificial intelligence and extend its capability. Bionics tries to build artificial systems based on the biological methods and systems found in nature.

Machine learning [547, 558–563] is the main branch of artificial intelligence which deals with the design and development of algorithms that allow the machine or computer to evolve behaviors based on example data or past experience. Machine learning algorithms can be organized into different categories: unsupervised learning, semi-supervised learning, supervised learning, transductive inference, active learning, transfer learning, reinforcement learning, and so on.

There are two basic models for machine learning. One is generative model [564] and the other is discriminative model [565]. A generative model can generate observable data given some hidden parameters. Examples of generative models include Gaussian mixture model, hidden Markov model, naive Bayes, Bayesian networks, Markov random fields,

---

Cognitive Radio Communications and Networking: Principles and Practice, First Edition.
Robert C. Qiu, Zhen Hu, Husheng Li and Michael C. Wicks.
© 2012 John Wiley & Sons, Ltd. Published 2012 by John Wiley & Sons, Ltd.

and so on. Hence, a generative model is a full probabilistic model of all variables and models the underlying process of how the data is generated [566]. A discriminative model only provides the dependence of the target variables on the observed variables which can be done directly by posterior probabilities or conditional probabilities. Hence, discriminative model can focus computational resources on given task and give better performance. However, a discriminative model looks like a black box and lacks explanatory power of the generative model. Examples of discriminative models include logistic regression, linear discriminant analysis, support vector machine, boosting, conditional random fields, linear regression, neural networks, and so on.

Artificial intelligence as well as machine learning can be generally applied to many different areas, for example, cognitive radio, cognitive radar, smart grid, computational transportation, data mining, robotics, web search engine, human computer interaction, manufacturing, bioengineering, and so on.

- **Cognitive Radio and Network.** Cognitive radio is a brand new concept for the wireless communication system. The idea of cognitive radio was first presented by Joseph Mitola III in a seminar at KTH, The Royal Institute of Technology, in 1998, and published later in an article [567] by Mitola and Gerald Q. Maguire, Jr in 1999. Software radio provides an ideal platform for the realization of cognitive radio [567], and cognitive radio makes software radio smart. Later Simon Haykin gave a review of cognitive radio and treated it as brain-empowered wireless communications [568]. The goal is to improve the utilization of a precious natural resource: the radio electromagnetic spectrum [568].

    Cognitive radio can be treated as one approach of implementing DSA on SDR platforms [525]. However cognitive radio is more than DSA. Cognition differentiates cognitive radio from any other radio system. Most of the research about cognitive radio focuses on the behaviors of one pair of cognitive radios. If multiple cognitive radios are taken into account or the network behaviors of cognitive radios are of interest, cognitive radio network will be the main research object. In cognitive radio network, cognition should cover from the physical layer through the application layer to reliably meet the requirements of the information system.

    Cognitive radio architecture and applications of machine learning to cognitive radio network have been presented in [569]. In cognitive radio engine, knowledge base, reasoning engine, and learning engine are three main components. Capacity maximization and DSA are used as examples to describe how cognitive radio works. Reasoning, learning, knowledge representation, and reconfiguration of cognitive radio have also been discussed in [570]. Learning is the basic function in cognitive radio network. The materialization of learning in cognitive radio network can be found in [571–578].
- **Cognitive Radar and Network.** A lot of algorithms which were infeasible decades ago are now coming possible. Such examples are common in machine learning and artificial intelligence. These algorithms revolutionize areas like robotics [579]. Radar is experiencing a similar revolution in the general direction of cognitive radar [580].

    The radar system evolves from the current adaptive radar and the radar with a function of waveform design to cognitive radar. The adaptive radar focuses more on the adaptivity at the receiver. The radar waveform design deals with the probing signal according to some optimization criterion. The dominant feature of cognitive radar is cognition, which means the radar can actively learn about the environment, and the whole radar system forms a dynamic closed feedback loop including the transmitter, environment, and receiver [580].

Cognitive radar only considers one pair of radar transceivers, and the cognition only focuses on the physical layer. In order to further enhance the capability of radar system, cognitive radar network is proposed. Cognitive radar network is not simply summation of multiple cognitive radars. Cognitive radar network itself at least integrates cognitive radio network, cognitive radar, MIMO radar, layered sensing, and so on. Cognition will run through physical layer to network layer and application layer.

With the support of cognition, radar network resource management will take care of operation, resource allocation, and maintenance of the networking system. Radar network resource management includes: (1) radio resource management; (2) network resource management; (3) radar task scheduling and prioritization.

Radio resource management is well studied in wireless communication. Similarly, DSA, spectrum management, power allocation, and so on are still very important for cognitive radar network. Network resource management focuses on the control strategy for the network behavior. Dynamic network configuration, adaptive routing, coordination, and competition should be taken into account.

Radar task scheduling and prioritization are application driven. Radar task scheduling and prioritization set the orders and priorities to all accepted radar tasks based on: (1) radio resource; (2) network resource; (3) the significance of radar task; (4) the urgency of radar task; (5) the condition of cognitive radar network. Radar task with higher priority will be scheduled first, and multiple radar tasks can be performed simultaneously. Thus radar task scheduling and prioritization should be executed dynamically and intelligently. Meanwhile, radar task admission control and radar task waiting list maintenance will also be taken into account under the framework of radar task scheduling and prioritization. If the capacity of cognitive radar network approaches its limitation or the heavy duties make the system unstable, the newest radar tasks cannot be admitted immediately. These tasks can be put in the waiting list for future service. The waiting list maintenance takes care of the order of radar tasks in the waiting list. Knowledge based resource management for multifunction radar takes a look at scheduling and task prioritization for adaptive radar in [581]. The analysis in [581] indicates that priorization is a key component to determining overall performance of radar system.

A partially observable Markov decision process (POMDP) is a well studied model and tool to solve decision making problem. POMDP is a generalization of a Markov decision process (MDP). A POMDP models a decision process in which it is assumed that the system dynamics are determined by an MDP, but the underlying state cannot be directly observed. Instead, it must maintain a probability distribution over the set of possible states based on observations and observation probabilities. Multivariate POMDPs are used for radar resource management in [582]. The problems of multitarget radar scheduling are formulated as multivariate POMDPs, the aim of which is to compute the scheduling policy to determine which target to choose and how long to continue with this choice so as to minimize a cost function [582]. Sensor scheduling for multiple target tracking and detection is discussed in [583]. The algorithm is also based on POMDP.

- **Smart Grid.** Smart grid explores and exploits two-way communication technology, advanced sensing, metering and measurement technology, modern control theory, network grid technology, and machine learning in the power and electricity system to make the power and electricity network stable, secure, efficient, flexible, economical and environmentally friendly.

  Novel control technology, information technology, and management technology should be effectively integrated to realize the smart information exchange within the power system from power generation, power transmission, power transformation, power distribution, power scheduling to power utilization. The goal of smart grid is to systematically optimize the cycle of power generation and utilization.

  Based on open system architecture and shared information mode, power flow, information flow and transaction flow can be syncretized. In this way, the operation performance of electric power enterprises can be increased. From electric power customer's perspective, demand response should be implemented. Customers would like to participate more activities in the power system and power market to reduce their electric power bill.

  Distributed energy resources, for example solar energy, wind energy, and so on, should also play an important role in smart grid. Versatile distributed energy resources can perform the peak power shaving and increase the stability of power system. However, distributed energy generation imposes a new challenge on the power system, especially on the distribution network. Power system planning, power quality, and so on should be reconsidered.

  To support smart grid, the infrastructure for the two-way communication should be set up dedicatedly for the power system only. In this way, secure, reliable, efficient communication and information exchange can be guaranteed. Meanwhile, the device, equipment, and facility of the current power system should also be updated and renovated. Novel technology for power electronics should be used to build advanced power devices, for example, transformer, relay, switch, storage, and so on.

  Machine learning for the New York City power grid has been presented in [584]. A general process for transforming historical electrical grid data into models that aim to predict the risk of failures for components and systems in the power grid is given [584]. These models can be used directly by power companies for the scheduling of maintenance work [584].

- **Computational Transportation.** Computational transportation [585, 586] or intelligent transportation [587–592] studies how to improve the safety, mobility, efficiency, and sustainability of transportation system by taking advantage of computer science, communication technology, information technology, sensing technology, computing technology, and control theory. Modeling, planning, and economic aspects of transportation are taken into account. The research topics and enabling solutions to transportation problems range from ride-sharing [593], routing, scheduling, and navigation, to autonomous/assisted driving, travel pattern analysis, and so on. More related information can be found in Computational Transportation Science at University of Illinois at Chicago.

- **Data Mining.** Data mining [561, 594–598] tries to discover new patterns and extract knowledge or intelligence from large scale data using methods at the intersection of artificial intelligence, statistics, and database system. Data mining can be widely used for science and engineering. Bioinformatics exploits data mining to generate new knowledge of biology and medicine, and discover new models of biological computation, for example, DNA computing, neural computing, evolutionary computing, and so on. Data mining is also useful for business applications. Take Internet advertising as an example, by data mining, more relevant advertisements can be sent to the right Internet audience at the right time.

- **Computer Vision.** Computer vision tries to obtain, process, analyze, and understand the real-world images or videos [599]. Information and intelligence can be extracted from the large scale data by computer vision. Machine learning is widely used in computer vision for detection, classification, recognition, tracking, and so on [600].

- **Robotics.** Robot is a virtual intelligent agent which can perform a variety of duties automatically or under guidance [601]. These duties can be part handling, assembly, painting, transport, surveillance, security, home help, and so on. The intelligence of robot is realized in software. Artificial intelligence gives robot the functions of perception, localization, modeling, reasoning, interaction, learning, planning, and so on. UAV can be treated as one kind of mobile robots.

- **Web Search Engine.** A web search engine [602] is mainly used to search for information on the website. Google, Yahoo, Bing, and so on are widely used web search engines. Machine learning is the powerful tool for web search engine. Commercial web search engines began to use machine learned ranking systems since the past decade. A ranking model is automatically constructed from training data by supervised learning or semisupervised learning. This ranking model for web search engine can reflect the importance of a particular web page.

- **Human Computer Interaction.** Human computer interaction [603] tries to design the interaction between people and computers. The researches about human computer interaction include cognitive models, speech recognition, natural language understanding, gesture recognition, data visualization, and so on. iPhone 4S can be treated as one kind of human computer interaction devices. iPhone 4s includes a new automated voice control system called Siri. Siri can allow the user to give the iPhone commands.

- **Social Network.** Social network is a network of social structure, social interdependency, or social relationships of human beings. Friendship, common interest, common belief, financial exchange, and so on are considered in the social network. Data related to social network have exploded recently due to the fast development of information technology. Thus, machine learning is a powerful tool to analyze social network for learning and inference [604–607].

- **Manufacturing.** Machine learning can be used in manufacturing to perform automatic and intelligent operations. In this way, the efficiency of manufacturing can be improved, especially for the dark factory with no involvement of human labor. The novel developments in machine learning and substantial applications of machine learning in modern industrial engineering and mass production have been presented in [608]. The analysis of data from simulations and experiments in the development phase and measurements during mass production plays a crucial role in modern manufacturing [608]. For example, various machine learning algorithms are applied to detection and recognition of spatial defect patterns in semiconductor fabrication processes [609–612]. These spatial defect patterns generated during integrated circuit (IC) manufacturing processes contain information about potential problems in the processes [612].

- **Bioengineering.** Bioengineering [613] tries to exploit both concepts of biology and engineering's analytical methodologies to deal with problems in life science. With the developments of mathematics and computer science, machine learning can be used in bioinformatics, medical innovations, biomedical image analysis, and so on.

7.1   Unsupervised Learning

Unsupervised learning [614–616] tries to find hidden or underlying structure from the unlabeled data. The key feature of unsupervised learning is that the data or examples given to the learner are unlabeled.

Clustering and blind signal separation are two categories of unsupervised learning algorithms [616]. Clustering assigns a set of objects into different groups or clusters such that the objects in the same group are similar [617]. The clustering algorithms include:

- k-means or centroid-based clustering [618–621];
- k-nearest neighbors [622, 623];
- hierarchical clustering or connectivity-based clustering;
- distribution-based clustering;
- density-based clustering.

Blind signal separation or blind source separation tries to separate a set of signals from a set of mixed signals without the information about the source signals or the mixing process [624]. The approaches for blind signal separation include:

- principal component analysis [625, 626];
- singular value decomposition [627];
- independent component analysis (ICA) [628–630];
- nonnegative matrix factorization [631–633].

Robust signal classification using unsupervised learning has been discussed in [634]. k-means clustering and the self-organizing map (SOM) are used as unsupervised classifiers. Meanwhile, the countermeasures to the class manipulation attacks are developed [634].

7.1.1   Centroid-Based Clustering

In centroid-based clustering [635], the whole data set is partitioned into different clusters. Each cluster is represented by a central vector. This central vector is not necessarily a member of the data set. Meanwhile, each member in the cluster has the smallest distance from the corresponding mean. If the number of the clusters is k, k-means clustering gives a corresponding optimization problem for centroid-based clustering.

Given a set of data $X = \{\mathbf{x}_1, \mathbf{x}_2, \ldots, \mathbf{x}_n\}$, k-means clustering attempts to partition the data set X into $k(k \leq n)$ sets $S_1, S_2, \ldots, S_k$ such that the sum of squared distances within the cluster is minimized [618–621]

$$\text{minimize} \sum_{i=1}^{k} \sum_{\mathbf{x}_l \in S_i} \|\mathbf{x}_l - \mathbf{y}_i\|^2, \tag{7.1}$$

where $\mathbf{y}_i$ is the mean of the cluster related to data set S_i.

7.1.2 k-Nearest Neighbors

The k-nearest neighbor (k-NN) algorithm assigns a class to an object by a majority vote of its k-nearest neighbors. Genetic programming with k-NN has been used in [636] to perform automatic digital modulation classification.

7.1.3 Principal Component Analysis

PCA is also called Karhunen-Loeve transform, Hotelling transform, or proper orthogonal decomposition [637]. PCA uses an orthogonal transformation to transform a set of correlated variables into a set of uncorrelated variables [637]. These uncorrelated variables are linear combinations of the original variables. They are called principal components. The number of principal components is less than or equal to the number of original variables. Thus, PCA is a widely used linear transformation for dimensionality reduction.

The goal of PCA is to ensure that the first principal component bears the largest variance and the second principal component has the second largest variance. Meanwhile, the directions of different principal components are orthogonal. Generally, PCA can be executed by eigenvalue decomposition of covariance matrix.

Given a set of high-dimensional real data $\mathbf{x}_1, \mathbf{x}_2, \ldots, \mathbf{x}_M$ where $\mathbf{x}_m \in \mathbf{R}^N$, PCA can be performed as:

1. $\bar{\mathbf{x}} = \frac{1}{M} \sum_{m=1}^{M} \mathbf{x}_m$.
2. $\tilde{\mathbf{x}}_m = \mathbf{x}_m - \bar{\mathbf{x}}$.
3. $\mathbf{C} = \frac{1}{M} \sum_{m=1}^{M} \tilde{\mathbf{x}}_M \tilde{\mathbf{x}}_M^T$.
4. Compute the eigenvalues $\lambda_1, \lambda_2, \ldots, \lambda_N$ of $\mathbf{C}$ and the corresponding eigenvectors $\mathbf{u}_1, \mathbf{u}_2, \ldots, \mathbf{u}_N$ where $\lambda_1 \geq \lambda_2 \geq \cdots \geq \lambda_N$.
5. Obtain the linear transformation matrix,

$$\mathbf{U} = [\mathbf{u}_1 \ \mathbf{u}_2 \ \cdots \ \mathbf{u}_K], \tag{7.2}$$

   where $K \ll N$.
6. Perform dimensionality reduction,

$$\mathbf{y} = \mathbf{U}^T \tilde{\mathbf{x}} \tag{7.3}$$

   and PCA approximation $\tilde{\mathbf{x}} = \mathbf{U}\mathbf{y}$.

In sum, PCA projects the data from the original directions or bases to the new directions or bases. Meanwhile, the data varies the most along the new directions. These directions can be determined by the eigenvectors of the covariance matrix corresponding to the largest eigenvalues. The eigenvalues relate to the variances of the data along the new directions. PCA gives a way to construct the linear subspace spanned by the new bases from the data.

$\tilde{\mathbf{x}} = \mathbf{U}\mathbf{y}$ is extended to $\tilde{\mathbf{X}} = \mathbf{U}\mathbf{Y}$. If $\mathbf{Y}$ contains as many zeros as possible, this problem is called sparse component analysis [638].

PCA can also be extended to its robust version. The background of robust PCA [639, 640] is to decompose a given large data matrix $\mathbf{M}$ as a low rank matrix $\mathbf{L}$ plus a sparse matrix $\mathbf{S}$, that is,

$$\mathbf{M} = \mathbf{L} + \mathbf{S}. \tag{7.4}$$

Specifically speaking, PCA finds a rank-r approximation of the given data matrix $\mathbf{M}$ in an l^2 sense by solving the following optimization problem,

$$\begin{array}{l} \text{minimize} \\ \|\mathbf{M} - \mathbf{L}\| \\ \text{subject to} \\ \text{rank}(\mathbf{L}) \leq r. \end{array} \tag{7.5}$$

This problem can be easily solved by SVD. An intrinsic drawback of PCA is that it can work efficiently only when the low rank matrix is corrupted with small and i.i.d. Gaussian noise. That is PCA is suitable for the model of $\mathbf{M} = \mathbf{L} + \mathbf{N}$ where $\mathbf{N}$ is the i.i.d. Gaussian noise matrix. PCA will fail when some of the entries in $\mathbf{L}$ are strongly corrupted as shown in Equation (7.4) in which the matrix $\mathbf{S}$ is a sparse matrix with arbitrarily large magnitude.

In order to find $\mathbf{L}$ and $\mathbf{S}$ from $\mathbf{M}$, robust PCA tries to solve the following optimization problem,

$$\begin{array}{l} \text{minimize} \\ \text{rank}(\mathbf{L}) + \lambda \|\mathbf{S}\|_1 \\ \text{subject to} \\ \mathbf{M} = \mathbf{L} + \mathbf{S}. \end{array} \tag{7.6}$$

From the convex optimization point of view, the rank function is a nonconvex function. Solving the optimization problem with a rank objective or rank constraint is NP-hard. However, it is known that the convex envelope of rank$(\mathbf{L})$ on the set $\{\mathbf{L} : \|\mathbf{L}\| \leq 1\}$ is the nuclear norm $\|\mathbf{L}\|_*$ [641]. Hence, the rank minimization can be relaxed to a nuclear norm minimization problem which is a convex objective function. In this regard, there are a series of papers that have studied the conditions required for successfully applying the nuclear norm heuristic to rank minimization from different perspectives [641–643]. Hence, the optimization problem (7.6) can be relaxed to

$$\begin{array}{l} \text{minimize} \\ \|\mathbf{L}\|_* + \lambda \|\mathbf{S}\|_1 \\ \text{subject to} \\ \mathbf{M} = \mathbf{L} + \mathbf{S}. \end{array} \tag{7.7}$$

In this way, $\mathbf{L}$ and $\mathbf{S}$ can be recovered.

Robust PCA is widely used in video surveillance, image processing, face recognition, latent semantic indexing, ranking, and collaborative filtering [639].

7.1.4 Independent Component Analysis

ICA tries to separate a mixed multivariate signal and identify the underlying non-Gaussian source signals or components that are statistically independent or as independent as possible [562, 629, 644]. Even though the source signals are independent, the observed signals are not independent due to the mixture operation. Meanwhile, the observed signals look like normal distributions [562]. A simple application of ICA is the "cocktail party problem." Assume in the cocktail party, there are two speakers denoted by $s_1(t)$ and $s_2(t)$ and there are two microphones recording time signals denoted by $x_1(t)$ and $x_2(t)$. Thus,

$$
\begin{aligned}
x_1(t) &= a_{11}s_1(t) + a_{12}s_1(t) \\
x_2(t) &= a_{21}s_1(t) + a_{22}s_1(t),
\end{aligned}
\tag{7.8}
$$

where a_{11}, a_{12}, a_{21}, and a_{22} are some unknown parameters that depend on the distances between the microphones and the speakers [629]. We would like to estimate two source signals $s_1(t)$ and $s_2(t)$ using only the recorded signals $x_1(t)$ and $x_2(t)$.

Using the matrix notation, the linear noiseless ICA can be written as

$$
\mathbf{X} = \mathbf{AS}, \tag{7.9}
$$

where the rows of $\mathbf{S}$ should be statistically independent. Due to the unknown $\mathbf{A}$ and $\mathbf{S}$, the variances and the order of the independent components cannot be determined [629]. In order to solve ICA problems, minimization of mutual information and maximization of non-Gaussianity are often used to achieve the independence of the latent sources.

The applications of ICA include separation of artifacts in magnetoencephalography (MEG) data, finding hidden factors in financial data, reducing noise in natural images, blind source separation for telecommunication [629]. ICA can also be used for chemical and biological sensing to extract the intrinsic surface-enhanced Raman scattering spectrum [645].

7.1.5 Nonnegative Matrix Factorization

Matrix decomposition has long been studied. A matrix decomposition is a factorization of a matrix into some canonical form. There are many different matrix decompositions, for example, LU factorization, LDU decomposition, Cholesky decomposition, rank factorization, QR decomposition, rank-revealing QR factorization, SVD, eigen-decomposition, Jordan decomposition, Schur decomposition, and so on.

Nonnegative matrix factorization [633, 646] is one kind of matrix decomposition with the nonnegative constraint on the factors. Mathematically speaking, a matrix $\mathbf{X}$ is factorized into two matrices or factors $\mathbf{W}$ and $\mathbf{H}$ such that

$$
\mathbf{X} = \mathbf{WH} + \mathbf{E} \tag{7.10}
$$

and all entries in $\mathbf{W}$ and $\mathbf{H}$ must be equal to or greater than zero where $\mathbf{E}$ represents approximation error.

There are many useful variants based on nonnegative matrix factorization [633]:

- Symmetric nonnegative matrix factorization,

$$
\mathbf{X} = \mathbf{WW}^T + \mathbf{E}. \tag{7.11}
$$

- Semi-orthogonal nonnegative matrix factorization,

$$
\mathbf{X} = \mathbf{WH} + \mathbf{E} \tag{7.12}
$$

and $\mathbf{W}^T\mathbf{W} = \mathbf{I}$ and $\mathbf{HH}^T = \mathbf{I}$.

- Three-factor nonnegative matrix factorization,

$$\mathbf{X} = \mathbf{WSH} + \mathbf{E}. \tag{7.13}$$

- Affine nonnegative matrix factorization,

$$\mathbf{X} = \mathbf{WH} + \mathbf{a1}^T + \mathbf{E}. \tag{7.14}$$

- Multilayer nonnegative matrix factorization,

$$\mathbf{X} = \mathbf{W}_1 \mathbf{W}_2 \cdots \mathbf{W}_L \mathbf{H} + \mathbf{E}. \tag{7.15}$$

- Simultaneous nonnegative matrix factorization,

$$\begin{aligned} \mathbf{X}_1 &= \mathbf{W}_1 \mathbf{H} + \mathbf{E}_1 \\ \mathbf{X}_2 &= \mathbf{W}_2 \mathbf{H} + \mathbf{E}_2. \end{aligned} \tag{7.16}$$

- Nonnegative matrix factorization with sparseness constraints on the each column of $\mathbf{W}$ and $\mathbf{H}$ [647].

Two-dimensional nonnegative matrix factorization can be extended to n-dimensional nonnegative tensor factorization [633, 648, 649]. Various algorithms for nonnegative matrix and tensor factorization are mentioned in [633]. In [650], Bregman divergences are used for generalized nonnegative matrix approximation.

Similar to robust PCA, the robust version of nonnegative matrix factorization is expressed as

$$\mathbf{X} = \mathbf{WH} + \mathbf{S} + \mathbf{E}, \tag{7.17}$$

where $\mathbf{S}$ is the sparse matrix. The optimization problem for robust nonnegative matrix factorization can be represented as

$$\text{minimize } \|\mathbf{X} - \mathbf{WH} - \mathbf{S}\|_F^2 + \lambda \|\mathbf{S}\|_1 \tag{7.18}$$

such that $\mathbf{W}$ and $\mathbf{H}$ are both the nonnegative matrices. The optimization problem (7.18) is not convex in $\mathbf{W}$, $\mathbf{H}$, and $\mathbf{S}$ jointly. Thus, we need to solve them separately [651]:

1. Solve the nonnegative matrix factorization problem for fixed $\mathbf{S}$.
2. Optimize $\mathbf{S}$ for fixed $\mathbf{W}$ and $\mathbf{H}$.

This procedure will be repeated until the algorithm converges.

Nonnegative matrix factorization is a special case of general matrix factorization. Probabilistic algorithms for constructing approximate matrix factorization have been comprehensively discussed in [652]. The core idea is to find structure with randomness [652]. Compared with standard deterministic algorithms for matrix factorization, the randomized methods are often faster and more robust [652].

7.1.6 Self-Organizing Map

An SOM is one kind of ANN within the category of unsupervised learning. SOM attempts to create spatially organized low-dimensional or internal representation (usually two-dimensional grid) of input signals and their abstractions which is called map [653–655]. SOM is different from other ANNs because a neighborhood function is used to preserve the topology of the input space. Thus, the nearby locations in the map represent the inputs with similar properties.

The training algorithms of SOM are based on the principle of competitive learning [653, 656] which is also used for the well-known vector quantization [657–659].

7.2 Supervised Learning

Supervised learning learns a function from supervised labeled training data [660]. In supervised learning, the training data consist of a set of training examples. Each training example includes an input object together with a desired output value. If the output value is discrete, the learned function is called a classifier. If the output value is continuous, the learned function is called a regression function. Algorithms for supervised learning generalize from the training data to the unseen data. The popular algorithms for supervised learning are:

- linear regression [661–664];
- logistic regression [665, 666];
- artificial neural network [667, 668];
- decision tree learning [669];
- random forests [670];
- naive Bayes classifier [671];
- support vector machines [672–674].

7.2.1 Linear Regression

Linear regression tries to model the relationship between a scalar dependent variable y and one or more explanatory (independent) variables $\mathbf{x}$. Mathematically speaking,

$$y = \mathbf{x}^T \mathbf{a} + \varepsilon, \tag{7.19}$$

where

$$\mathbf{x} = [x_1 \ x_2 \ \cdots \ x_P]^T; \tag{7.20}$$

$\mathbf{a}$ is called the parameter vector or regression coefficients

$$\mathbf{a} = [a_1 \ a_2 \ \cdots \ a_P]^T; \tag{7.21}$$

and ε is the noise or error.

If there are N dependent variables, Equation (7.19) can be extended to

$$\mathbf{y} = \mathbf{X}^T \mathbf{a} + \varepsilon, \tag{7.22}$$

where

$$\mathbf{y} = [y_1 \ y_2 \ \cdots \ y_N]^T \tag{7.23}$$

$$\mathbf{X} = [\mathbf{x}_1 \ \mathbf{x}_2 \ \cdots \ \mathbf{x}_N] \tag{7.24}$$

$$\varepsilon = [\varepsilon_1 \ \varepsilon_2 \ \cdots \ \varepsilon_N]^T. \tag{7.25}$$

7.2.2 Logistic Regression

Logistic regression [675] is a nonlinear regression which can predict the probability of an event occurrence using a logistic function. A simple logistic function is defined as

$$f(t) = \frac{1}{1 + \exp(-t)},$$ (7.26)

which always takes on values between zero and one. Thus, logistic regression can be expressed as

$$y = f(\mathbf{x}) = \frac{1}{1 + \exp(-(a_0 + a_1 x_1 + \cdots + a_P x_P))},$$ (7.27)

where a_0 is called intercept and $a_1, a_2, \ldots, a_P$ are called regression coefficients of $x_1, x_2, \ldots, x_P$.

Logistic regression is a popular way to model and analyze binary phenomena, which means the dependent variable or the response variable is a two-valued variable [676].

7.2.3 Artificial Neural Network

The idea of artificial neural network [677] is borrowed from biological neural network to mimic the real life behavior of neurons. Artificial neural network is an adaptive system used to model relationship between inputs and outputs. The mathematical expression of the simplest artificial neural network is

$$o = f\left(\sum_{n=1}^{N} w_n x_n\right),$$ (7.28)

where $x_1, x_2, \ldots, x_N$ are inputs; $w_1, w_2, \ldots, w_N$ are the corresponding weights; o is output; and f is an activation (transfer) function.

Perceptron, one type of artificial neural network, is a binary classifier which maps its inputs to an output with binary values. Given threshold θ, if $\sum_{n=1}^{N} w_n x_n \geq \theta$, then $o = 1$; otherwise $o = 0$. This single-layer perceptron has no hidden layer. The single-layer perceptron can be extended to the multilayer perceptron which consists of multiple layers of nodes in a directed graph. The multilayer perceptron can use backpropagation algorithm to learn the network.

7.2.4 Decision Tree Learning

A decision tree [678] is a tree-like graph or model for decision, prediction, classification, and so on. In a decision tree, each internal node tests an attribute. Each branch corresponds to one possible value of attribute. Each leaf node assigns a classification for the observation. Decision tree learning tries to learn a function which can be represented as a decision tree [679]. A number of decision trees can be used together to form a random forest classifier which is an ensemble classifier [680]. In random forest, each tree is grown at least partially at random. Bootstrap aggregation is used for parallel combination of learners which is independently trained on distinct bootstrap samples. Final result is the mean prediction or class with maximum votes. Random forest can increase accuracy by reducing prediction variance.

7.2.5 Naive Bayes Classifier

A naive Bayes classifier [681] is a probabilistic classifier based on Bayes' theorem with independence assumptions.

Based on Bayes' theorem, the naive Bayes probabilistic model can be expressed as,

$$p(C \mid X_1, X_2, \ldots, X_N) \propto p(C) \prod_{n=1}^{N} p(X_n \mid C), \tag{7.29}$$

where C is a dependent class variable and $X_1, X_2, \ldots, X_N$ are the feature variables. $p(C \mid X_1, X_2, \ldots, X_N)$ is the posterior probability. $p(C)$ is the prior probability. $p(X_n \mid C)$ is the likelihood probability.

According to the maximum a posteriori (MAP) decision rule, A naive Bayes classifier can be written as [671]

$$c = f(x_1, x_2, \ldots, x_N) = \arg\max p(C = c) \prod_{n=1}^{N} p(X_n = x_n \mid C = c). \tag{7.30}$$

7.2.6 Support Vector Machines

SVM [682] is a set of the supervised learning algorithms used for classification and regression. SVM includes linear SVM, kernel SVM [683, 684], multiclass SVM [685–692], support vector regression [693–697]. Design of learning engine based on SVM in cognitive radio has been mentioned in [698]. Both classification and regression results of SVM for eight kinds of modulation modes are demonstrated. The experimental data come from 802.11a protocol platform. SVM is used for MAC protocol classification in a cognitive radio network [699]. The received power mean and variance are chosen as two features for SVM. Two MAC protocols, time division multiple access and slotted Aloha, are classified.

Let's study SVM from the linear two-class SVM [674]. Given the training data set having M pairs of inputs and outputs,

$$(\mathbf{x}_i, l_i), i = 1, 2, \ldots, M \tag{7.31}$$

and

$$l_i \in \{-1, 1\}. \tag{7.32}$$

SVM attempts to find the separating hyperplane

$$\mathbf{w} \cdot \mathbf{x} - b = 0 \tag{7.33}$$

with the largest margin satisfying the following constraints:

$$\begin{aligned} \mathbf{w} \cdot \mathbf{x}_i - b &\geq 1, \text{ for } l_i = 1 \\ \mathbf{w} \cdot \mathbf{x}_i - b &\leq -1, \text{ for } l_i = -1 \end{aligned} \tag{7.34}$$

in which $\mathbf{w}$ is the normal vector of the hyperplane and $\cdot$ stands for inner product. The constraint (7.34) can be combined into:

$$l_i(\mathbf{w} \cdot \mathbf{x}_i - b) \geq 1. \tag{7.35}$$

The distance between two hyperplanes $\mathbf{w} \cdot \mathbf{x}_i - b = 1$ and $\mathbf{w} \cdot \mathbf{x}_i - b = -1$ is $\frac{2}{\|\mathbf{w}\|}$. In order to obtain the largest margin, the following optimization is used [674]

$$\begin{aligned}
&\text{minimize} \\
&\tfrac{1}{2}\|\mathbf{w}\|^2 \\
&\text{subject to} \\
&l_i(\mathbf{w} \cdot \mathbf{x}_i - b) \geq 1, i = 1, 2, \ldots, M.
\end{aligned} \tag{7.36}$$

The dual form of the optimization problem (7.36) by introducing Lagrange multipliers $\alpha_i \geq 0, i = 1, 2, \ldots, M$ is [674]

$$\begin{aligned}
&\text{maximize} \\
&\sum_i \alpha_i - \tfrac{1}{2}\sum_{i,j} \alpha_i \alpha_j l_i l_j \mathbf{x}_i \cdot \mathbf{x}_j \\
&\text{subject to} \\
&\sum_i \alpha_i l_i = 0 \\
&\alpha_i \geq 0, , i = 1, 2, \ldots, M.
\end{aligned} \tag{7.37}$$

The solution to $\mathbf{w}$ can be expressed in terms of a linear combination of the training vectors as

$$\mathbf{w} = \sum_{i=1}^{M} \alpha_i l_i \mathbf{x}_i. \tag{7.38}$$

Those $\mathbf{x}_i, i = 1, 2, \ldots, M_{\mathrm{SV}}$ with $\alpha_i > 0$ are called support vectors which lie on the margin and satisfy $l_i(\mathbf{w} \cdot \mathbf{x}_i - b) = 1$. Thus, b can be obtained as

$$b = \frac{1}{M_{\mathrm{SV}}} \sum_{i=1}^{M_{\mathrm{SV}}} (\mathbf{w} \cdot \mathbf{x}_i - l_i). \tag{7.39}$$

Thus, a classifier based on SVM can be written as [674]

$$f(\mathbf{x}) = \mathrm{sign}\left(\sum_{i=1}^{M} \alpha_i l_i \mathbf{x}_i \cdot \mathbf{x} - b\right). \tag{7.40}$$

When the number of classes for outputs is more than two, multiclass SVM can be used to perform multiclass classification. The common approach for multiclass SVM is to decompose the single multiclass classification problem into multiple two-class classification problems. Each two-class classification problem can be addressed by the well known two-class SVM. Within this framework, one-against-all and one-against-one are widely used [692]. Besides, a pairwise coupling strategy can be exploited to combine the probabilistic outcomes of all the one-against-one two-class classifiers to obtain the estimates of the posterior probabilities for the test input [692, 700].

Based on the idea of SVM, the multiclass classification problem can also be handled by solving one single optimization problem [685, 688, 701].

The linear two-class SVM can be modified to tolerate some misclassification inputs, which is called soft margin SVM [674]. Soft margin SVM can be done by introducing a nonnegative slack variable $\xi_i, i = 1, 2, \ldots, M$ which measures the degree of misclassification for the input $\mathbf{x}_i$. Hence, the constraint (7.34) should be modified as

$$
\begin{aligned}
\mathbf{w} \cdot \mathbf{x}_i - b &\geq +1 - \xi_i, \text{ for } l_i = 1 \\
\mathbf{w} \cdot \mathbf{x}_i - b &\leq -1 + \xi_i, \text{ for } l_i = -1,
\end{aligned}
\tag{7.41}
$$

which can be combined into

$$
l_i(\mathbf{w} \cdot \mathbf{x}_i - b) \geq 1 - \xi_i
\tag{7.42}
$$

and $\xi_i \geq 0, i = 1, 2, \ldots, M$.

The optimization problem for soft margin SVM is expressed as [674]

$$
\begin{aligned}
&\text{minimize} \\
&\tfrac{1}{2}\|\mathbf{w}\|^2 + C \sum_{i=1}^{M} \xi_i \\
&\text{subject to} \\
&l_i(\mathbf{w} \cdot \mathbf{x}_i - b) \geq 1 - \xi_i, \xi_i \geq 0, i = 1, 2, \ldots, M,
\end{aligned}
\tag{7.43}
$$

where C is a trade-off parameter to compromise the slack variable penalty and the size of margin. The dual form of the optimization problem (7.43) is [674]

$$
\begin{aligned}
&\text{maximize} \\
&\sum_i \alpha_i - \tfrac{1}{2} \sum_{i,j} \alpha_i \alpha_j l_i l_j \mathbf{x}_i \cdot \mathbf{x}_j \\
&\text{subject to} \\
&\sum_i \alpha_i l_i = 0 \\
&0 \leq \alpha_i \leq C, , i = 1, 2, \ldots, M.
\end{aligned}
\tag{7.44}
$$

If the value of output l_i is continuous, the learned function is called a regression function. SVM can be extended to supoort vector regression. Analogously to the soft margin SVM, the optimization problem for support vector regression can be written as [697]

$$
\begin{aligned}
&\text{minimize} \\
&\tfrac{1}{2}\|\mathbf{w}\|^2 + C \sum_{i=1}^{M} (\xi_i^+ + \xi_i^-) \\
&\text{subject to} \\
&l_i - (\mathbf{w} \cdot \mathbf{x}_i - b) \leq \varepsilon + \xi_i^+, i = 1, 2, \ldots, M \\
&(\mathbf{w} \cdot \mathbf{x}_i - b) - l_i \leq \varepsilon + \xi_i^-, i = 1, 2, \ldots, M \\
&\xi_i^+ \geq 0, \xi_i^- \geq 0, i = 1, 2, \ldots, M.
\end{aligned}
\tag{7.45}
$$

where ε is a parameter to determine the region bounded by $l_i \pm \varepsilon, i = 1, 2, \ldots, M$ which is call $l_i \pm \varepsilon$-insensitive tube. The dual form of the optimization problem (7.45) is [697]

$$
\begin{aligned}
&\text{maximize} \\
&\sum_{i=1}^{M} l_i(\alpha_i^+ - \alpha_i^-) - \varepsilon \sum_{i=1}^{M} (\alpha_i^+ + \alpha_i^-) - \tfrac{1}{2} \sum_{i,j} (\alpha_i^+ - \alpha_i^-)(\alpha_j^+ - \alpha_j^-)\mathbf{x}_i \cdot \mathbf{x}_j \\
&\text{subject to} \\
&\sum_{i=1}^{M} (\alpha_i^+ - \alpha_i^-) = 0, i = 1, 2, \ldots, M \\
&0 \leq \alpha_i^+ \leq C, i = 1, 2, \ldots, M \\
&0 \leq \alpha_i^- \leq C, i = 1, 2, \ldots, M.
\end{aligned}
\tag{7.46}
$$

Thus,

$$\mathbf{w} = \sum_{i=1}^{M} (\alpha_i^+ - \alpha_i^-) \mathbf{x}_i \qquad (7.47)$$

and

$$f(\mathbf{x}) = \text{sign} \left(\sum_{i=1}^{M} (\alpha_i^+ - \alpha_i^-) \mathbf{x}_i \cdot \mathbf{x} + b \right). \qquad (7.48)$$

7.3 Semisupervised Learning

Supervised learning exploits the labeled data for training to learn the function. However, the labeled data, sometimes, are hard or expensive to obtain and generate. While the unlabeled data are more plentiful than the labeled data [702]. In order to make use of both labeled data and unlabeled data for training, semisupervised [703] learning can be explored. Semisupervised learning falls between unsupervised learning and supervised learning. The underlying phenomenon behind semisupervised learning is that a large amount of unlabeled data used together with a small amount of labeled data for training can improve machine learning accuracy [704, 705].

7.3.1 Constrained Clustering

Constrained clustering [706–708] can be treated as clustering with side information or additional constraints. These constraints include pairwise must-link constraints and cannot-link constraints. The must-link constraints mean two members or data points must be in the same cluster while the cannot-link constraints mean two data points cannot be in the same cluster. Take the k-means clustering as an example. The penalty cost function related to the must-link constraints and the cannot-link constraints can be added to the optimization problem (7.1) to form the optimization problem for the constrained k-means clustering. Besides, if the partial label information is given, a small amount of labeled data can aid the clustering of unlabeled data [709]. In [709], the seed clustering is used to initialize the k-means algorithm.

7.3.2 Co-Training

Co-training is also a semisupervised learning technique [702, 710]. In co-training, the features of each input are divided into two different feature sets. These two feature sets should be conditionally independent given the class of the input. Meanwhile, the class of the input can be accurately predicted from each feature set alone. In other words, each feature set contains sufficient information to determine the class of the input [702]. Co-training first learns two different classifiers based on two different feature sets using the labeled training data. Then, each classifier will label several unlabeled data with more confidence. These data will be used to construct the additional labeled training data. This procedure will be repeated until convergence.

7.3.3   Graph-Based Methods

Recently, graph-based methods for semisupervised learning have become popular [711]. Graph-based methods for semisupervised learning are nonparametric, discriminative, and transductive in nature [711]. The first step of graph-based methods is to create the graph based on both labeled data and unlabeled data. The data correspond to the nodes on the graph. The edge together with the weight between two nodes is determined by the inputs of the corresponding data. The weight of the edge reflects the similarity of two data inputs. The second step of graph-based methods is to estimate a smooth function on the graph. This function can predict the classes for all the nodes on the graph. Meanwhile, the predicted classes of the labeled nodes should be close to the given classes. Thus, how to estimate this function can be expressed as the optimization problem with two terms [711]. The first term is a loss function and the second term is a regularizer [711].

7.4   Transductive Inference

Transductive inference [712–714] is similar to semisupervised learning. Transductive inference tries to predict outputs for the test inputs based on the training data and test inputs. Transduction is different from the well-known induction. In induction, general rules are first obtained from the observed cases; then these general rules are applied to the test cases. Thus, the performances transductive inference are inconsistent on different test cases.

7.5   Transfer Learning

Transfer learning or inductive transfer [715,716] focuses on gaining knowledge from solving one problem or previous experience and applying it to a different but related problem. Markov logic networks [717] and Bayesian networks [718] have already been exploited for transfer learning.

Multitask learning or learning to learn is one kind of transfer learning [719]. Multitask learning tries to learn a problem together with other related problems simultaneously, with consideration of the commonality among the problems.

7.6   Active Learning

Active learning is also called optimal experimental design [720,721]. Active learning is a form of supervised learning in which the learner can interactively ask for information. Specifically speaking, the learner actively queries the user, teacher, or expert to label the unlabeled data. And then, supervised learning is exploited. Since the learner can select the training examples, the number of examples to learn a function can often be smaller than the number needed in common supervised learning. However, there is a risk for active learning. Unimportant or even invalid examples may be chosed. The basic experimental design types for active learning include A-optimal design which minimizes the trace of the matrix, D-optimal design which minimizes the log-determinant of the matrix, and E-optimal design which minimizes the maximum eigenvalue of the matrix. All these design problems can be solved by convex optimization. Active learning based on locally linear reconstruction is presented in [722] where local structure of the data space is taken into account.

7.7 Reinforcement Learning

Reinforcement learning [723–725] is a very useful and fruitful area of machine learning. Reinforcement learning tries to learn how to act in response to an observation of the world in order to maximize some kind of cumulative reward. Every action taken has some influence on the environment. The environment will give its feedback through rewards to the learner. This feedback can guide the learner to make the decision for the next action. Reinforcement learning is widely studied in control theory, operation research, information theory, economics, and so on. Many algorithms for reinforcement learning are highly related to dynamic programming [726, 727]. Reinforcement learning is a dynamic and life-long learning with focus on the online performance. Thus, there is a trade-off between exploration and exploitation in reinforcement learning [728, 729]. The basic components in reinforcement learning should include environment states, possible actions, possible observations, transitions between states, and rewards.

Reinforcement learning is widely used in cognitive radio network for exploration and exploitation [730–746]. Three learning strategies will be presented in detail:

- Q-learning;
- Markov decision process;
- partially observable Markov decision process.

7.7.1 Q-Learning

Q-learning is a simple but useful reinforcement learning technique [747, 748]. Q-learning learns a utility function of a given action in a given state. Q-learning follows a fixed state transition and does not require the environment information.

Given the current state s_t and the action a_t, the utility function $Q(s_t, a_t)$ is learned or updated as

$$Q(s_t, a_t) = Q_{old}(s_t, a_t) + \alpha_t(s_t, a_t)\left(r(s_t, a_t) + \gamma \max_{a_{t+1}} Q(s_{t+1}, a_{t+1}) - Q_{old}(s_t, a_t)\right),$$
(7.49)

where $\alpha_t(s_t, a_t) \in (0, 1]$ is the learning rate; $r(s_t, a_t)$ is the immediate reward; $\gamma \in [0, 1)$ is the discount factor; s_{t+1} is the next state due to the state transition from the current state s_t by the action a_t. If α_t is equal to 1 for all the states and all the actions, Equation (7.49) can be reduced to

$$Q(s_t, a_t) = r(s_t, a_t) + \gamma \max_{a_{t+1}} Q(s_{t+1}, a_{t+1}).$$
(7.50)

Finally, the utility function can be learned through iteration. For each state, the selected action should be

$$\pi(s) = \arg\max_a Q(s, a).$$
(7.51)

Q-learning and its variants are widely used in cognitive radio network [734, 749–762].

7.7.2 Markov Decision Process

MDP [763] is a mathematical framework for studying the decision-making problem. MDP can be treated as an extension of Markov chain. Mathematically speaking, a Markov chain is a sequence of random variables $X_1, X_2, X_3, \ldots, X_t, \ldots$ with the Markov property, that is, the memoryless property of a stochastic process,

$$
\Pr(X_{t+1} = x \mid X_t = x_t, X_{t-1} = x_{t-1}, \ldots, X_2 = x_2, X_1 = x_1)
$$
$$
= \Pr(X_{t+1} = x \mid X_t = x_t),
$$

(7.52)

which means the following states and the previous states are independent given the current state.

An MDP consists of [723]:

- a set of states S;
- a set of actions A;
- a reward function $R(s, a)$;
- a state transition function $T(s, a, s') = \Pr(s_{t+1} = s' \mid s_t = s, a_t = a)$.

The goal of MDP is to find a policy $a = \pi(s)$ for the decision maker. When the police is fixed, MDP behaves like a Markov chain. Typically, the optimization problem of MDP can be formulated as

$$
\text{maximize} \sum_{t=0}^{\infty} \gamma^t R(s_t, \pi(s_t)).
$$

(7.53)

There are three basic methods to solve MDP:

- value Iteration;
- policy Iteration;
- linear programming [764–767].

For value iteration and policy iteration, the optimal value function is defined as [723],

$$
V^*(s) = \max_a \left(R(s, a) + \gamma \sum_{s' \in S} T(s, a, s') V^*(s') \right), \forall s \in S
$$

(7.54)

and given the optimal value function, the optimal policy can be obtained as [723],

$$
\pi(s) = \arg\max_a \left(R(s, a) + \gamma \sum_{s' \in S} T(s, a, s') V^*(s') \right).
$$

(7.55)

Value iteration tries to find the optimal value function and then obtain the optimal policy. The core part of value iteration is [723]:

1. Initialize $V(s)$ arbitrarily.
2. Let $V'(s)$ be equal to $V(s)$.

3. For $\forall s \in S$, calculate

$$U(s, a) = R(s, a) + \gamma \sum_{s' \in S} T(s, a, s') V'(s') \tag{7.56}$$

and

$$V(s) = \max_a U(s, a). \tag{7.57}$$

4. If $\max_s |V'(s) - V(s)|$ is less than the pre-defined threshold, the optimal value function $V(s)$ is obtained; otherwise go to step 2.

Policy iteration updates the policy directly. The core part of policy iteration is [723]:

1. Initialize $\pi(s)$ arbitrarily.
2. Let $\pi'(s)$ be equal to $\pi(s)$.
3. Solve the linear equations,

$$V(s) = R(s, \pi'(s)) + \gamma \sum_{s' \in S} T(s, \pi'(s), s') V(s') \tag{7.58}$$

and improve the policy,

$$\pi(s) = \arg \max_a \left(R(s, a) + \gamma \sum_{s' \in S} T(s, a, s') V(s') \right). \tag{7.59}$$

4. If $\pi'(s)$ is the same as $\pi(s)$, then the optimal policy is obtained; otherwise go to step 2.

MDP and its variants can be exploited in cognitive radio network [736, 737, 740, 768–782].

7.7.3 Partially Observable MDPs

POMDP is an extension of MDP. The system dynamics are modeled by MDP. However, the underlying state cannot be fully observed. POMDP models the interaction procedure of an agent with the outside world [783]. An agent first observes the outside world, then it tries to estimate the belief state using the current observation. The solution of POMDP is the optimal policy for choosing actions.

An POMDP consists of

- a set of states S;
- a set of actions A;
- a set of observations O;
- a reward function $R(s, a)$;
- a state transition function $T(s, a, s') = \Pr(s_{t+1} = s' \mid s_t = s, a_t = a)$;
- an observation function $\Omega(o, s', a) = \Pr(o_{t+1} = o \mid s_{t+1} = s', a_t = a)$.

Define a belief state over the states as

$$\mathbf{b}_t = \begin{bmatrix} b_t(s_1) \\ b_t(s_2) \\ \vdots \end{bmatrix}, \tag{7.60}$$

where $b_t(s) \geq 0, \forall s \in S$ and $\sum_{s \in S} b(s) = 1$. There are uncountably infinite number of belief states.

Given $\mathbf{b}_t$ and a_t, if $o \in O$ is observed with probability $\Omega(o, s', a)$, $\mathbf{b}_{t+1}$ can be obtained as [784]

$$b_{t+1}(s') = \frac{\Omega(o, s', a) \sum_{s \in S} T(s, a, s') b_t(s)}{\Pr(o \mid a, \mathbf{b}_t)}, \tag{7.61}$$

where

$$\Pr(o \mid a, \mathbf{b}_t) = \sum_{s' \in S} \Omega(o, s', a) \sum_{s \in S} T(s, a, s') b_t(s). \tag{7.62}$$

Define the belief state transition function as [784]

$$\tau(\mathbf{b}, a, \mathbf{b}') = \Pr(\mathbf{b}' \mid \mathbf{b}, a) \tag{7.63}$$

and $\tau(\mathbf{b}, a, \mathbf{b}') = \Pr(o \mid a, \mathbf{b})$ if $\mathbf{b}'$, $\mathbf{b}$, a, and o follows Equation (7.61); otherwise $\tau(\mathbf{b}, a, \mathbf{b}')$ is equal to zero. Thus, POMDP can be treated as infinite state MDP with [784, 785]

- a set of belief states B;
- a set of actions A;
- a belief state transition function shown in Equation (7.63);
- a reward function $\rho(\mathbf{b}, a) = \sum_{s \in S} b(s) R(s, a)$.

Solving a POMDP is not easy. The first detailed algorithms for finding exact solutions of POMDP were introduced in [786]. There exist some software tools for solving POMDP, such as pomdp-solve [787], MADP [788], ZMDP [789], APPL [790], and Perseus [791]. Among them, APPL is the fastest one in most cases [790].

POMDP and its variants are widely used in cognitive radio network [792–818].

7.8 Kernel-Based Learning

Kernel-based learning [819] is the great extension of machine learning by different kernel functions. Kernel SVM [683, 684], kernel PCA [820–823], and kernel Fisher discriminant analysis [824, 825] are widely used. Kernel functions can implicitly map the data from original low-dimensional linear space $\mathbf{x}$ to high-dimensional feature nonlinear space $\Phi(\mathbf{x})$.

Kernel function $K(\mathbf{x}, \mathbf{y})$ is defined as the inner product of $\Phi(\mathbf{x})$ and $\Phi(\mathbf{y})$. If we know the analytic expression of kernel function and we only care about the inner product of $\Phi(\mathbf{x})$ and $\Phi(\mathbf{y})$, then we do not need to know the mapping nonlinear function Φ explicitly. This is called the kernel trick. The commonly used kernel functions are:

- Gaussian kernels: $K(\mathbf{x}, \mathbf{y}) = \exp\left(-\frac{\|\mathbf{x}-\mathbf{y}\|}{2\sigma^2}\right)$;
- homogeneous polynomial kernels: $K(\mathbf{x}, \mathbf{y}) = (\mathbf{x} \cdot \mathbf{y})^d$;
- inhomogeneous polynomial kernels: $K(\mathbf{x}, \mathbf{y}) = (\mathbf{x} \cdot \mathbf{y} + 1)^d$;
- sigmoid kernels: $K(\mathbf{x}, \mathbf{y}) = \tanh(a\mathbf{x} \cdot \mathbf{y} + b)$.

Gaussian kernels, polynomial kernels, and sigmoid kernels are all data independent. Given kernel functions and training data, we can get kernel matrix. However, kernel matrix can also be learned and optimized from data [388, 826–828]. In [829], Bregman matrix divergences are used to learn the low-rank kernel matrix.

The brilliance of the optimization problem (7.37) and Equation (7.40) is that the inner product between inputs is used. By applying the kernel trick, linear two-class SVM can be easily extended to the nonlinear kernel SVM. In the feature space [683, 684],

$$\mathbf{w} = \sum_{i=1}^{M} \alpha_i l_i \Phi(\mathbf{x}_i) \tag{7.64}$$

and

$$\mathbf{w} \cdot \Phi(\mathbf{x}) = \sum_{i=1}^{M} \alpha_i l_i \langle \Phi(\mathbf{x}_i), \Phi(\mathbf{x}) \rangle = \sum_{i=1}^{M} \alpha_i l_i K(\mathbf{x}_i, \mathbf{x}). \tag{7.65}$$

Thus, a classifier based on kernel SVM can be written as

$$f(\mathbf{x}) = \text{sign}\left(\sum_{i=1}^{M} \alpha_i l_i K(\mathbf{x}_i, \mathbf{x}) - b\right). \tag{7.66}$$

Besides, kernel principal angles are explored for machine learning related tasks [830, 831]. The principal angles, also called canonical angles, give information about the relative position of two subspaces of a Euclidean space [832–835].

7.9 Dimensionality Reduction

In large scale cognitive radio networks, there is a significant amount of data. However, in practice, the data is highly correlated. This redundancy in the data increases the overhead of cognitive radio networks for data transmission and data processing. In addition, the number of degrees of freedom (DoF) in large scale cognitive radio networks is limited. The DoF of a K user $M \times N$ MIMO interference channel has been discussed in [836]. The total number of DoF is equal to $\min(M, N) * K$ if $K \leq R$, and $\min(M, N) * \frac{R}{R+1} * K$ if $K > R$, where $R = \frac{\max(M,N)}{\min(M,N)}$. This is achieved based on interference alignment [541, 837, 838]. Theoretical analysis about DoF in cognitive radio has been presented in [839, 840]. The DoF corresponds to the key variables or key features in the network. Processing the high-dimensional data instead of the key variables will not enhance the performance of the network. In some cases, this could even degrade the performance. Hence, compact representations of the data using dimensionality reduction is critical in cognitive radio networks.

Due to the curse of dimensionality and the inherent correlation of data, dimensionality reduction [841] is very important for machine learning. Meanwhile, machine learning provides the powerful tools for dimensionality reduction. Dimensionality reduction tries to reduce the number of random variables or equivalently the dimension of the data under consideration. Dimensionality reduction can be divided into feature selection and feature extraction [842]. Feature selection tries to find a subset of the original variables or features. Feature extraction transforms the data from the high-dimensional space to low-dimensional space. PCA is a widely used linear transformation for feature extraction. However, there are also many powerful nonlinear dimensionality reduction techniques.

Many nonlinear dimensionality reduction methods are related to manifold learning algorithms [843–846]. The data set most likely lies along a low-dimensional manifold embedded in a high-dimensional space [847]. Manifold learning attempts to uncover the underlying manifold structure in a data set. These methods include:

- kernel principal component analysis [820–823];
- multidimensional scaling [848–850];
- isomap [843, 851–853];
- locally-linear embedding [854–856];
- Laplacian eigenmaps [857, 858];
- diffusion maps [859, 860];
- maximum variance unfolding or semidefinite embedding [861–864].

7.9.1 Kernel Principal Component Analysis

Kernel PCA is a kernel-based machine learning algorithm. It uses the kernel function to implicitly map the original data to a feature space, where PCA can be applied. Assuming the original dimensionality data are a set of M samples $\mathbf{x}_i \in \mathbf{R}^N$, $i = 1, 2, \ldots, M$, the reduced dimensionality samples of $\mathbf{x}_i$ are $\mathbf{y}_i \in \mathbf{R}^K$, $i = 1, 2, \ldots, M$, where $K \ll N$. x_{ij} and y_{ij} are componentwise elements in $\mathbf{x}_i$ and $\mathbf{y}_i$, respectively.

Kernel PCA uses the kernel function

$$K(\mathbf{x}_i, \mathbf{x}_j) = \varphi(\mathbf{x}_i) \cdot \varphi(\mathbf{x}_j) \tag{7.67}$$

to implicitly map the original data into a feature space $\mathbf{F}$, where φ is the mapping from original space to feature space and $\cdot$ represents inner product. In $\mathbf{F}$, PCA algorithm can work well.

A function is a valid kernel if there exists a mapping φ satisfying Equation (7.67). Mercer's condition [683] gives us the condition about what kind of functions are valid kernels.

If $K(\cdot, \cdot)$ is a valid kernel function, the matrix

$$\mathbf{K} = \begin{bmatrix} K(\mathbf{x}_1, \mathbf{x}_1) & K(\mathbf{x}_1, \mathbf{x}_2) & \cdots & K(\mathbf{x}_1, \mathbf{x}_M) \\ K(\mathbf{x}_2, \mathbf{x}_1) & K(\mathbf{x}_2, \mathbf{x}_2) & \cdots & K(\mathbf{x}_2, \mathbf{x}_M) \\ \vdots & \vdots & \vdots & \vdots \\ K(\mathbf{x}_M, \mathbf{x}_1) & K(\mathbf{x}_M, \mathbf{x}_2) & \cdots & K(\mathbf{x}_M, \mathbf{x}_M) \end{bmatrix} \tag{7.68}$$

must be positive semidefinite [865]. The matrix $\mathbf{K}$ is the so-called kernel matrix.

Assuming the mean of feature space data $\varphi(\mathbf{x}_i), i = 1, 2, \ldots, M$ is zero, that is,

$$\frac{1}{M} \sum_{i=1}^{M} \varphi(\mathbf{x}_i) = 0. \tag{7.69}$$

The covariance matrix in $\mathbf{F}$ is

$$\mathbf{C}_F = \frac{1}{M} \sum_{i=1}^{M} \varphi(\mathbf{x}_i)\varphi(\mathbf{x}_i)^T. \tag{7.70}$$

In order to apply PCA in $\mathbf{F}$, the eigenvectors $\mathbf{v}_i^F$ of $\mathbf{C}_F$ are needed. As we know that the mapping φ is not explicitly known, thus the eigenvectors of $\mathbf{C}_F$ can not be as easily derived as PCA. However, the eigenvectors $\mathbf{v}_i^F$ of $\mathbf{C}_F$ must lie in the span [86] of $\varphi(\mathbf{x}_j), j = 1, 2, \ldots, M$, that is,

$$\mathbf{v}_i^F = \sum_{j=1}^{M} \alpha_{ij} \varphi(\mathbf{x}_j). \tag{7.71}$$

It has been proved that $\boldsymbol{\alpha}_i, i = 1, 2, \ldots, M$ are eigenvectors of kernel matrix $\mathbf{K}$ [86]. In which α_{ij} are component-wise elements of $\boldsymbol{\alpha}_i$.

Then the procedure of kernel PCA can be summarized in the following six steps:

1. Choose a kernel function $K(\cdot, \cdot)$.
2. Compute kernel matrix $\mathbf{K}$ based on Equation (7.67).
3. Obtain the eigenvalues $\lambda_1^{\mathbf{K}} \geq \lambda_2^{\mathbf{K}} \geq \cdots \geq \lambda_M^{\mathbf{K}}$ and the corresponding eigenvectors $\boldsymbol{\alpha}_1, \boldsymbol{\alpha}_2, \ldots, \boldsymbol{\alpha}_M$ by diagonalizing $\mathbf{K}$.
4. Normalize $\mathbf{v}_j^F$ by [86]

$$\boldsymbol{\alpha}_j = \frac{\boldsymbol{\alpha}_j}{\sqrt{\lambda_j^{\mathbf{K}}}}. \tag{7.72}$$

5. Constitute the basis of a subspace in $\mathbf{F}$ from the normalized eigenvectors $\mathbf{v}_j^F, j = 1, 2, \ldots, K$.
6. Compute the projection of a training point $\mathbf{x}_i$ on $\mathbf{v}_j^F, j = 1, 2, \ldots, K$ by

$$y_{ij} = (\mathbf{v}_j^F, \varphi(\mathbf{x}_i)) = \sum_{n=1}^{M} \alpha_{jn} K(\mathbf{x}_n, \mathbf{x}_i) \tag{7.73}$$

in which the reduced dimensionality data in feature space corresponding to $\mathbf{x}_i$ is $\mathbf{y}_i = (y_{i1}, y_{i2}, \ldots, y_{iK})$.

So far the mean of $\varphi(\mathbf{x}_i), i = 1, 2, \ldots, M$ has been assumed to be zero. In fact, the zero mean data in the feature space are

$$\varphi(\mathbf{x}_i) - \frac{1}{M} \sum_{i=1}^{M} \varphi(\mathbf{x}_i). \tag{7.74}$$

The kernel matrix for this centering or zero mean data can be derived by [86]

$$\tilde{\mathbf{K}} = \mathbf{HKH} \tag{7.75}$$

in which $\mathbf{H} = \mathbf{I} - \frac{1}{N}\mathbf{1}\mathbf{1}^T$ is the so-called centering matrix, $\mathbf{I}$ is an identity matrix, $\mathbf{1}$ is all-one vector.

Kernel PCA can be used for noise reduction which is a nontrivial task. S. Mika and co-workers have proposed an iterative scheme on noise reduction for Gaussian kernels [821]. This method needs to rely on the nonlinear optimization. However, a distance-constraint based method has been proposed by J. Kwok and I. Tsang which just relies on linear algebra [823]. In order to apply kernel PCA for noise reduction, the pre-image $\tilde{\mathbf{x}}_i$ (in original space) of $\mathbf{y}_i$ (in feature space) is needed. The distance-constraint based method for noise reduction makes use of the distance relationship [866] found by Williams between original space and feature space for some specific kernels. It tries to find the distance between $\tilde{\mathbf{x}}_i$ and $\mathbf{x}_j$ once the distance between $\mathbf{y}_i$ and $\varphi(\mathbf{x}_j)$ is known. $d(\mathbf{x}_i, \mathbf{x}_j)$ is used to represent distance between two vectors $\mathbf{x}_i$ and $\mathbf{x}_j$.

It has been proved that the squared distance between $\mathbf{y}_i$ and $\varphi(\mathbf{x}_j)$ can be derived by [823]

$$d^2(\mathbf{y}_i, \varphi(\mathbf{x}_j))$$

$$= (\mathbf{k}_{\mathbf{x}_i} + \tfrac{1}{N}\mathbf{K}\mathbf{1} - 2\mathbf{k}_{\mathbf{x}_j})^T \mathbf{H}^T \mathbf{M} \mathbf{H}(\mathbf{k}_{\mathbf{x}_i} - \tfrac{1}{N}\mathbf{K}\mathbf{1}) \tag{7.76}$$

$$+ \tfrac{1}{N^2}\mathbf{1}^T\mathbf{K}\mathbf{1} + \mathbf{K}_{ii} - \tfrac{2}{N}\mathbf{1}^T\mathbf{k}_{\mathbf{x}_j},$$

where $\mathbf{k}_{\mathbf{x}_i} = (K(\mathbf{x}_i, \mathbf{x}_1), K(\mathbf{x}_i, \mathbf{x}_2), \ldots, K(\mathbf{x}_i, \mathbf{x}_M))^T$ and $\mathbf{M} = \sum_{k=1}^{K} \frac{1}{\tilde{\lambda}_k}\tilde{\boldsymbol{\alpha}}_k\tilde{\boldsymbol{\alpha}}_k^T$ in which $\tilde{\lambda}_k$ and $\tilde{\boldsymbol{\alpha}}_k$ are the k-th largest eigenvalues and corresponding column eigenvectors of $\tilde{\mathbf{K}}$.

By making use of the distance relationship [866] between original space and feature space, if the kernel is the radial basis kernel, then

$$d^2(\tilde{\mathbf{x}}_i, \mathbf{x}_j) = -\tfrac{1}{\gamma} \log(0.5(\mathbf{K}_{ii} + \mathbf{K}_{jj} - d^2(\mathbf{y}_i, \varphi(\mathbf{x}_j)))). \tag{7.77}$$

Once the above distances are derived, $\tilde{\mathbf{x}}_i$ can be reconstructed [823].

7.9.2 Multidimensional Scaling

Multidimensional scaling (MDS) is a set of data analysis techniques used to explore the structure of similarities or dissimilarities in data [867]. The high-dimensional data can be displayed in 2-D or 3-D visualization.

Given a set of the high-dimensional data $\{\mathbf{x}_1, \mathbf{x}_2, \ldots, \mathbf{x}_M\}$, the distance between $\mathbf{x}_i$ and $\mathbf{x}_j$ is δ_{ij}. Arbitrary distance function can be used to define the similarity between $\mathbf{x}_i$ and $\mathbf{x}_j$. Take Euclidean distance as an example, the goal of MDS is to find a set of the low-dimensional data $\{\mathbf{y}_1, \mathbf{y}_2, \ldots, \mathbf{y}_M\}$ such that

$$\|\mathbf{y}_i - \mathbf{y}_j\| \approx \delta_{ij} \tag{7.78}$$

for all $i = 1, 2, \ldots, M$ and $j = 1, 2, \ldots, M$. The low dimensional embedding can preserve pairwise distances. Thus, MDS can be expressed as an optimization problem [848–850]

$$\text{minimize} \sum_{i<j} (\|\mathbf{y}_i - \mathbf{y}_j\| - \delta_{ij})^2, \tag{7.79}$$

where the sum of the squared differences between ideal distances in the original space and actual distances in the low-dimensional space is used as the cost function. Stress majorization can be used as a solver. It is well known that classical MDS is equivalent to PCA when Euclidean distance is used for some particularly selected cost functions [843] which simplifies the algorithm for MDS.

Local MDS is a technique for the nonlinear dimensionality reduction [850, 868]. MDS is executed locally instead of globally. Mathematically speaking, the optimization problem of local MDS can be expressed as

$$
\text{minimize} \quad \sum_{(i,j)\in\Omega}(\|\mathbf{y}_i - \mathbf{y}_j\| - \delta_{ij})^2 + \sum_{(i,j)\notin\Omega} w(\|\mathbf{y}_i - \mathbf{y}_j\| - \delta_\infty)^2, \quad (7.80)
$$

where Ω is a symmetric set of nearby pairs (i, j) which describes the local fabric of a high-dimensional manifold [868]; δ_∞ is a very large value of dissimilarity and w is a small weight. If δ_∞ goes to infinity and $w = \frac{t}{2\delta_\infty}$, the optimization problem (7.80) can be reduced to [868]

$$
\text{minimize} \quad \sum_{(i,j)\in\Omega}(\|\mathbf{y}_i - \mathbf{y}_j\| - \delta_{ij})^2 - t \sum_{(i,j)\notin\Omega} \|\mathbf{y}_i - \mathbf{y}_j\|, \quad (7.81)
$$

where the first term forces $\|\mathbf{y}_i - \mathbf{y}_j\|$ to approach δ_{ij} locally and the second term pushes nonlocal data far away from each other [868].

7.9.3 Isomap

Isomap is classical MDS where small pairwise distances between neighboring data are preserved while large pairwise distances between faraway data are replaced by geodesic distances which can be estimated by computing the shortest path distances along the neighborhood graph [843, 868, 869]. There are three steps to perform Isomap [843]. The first step is to construct neighborhood graph. The neighborhood graph can be determined by ε-neighborhoods or k-nearest neighbors. The second step is to compute shortest paths to estimate geodesic distances. Floyd-Warshall algorithm can be applied. The third step is to apply classical MDS to the matrix of graph distances and obtain the low-dimensional embedding.

7.9.4 Locally-Linear Embedding

Locally linear embedding (LLE) tries to discover low-dimensional, neighborhood preserving embedding of the high-dimensional data by using a locally linear approximation of the data manifold. Hence, data can be represented as the linear combinations of their neighbors. The first step for LLE is to calculate the weight matrix $\mathbf{W}$ based on the following optimization problem [854],

$$
\begin{aligned}
&\text{minimize} \\
&\sum_i (\mathbf{x}_i - \sum_j (\mathbf{W})_{i,j}\mathbf{x}_j)^2 \\
&\text{subject to} \\
&\sum_j (\mathbf{W})_{i,j} = 1.
\end{aligned} \quad (7.82)
$$

where $\mathbf{x}_i$ can only be reconstructed from its neighbors [854]. Hence, $(\mathbf{W})_{i,j}$ will be equal to zero if $\mathbf{x}_i$ and $\mathbf{x}_j$ are not neighbors.

The second step of LLE is to perform dimensionality reduction by solving the optimization problem shown below [854]:

$$\text{minimize}$$
$$\sum_i (\mathbf{y}_i - \sum_j (\mathbf{W})_{i,j} \mathbf{y}_j)^2, \tag{7.83}$$

where $\mathbf{W}$ is the solution to the optimization problem (7.82). Meanwhile, the local affine structure is preserved.

7.9.5 Laplacian Eigenmaps

Laplacian eigenmaps use the notion of the Laplacian of the graph to compute a low-dimensional representation of the high-dimensional data that can optimally preserve local neighborhood information [857]. Similar to LLE, the first step of Laplacian eigenmaps is to construct the neighborhood graph. The second step is to get weight matrix based on the neighborhood graph. If $\mathbf{x}_i$ and $\mathbf{x}_j$ are neighbors, then $(\mathbf{W})_{i,j} = 1$ and $(\mathbf{W})_{j,i} = 1$; otherwise $(\mathbf{W})_{i,j} = 0$. Thus, $\mathbf{W}$ is the symmetric matrix. The third step is to perform dimensionality reduction by computing eigenvalues and eigenvectors for the generalized eigen-decomposition problem [857],

$$\mathbf{Lu} = \lambda \mathbf{Du}, \tag{7.84}$$

where $\mathbf{D}$ is a diagonal matrix and $(\mathbf{D})_{i,i} = \sum_j (\mathbf{D})_{i,j}$; $\mathbf{L} = \mathbf{D} - \mathbf{W}$ is the Laplacian matrix which is a positive semidefinite matrix. The embedding of the low-dimensional data is given by the eigenvectors corresponding to the smallest nonzero eigenvalues.

7.9.6 Semidefinite Embedding

Within the framework of manifold learning, the current trend is to learn the kernel using SDP [8, 388] instead of defining a fixed kernel. The most prominent example of such a technique is semidefinite embedding (SDE) or maximum variance unfolding (MVU) [861]. MVU can learn the inner product matrix of $\mathbf{y}_i$ automatically by maximizing their variance, subject to the constraints that $\mathbf{y}_i$ are centered, and local distances of $\mathbf{y}_i$ are equal to the local distances of $\mathbf{x}_i$. Here, the local distances represent the distances between $\mathbf{y}_i$ ($\mathbf{x}_i$) and its k-nearest neighbors, in which k is a parameter.

The intuitive explanation of this approach is that when an object such as string is unfolded optimally, the Euclidean distances between its two ends must be maximized. Thus, the optimization objective function can be written as [861–864]

$$\text{maximize} \sum_{ij} \|\mathbf{y}_i - \mathbf{y}_j\|^2, \tag{7.85}$$

subject to the constraints,

$$\sum_i \mathbf{y}_i = 0$$
$$\|\mathbf{y}_i - \mathbf{y}_j\|^2 = \|\mathbf{x}_i - \mathbf{x}_j\|^2 \text{ when } \eta_{ij} = 1 \tag{7.86}$$

in which $\eta_{ij} = 1$ means $\mathbf{x}_i$ and $\mathbf{x}_j$ are k-nearest neighbors otherwise $\eta_{ij} = 0$.

Apply inner product matrix

$$\mathbf{I} = (\mathbf{y}_i \cdot \mathbf{y}_j)_{i,j=1}^{M} \tag{7.87}$$

of $\mathbf{y}_i$ to the above optimization can make the model simpler. The procedure of MVU can be summarized as follows:

1. Optimization step: because $\mathbf{I}$ is an inner product matrix, it must be positive semidefinite. Thus the above optimization can be reformulated into the following form [861]

$$
\begin{aligned}
&\text{maximize} \\
&\text{trace}(\mathbf{I}) \\
&\text{subject to} \\
&\mathbf{I} \succ 0 \\
&\sum_{ij} \mathbf{I}_{ij} = 0 \\
&\mathbf{I}_{ii} - 2\mathbf{I}_{ij} + \mathbf{I}_{jj} = D_{ij}, \text{when } \eta_{ij} = 1
\end{aligned}
\tag{7.88}
$$

where $D_{ij} = \|\mathbf{x}_i - \mathbf{x}_j\|^2$, and $\mathbf{I} \succ 0$ represents $\mathbf{I}$ is positive semidefinite.
2. The eigenvalues $\lambda_1^y \geq \lambda_2^y \geq \cdots \geq \lambda_M^y$ and the corresponding eigenvectors $\mathbf{v}_1^y, \mathbf{v}_2^y, \ldots,$ $\mathbf{v}_M^y$ are obtained by diagonalizing $\mathbf{I}$.
3. Dimensionality reduction by

$$y_{ij} = \sqrt{\lambda_j^y} v_{ij}^y \tag{7.89}$$

in which v_{ij}^y are componentwise elements of $\mathbf{v}_i^y$.

Landmark-MVU (LMVU) [870] is a modified version of MVU which aims at solving larger-scale problems than MVU. It works by using the inner product matrix $\mathbf{A}$ of randomly chosen landmarks from $\mathbf{x}_i$ to approximate the full matrix $\mathbf{I}$, in which the size of $\mathbf{A}$ is much smaller than $\mathbf{I}$.

Assuming the number of landmarks is m which are $\mathbf{a}_1, \mathbf{a}_2, \ldots, \mathbf{a}_m$, respectively. Let $\mathbf{Q}$ [870] denote a linear transformation between landmarks and original dimensional data $\mathbf{x}_i \in \mathbf{R}^N, i = 1, 2, \ldots, M$, accordingly,

$$
\begin{pmatrix} \mathbf{x}_1 \\ \mathbf{x}_2 \\ \vdots \\ \mathbf{x}_M \end{pmatrix} \approx \mathbf{Q} \cdot \begin{pmatrix} \mathbf{a}_1 \\ \mathbf{a}_2 \\ \vdots \\ \mathbf{a}_m \end{pmatrix}
\tag{7.90}
$$

in which

$$\mathbf{x}_i \approx \sum_j \mathbf{Q}_{ij} \mathbf{a}_j. \tag{7.91}$$

Assuming the reduced dimensionality landmarks of $\mathbf{a}_1, \mathbf{a}_2, \ldots, \mathbf{a}_m$ are $\tilde{\mathbf{y}}_1, \tilde{\mathbf{y}}_2, \ldots, \tilde{\mathbf{y}}_m$, and the reduced dimensionality samples of $\mathbf{x}_1, \mathbf{x}_2, \ldots, \mathbf{x}_M$ are $\mathbf{y}_1, \mathbf{y}_2, \ldots, \mathbf{y}_M$, then the linear transformation between $\mathbf{y}_1, \mathbf{y}_2, \ldots, \mathbf{y}_M$ and $\tilde{\mathbf{y}}_1, \tilde{\mathbf{y}}_2, \ldots, \tilde{\mathbf{y}}_m$ is $\mathbf{Q}$ as well [870], consequently,

$$\begin{pmatrix} \mathbf{y}_1 \\ \mathbf{y}_2 \\ \vdots \\ \mathbf{y}_M \end{pmatrix} \approx \mathbf{Q} \cdot \begin{pmatrix} \tilde{\mathbf{y}}_1 \\ \tilde{\mathbf{y}}_2 \\ \vdots \\ \tilde{\mathbf{y}}_m \end{pmatrix}. \tag{7.92}$$

Matrix $\mathbf{A}$ is the inner-product matrix of $\mathbf{a}_1, \mathbf{a}_2, \ldots, \mathbf{a}_m$,

$$\mathbf{A} = (\tilde{\mathbf{y}}_i \cdot \tilde{\mathbf{y}}_j)_{i,j=1}^m. \tag{7.93}$$

Hence the relationship between $\mathbf{I}$ and $\mathbf{A}$ is

$$\mathbf{I} \approx \mathbf{Q}\mathbf{A}\mathbf{Q}^T. \tag{7.94}$$

The optimization problem (7.88) can be reformulated into the following form:

$$\begin{aligned} &\text{maximize} \\ &\text{trace}\,(\mathbf{Q}\mathbf{A}\mathbf{Q}^T) \\ &\text{subject to} \\ &\mathbf{A} \succ 0 \\ &\textstyle\sum_{ij}(\mathbf{Q}\mathbf{A}\mathbf{Q}^T)_{ij} = 0 \\ &D_{ij}^{\mathbf{y}} \le D_{ij}, \text{ when } \eta_{ij} = 1, \end{aligned} \tag{7.95}$$

where

$$D_{ij} = \|\mathbf{x}_i - \mathbf{x}_j\|^2 \tag{7.96}$$

$$D_{ij}^{\mathbf{y}} = (\mathbf{Q}\mathbf{A}\mathbf{Q}^T)_{ii} - 2(\mathbf{Q}\mathbf{A}\mathbf{Q}^T)_{ij} + (\mathbf{Q}\mathbf{A}\mathbf{Q}^T)_{jj} \tag{7.97}$$

and $\mathbf{A} \succ 0$ represents $\mathbf{A}$ is positive semidefinite. The optimization problem (7.95) differs from the optimization problem (7.88) in that equality constraints for nearby distances are relaxed to inequality constraints in order to guarantee the feasibility of the simplified optimization model.

LMVU can increase the speed of programming but at the cost of accuracy.

7.10 Ensemble Learning

Ensemble learning tries to use multiple models to obtain better predictive performance which means a target function is learned by training a finite set of individual learners and combining their predictions [871]. Ideally, if there are M models with uncorrelated errors, simply by averaging them the average error of a model can be reduced by a factor of M [872]. The common combination schemes include [873]:

- voting;
- sum, mean, median;
- generalized ensemble;

- adaptive weighting;
- stacking;
- borda count;
- logistic regression;
- class set reduction;
- Dempster-Shafer;
- fuzzy integrals;
- mixture of local experts;
- hierarchical mixture of local experts;
- associative switch;
- bagging;
- boosting;
- random subspace;
- neural tree;
- error-correcting output codes [874].

Sometimes, the more general concept than ensemble learning is meta learning. Meta learning [875] tries to learn the interaction between the mechanism of learning and the concrete contexts in which that mechanism is applicable based on meta data [876].

7.11 Markov Chain Monte Carlo

MCMC methods [877–879] are a class of sampling algorithms. A Markov chain is constructed such that the equilibrium distribution of Markov chain is the same as the desired density of the sampled probability distribution.

The key point of Monte Carlo principle is to draw a set of i.i.d. samples $x_n, n = 1, 2, \ldots, N$ from the PDF $p(x)$ defined on a high-dimensional space [878]. These N samples can be exploited to approximate the PDF $p(x)$ as [878]

$$p_N(x) = \frac{1}{N} \sum_{n=1}^{N} \delta(x - x_n). \tag{7.98}$$

Based on Equation (7.98), Monte Carlo integration tries to compute integral using large randomly-generated numbers,

$$I_N(f) = \frac{1}{N} \sum_{n=1}^{N} f(x_n). \tag{7.99}$$

As $N \to \infty$, then,

$$I(f) = \int f(x) p(x) dx. \tag{7.100}$$

Suppose we want to calculate the integral $\int f(x) q(x) dx$. However, the samples from the PDF $q(x)$ are hard to generate. But $\frac{q(x)}{p(x)}$ is easy to evaluate. Thus,

$$\int f(x) q(x) dx \approx \frac{1}{N} \sum_{n=1}^{N} f(x_n) \left(\frac{q(x_n)}{p(x_n)} \right), \tag{7.101}$$

where x_n is drawn from the PDF $p(x)$. This method is called importance sampling.

The Metropolis-Hastings algorithm is one of the most popular MCMC methods [878]. In order to get samples from the PDF $p(x)$, the Metropolis-Hastings algorithm is performed as [878]:

1. Start with any initial sample x_0 such that $p(x_0) > 0$.
2. Sample u from the uniform distribution between 0 and 1.
3. Sample x^* from the proposal distribution $q(x_* \mid x_n)$.
4. Calculate

$$\alpha = \min\left\{1, \frac{p(x^*)q(x_n \mid x^*)}{p(x_n)q(x^* \mid x_n)}\right\}. \tag{7.102}$$

5. Accept x^* as $x_{n+1} = x^*$ if $u < \alpha$; otherwise, reject x^* as $x_{n+1} = x_n$.
6. Go to step 2.

The Metropolis-Hastings algorithm can be reduced to the Metropolis algorithm if the proposal distribution is symmetric, that is,

$$q(x^* \mid x_n) = q(x_n \mid x^*), n = 1, 2, \ldots \tag{7.103}$$

and the calculation of α in Equation (7.102) can be simplified as

$$\alpha = \min\left\{1, \frac{p(x^*)}{p(x_n)}\right\}. \tag{7.104}$$

MCMC can work for various algorithms and applications within the framework of machine learning [878, 880–898]. MCMC can be explored for various optimization problems, especially for large-scale or stochastic optimization [899–907].

Gibbs sampling is a special case of the Metropolis-Hastings algorithm. Gibbs sampling gets samples from the simple conditional distributions instead of the complex joint distributions. If the joint distribution of $\{X_1, X_2, \ldots, X_N\}$ is $p(x_1, x_2, \ldots, x_N)$, the k-th sample $\{x_1^{(k)}, x_2^{(k)}, \ldots, x_N^{(k)}\}$ can be obtained sequentially as follows [908]:

1. Initialize $\{X_1, X_2, \ldots, X_N\}$ as $\{x_1^{(0)}, x_2^{(0)}, \ldots, x_N^{(0)}\}$.
2. Sample $x_n^{(k)}$ from the conditional distribution

$$x_n^{(k)} \sim$$
$$\Pr(X_n = x_n \mid X_1 = x_1^{(k)}, \ldots, X_{n-1} = x_{n-1}^{(k)}, X_{n+1} = x_{n+1}^{(k-1)}, \ldots, X_N = x_N^{(k-1)}). \tag{7.105}$$

MCMC has also been applied to cognitive radio network [909–912].

7.12 Filtering Technique

Filtering is the common approach in signal processing. For communication or radar, filtering can be used to perform frequency band selection, interference suppression, noise reduction, and so on. In machine learning, the meaning of filtering is greatly extended. Kalman filtering and particle filtering are explored to deal with the sequential data, for example, time series data. Collaborative filtering are exploited to perform recommendations or predictions.

7.12.1 Kalman Filtering

Kalman filtering is a set of mathematical equations that provides an efficient computational and recursive strategy to estimate the state of a process such that the mean of squared error can be minimized [913, 914]. Kalman filtering is very popular in the areas of autonomous or assisted navigation and moving target tracking.

Let's start from the simple linear discrete Kalman filtering to understand how Kalman filtering works. There are two basic equations in Kalman filtering. One is the state equation to represent the state transition. The other is the measurement equation to obtain the observation. The linear state equation can be expressed as

$$\mathbf{x}_n = \mathbf{A}_n \mathbf{x}_{n-1} + \mathbf{B}_n \mathbf{u}_n + \mathbf{w}_n, \tag{7.106}$$

where

- $\mathbf{x}_n$ is the current of a process or a dynamic system and $\mathbf{x}_{n-1}$ is the previous state;
- $\mathbf{A}_n$ represents the current state transition model;
- $\mathbf{u}_n$ is the current system input;
- $\mathbf{B}_n$ is the current control model;
- $\mathbf{w}_n$ is the current state noise which follows a zero mean multivariate normal distribution with covariance $\mathbf{W}_n$.

The linear measurement equation is represented as

$$\mathbf{z}_n = \mathbf{H}_n \mathbf{x}_n + \mathbf{v}_n, \tag{7.107}$$

where

- $\mathbf{z}_n$ is the measurement of the current state $\mathbf{x}_n$;
- $\mathbf{H}_n$ is the current observation model;
- $\mathbf{v}_n$ is the current measurement noise which follows a zero mean multivariate normal distribution with covariance $\mathbf{V}_n$.

The goal of Kalman filtering is to estimate $\tilde{\mathbf{x}}_n, n = 1, 2, \ldots$ given $\mathbf{x}_0$. Meanwhile, $\mathbf{A}_n$, $\mathbf{B}_n$, $\mathbf{H}_n$, $\mathbf{W}_n$, and $\mathbf{V}_n$ are all known. State noises and measurement noises are all mutually independent.

There are two main steps in Kalman filtering [913]. One is the predict step and the other is update step. These two steps are performed iteratively.

The procedure of the predict step is [913]:

1. Predict a priori current state $\tilde{\mathbf{x}}_{n|n-1}$

$$\tilde{\mathbf{x}}_{n|n-1} = \mathbf{A}_n \tilde{\mathbf{x}}_{n-1|n-1} + \mathbf{B}_n \mathbf{u}_n. \tag{7.108}$$

2. Predict a priori current error covariance of state estimation

$$\mathbf{P}_{n|n-1} = \mathbf{A}_n \mathbf{P}_{n-1|n-1} \mathbf{A}_n^T + \mathbf{W}_n. \tag{7.109}$$

The procedure of the update step is [913]

1. Obtain the current measurement residual $\mathbf{r}_n$

$$\mathbf{r}_n = \mathbf{z}_n - \mathbf{H}_n \tilde{\mathbf{x}}_{n|n-1}. \tag{7.110}$$

2. Obtain the current residual covariance $\mathbf{R}_n$

$$\mathbf{R}_n = \mathbf{H}_n \mathbf{P}_{n|n-1} \mathbf{H}_n^T + \mathbf{V}_n. \tag{7.111}$$

3. Get the current gain of Kalman filtering $\mathbf{G}_n$

$$\mathbf{G}_n = \mathbf{P}_{n|n-1} \mathbf{H}_n^T \mathbf{R}_n^{-1}. \tag{7.112}$$

4. Update a posteriori current state $\tilde{\mathbf{x}}_{n|n}$ which can be treated as $\tilde{\mathbf{x}}_n$

$$\tilde{\mathbf{x}}_{n|n} = \tilde{\mathbf{x}}_{n|n-1} + \mathbf{G}_n \mathbf{r}_n. \tag{7.113}$$

5. Update a posteriori current error covariance of state estimation $\mathbf{P}_{n|n}$

$$\mathbf{P}_{n|n} = (\mathbf{I} - \mathbf{G}_n \mathbf{H}_n) \mathbf{P}_{n|n-1}. \tag{7.114}$$

Linear Kalman filtering can be extended to the extended Kalman filtering and the unscented Kalman filtering to deal with the general nonlinear dynamic system. In the nonlinear Kalman filtering, the state transition function and state measurement function can be the nonlinear functions shown as

$$\mathbf{x}_n = f(\mathbf{x}_{n-1}, \mathbf{u}_n) + \mathbf{w}_n \tag{7.115}$$

and

$$\mathbf{z}_n = h(\mathbf{x}_n) + \mathbf{v}_n. \tag{7.116}$$

If the nonlinear functions are differentiable, the extended Kalman filtering computes the Jacobian matrix to linearize the nonlinear functions [913]. The state transition model $\mathbf{A}_n$ can be represented as

$$\mathbf{A}_n = \frac{\partial f}{\partial \mathbf{x}} |_{\tilde{\mathbf{x}}_{n-1|n-1}, \mathbf{u}_n} \tag{7.117}$$

and the state observation model $\mathbf{H}_n$ can be represented as

$$\mathbf{H}_n = \frac{\partial h}{\partial \mathbf{x}} |_{\tilde{\mathbf{x}}_{n|n-1}}. \tag{7.118}$$

If the functions f and g are highly nonlinear and the state noise and measurement noise are involved in the nonlinear functions, the performance of the extended Kalman filtering is poor. We need to resort to the unscented Kalman filtering to handle this tough situation [915, 916].

The unscented transformation is the basis of the unscented Kalman filtering. The unscented transformation can calculate the statistics of a random variable which goes through a nonlinear function [915]. Given an L-dimensional random variable $\mathbf{x}$ with mean $\bar{\mathbf{x}}$ and covariance $\mathbf{C_x}$, we would like to calculate the statistics of $\mathbf{y}$ which satisfies $\mathbf{y} = f(\mathbf{x})$. Based on the unscented transformation, $2L + 1$ sigma vectors are sampled according to the following rule,

$$
\begin{aligned}
\mathbf{x}_0 &= \bar{\mathbf{x}} \\
\mathbf{x}_l &= \bar{\mathbf{x}} + (\sqrt{(L+\lambda)\mathbf{C_x}})_l, l = 1, 2, \ldots, L \\
\mathbf{x}_l &= \bar{\mathbf{x}} - (\sqrt{(L+\lambda)\mathbf{C_x}})_{l-L}, l = L+1, L+2, \ldots, 2L,
\end{aligned}
\tag{7.119}
$$

where λ is a scaling parameter and $(\sqrt{(L+\lambda)\mathbf{C_x}})_l$ is the l-th column of the matrix square root [916]. These sigma vectors go through the nonlinear function to get the samples of $\mathbf{y}$,

$$
\mathbf{y}_l = f(\mathbf{x}_l), l = 0, 1, 2, \ldots, 2L.
\tag{7.120}
$$

Thus, the mean and covariance of $\mathbf{y}$ can be approximated by the weighted sample mean and the weighted sample covariance [916],

$$
\bar{\mathbf{y}} \approx \sum_{l=0}^{2L} w_l^{(m)} \mathbf{y}_l
\tag{7.121}
$$

and

$$
\mathbf{C_y} \approx \sum_{l=0}^{2L} w_l^{(c)} (\mathbf{y}_l - \bar{\mathbf{y}})(\mathbf{y}_l - \bar{\mathbf{y}})^T,
\tag{7.122}
$$

where $w_l^{(m)}$ and $w_l^{(c)}$ are given deterministically [916].

The state transition function and the state measurement function for the unscented Kalman filtering can be written as [916]

$$
\mathbf{x}_n = f(\mathbf{x}_{n-1}, \mathbf{u}_{n-1}, \mathbf{w}_{n-1})
\tag{7.123}
$$

and

$$
\mathbf{z}_n = h(\mathbf{x}_n, \mathbf{v}_n).
\tag{7.124}
$$

The state vector for the unscented Kalman filtering is redefined as the concatenation of $\mathbf{x}_n$, $\mathbf{w}_n$, and $\mathbf{v}_n$ [916],

$$
\mathbf{x}_n^{UKF} = \begin{bmatrix} \mathbf{x}_n \\ \mathbf{w}_n \\ \mathbf{v}_{n+1} \end{bmatrix}.
\tag{7.125}
$$

There are three steps in the unscented Kalman filtering. The first step is to calculate sigma vectors. The second step is the predict step and the third is the update step. These three steps are performed iteratively. The whole procedure of the unscented Kalman filtering is [916],

1. Calculate sigma vectors

$$\mathbf{x}_{0,n-1}^{\text{UKF}} = \tilde{\mathbf{x}}_{n-1|n-1}^{\text{UKF}}$$

$$\mathbf{x}_{l,n-1}^{\text{UKF}} = \tilde{\mathbf{x}}_{n-1|n-1}^{\text{UKF}} + \left(\sqrt{(L+\lambda)\mathbf{P}_{n-1}^{\text{UKF}}} \right)_l, l = 1, 2, \ldots, L$$

$$\mathbf{x}_{l,n-1}^{\text{UKF}} = \tilde{\mathbf{x}}_{n-1|n-1}^{\text{UKF}} - \left(\sqrt{(L+\lambda)\mathbf{P}_{n-1}^{\text{UKF}}} \right)_{l-L}, l = L+1, L+2, \ldots, 2L, \tag{7.126}$$

where $\mathbf{P}_{n-1}^{\text{UKF}}$ is equal to

$$\mathbf{P}_{n-1}^{\text{UKF}} = \begin{bmatrix} \mathbf{P}_{n-1|n-1} & \mathbf{0} & \mathbf{0} \\ \mathbf{0} & \mathbf{W}_{n-1} & \mathbf{0} \\ \mathbf{0} & \mathbf{0} & \mathbf{V}_n \end{bmatrix}. \tag{7.127}$$

2. The predict step

$$\mathbf{x}_{l,n|n-1}^{\text{x}} = f(\mathbf{x}_{l,n-1}^{\text{x}}, \mathbf{u}_{n-1}, \mathbf{x}_{l,n-1}^{\text{w}}), l = 0, 1, 2, \ldots, 2L \tag{7.128}$$

$$\tilde{\mathbf{x}}_{n|n-1} = \sum_{l=0}^{2L} w_l^{(m)} \mathbf{x}_{l,n|n-1}^{\text{x}} \tag{7.129}$$

$$\mathbf{P}_{n|n-1} = \sum_{l=0}^{2L} w_l^{(c)} (\mathbf{x}_{l,n|n-1}^{\text{x}} - \tilde{\mathbf{x}}_{n|n-1})(\mathbf{x}_{l,n|n-1}^{\text{x}} - \tilde{\mathbf{x}}_{n|n-1})^T. \tag{7.130}$$

3. The update step

$$\mathbf{z}_{l,n|n-1} = h(\mathbf{x}_{l,n|n-1}^{\text{x}}, \mathbf{x}_{l,n|n-1}^{\text{v}}) \tag{7.131}$$

$$\tilde{\mathbf{z}}_{n|n-1} = \sum_{l=0}^{2L} w_l^{(m)} \mathbf{z}_{l,n|n-1} \tag{7.132}$$

$$\mathbf{r}_n = \mathbf{z}_n - \tilde{\mathbf{z}}_{n|n-1} \tag{7.133}$$

$$\mathbf{P}_{n|n-1}^{\text{z z}} = \sum_{l=0}^{2L} w_l^{(c)} (\mathbf{z}_{l,n|n-1} - \tilde{\mathbf{z}}_{n|n-1})(\mathbf{z}_{l,n|n-1} - \tilde{\mathbf{z}}_{n|n-1})^T \tag{7.134}$$

$$\mathbf{P}_{n|n-1}^{\text{x z}} = \sum_{l=0}^{2L} w_l^{(c)} (\mathbf{x}_{n|n-1}^{\text{x}} - \tilde{\mathbf{x}}_{n|n-1})(\mathbf{z}_{l,n|n-1} - \tilde{\mathbf{z}}_{n|n-1})^T \tag{7.135}$$

$$\mathbf{G}_n = \mathbf{P}_{n|n-1}^{\text{x z}}(\mathbf{P}_{n|n-1}^{\text{z z}})^{-1} \tag{7.136}$$

$$\tilde{\mathbf{x}}_{n|n} = \tilde{\mathbf{x}}_{n|n-1} + \mathbf{G}_n \mathbf{r}_n \qquad (7.137)$$

$$\mathbf{P}_{n|n} = \mathbf{P}_{n|n-1} - \mathbf{G}_n \mathbf{P}^{z\,z}_{n|n-1} \mathbf{G}_n^T. \qquad (7.138)$$

Kalman filtering and its variants have been used in cognitive radio network [917–927].

7.12.2 Particle Filtering

Particle filtering is also called sequential Monte Carlo method [878, 928, 929]. Particle filtering is the sophisticated model estimation technique based on simulation [928]. Particle filtering can perform the on-line approximation of probability distribution using a set of randomly chosen weighted samples or particles [878]. Similar to PSO, multiple particles are generated in particle filtering and these particles can evolve.

Similar to Kalman filtering, particle filtering also has an initial distribution, a dynamic state transition model, and the state measurement model [878]. Assume $\mathbf{x}_0, \mathbf{x}_1, \mathbf{x}_2, \ldots, \mathbf{x}_n, \ldots$ are the underlying or latent states and $\mathbf{y}_1, \mathbf{y}_2, \ldots, \mathbf{y}_n, \ldots$ are the corresponding measurements.

- The initial distribution of $\mathbf{x}$ is $p(\mathbf{x}_0)$.
- The dynamic state transition model is $p(\mathbf{x}_n \mid \mathbf{x}_{0:n-1}, \mathbf{y}_{1:n-1}), n = 1, 2, \ldots$.
- The state measurement model is $p(\mathbf{y}_n \mid \mathbf{x}_{0:n}, \mathbf{y}_{1:n-1}), n = 1, 2, \ldots$.

$\mathbf{x}_{0:n} = \{\mathbf{x}_0, \mathbf{x}_1, \ldots, \mathbf{x}_n\}$ and $\mathbf{y}_{1:n} = \{\mathbf{y}_1, \mathbf{y}_2, \ldots, \mathbf{y}_n\}$. Markov conditional independence can be used to simplify the models as $p(\mathbf{x}_n \mid \mathbf{x}_{0:n-1}, \mathbf{y}_{1:n-1}) = p(\mathbf{x}_n \mid \mathbf{x}_{n-1})$ and $p(\mathbf{y}_n \mid \mathbf{x}_{0:n}, \mathbf{y}_{1:n-1}) = p(\mathbf{y}_n \mid \mathbf{x}_n)$ [878].

The basic goal of particle filtering is to approximate the posterior $p(\mathbf{x}_{0:n} \mid \mathbf{y}_{1:n})$ as

$$p(\mathbf{x}_{0:n} \mid \mathbf{y}_{1:n}) \approx \sum_{l=1}^{L} w_{l,n} \delta(\mathbf{x}_{0:n} - \mathbf{x}_{l,0:n}), \qquad (7.139)$$

where L is the number of particles used in particle filtering; $\mathbf{x}_{l,0:n}$ is the l-th particle maintaining the whole trajectory; $w_{l,n}$ is the corresponding weight which should be normalized,

$$\sum_{l=1}^{L} w_{l,n} = 1, n = 1, 2, \ldots. \qquad (7.140)$$

Based on the concept of importance sampling, sequential importance sampling is used to generate the particle and the associated weight [930]. The importance function is chosen such that [930]

$$q(\mathbf{x}_{0:n} \mid \mathbf{y}_{1:n}) = q(\mathbf{x}_n \mid \mathbf{x}_{0:n-1}, \mathbf{y}_{1:n}) q(\mathbf{x}_{0:n-1} \mid \mathbf{y}_{1:n-1}). \qquad (7.141)$$

Given $\mathbf{x}_{l,0:n-1}, l = 1, 2, \ldots, L$ and $w_{l,n-1}, l = 1, 2, \ldots, L$, particle filtering updates the particle and the weight as follows [878, 930]:

1. Sample $\mathbf{x}_n$ as,

$$\mathbf{x}_{l,n} \sim q(\mathbf{x}_n \mid \mathbf{x}_{l,0:n-1}, \mathbf{y}_{1:n}), l = 1, 2, \ldots, L. \tag{7.142}$$

2. Augment the old particle $\mathbf{x}_{l,0:n-1}$ to the new particle $\mathbf{x}_{l,0:n}$ with $\mathbf{x}_{l,n}$.
3. Update the weight as,

$$w_{l,n} = w_{l,n-1} \frac{p(\mathbf{y}_n \mid \mathbf{x}_{l,n}) p(\mathbf{x}_{l,n} \mid \mathbf{x}_{l,n-1})}{q(\mathbf{x}_{l,n} \mid \mathbf{x}_{l,0:n-1}, \mathbf{y}_{1:n})}. \tag{7.143}$$

4. Normalize the weight as shown in Equation (7.140).

If the importance function is simply given by $p(\mathbf{x}_n \mid \mathbf{x}_{l,n-1})$, then the weight can be updated as [928, 930],

$$w_{l,n} = w_{l,n-1} p(\mathbf{y}_n \mid \mathbf{x}_{l,n}). \tag{7.144}$$

However, after a few iterations for sequential importance sampling, most of the particles have a very small weight. The particles fail to represent the probability distribution accurately [928]. In order to avoid this degeneracy problem, sequential importance resampling is exploited. Resampling is a method that gets rid of particles with small weights and replicates particles with large weights [930]. Meanwhile, the equal weight is assigned to each particle.

Particle filtering and its variants have been used in cognitive radio network [931–933].

7.12.3 Collaborative Filtering

Collaborative filtering [934] is the filtering process of information or pattern. Collaborative filtering deals with large scale data involving collaboration among multiple data sources. Collaborative filtering is a method to build the recommender system, which exploits the known preferences of some users to make recommendation or prediction of the unknown preferences for other users [935]. Item-to-item collaborative filtering is used by Amazon.com to match each of the user's purchased and rated items to the similar items which are combined into a recommendation list [936]. Netflix, an American provider of on-demand Internet streaming media, held an open competition for the best collaborative filtering algorithm. A large scale industrial dataset with 480,000 users and 17,770 movies was used for the competition [935].

The challenges of collaborative filtering are [935]:

- data sparsity;
- scalability;
- synonymy;
- gray sheep and black sheep;
- shilling attacks;
- personal privacy.

Matrix completion [643, 937] can be used to address the issue of data sparsity in collaborative filtering. The data matrix for collaborative filtering can be recovered even if this matrix is extremely sparse as long as the matrix is well approximated by a low-rank matrix [937, 938].

There are three categories for collaborative filtering algorithms [935]:

- memory-based collaborative filtering, for example, neighborhood-based collaborative filtering, top-N recommendation, and so on;
- model-based collaborative filtering, for example, Bayesian network collaborative filtering, clustering collaborative filtering, regression-based collaborative filtering, MDP-based collaborative filtering, latent semantic collaborative filtering, and so on;
- hybrid collaborative filtering.

Collaborative filtering can be explored for cognitive radio network [939–942].

7.13 Bayesian Network

Bayesian network [943, 944] is also called belief network or directed acyclic graphical model. Bayesian network explicitly uncovers the probabilistic structure of dependency in a set of random variables. It uses a directed acyclic graph to represent the dependency structure, in which each node denotes a random variable and each edge denotes the relation of dependency. Bayesian network can be extended to dynamic Bayesian network to model the sequential data or the dynamic system. The sequential data can be anywhere. Speech recognition, visual tracking, motion tracking, financial forecasting and prediction, and so on are the temporal sequential data [945].

The well-known hidden Markov model (HMM) can be treated as one simple dynamic Bayesian network. Meanwhile, the variants of HMM can be modeled as dynamic Bayesian networks. These variants include [945]:

- auto-regressive HMM;
- HMM with mixture-of-Gaussians output;
- input-output HMM;
- factorial HMM;
- coupled HMM;
- hierarchical HMM;
- mixtures of HMM;
- semi-Markov HMM;
- segment HMM.

State space model and its variants can also be modeled as dynamic Bayesian networks [945]. The basic state space model is also known as dynamic linear model or Kalman filter model [945].

Bayesian network is a powerful tool for learning and inference in cognitive radio network. Various applications of Bayesian network in cognitive radio network can be found in [536, 946–949].

7.14 Summary

In this chapter, machine learning has been presented. Machine learning can be applied everywhere to make the system intelligent. In order to give readers the whole picture of machine learning, almost all the topics related to machine learning have been covered in this chapter which include unsupervised learning, supervised learning, semi-supervised learning, transductive inference, transfer learning, active learning, reinforcement learning, kernel-based learning, dimensionality reduction, ensemble learning, meta learning, MCMC, Kalman filtering, particle filtering, collaborative filtering, Bayesian network, HMM, and so on. Meanwhile, the references about applications of machine learning in cognitive radio network have been given. Machine learning will be the basic tool to make cognitive radio network practical.

7.10 Summary

8

Agile Transmission Techniques (I): Multiple Input Multiple Output

MIMO [68, 950–954] in wireless communication tries to exploit multiple antennas at both the transmitter and the receiver to improve the performances of wireless communication without additional radio bandwidth. These performances can be spectral efficiency, data throughput, link range, link reliability, QoS of multiuser situation, and so on. MIMO is the core technology of modern wireless communication. MIMO is widely adopted as radio communication standards by IEEE 802.11, IEEE 802.16, and 3GPP LTE.

8.1 Benefits of MIMO

The benefits of MIMO can be generally summarized as three different gains:

- array gain;
- diversity gain;
- multiplexing gain.

8.1.1 Array Gain

Array gain means the average SNR increase at the receiver due to the signal coherent combination by using multiple-antennas at transmitter and/or receiver [953, 955]. Array gain can also be called power gain which can increase energy efficiency. In order to exploit array gain, channel knowledge or channel state information is required at both transmitter and receiver. Beamforming is the signal processing technique which brings array gain.

8.1.2 Diversity Gain

Diversity is used to combat fading in wireless communication [953, 956]. Fading will cause the signal power to drop significantly and degrade the communication perfor-

Cognitive Radio Communications and Networking: Principles and Practice, First Edition.
Robert C. Qiu, Zhen Hu, Husheng Li and Michael C. Wicks.
© 2012 John Wiley & Sons, Ltd. Published 2012 by John Wiley & Sons, Ltd.

mance [953]. Thus, multiple copies of the same signal can be transmitted through two or more different communication channels. Diversity gain can also increase SNR. The commonly used diversity schemes include:

- time diversity;
- frequency diversity;
- space diversity;
- polarization diversity;
- multi-user diversity.

In order to combine the signals from multiple communication channels at the receiver, diversity combining techniques are needed which include:

- selection combining;
- switched combining;
- equal-gain combining;
- maximal-ratio combining.

If there are multiple antennas at the transmitter, transmit diversity is applied. Transmit diversity can be extracted with or without channel knowledge at the transmitter [953].

We can also simply understand and differentiate array gain and diversity gain from the perspective of random process. For the superposition of several random processes, array gain can increase the mean and diversity gain can reduce the variance compared with the single random process.

8.1.3 Multiplexing Gain

Multiplexing gain refers to the increase in capacity due to the simultaneous transmission of different data streams on multiple spatial dimensions without additional power and radio bandwidth [957]. Multiplexing gain can be achieved with or without channel knowledge at the transmitter.

There is a fundamental tradeoff between diversity gain and multiplexing gain when MIMO is explored [958, 959].

8.2 Space Time Coding

Space time coding tries to improve the reliability or link quality of data transmission by using multiple antennas at the transmitter [960]. The two main types of space time coding are [960]:

- space time block coding (STBC);
- space time trellis coding (STTC).

The coding is performed jointly in both time and space domains. The transmitted signals in different time slots and from different antennas have some levels of correlation, which leads to information redundancy. However, in order to provide multiplexing gain, layered space time coding is explored. All these three space time codes require the receiver to have channel state information.

8.2.1 Space Time Block Coding

Space time block coding [961] is based on a linear code which is an error-correcting code. A space time block code can be represented as

$$
\begin{bmatrix}
x_{11} & x_{12} & \cdots & x_{1M} \\
x_{21} & x_{22} & \cdots & x_{2M} \\
\vdots & \vdots & \vdots & \vdots \\
x_{T1} & x_{T2} & \cdots & x_{TM}
\end{bmatrix},
\tag{8.1}
$$

where x_{tm} is the modulated symbol which will be transmitted in the t-th time slot from the m-th antenna. M is the number of antennas at the transmitter and T is the number of total time slots. If K symbols are encoded within T time slots by space time block code, then the code rate of space time block code is

$$
r = \frac{K}{T}.
\tag{8.2}
$$

Alamouti code is the simplest and the most well-known space time block code [962]. Alamouti code was originally designed for the system with two antennas at the transmitter and the coding matrix is expressed as [962]

$$
\mathbf{C}_2 = \begin{bmatrix} x_1 & x_2 \\ -x_2^H & x_1^H \end{bmatrix}.
\tag{8.3}
$$

Alamouti code does not require channel knowledge at the transmitter and obtains the gain of transmit diversity. Alamouti code can achieve full code rate of one. Alamouti code is also an orthogonal space time block code [963, 964], which means for Alamouti code, the product of its coding matrix with its Hermitian transpose is equal to the 2×2 identity matrix,

$$
\mathbf{C}_2 \mathbf{C}_2^H = \begin{bmatrix} |x_1|^2 + |x_2|^2 & 0 \\ 0 & |x_1|^2 + |x_2|^2 \end{bmatrix}.
\tag{8.4}
$$

Generally, the orthogonal space time block coding is performed such that any pair of columns from the coding matrix is orthogonal. In other words, the data vectors for different antennas are mutually orthogonal. This orthogonality will make the decoding at the receiver simple, linear, and optimal.

If there are three antennas at the transmitter, the coding matrix with the code rate of $\frac{1}{2}$ is [964]

$$
\mathbf{C}_{3,\frac{1}{2}} = \begin{bmatrix}
x_1 & x_2 & x_3 \\
-x_2 & x_1 & -x_4 \\
-x_3 & x_4 & x_1 \\
-x_4 & -x_3 & x_2 \\
x_1^H & x_2^H & x_3^H \\
-x_2^H & x_1^H & -x_4^H \\
-x_3^H & x_4^H & x_1^H \\
-x_4^H & -x_3^H & x_2^H
\end{bmatrix}
\tag{8.5}
$$

and the coding matrix with the code rate of $\frac{3}{4}$ is [964]

$$
C_{3,\frac{3}{4}} = \begin{bmatrix} x_1 & x_2 & \frac{x_3}{\sqrt{2}} \\ -x_2^H & x_1^H & \frac{x_3}{\sqrt{2}} \\ \frac{x_3^H}{\sqrt{2}} & \frac{x_3^H}{\sqrt{2}} & \frac{-x_1-x_1^H+x_2-x_2^H}{2} \\ \frac{x_3^H}{\sqrt{2}} & -\frac{x_3^H}{\sqrt{2}} & \frac{x_1-x_1^H+x_2+x_2^H}{2} \end{bmatrix}.
\tag{8.6}
$$

If there are four antennas at the transmitter, the coding matrix with the code rate of $\frac{1}{2}$ is [964]

$$
C_{4,\frac{1}{2}} = \begin{bmatrix} x_1 & x_2 & x_3 & x_4 \\ -x_2 & x_1 & -x_4 & x_3 \\ -x_3 & x_4 & x_1 & -x_2 \\ -x_4 & -x_3 & x_2 & x_1 \\ x_1^H & x_2^H & x_3^H & x_4^H \\ -x_2^H & x_1^H & -x_4^H & x_3^H \\ -x_3^H & x_4^H & x_1^H & -x_2^H \\ -x_4^H & -x_3^H & x_2^H & x_1^H \end{bmatrix}
\tag{8.7}
$$

and the coding matrix with the code rate of $\frac{3}{4}$ is [964]

$$
C_{4,\frac{3}{4}} = \begin{bmatrix} x_1 & x_2 & \frac{x_3}{\sqrt{2}} & \frac{x_3}{\sqrt{2}} \\ -x_2^H & x_1^H & \frac{x_3}{\sqrt{2}} & -\frac{x_3}{\sqrt{2}} \\ \frac{x_3^H}{\sqrt{2}} & \frac{x_3^H}{\sqrt{2}} & \frac{-x_1-x_1^H+x_2-x_2^H}{2} & \frac{x_1-x_1^H-x_2-x_2^H}{2} \\ \frac{x_3^H}{\sqrt{2}} & -\frac{x_3^H}{\sqrt{2}} & \frac{x_1-x_1^H+x_2+x_2^H}{2} & \frac{-x_1-x_1^H-x_2+x_2^H}{2} \end{bmatrix}.
\tag{8.8}
$$

8.2.2 Space Time Trellis Coding

Space time block code can only provide diversity gain. In order to exploit both diversity gain and coding gain, we need to explore space time trellis coding [965, 966]. Space time trellis coding combines transmit diversity and trellis coded modulation to improve the BER performance. The encoding and decoding of space time trellis code are more complex than the counterparts of space time block code due to the utilization of convolutional coding.

8.2.3 Layered Space Time Coding

Layered space time coding can provide multiplexing gain [967]. Meanwhile, diversity gain and coding gain can still be achieved dependent on code structure. Based on layered space time coding, Bell laboratories layered space time (BLAST) is the well known transceiver architecture for achieving spatial multiplexing over MIMO wireless communication system [952, 968–970]. BLAST is an extraordinarily bandwidth-efficient approach to wireless communication which takes advantage of the spatial dimension by transmitting and detecting a number of independent co-channel data streams using multiple, essentially co-located, antennas. At the transmitter, several independent data streams are sent from multiple antennas on the same bandwidth. The encoding formats of BLAST include:

- Diagonal BLAST (D-BLAST);
- Horizontal BLAST (H-BLAST);
- Vertical BLAST (V-BLAST);
- Turbo BLAST [971].

At the receiver, each receive antenna sees all of the transmitted data streams superimposed. There are three main decoding strategies:

- ML decoding;
- linear decoding includes zero-forcing criterion and MMSE criterion;
- nonlinear decoding called successive interference cancellation [972].

In successive interference cancellation, ordering plays an important role [967, 973, 974]. The received symbol with the highest SINR among all the undetected symbols should be detected first. Then, this symbol will be canceled as the interference for the following procedure.

8.3 Multi-User MIMO

Multiuser MIMO [975] can be treated as advanced MIMO which extends MIMO techniques from a single wireless communication link to multiple users.

8.3.1 Space-Division Multiple Access

In wireless communication, there are four main multiple access methods which allow multiple users to share the same transmission channel using different radio resources:

- Frequency-division multiple access (FDMA). FDMA is based on the frequency-division multiplexing scheme. Different nonoverlapping frequency bands are allocated to different users or different data streams. An example of an FDMA system is the first-generation cellular network. In order to increase spectral efficiency of FDMA, OFDMA is used based on the well-known OFDM scheme.
- Time-division multiple access (TDMA). TDMA is based on the time-division multiplexing scheme. Different time slots are allocated to different users or different data streams. An example of TDMA system is the second-generation cellular network.
- Code-division multiple access (CDMA). CDMA is based on the code-division multiplexing scheme. CDMA is also called spread spectrum multiple access. Different codes are allocated to different users or different data streams. An example of CDMA system is the third-generation cellular network.

- Space-division multiple access (SDMA). In SDMA, different spatial subchannels or spatial pipes are allocated to different users through spatial multiplexing or spatial diversity. The cellular network deployed with multiple antennas can explore SDMA to support multiuser wireless communication.

In order to implement SDMA, smart antenna, beamforming or phase array technique can be used for directional signal transmission or reception. In this way, power is saved and interference is avoided. The researches related to SDMA, smart antenna, beamforming, and phase array technique can be found in [976–988].

8.3.2 MIMO Broadcast Channel

MIMO broadcast channel [68, 953, 989, 990] is the multiuser downlink channel. In MIMO broadcast channel, the joint signal processing is allowed at the transmitter.

In MIMO broadcast channel, there is one transmitter with $M > 1$ antennas and there are K users to receive the signals. There are $N_k \geq 1$ antennas at the k-th user. Assume $\mathbf{x} \in C^{M \times 1}$ is the transmitted signal which contains the independent information for each of the users [989]. The covariance matrix of $\mathbf{x}$ is $\mathbf{C_x}$. The average transmitted power should be bounded which means trace$(\mathbf{C_x}) \leq P$ [989].

$\mathbf{H}_k \in C^{N_k \times M}$ is the channel state matrix from the transmitter to the k-th user. $\mathbf{n}_k \in C^{N_k \times 1}$ represents the circularly symmetric complex Gaussian noise at the k-th user which follows normal distribution with zero mean and unit variance on each vector component [989]. Let $\mathbf{y}_k \in C^{N_k \times 1}$ be the received signal at the k-th user which can be expressed as

$$\mathbf{y}_k = \mathbf{H}_k \mathbf{x} + \mathbf{n}_k, k = 1, 2, \ldots, K. \tag{8.9}$$

Let

$$\mathbf{y} = \begin{bmatrix} \mathbf{y}_1 \\ \mathbf{y}_2 \\ \vdots \\ \mathbf{y}_K \end{bmatrix} \tag{8.10}$$

$$\mathbf{H} = \begin{bmatrix} \mathbf{H}_1 \\ \mathbf{H}_2 \\ \vdots \\ \mathbf{H}_K \end{bmatrix} \tag{8.11}$$

and

$$\mathbf{n} = \begin{bmatrix} \mathbf{n}_1 \\ \mathbf{n}_2 \\ \vdots \\ \mathbf{n}_K \end{bmatrix} \tag{8.12}$$

then the mathematical model for MIMO broadcast channel is

$$\mathbf{y} = \mathbf{Hx} + \mathbf{n}. \tag{8.13}$$

Different from single user wireless communication system, for multiuser MIMO, the sum-rate capacity and the achievable rate region are used to evaluate the performance of potential algorithms or schemes. In MIMO broadcast channel, dirty paper coding [991–993], a precoding technique, is exploited to achieve the sum-rate capacity. The idea of dirty paper coding is if the interference is known, the interference can be presubtracted at the transmitter. In this way, the performance remains the same even with the interference. The sum-rate capacity of MIMO broadcast channel can be achieved by solving the following optimization problem [989, 992–995]:

$$
\begin{aligned}
&\text{maximize} \\
&\sum_{k=1}^{K} \log \frac{|\mathbf{I}+\mathbf{H}_k(\sum_{j \leq k}\mathbf{C}_j)\mathbf{H}_k^{H}|}{|\mathbf{I}+\mathbf{H}_k(\sum_{j < k}\mathbf{C}_j)\mathbf{H}_k^{H}|} \\
&\text{subject to} \\
&\sum_{k=1}^{K} \text{trace}(\mathbf{C}_k) \leq P \\
&\mathbf{C}_k \geq 0, k = 1, 2, \ldots, K,
\end{aligned}
\tag{8.14}
$$

where the maximization is over the $M \times M$ positive semidefinite covariance matrices $\mathbf{C}_1, \mathbf{C}_2, \ldots, \mathbf{C}_K$. The optimization problem (8.14) is not a concave optimization problem which is hard to solve. Meanwhile, both all the channel state information and the additive interference information should be known. Ordering of users for precoding is also very important. Because by using dirty paper coding, the interference from the unintended signal can be reduced or completely eliminated [989].

Due to the duality of MIMO broadcast channel and MIMO multiple access channel, the sum-rate capacity of MIMO broadcast channel is equal to the sum-rate capacity of the dual MIMO multiple access channel which gives a beautiful solution to the optimization problem (8.14) [989, 993].

Although the sum-rate capacity of MIMO broadcast channel can be achieved by dirty paper coding, it is hard to implement dirty paper coding with high computational complexity in practice [996]. Hence, the pre-equalizer can be explored. Zero-forcing precoding is a transmission method in MIMO broadcast channel [997, 998]. Zero-forcing beamforming has been presented in [996] together with the user selection scheme and scheduling scheme. Low-complexity linear zero-forcing has been proposed for MIMO broadcast channel in [999]. Random matrix theory has been used to analyze the zero-forcing precoding in MISO broadcast channel with limited feedback [1000]. Zero-forcing precoding is also used together with nonlinear Tomlinson-Harashima precoding to improve the performance of MIMO broadcast system [1001]. Zero-forcing dirty paper coding is proposed in [992]. Besides zero-forcing equalizer, the other well-known linear equalizer is MMSE equalizer. Error performance has been analyzed for linear zero-forcing and MMSE precoders in MIMO broadcast channel [1002]. If imperfect channel knowledge is assumed at the transmitter, robust MMSE precoding is presented in [1003]. Precoding for point-to-multipoint transmission over MIMO ISI channels has been presented in

[1004]. Both intersymbol interference and multiuser interference are taken into account. The nonlinear spatial/temporal Tomlinson-Harashima precoding is explored [1004, 1005].

Block diagonalization is a popular linear precoding for MIMO broadcast channel [1006–1008]. The signal of each user is multiplied by the precoding matrix before the signal is transmitted. In order to eliminate the mutual interference, the precoding matrix for each user should be designed to lie in the null space of channel matrix of all the other users. Hence, the number of users supported by block diagonalization is dependent on transmitter antennas, receiver antennas of each user, and channel state information. Block diagonalization can be treated as the generalized zero-forcing or channel inversion to deal with MIMO broadcast channel when users have more than one antenna [1009]. MMSE based block diagonalization can also be applied [1010]. The achievable throughput for the optimal strategy of dirty paper coding has been compared to that achieved with suboptimal and lower complexity linear precoding, for example, zero-forcing and block diagonalization, in high SNR for MIMO broadcast channel [1011]. Both strategies exploit all available spatial dimensions and therefore have the same multiplexing gain, but an absolute difference in terms of throughput does exist [1011].

Most of the precoding schemes require channel state information at the transmitter. However, it is difficult for the transmitter to have perfect channel knowledge. Meanwhile, in order to reduce the overhead of the system, finite rate feedback is practical. MIMO broadcast channels with imperfect channel state information, partial side information, limited feedback, or finite rate feedback have been considered in [1003, 1008, 1012–1017].

Multiuser diversity is one form of selection diversity among users when the number of users is large [996]. Multi-user diversity can be achieved by user selection and scheduling. In MIMO broadcast system with large number of users, the transmitter cannot serve all the users simultaneously. Multiuser selection and scheduling should be used to choose a group of users to be served. The selection criteria can be the channel conditions of users, fairness, sum-rate capacity of the system, and so on. Multiuser selection and scheduling in MIMO broadcast channel can be found in [996, 1006, 1014–1016, 1014–1016, 1018–1020].

The work about MIMO broadcast channel or MIMO downlink system in cognitive radio network can be found in [1021–1025].

8.3.3 MIMO Multiple Access Channel

MIMO multiple access channel [990, 993, 1026] is the multiuser uplink channel. In MIMO multiple access channel, the joint signal processing is allowed at the receiver.

In MIMO multiple access channel, there is one receiver with $M > 1$ antennas and there are K users to transmit the signals. There are $N_k \geq 1$ antennas at the k-th user. Assume $\mathbf{x}_k \in C^{N_k \times 1}$ is the transmitted signal from k-th user. The covariance matrix of $\mathbf{x}_k$ is $\mathbf{Q}_k$. If there is a sum-power constraint, then $\sum_{k=1}^{K} \text{trace}(\mathbf{Q}_k) \leq P$.

$\mathbf{H}_k^H \in C^{M \times N_k}$ is the channel state matrix from the k-th user to the receiver. $\mathbf{n} \in C^{M \times 1}$ represents the noise at the receiver. Hence, the mathematical model of MIMO multiple access channel can be expressed as

$$\mathbf{y}_{\text{MAC}} = \sum_{k=1}^{K} \mathbf{H}_k^H \mathbf{x}_k + \mathbf{n}. \tag{8.15}$$

The sum-rate capacity of MIMO multiple access channel can be achieved by solving the following optimization problem [989, 990]:

$$
\begin{aligned}
&\text{maximize} \\
&\log |\mathbf{I} + \textstyle\sum_{k=1}^{K} \mathbf{H}_k^H \mathbf{Q}_k \mathbf{H}_k| \\
&\text{subject to} \\
&\textstyle\sum_{k=1}^{K} \mathrm{trace}(\mathbf{Q}_k) \leq P \\
&\mathbf{Q}_k \geq 0, k = 1, 2, \ldots, K,
\end{aligned}
\tag{8.16}
$$

where the maximization is over the $N_k \times N_k$ positive semidefinite covariance matrices $\mathbf{Q}_k, k = 1, 2, \ldots, K$. The objective function in the optimization problem (8.16) is a concave function of the covariance matrices. The efficient numerical algorithms exist to solve the optimization problem (8.16), for example, iterative water filling algorithm [990, 1027]. Meanwhile, it is well known that the dirty paper rate region for MIMO broadcast channel is equal to the capacity region of the dual MIMO multiple access channel with sum-power constraint P [993, 989, 990]. Meanwhile, there is a deterministic transformation which maps from uplink covariance matrices $\mathbf{Q}_1, \mathbf{Q}_2, \ldots, \mathbf{Q}_K$ to downlink covariance matrices $\mathbf{C}_1, \mathbf{C}_2, \ldots, \mathbf{C}_K$ that achieve the same rate and use the same power [990].

8.3.4 MIMO Interference Channel

MIMO interference channel [836, 1028, 1029] includes more than one transmitter and more than one receiver. In MIMO interference channel, neither transmitters nor receivers directly involve joint signal processing.

Assume there are K transmitter-receiver communication links in MIMO interference channel [1029]. There are M_k antennas at the k-th transmitter and there are N_k antennas at the corresponding receiver. $\mathbf{x}_k \in C^{M_k \times 1}$ is the transmitted signal vector for the k-th user. $\mathbf{H}_{kl} \in C^{N_k \times M_l}$ represents the channel state matrix from the l-th transmitter to the k-th receiver. Hence, the received signal vector $\mathbf{y}_k \in C^{N_k \times 1}$ for the k-th receiver is [1029]

$$
\mathbf{y}_k = \mathbf{H}_{kk}\mathbf{x}_k + \sum_{l=1,l\neq k}^{K} \mathbf{H}_{kl}\mathbf{x}_l + \mathbf{n}_k,
\tag{8.17}
$$

where $\mathbf{n}_k$ is the AWGN vector at the k-th receiver with zero mean and covariance matrix $\mathbf{C}_{\mathbf{n}_k}$. $\sum_{l=1,l\neq k}^{K} \mathbf{H}_{kl}\mathbf{x}_l$ is the interference to the k-th receiver.

The straightforward way to handle MIMO interference channel is to exploit precoding matrix $\mathbf{V} \in C^{M_k \times d_k}$ and filtering matrix $\mathbf{U} \in C^{d_k \times N_k}$ to suppress the inference which can be expressed as [1029]

$$
\mathbf{r}_k = \mathbf{U}_k\mathbf{H}_{kk}\mathbf{V}_k\mathbf{s}_k + \sum_{l=1,l\neq k}^{K} \mathbf{U}_k\mathbf{H}_{kl}\mathbf{V}_l\mathbf{s}_l + \mathbf{U}_k\mathbf{n}_k,
\tag{8.18}
$$

where d_k is the number of independent data streams $\mathbf{s}_k$ for the k-th user. $\mathbf{U}_k$ performs the linear dimensionality reduction from $\mathbf{y}_k \in C^{N_k \times 1}$ to $\mathbf{r}_k \in C^{d_k \times 1}$ [1029]. $\mathbf{r}_k$ can be further processed to extract the transmitted signals.

It is well known that the capacity in AWGN channel is proportional to $\log(\text{SNR})$ at high SNR. Hence, we can use the simple concept of spatial degrees of freedom to quantify the maximum multiplexing gain of the MIMO system [1028]. The spatial degrees of freedom can be defined as [1028]

$$\eta = \lim_{\rho \to \infty} \frac{C_{\Sigma}(\rho)}{\log(\rho)}, \tag{8.19}$$

where ρ represents SNR and $C_{\Sigma}(\rho)$ is the corresponding sum capacity.

For a single user MIMO system with M transmitters and N receivers, $\eta = \min\{M, N\}$ [1028]. For MIMO broadcast channel with two users, $\eta = \min\{M, N_1 + N_2\}$ where M is the number of antennas at the transmitter and $N_k, k = 1, 2$ is the number of antennas at the k-th receiver [1028]. Similar result is obtained for MIMO multiple access channel. For two-user MIMO Gaussian interference channel with full rank channel state matrices, if perfect channel knowledge is known at all transmitters and receivers,

$$\eta = \min\{M_1 + M_2, N_1 + N_2, \max\{M_1, N_2\}, \max\{M_2, N_1\}\}, \tag{8.20}$$

where $M_k, k = 1, 2$ is the number of antennas at the k-th transmitter and $N_k, k = 1, 2$ is the number of antennas at the k-th corresponding receiver [1028]. The zero-forcing scheme is sufficient to obtain all the available degrees of freedom [1028].

Furthermore, degrees of freedom of MIMO Gaussian interference channel with K users have been discussed in [836]. Assume there are M antennas for each transmitter and there are N antennas for each receiver. For the outer bound of degrees of freedom [836],

$$\eta \le K \min\{M, N\}, \text{if } K \le R \tag{8.21}$$

and

$$\eta \le K \frac{\max\{M, N\}}{R + 1}, \text{if } K > R, \tag{8.22}$$

where

$$R = \lfloor \frac{\max\{M, N\}}{\min\{M, N\}} \rfloor. \tag{8.23}$$

The achievable inner bound of degrees of freedom is obtained under the assumption that the channel coefficients are time-varying and drawn from a continuous distribution [836]. For the inner bound of degrees of freedom [836],

$$\eta \ge K \min\{M, N\}, \text{if } K \le R \tag{8.24}$$

and

$$\eta \ge \frac{R}{R + 1} \min\{M, N\}, \text{if } K > R. \tag{8.25}$$

When R defined in Equation (8.23) is equal to an integer, then the bound is tight which means [836]

$$\eta = K \min\{M, N\}, \text{if } K \le R \tag{8.26}$$

and

$$\eta = \frac{R}{R + 1} \min\{M, N\}, \text{if } K > R, \tag{8.27}$$

where the results of the achievable degrees of freedom is based on interference alignment [836]. If channel coefficients of MIMO interference channel are constant, using interference alignment together with zero-forcing can achieve more degrees of freedom than using only zero-forcing for some situations [836]. For example, if MIMO Gaussian interference channel with constant channel coefficients has $R + 2$ users where each transmitter has $M > 1$ antennas and each receiver has RM, $R = 2, 3, \ldots$ antennas, then $RM + \left\lfloor \frac{RM}{R^2 + 2R - 1} \right\rfloor$ degrees of freedom can be obtained [836]. RM degrees of freedom can be achieved using zero-forcing [836]. Hence, if $M \geq R + 2$, then $\lfloor \frac{RM}{R^2 + 2R - 1} \rfloor > 0$ and more than RM degrees of freedom can be achieved [836].

There are three general approaches to deal with interference management:

- decode interference;
- treat interference as noise;
- orthogonalize interference and signal in time, frequency, code, and space, for example, interference alignment.

Interference alignment is the core technique used in MIMO interference channel. Interference alignment refers to a construction of signals such that they cast maximized overlapping shadows at the receivers where they constitute interference while they retain distinguishable at the receivers where they are desired [541]. Hence, we need to restrict interference into some subspaces and remain other subspaces for interference free communication. The challenge of interference alignment is that the global channel knowledge is required. Distributed interference alignment has been presented using only local channel knowledge instead of global channel knowledge [1030].

The benefits of user cooperation and cognitive message sharing for degrees of freedom of a two-user MIMO interference channel have been explored in [839]. The term "cognitive message sharing" refers to the genie-aided type of message sharing among cognitive radios [839]. Cognitive message sharing can increase the sum degrees of freedom [839]. Meanwhile, a cognitive transmitter may be more beneficial than a cognitive receiver [839]. Constrained interference alignment and the spatial degrees of freedom of MIMO cognitive networks have been studied explicitly in [1031]. Cognitive radios can align their transmitted signals to produce a number of interference-free channels at each secondary receiver without causing any interference to the primary user [1031].

Opportunistic interference alignment in MIMO cognitive networks has been presented in [1032]. Power limitations lead the primary user to leave some of its spatial directions unused [1032]. The opportunistic link of cognitive radio can be used to transmit data if it is possible to align the interference produced on the primary link into unused spatial directions [1032]. Similarly, opportunistic spatial orthogonalization has been proposed to allow the existence of secondary users even if the primary user occupies all the frequency bands all the time [1033]. Opportunistic spatial orthogonalization can be interpreted as an opportunistic interference alignment scheme where the interference from multiple secondary users is opportunistically aligned at the direction that is orthogonal to the primary user's signal space [1033].

8.4   MIMO Network

Traditionally, MIMO is a physical-layer technique. However, we cannot ignore the great impact of MIMO on the performance of the whole network. Hence, cross-layer MIMO, cooperative MIMO, MIMO routing, and so on have attracted much attention recently [954, 975].

Cross-layer MIMO explores cross-layer optimization for the networking system using MIMO technique and configuration. Cross-layer optimization breaks virtually strict boundaries between layers and jointly designs and optimizes the whole communication architecture from physical layer to application layer [1034]. Cross-layer optimization for MIMO-based wireless ad hoc network has been studied in [1035]. Multihop/multipath routing optimization, power allocation, and bandwidth allocation are considered jointly. Cross-layer optimization for MIMO-based mesh network with Gaussian vector broadcast channel has been presented in [1036]. Jointly optimizing power allocation for dirty paper coding and multihop/multipath routing in MIMO-based mesh network is considered. Distributed link scheduling, power control, and routing for multihop wireless MIMO network have been developed in [1037].

A cross-layer optimization framework for effective interference management has been developed to understand fundamental tradeoffs among possible MIMO gains in multihop networks [1038]. Network utility maximization is also used for cross-layer design of MIMO-enabled wireless local area network(WLAN) [1039]. A cross-layer framework to determine the user selection, rate selection, power selection as well as diversity/multiplexing selection has been studied for multiuser MIMO system [1040]. A cross-layer transmission control protocol (TCP) modeling framework for MIMO wireless system has been presented and the TCP performances of two representative MIMO systems, namely, the BLAST system and the orthogonal STBC system, have been analyzed [1041]. For service-differentiated multiuser MIMO systems, joint feedback and scheduling scheme is used to meet both average and instantaneous delay constraints of delay sensitive applications [1042]. A cross-layer design approach to multicast service in wireless network with MIMO links has been shown in [1043].

Cooperative MIMO explores the distributed MIMO techniques in the coordinated fashion [1044]. In cooperative MIMO, antennas belong to different users, terminals, or base stations with different geo-locations. Diversity gain, especially cooperative diversity, and multiplexing gain can still be achieved in cooperative MIMO. Simulation results shown in [1044] indicate that distributed MIMO systems can provide large spatial diversity, and the data rate in cooperative networks can be significantly increased. Relay is one realization of cooperative MIMO. The basic relay strategies are:

- amplify-and-forward relay;
- decode-and-forward relay;
- compress-and-forward relay.

Infrastructure relay transmission with cooperative MIMO has been studied in [1045]. A signal is transmitted from a base station to a randomly located mobile station using fixed-location relay stations [1045]. Compress-and-forward cooperative MIMO relaying with full channel knowledge has been studied in [1046]. An achievable rate on the

Gaussian MIMO relay channel is derived by using distributed vector compression techniques [1046]. Throughput maximization of ad-hoc wireless networks using adaptive cooperative diversity and truncated ARQ has been presented in [1047]. The relay nodes are not fixed and are selected according to the channel conditions [1047]. Transmitter antenna selection strategies of cooperative MIMO amplify-and-forward relay network have been analyzed in [1048]. One optimal strategy and two suboptimal strategies are considered. Optimized distributed MIMO for cooperative relay network has been introduced in [1049]. An optimization criterion has been derived for the decision of signature vector used in the optimized distributed MIMO. Discrete stochastic algorithms have been exploited for joint transmit diversity optimization and relay selection for multirelay cooperative MIMO system [1050]. The performance is shown to converge to the optimum exhaustive solution [1050].

Joint source and relay optimization for two-way MIMO multirelay networks has been investigated in [1051]. An iterative algorithm is developed to jointly optimize the source, relay, and receive matrices such that the two-way sum MSE of the signal waveform estimation is minimized [1051]. Cooperative power scheduling for a network of MIMO links has been presented in [1052]. Price-based iterative water filling algorithm is a distributed algorithm by which each link computes its power scheduling through an iterative and cooperative process [1052]. Cooperative and constrained MIMO communications in wireless ad hoc/sensor networks have been investigated in [1053]. Given constraints of energy, delay, and data rate, a source node dynamically selects its cooperating nodes from its neighbors to form a virtual MIMO system with communication to the destination node [1053]. Similarly in a wireless sensor network, it is possible to group several sensors to form a virtual MIMO link [1054]. A distributed MIMO-adaptive energy-efficient clustering/routing scheme has been proposed in [1054], which tries to reduce energy consumption in a multihop wireless sensor network.

Multicell MIMO cooperative networks have been analyzed in [1055]. Multicell cooperation can dramatically improve the system performance by exploiting intercell interference [1055]. A linear precoding technique called soft interference nulling has been explored in cooperative MIMO cellular networks for low-complexity implementation [1056]. Cooperation among base stations allows the joint encoding of user signals, which can successfully handle the interference [1056]. The idea of cooperation is also used in LTE-Advanced called coordinated multipoint transmission/reception [1057–1059]. Cooperative cellular networks using multiuser MIMO have been considered in [1060]. The impacts of the scheduling criterion, cell density, and coordination on the average and cell edge user rates across different designs have been analyzed [1060]. QoS-aware base station selections for distributed MIMO links in broadband wireless networks have been presented in [1061]. The BS usages can be minimized and the interfering range can be reduced [1061]. Capacity of a multicell multi-antenna cooperative cellular network with co-channel interference has been analyzed in [1062]. Simulation results shown in [1062] indicate that cooperative transmission can improve the capacity performance of multicell multi-antenna cooperative cellular networks. Capacity of MIMO cellular systems with base station cooperation has been comprehensively studied in [1063]. Several bounds are derived for the minimized transmitted power of the rate-constrained MIMO cellular systems with various base station cooperation strategies [1063].

Wideband waveform design for relay cognitive network has been presented in [1064]. Among the basic relay strategies, amplify-and-forward is the simplest scheme. The received signal at the relay node is multiplied by the complex value and then retransmitted to the destination. Extending from narrowband relay network to wideband relay network, amplify-and-forward can be replaced by its wideband counterpart, called filter-and-forward [1065]. FIR filter is implemented at the relay node. The received signal is filtered and then re-transmitted to the receiver. Besides, in order to improve the performance, the approach based on multiple relay nodes is also proposed in [1065] to perform distributed cooperative beamforming. In relay cognitive networks [1064], there is one transmitter, several relay nodes and one or several receivers. Assume there is no direct communication link between the transmitter and the receiver. All the channel knowledge is perfectly known. The transmitted waveform and the FIR filters at the relay nodes are jointly designed such that the received SNR can beat the threshold derived from QoS. However, the general formulation of this problem is not convex. It is hard to find the global optimal solutions for the transmitted waveform plus the relay FIR filters. Thus a new iterative method is proposed in [1064] to obtain the suboptimal solution and the received SNR can still be guaranteed. With any initial transmitted waveform, the relay FIR filters are jointly optimized similar to distributed cooperative beamforming. Then the transmitted waveform is optimized with fixed relay FIR filters. Afterward, the relay FIR filters are redesigned based on the previous optimized transmitted waveform. This process continues until the transmitted waveform converges to a stable waveform. Because cognitive network is taken into account, there are spectral mask constraints imposed on the transmitted waveform and the forwarded signal from each relay node, which will make the optimization issue more complex [1064].

8.5 MIMO Cognitive Radio Network

MIMO can be fully explored in cognitive radio network. Spatial diversity, spatial multiplexing, beamforming, smart antenna, and so on can help cognitive radio network to access the valuable spectrum and increase the link quality as well as spectrum efficiency. Optimal spectrum sharing in MIMO cognitive radio networks via SDP has been presented in [815]. A unified homogeneous QCQP formulation is applied to three scenarios in which the cognitive radio has complete, partial, or no knowledge about the channels to the primary users [815]. The homogeneous QCQP formulation, though nonconvex can be relaxed to SDP [815]. Similarly, dynamic spectrum management for multi-antenna cognitive radio system with imperfect channel state information has been studied in [1066]. A linear matrix inequality (LMI) transformation is used to facilitate the proper treatment of channel uncertainty at the transmitter of cognitive radio [1066]. Opportunistic spectrum sharing has also been exploited for multi-antenna cognitive radio network [1067]. The channel capacity of cognitive radio is characterized under both its own transmitted power constraint and interference power constraints imposed by primary users [1067]. Similarly, interference minimization approach in MIMO-based cognitive radio networks has been studied in [1068]. The proposed precoder tries to maximize the sum capacity through the minimization of interference power [1068].

Antenna selection in MIMO cognitive radios has been addressed in [1069]. Two solutions are given to the problem of joint transmit-receive antenna selection which aims at maximizing cognitive radio data rates and satisfying interference constraints at the primary users [1069]. Cross-layer antenna selection and learning-based channel allocation for MIMO cognitive radios have been proposed in [1070]. The spectrum efficiency and fairness issue are considered [1070]. Optimal resource allocation for MIMO ad hoc cognitive radio networks has been discussed in [1071]. A semidistributed algorithm is introduced for the maximization of the weighted sum-rate of secondary users [1071]. The throughput of MIMO-empowered multihop cognitive radio networks has been investigated in [1072]. The goal is to achieve the ultimate flexibility and efficiency in DSA and spectrum utilization [1072]. A tractable mathematical model is developed to capture the essence of channel assignment of cognitive radio and degree-of-freedom allocation for MIMO within a channel [1072].

Game theory is widely used in MIMO cognitive radio network. A competitive optimality principle based on game theory is explored in cognitive MIMO radio [1073]. Similarly, a novel game theoretical formulation is proposed to solve one of the challenging and unsolved resource allocation problems in MIMO cognitive radio network: how to allow the concurrent communication over MIMO channels among cognitive radios in a decentralized way, under different interference constraints [1074]. Robust MIMO cognitive radio via game theory has been presented in [1075]. Cognitive radios compete with each other over the resources made available by primary users, by maximizing their own information rates subject to the transmitted power and robust interference constraints [1075].

The work related to applications of beamforming in cognitive radio network can be found in [575, 1076–1094]. Some of the efforts are put into the robust beamforming design for reliable communication in cognitive radio network. Beamforming can also be used for routing due to its directional transmission. Directional transmission can reduce the interference area and directional reception can avoid interference from other radios. Hence, beamforming can increase the efficiency of spectrum sharing. Beamforming supported routing has been exploited in ad hoc network and wireless mesh network which can be straightforwardly extended to cognitive radio network [542, 1095, 1096].

8.6 Summary

MIMO transmission techniques have been presented in this chapter. MIMO can bring array gain, diversity gain, and multiplexing gain by taking advantage of multiple antennas at the transceivers. The basic topics about MIMO have been covered which include space time coding, multiuser MIMO, MIMO network, and so on. The references about applications of MIMO in cognitive radio network are also given. MIMO explores the spatial radio resources to support spectrum access and spectrum sharing in cognitive radio network.

9

Agile Transmission Techniques (II): Orthogonal Frequency Division Multiplexing

OFDM is a technique of digital data transmission based on multicarrier modulation [68, 1097–1099]. The history of OFDM can be traced back to the middle 1960s [1097, 1100–1103]. OFDM is the extension of the frequency division multiplexing scheme. In frequency domain, though the signals of subchannels or subcarriers are overlapped, they are orthogonal after demodulation. Hence, OFDM improves efficiency of spectrum utilization compared to the frequency division scheme which assigns nonoverlapping frequency bands to different signals [68].

OFDM is an effective tool to handle ISI without using equalization at the receiver. The high-data-rate data stream is divided into many low-data-rate substreams and these substreams are sent over many different subchannels [68]. The bandwidth of each subchannel is much smaller than the total system bandwidth [68]. Meanwhile, the bandwidth of each subchannel is less than the coherent bandwidth of the radio channel. Hence, the effect of ISI on each sub-channel is small and flat fading can be assumed for each subchannel [68].

OFDM is the core technology in the current wireless and wired communications. OFDM is widely used by 3GPP LTE, WLAN, WiMAX, digital TV [1104], power line communication, ADSL, VDSL, and HDSL [1105].

9.1 OFDM Implementation

OFDM can use DFT or FFT for efficient implementation. If N subcarriers are used in OFDM, the continuous-time baseband OFDM signal can be expressed as

$$x(t) = \sum_{k=0}^{N-1} X[k] \exp(j2\pi kt/T), 0 \le t < T, \tag{9.1}$$

Cognitive Radio Communications and Networking: Principles and Practice, First Edition.
Robert C. Qiu, Zhen Hu, Husheng Li and Michael C. Wicks.
© 2012 John Wiley & Sons, Ltd. Published 2012 by John Wiley & Sons, Ltd.

where T is the time duration of one OFDM symbol and $X[k]$ is the complex data symbol assigned to the $(k+1)$-th subcarrier with the central frequency of $\frac{k}{T}$. The frequency space of the adjacent subcarriers is $\frac{1}{T}$. N subcarriers are mutually orthogonal over T.

$x(t)$ is sampled with sampling interval of $\frac{T}{N}$. If $x[n] = x(\frac{nT}{N}), n = 1, 2, \ldots, N-1$, then the discrete-time OFDM signal is

$$x[n] = \sum_{k=0}^{N-1} X[k] \exp\left(\frac{j2\pi kn}{N}\right), 0 \leq n \leq N-1, \qquad (9.2)$$

where Equation (9.2) can be implemented by IDFT/IFFT which means IDFT generates the time-domain OFDM symbol consisting of the sequence $x, x[1], \ldots, x[n-1]$ from a complex data symbol stream $X, X[1], \ldots, X[n-1]$ which can be treated as the frequency-domain data. The following discussion about OFDM implementation will be based on the discrete-time model. DFT and IDFT operations are represented simply as

$$X[n] = \text{DFT}\{x[n]\} \qquad (9.3)$$

and

$$x[n] = \text{IDFT}\{X[n]\}. \qquad (9.4)$$

One property of DFT used in OFDM implementation is circular convolution in time domain leads to multiplication in frequency domain [68]. The N-point circular convolution of $x[n]$ and $h[n]$ is defined as [68]

$$y[n] = x[n] \otimes h[n]$$
$$= \sum_k h[k]x[n-k]_N, \qquad (9.5)$$

where $[n-k]_N$ denotes $n-k$ modulo N. $x[n-k]_N$ is a periodic version of $x[n-k]$ with period of N [68]. Thus,

$$Y[n] = \text{DFT}\{y[n]\}$$
$$= \text{DFT}\{x[n] \otimes h[n]\}$$
$$= \text{DFT}\{x[n]\}\text{DFT}\{h[n]\}$$
$$= X[n]H[n], n = 0, 1, 2, \ldots, N-1. \qquad (9.6)$$

If $h[n], 0 \leq n \leq L$ represents a discrete time channel impulse response, $y[n], 0 \leq n \leq N$ can be expressed as [68]

$$\begin{bmatrix} y[N-1] \\ y[N-2] \\ \vdots \\ y \end{bmatrix} = \begin{bmatrix} h & h[1] & \cdots & h[L] & 0 & \cdots & 0 \\ 0 & h & \cdots & h[L-1] & h[L] & \cdots & 0 \\ \vdots & \vdots & \vdots & \vdots & \vdots & \vdots & \vdots \\ 0 & \cdots & 0 & h[0] & \cdots & h[L-1] & h[L] \end{bmatrix}$$

$$
\times \begin{bmatrix} x[N-1] \\ x[N-2] \\ \vdots \\ x \\ x[-1] \\ x[-2] \\ \vdots \\ x[-L] \end{bmatrix} + \begin{bmatrix} n[N-1] \\ n[N-2] \\ \vdots \\ n \end{bmatrix}, \tag{9.7}
$$

where $n, n[1], \ldots, n[N-1]$ are AWGNs.

In order to eliminate ISI from the previous symbol and explore the property of circular convolution mentioned in Equation (9.6), a guard interval with cyclic prefix is exploited [68]. Cyclic prefix adds prefix of a symbol using a repetition of the end. Thus, $x[-1] = x[N-1], x[-2] = x[N-2], \ldots, x[-L] = x[N-L]$. In this way, OFDM implemented by IDFT and DFT can decompose ISI channel into N orthogonal subchannels [68].

9.2 Synchronization

One of the challenging problems in OFDM is synchronization [1106]. OFDM system is very sensitive to synchronization errors, especially carrier frequency offsets [1107, 1108]. In OFDM system, there are four synchronizations involved:

- carrier frequency synchronization;
- sampling timing synchronization;
- sampling frequency synchronization;
- symbol synchronization or frame synchronization.

Carrier frequency offset can destroy the orthogonality among subcarriers and cause intercarrier interference (ICI) which will greatly degrade the system performance. Thus, carrier frequency synchronization tries to remove carrier frequency offset and the corresponding phase offset. Carrier frequency synchronization is usually performed in two steps. The first step is coarse synchronization which usually reduces carrier frequency offset to within one-half of the subcarrier spacing [1109, 1110]. Then, the second step is the fine carrier synchronization which further estimates and reduces the residual carrier frequency offset [1109, 1110]. Carrier frequency synchronization algorithms can be:

- time domain correlation algorithm based on training symbol;
- frequency domain correlation algorithm based on training symbol;
- ML estimator based on training symbol [1111];
- ML estimator based on cyclic prefix [1112, 1113];
- blind synchronization [1114, 1115].

For OFDM blind carrier offset estimation, the method called ESPRIT does not need training symbols, pilot tones, or excess cyclic prefix [1114]. The inherent structure of OFDM signals can be used to provide an accurate carrier frequency offset estimate [1114].

Oversampling is exploited for blind estimation of OFDM carrier frequency offset [1115]. The intrinsic phase shift of neighboring sample points caused by carrier frequency offset should be common among all subcarriers [1115]. Only a single OFDM symbol is required to achieve reliable estimation which makes the blind method data efficient [1115]. The second-order cyclostationarity of OFDM signals has been exploited for blind estimation of symbol timing and carrier frequency offset [101]. A cyclic prefix is not needed necessarily [101]. The similar idea has also been explored in [1116]. The blind estimator exploits the second-order cyclostationarity of received signals and then uses the symbol-timing and carrier frequency offset information appearing in the cyclic correlation [1116]. No channel state information is required [1116].

A blind synchronizer based on SINR maximization for OFDM systems has been developed in [1117]. Besides, the synchronization algorithms taking advantage of the redundancy introduced by cyclic prefix are still treated as the blind algorithms. The joint ML symbol timing and carrier frequency offset estimator has originally presented in [1112]. Redundancy information included in cyclic prefix is utilized without additional training symbols or pilot tones [1112]. Furthermore, a new class of blind cyclic-based estimators for carrier frequency offset and symbol timing estimation have been developed in [1113]. A new likelihood function is derived for joint estimation [1113]. The resulting probabilistic measure is used to develop three unbiased estimators, that is, ML estimator, minimum variance unbiased estimator, and moment estimator [1113]. Virtual carriers are used as intrinsic structure information of OFDM signals for blind estimation of OFDM carrier frequency offset [1118–1120]. MUSIC-like estimation algorithm and ML estimation are explored.

Sampling frequency offset can also cause ICI due to the loss of orthogonality between the subcarriers. Pilot symbol, training symbol, or reference symbol can be used for sampling timing synchronization and sampling frequency synchronization [1121, 1122]. Besides, a novel blind estimation algorithm for sampling clock offset based on second order statistics of the received OFDM samples is devised in [1123], which can be used successfully for noncooperative communications.

Joint synchronization can also be applied to OFDM systems [101, 1112, 1113, 1116, 1117, 1124–1126]. In this way, we do not need to do different synchronizations separately. The different synchronizations are considered simultaneously.

Robust frequency synchronization for OFDM-based cognitive radio systems has been discussed in [1127]. Carrier frequency offset is estimated in the presence of narrow-band interference [1127]. The carrier frequency offset and interference power on each subcarrier are jointly estimated through ML method [1127].

Carrier frequency synchronization and sampling frequency synchronization can reduce ICI. Sphere decoding together with a new search strategy is developed to reduce ICI for OFDM systems [1128]. The suppression of ICI in OFDM systems has also been mentioned in [1129]. The time variations of the channel during one OFDM frame destroy the orthogonality of different subcarriers and result in power leakage among the subcarriers [1129]. A simple and efficient polynomial surface channel estimation technique is proposed to obtain the necessary channel state information first [1129]. Based on the estimated channel state information, a MMSE based OFDM detection technique is used to reduce the performance degradation caused by ICI distortion [1129]. Iterative methods for cancellation of ICI in OFDM Systems have been suggested in [1130]. Operator-perturbation technique is used for the inversion of a linear system of equations [1130]. Furthermore, serial and parallel interference cancellations are proposed to drastically reduce the error floor caused by ICI [1130]. Similarly, an iterative method for frequency domain estimation and compensation of ICI in OFDM systems has been presented in [1131]. There are two steps in the proposed iterative method. Firstly, correlation between received signal and estimated transmitted signal is used to estimate the channel matrix, and the second step estimates the actual transmitted data by means of MMSE equalization [1131]. An iterative algorithm for estimating multipath complex gains with ICI mitigation has also been proposed in [1132]. The ICI self-cancellation schemes have been analyzed in [1133–1139] which include:

- time domain windowing techniques;
- precoding techniques.

Similarly, two-path parallel cancellation schemes can also be used for ICI cancellation [1140–1143]. Reducing ICI in OFDM systems by partial transmit sequence and selected mapping has been proposed in [1144, 1145]. In partial transmit sequence, each block of subcarriers is multiplied by a constant phase factor and these phase factors are optimized to minimize the peak interference to carrier ratio [1144]. In selected mapping, several independent OFDM symbols representing the same information are generated and the OFDM symbol with lowest peak interference to carrier ratio is chosen for transmission [1144]. A novel ICI mitigation method for OFDM by taking advantage of a planar extended Kalman filter has been developed in [1146]. Kalman filter algorithm can estimate and track the frequency offset caused by Doppler in high mobility [1146, 1147]. The Doppler frequency drift information can be updated at each step to get a more accurate result [1146]. Estimation and suppression of ICI due to phase noise in OFDM systems have been discussed in [1148–1151].

9.3   Channel Estimation

Channel estimation in wireless communication system tries to find the time domain characteristics and the frequency domain characteristics of radio channel. For OFDM system, channel estimation identifies the channel gains for different subchannels at different time slots which can be viewed as a two-dimensional lattice in a time-frequency plane [1152]. The two main challenges for channel estimation are:

- how to design pilot symbol pattern;
- how to design an estimator with both low complexity and good channel tracking ability.

Pilot information is the transmitted data or signals known at the receiver. Pilot information can be used as a reference for channel estimation. Locations, powers, and phases of pilot symbols play important roles in the channel estimation [1153]. The basic pilot patterns are [1152]:

- block-type pilot symbol pattern;
- comb-type pilot symbol pattern [1154, 1155].

Block-type pilot symbol patterns are used in slow fading wireless channel. The pilot symbols are inserted into all subcarriers of one OFDM symbol within a specific period. There is no need for interpolation in frequency domain. The estimated channel state will be used to decode the received data inside the block until the next OFDM symbol with pilot information arrives [1152]. Comb-type pilot symbol pattern are mostly used in fast fading wireless channel. Pilot symbols are inserted into certain subcarriers of every OFDM symbol. The interpolation is needed in frequency domain. The one-dimensional interpolation methods can be [1152]:

- linear interpolation;
- second-order interpolation;
- low-pass interpolation;
- spline cubic interpolation.

The basic estimators for channel estimation include [1152, 1156–1158]:

- least square estimator;
- MMSE estimator;
- ML estimator;
- parametric channel modeling estimator;
- filter-based estimator.

Two-dimensional interpolation can also be performed. The optimal solution in terms of MMSE is two-dimensional Wiener filter interpolation [1152]. However, two-dimensional interpolation requires a huge computational complexity. Hence, two-dimensional interpolation can be simplified to two concatenated one-dimensional interpolations in frequency domain and time domain sequentially [1152]. In this way, the system complexity is reduced.

In OFDM-based cognitive radio, due to the noncontiguous positions of the available subcarriers for the secondary users, the conventional pilot design methods are no longer effective [1159]. To obtain satisfactory channel estimation performance, a shift pilot scheme is proposed in [1160–1162]. After deactivating some of pilot tones according to the spectrum sensing result, the shift pilot scheme chooses some nearest activated data subcarriers as the new pilot tones. However, the positions of pilot tones are not optimized [1159]. An efficient pilot design method for OFDM-based cognitive radio systems has been proposed in [1159]. The pilot design including placement is formulated as a optimization problem. Besides, OFDM pilot design for channel estimation with null edge subcarriers has been presented in [1163]. Null subcarriers on band edges can reduce adjacent channel interference [1163]. An arbitrary-order polynomial parameterization of the pilot subcarrier indices is exploited [1163].

Blind channel estimation [1164–1166] and semiblind channel estimation [1167–1169] are also used in the OFDM system. Joint channel estimation and synchronization in OFDM systems can be found in [1170–1175].

9.4 Peak Power Problem

One OFDM symbol is the superposition of many independent modulated subcarriers. If this addition is executed coherently, the instantaneous power of OFDM signal will be big, which leads to high PAPR. High PAPR will reduce the efficiency of linear power amplifier at the transmitter. If the power is beyond the linear region of power amplifier, the OFDM signal will be distorted. Meanwhile, high PAPR require thigh resolution ADC with high dynamic range at the receiver [68]. Hence, we need to reduce PAPR for OFDM communication system. The PAPR of a continuous-time signal is [68]

$$\text{PAPR} = \frac{\max\{|x(t)|^2\}}{E\{|x(t)|^2\}} \tag{9.8}$$

and the counterpart of a discrete-time signal is

$$\text{PAPR} = \frac{\max\{|x[n]|^2\}}{E\{|x[n]|^2\}}. \tag{9.9}$$

There arc many ways to reduce PAPR of OFDM signals [68, 1176–1178]:

- clipping and windowing [1179, 1180];
- adaptive symbol selection scheme [1179];
- selective mapping [1181–1183];
- partial transmission sequence [1183–1187];
- phase optimization [1188];
- nonlinear companding transformation [1189];
- special coding techniques [1190];
- constellation shaping [1191, 1192];
- pulse shaping [1193].

9.5 Adaptive Transmission

Adaptive transmission can adapt the coding and modulation scheme and other signal and protocol parameters, for example, transmitted power, signaling bandwidth based on prevailing channel conditions in order to increase spectrum efficiency. Adaptive transmission requires some channel state information at the transmitter. There are various metrics which can be used as channel state information, for example, SNR, SINR, BER, and packet error rate [1194]. Adaptive transmission can be exploited over a fading channel to improve the energy efficiency and increase the data rate [1195]. Meanwhile, adaptive transmission can modify transmission scheme according to the radio interference.

Adaptive transmission can be applied to MIMO system where there are multiple spatial subchannels with different channel gains. We can dynamically determine the coding and modulation scheme as well as transmitted power for each subchannel. Adaptive modulation and coding in MIMO WiMAX with limited feedback has been experimentally evaluated in [1196]. The condition number of the spatial correlation matrix is used as an indicator of the spatial selectivity of the MIMO channel for adaptive MIMO transmission

[1197]. The adaptive algorithm selects the MIMO transmission method, among spatial multiplexing, double space-time transmit diversity, and beamforming, to enhance the spectral efficiency for a target error rate performance and transmitted power [1197].

Adaptive transmission can be exploited in OFDM system where there are multiple parallel subchannels in frequency domain. The independent coding and modulation scheme will be selected for each subchannel [1194]. High-level modulation and high-rate coding will be used on subchannel with good channel condition [1194]. In order to support adaptive transmission, adaptation threshold and adaptation rate should be determined [1194]. Meanwhile, feedback overhead and computation load should also be taken into account when adaptive transmission is explored. For example, adaptive transmission based on sub-band instead of subcarrier for OFDM system can reduce the demanding overhead [1194]. Adaptive transmission for OFDM system can be found in [1198–1201].

Take adaptive modulation as an example to show how adaptive transmission works in OFDM system. Assume there are N subchannels without consideration of ISI and ICI. The power allocated to the n-th subchannel is P_{tn}. The number of bits transmitted over the n-th subchannel is b_n which corresponds to the modulation scheme. Based on the transmitted power, the gain of subchannel, and the modulation scheme, BER can be obtained. BER for the n-th sub-channel is P_{en}. If we would like to minimize the total transmitted power given the constraints of data rate and BER, the optimization problem of adaptive modulation with bit loading can be expressed as

$$\text{minimize}$$
$$\sum_{n-1}^{N} P_{tn}$$
$$\text{subject to} \qquad\qquad (9.10)$$
$$\sum_{n-1}^{N} b_n = b_{\text{target}}$$
$$P_{en} \leq P_{e\text{target}}, n = 1, 2, \ldots, N,$$

where b_{target} is the minimum number of bits transmitted over one OFDM symbol and $P_{e\text{target}}$ is the maximum tolerable BER. If we would like to maximize the data rate given the constraints of transmitted power and BER, the corresponding optimization problem is,

$$\text{maximize}$$
$$\sum_{n=1}^{N} b_n$$
$$\text{subject to} \qquad\qquad (9.11)$$
$$\sum_{n=1}^{N} P_{tn} \leq P_{t\text{target}}$$
$$P_{en} \leq P_{e\text{target}}, n = 1, 2, \ldots, N$$

where $P_{t\text{target}}$ is the maximum total transmitted power for one OFDM symbol.

There are three basic algorithms to solve the optimization problem for adaptive modulation:

- Hughes-Hartogs algorithm [1202, 1203];
- Chow algorithm;
- Fischer algorithm [1204, 1205].

The Hughes-Hartogs algorithm is a greedy algorithm based on gradient allocation. Every incremental power to transmit one additional bit over each subchannels is compared.

One bit will be added to the subchannel with the least incremental power. The whole procedure is repeated until b_{target} is achieved. The Chow algorithm loads the bit based on the capacity of subchannel while the Fischer algorithm allocates bit from the minimization of BER.

Generally, adaptive transmission also includes dynamic radio resource allocation for various OFDM systems [506, 1206, 1207].

9.6 Spectrum Shaping

Due to the easy power control, adaptive transmission, and pulse shaping of each subcarrier in OFDM signal, spectrum shaping can be performed for OFDM-based broadband wireless communication. Spectrum shaping plays an important role for interference management, DSA, and so on. Thus, OFDM is the key technology used in cognitive radio network for spectrum access and spectrum sharing.

Spectrum shaping of OFDM-based cognitive radio signals has been presented in [1208]. Modulated OFDM sub-carriers suffer from high side-lobes which result in adjacent channel interference [1208]. Hence, active cancellation carrier and raised cosine windowing are used to reduce adjacent channel interference [1208]. Sidelobe suppression in OFDM-based spectrum sharing systems using adaptive symbol transition has been proposed in [1209]. An extension is added to OFDM symbol that is calculated using the optimization method to minimize adjacent channel interference [1209]. Similarly, reduction of out-of-band radiation in OFDM systems by insertion of cancellation carriers has been investigated in [1210]. Spectral sculpting for OFDM-based spectrum access has been studied in [1211]. The idea is also to add a cancellation signal to the OFDM signal to cancel interference in the target spectrum band caused by data tones, so that interference received by primary user can be limited [1211]. The researches about active interference cancellation can also be found in [1212–1214]. Dynamic spectral shaping has been used in cognitive radio to avoid spectral bands used by licensed users and maintain specified target SINR at the receiver of cognitive radio [1215].

NC-OFDM can be used for spectrum shaping in cognitive radio by deactivating subcarriers located in the spectrum band occupied by primary user. An efficient implementation of NC-OFDM transceivers for cognitive radio has been presented in [539]. The main idea is to prune the FFT efficiently and quickly [539]. Similarly, low-power FFT design for NC-OFDM in cognitive radio systems has been introduced in [1216]. The resource allocation in NC-OFDM based cognitive radios can be found in [1217]. Portfolio optimization is used to achieve QoS maintenance [1217].

9.7 Orthogonal Frequency Division Multiple Access

OFDMA is the multiple access technique based on the popular OFDM digital modulation scheme [1218]. Different subcarriers or different subsets/groups of subcarriers are chosen to different users. Multiple users can be served simultaneously. OFDMA is a promising technique to improve the transmission reliability and efficiency of multiuser wireless communications [1219]. The fundamental relationship between multiplexing and diversity in OFDMA systems has been investigation in [1219]. The proposed H-matching method achieves the optimal outage performance at a given target multiplexing gain, which shows that the optimal diversity multiplexing tradeoff can be achieved by only allocating subcarriers [1219]. For green communication, energy efficiency and spectrum efficiency tradeoff in downlink OFDMA networks has been discussed in [1220]. Under the general tradeoff framework between energy efficiency and spectrum efficiency, energy efficiency is strictly quasiconcave in spectrum efficiency [1220]. Uplink synchronization in OFDMA spectrum-sharing systems has been considered in [1221]. The frequency and timing errors of multiple unsynchronized users are estimated [1221].

Optimal radio resource allocation can improve the performances of OFDMA downlink systems [1222]. Weighted sum rate maximization and weighted sum power minimization problems are considered with the assumption that each tone is taken by at most one user [1222]. Lagrange dual decomposition method is exploited due to the non-convex property of the original resource allocation problems [1222]. Resource allocation for OFDMA-based cognitive radio network with application to H.264 scalable video transmission has been presented in [1223]. Minimum and maximum rate constraints are considered for the transmission of scalable video sequences [1223]. Integer programming is used to determine how to allocate radio resources to different cognitive radio users with consideration of interference tolerances of primary users [1223]. Radio resource allocation in OFDMA cognitive radio systems has also been considered in [1224]. A novel three-step cross-layer optimization of OFDMA radio resource allocation has been developed to keep fairness among users and maximize total capacity [1224]. For multicast service in cognitive radio network, taking the maximization of the expected sum rate of cognitive multicast groups as the design objective, an efficient joint subcarrier and power allocation scheme is proposed in [1225].

Joint cross-layer scheduling and spectrum sensing have been explored in the downlink transmission of an OFDMA-based cognitive radio system [1226]. The power allocation and the subcarrier assignment across the secondary users are adjusted to optimize a system utility [1226]. Meanwhile, distributed implementation for the cross-layer sensing and scheduling design is given using the primal-dual decomposition approach [1226]. Distributed resource allocation for OFDMA-based relay networks has been investigated in [1227]. The data rate and user fairness can be improved by cognitive radio techniques used at the relay nodes [1227]. Iterative waterfilling and its variants are exploited for resource allocation [1227]. A novel subchannel and power allocation scheme for multicell OFDMA networks with cognitive radio functionality has been proposed in [1228]. Intercell interference together with the interference to the primary user is considered [1228]. Dual decomposition method is exploited to derive a distributed algorithm [1228]. Similarly, coexistence and optimization of a multicell OFDMA-based cognitive radio network which is overlaid with a multicell primary radio network have been studied in [1229]. A Lagrange duality based technique is used to optimize the weighted sum rate of secondary users over multiple cells [1229]. Interference-aware radio resource allocation in OFDMA-based cognitive radio networks has been presented in [1230]. Out-of-band emissions from cognitive radio and the interference that arises as a result of imperfect spectrum sensing are explicitly considered [1230]. The resource allocation problem is formulated as a mixed-integer nonlinear programming problem [1230]. In order to combat a passive multi-antenna eavesdropper and the effects of imperfect channel state information at the transmitter, secure resource allocation and scheduling for OFDMA decode-and-forward relay networks have been explored [1231]. The packet data rate, secrecy data rate, power, and subcarrier allocation policies are optimized to maximize the average secrecy outage capacity [1231]. Distributed energy efficient spectrum access in cognitive radio ad hoc networks has been considered in [1232]. A fully distributed subcarrier selection and power allocation algorithm is proposed by combining an unconstrained optimization method with a constrained partitioning procedure [1232]. Distributed resource allocation for cognitive radio ad hoc networks with spectrum-sharing constraints has also been discussed in [1233]. A dual decomposition framework is explored for the realization of distributed solutions [1233].

A novel spectrum trading model for OFDMA-based cognitive radio systems has been introduced in [1234]. Primary users can trade their spare subcarriers with secondary users for better utilities [1234]. Pricing policies and market equilibrium are also considered [1234]. Similarly, joint pricing and resource allocation for OFDMA-based cognitive radio systems has been presented in [1235]. The secondary users try to maximize their capacity under three different constraints: total transmitted power, a given budget for sharing subchannels, and tolerable interference to the primary users [1235]. Nash bargaining is explored for efficient resource allocation with flexible channel cooperation in OFDMA cognitive radio networks [1236]. In cooperative cognitive radio networks, secondary users cooperatively relay data for primary users in order to access the spectrum [1236]. A novel design of flexible channel cooperation is proposed, which allows secondary users to freely optimize the utilization of channels for transmitting data of primary users along with their own data [1236].

9.8 MIMO OFDM

MIMO and OFDM can be exploited together for high-capacity, high-reliability wireless connectivity [1237–1241].

The researches about synchronization for MIMO OFDM system can be found in [1242–1244]. Channel estimation for MIMO OFDM system has been discussed in [1245–1252]. Adaptive transmission can also be extended to MIMO OFDM system [1253–1255]. Various radio resource allocation and management strategies for MIMO OFDMA systems have also been developed [1256–1262] to optimize the performance of the multiuser system.

9.9 OFDM Cognitive Radio Network

The underlying sensing and spectrum shaping capabilities of OFDM, together with its flexibility and adaptivity, probably make it the best transmission technology for cognitive radio network to perform DSA and spectrum sharing [1263, 1264].

Optimal and suboptimal power allocation schemes for OFDM-based cognitive radio systems have been presented in [1265]. Furthermore, adaptive power loading with statistical interference constraint is developed in [1266]. Cognitive radio transmitter does not require the instantaneous channel quality feedback from the receivers of primary users [1266]. An efficient power loading scheme is also studied in [1267]. Cognitive radios may use both active and nonactive bands of primary users as long as the generated interference is within the interference temperature limits of primary users [1267]. A risk-return model is explored to perform energy-efficient power allocation in OFDM-based cognitive radio systems [1268]. Based on a risk-return model, a convex optimization problem is formulated [1268]. A fast power allocation algorithm is mentioned in [1269]. Resource allocation in an OFDM-based cognitive radio system has been formulated as a multidimensional knapsack problem with consideration of sub-carrier, bit, and power [1270]. A low-complexity, greedy, max-min algorithm is proposed to give the near-optimal solution [1270]. Cross-layer resource allocation is also explored for multiuser OFDM-based cognitive radio systems with consideration of both real-time and non-real-time applications [1271]. Due to the dynamic nature of the available spectrum, two challenges, that is, problem feasibility and false urgency, are explicitly addressed [1271].

A distributed resource allocation algorithm is proposed for OFDM cognitive radio systems to provide good fairness among users [1272]. Queue-aware subchannel and power allocation for downlink OFDM-based cognitive radio networks has been investigated in [1273]. Secondary users with small queue backlogs are only given sufficient rates to support their demands and the remaining radio resources are shared among highly backlogged users [1273]. The work in [1273] is extended to downlink OFDMA cognitive radio networks in [1274]. The achievable rate of an OFDM-based cognitive radio system sharing the spectrum with an OFMDA-based primary system has been studied in [1275]. Rate loss constraint is used for primary transmission protection [1275]. Relay and power allocation schemes for OFDM-based cognitive radio systems have been developed in [1276]. The capacity of cognitive radio using relay is optimized while total transmitted power is bounded and the interference introduced to the primary user is kept within a prescribed threshold [1276]. The corresponding optimization problem is a mixed-integer problem, which is NP-hard [1276].

Robust transmit power control for cognitive radio network based on OFDM technology has been discussed in [397]. Robust optimization problem for multiuser dynamic power control is given [397]. Robust optimization can guarantee the acceptable performance under the worst case conditions. Robust optimization is a conservative approach, but it can provide seamless communication [397]. Due to the dynamic nature of cognitive radio network and the delay introduced by the feedback channel, it is hard to obtain the accurate and real-time information for interference [397]. Hence, robust optimization gives us a way to address this issue by taking into account the worst case uncertainty in the interference and noise [397]. Multiuser radio resource allocation is a game problem. Robust iterative water filling algorithm is exploited to solve the robust game [397].

OFDM can be exploited together with MIMO to support wireless transmission in cognitive radio network. The researches about MIMO-OFDM based cognitive radio network can be found in [1277–1283]. Most of these efforts are related to radio resource management.

9.10   Summary

OFDM transmission techniques have been presented in this chapter. The critical issues in OFDM systems including OFDM implementation, synchronization, ICI, channel estimation, peak power problem, adaptive transmission, spectrum shaping, OFDMA, and so on have been discussed. OFDM is the basic transmission technique used in cognitive radio network. DSA and spectrum sharing can be well supported by OFDM.

10

Game Theory

10.1 Basic Concepts of Games

In cognitive radio networks, there are many secondary users, as well as possible attackers; each has its own action and payoff and can be considered as a rational agent, as long as secondary users are usually equipped with powerful computing devices. Hence, it is natural to introduce game theory, which traces back to early 1900s [1284] and analyzes the possible conflict or collaboration and the corresponding strategies of rational players, to study the interactions among the agents in the cognitive radio network. In this section, we introduce the basic concepts in game theory. There are many types of games, such as the strategic-form games, repeated games, stochastic games, and differential games, etc. A comprehensive introduction of game theory can be found in [1285]. A more modern introduction to game theory, from the computer science perspective, can be found in [1286]. Games in dynamical systems, such as Markov processes or continuous time systems, are discussed in [1287] and [1288]. As a preliminary introduction to game theory, we focus on the simplest strategic-form game in this book, which provides a starting point for studying more complicated games, and then provide a brief introduction to Bayesian games and stochastic games which are carried out in multiple stages.

10.1.1 Elements of Games

For simplicity, we assume that there are two players in the game. It is not difficult to extend the two-player game to the general case with multiple players. Before we step into the formal formulation of a game, we first explain a famous example of game, the *prisoner's dilemma*, which will be used as the example throughout the introduction. In this game, there are two prisoners being accused for a crime which can be convicted only when one or more confesses. If one prisoner confesses while the other does not, the latter one will be sentenced to 6 years while the former can go free. If both confess, both will be sentenced to 5 years. If both do not confess, they are both sentenced to 1 year. As will be seen, this game will yield a very surprising result.

Cognitive Radio Communications and Networking: Principles and Practice, First Edition.
Robert C. Qiu, Zhen Hu, Husheng Li and Michael C. Wicks.
© 2012 John Wiley & Sons, Ltd. Published 2012 by John Wiley & Sons, Ltd.

In a two-player strategic-form game, we assume that each player has the following elements:

- Action: Each player can take a finite number of actions. We denote by $A_i = \{a_{i1}, \ldots, a_{in_i}\}$, where a_{ij} stands for an action and n_i is the total number of actions, the set of actions for player i. Take the prisoner's dilemma as an example, we have $a_{11} = a_{21} =$ confess and $a_{12} = a_{22} =$ not confess. Obviously $n_1 = n_2 = 1$.
- Strategy: Each player can choose the action in a random manner. We denote by π_{ij} the probability that player i chooses action j. The probabilities of action are called strategies, denoted by π_i for player i. When there exists a j such that $\pi_{ij} = 1$, we call it pure strategy, that is, player i only chooses action j; otherwise, we call it mixed strategy since the player has more than one options. In the example of prisoner's dilemma, a strategy of prisoner 1 is to confessor with probability 0.6 and not to confess with probability 0.4. The prisoner can flip a asymmetric coin to make the decision.
- Payoff: After the players choose their actions, they will receive payoffs which are functions of the actions. We denote by $r_i(a_{1m}, a_{2n})$ the payoff of player i if the actions are a_{1m} and a_{2n}. For example, the prisoner receives payoff -6 if he chooses not to confessor but the other prisoner chooses to confess. The payoffs can be represented by the following table:

$$\begin{pmatrix} (-1, -1) & (-6, 0) \\ (0, -6) & (-5, -5) \end{pmatrix},$$ (10.1)

where the rows and columns represent the actions of players 1 and 2, respectively. One special type of the payoff is $r_1(a_{1m}, a_{2n}) = -r_2(a_{1m}, a_{2n})$, that is, the sum of the payoffs of the two players is 0. We call it zero-sum game. Usually, we use it to model the game between two players with completely conflicting interest. In zero-sum games, we need to specify only the payoff of one player. The payoff of the other player is obtained correspondingly. Obviously, the game of the two prisoners is not a zero-sum one.

10.1.2 Nash Equilibrium: Definition and Existence

Now we discuss the Nash equilibrium of the game, which is the key concept in game theory and a corner stone of modern economics. It was named after the legendary mathematician John. F. Nash, Jr [1289]. First, the expected payoff the each player is given by

$$\bar{R}_i(\pi_1, \pi_2) = \sum_{j=1}^{n_1} \sum_{k=1}^{n_2} \pi_{1j}\pi_{2k}r_i(a_{1j}, a_{2k}), \qquad i = 1, 2.$$ (10.2)

We assume that both players are rational. Hence, they want to choose strategies to maximize their own expected payoffs. However, when player 1 fixes a strategy, player 2 may change its strategy such that the expected payoff of player 1 is reduced. Hence, each player must consider the possible strategy of each other and make corresponding decisions. So how can they decide their strategy in such a bilateral decision scenario? As we will see, the players can only choose strategies at Nash equilibrium if they are rational.

We call a pair of strategy, denoted by π_1^* and π_2^* a Nash equilibrium if

$$\begin{cases} \bar{R}_1(\pi_1^*, \pi_2^*) \geq \bar{R}_1(\pi_1, \pi_2^*), \forall \pi_1 \\ \bar{R}_2(\pi_1^*, \pi_2^*) \geq \bar{R}_2(\pi_1^*, \pi_2^*), \forall \pi_2 \end{cases} . \tag{10.3}$$

An intuitive explanation of (10.3) is that, at the Nash equilibrium, if one players changes its own strategy unilaterally, it will receive less expected payoff. Hence, at the Nash equilibrium, both players do not want to change their strategies; thus the game reaches an equilibrium.

We can examine this concept using the example of prisoner's dilemma. We examine only the pure strategies now and will check mixed strategies in another example. Obviously, the pure strategies of (confess, confess) is not a Nash equilibrium since, if player 1 changes his action to not confess, he can improve his payoff from -1 to 0, which contradicts the definition of Nash equilibrium. Similarly, the pure strategies of (confess, not confess) is also not a Nash equilibrium since the prisoner who takes the action of not confess can switch his action to confess such that his payoff is improve from -6 to -5. Finally, the pure strategy (confess, confess) is a Nash equilibrium since, if a prisoner changes to not confess, his payoff will be decreased from -5 to -6. The analysis shows that both prisoners will choose not to confess and then be sentenced to 5 years, which is a much worse result than the situation in which both do not confess (both sentenced to 1 year).

We have seen that, in the example of prisoner's dilemma, there is a pure strategy Nash equilibrium. However, not every game has a pure strategy Nash equilibrium. Consider the following zero-sum game with two actions for each player and the following payoff table:

$$\begin{pmatrix} (2, -2) & (1, -1) \\ (1, -1) & (3, -3) \end{pmatrix} . \tag{10.4}$$

We can verify that all four possible combinations of pure strategies are not Nash equilibrium. For example, when the pure strategies are (a_{11}, a_{21}), the payoffs are 2 and -2. Then, player 2 desires to switch to action a_{22} such that his payoff will be increased from -2 to -1. Then, does this mean that this game has no Nash equilibrium? No. We have not checked the mixed strategies yet. Actually, it is easy to verify that the following mixed strategy:

$$\begin{cases} \pi_{11} = \frac{2}{3}, & \pi_{12} = \frac{1}{3} \\ \pi_{21} = \frac{2}{3}, & \pi_{22} = \frac{1}{3} \end{cases} . \tag{10.5}$$

The verification is left as an exercise (Problem 1).

Now, we have examined the Nash equilibria of two games. A question arises naturally: does every strategy-form game have Nash equilibrium? The answer is *yes*: each finite[1] strategic game has at least one Nash equilibrium. A rigorous proof can be found in Section 3.12 of [1285]. We omit it here due to the limited length.

_____

[1] Here finite means that the set of actions is finite.

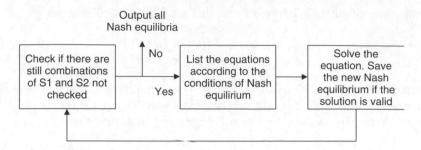

Figure 10.1   Procedure of computing Nash equilibrium.

10.1.3   Nash Equilibrium: Computation

Once we solved the problem of the existence of Nash equilibrium, the next question is how to find the Nash equilibrium. A detailed rigorous discussion about the computational complexity of finding Nash equilibrium can be found in Chapter 2 of [1286]. In this book, we provide some working techniques for computing Nash equilibrium, which is illustrated in Figure 10.1. We first consider the general case of payoffs. We denote by S_i the set of actions that player i will take with nonzero probabilities. Then, if there exists a Nash equilibrium (π_1, π_2) over the action sets S_1 and S_2, that is, at the Nash equilibrium both players will confine their actions within these two sets, there must exist numbers w_1 and w_2 such that the following conditions are satisfied:

$$\begin{cases} \sum_{y \in S_2} \pi_{2y} r_1(x, y) = w_1, & x \in S_1 \\ \sum_{x \in S_1} \pi_{1x} r_2(x, y) = w_2, & y \in S_2 \\ \pi_{ij} = 0, & i = 1, 2, j \notin S_i \\ \sum_{j \in S_i} \pi_{ij} = 1, & i = 1, 2 \end{cases}. \tag{10.6}$$

It is quite easy to understand the last two equations since they are simply the definition of S_i and the normalization condition of probability. The first two equations are more essential to the Nash equilibrium. An intuitive explanation for the first equation is that, at the Nash equilibrium, the expected payoff of taking an action x, namely $\sum_{y \in S_2} \pi_{2y} r_1(x, y)$, is the same as that of taking any other action in S_1. Otherwise, say player 1 receives more expected payoff when taking action 1, it will put more probability to action 1, thus breaking the Nash equilibrium. So is the explanation for the second equation. By solving the above linear equations, we can obtain the strategies at the Nash equilibrium, which is quite straightforward. The real challenges is how to choose sets S_1 and S_2. Unfortunately, there is no systematic approach to find S_1 and S_2. One approach is to exhaustively search all possible combinations of S_1 and S_2. In some cases, we can also incorporate some *a priori* information about the Nash equilibrium.

We use the game with payoffs defined in (10.3) to illustrate the computation of Nash equilibrium. First, we assume $S_1 = \{a_{11}, a_{12}\}$ and $S_2 = \{a_{21}, a_{22}\}$, that is, both players will take each action with nonzero probability. Then, we have

$$\begin{cases} 2\pi_{21} + (1 - \pi_{21}) = \pi_{21} + 3(1 - \pi_{21}) \\ -2\pi_{11} - (1 - \pi_{11}) = -\pi_{11} - 3(1 - \pi_{11}) \end{cases}, \tag{10.7}$$

according to the first two equations in (10.6). Note that the last condition in (10.6) has been implicitly incorporated since we assume $\pi_{12} = 1 - \pi_{11}$ and $\pi_{22} = 1 - \pi_{21}$. Solving the equations results in $\pi_{11} = 1/3$ and $\pi_{21} = 1/3$.

We can also take the Prisoner's dilemma as another example. First, we assume $S_1 = \{a_{11}, a_{12}\}$ and $S_2 = \{a_{21}, a_{22}\}$. Then, according to the conditions in (10.6), we have

$$\begin{cases} -\pi_{21} - 6(1 - \pi_{21}) = -5(1 - \pi_{21}) \\ -\pi_{11} - 6(1 - \pi_{11}) = -5(1 - \pi_{11}) \end{cases}. \tag{10.8}$$

Obviously, there is no solution to these equations. Hence, the assumption that $S_1 = \{a_{11}, a_{12}\}$ and $S_2 = \{a_{21}, a_{22}\}$ is incorrect. Hence, we should check other possible combinations of S_1 and S_2. Finally, we will find that (confess, confess) is the only Nash equilibrium.

10.1.4 Nash Equilibrium: Zero-Sum Games

In the previous discussions on the Nash equilibrium, the payoffs are of general form. As we have introduced, zero-sum game is an important type of games. The Nash equilibrium of zero-sum game has a special structure which is worthy of special introduction. It has been shown that the Nash Equilibrium of a zero-sum game, denoted by (π_1^*, π_2^*), is given by

$$\begin{cases} \pi_1^* = \max_{\pi_1} \min_{\pi_2} R_1(\pi_1, \pi_2) \\ \pi_2^* = \min_{\pi_2} \max_{\pi_1} R_1(\pi_1, \pi_2) \end{cases}. \tag{10.9}$$

The above equations are very intuitive. At the Nash equilibrium, player 1 wants to maximize the expected payoff minimized by player 2, while player 2 wants to minimize the expected payoff maximized by player 1. Such a maxmin and minimax structure embodies the conflicting nature of zero-sum game. We call the maxmin value the *value* of the zero-sum game.

10.1.5 Nash Equilibrium: Bayesian Case

In the previous discussions, we assume that the players know the payoffs of each other perfectly. However, in many cases, this assumption may not be true. As will be discussed, in the collaborative spectrum sensing, a secondary user may not be sure about the trust-worthiness of a collaborator: it could be an honest collaborator, or a malicious attacker. For the two different possibilities, the other player may have different payoffs. When the collaborator is an attacker (honest secondary user), it received positive (negative) payoff when the spectrum sensing fails. In this case, we say the game has incomplete information.

To better describe the game, we define the type of each player, denoted by t_i for player i, and denote by T_i the set of possible types. For example, the collaborator in the collaborative spectrum sensing may have two types: honest or malicious. Each type means a set of payoffs. The payoff of player i is not only determined by the actions of both players but also its type. Hence, the payoff of player is given by $r_i(t_i, a_1, a_2)$.

The players have conjectures on the type of each other. We denote by $p_i(t_{-i}|t_i)$ the player i's conjecture on the probability that the other player (here $-i$ means the player other than player i) has type t_{-i}, when player i's own type is t_i. We assume that the probabilities $\{p_i(t_{-i}|t_i)\}_{i,t_i,t_{-i}}$ are perfectly known to both players. The strategy of each player is also dependent on its own type. Take the collaborative spectrum sensing as an example, a malicious collaborator will be more likely to send out a false report while an honest collaborator always sends its own observation. Hence, the strategy can be written as $\pi_i(\cdot|t_i)$. We denote by $\pi_i = \{\pi_i(\cdot|t_i)\}_{t_i \in T_i}$. Then, when player i has type j, its expected payoff will be

$$\bar{R}_i(\pi_1, \pi_2|t_i) = \sum_{t_{-i} \in T_{-i}} p_i(t_{-i}|t_i) \sum_{j=1}^{n_1} \sum_{k=1}^{n_2} \pi_{1j}(a_{1j|t_1}) \pi_{2k}(a_{2k}|t_2) r_i(a_{1j}, a_{2k}). \quad (10.10)$$

In contrast to the strategic form game with perfect information, the players with imperfect information must consider the different possible types of the other player, as the type of the opponent can affects its payoff.

For the Bayesian game, we can define a Bayesian equilibrium, in which for every player i and every type t_i of itself, the following equation is satisfied:

$$\begin{cases} \pi_1^*(\cdot|t_1) = \arg\max_\pi \bar{R}_1(\pi, \pi_2^*|t_1) \\ \pi_2^*(\cdot|t_2) = \arg\max_\pi \bar{R}_2(\pi_1^*, \pi|t_2) \end{cases}. \quad (10.11)$$

Similarly to the Nash equilibrium, at the Bayesian equilibrium, unilaterally changing the strategy does not increase the expected payoff according to the belief on the opponent's type. Obviously the computation of Bayesian equilibrium is more complicated.

10.1.6 Nash Equilibrium: Stochastic Games

In strategic-form games and Bayesian games, the game lasts for only one snapshot. However, in many problems, the game may last for many stages. Moreover, there may exists a system state evolving with time (usually impacted by the actions of both players) and the payoffs are dependent on the system state. For example, in cognitive radio networks, the queue lengths can be considered as system state. The payoff of each secondary user may be dependent on the system state. For example, it may be more important for a secondary user with more packets in its queue to access the spectrum than for a secondary user with less packets. The evolution of queue lengths is also dependent on the actions taken by the secondary users, namely the channels to access. We can such multistage and state dependent games *stochastic games*. Note that a stochastic game can be considered as an extension of one-stage games to multiple-stage ones. Meanwhile, we can also consider it as an extension from the one-decider optimization problems discussed in Chapter 8 to the case of multiple rational deciders.

To describe stochastic games, in addition to the elements defined for strategic-form games, we have the following elements:

- System State: For simplicity, we assume that there are finitely many system states, denoted by $s_1, \ldots, s_M$. The set of possible states is denoted by $\mathcal{S}$.
- State Transitions: We assume that the state transitions are Markovian; that is, the state transition is dependent on only the current system state and the players' actions. We

denote by $Q(s_m|s_n, a_1, a_2)$ the probability that the system state transits from s_n to s_m when the players' actions are a_1 and a_2, respectively.

- Payoff: In each stage, each player receives some payoff which is determined by the actions and current system state. We denote by $r_i(a_1, a_2, s_m)$ the payoff of player i when the actions are a_1 and a_2, and the system state is s_m. Then, the total payoff is accumulated through the stages. There are two possible definitions for the total payoff. One is the discounted sum of rewards, which is given by

$$R_i = \sum_{t=0}^{\infty} \beta^t r_i(a_1(t), a_2(t), s_m(t)), \tag{10.12}$$

where $0 < \beta < 1$ is a discounting factor. The other definition of total payoff is the average one, which is given by

$$R_i = \lim_{T \to \infty} \frac{1}{T} \sum_{t=0}^{T} r_i(a_1(t), a_2(t), s_m(t)). \tag{10.13}$$

For simplicity, we consider only the discounted sum in (10.12) due to the simplicity of analysis. The analysis for the average payoff is much more complicated, which can be found in [1288].

- State Dependent Strategy: Now the strategy of each player should be dependent on the current system state since it needs to consider the payoff subject to the current state as well as the future system state evolution. We denote by $\pi_i(\cdot|s)$ the strategy of player i when the current system state is s and by π_i the set $\{\pi_i(\cdot|s)\}_i$.

For the stochastic game, we define the Nash equilibrium as the strategy pair π_1^* and π_2^* such that

$$\begin{cases} E[R_1](\pi_1, \pi_2^*) \le E[R_1](\pi_1^*, \pi_2^*) \\ E[R_2](\pi_1^*, \pi_2) \le E[R_2](\pi_1^*, \pi_2^*) \end{cases}. \tag{10.14}$$

Again, at the Nash equilibrium, unilaterally changing the strategy does not increase the expected payoff. The only difference from the one-stage game is that the stochastic game needs to consider the rewards along infinitely many stages.

Next we will study the Nash equilibrium of stochastic games. For simplicity of analysis, we assume that the game is a zero-rum one, namely $r_2 = -r_1$. The discussion of general-sum games can be found in [1288]. First, we define matrices $\mathbf{R}(s)$, $\forall s \in \mathcal{S}$. For a given system state s, $\mathbf{R}(s)$ contains the payoffs of player 1 with respect to different action pairs, namely

$$(\mathbf{R}(s))_{mn} = r_1(s, a_{1m}, a_{2n}). \tag{10.15}$$

Since we consider only zero-sum games, the payoff information for each state can be summarized in a matrix. Then, we define a value vector $\mathbf{v}$, each element of which corresponds to the expected payoff of a state. For example, $\mathbf{v}_1$ is the expected future payoff of player 1 when the current state is s_{11}. Based on the definitions of $\mathbf{R}(s)$ and $\mathbf{v}$, we define another matrix $\tilde{\mathbf{R}}(s, \mathbf{v})$ as

$$(\tilde{\mathbf{R}}(s, \mathbf{v}))_{mn} = r(s, a_{1m}, a_{2n}) + \beta \sum_{s' \in \mathcal{S}} p(s'|s, a_{1m}, a_{2n}) \mathbf{v}(s'), \tag{10.16}$$

where $\mathbf{v}(s')$ is the element of $\mathbf{v}$ corresponding to state s'. Readers who have read Chapter 8 carefully can feel familiar with (10.16). Yes, it is very similar to the Bellman's equation for dynamic programming. We can take a closer look at the two terms on the right hand side of (10.16): the first term is the instantaneous payoff of player 1 when the actions are a_{1m} and a_{2n}; the second term is the expected payoff in the future since all possible state transitions are considered and $\mathbf{v}(s')$ means the future payoff when the next system state is s'. Obviously, the left hand side of (10.16) is the expected payoff when the actions are a_{1m} and a_{2n}, and the current system state is s.

Obviously, $\mathbf{R}$ is known since the instantaneous payoffs are assumed to be known. If $\mathbf{v}$ is known, it is also easy to obtain $\tilde{\mathbf{R}}$. However, $\mathbf{v}$ is still unknown. Without $\mathbf{v}$, we are unable to evaluate the future rewards when certain actions are taken. Fortunately, Shapley showed that the value vector $\mathbf{v}$ can be determined by $\tilde{\mathbf{R}}$ via the following equation [1290]:

$$\mathbf{v}(s) = val\,[\tilde{\mathbf{R}}(s, \mathbf{v})], \tag{10.17}$$

where val means the value of the zero-sum game determined by matrix $\tilde{\mathbf{R}}(s, \mathbf{v})$.

Equation (10.17) looks surprisingly elegant; actually it is very intuitive. We can understand (10.17) in the following way. First, we assume that $\mathbf{v}$ has been given by a genie. Hence, when the players take certain actions, the instantaneous payoff and the expected future payoff can be determined from $\mathbf{R}$ and $\tilde{\mathbf{R}}$, respectively. Then, given the current system state, we can simplify the multiple-stage game into a single-stage one by incorporating the expected future payoff into the instantaneous one, thus obtaining a zero-sum game with the payoff matrix $\tilde{\mathbf{R}}$. Note that we have multiple one-stage zero-sum games since each system state corresponds to a game. Finally, when the players take actions, they will choose Nash equilibrium ones corresponding to the equivalent zero-sum game. The value of the zero-sum game is then equal to the corresponding element in the value vector $\mathbf{v}$. Once $\mathbf{v}$ is determined, the strategies at the Nash equilibrium, given the system state, can also be obtained from analyzing the zero-sum game with $\tilde{\mathbf{R}}$.

Although we have found the equation describing the value vector $\mathbf{v}$, it is still unclear how to compute $\mathbf{v}$ since we do not have an explicit expression for the functional val. Fortunately, we have some efficient algorithms for computing the Nash equilibrium of some special stochastic games, which will be explained below.

First, we consider the special case in which the system state is controlled by only one player. Without loss of generality, we assume that this controlling player is player 1. Then, the value vector and the strategy π_2 can be obtained by solving the following linear programming problem:

$$\min_{\mathbf{v}, \pi_2} \sum_{s \in \mathcal{S}} \mathbf{v}(s)$$

$$s.t. \quad v(s) \geq \sum_{a_2} \mathbf{R}(s)_{a_1, a_2} \pi_2(a_2|s) + \beta \sum_{s' \in \mathcal{S}} p(s'|s, a_1)\, \mathbf{v}(s'), \qquad s \in \mathcal{S}, \forall a_1$$

$$\sum_{a_2} \pi_2(a_2|s) = 1, \qquad \forall s \in \mathcal{S}$$

$$\pi_2(a_2|s) \geq 0, \qquad s \in \mathcal{S}, \forall a_2. \tag{10.18}$$

There are many efficient algorithms to solve the above linear programming problem, for example, the simplex method or the interior-point method. The rigorous proof of the conclusion that the above linear programming results in the Nash equilibrium of the single-controller stochastic game is omitted due to the limited space. Here we can provide some intuitions for the linear programming problem. The second and the third constraints are obvious since they are simply the requirements of normalization and nonnegativity of probabilities. The first constraint means that, given the action of player 1, the strategy of player 2 can always try to reduce the payoff of player 1; hence, the right hand side of the constraint is always less than or equal to the actual value. In the objective function, the average value is minimized, which represents the impact of player 2's action. Note that the dual form of the linear programming can also be used to compute the value vector and the strategy of player 1 at the Nash equilibrium can be obtained correspondingly. The details can be found in [1288, pp. 94–95].

We notice that the assumption of single state controller is implicitly embedded in the first constraint in (10.18), where the state transition is independent of the strategy of player 2 since the state is dependent on only player 1. When the system state is dependent on the actions of both players, the optimization problem can be rewritten as

$$\min_{\mathbf{v}, \pi_2} \sum_{s \in \mathcal{S}} \mathbf{v}(s)$$

$$s.t. \quad v(s) \geq \sum_{a_2} \mathbf{R}(s)_{a_1, a_2} \pi_2(a_2|s) + \beta \sum_{a_2} \sum_{s' \in \mathcal{S}} p(s'|s, a_1, a_2) \mathbf{v}(s') \pi_2(a_2|s), \qquad s \in \mathcal{S}, \forall a_1$$

$$\sum_{a_2} \pi_2(a_2|s) = 1, \qquad \forall s \in \mathcal{S}$$

$$\pi_2(a_2|s) \geq 0, \qquad s \in \mathcal{S}, \forall a_2, \tag{10.19}$$

where we added the impact of the strategy of player 2 to the first constraint. The involvement of the strategy of player 2 also makes the optimization problem nonlinear since there exist products of value $\mathbf{v}(s)$ and probability $\pi_2(a_2|s)$. Hence, we are no longer able to solve the optimization problem using linear programming approaches. Many other approaches like Newton's method can be used to solve the optimization problem. More details can be found in Section 3.7 of [1288].

In the remainder of this chapter, we will use three typical games in cognitive radio, namely the primary user emulation attack, channel synchronization and collaborative spectrum sensing, to illustrate the above explanations of game theory. The types of the three games are summarized in Table 10.1.

Table 10.1 Optimal Strategies for Cases 1 and 2

Game Type	PUE Attack	Chan. Synch.	Spectrum Sensing
Strategic-form game	X	—	—
Bayesian game	—	X	X
Stochastic game	X	—	—
Zero-sum game	X	—	X
Collaboration game	—	X	—

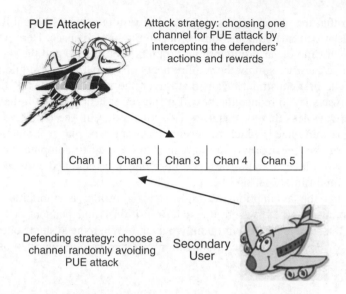

PUE Attacker   Attack strategy: choosing one
                          channel for PUE attack by
                          intercepting the defenders'
                          actions and rewards

| Chan 1 | Chan 2 | Chan 3 | Chan 4 | Chan 5 |

Defending strategy: choose a   Secondary
channel randomly avoiding   User
PUE attack

Figure 10.2   Illustration of the dogfight in spectrum.

10.2   Primary User Emulation Attack Games

In this section, we consider another type of game in cognitive radio network, namely the primary user emulation (PUE) attack game. PUE attack is a serious threat to cognitive radio networks; hence, it is important to analyze the game between secondary users and PUE attackers. Meanwhile, it is also a good example to illustrate how to analyze stochastic games.

10.2.1   PUE Attack

The dynamical spectrum access in cognitive radio, particularly the spectrum sensing mechanism, also incurs vulnerabilities for the communication system. One is the false report attack in collaborative spectrum sensing, which has been discussed in the previous section. Another serious threat is the primary user emulation (PUE) attack, originally proposed in [1291]. As illustrated in Figure 10.2, in PUE attacks, the attacker sends out signal emulating that of primary users during the spectrum sensing period, such that the secondary users will be "scared" away even if the spectrum is actually idle, based on the assumption that it is difficult for secondary users to distinguish the signals of primary user and the attacker. This assumption is usually true, especially when energy detection is used in spectrum sensing. Such a PUE attack is very efficient for the attacker since only very weak power is consumed due to the high requirement on the spectrum sensing sensitivity of secondary users; hence, it is much more power efficient (for the attacker side) than traditional jamming attackers which use high power to suppress the legitimate signals.

There are usually two approaches to combat the PUE attack:

- Proactive Approach [1292, 1293]: In this approach, the secondary users detect the attacker in a proactive manner. Although the secondary users cannot distinguish the signal structures of primary users and attackers, they can collaboratively estimate the transmit power of the radio emitter. It is assumed that the attacker has much less transmit power than the primary user, which is reasonable if the primary user is TV station. Then, the radio emitter with low transmit power will be considered as an attacker. Such an approach is very effective for cognitive radio in the TV band. However, if primary users can also have low transmit power, it is impossible to distinguish the attacker from primary user by merely considering the signal power.

- Passive Approach [1294, 1295]: If the proactive approach does not work, we can only carry out passive approach. We assume that there are multiple channels in the licensed spectrum, which is true for practical systems. Furthermore, we assume that the attacker cannot cover all channels since it requires expensive equipments for a wideband transmission. Hence, the secondary users can sense/access channels in a random manner such that the attacker is unable to always block the transmission (of course the probability that a secondary user happens to sense the channel that the attacker is attacking is nonzero). This is similar to the frequency hopping in jamming and anti-jamming; hence it is called dogfight in spectrum in [1294].

In this section, we adopt the passive approach and model it as a game between the cognitive radio network and the PUE attacker.

10.2.2 Two-Player Case: A Strategic-Form Game

We first consider the simplest case, in which there is one attacker and one cognitive radio transmitter. We consider a cognitive radio system having N licensed channels. We denote by p_{nI} the idle probability of channel n. Without loss of generality, we assume that $p_{1I} < p_{2I} < \ldots < p_{NI}$. For simplicity of analysis, we assume that the attacker (secondary user) can attack (sense) one channel at a time. If they happen to choose the same channel, the secondary user will be unable to use this channel; otherwise, whether the secondary user can use this channel depends on only the activity of primary user. We consider only one stage and will extend it to multiple stages later.

Then, we can model the dogfight as a strategic-form game. Below are the elements of the game:

- Player: We assume that there are two players: player 1 is the secondary user (the transmitter) and player 2 is the PUE attacker.
- Action and Strategy: In such a dogfight game, the action space of the secondary user (attacker) contains the choice of channel to sense (to jam). The strategies of the secondary user and attacker are the probabilities to sense and jam different channels, denoted by $\{u_i\}_{i=1,\ldots,N}$ and $\{v_i\}_{i=1,\ldots,N}$, respectively.
- Reward: For player 1 (the secondary user), we assume that it receives a reward p_{iI} when it chooses channel i to sense and the channel is not attacked by the PUE attacker. Note that this reward is the idle probability of channel i. Hence, it is actually the expectation of the actual reward if we define the actual reward as 1 when the secondary user finds an idle channel to transmit. The definition of the reward simplifies the analysis since it does not involve the actual state of primary user. When channel i is being attacked by player 2, player 1 receives reward 0. We assume that this is a zero-sum game since

the two players have completely conflicting interests. Then, the reward of player 2 is also determined.

The Nash equilibrium of the dogfight game is disclosed in the following theorem [1294]. An interesting observation is that some channels exist with bad qualities (that is, small idle probabilities) that both players will not access. The probabilities of the corresponding actions are then equal to zero.

Theorem 10.1 *Define K as*

$$K = \max \left\{ k \left| \frac{\frac{k-1}{P_{N-k+1,I}}}{\sum_{j=N-k+1}^{N} \frac{1}{p_{jI}}} < 1 \right. \right\}. \tag{10.20}$$

Then, there is a unique Nash equilibrium point in the game of spectrum dogfight, which is given by

$$u_i = \begin{cases} \frac{\frac{1}{p_{iI}}}{\sum_{j=N-K+1}^{N} \frac{1}{p_{jI}}}, & i = N - K + 1, \ldots, N, \\ 0, & i = 1, \ldots, N - K \end{cases} \tag{10.21}$$

and

$$v_i = \begin{cases} 1 - \frac{\frac{K-1}{p_{iI}}}{\sum_{j=N-K+1}^{N} \frac{1}{p_{jI}}}, & i = N - K + 1, \ldots, N \\ 0, & i = 1, \ldots, N - K \end{cases} \tag{10.22}$$

The above Nash equilibrium is illustrated using real measurement data using the system shown in Figure 10.3. An E4407B-COM ESA-E spectrum analyzer is used to collect the spectrum activity. The range of 2.4 GHz to 2.5 GHz is divided into 20 channels, each spanning 5 MHz. The measurement is carried out for both inside and outside the Ferris Hall of the University of Tennessee. The busy probabilities of the 20 channels are shown in Figure 10.4.

For both the indoor and outdoor measurements, we show the sensing/jamming probabilities at the Nash equilibrium in Figure 10.5. We observe that, for the indoor case, $K = 19$, that is, only one channel will not get involved in the game; while $K = 18$ for the outdoor environment. Note that this conclusion is valid for only the set of measurements used in the simulation. It may not be true for other spectrum environments.

10.2.3 Game in Queuing Dynamics: A Stochastic Game

In the previous discussion, we consider only two players. However, in practice, there could be multiple PUE attackers while the defender could be the whole cognitive radio network. Moreover, the goal of the strategic game in the previous discussion is essentially increasing/decreasing the throughput of the secondary user transmitter. However, this goal may not be reasonable in a network with delay-tolerant traffics. For example, if the goal of the secondary user is to deliver all packets to the destination regardless of the delay, the

Figure 10.3   A picture of the spectrum measurement system.

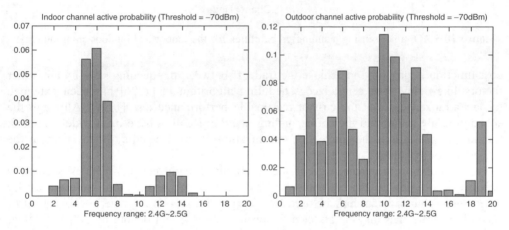

Figure 10.4   The probabilities of different channels of both indoor (upper figure) and outdoor (lower figure) environments.

attacker has no reward if all packets are eventually delivered. Hence, here we extend the strategic game of the PUE attack to the more interesting scenario of network-wide game for the queuing dynamics in the cognitive radio network. Since the reward is dependent on the system state, namely the queue lengths of each node, the game is a stochastic one over multiple stages. Briefly speaking, the goals of the players are to stabilize (the cognitive radio network side) or destabilize (the attacker side) the queuing dynamics in the cognitive radio network.

Note that the queuing dynamics have been widely studied in wireless communication system. In their seminal work [1296], Tassiulas and Ephremides proposed a scheduling algorithm for wireless communication networks that achieves the maximal throughput region. In the context of cognitive radio network, the scheduling algorithm is extended [1297], which will be explained in details in Chapter 10. In [1298], a "drift-plus-penalty"

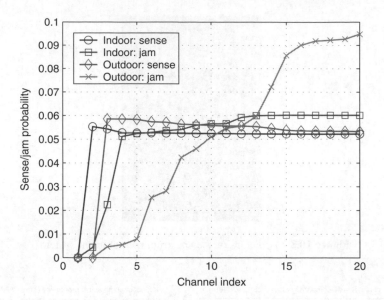

Figure 10.5 Optimal sensing/jamming probabilities for the indoor and outdoor environments.

cost function is proposed to achieve the tradeoff between the queuing stability and other factors like delay. The centralized scheduling algorithm in [1296] has been extended to decentralized cases at the cost of reasonable performance loss [1299]. Although the scheduling algorithm and the corresponding queuing stability have been widely studied, they are almost all on the single side optimization of the scheduling policy without the consideration of attacks.

For formulating the game, we consider a cognitive radio network with N secondary users, whose topology can be represented by a graph. We assume that there are totally M licensed channels that may be used by K primary users. We denote by $\mathcal{N}_k$ the set of secondary users that may be affected by primary user k and denote by $\mathcal{M}_k$ the set of channels that primary user k occupies when it is active. For simplicity, we assume that the activities in different time slots of each primary user are mutually independent, and the probability of being active is denoted by p_k for primary user k. At time slot t, the status of channel m is denoted by s_m; that is, $s_m = 0$ when the channel is not being used by primary users and $s_m = 1$ otherwise. Due to the limited capability of spectrum sensing, we assume that each secondary user can sense only one channel during the spectrum sensing period.

We assume that there are totally F data flows in the cognitive radio network. We denote by S_f and D_f the source and destination nodes of flow f, respectively. We assume that the packet arrival at the source node of data flow f is Poisson with expectation a_f. The routing paths of the F data flows can be represented by an $F \times N$ matrix $\mathbf{R}$, in which $R_{fn} = 1$ if data flow f passes through secondary user n and $R_{fn} = 0$ otherwise. We denote by $\mathcal{I}_n$ the set of data flows passing through secondary user n.

For each flow, the data are packetized using the same packet length. Each secondary user has one buffer for each data flow passing through it. In each time slot, the secondary users will choose one packet from its buffer(s), if there is any, for the opportunistic spectrum access. Suppose that one channel can support only one data flow in one time slot. We assume that there are sufficiently many channels such that any set of interfering secondary users can be assigned to different channels. Thus, all secondary users can transmit simultaneously (if there is no primary user) by appropriately allocating the channels.

When secondary user n decides to transmit to a neighbor j, and an idle channel, say channel m, is assigned to secondary user n, the packet can be delivered successfully with probability p_{njm} which is determined by the channel quality. Hence, the probability that a packet can be delivered is given by

$$\mu_{njm} = p_{njm} \prod_{k:n\in\mathcal{N}_k,m\in\mathcal{M}_k} (1 - p_k). \tag{10.23}$$

We assume that there are totally L PUE attackers. In each time slot, each attacker chooses Q $(Q \leq M)$ channels to attack. We denote by $\mathcal{V}_l$ the set of potential secondary user victims that are jammed by attacker l. We assume that the attackers have perfect knowledge about the current state of the cognitive radio network. Such an assumption can make the game theoretic analysis easier. It provides a starting point for more complicated cases in which the attackers have only partial observations on the network state. Moreover, this assumption is reasonable if any node in the cognitive radio network is compromised, or the attackers have acquired the secrecy key and can decode/decypher the messages like current queue lengths.

Then, we formulate the game between the cognitive radio network and the attackers in the following way:

- Players: We consider only two players, namely the cognitive radio network and the attackers. This implicitly assumes that there are two centralized controllers making decisions for the network and the attackers, respectively.
- System State: We denote by $\mathbf{s}$ the system state, which is composed by all queue lengths (denoted by $\{q_{fn}\}_{f=1,\ldots,F,n=1,\ldots,N}$). The state space is denoted by $\mathcal{S}$.
- Actions: The set of actions of the attackers and secondary users are denoted by $\mathcal{A}_a$ and $\mathcal{A}_s$, respectively. The actions of the attackers, denoted by $\mathbf{a}_a$, include the channels to jam of each attack, which are denoted by $\{\mathbf{c}_l^a\}_{l=1,\ldots,L}$ ($\mathbf{c}_l$ is a vector containing the Q channels to jam). The actions of the secondary users, denoted by $\mathbf{a}_s$, have more elements. It consists of the channel assignment, as well as the flow schedule (which flow to choose the packet if there are multiple flows?). We denote by $c_n(t)$ and $f_n(t)$ the assigned channel and scheduled flow at secondary user n at time slot t.
- Reward: This is the key element in this PUE game for queuing dynamics. Recall that the goals of the players are to stabilize and destabilize the queuing dynamics, respectively. Hence, we need a quantity to characterize the system stability. We follow the analysis in [1296], we define the following Lyapunov function, which is given by

$$V(\mathbf{s}(t)) = \sum_{f=1}^{F} \sum_{n=1}^{N} q_{fn}^2(t), \tag{10.24}$$

that is, the square sum of all queue lengths. Obviously, a larger Lyapunov function implies more instability in the queuing dynamics. As illustrated in Figure 10.6, the

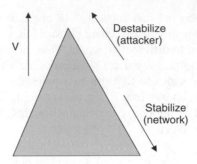

Figure 10.6   An illustration of the Lyapunov function.

Lyapunov function means the energy of the queuing system. The attacker wants to increase it while the network wants to decrease it. We can rewrite $V(\mathbf{s}(t))$ as

$$V(\mathbf{s}(t)) = V(\mathbf{s}(0)) + \sum_{r=1}^{t} V(\mathbf{s}(r)) - V(\mathbf{s}(r-1)). \tag{10.25}$$

Then, we observe that the Lyapnov function is the sum of the Lyapunov drift of each time slot, namely the increase of the Lyapunov function $d(t) = E[V(\mathbf{s}(t)) - V(\mathbf{s}(t-1))]$ [1296, 1298]. Hence, we can define the Lyapunov drift $d(t)$ as the reward of the attacker. For a positive $d(t)$, the system becomes more unstable, thus benefiting the attackers. We model the game as a zero-sum one, thus defining the payoff of the network. We add a discounting factor $0 < \beta < 1$ to the reward such that the total payoff of the attacker is given by

$$R = \sum_{t=0}^{\infty} \beta^t d(t), \tag{10.26}$$

which simplifies the analysis since it is much easier to analysis the game with a discounted sum of rewards. It is also possible to consider the average of the Lyapunov drift, which makes the analysis more complicated. Note that this definition of reward is motivated by the classical works on scheduling queuing network in which the scheduling algorithm tries to minimize the Lyapnov drift in order to stabilize the queuing dynamics [1296, 1298].

We applied the Shapley's theorem and the corresponding numerical approach, which are introduced in Section 10.1, to compute the Nash equilibrium defined above. The example is illustrated in Figure 10.7, in which there is one attacker and three secondary users. We assume that there are only two channels over which two data flows are sent from secondary user 3 to secondary users 1 and 2, respectively. The attacker can only interfere secondary user 3. For simplicity, we assume that secondary user 3 can sense and transmit over both channels simultaneously; hence, there are only two possible actions for secondary user 3. When computing the Nash equilibrium, we assume that the strategy is the same for the state of more than 10 packets in either buffer; otherwise, there will be infinitely many system states. Note that the computation is prohibitively difficult to large networks. Hence, we consider only the small scale network.

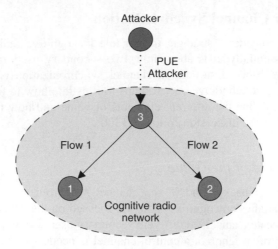

Figure 10.7 An illustration of the example.

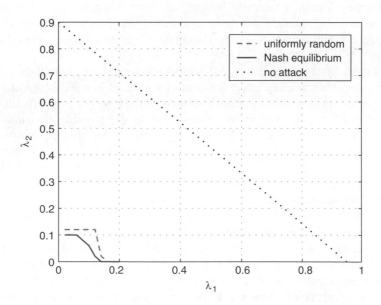

Figure 10.8 Rate region subject to PUE attacks in Figure 10.7.

In Figure 10.8, we show the rate region subject to PUE attacks for the network in Figure 10.7. We judge whether a given set of rates is stable by carrying out the simulation for the queuing dynamics; if one of the queues has more than 50 packets after 2000 time slots, we claim that the rates are unstable. We tested the case of Nash equilibrium, uniformly choosing the actions and no PUE attack. The region of each case is the area below the corresponding curve. We observe that the PUE attack can cause a significant reduction of the rate region.

10.3   Games in Channel Synchronization

In this section, we consider a strategic form game in cognitive radio with the essence of collaboration. Essentially, it is about how two secondary users can synchronize the channel information. Since a success in channel synchronization is beneficial to both players, the game is a collaborative one. It helps to illustrate how to formulate a problem into a game, how to define the different elements of game and how to analyze the Nash equilibrium. Note that the discussion follows [1300].

10.3.1   Background of the Game

Now, we introduce the background of the channel synchronization game in cognitive radio. Most current studies on cognitive radio are focused on the data communication, for example, how to allocate channels to different secondary user during the data transmission period. However, as often ignored, a control channel is needed to convey control signalings, for example, ACK/NACK messages, routing tables or SYN messages in transport layer. Uniquely to cognitive radio, control channel is needed for channel synchronization; that is, the transmitter needs to inform the receiver which channel it will use for its data transmission, since the available channels could be dynamic and change with time. Therefore, control channel, as the backbone of cognitive radio networks, is a key design issue.

Many wireless systems use dedicated channels, for example, a frequency channel, a set of time slots or a spreading code, for control signaling. For reliability, the dedicated resource is predetermined since the whole band is fixed. However, it is difficult to allocate a reliable channel in cognitive radio systems due to the possibly dynamic spectrum. One possible approach is to use UWB signal, which can overlay on existing wireless systems and does not need dedicated channel. However, UWB signal is limited by its short transmission range (around 10 to 15 meters) and by the fact that it is typically used in indoor environments. Another approach is to use unlicensed frequency band such as the industrial, scientific and medical (ISM) band. However, it has to compete with WLAN such as IEEE 802.11 which also use this band and may cause substantial interference and damage to the control signals in cognitive radio systems.

In this book, we assume that a set of licensed channels are used as the control channel. Similar schemes are also considered in [1301, 1302]. In these studies, symmetric environments are assumed at both the transmitter and receiver; that is, they share the same spectrum occupancies. However, in practical systems, the transmitter may not have the information that some channels have been strongly interfered by primary users at the receiver's location. Therefore, if the transmitter uses only one frequency channel for the control signaling, the receiver may never receive it and then loses connection. Hence, the receiver should not monitor only one channel; we need to intelligently synchronize frequency channels for transmission in the control channel.

10.3.2   System Model

The system is illustrated in Figure 10.9. We consider two secondary users, one planning to transmit a message to the other in the control channel. Suppose that the control channel

Figure 10.9   Illustration of the channel synchronization.

contains N licensed frequency channels while the transmitter/receiver can transmit/receive over only one channel. We denote by p_n and q_n the probabilities that channel n is idle at the transmitter and receiver sides, respectively. Stacking them into vectors, we define $\mathbf{p} = (p_1, \ldots, p_N)$ and $\mathbf{q} = (q_1, \ldots, q_N)$. For simplicity, we assume that all these probabilities are nonzero.

For each frequency channel, we define the state as whether it is occupied by primary users (B: Busy) or not (I: Idle). We have the following assumptions:

- The transmitter (receiver) knows the states of all channels at its own location before each time slot, which can be accomplished by spectrum sensing. However, they do not know the spectrum state of their partners.
- We consider a perfect spectrum sensing; that is, there is no spectrum sensing error.
- We assume that the occupancies of different frequency channels are mutually independent. The spectrum situations at the transmitter and receiver are also mutually independent.
- Both transmitter and receiver know the beginning of channel synchronization, which can be achieved by a perfect time synchronization and a uniform timing structure.

For simplicity, we consider only one-stage synchronization; that is, if the receiver fails in choosing the same channel as the transmitter does, then the synchronization fails and does not continue. The discussion on multistage synchronization can be found in [1300]. We denote by $\mu_n^{\mathbf{T}}$ the probability of choosing channel n for transmission when the set of idle bands is $\mathbf{T}$ at the transmitter. Similarly, we denote by by $\nu_n^{\mathbf{R}}$ the probability of choosing band n to receive when the set of idle bands is $\mathbf{R}$ at the receiver.

10.3.3   Game Formulation

Based on the above mechanism of channel synchronization, we can formulate it as a game and define the following elements of the game:

- Player: Obviously, the transmitter and the receiver are the two players.
- Action: For each player, the action space is the selection of frequency channel to access. The mixed strategies for the transmitter and receiver are the probabilities of transmitting and receiving over different channels, that is, $\{\mu_n^{\mathbf{T}}\}_{n,\mathbf{T}}$ and $\{\nu_n^{\mathbf{R}}\}_{n,\mathbf{R}}$, respectively. We denote by $S_{\mathbf{T}}^{TX}$ and $S_{\mathbf{R}}^{RX}$ the mixed strategies of the transmitter and receiver when the corresponding sets of usable bands are $\mathbf{T}$ and $\mathbf{R}$, respectively.

- Type: Since the two players do not know the spectrum situation of each other, this collaboration game has incomplete information and can be modeled as a Bayesian game. Each player has a type, namely the set of usable frequency channels **T** or **R**, which is known to itself but unknown to its collaborator. Since the statistics of frequency channels are assumed to be perfectly known to both players, the *a priori* probability of the receiver's type can be computed by the transmitter, namely

$$P(\mathbf{R}) = \prod_{n \in \mathbf{R}} q_n \prod_{m \notin \mathbf{R}} (1 - q_m). \tag{10.27}$$

Similarly, the *a priori* probability of the transmitter's type can also be computed by the receiver, which is given by

$$P(\mathbf{T}) = \prod_{n \in \mathbf{T}} p_n \prod_{m \notin \mathbf{T}} (1 - p_m). \tag{10.28}$$

- Reward: If the transmitter and receiver choose the same frequency band, the reward is 1; otherwise, the reward is 0. Hence, in this game, both players share the same reward.

10.3.4 Bayesian Equilibrium

According to the definition of Bayesian equilibrium, if $\{\mu_n^{\mathbf{T}}\}_n$ and $\{v_n^{\mathbf{R}}\}_n$ are Bayesian equilibrium strategies, they should satisfy the following equations:

$$S_{\mathbf{T}}^{TX} = \arg\max \sum_{\mathbf{R}} P(\mathbf{R}) \left(\sum_{j \in \mathbf{R}, \mathbf{T}} v_j^{\mathbf{R}} \mu_j^{\mathbf{T}} \right) \tag{10.29}$$

and

$$S_{\mathbf{R}}^{RX} = \arg\max \sum_{\mathbf{T}} P(\mathbf{T}) \left(\sum_{i \in \mathbf{R}, \mathbf{T}} \mu_i^{\mathbf{T}} v_i^{\mathbf{R}} \right). \tag{10.30}$$

We can search for equilibrium points using the following procedure: let $D_{\mathbf{T}}^{Tx}$ and $D_{\mathbf{R}}^{Rx}$ denote the sets of bands whose sensing probabilities are nonzero when the sets of usable bands at the transmitter and receiver are **T** and **R**, respectively, that is,

$$D_{\mathbf{T}}^{Tx} = \{n | \mu_n^{\mathbf{T}} > 0, n \in \mathbf{T}\}, \tag{10.31}$$

and

$$D_{\mathbf{R}}^{Rx} = \{n | v_n^{\mathbf{R}} > 0, n \in \mathbf{R}\}. \tag{10.32}$$

For all possible combinations of $D_{\mathbf{T}}^{Tx}$ and $D_{\mathbf{R}}^{Rx}$ and for all possible **T** and **R**, we have the following equations:

$$\sum_{i \in \mathbf{T}} P(\mathbf{T}) \mu_i^{\mathbf{T}} = C_1(D_{\mathbf{R}}^{Rx}), \qquad \forall i \in D_{\mathbf{R}}^{RX}, \forall \mathbf{R}, \tag{10.33}$$

and

$$\sum_{j \in R} P(\mathbf{R}) v_j^{\mathbf{T}} = C_2(D_{\mathbf{T}}^{Tx}), \qquad \forall j \in D_{\mathbf{T}}^{Tx}, \forall \mathbf{T}, \tag{10.34}$$

as well as the constraints

$$\sum_{i \in D_{\mathbf{T}}^{Tx}} \mu_i^{\mathbf{T}} = 1, \tag{10.35}$$

and

$$\sum_{j \in D_{\mathbf{R}}^{Rx}} v_j^{\mathbf{R}} = 1. \tag{10.36}$$

Note that $C_1(D_{\mathbf{R}}^{Rx})$ and $C_2(D_{\mathbf{T}}^{Tx})$ are constants (to be determined) independent of i and j but dependent on $D_{\mathbf{R}}^{Rx}$ and $D_{\mathbf{T}}^{Tx}$, respectively. Intuitively, Equations (10.33) and (10.34) mean that the players are indifferent among the selections of bands whose probabilities are positive.

For all combinations of $\{D_{\mathbf{T}}^{Tx}\}_{\mathbf{T}}$ and $\{D_{\mathbf{R}}^{Rx}\}_{\mathbf{R}}$, we list all Equations (10.33) and (10.34), as well as the constraints (10.35) and (10.36, and then solve it. It is possible that, for some combinations of $D_{\mathbf{T}}^{Tx}$ and $D_{\mathbf{R}}^{Rx}$, there is no solution for the above equations. It is also possible that there are multiple (maybe uncountable) solutions. In this case, we choose only the solution that maximizes a certain linear object function, thus converting the procedure of solving equations to an optimization problem which can be efficiently solved by linear programming. In this chapter we consider the following linear programming problem:

$$\max_{\{\mu_n^{\mathbf{T}}\}_{n,\mathbf{T}}, \{\mu_n^{\mathbf{R}}\}_{n,\mathbf{R}}} \sum_{\mathbf{T},\mathbf{R}} P(\mathbf{T}) C_2(D_{\mathbf{T}}^{Tx}) + P(\mathbf{R}) C_1(D_{\mathbf{R}}^{Rx}), \tag{10.37}$$

subject to constraints (10.34), (10.33), (10.35) and (10.36), as well as $0 \leq \mu_i^{\mathbf{T}} \leq 1$ and $0 \leq v_i^{\mathbf{R}} \leq 1$.

After exhaustive search for all possibilities of $\{D_{\mathbf{T}}^{Tx}\}_{\mathbf{T} \subset \{1,\dots,n\}}$ and $\{D_{\mathbf{R}}^{Rx}\}_{\mathbf{R} \subset \{1,\dots,n\}}$, we can obtain all equilibrium points. It is easy to check that we need to verify

$$\prod_{t=1}^{N} (2^t - 1)^{\binom{N}{t}} \tag{10.38}$$

possibilities for $\{D_{\mathbf{T}}^{Tx}\}_{\mathbf{T} \subset \{1,\dots,n\}}$ and the same for $\{D_{\mathbf{R}}^{Rx}\}_{\mathbf{R} \subset \{1,\dots,n\}}$. When N is not small, we need to search prohibitively large number of possibilities, which is infeasible for numerical computations. Therefore, in numerical simulations, we consider only the case of small N. For large N, it is still an open problem to find an efficient algorithm to compute the equilibria according to the features of the collaboration game. The application of the above procedure in the simplest case of $N = 2$ is left as an exercise problem.

10.3.5 Numerical Results

We consider the case $N = 3$, that is, there are three available bands. According to (10.38), there are 189 possible configurations of nonzero input distributions for the transmitter

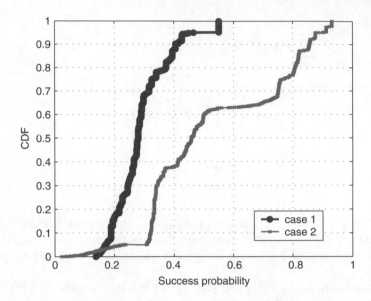

Figure 10.10   CDF of the success probability for different equilibrium points.

Table 10.2   Optimal Strategies for Cases 1 and 2

available bands	Tx Case1	Rx Case1	Tx Case2	Rx Case2
1,2	2	1	1	1
1,3	1	1	1	1
2,3	2	2	2	2
1,2,3	2	1	1	1

(same for the receiver). Therefore, we need to check 35721 possible joint configurations. We consider two cases. In Case 1, $\mathbf{p} = (0.8, 0.6, 0.3)$ and $\mathbf{q} = (0.3, 0.9, 0.7)$. We found totally 975 equilibrium points and the highest successful synchronization probability is 0.5508. In Case 2, $\mathbf{p} = (0.99, 0.93, 0.97)$ and $\mathbf{q} = (0.94, 0.98, 0.90)$, in which 2237 equilibrium points are found and the highest successful synchronization probability is 0.9311. The cumulative distribution functions (CDFs) of the probabilities of successful synchronization of different equilibrium points are provided in Figure 10.10. We observe that different equilibrium points may yield considerably different performances.

By examining the obtained equilibrium points, we found the optimal strategies yielding the maximal success probabilities are pure strategies for both cases and the corresponding band selections are provided in Table 10.2 (we do not list the cases when there is only one available band since the corresponding band synchronization is trivial), where Tx and Rx stand for transmitter and receiver, respectively.

10.4   Games in Collaborative Spectrum Sensing

In this section, we consider the game in collaborative spectrum sensing by formulating a Bayesian game. Hence, on one hand, we can better understand the possible attack/defense in collaborative spectrum sensing; on the other hand, we can use it as an example to illustrate how to formulate and analyze Bayesian games. Note that the discussion mainly follows that in [1303].

10.4.1 False Report Attack

Collaborative spectrum sensing has been introduced in Chapter 3. Hence, the readers can check Chapter 3 for the details. In the collaborative spectrum sensing in Chapter 3, we implicitly assume that every secondary user is honest; that is, they exchange their true observations or decisions. However, this assumption may not be true in the following situations:

- When a collaborator is malicious, it may send out false reports in order to ruin the collaborative spectrum sensing.
- When a collaborator is selfish, it may report a busy channel although the channel is actually idle, such that it can use this channel by itself.
- When a collaborator is malfunctioning, for example, the spectrum sensor configuration is incorrect, the reports may not be true, although it does not intend to do so.

In all these cases, the incorrect reports may incur spectrum sensing errors, thus degrading the system performance. For simplicity, we consider only the malicious attacker case and the corresponding false report attack. Many studies have been paid to detect such false report attacks [1304–1307]. Usually these schemes are based on a centralized collaborative spectrum sensing in which a center will collect the reports and make decision on whether an attacker exists and who the attacker is. In the decentralized spectrum sensing, each secondary user needs to make its own decision. Since the type, honest or malicious, is unknown, we can model it as a Bayesian game and analyze the Bayesian equilibrium therein.

10.4.2 Game Formulation

For simplicity, we consider the spectrum sensing over only one channel. We denote by B and I the busy and idle states of primary user, respectively. For general case, we denote by S the primary user state. The corresponding *a priori* probabilities are denoted by P_B and P_I, respectively.

We consider two secondary users in the collaborative spectrum sensing, in which they exchange messages. One secondary user is honest (player 1) while the other one secondary user is malicious (player 2). Player 1 does not know the type of player 2 while player 2 knows that player 1 is honest. We denote by X_i the local observation of player i during the spectrum sensing. We assume that N possible observations during the spectrum sensing, denoted by $O_1, \ldots, O_N$, which is reasonable for practical systems due to the quantization of measurements in spectrum sensing. The observations are mutually independent for different players due to independent noise. We denote by $P(X|S)$ the probability of observing X conditioned on the channel state S, which is common for both players. It is assumed that the probability $P(X|S)$ is perfectly known to both player.

The two secondary users exchange their local observations to each other. Player 1 sends its local observation, X_1, to player 2 first. In the viewpoint of player 1, if player 2 is malicious, it sends a fake value X_2' back to player 1. Note that the fake value X_2' is determined by X_1, X_2 and its attacking strategy; if player 2 is honest, it sends the original observation X_2 to secondary user 1. Then, player 1 makes a decision on whether primary user exists, based on X_1, X_2' and its own strategy.

10.4.3 Elements of Game

For simplicity we consider only a single round of the game. It is more interesting to study multiple stages to see how the honest player accumulates its belief on its collaborator and how the malicious attacker sometimes pretends to be innocent in order to elude player 1. However, this is beyond the scope of this book.

Note that this game is slightly different from the one introduced in Section 10.1 since the actions of the two players are not taken simultaneously (recall that player 2 decides its report X_2' first and then player 1 makes a decision on the sensing result). On the other hand, the information is also asymmetric since player 2 knows that player 1 is honest while player 1 does not know the type of player 2. Hence, it is essentially a signaling game [1308], a special type of Bayesian game, in which one player (leader) has a private type while the other player (follower) has a public type. The leader takes an action first. Then, the follower decides its own action by guessing the type of the leader from the leader's action.

Below, we define the elements of the signaling game, namely type, action, strategy and reward, as illustrated in Figure 10.11:

- Type: The type of player i consists of whether it is honest (H) or malicious (M) and is denoted by c_i. Note that the observation X_i is also a part of type if it could be private. The type of player 1, that is, being honest and observation X_1, is known to both players. However, the type of player 2, namely being malicious and having observation X_2, is unknown to player 1. Player 1 has an *a priori* probability (or belief) of player 2 being malicious, which is denoted by π_M. Summarizing the above discussion, we denote by $T_i = (C_i, X_i)$ the type of player i.
- Action: The action of player 1 is the decision on the spectrum sensing result, given the report from player 2 and its own observation. The action of player 2 is how to fake the

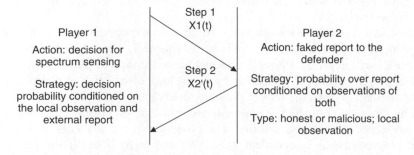

Figure 10.11 Illustration of the game elements.

report, X_2': should it send the true observation or not? If not, what observation should it report to player 1?

- Strategy: For player 1, its strategy is the probability of claiming that the channel is busy, given its own observation and the report from player 2. We denote this probability by $\pi_1(B|i, j)$ if $X_1 = O_i$ and $X_2' = O_j$. The strategy of player 2 is the probability of reporting observation $O_n, n = 1, \ldots, N$, which is denoted by $\pi_2(n|i, j)$, given $X_1 = O_i$ and $X_2 = O_j$. Recall that we use π_1 and π_2 the overall strategies of players 1 and 2, respectively.

- Reward: There are two types of costs in the spectrum sensing, namely the missed detection and false alarm. We denote by C_M the cost incurred by missed detection, that is, claiming an idle channel while primary users actually exist. Similarly, we denote by C_F the cost caused by false alarm, that is, claiming a busy channel while no primary user exists. Then, the reward of player 1 is $-C_M$ upon a missed detection and $-C_F$ upon a false alarm. The reward is 0 for a correct detection. Given observation X_1, report X_2', type T_2 and player 1's strategy π_1, the expected reward of player 1 is given by

$$r_1(\pi_1, X_1, X_2', T_2) = -C_F P(I|X_1, T_2)\pi_1(B|X_1, X_2')$$
$$- C_M P(B|X_1, T_2)(1 - \pi_1(B|X_1, X_2')), \qquad (10.39)$$

where $P(I|X_1, T_2)$ and $P(B|X_1, T_2)$ are the actual *a posteriori* probabilities of the channel being idle and busy, respectively, given the observations X_1 and X_2. It is easy to verify that $P(B|X_1, T_2)$ is given by

$$P(B|X_1, T_2) = \frac{P(X_1|B)P(X_2|B)P_B}{P(X_1|B)P(X_2|B)P_B + P(X_1|I)P(X_2|I)P_I}, \qquad (10.40)$$

where X_2 is the actual observation of secondary user 2, as a part of T_2.

We model the game as a zero-sum one since the goal of the malicious user is to ruin the collaborative spectrum sensing. When the true observations are X_1 and X_2, the expected reward of player 2 is equal to

$$r_2(\pi_1, \pi_2, X_1, X_2) = \sum_{n=1}^{N} C_M P(B|X_1, X_2)\pi_2(O_n|X_1, X_2)(1 - \pi_1(B|X_1, O_n))$$

$$+ \sum_{n=1}^{N} C_F P(I|X_1, X_2)\pi_2(O_n|X_1, X_2)\pi_1 B|X_1, O_n. \qquad (10.41)$$

Note that, although the true cost is modeled as zero-sum, the sum of the expected rewards in (10.39) and (10.41) may not be zero. The reason is that the two players have different sets of information, thus having different capabilities of predicting the rewards.

10.4.4 Bayesian Equilibrium

Now we begin to analyze the Bayesian equilibrium of the game. According to the definition of Bayesian equilibrium, the following conditions should be satisfied for the equilibrium strategies π_1^* and π_2^*:

- Player 2: For any types T_1 and T_2 (both are known to player 2), we have

$$\pi_2^*(\cdot|T_1, T_2) \in \arg\max_{\pi_2} r_2(\pi_1^*, \pi_2, X_1, X_2), \tag{10.42}$$

where the observations X_1 and X_2 are parts of the types T_1 and T_2, respectively. This condition means that the equilibrium strategy of player 2 should be optimal for any type pairs T_1 and T_2, given the strategy of player 1 π_1^*.
- Player 1: For any report X_2' and type T_1, which are the observations of player 1, we have

$$\pi_1^*(\cdot|T_1, X_2') \in \arg\max_{\pi_1} \sum_{T_2} \mu(T_2|X_2', T_1) r_1(\pi_1, X_1, X_2', T_2), \tag{10.43}$$

where $\mu(T_2|X_2', T_1)$ is the conjecture on the real type of player 2 when player 1 has type T_1 and receives a report X_2' from player 2.
- Conjecture of Type 2: The *a posteriori* probability of type T_2 given report X_2' and type T_1, $\mu(T_2|X_2', T_1)$, is given by (the derivations of the equations will be left as exercises)

$$\mu(T_2|X_2', T_1) = \frac{P(T_2|T_1)P(X_2'|T_2, T_1)}{\sum_{\tilde{T}_2} P(\tilde{T}_2|T_1)P(X_2'|\tilde{T}_2, T_1)}, \tag{10.44}$$

where the conditional probability $P(T_2|T_1)$ is given by

$$P(T_2|T_1) = \frac{P(T_1, T_2)}{P(X_1|S = B)P_B + P(X_1|S = I)P_I}, \tag{10.45}$$

and the joint probability $P(T_1, T_2)$ is given by

$$\begin{aligned} P(T_1, T_2) = &P(C_2)P_B P(X_1|S = B)P(X_2|S = B) \\ &+ P(C_2)P_I P(X_1|S = I)P(X_2|S = I), \end{aligned} \tag{10.46}$$

where

$$P(C_2) = \begin{cases} \pi_M, & \text{if } C_2 = M \\ 1 - \pi_M, & \text{if } C_2 = H \end{cases}. \tag{10.47}$$

Note that $P(X_2'|T_2, T_1)$ in (10.44) is given by

$$\begin{aligned} &P(X_2'|T_2, T_1) \\ &= \begin{cases} 1 &, \quad \text{if } C_2 = H, X_2 = X_2' \\ 0 &, \quad \text{if } C_2 = H, X_2 \neq X_2' \\ \pi_2(X_2'|X_1, X_2) &, \quad \text{if } C_2 = M \end{cases}. \end{aligned} \tag{10.48}$$

Now, we discuss the computation of Bayesian equilibrium by optimizing the strategy of player 1 first and then optimizing that of player 2, based on the above conditions of the Bayesian equilibrium.

- Player 1: We first fix the strategy of player 2 and substitute (10.39) into (10.43), thus obtaining

$$\sum_{T_2} \mu(T_2|X_2', T_1) r_1(\sigma_1, X_1, X_2', T_2)$$

$$= -C_F \left(\sum_{T_2} \mu(T_2|X_2', T_1) P(I|X_1, T_2) \right) \pi_1(B|X_1, X_2')$$

$$- C_M \left(\sum_{T_2} \mu(T_2|X_2', T_1) P(B|X_1, T_2) \right) (1 - \pi_1(B|X_1, X_2'))$$

$$= R_F(T_1, X_2') \pi_1(B|X_1, X_2') + R_M(T_1, X_2')(1 - \pi_1(B|X_1, X_2')), \qquad (10.49)$$

where $R_F(T_1, X_2')$ is the expected reward when false alarm occurs, which is negative and is defined as

$$R_F(T_1, X_2') = -C_F \sum_{T_2} \mu(T_2|X_2', T_1) P(I|X_1, T_2), \qquad (10.50)$$

and $R_M(T_1, X_2')$ is the expected reward when missed detection occurs, which is defined as

$$R_M(T_1, X_2') = -C_M \sum_{T_2} \mu(T_2|X_2', T_1) P(B|X_1, T_2). \qquad (10.51)$$

Obviously, the optimal strategy of secondary user 1, given T_1 and X_2', is given by the following pure strategy:

$$\pi_1(B|X_1, X_2') = \begin{cases} 1, & \text{if } R_M(T_1, X_2') > R_F(T_1, X_2') \\ 0, & \text{if } R_M(T_1, X_2') \le R_F(T_1, X_2') \end{cases}. \qquad (10.52)$$

Upon the event $R_M(T_1, X_2') = R_F(T_1, X_2')$, we simply assign zero to $\pi_1(B|X_1, X_2')$, since it is an event with zero probability. Intuitively, the optimal strategy of player 1 in (10.52) is to choose the decision corresponding to the minimum of the expected risks of miss detection and false alarm. Note that the current optimal strategy of player 1 is still dependent on the strategy of player 2 since (10.48) is dependent on $\pi_2(X_2'|X_1, X_2)$.

- Player 2: Now, we derive the optimal strategy of player 2. Substituting the optimal strategy of secondary user 1 into the reward of secondary user 2 in (10.41), we have

$$r_2(\pi_1^*, \pi_2, X_1, X_2) = \sum_{n=1}^{N} C_M P(B|X_1, X_2)\pi_2(O_n|X_1, X_2)$$

$$\times I(R_M(T_1, n) \le R_F(T_1, n))$$

$$+ \sum_{n=1}^{N} C_F P(I|X_1, X_2)\pi_2(O_n|X_1, X_2)$$

$$\times I(R_F(T_1, n) > R_F(T_1, n)), \tag{10.53}$$

where $I(\cdot)$ is the characteristic function of the corresponding event. Then, the equilibrium strategy π_2^* can be obtained by optimizing (10.53). We observe that the strategies corresponding to different X_1's do not couple with each other; hence, we can optimize (10.53) for different X_1's separately.

Note that the same strategy π_2 must optimize the N rewards, each corresponding to a given X_2, simultaneously, thus resulting in multiple objectives for the optimization. Although we can optimize the N rewards separately (e.g., for given X_2, we optimize $\pi_2(X_1'|X_1, X_2)$), we can convert the multiple-objective optimization into a single-objective one in the following manner:

$$\bar{r}_2(\pi_1^*, \pi_2, X_1) = \sum_{X_2} P(X_2|X_1)r_2(\pi_1^*, \pi_2, X_1, X_2), \tag{10.54}$$

whose solution must be an equilibrium point since it must maximize all individual rewards (each corresponding to an X_2) in (10.53). Summarizing (10.53) and (10.54), we have the following objective function for the strategy of player 2, which is given by

$$\sum_{X_2} P(X_2|X_1) \sum_{n=1}^{N} C_M P(B|X_1, X_2)\pi_2(O_n|X_1, X_2)$$

$$\times I(R_M(T_1, n|\pi_2) \le R_F(T_1, n|\pi_2))$$

$$+ \sum_{n=1}^{N} C_F P(I|X_1, X_2)\pi_2(O_n|X_1, X_2)$$

$$\times I(R_F(T_1, n|\pi_2) > R_F(T_1, n|\pi_2)), \tag{10.55}$$

where we added π_2 to the arguments of function R_M and R_F since both functions are dependent on the strategy of player 2.

We summarize the procedure of computation for the Bayesian equilibrium in Procedures 1 and 2, respectively. Procedure 2 is the main procedure, in which the strategy of player 2 is computed, while Procedure 1 is a subfunction of Procedure 2, which is used to compute the optimal strategy of player 1 and the functions R_M and R_F.

Procedure 1 Procedure of optimal strategy of player 1

1: Input: the strategy of player 2 π_2
2: **for** Each observation X_1 **do**
3:   **for** Each report X_2' **do**
4:     Use Equations (10.44) to (10.48) to compute the *a posteriori* probability $\mu(T_2|X_2', T_1)$.
5:     Use (10.40) to compute $P(I|X_1, T_2)$ and $P(B|X_1, T_2)$.
6:     Use (10.50) and (10.51) to compute the risks $R_F(T_1, X_2')$ and $R_M(T_1, X_2')$.
7:     Choose the decision using (10.52).
8:   **end for**
9: **end for**
10: Output: The functions R_F and $R - M$.

Procedure 2 Procedure of optimal strategy of player 2

---

1. **for** Each observation X_1 **do**
2.   Optimize π_2 using the optimization in (10.55); the functions R_M and R_F are evaluated using Procedure 1.
3. **end for**

Note that it is very difficult carry out an analytic optimization (e.g., using the KKT condition) for (10.55). Hence we can use the function of constrained optimization in the optimization toolbox of Matlab for the optimization.

10.4.5 Numerical Results

The analysis on the Bayesian equilibrium can be demonstrated by numerical simulations [1303]. For simplicity of computation, we assume that $N = 4$, that is, there are 4 possible observations (e.g., very high energy, high energy, medium energy and low energy). The discrete observation is reasonable since continuous observations can be discretized. The observation distribution for different primary user state is given by the following matrix:

$$\begin{pmatrix} 0.5 & 0.2 & 0.17 & 0.13 \\ 0.13 & 0.17 & 0.2 & 0.5 \end{pmatrix}, \tag{10.56}$$

X2 =	X1 = O2				X1 = O3			
	O1	O2	O3	O4	O1	O2	O3	O4
P(X2') = O1	0	0	0.5	0.5	0	0.33	0.33	0.33
P(X2') = O2	0	0	0.5	0.5	1	0	0	0
P(X2') = O3	0.56	0.44	0	0	1	0	0	0
P(X2') = O4	0.95	0.05	0	0	1	0	0	0

Figure 10.12 Examples of attacking strategies in typical situations.

where the first row means the probabilities of observation when the channel is idle while the second row is the probabilities when the channel is busy. The four columns in the matrix, indexed from left to right, represent the observations O_1 to O_4, respectively. When primary user is not present, it is more possible to receive observations with lower indices (e.g., O_1); on the other hand, when primary users emerge, observations with higher indices (e.g., O_4) will have high probabilities. We set $C_M = 2$ and $C_F = 1$ since missed detections in spectrum sensing typically incur more damage than false alarms.

The Bayesian equilibrium is computed using the algorithms in Procedures 1 and 2. The optimization is achieved using the optimization toolbox in Matlab. In Figure 10.12, two examples of the attacking strategies are shown when $X_1 = O_2$ and $X_1 = O_3$, respectively. We set $\pi_M = 0.1$, that is, player 1 has the belief that player 2 is a malicious one with probability 0.1. We observe that, when $X_1 = O_2$ and $X_2 = O_1, O_2$, that is, the channel is more likely to be idle, player 2 intends to send reports indicating the existence of primary users in order to incur false alarms. On the other hand, when $X_1 = O_2$ and $X_2 = O_3, O_4$, that is, the channel is more likely to be busy, the attacker believes that the channel is actually busy and thus sends reports indicating an idle channel. We can also observe similar strategies for the case $X_1 = O_3$.

11

Cognitive Radio Network

In all our previous discussions, we focused on point to point communications using cognitive radio technology. Having solved the problem of two-party communications, we can now focus on using the cognitive radio link to form a network. Wireless networking has been widely studied for decades. However, the revolutionary new spectrum access mechanism in cognitive radio incurs substantial challenges for the design of networks. In this chapter, we provide a brief introduction to the basics of general networks; more details about networks can be found in [1309]. Then, we will study the special design suitable for cognitive radio in different layers in a bottom-up manner.

11.1 Basic Concepts of Networks

Intuitively, a network is an ensemble of parties that can communicate with each other, directly or indirectly. Usually, a network can be represented by a graph, in which each node represents a communication party while each edge means that the two incident nodes can communicate with each other.

11.1.1 Network Architecture

There are typically two types of architectures for networks, as illustrated in Figure 11.1, both having plenty of applications.

- Cellular networks: This can also be called server-client architecture. In such a network, there exist multiple base stations and many mobile stations. Two mobile stations cannot communicate directly even if they are within a communication range. Their information transmission must take route through a base station. Our cellular phones fall in this category.
- Peer-to-peer networks: This architecture can also be called an ad hoc one. In a peer-to-peer network, there is no centralized base station. Nodes within communication range can talk to each other directly. If two nodes are too far away from each other, they can communicate with an intermediate relay node. Such an architecture is of particular use in sensor networks or battlefield communication networks.

Cognitive Radio Communications and Networking: Principles and Practice, First Edition.
Robert C. Qiu, Zhen Hu, Husheng Li and Michael C. Wicks.
© 2012 John Wiley & Sons, Ltd. Published 2012 by John Wiley & Sons, Ltd.

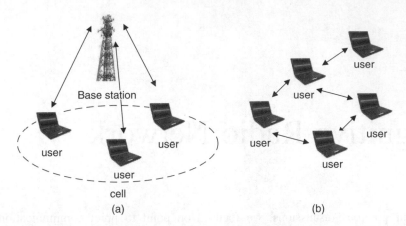

Figure 11.1   An illustration of the two types of architectures. (a) Cellular architecture. (b) Peer-to-peer architecture.

Cognitive radio can adopt both architectures. An advantage of the cellular system is that the base station has a powerful sensing and data processing capability, which can fuse the spectrum sensing results of different secondary users and schedule the transmissions. On the other side, the cost of the base station may not justify the risk that there emerge many primary users such that the secondary users within the cell cannot find enough spectrum for data transmission. The peer-to-peer architecture has the opposite advantages and disadvantages.

11.1.2   Network Layers

To facilitate the design of communication networks, the functionalities of networks can be organized into a stack of layers. Each layer takes charge of different tasks and communicate with adjacent layers. Certain protocols are also designed for the interfaces between two adjacent layers. The most popular definition of network layers is the OSI reference model developed by the International Standards Organization (ISO). Another typical model is the TCP/IP reference model. Both are illustrated in Figure 11.2. The details of the models are explained as follows.

We first introduce the OSI model, in which the network is divided into seven layers. Since the layers of session and presentation are not used in most network designs, we focus on the remaining five layers which have been widely used in the design and analysis of communication networks.

- Physical Layer: The physical layer concerns how to transmit information bits from a transmitter to a receiver. It mainly concerns the modulation/demodulation, coding/decoding,[1] and signal processing for transmission and reception. Most of our previous discussions fall in the physical layer.

---

[1] In the original definition of OSI model, the coding and decoding are issues in the data link layer; however, in practice, people usually consider them as physical layer issues.

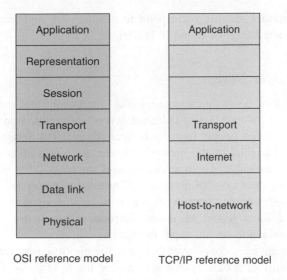

Figure 11.2   An illustration of layers in OSI and TCP/IP.

- Data Link Layer: This layer takes charge of tasks such as frame acknowledgment, flow regulation and channel sharing. The last one, called MAC is the most important for wireless networks due to the broadcast nature of wireless transmissions, thus usually considered as an independent layer. Essentially, the MAC layer addresses the resource allocation, for example, how to allocate different communication channels to different users, and scheduling, for example, when there is a competition among users, which user should obtain the priority to transmit. We have mentioned some MAC layer issues in cognitive radio before. In this chapter, we will explain more details about it.
- Network Layer: This layer determines how to find a route in the network from the source to the destination. For example, we need to design an addressing mechanism to for the routing. Moreover, when the addresses of the source and destination are known, we need to design algorithms for the network to find a path with the minimum cost (e.g., number of hops) to the destination. When a path is broken by emergency, the network layer needs to find a new path for the data flow.
- Transport Layer: This layer receives data from the application layer, splits it into smaller units if needed, and then pass to the network layer. The main job of the transport layer is congestion control, that is, how to control the source rate according to the congestion situation in the network.
- Application Layer: It provides various protocols for different applications. For example the HyperText Transfer Protocol (HTTP) is used for websites.

Note that the physical, data link and network layers concern the intermediate nodes in the network, while the transport and application layers concern only the two ends of data flow, namely the source and destination.

In the TCP/IP reference model, the physical and data link layers are not well specified. They are considered as the host-to-network layer. The Internet and TCP layers in the

TCP/IP reference model roughly correspond to the network and transport layers in the OSI model. More details can be found in [1309].

11.1.3   Cross-Layer Design

In traditional design of communication networks, the design and operation of different layers are carried out independently. Different layers are coupled only through the inter layer interfaces. However, people have found that the isolated design of different layers may decrease the efficiency of the network. Hence, the cross-layer design is proposed for communication networks. An excellent tutorial on cross-layer design can be found in [1034]. Moreover, the theory of network utility maximization and optimization based decomposition provides unified framework for cross-layer designs in communication network. Due to the limited space, we do not introduce this theory. Readers can find a comprehensive introduction on this topic in [528].

A motivating example for cross-layer design is the opportunistic scheduling in cellular systems [1310]. Consider a base station serving for multiple users. Different users may have different channel gains. The base station needs to schedule the transmission of the multiple users. Recall that scheduling is a MAC layer issue while the channel gain is a physical layer quantity. In traditional layered design, the scheduling algorithm does not take channel gains into account. However, it has been shown that, to maximize the sum capacity, it is optimal to schedule only the user having the largest channel gain. Hence, we see that, if the sum capacity is the performance metric, it is more desirable to take the physical layer issues into the scheduling algorithm in the MAC layer, thus resulting in a cross-layer design.

Note that the frequency spectrum is actually a physical layer concept. Hence, if we adopt the layered design for cognitive radio networks, only the physical layer issues like spectrum sensing or transmission schemes for non-contiguous spectrum bands are concerned; the networking issues like scheduling and routing still follow traditional designs. However, such a layered may result in a low efficiency of spectrum utilization. Take the above opportunistic scheduling for instance. We can replace the channel gain with the channel availability in the context of cognitive radio. Then, it is obvious that the scheduling should take the spectrum situation into account since only secondary users having available channels should be scheduled. Similarly, routing in the network layer should also consider the spectrum situations such that it can circumvent the region with more frequent emergences of primary users. Therefore, the cross-layer design is a must for high performance cognitive radio networks and each layer should be aware of the spectrum in the physical layer.

11.1.4   Main Challenges in Cognitive Radio Networks

Since it is necessary to incorporate the spectrum situation in the physical layer into account when designing the upper layers of cognitive radio networks, the novel mechanism of spectrum access brings many challenges to the design and performance. The major challenge is the spectrum dynamics. The corresponding impact on all upper layers is explained as follows.

- MAC Layer: As we have explained, the main taks of MAC layer is to allocate communication resources to different transmitters. However, in cognitive radio, the available communication resource could be dynamical. Hence, the MAC layer must be adaptive to the current spectrum situation, which demands high processing speed and fast information collection.
- Network Layer: In traditional wireless communication networks, once a path is found, it will be used for a long period of time. However, in cognitive radio, the data path should be adaptive to the spectrum situation. For example, when a primary user emerges and blocks a data path, the data path should either be re-routed or should wait for the recovery of the original path, which is a decision problem. Since the spectrum is random, the data path could also be random.
- Transport Layer: The congestion control mechanism used for the Internet cannot be applied directly in cognitive radio networks. A key difficulty is that it is difficult to distinguish the packet drop due to congestion and the packet blockage due to the emergence of primary users. Hence, an explicit mechanism should be designed to inform the source about the primary user emergence such that the source node can better control the data traffic rate.

In the subsequent sections, we will explain how to address the above challenges for different layers. For each layer, we will explain the general principle of the spectrum-aware design and use a typical example of algorithm or protocol to illustrate the principle.

11.1.5   Complex Networks

Another topic that we will mention is the complex network phenomenon in cognitive radio networks. Although both studies on network design and complex networks concern communication networks, the latter is more focused on the interesting properties when the size of network becomes very large.

An important complex network phenomenon in communication networks is the phase transition of the network connectivity [1311]. Suppose that there are sufficiently many nodes within a network, there exists a communication link between two nodes with a certain probability p. Then, there exists a critical value denoted by p_c such that, when $p < p_c$, most nodes are separated; when $p > p_c$, most nodes are connected. Hence, the network connectivity experiences a sudden change at $p = p_c$, similarly to the conversion from liquid water to vapor at $100\,°C$. This phenomenon occurs only when the network is sufficiently large and complex.

Below are some other examples of complex network phenomenon.

- The epidemic propagation in social networks [1312], that is, how the network topology affects the propagation of certain behavior in complex networks.
- The vulnerability of large power networks [1313], that is, how failures can propagate within a large power network.
- The small world phenomenon in complex networks [1314], that is, how many intermediate acquaintances are needed for any two people to get connect.
- The synchronization phenomenon in many complex systems [1315], that is, how a network of oscillators become synchronized or lose synchronization.

Since there exist interactions among secondary users in cognitive radio networks, such as collaborations or recommendations, it is interesting to study the complex network phenomenon in cognitive radio networks, which provides insights for the design and analysis. Hence, we will also discuss the complex network analysis for cognitive radio networks.

11.2 Channel Allocation in MAC Layer

Now we begin to study the networking in cognitive radio systems from the MAC layer. As we have explained, the major task of MAC layer is to allocate communication resource to different secondary users. In this section, we consider the resource allocation for elastic data traffics. The discussion on data traffics having constant source rate will be given in the next section. Since the major challenging for networking with cognitive radio is the dynamic communication resource, we will see how an effective algorithm addresses this challenge.

11.2.1 Problem Formulation

There have been many studies on the channel allocation in cognitive radio [1316, 1317]. In this chapter, we focused on the study in [1318] (best paper award of IEEE Globecom2010). In the study, a set of L licensed channels are considered. We assume that there are $2N$ secondary users, namely source nodes $1, 2, \ldots, N$ and destination nodes $1, 2, \ldots, N$, forming N source-destination pairs. We denote by $a_{s,i}^l$ and $a_{d,i}^l$ the availability of channel l at source node s and destination node d, respectively. Value 1 for $a_{s,i}^l$ or $a_{d,i}^l$ means the corresponding channel is available; otherwise, the value equals 0.

An example is shown in Figure 11.3. Three transmission pairs are shown in the cognitive radio network. When the transmitter and receiver have common available channels, they are able to communicate. Otherwise, like S3 and D3, the communication is unfeasible because they cannot find a common channel.

We consider the following protocol for the cognitive radio network. Each time slot is divided into three periods: sensing, access and data transmission.

- Sensing period: Each secondary user senses the spectrum and determines the available channels. Then, each secondary user chooses one available channel as its working channel. For a transmitter, the working channel is for transmission, while, for a receiver, the working channel is for listening.
- Access period: This period mainly solves the possible collisions among multiple secondary users within the same channel. This period is divided into K mini-slots. Each source node chooses a random number for the time of sensing a request-to-send (RTS). Before sending the RTS, the source node listens to the channel; once any spectrum activity is detected, the source nodes keeps silent. When the destination node receives an RTS, it sends back a clear-to-send (CTS). Note that such a mechanism is very similar to a carrier sense multiple access (CSMA) one.
- Data transmission: Once a source node and a destination node are hooked up via RTS and CTS, they can begin the data transmission during the remainder of the time slot. The transmission approach can be the same as traditional communication systems.

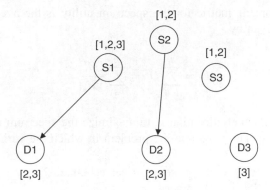

Figure 11.3   An example of channel situation.

Obviously, the spectrum sensing period and the data transmission period are not the focus of the study. We are focused on the access period, particularly, on the problem of which channel to access. This problem is essentially a resource allocation one.

11.2.2 Scheduling Algorithm

In [1318], the scheduling of the channels is formulated as an optimization problem. The essential purpose of the optimization is to maximize the utilization of the licensed spectrum, which coincides with the purpose of cognitive radio.

First, we denote by M_s^l and M_d^l the sets of source nodes and destination nodes that choose channel l during the spectrum sensing period. We define $M^l = M_s^l \cap M_d^l$, namely the set of transmission pairs that can use channel l for the data transmission.

We denote by $\{X_k\}_{k=1,\dots,|M^l|}$ the random backoff values during the access period of the source nodes choosing channel l. We define $W = \min\{X_k\}_{k=1,\dots,|M^l|}$. Then, the utility of channel l, defined as the probability that a successful transmission occurs over channel l, is given by

$$U^l = \sum_{i=1}^{|M^l|} P(W = X_i) \sum_{x=1}^{K} (p(X_i = x) P(\cap_{j \neq i}\{X_j > x\}))$$

$$= \frac{|M^l|}{K|M_s^l|} \sum_{x=1}^{K} \left(\frac{K-x}{K} \right)^{|M_s^l|-1}. \tag{11.1}$$

We focus on the case $K \to \infty$. Then, we have

$$\sum_{x=1}^{K} \left(\frac{K-x}{K} \right)^{|M_s^l|-1} = 1, \tag{11.2}$$

and

$$U^l = \frac{|M^l|}{|M_s^l|}. \tag{11.3}$$

Obviously, a reasonable metric for the spectrum utility is the average utility over all channels, which is given by

$$U = \sum_{l=1}^{L} U^l = \sum_{l=1}^{L} \frac{|M^l|}{|M_s^l|}. \tag{11.4}$$

Then, the goal of channel allocation is to maximize the spectrum utilization U. It can be formulated as an integer programming problem in which the variables are

$$x_{s,i} = \begin{cases} 1, & \text{if source } i \text{ selects channel } l \\ 0, & \text{otherwise} \end{cases}, \tag{11.5}$$

and

$$x_{d,i} = \begin{cases} 1, & \text{if destination } i \text{ selects channel } l \\ 0, & \text{otherwise} \end{cases}. \tag{11.6}$$

Based on the above definitions, we have

$$U^l = \frac{\sum_{i=1}^{N} x_{s,i}^l x_{d,i}^l}{\sum_{i=1}^{N} x_{s,i}^l}. \tag{11.7}$$

Then, the optimization problem is formulated as

$$\max_{\{x_{d,i}\},\{x_{s,i}\}} \sum_{l=1}^{L} \frac{\sum_{i=1}^{N} x_{s,i}^l x_{d,i}^l}{\sum_{i=1}^{N} x_{s,i}^l}$$

$$s.t. \ x_{s,i}^l = 0 \text{ or } 1, \quad \forall l, i$$

$$x_{d,i}^l = 0 \text{ or } 1, \quad \forall l, i$$

$$\sum_{l=1}^{L} x_{s,i}^l = 1$$

$$\sum_{l=1}^{L} x_{d,i}^l = 1$$

$$x_{s,i}^l = 0, \quad \text{if } a_{s,i}^l = 0$$

$$x_{d,i}^l = 0, \quad \text{if } a_{d,i}^l = 0. \tag{11.8}$$

Obviously, the objective function is the spectrum utility. The first two constraints are the binary values of the variables. The third and the fourth constraints mean that one source or destination can choose only one channel. The last two constraints mean that a secondary user should not choose a channel that has been occupied by primary users.

11.2.3   *Solution*

Although the channel allocation has been formulated as an optimization problem, it is challenging to solve it. Unfortunately, it has been shown in [1318] that the optimization problem in (11.8) is an NP-hard one. Hence, it is impossible to find a polynomial time algorithm for the optimization. For a large cognitive radio network, we have to use some heuristic approaches. In [1318], both centralized and decentralized greedy algorithms are introduced and achieve good performance. Below we provide a brief introduction on both approaches.

Centralized Algorithm: In this approach, we divide the transmission pairs into two groups $\mathcal{G}_{wc}$ and $\mathcal{G}_{oc}$, which have at least one common channel or have no common channel between the transmitter and receiver, respectively. We also denote by $\mathcal{C}_{sd,i}$ the set of common channels for transmission pair i and define $\mathcal{C}_{sd} = \cup_i \mathcal{C}_{sd,i}$. Then, the following steps are used for the channel allocation. The detailed description and pseudocode can be found in [1318].

1. Initialization: We initialize $\mathcal{G}_{wc}$, $\mathcal{G}_{oc}$, $\mathcal{C}_{sd,i}$ and $\mathcal{C}_{sd}$.
2. Constructing a bipartite graph: We use the sets $\mathcal{G}_{wc}$ and $\mathcal{C}_{sd}$ to construct a bipartite graph, which is denoted by G, where an edge points from a transmission pair to a channel if the channel is a commonly available one for the transmission pair.

    An example is shown in Figure 11.4, in which there are three transmission pairs and four channels. From the bipartite graph, we can see that transmission pair 2 has common channels 2 and 3, while channel 2 is also available for the transmitter and receiver in pair 1; hence, pairs 1 and 2 may conflict over channel 2. Obviously, the bipartite graph provides the information about available channels for different transmission pairs, as well as possible collisions among the transmission pairs.

3. Matching maximization: With the aid of the bipartite graph, we can formulate the optimization problem in (11.8) as a maximum bipartite matching problem in graph theory. In a bipartite graph, a matching means a set of edges, in which any two edges do not share a node. In the context of channel allocation, if an edge is within the matching, we say that the corresponding transmission pair uses the corresponding channel. The requirement of no common nodes in the matching is due to (a) any two transmission pairs cannot use the same channel which may incur collision; (b) one transmission pair can use only at most one channel. A maximum matching means that adding any more edge to the set will make the set no longer a matching.

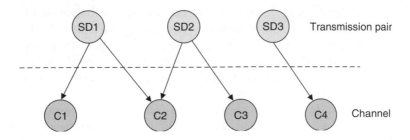

Figure 11.4   An example of the bipartite graph used in channel allocation.

For example, in Figure 11.4, the edges $\{SD1 \rightarrow C2, SD2 \rightarrow C3, SD3 \rightarrow C4\}$ is a maximum matching. When we add one more edge, say $SD2 \rightarrow C2$, the definition of matching will be violated.

Hence, when we maximize the matching in the bipartite graph, it means that we can no longer add one more channel to the transmission, thus maximizing (could be locally) the spectrum utility. Many algorithms can be used for the matching maximization, for example the greedy algorithm.

Decentralized Algorithm: In the previous discussion, the scheduling is carried out by a centralized scheduler. In many cases, for example, in ad hoc cognitive radio networks, there is no such a center that can carry out the optimization for the channel allocation. The channel allocation has to be carried out in a distributed manner.

A heuristic algorithm is proposed for the decentralized channel allocation, based on a predetermined priority order. In each time slot, the secondary users sort the channels using a common order (which could change with time). Then, each secondary user choose the channel that has the highest priority among all its available channels. In [1318], the order is simply defined as a round-robin manner; that is, the order at time t is given by

$$p_{\text{mod } (h+t-1,L)+1} > p_{\text{mod } (h+t,L)+1} > \ldots > p_{\text{mod } (h+t+L-2,L)+1}, \qquad (11.9)$$

where p_c is the priority of channel c. It is easy to see that the secondary users can keep synchronized in the priority order. The rationale of this round-robin priority is to guarantee the fairness among different secondary users and channels. The performance analysis of the decentralized algorithm is left as an exercise problem.

11.2.4 Discussion

Note that the introduced scheduling algorithm is far from solving the scheduling problem in cognitive radio networks. There are still many problems to be solved, such as

- QoS scheduling: The scheduling algorithm proposed in this section aims at maximizing the spectrum utilization. However, this may starve some unfortunate nodes, thus not being able to guarantee the QoS. Hence, it is necessary to study the scheduling for the QoS of secondary users. In the next section, we will consider the QoS of source data rate and study the corresponding scheduling algorithm to stabilize the queuing dynamics.
- Scheduling with limited communications: In the decentralized algorithm discussed in this section, we assume that the priorities have been predetermined. However, this may not result in the optimal scheduling. A better approach is to let the secondary users exchange a limited amount of messages such that the scheduling can be better adapted to the spectrum environment. This is equivalent to carrying out a distributed maximization of discrete objective function.
- Partial observability: In this section, we have assumed that each secondary user can perfectly detect the primary user activity in all licensed spectrum band. However, this assumption may not be true if the licensed band is wide, since it requires a significantly high sampling rate to perfectly reconstruct a wideband signal. In this case, a practical

approach is to sense only a few channels within the sampling capability. Then, the scheduling becomes how to allocate different channels to different secondary users to sense. Due to the randomness of spectrum, the objective function could be chosen as the expectation of the spectrum utility or other metrics.

11.3   Scheduling in MAC Layer

In the previous section, the criterion for scheduling the channels is to maximizing the channel utility, or equivalently maximizing the throughput. However, this is not suitable for traffics with tight QoS requirements, for example, with fixed source data rates. Hence, in this section, we consider the case with strict data rates and study the scheduling algorithm to stabilize the queuing dynamics in the cognitive radio network. Note that we follow the argument in [1319].

11.3.1   Network Model

We consider a cognitive radio network with N secondary users and M primary users, which are illustrated in Figure 11.5. Each primary user uses a single licensed channel. We denote by $\mathcal{I}_{nm}$ the set of channels that secondary user n interferes when it is accessing channel m. For simplicity, it is assume that each primary user can interfere all secondary users using the corresponding channel. Note that there is no channels never used by any primary user; hence, the total number of channels is also M. We denote by $s_m(t)$ the state of channel m at time slot t: channel m is available when $s_m(t) = 1$ and is not available otherwise. As in many studies on cognitive radio networks, we assume that the primary user state follows a Markov chain. Given the primary user state $\mathbf{s}(t-1) = (s_1(t), \ldots, s_M(t))$, the probability that channel m is idle is given by $P_m(t)$, namely

$$P_m(t) = E[s_m(t)|\mathbf{s}(t-1)]. \tag{11.10}$$

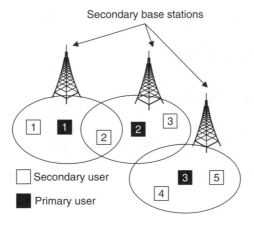

Figure 11.5   An illustration of the primary and secondary networks.

Obviously, a secondary user can successfully transmit a packet through channel m only if the following conditions hold

- $s_m(t) = 1$, that is, primary users are not using channel m.
- The transmissions of all other secondary users i do not interfere channel m at the secondary user.

We assume that each secondary user receives external data in an i.i.d. manner. The packet arrival process is denoted by $A_n(t)$ for secondary user n. The average arrival rate is denoted by λ_n packets/time slot. Unlike the frequently used Poisson arrival model, we put an upper bound for the number of packet arrivals, which is denoted by $A_{\max}$. The backlog in the queue of secondary user n is denoted by $U_n(t)$ at time slot t. The number of new packets admitted into the queue is denoted by $R_n(t)$ (note that $R_n(t) \leq A_n(t)$ since a newly arriving packet is not necessarily admitted into the queue). The number of attempted packet transmissions in channel m by secondary user n is denoted by $\mu_{nm}(t)$ at time slot t. For simplicity, we assume that μ_{nm} is either 0 or 1; that is, a secondary user either transmit one packet in a given channel or transmit nothing at all. It is interesting to study the case in which a secondary user can transmit two or more packets in a channel according to the channel quality; however, this will be much more complicated.

Then, the queuing dynamics are given by

$$U_n(t + 1) = \max \left[U_n(t) - \sum_{m=1}^{M} u_{nm}(t)s_m(t), 0 \right] + R_n(t), \qquad (11.11)$$

where the constraints are

$$u_{nm} = 0 \text{ or } 1, \qquad \forall m, n \qquad (11.12)$$

$$u_{nm}(t) \leq h_{nm}(t), \qquad \forall m, n$$

$$0 \leq \sum_{m=1}^{M} u_{nm}(t) \leq 1, \forall n$$

$$u_{nm}(t) = 1 \Leftrightarrow \sum_{j=1}^{M} \sum_{i=1, i \neq n}^{N} I_{ij}^{m} u_{ij}(t) = 0, \qquad m, n$$

$$0 \leq R_n(t) \leq A_n(t). \qquad (11.13)$$

The physical meaning of the constraints is:

1. The transmission is either successful ($u_{nm=1}$) or not ($u_{nm} = 0$).
2. The transmission is limited by the spectrum occupancy.
3. Each secondary user can transmit over only one channel.
4. If secondary user transmits within a channel, all channels that will be interfered will not be used by other secondary users.
5. The number of packets admitted into the queue is limited by the number of arriving packets.

11.3.2 Goal of Scheduling

In [1319], it is assumed that the goal of scheduling is to maximize the weighted throughput. To that end, we define

$$r_n = \lim_{t \to \infty} \frac{1}{t} \sum_{s=0}^{t-1} R_n(s), \tag{11.14}$$

which means the throughput of secondary user n. We also define a metric to measure how much interference the secondary users cause on the primary users, which is given by

$$c_m(t) = I(\text{a collision with primary user occurs in channel } m \text{ in time slot } t). \tag{11.15}$$

Then, the average collision with primary users in channel m is defined as

$$c_m = \lim_{t \to \infty} \frac{1}{t} \sum_{s=0}^{t-1} c_m(s). \tag{11.16}$$

Then, the goal of scheduling is to carry out the following optimization:

$$\max \sum_{n=1}^{N} w_n r_n$$
$$s.t. \ 0 \le r_n \le \lambda_n, \qquad n$$
$$c_m \le \rho_m, \qquad m$$
$$\mathbf{r} \in \Lambda, \tag{11.17}$$

where w_n is the weight for secondary user n, $\mathbf{r} = (r_1, \ldots, r_N)$ and Λ is the network capacity region within which the queuing dynamics are stable.

Obviously the objective function is the weighted sum of throughputs. The meanings of constraints are given below:

1. The throughput cannot be larger than the packet arrival rate.
2. The collision with primary users should be confined within a certain range; otherwise, the secondary users will cause too much interference to the primary system.
3. The throughput should be within the capacity of the cognitive radio network.

11.3.3 Scheduling Algorithm

Now, we begin to study how to scheduling the data traffic in cognitive radio in order to optimize the constrained objective function in (11.17). There are two components in the scheduling algorithm:

- Flow control, that is, how to admit the packets into the queues. Here, the control variable is $R_n(t)$, that is, how much packet to admit into the queues.
- Resource allocation, that is, how to allocate different channels to different users. Here, the control variable is $u_{nm}(t)$, that is, which secondary user should be scheduled over which channel.

The two components are formulated as the following two separate optimization problems. The integration of the two components is called Cognitive Network Control (CNC) algorithm [1319].

• Flow control: The packet admittance is carried out according to the following optimization problem:

$$\min_{R_n(t)} R_n(t)(U_n(t) - V w_n)$$

$$s.t. \ 0 \le R_n(t) \le \Lambda_n(t). \tag{11.18}$$

Here V is a predetermined constant which controls the tradeoff between throughput and delay (when the queue length is too large, the delay will be increased).

The solution is obvious:

$$R_n(t) = \begin{cases} \Lambda_n(t), & \text{if } U_n(t) \le V w_n \\ 0, & \text{if } U_n(t) > V w_n \end{cases},$$

that is, if the queue length is not large, all arriving packets are admitted; otherwise, no packet admittance.

• Resource allocation: The resource allocation can be formulated as another optimization problem. Before formulating the optimization problem, we need to define a 'virtual' queue which represents the amount of collisions with primary users. The dynamics of queue are defined as (recall that $c_m(t)$ is the notation for the account of collisions with primary users)

$$X_m(t+1) = \max[X_m(t) - \rho_m, 0] + c_m(t), \tag{11.19}$$

where X_m is the backlog of the virtual queue in channel m. Recall that ρ_m is the average number of collisions at channel m. An intuitive explanation for the virtual queue is that, when the number of collisions is more than the average one, the queue accumulates; otherwise the queue depletes such as that we allow more collisions. The incorporation of this virtual queue is used to prevent too many collisions with primary users and avoid the violation of the game rule of cognitive radio.

Based on the definition of the virtual queue, we can define the optimization problem for the resource allocation, which is given by

$$\max_{\{u_{nm}\}} \sum_{n,m} u_{nm}(t) \left(U_n(t) P_m(t) - \sum_{k=1}^{M} X_k(t)(1 - P_k(t)) I_{nm}^k \right)$$

$$s.t. \text{ The constraints in (11.12).} \tag{11.20}$$

An intuitive explanation is given as follows: $U_n(t) P_m(t)$ means the desire of transmitting the packet in secondary user n using channel m, since a larger queue length or a larger channel idle probability motivates the system to assign $u_{nm} = 1$; meanwhile the term $\sum_{k=1}^{M} X_k(t)(1 - P_k(t)) I_{nm}^k$ provides a penalty on the possible collisions with primary users (here I_{nm}^k takes all possibly interfered channels into account); then, the scheduling should achieve a tradeoff between both the queue depletion and the collisions with primary users. The solution of the above optimization will be discussed in the next subsection.

When all channels are orthogonal to each other (that is, there is no cross-channel interference), the optimization problem in (11.20) can be simplified to

$$\max_{\{u_{nm}\}} \sum_{n,m} u_{nm}(t)(U_n(t)P_m(t) - X_m(t)(1 - P_k(t)))$$

s.t. The constraints in (11.12).   (11.21)

Note that the above simplified problem is a maximum weight match (MWM) problem on an $N \times M$ bipartite graph. If a centralized scheduling is used, this can be achieved within a polynomial time.

11.3.4 Performance of the CNC Algorithm

Now, we begin to study the the performance of the CNC algorithm. It follows the framework of Lyapunov function which Tassiulus and Emphremides have exploited for the stability of controlled queueing dynamics [1296]. Recall that this is also discussed, when we discuss the game between PUE attackers and cognitive radio network.

To that end, we let $L(\mathbf{q})$ be a function of queue lengths $\mathbf{q}$, which is scalar and nonnegative. We call it Lyapunov function, which can be considered as the energy of the queuing system. The Lyapunov drift, intuitively meaning the decrease of the system energy, is defined as

$$\Delta L(t) = E[L(\mathbf{q}(t+1)) - L(\mathbf{q}(t))].$$ (11.22)

Obviously, the smaller the Lyapunov drift is, the more stable the queuing dynamics are. It has been shown in [1319] that

$$\Delta L(t) - V E\left[\sum_{n=1}^{N} w_n R_n(t)\right] \le B - E\left[\sum_{n=1}^{N} U_n(t)\left(\sum_{m=1}^{M} u_{nm}(t) S_m(t) - R_n(t)\right)\right]$$

$$- E\left[\sum_{m=1}^{M} X_m(t)(\rho_m - \hat{c}_m(t))\right]$$

$$- V E\left[\sum_{n=1}^{N} w_n R_n(t)\right],$$ (11.23)

where V is a control parameter.

In [1319], the Lyapunov function is defined as

$$L(\mathbf{q}(t)) = \frac{1}{2}\left[\sum_{n=1}^{N} U_n^2(t) + \sum_{m=1}^{M} X_m^2(t)\right].$$ (11.24)

In contrast to the stability analysis for traditional queuing systems, in which the Lyapunov function is defined as the square sum of the queue lengths, the Lyapunov function defined for the cognitive radio network has an extra term, namely the square sum of the lengths of the virtual queues, which addresses the unique issue of collisions with primary users. If we consider the virtual queues as normal queues, the analysis falls in the same framework as traditional networks.

Then, we have the following fact, which is called the *Optimal Stationary Randomized Policy* in [1319]. Consider any rate vector denoted by $(\lambda_1, \ldots, \lambda_N)$ (recall that λ_n is the average arriving rate of secondary user n). We can always find a stationary randomized scheduling policy *STAT*. It selects feasible solutions $R_n^{STAT}(t)$, $u_{nm}^{STAT}(t)$ as a function of the channel state $\mathbf{P}(t)$ and $\mathbf{H}(t)$. They can yield the following equations:

$$E[R_n^{STAT}(t)] = r_n^*, \tag{11.25}$$

and

$$u_n^{STAT} \equiv \lim_{t \to \infty} \frac{1}{t} \sum_{s=0}^{t-1} E\left[\sum_{m=1}^{M} u_{nm}^{STAT}(s)s_m(s)\right] \geq r_n^*, \tag{11.26}$$

and

$$\hat{c} \equiv \lim_{t \to \infty} \frac{1}{t} \sum_{s=0}^{t-1} E[\hat{c}_m^{STAT}(t)] \leq \rho_m. \tag{11.27}$$

This claim can be proved by using the approaches in [1320]. Then, it is shown in [1319] that the CNC algorithm can minimize the right hand side of (11.23) over all feasible actions that can be made in time slot t.

11.3.5 Distributed Scheduling Algorithm

For the distributed case, we consider the case of orthogonal channels. Hence the optimization problem is formulated as (11.21), which is essentially an MWM problem. To achieve a distributed version (also constant time), a greedy maximum matching algorithm is proposed in [1319].

The algorithm is described as follows:

- Step 0: Assign weights to each communication link. The weights are obtained from (11.21).
- Step 1: Choose the link having the largest weight. Activate it.
- Step k $(k > 1)$: Choose the link having the largest weight among the links that have not been activated and do not have conflict with the activated links. If there is no more feasible link, stop.

The greedy algorithm tries to increase the objective function as much as possible in each step. However, once a link is activated, it does not remove it. Hence, the algorithm could be suboptimal. It has been shown in [1319] that the throughput utility achieved by the CNC algorithm is within a bounded range of the optimal performance.

11.4   Routing in Network Layer

In this section, we focus on the network layer and study the routing issues in cognitive radio networks.

11.4.1   Challenges of Routing in Cognitive Radio

The key challenge for routing in cognitive radio networks is the dynamical spectrum occupancies (or equivalently the dynamical transmission opportunities). When the channel being used by a secondary user is occupied by primary users, the corresponding link is broken; thus the data route no longer works. Then, there are three possible actions the secondary user can take:

- Wait: If the primary users will leave the channel soon, the secondary user can wait until the channel is cleared.
- Switching channels: If there are multiple channels, the secondary user can also try to sense other channels and resume the transmission until an available channel is found.
- Re-routing: If there is only one channel and the primary user does not leave quickly, a new path has to be found in order to resume the data traffic.

Note that all the above actions incur cost. It incurs packet delay when waiting for the primary user to leave. In most wireless hardware, it takes time to switch to a new channel. It also incurs significant overhead to carry out re-routing since the secondary users need to exchange messages on the path information. Then, it is determined by the dominating factor to choose the corresponding action. This results in two type of routing schemes:

- Stationary Routing: The data traffic has a fixed route. This type routing requires a quick channel switch mechanism or the primary user can leave quickly (that is, the spectrum is highly dynamical). Or when the spectrum is highly stationary (in this case, the network is similar to a traditional one), the route can also be stationary, since the primary users seldom break the data path; however, in this case, a re-routing mechanism may be needed in case the emergence of primary users.
- Dynamical Routing: In this case, the data traffic does not have a fixed route. The packet forwarding is adaptive to the spectrum situation or is random. The dynamical routing is suitable for the case in which the spectrum is moderately dynamical and the channel switching incurs significant overhead; otherwise, the secondary users can either wait for the leave of primary users or quickly switch to another channel.

The change of routing strategy with respect to the level of dynamical spectrum is illustrated in Figure 11.6. Hence, the selection of routing strategy should be highly dependent on the spectrum environment and the hardware specifications of the secondary users.

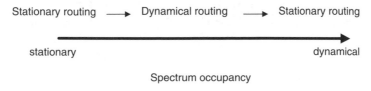

Figure 11.6   An illustration of the change of routing strategy with respect to the spectrum dynamics.

11.4.2   Stationary Routing

In this subsection, we study the stationary routing in cognitive radio network, that is, the data path is fixed throughout the operation of the network. There have been many corresponding studies. We follow the one in [1321], which takes the following unique challenges in cognitive radio networks as summarized in Section I of [1321]:

- The routing can protect the primary users explicitly, as illustrated in Figure 11.7.
- There are two classes of secondary user services: class I assigns more significance to end-to-end latency under the constraint of primary user interference; class II puts more priority on the protection of primary users at the cost of certain performance loss for cognitive radio network.
- The routing algorithm must be scalable and takes the routing and spectrum selection into account jointly.

The algorithm proposed in [1321] is called Cognitive Radio Routing Protocol (CRP). Briefly speaking, CRP has the following stages:

- Stage 1. Spectrum Band Selection: In this stage, the secondary users choose the best spectrum band according to their local observations. An optimization problem is formulated for each class of secondary users. Each secondary user develops an initiative for the routing procedure.
- Stage 2. Next Hop Selection: In this stage, the initiative of each secondary user is mapped to a delay function when the Route Requests (RRQE) messages are exchanged. A ranking of neighboring nodes are established by the delay function. And finally a routing path is formed by the destination.

Moreover, a mechanism for route maintenance is also proposed in [1321] in case that any cognitive radio link is broken due to the emergence of new primary users. The details of CRP are described in the remainder of this subsection.

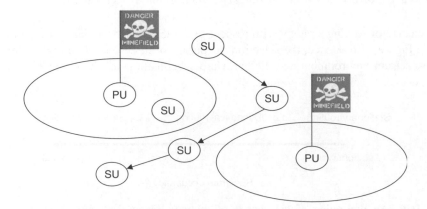

Figure 11.7   An illustration of the avoidance of primary user regions.

11.4.2.1 Stage 1. Spectrum Selection

To select the spectrum (also for the route selection), we need metrics to measure the usefulness of different spectrum channels. In [1321], the following five metrics are selected:

- Probability of bandwidth availability. Essentially this metric measure the possibility that the channel is not occupied by primary users and can be used for data transmission. For channel i in band k, this probability is estimated as

$$p_i^k = \frac{\alpha_i^k}{\alpha_i^k + \beta_i^k}, \tag{11.28}$$

where $\frac{1}{\alpha_i^k}$ ($\frac{1}{\beta_i^k}$) is the average time that channel i in band k is idle (busy). Obviously, it is more desirable to choose a channel with a higher available probability. Then, the available probability of the channels selected within band k is given by

$$M_B^k = \prod_{i \in C^k} p_i^k, \tag{11.29}$$

where C^k is the set of channels selected to use within band k. Obviously, this expression is based on the assumption that the spectrum occupancies at different channels are mutually independent.

- Variance of capacity. This metric measures the variance of the capacity that can be provided by a band. The larger the variance is, the more jitter the transmission will incur. Particularly, it has been found that a large variance of capacity will cause a significant difficulty in the design of transport layer [1322], which will be explained in detailed later. To compute this metric, we define the metric for channel i in band k as

$$\xi_i^k = \frac{1}{N_v} \sum_{s=1}^{N_v} \left[\frac{1}{\beta_i^k} - t_s^{OFF} \right]^2, \tag{11.30}$$

where N_v is the number of previous busy periods (which can be considered as a window for estimating the variance) and t_s^{OFF} is the length of the s-th busy period. Obviously, ξ_i^k is the estimation of the variance of busy period within the time window specified by N_v. Then, for band k, we can define the metric as

$$V_B^k = \psi_k \sum_{i \in C^k} \xi_i^k, \tag{11.31}$$

where ψ_k is the bandwidth of each channel in band k. The purpose of scaling by ψ_k is to compute the variance of the number of bits that can be conveyed.

- Spectrum propagation characteristics. Different spectrum bands may have different characteristics of radio propagation. Particularly, the lower frequency bands may have better performance. In [1321], this metric is defined as

$$D_k = \left[\left(\frac{c}{4\pi f_k} \right)^2 \frac{P_{tx}^{CR}}{P_{rx}^{CR}} \right]^{\frac{1}{\beta}}, \tag{11.32}$$

where c is the light speed, P_{tx}^{CR} is the maximum transmit power of secondary user, P_{rx}^{CR} is the threshold of receive power at the receiver, f_k is the frequency of the band and β is the attenuation factor.

- Protection for primary users: A key consideration of cognitive radio network is to avoid interference to primary users. Hence, this should be considered in the spectrum selection and routing. Areas with primary users can be considered as minefields and secondary users should try to avoid them.

When a secondary user forwards its packets to a neighbor, it needs to consider the overlap between the transmission ranges of the next hop neighbor and primary user. For secondary user i and primary user j, this overlap can be computed in the following way:

$$
A_{i,j} = D_K^2 \cos^{-1} \left\{ \frac{D_{i,j}^2 + D_k^2 - (r_k^j)^2}{2D_{i,j}D_k} \right\}
$$

$$
+ (r_k^j)^2 \cos^{-1} \left\{ \frac{D_{i,j}^2 - D_k^2 + (r_k^j)^2}{2D_{i,j}r_k^j} \right\}
$$

$$
- \frac{1}{2}\sqrt{s(s - 2D_{i,j})(s - 2D_k)(s - 2r_k^j)}, \tag{11.33}
$$

where $s = D_{i,j}^2 + D_k^2 + r_k^j$, $D_{i,j}$ is the distance between secondary user i and primary user j, D_k is the propagation distance in (11.32) and r_k^j is the transmission range of primary user j. The detailed derivation can be found in [1321]. Then, for secondary user i and spectrum band k, the metric for measuring the protection of primary users is defined as

$$
A_i^k = \frac{\cup_{j=1,\ldots,N_p} A_{i,j}}{\pi D_k^2}, \qquad \text{if } D_{i,j} < D_k + r_k^j, \tag{11.34}
$$

where N_p is the number of primary users.

- Consideration on spectrum sensing: Since it is difficult to distinguish the signal of primary users from that of secondary users (particularly when the energy detection is used), when a secondary user is carrying out spectrum sensing, nearby secondary users are forbidden to transmit in order to avoid interfering the spectrum sensing procedure. Hence, the secondary users close to a selected data path are substantially affected by this requirement from spectrum sensing.

To formulate the metric representing the requirement of spectrum sensing, we denote by T_s^z and T_t^z the times of spectrum sensing and transmission for secondary user z, which lies within the carrier-sensing time interval of secondary user y. When we consider the requirement of spectrum sensing of both users x and z, the available time for data transmission of user y is then illustrated in Figure 11.8. Hence the time available for the data transmission of user y is given by

$$
T_a^y = \max_j \{T_s^j + T_t^j\} - \cup \{T_s^i\}, \qquad i \in I_y, \tag{11.35}
$$

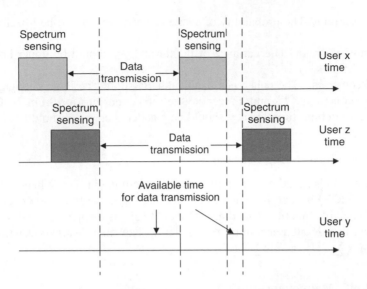

Figure 11.8 An illustration of the impact of spectrum sensing on the data transmission of nearby secondary users.

where I_y is the set of secondary users that can be interfered by user y. Based on the above definition and discussion, we define the metric as

$$T_f^y = \frac{T_a^y}{\max\{T_s^j + T_t^j\}}, \qquad i \in I_y. \tag{11.36}$$

Based on the above definitions of metrics, the spectrum selection problem for secondary user x can be modeled as the following optimization problem:

$$\max_k O_I = D_k T_f^x$$

$$\text{or} \quad \min_k O_{II} = D_k A_x^k$$

$$s.t. \quad M_B^k > P_B^{|C^k|}$$

$$V_B^k < J_T$$

$$B_c^k > \psi_k |C^k|$$

$$T_\Delta^s + T_\Delta^c (1 - M_B^k) < T_{th}. \tag{11.37}$$

The explanation of the above optimization problem is given as follows:

- Objective function: There could be two types of objective functions, namely O_I and O_{II}. For O_I, the goal is to maximize the transmission time such that the end-to-end delay can be minimized; the avoidance of interference to primary users is of the secondary priority. For O_{II}, the goal is to minimize the interference caused to primary users.

- The first constraint: The probability of available spectrum should be lower bounded by a threshold.
- The second constraint: The variance of the throughput should be upper bounded by a threshold.
- The third constraint: The coherence bandwidth B_c^k should be large enough.
- The fourth constraint: T_Δ^s and T_Δ^c denote the inter- and intraspectrum switching time. Hence, the switching time latency should be smaller than a threshold.

11.4.2.2   Stage 2. Next Hop Selection

The route selection is based on the optimization problem in (11.37). When each secondary x receives an RREQ packet, it computes the optimal objective function O_I or O_{II}. Then, the optimal objective function, called initiative in [1321], is mapped to a delay. The RREQ packet is then broadcasted according to this delay. The final destination will choose the route with $\max \sum_j O_I^j$ or $\min \sum_j O_{II}^j$.

11.4.2.3   Route Maintenance

We assume that a route is broken due to the mobility of secondary users. For example, when a secondary user in a data path moves to an area with many primary users, this route may be damaged according to the protection of primary users. At this time, a new route should be found to resume the data traffic.

A simple scheme for route maintenance is discussed in [1321]. In such a scheme, it is supposed that each secondary user knows its own location (e.g., using a GPS), as well as the locations of primary users. When a secondary user finds that it is close to primary users by comparing with a threshold, it signals its previous hop user that a new route discovery procedure should be initiated. Then, the previous hop user sends out RREQ packets and finds a new route.

11.4.3   Dynamic Routing

Now, we use the proposed scheme in [755] to illustrate how to carry out dynamic routing. It is assumed that the secondary user do not have prior information about the spectrum situation. Hence, the reinforcement learning (introduced in Chapter 7) is applied for the secondary users to learn how to carry out the routing.

First, we describe how the MAC layers works in the network studied in [755]. Suppose that secondary user a plans to send a packet to secondary user b. It first choose a channel that is idle by carrying out spectrum sensing. Then, it sends a request-to-send (RTS) over this available channels. If no clear-to-send (CTS) from secondary user b is received in a predetermined period of time, a sends the RTS again until it receives the CTS. When b sends back the CTS, it also knows that a will transmit over this channel; thus, b will wait for the packets from a in this channel. Note that the RTS-CTS mechanism is the same as many traditional communications systems, such as the systems with CSMA-CA mechanism. However, the novelty of this approach is that the channel information is implicitly conveyed in the RTS message.

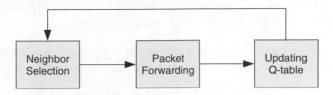

Figure 11.9 The Q-learning procedure for routing.

Then, we describe how the reinforcement learning is applied in the routing procedure, as illustrated in Figure 11.9. Since we assume that the secondary users do not any prior information about the spectrum activities, each secondary user (say user a) keeps a Q-table in which each element is a 2-tuple, denoted by $Q_a(b, d)$ where b is the ID of a neighbor and d is the destination. Intuitively, $Q_a(b, d)$ means the metric of routing when secondary user a sends a packet to destination d via user b. A larger $Q_a(b, d)$ implies that secondary user a should choose to forward packets to neighbor b more frequently.

There are two approaches for selecting the next-hop neighbor when secondary user a has a data packet for destination d:

- Deterministic forwarding: In this approach, secondary user a selects the neighbor having the highest Q-value to forward, that is,

$$t = \arg\max_{b \sim a} Q_a(b, d), \tag{11.38}$$

where $b \sim a$ means that b is a neighbor of a. Note that [755] adopted this approach.
- Stochastic forwarding: In the stochastic forwarding, secondary user chooses the forwarding neighbor in a random manner. There are also two approaches for the stochastic forwarding. In the first approach, user a forwards packets to t obtained in (11.38) with a predetermined probability p or, with probability $1 - p$, randomly selects another neighbor other than t. Usually $p > 0.5$. In the second approach, the probability for choosing a neighbor b is given by the following Boltzman distribution:

$$P(\text{choose } b) = \frac{\exp(\beta Q_a(b, d))}{\sum_{c \sim a} \exp(\beta Q_a(c, d))}, \tag{11.39}$$

where β is a parameter called inverse temperature. Obviously, when $Q_a(b, d)$ is larger (that is, the reward of choosing neighbor b is larger), the probability of choosing neighbor b is also larger. The parameter β is used to control the level of concentration on the neighbors with good Q values. When β is large (that is, low temperature), the selection will be more focused on neighbors having large Q values. When $\beta \to \infty$, the rule converges to the deterministic one in (11.38). When β is small (that is, the temperature is large), the probability of selection is more dispersed among all neighbors.

An advantage of the stochastic forwarding over the deterministic one is as follows. When the forwarding neighbor is chosen randomly, each neighbor has a chance to be selected. Hence, a good candidate neighbor having a bad initial Q-value can get chance to improve its Q-value such that it will receive more selections in the future. In the deterministic case, if a good candidate neighbor has a bad initial Q-value, it may never be selected since another neighbor may dominate it forever.

After selecting the forwarding neighbor, the secondary user needs to update the Q-value according to the experience of this action. Usually, the updating rule is given by

$$Q_a^{t+1}(b, d) = (1 - \alpha)Q_a^t(b, d) + \alpha r_t, \qquad (11.40)$$

where t is the index of time, r_t is the reward in the t-th action and $0 < \alpha < 1$ is the factor controlling the speed of learning. When α is too large, it may be considerably impacted by the randomness in the reward; when α is too small, the learning could be too slow. Hence, it is important to choose a good value for α.

When secondary user a is able to know the spectrum occupancies of nearby secondary users, it can add one more item to the Q-value, that is, the local spectrum state s. Hence, the Q-value is changed to $Q_a(b, d, s)$. The local spectrum state s can refine the Q-values and make the learning adaptive to instantaneous spectrum situations. In the previous mechanism of Q-value, the forwarding neighbor is unaware of the local spectrum situation. Even when a neighbor, say b, is unable to transmit due to the emergence of primary user, secondary user a still forward its packet to b, if $Q_a(b, d)$ is large. The addition of the local spectrum state s can prevent such a scenario. However, it requires the information exchange among neighbors; moreover, the learning speed will be much slower since now we have much more Q-values to learn.

The learning based dynamical routing can track the dynamics of spectrum occupancies. Even if new primary users emerge or some primary users leave, the Q-learning based routing can still adapt to the new spectrum environment since the Q-values can be updated in time.

11.5   Congestion Control in Transport Layer

As we have explained, the main functionality of the transport layer is the congestion control, that is, controlling the source traffic rate to avoid congestions at bottlenecks in the network. Due to the new mechanism of spectrum access, new challenges arise for the design of congestion control in cognitive radio networks. In this section, we will first introduce the mechanism of congestion control in Internet. Then, we will explain the new challenges in cognitive radio networks and discuss various schemes addressing these challenges.

11.5.1   Congestion Control in Internet

On the Internet, congestion control, usually in TCP, is carried out by locally estimating the possible congestions (not by explicit congestion notifications (ECNs)). The details of TCP can be found in [1323]. In TCP, each source node maintains a sliding window, which allows the source node to transmit multiple packets before sending an acknowledgement. A key mechanism in TCP is how to determine the window size. A general principle of TCP[2] is: before a congestion occurs, the window size increases in an additive manner;

---

[2] There are many versions of TCP. Each version has different details for the window size control.

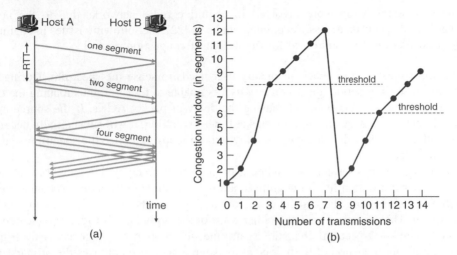

Figure 11.10 An illustration of the evolution of window size. (a) Increase of window size. (b) Evoluation of window size.

when a congestion occurs, the window size decreases exponentially. Since there is no explicit notification on the congestion, the source node claims a congestion when the timer for the acknowledgement (ACK) of a packet expires (at this time, the source node believes that the packet is lost or seriously delayed due to the congestion).

An illustration of the window size increase is shown in Figure 11.10 (a). At the very beginning, the window size is 1. When an ACK is received, the window size is increased by 1. Hence, after 3 batches of transmission, the window size becomes 4. As shown in Figure 11.10, when the window size is increased to a threshold (equaling 8 at time slot 3), the increase of the window size becomes linear. At time slot 7, a packet drop occurs because the timer for ACK expires. Hence, the window size is decreased to 1. The increase procedure is repeated again and the threshold is set to 6, that is, half of the peak window size in the previous increase stage.

11.5.2 Challenges in Cognitive Radio

Note that the traditional congestion control mechanism is designed for wired networks on the Internet, in which the main reason for packet drop is congestion. Hence, the congestion is determined by the drop of packets. However, this assumption is incorrect in wireless communication networks, where the packet drop can also be incurred by bad channel conditions (e.g., deep fading situation). Hence, if we still apply the traditional congestion control mechanism in wireless network, it may cause false action of the source node. For example, an intermediate communication link experiences a temporary deep fading and thus causes a packet drop. Actually, the bad channel condition may be recovered soon and there is no congestion; however, the source node will significantly decrease its traffic rate, thus causing the underutilization of the wireless spectrum. Hence, many approaches have been proposed to address the congestion control in wireless communication network.

The situation is even more involved in cognitive radio networks due to the novel mechanism of spectrum access. As pointed out by [1322], the following issues in cognitive radio networks must be addressed for the design of congestion control:

- Spectrum sensing state: Each secondary user needs to sense the spectrum. During the spectrum sensing period, the secondary user is unable to transmit, thus making the data path virtually disconnected and causing the delay of packet ACKs. If the source node simply claims a congestion upon a packet timer expiration, it will substantially decrease its traffic rate while the data path actually will be recovered once the spectrum sensing is completed.
- Effect of primary user activity: When a primary user emerges, it is possible to interrupt a communication link in cognitive radio. Then, the corresponding secondary user will temporarily lose the capability of forwarding packets, as well as sending ACKs upstream. The secondary user can either wait for the primary user to leave or search for a new available spectrum channel. Again, the interruption may unnecessarily trigger the significant reduction of traffic rate at the source node if it confuses the primary user interruption with a real network congestion.
- Channel switching: When a secondary user changes its current working channel (e.g., because of a primary user emerging in this channel), it is unable to transmit. However, after a certain channel switching time, the transmission can be recovered. Hence, there is not need to decrease the transmission window size.

11.5.3 TP-CRAHN

In [1322], a protocol of congestion control has been proposed to address the above challenges in the transport layer of cognitive radio networks. The key feature of the new protocol is that multiple new states are added to the transport layer, which represent the new features in cognitive radio, as illustrated in Figure 11.11. Then, we will discuss these states and the corresponding state transitions.

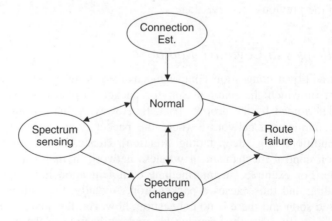

Figure 11.11 An illustration of state transmission in TP-CRAHN.

11.5.3.1 States

There are multiple states in the TP-CRAHN (here we omit the state of Mobility Predicted proposed in [1322] since it is less relevant to the cognitive radio mechanism):

- Normal: This state is the same as the state in traditional TCP. Within this state, the connection operates as a traditional wireless network. When the connection is established, there is no primary user and the secondary users are not interrupted by the spectrum sensing period, the connection stays in the normal state. The normal state of TP-CRAHN is different from traditional TCP in the congestion avoidance and feedback information, which will be explained later.
- Connection Establishment: The connection is established during this state. The procedure of the connection establishment is very similar to the traditional three-way handshake in TCP newReno. The key difference is that some basic information for the operation of cognitive radio is also established during this procedure. First, a synchronization packet (SYN) is sent by the source node to the destination. The intermediate nodes attach their information including their IDs, time stamps and tuples (t_i^1, t_i^2, t_i^s), where i is the ID, t_i^1 is the time to go before the next round of spectrum sensing, t_i^2 is the time interval between two spectrum sensing actions and t_i^3 is the time needed for spectrum sensing.
- Spectrum Sensing: In this state, one (or more) secondary user is in the state of spectrum sensing and cannot forward data packets. During this period, the connection needs to do two things:
  - (a) Flow Control: Since the secondary user in spectrum sensing (say, user i) is unable to forward or receive packets, user $i - 1$, which is the node right before user i may be overwhelmed by the incoming packets. Hence, the effective transmission window size can be changed to

    $$ewnd = \min\{cwnd, rwnd, B_{i-1}^f\}, \tag{11.41}$$

    where B_{i-1}^f is the free buffer space of user $i - 1$. Then, the buffer of user $i - 1$ will not overflow. More details of the operations can be found in [1322].
  - (b) Sensing Time Regulation: If no (or limited) primary user activities have been detected in a channel, the spectrum sensing time over this channel can be reduced in order to reduce the end-to-end throughput. The detailed mechanism for reducing the spectrum sensing time is given in [1322].
- Spectrum Change State: In this state, a communication link is interrupted by primary user(s) and thus cannot forward packets. An illustration is given in Figure 11.12. Suppose that secondary user i is affected by a primary user. Upon detecting the primary user, it sends an explicit pause notification (EPN) to the source node to freeze the whole connection, since the connection is temporarily in outage. Then, user i sends a list of available channels to user $i - 1$, which in turn feeds back a channel selection to user i. After this handshake, a new spectrum channel is established between user i and user $i - 1$. The next task is to measure the capacity of the new channel. User i sends back a probe packet and receives an ACK from user $i - 1$. From the probe packet and ACK, the bandwidth of the new channel can be estimated to update the round trip time. Details can be found in [1322].

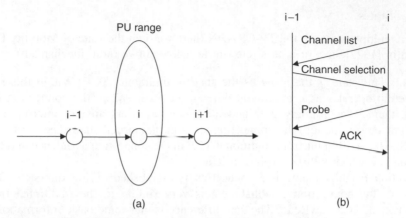

(a)   (b)

Figure 11.12   An illustration of spectrum change. (a) Illustration of the primary user interruption. (b) Information exchange for spectrum change.

As we have mentioned, the operation in the normal state is different from the traditional TCP in the following two aspects:

- Explicit Congestion Notification (ECN): In traditional communication networks, the congestion is detected by packet losses or the timeout of ACK. This is reasonable in wired communication networks since the main reason of packet loss or long delay is congestion. However, as we have explained, there could be many other reasons for packet loss or long ACK latency such as fading in general wireless communication networks or primary user emergence in cognitive radio networks. Hence, it is more reasonable to use ECN in cognitive radio networks in order to differentiate different reasons of congestions. The ECN packet is sent from the affected node to the source node in two ways: (a) an independent ECN packet is sent to the source node directly; (b) the ECN message is also piggybacked to data packets that are sent to the destination; then the destination node will send the ECN message to the source node in an ACK packet. The procedure is illustrated in Figure 11.13. Upon receiving an ECN message, the source node make an assessment on the timeliness of the ECN. If the time stamp of the ECN is within a certain range, the source node will decrease its window size.
- Feedback through ACK: Each intermediate node along the path will piggyback the following information to the data packets to the destination node, which will be sent back to the source node in ACK packets: (a) residual buffer space; (b) observed link bandwidth; (c) total link latency.

11.5.4   Early Start Scheme

As we have seen above, it is reasonable to freeze or slow the whole connection or some upstream nodes when an intermediate secondary user is interrupted by primary users. Only when the affected secondary user sends back a notification to clear the freezing, can a packet before the affected node resume its journey along the data path. We call such a strategy the 'slow-and-start' one, which is similar to what drivers do when meeting

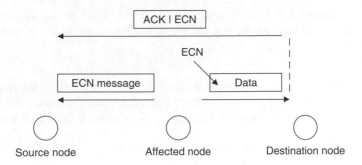

Figure 11.13   An illustration of the ECN transmission.

perturbations on the highway. Hence, this will bring interruptions to the data traffic. As we will see, even small perturbations to the data traffic may cause a serious damage. We will find that the reason for the damage is due to the 'slow-and-start' policy. Hence, we will proposed an early start policy within the framework of network utility maximization.

11.5.4.1 Traffic Perturbation

To analyze the impact of the perturbation on the data traffic, we regard the packets as vehicles running on a highway due to the similarity between data and vehicle traffics. For simplicity, only one data flow along a single path is considered; otherwise, the analysis for multiple traffics will be much more involved. We consider a fluid model; that is, the time and space in the data traffic are continuous, thus facilitating the analysis of stability. Although being different from the real data traffic in cognitive radio networks, the continuous model will reveal much insight.

We assume that many packets are distributed along a path with ascending indices; that is, packet n immediately follows packet $n - 1$. We denote by x_n the location of the n-th packet, which is a continuous variable. The dynamics of packet n are assumed to satisfy

$$\frac{d^2 x_n(t)}{dt^2} = f(x_{n-1}(t) - x_n(t)) - b\frac{dx_n(t)}{dt}, \tag{11.42}$$

where $\frac{d^2 x_n(t)}{dt^2}$ is the acceleration rate, $\frac{dx_n(t)}{dt}$ is the speed and $f(x_{n-1}(t) - x_n(t))$ is a monotonically increasing function representing the impact of interpacket interval on the acceleration rate. This model is similar to the one proposed for the vehicle traffics in [1324]. The mode of the packet dynamics in (11.42) are justified by the following features:

- When the distance between packets becomes small, the behind packet will slow down in order to avoid congestion/collision. When primary user interruption occurs and causes a packet to stop, the succeeding packets will slow down more radically and the speed will decrease to zero asymptotically.
- When the speed is high, the acceleration rate may become negative and bring the speed down, thus avoiding an infinite speed. This is reasonable due to the control mechanism of transmission window.

We assume that the initial condition satisfies $x_{n-1}(0) - x_n(0) = \Delta x$; that is, the packets are evenly distributed. We assume that there exists a positive number v such that

$$f(\Delta x) = bv, \tag{11.43}$$

which implies that $x_n(t) = x_n(0) + vt$ is a solution to the differential equation in (11.42).

It is very difficult to analyze the stability of the nonlinear dynamics. To linearize the dynamics, we denote by $y_n(t)$ the disturbance of the stationary state $x_n(0) + vt$, that is,

$$x_n(t) = x_n(0) + vt + y_n(t). \tag{11.44}$$

On assuming that $y_n(t)$ is sufficiently small and by linearizing the dynamics in (11.42), we have

$$\frac{d^2 y_n(t)}{dt^2} = f_0(y_{n-1}(t) - y_n(t)) - b\frac{dy_n(t)}{dt} + o(\mathbf{y}(t)), \tag{11.45}$$

where f_0 is the derivative of function f at Δx and $\mathbf{y} = (y_1, y_2, \ldots, y(N))$. In the subsequence, we ignore the higher order term $o(\mathbf{y}(t))$ and consider only the linear dynamics.

By defining $\frac{dy_n}{dt} = \theta_n$ and $\mathbf{z} = (y_1, y_2, \ldots, y_N, \theta_1, \ldots, \theta_N)$, it is easy to verify that the second-order dynamics in (11.45) can be rewritten in a vector form, which is given by

$$\frac{d\mathbf{z}}{dt} = \mathbf{A}\mathbf{z}(t), \tag{11.46}$$

where

$$\mathbf{A} = \begin{pmatrix} \mathbf{0} & \mathbf{I} \\ \mathbf{F} & -b\mathbf{I} \end{pmatrix}, \tag{11.47}$$

where $\mathbf{0}$ is the zero matrix, $\mathbf{I}$ is the identity matrix and $\mathbf{F}$ is a circulant matrix with the first row being $(-1, 0, \ldots, 0, 1)$. For example, when $N = 4$, we have

$$\mathbf{F} = f_0 \begin{pmatrix} -1 & 0 & 0 & 1 \\ 1 & -1 & 0 & 0 \\ 0 & 1 & -1 & 0 \\ 0 & 0 & 1 & -1 \end{pmatrix}. \tag{11.48}$$

Then, we can prove the following conclusion for the system dynamics, whose detailed proof is given in [1325].

Proposition 11.1 *The traffic dynamics are stable for all N if and only if the following condition holds:*

$$f_0 \geq \frac{b^2}{2}. \tag{11.49}$$

Some numerical simulation results related to this conclusion can be found in [1325].

Now, we discuss the implications of the conclusion in Proposition 11.1 in the context of cognitive radio networks. According to Proposition 11.1, when f_0 is not large enough, the traffic is unstable at the equilibrium point; hence, a small perturbation may result in the traffic congestion, which is called *phantom traffic jam* [1326] in the community of transport traffic analysis, even if the highway can provide sufficient space when there is no perturbation. People have found that the phantom jam is caused by the following two reasons [1326]:

- *Overreaction of drivers*: that is, the drivers may be too sensitive to its speed, or equivalently b is too large. When b is large, the driver is more included to slow down when its speed is high.
- *Chain reaction of followers*: that is, the reaction to the preceding vehicle. For example, suppose that three cars A, B and C are running on the highway in the order of $C \rightarrow B \rightarrow A$. When vehicle A slows down or stops due to some interruption, vehicle B will decelerate more quickly; on the other hand, if vehicle B accelerates, the acceleration of vehicle C will be delayed since the driver of vehicle C needs some time to react. A small f_0 implies the slow reaction to the preceding vehicle, which may result in $f_0 < b^2/2$ and thus the instability.

These conclusions in the community of transport traffic analysis provide substantial insight for cognitive radio networks. Correspondingly, in the design of cognitive radio networks, one needs to control the parameters such that f_0 is increased and b is decreased, in order to avoid the overreaction and chain reaction of followers. It is difficult to control b in the transport layer. However, we can control f_0 in cognitive radio networks. For example, if a follower, that is, a upstream neighbor node, can respond very quickly to the notification from the preceding secondary user that the interruption due to primary user emergence has been mitigated (either due to finding a new channel or the leave of primary user), then f_0 is increased. Another possibility is that the follower secondary user can predict the time when the interruption is alleviated and then accelerate its packet transmission in advance (imagine a driver who can predict the time of interruption clearance). As has been demonstrated in many studies, the future activities of spectrum can be predicted to some extent. This motivates us the subsequent work.

11.5.4.2 Network Utility Maximization

As we have discussed above, small perturbations due to primary user emergence may significantly affect the data traffic in cognitive radio networks. To address the primary user interruptions, the key is to improve the reaction speed of upstream nodes. Now, we propose to use the framework of NUM to study it.

We assume that there are totally N data flows and M cognitive radio links in the cognitive radio network. The licensed frequency band is divided into multiple channels and the time is divided into time slots. Each time slot lasts T_s seconds and consists of a spectrum sensing period and then a data transmission period. For simplicity, we assume that each secondary user can access one channel at a time. It is straightforward to extend the results in this paper to the general case that each secondary user can use multiple channels simultaneously. When an idle licensed channel is found, the secondary user will keep transmitting over this channel until a primary user emerges over this channel. Upon the emergence of primary user, the secondary user will switch to sense other channels until a new available channel is found to resume the transmission.

To simplify the analysis, we assume that the secondary users are full-duplex, that is, they are able to receive and transmit simultaneously, by employing two radios. There are sufficiently many idle licensed channels such that different cognitive radio links use different channels for transmission, thus avoiding the co-channel interference. Moreover, we ignore the problem of multiple access and assume that each secondary user can perfectly receive all the incoming data flows by employing one radio for one incoming data flow. These assumptions will be relaxed in our future study.

We denote by $\mathbf{R}$ the routing matrix, in which the rows are links and the columns are the flows. If flow j passes through link i, $\mathbf{R}_{ij} = 1$; otherwise $\mathbf{R}_{ij} = 0$. We denote by $c_i(t)$ the channel capacity of link i at time slot t. A block fading model is considered for the cognitive radio link, that is, the corresponding channel gain is constant during the time slot and changes at the beginning of the next time slot. If there are more than one data flows passing through a secondary user, we assume that the secondary user uses a round-robin scheduling algorithm, that is, allocating the same transmission time for different flows.

Now we begin to introduce the theory of network utilization maximization. In NUM, each data flow, say flow i, is associated with a utility function $U_i(x_i)$ where x_i is the data generation rate of flow i. Then, the congestion can be formulated as the optimization of the total utility within the constraint of channel capacities, which is given by

$$\max_{\mathbf{x}} \sum_{n=1}^{N} U_n(x_n)$$

$$s.t. \ \mathbf{Rx} \leq \mathbf{c}, \tag{11.50}$$

where $\mathbf{x} = (x_1, \ldots, x_N)$ is the vector containing the data rates of all flows and channel capacity vector $\mathbf{c} = (c_1, \ldots, c_M)$ is assumed to be constant. It is well known that this optimization can be decomposed as local optimizations via a pricing mechanism, that is, a price is set at each link and each flow source determines its data generation rate by maximizing its utility minus the cost due to the price, that is,

$$\mathbf{x}_n^*(t) = \arg\max_{x} \left(U(x) - x \sum_{j:\mathbf{R}_{nj}=1} \lambda_j(t) \right), \tag{11.51}$$

where $\lambda_j(t)$ is the price of link j at time slot t. In practice, the source rate can be smoothed using a weighting factor, that is,

$$\mathbf{x}_n(t+1) = (1-w)\mathbf{x}_n(t) + w\mathbf{x}_n^*(t+1), \tag{11.52}$$

where $0 < w < 1$ is the weighting factor.

Then, the evolution of the data source rates and prices can be written as

$$\begin{cases} x_n^*(t+1) = F(x_n(t), q_n(t)), & n = 1, \ldots, N \\ \lambda_m(t+1) = G(\lambda_m(t), y_m(t)), & m = 1, \ldots, M, \end{cases} \tag{11.53}$$

where $q_n(t)$ is the n-th element of vector $\mathbf{q} = \mathbf{R}^T \lambda(t)$ and means the sum of prices along the path of data flow n; $y_m(t)$ is the m-th element of the vector $\mathbf{Rx}$ and means the total throughput in link m; F and G are the functions for updating the source rates and prices. When the evolution functions F and G are properly chosen, the source rate and price dynamics will converge to the optimal solution of the NUM problem in (11.50). Throughout the discussion, we assume that every flow source can receive the current price information immediately, which can be realized by using a broadcast mechanism. It is interesting to study the case with a delayed price, which is beyond the scope of this book.

The traditional NUM formulation assumes that each flow source is tolled by the prices immediately. This may not be reasonable in cognitive radio networks with dynamic spectrum activities, since there is still a period of time before the newly generated packets arrive at the tolling links. For example, at time t, the flow source using TCP receives a high price from a bottleneck link which is interrupted by primary users and then immediately decreases its window size or even freezes the data flow; however, when the packets in the small window arrive at the bottleneck node, the price at this link may have become lower because the primary user has left or the link has found a new licensed channel. Hence, the traditional NUM formulation may make the sources in cognitive radio networks too conservative. To alleviate this problem, we introduce the real time network utility maximization (RT-NUM), in which each packet is tolled by each link using the price when the packet passes through the link, instead of the price when the packet is generate.

To formulate the RT-NUM mechanism, we assume that the data rate of a generic flow m at time slot $x_m(t)$ is proportional to the number of packets generated at this time slot; these packets will arrive at the cognitive radio link of the h-th hop and thus receive the tolls of this link at time slot $t + h^3$. Hence, the source rate in RT-NUM is given by

$$\mathbf{x}_n(t) = \arg\max_x \left(U(x) - x \sum_{j:\mathbf{R}_{nj}=1} \lambda_j(t + h_{nj}) \right), \tag{11.54}$$

where h_{nj} is the number of hops from the source to link j.

The difficulty in analyzing the price dynamics with time varying channel conditions is that there is no simple and explicit expression for describing the channel condition dynamics. Moreover, there are many communication links, which involve the high dimensional integral in terms of many channel gains. Hence, we use phenomenological models for describing the price, similarly to the price models used in economics or financial market. We will consider both the Geometric Brownian Motion and Jump Diffusion Process models.[4] Note that both models are based on the continuous time while the timing in cognitive radio is discrete; however, the continuous time model can significantly simplify the analysis.

The Geometric Brownian Motion (GBM) model has been widely used to model the price evolutions like stock. The use of GBM for price modeling traces back to the studies of P. A. Samuelson in 1965 [1328]. In GBM, the price dynamics are described using the following stochastic differential equation (SDE):

$$d\lambda(t) = \mu\lambda(t)dt + \sigma\lambda(t)dW(t), \tag{11.55}$$

---

[3] This assumption may not be true in practical systems since the packets could be blocked or lost before reaching the h-th hop. However, this assumption simplifies the analysis and will achieve a good performance, as will be shown in the numerical simulations.

[4] There could be many other models like the Lévy process [1327]. We will study them in the future.

where μ and σ are the parameters of drift and volatility, and $W(t)$ is the standard Brownian motion. Note that the drift means the average shift of the price while the volatility measures the variance of the price with respect to the time. We assume that there is no drift in the price; hence we can focus on only the volatility σ.

An equivalent description of GBM is given by

$$P(\lambda(t + \tau) = x | \lambda(t)) = \frac{1}{\sqrt{2\pi\tau}\sigma x}$$

$$\times \exp\left(-\frac{1}{2}\left(\frac{\log x - \log \lambda(t) - \mu\tau}{\sigma\tau}\right)^2\right) \tag{11.56}$$

We consider the discrete time prices $\{\lambda(t)\}_{t=0,1,2,...}$ as samples of the underlying continuous time price. Suppose that a source node observes T $(T > 1)$ consecutive price samples. Then, volatility parameter can be calibrated using the following estimation [1329]

$$\hat{\sigma} = \sqrt{\frac{1}{T-1}\sum_{t=1}^{T}\left(\frac{\log \lambda(t) - \log \lambda(t-1)}{\sqrt{T_s}} - \hat{\mu}\right)^2}, \tag{11.57}$$

where

$$\hat{\mu} = \frac{1}{T}\sum_{t=1}^{T}\frac{\log \lambda(t) - \log \lambda(t-1)}{\sqrt{T_s}}. \tag{11.58}$$

When the prices in cognitive radio evolve smoothly (that is, the channel conditions change slowly), the GBM model is reasonable. However, the abrupt emergence of primary users may cause a radical price change in the 'market' (imagine how fast the gas price increases when a major gas refinery plant is out of work). Hence, it is preferable to model the price jump caused by the primary user interruptions. Hence, we also consider the modeling of price using the Jump Diffusion Process (JDP) which traces back to the pioneering work of R. C. Merton [1330].

A JDP is a combination of the diffusion process in (11.55), which represents the channel condition variation, and a jump process, which represents the emergence and disappearance of primary users. On assuming that the jump process is a Poisson process with the intensity parameter ρ, the dynamics of a typical JDP can be written as [1329]

$$\frac{d\lambda(t)}{\lambda(t)} = (\mu - \rho k)dt + \sigma dW(t) + (Y(t) - 1)dq(t), \tag{11.59}$$

where μ and σ are the drift and volatility; $W(t)$ is the standard Brownian motion; $q(t)$ is the Poisson process; $Y(t) - 1$ is a random variable standing for the jump magnitude; and $k = E[Y(t) - 1]$. Note that the Poisson process satisfies

$$\begin{cases} P(dq(t) = 0) = 1 - \rho dt \\ P(dq(t) = 1) = \rho dt \\ P(dq(t) > 1) = o(dt) \end{cases}. \tag{11.60}$$

For simplicity, we assume that the jump magnitude can take two values, namely $+A$ or $-A$. When $Y(t) - 1 = A$, a primary user emerges, while $Y(t) - 1 = -A$ means the disappearance of the primary user or the blocked secondary user has found a new available channel. One drawback of the dynamics in (11.59) is that the parameter ρ results in the same expected periods of high price and low price, which does not satisfy the assumption that the activity of primary user is sparse. Hence, we propose to model the jump process as a heterogeneous Poisson process, that is, the intensity parameter equals ρ_l when the current price is low and ρ_h when the current price is high, respectively. Obviously, we have $\rho_h > \rho_l$ due to the assumption of sparse activity of primary users.

Below we propose a heuristic approach to estimate the parameters A, ρ, μ and σ, based on price observations $\lambda(0), \ldots, \lambda(T)$.

- Parameters of jump process: We divide the price samples into two classes, high price and low price. This can be accomplished by comparing the prices with a predetermined threshold or using a clustering approach. Denoting by $\mathcal{P}_h$ and $\mathcal{P}_l$ the index sets of high and low prices, respectively, we use the following simple estimation for A, which is given by

$$\hat{A} = \frac{1}{|\mathcal{P}_h|} \sum_{t \in \mathcal{P}_h} \lambda(t) - \frac{1}{|\mathcal{P}_l|} \sum_{t \in \mathcal{P}_l} \lambda(t). \tag{11.61}$$

The intensity parameter ρ_h is estimated by

$$\rho_h = \frac{\text{total number of jumps}}{|\mathcal{P}_h| T_s}. \tag{11.62}$$

The parameter ρ_l can be estimated similarly.

- Drift and volatility: Again, we assume that $\mu = 0$, that is, there is no nonzero drift in the pricing process. For estimating the volatility, we remove the offset due to the primary user emergence in the high price by

$$\tilde{\lambda}(t) = \begin{cases} \lambda(t), & t \in \mathcal{P}_l \\ \lambda(t) - \hat{A}, & t \in \mathcal{P}_h \end{cases}. \tag{11.63}$$

Then, we use the normalized price samples $\{\tilde{\lambda}(t)\}_{t=0,\ldots,T}$ to estimate the volatility σ.

The purpose of the pricing model is to study the congestion control in cognitive radio networks, that is, how the source controls the congestion window size, or equivalently, the amount of outgoing packets. We consider this problem as purchasing assets in the area of mathematical finance. We can consider the following two types of assets:

- Asset 1: not generating a packet.
- Asset 2: generating a packet.

Suppose that the source obtains W tokens in each time slot and the unit price of each asset is 1. Hence, the maximum number of packets that the source can send is given by

W. When $W_1 < W$ tokens is used to purchase asset 2, the source generates W_1 packets for the data traffic. Then, the problem is formulated as

$$\max_{W_1 \leq W} E \left[U(W_1) - \sum_{w=1}^{W_1} p(t + \tau_w) \right],$$ (11.64)

where τ_w is the time needed for packet w to reach the bottleneck node. Note that here we consider only the toll at the bottleneck node. It is straightforward to extend to the case of considering all the tolls along the path. The source can collect the historic data, compute the corresponding mathematical model and then optimize the asset purchase.

We randomly drop 50 secondary users and 10 primary users within a 1km×1km square. The activities of primary user satisfy a two-state Markov chain. The maximum distance for communication is 250 meters. We assume that there are 10 data flows within the cognitive radio network and the shortest path routing is used to establish the flow paths. The fading process of each wireless channel follows the 3GPP2 standard. Ten realizations of the network are used to collect the required statistics.

Several realizations of the price process are shown in Figure 11.14. The peaks are due to the primary user activities. When the channel is idle, the price fluctuation is due to the channel fading process.

We first consider the GBM model. Figure 11.15 shows the CDF of the difference of prices (in the logarithm scale), namely $\log \lambda(t + 1) - \log \lambda(t)$. For standard GBM, the difference is Gaussian. By comparing the empirical distribution of the difference with the Gaussian distribution with the same expectation and variance, we observe that there is still some gap, which means that the modeling is imperfect.

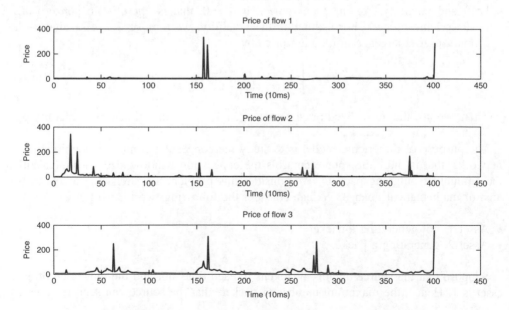

Figure 11.14 Samples of price evolutions at different links.

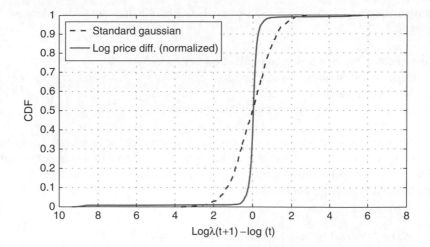

Figure 11.15 Comparison of the price change CDF.

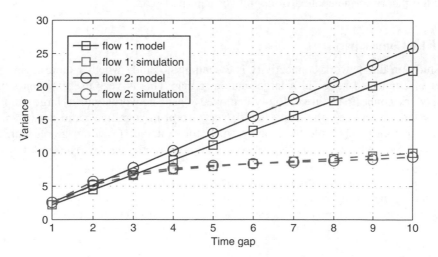

Figure 11.16 Comparison of the variance.

Figure 11.16 shows the variance of $\log \lambda(t)$ as a function of time for two data flows. In standard GBM model, the variance should be a linear function of time. As we have observed, the linearity holds for a short time; then, the actual variance becomes less than the standard linear function. Hence, the GBM cannot be used for modeling long-term price evolution, which is also true in stock price models. Similar conclusions also hold for the JDP model.

11.6   Complex Networks in Cognitive Radio

Complex network is a powerful mathematical tool to describe many interesting phenomena when the network size is sufficiently large, particularly when nodes in the network have mutual impacts. It has been used to analyze many behaviors of wireless networks, for example,

- the connectivity of wireless communication networks;
- the information flow in random networks;
- the navigation in random networks;
- the behavior propagation in cognitive radio networks.

In this section, we will focus on the connectivity and behavior propagation in cognitive radio networks, which are of significant importance for analyzing the performance of cognitive radio networks and can also help to understand how to disclose the complex network behaviors in large networks.

11.6.1 Brief Introduction to Complex Networks

We first provide a brief introduction to complex networks. In the next subsection, we will apply the theory to the context of cognitive radio networks.

11.6.1.1 Connectivity

The study on the network connectivity is essentially to study whether there is an infinite subset of connected nodes in a large random graph. This is very useful when studying the network connectivity in randomly deployed sensor networks with a large number of sensors. As we will see, the study is also very useful in cognitive radio networks. Note that the random graph could have many types, for example, a random graph with a grid topology or a stochastic geometric network with node randomly deployed in a plane. For simplicity, we introduce only the former case.

We consider a network with a grid topology, as illustrated in Figure 11.17. Two adjacent nodes are connected by an edge with probability p. Obviously, the larger p is, the more nodes are connected within the same subgraph. It has been shown that there exists a critical

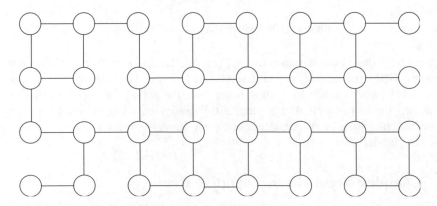

Figure 11.17   Illustration of bond percolation.

value for the probability, denoted by p_c. When $p > p_c$, with probability one, the graph contains a connected subgraph with infinite cardinality; when $p < p_c$, with probability one, the graph contains only connected subgraphs with finite sizes. We say that there is a phase change at p_c since the connectivity experiences a sudden change at p_c with probability, which is similar to phase change from liquid water to vapor. This is called bond percolation since the random factor is the existence of the edges.

Such a phase change of the connectivity is also valid for site percolation. In this case, each node could be open or closed, with probability q, and every neighbors nodes are connected by an edge. If a node is closed, an object cannot move through this node. Using almost the same argument, we can show that there exists a critical probability q_c such that, when $q > q_c$, an object can move through infinite nodes in the graph, and when $q < q_c$, an object can only move through finite nodes.

The phase change can also be extended to the case of stochastic geometric graphs. In such graphs, nodes are randomly distributed in the space. Two nodes are connected if and only if their distance is below a threshold. Then, the spatial density of the nodes experience a phase change for the network connectivity.

When discussing the connectivity in cognitive radio networks, the connectivity of two nodes depends on their distances and the positions of primary users. Hence, the densities of both primary and secondary users play important roles in the network connectivity. We will discuss this later.

11.6.1.2 Epidemic Propagation

Epidemic propagation is an important topic in the theory of complex networks. Intuitively, the theory of epidemic propagation studies how epidemic is propagated in the social network of population. It is very useful in studies on epidemics. However, here the epidemic is not confined to real disease. It can also be many other social behaviors such as rumors and habits. Such a study can use mathematics to describe how the social network structure affects the propagation of a certain behavior (either harmful or beneficial), thus helping researchers to understand or control such a propagation.

An illustration of epidemic propagation is given in Figure 11.18. The social network of population is represented by a random network in which each node is a person and each edge means that the two people corresponding to the two end nodes have contact with each other. There are three types of nodes in the network:

- Infected: Such a person is infected by the epidemic, thus being able to infect his neighbor in the social network. For example, in Figure 11.18, node 5 may propagate the epidemic to his neighbor 1 or 7.
- Susceptible: Such a person is not infected by the epidemic; however, he is susceptible to the epidemic if one or more neighbors of him is in the state of infected. For example, in Figure 11.18, node 1 can be infected from his infected neighbor 4 or 5.
- Recovered: Such a person has recovered from the epidemic and will no longer be infected; he also does not propagate the epidemic to his neighbors.

Given the above three types of nodes, there are three different models for the epidemic propagation with different assumptions:

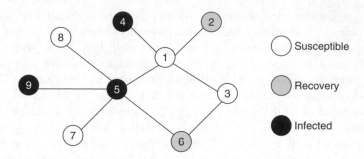

Figure 11.18 An illustration of epidemic propagation.

- SIR (susceptible-infected-recovered) model: In this model, each infected person may recover from the epidemic and then will never be infected. Hence, in a typical SIR model, all people will recover and then the epidemic extinguishes in the network.
- SIS (susceptible-infected-susceptible) model: In this model, each infected person may be cured and then becomes susceptible again. It is possible that all people become susceptible and then the epidemic extinguishes due to the lack of infection source. It is also possible that the epidemic never extinguish and there are always some people being infected.
- SI (susceptible-infected) model: In a contrast to the SIS model, each person in the SI model cannot be cured and stays in the state of infected. Hence, finally, all people will become infected.

For simplicity, we consider only the SIS model. We denote by i_k the proportion of the infected people having degree k, that is, $i_k = I_k/N_k$ where I_k is the number of infected people having degree k and N_k is the total number of people having degree k. It is assumed that, for a susceptible person, his infection from an infected neighbor is given by $\beta \delta t$ within a short time period δt, where β is called the spreading rate. We also assume that a infected node becomes susceptible again with probability $\mu \delta t$ for sufficiently small time δt. Then, the evolution of i_k is described the following ordinary differential equation:

$$\frac{i_k(t)}{dt} = \beta(1 - i_k(t))k\Theta_k(t) - \mu i_k(t). \tag{11.65}$$

The explanations of each term are given as follows:

- β is the spreading rate which determines the probability that a susceptible person is infected from a neighbor.
- $1 - i_k(t)$ is the proportion of the susceptible people having degree k.
- $\Theta_k(t)$ is the probability that one of the neighbor of a person with degree k is infected. It is scaled by k since the susceptible person has k neighbors.

For a general network topology, it is very difficult to analyze the term Θ_k. However, if the network has no degree correlations (that is, for a node with degree k, the probability that one of its neighbors has degree k', denoted by $P(k'|k)$ is independent of k), the analysis can be simplified. It is easy to show that the probability $P(k'|k)$ is given by

$$P(k'|k) = \frac{k'P(k')}{<k>}, \tag{11.66}$$

where $<k> = \sum_m m P(m)$ and $P(k')$ is the probability that a generic node has degree k'. Note that the no-degree-correlation property is true for small world and scale free networks. Then, for networks without degree correlation, one obtains

$$\Theta_k(t) = \frac{\sum_{k'} k'P(k')i_{k'}(t)}{<k>}, \tag{11.67}$$

which is independent of k. Hence, we can ignore the subscript k in Θ_k.

It is difficult to directly analyze the dynamics in (11.65) and (11.67) since there is no explicit expression for the solution of the ordinary differential equation. However, we can analyze the stationary solution as $t \to \infty$. We assume that the ordinary differential equation has a stable solution such that

$$\frac{di_k(t)}{dt} \to 0, \qquad \text{as } t \to \infty. \tag{11.68}$$

Then, $i_k(t)$ and $\Theta(t)$ converge to i_k and Θ, respectively. We have

$$i_k = \frac{k\beta\Theta}{\mu + k\beta\Theta}, \tag{11.69}$$

which implies

$$\Theta = \frac{1}{k>} \sum_k kP(k) \frac{\beta k\Theta}{\mu + \beta k\Theta}. \tag{11.70}$$

The stationary distribution of the infected and susceptible people can be obtained from (11.69) and (11.70).

The analysis of SI and SIR models is similar. The difference is

- for the SI model, the last term in (11.65) does not exist since any infected person cannot be recovered to susceptible;
- for the SIR model, the proportion of the recovered people should be added, thus increasing the number of differential equations.

11.6.2 Connectivity of Cognitive Radio Networks

The connectivity of cognitive radio network has been studied. In this book, we focus on the model and analysis.

11.6.2.1 Network Model

We consider a cognitive radio network distributed in an infinite plane as a two-dimensional Poisson process. A primary network is also distributed in a Poisson manner. The densities of the cognitive radio and primary networks are denoted by λ_S and λ_P. It is assumed that the primary system has a time slotted structure, in which each time slot lasts 1 time unit. In different time slots, the set of active primary users changes in an i.i.d. manner. This assumption is reasonable for networks with random data traffics or random scheduling of traffic. The secondary users within a certain range can communicate with each other, if they are not within a certain range of a primary user.

11.6.2.2 Conclusions on Connectivity

In summary, the network connectivity of cognitive radio system is illustrated in Figure 11.19. The observations are given below:

- There is a critical value for λ_S, denoted by λ_c, such that when $\lambda_S < \lambda_c$, the cognitive radio network is never connected, regardless of the primary user density.
- For each secondary user density larger than λ_c, there exists a critical value for λ_P (as a function of λ_S) such that when λ_P is larger than this critical value, the network is intermittently connected; otherwise, the network is connected.

Notice that the existence of λ_c is independent of primary users and is well known in traditional wireless communication networks. The impact of primary user density is unique for cognitive radio networks.

Note that the above conclusion concerns the definitions of network connectivity in the cognitive radio network. The network connectivity is defined by the minimum multihop delay (MMD):

- If the MMD between two randomly selected secondary users is infinite (or equivalently, there is no finite path between these two users) with probability 1, we say that the network is not connected.
- If the MMD between two randomly selected secondary users is finite with a positive probability, we say that the network is connected.

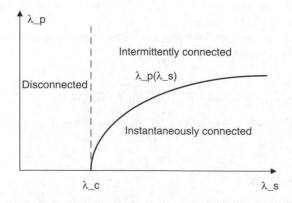

Figure 11.19 Different regions of the connectivity.

For the connected network, we distinguish intermittent connection from instantaneous connection:

- When we say that the cognitive radio network is intermittently connected, there is no infinite connected component in the network. However, a message can wait for some period of time for the recovery of spectrum and finally reach the destination.
- When we say that the cognitive radio network is instantaneously connected, there is an infinite connected component in the network.

To be more mathematical, we assume that the delay is determined by only the waiting time for spectrum opportunities. We denote by $t(s, d)$ the MMD from source node s to destination node d, and by $h(s, d)$ the distance between s and d. Then, we have

$$\lim_{h(s,d)\to\infty} \frac{t(s, d)}{h(s, d)} = \begin{cases} = 0, & \text{if instantaneously connected} \\ > 0, & \text{if intermittently connected,} \end{cases} \quad (11.71)$$

almost surely. Hence, if we ignore the propagation delay and the processing delay at each node, the cognitive radio network is almost the same as traditional ad hoc networks since the primary users do not affect the scaling law.

11.6.3 Behavior Propagation in Cognitive Radio Networks

In cognitive radio networks, each node can sense and compute, thus being able to 'see' and 'think'. Meanwhile, they can also communicate to each other, that is, they are able to 'talk'. This forms a social and the connections of the secondary users form a complex social network, as illustrated in Figure 11.20. The patterns of spectrum access can be considered as social behaviors that can be propagated in the network, since the secondary users can exchange information and shape their spectrum access patterns for better spectrum utilizations.

11.6.3.1 Network Model

We consider the behavior of 'favorite channel' in cognitive radio. We assume a cognitive radio network with N secondary users and K licensed channels. The locations of these N

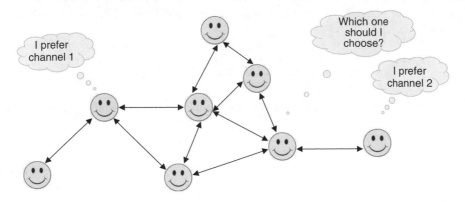

Figure 11.20 An illustration of social network in cognitive radio.

secondary users are denoted by $X_1, \ldots, X_N$. The network can be represented by a graph with multiple nodes, each of which represents a secondary user, and multiple edges, each of which represent a communication link. The secondary users adjacent in the graph can communication with each other. Primary users can emerge at any channel at any time slot. A secondary user cannot transmit over a channel that primary users are using. For facilitating the analysis, we assume that the secondary users are randomly distributed in the plane. We assume that the N secondary users are independently and uniformly distributed within a square S with area AN, that is, averagely each secondary user obtains an area of A^5. Formally, this means that, for any region $R \in \mathbb{R}^2$, the probability that a given secondary user n falls in region R is given by

$$P(X_n \in R) = \frac{|R \cap S|}{AN}. \tag{11.72}$$

As $N \to \infty$, the distribution of degree converges to a Poisson distribution with expectation λ, which is given by

$$\lambda = \frac{\pi d_{\max}^2}{A}. \tag{11.73}$$

As we have discussed in the introduction of SIS model of epidemic propagation, it is very important to obtain the conditional distribution of node degrees $P(k'|k)$, that is the probability that a node has a degree k' when one of his neighbors has degree k. We have explained that $P(k'|k)$ is independent of k (given $k > 0$) for small world and scale free networks. However, in the context of nodes randomly distributed in a plane, this independence no longer holds. The reason is intuitive. Suppose that node 1 has degree k and one of its neighbors, node 2, has degree k'. When k is large, there are many nodes around node 1. Then, with a large probability, many of these nodes also fall in the neighborhood of node 2. Hence, the probability that k' is also large is large. To be more mathematically precise, given a node with degree k, the probability that an arbitrary neighbor of it has degree k' is given by

$$P(k'|k) = \int_0^{d_{\max}} P(k'|r) \frac{2r}{d_{\max}^2} dr, \tag{11.74}$$

where $P(k'|r)$ is the distribution of the sum of two independent random variables r_1 and $r_2 + 1$. The random variable r_1 has a binomial distribution $B(n, \rho(r))$, where $n = k - 1$ and

$$\rho(r) = \frac{2d_{\max}^2 \cos^{-1}\left(\frac{r}{2d_{\max}}\right) + \frac{r}{2}\sqrt{d_{\max}^2 - \left(\frac{r}{2}\right)^2}}{\pi d_{\max}^2}. \tag{11.75}$$

The random variable r_2 has a Poisson distribution with expectation $\lambda'(r)$, which is given by

$$\lambda'(r) = \frac{d_{\max}^2 \left(\pi - 2\cos^{-1}\left(\frac{r}{2d_{\max}}\right)\right) - \frac{r}{2}\sqrt{d_{\max}^2 - \left(\frac{r}{2}\right)^2}}{A}. \tag{11.76}$$

---

[5] It is easy to extend to general nonuniform distribution case.

11.6.3.2 Favorite Channel

Now, we consider the social behavior of secondary users. We consider a certain channel, say channel 1. Each secondary user keeps a channel that it prefers to sensing, which is coined *favorite channel*. Each secondary user must be in one of two possible channels, namely 0 (the favorite channel of this secondary user is not channel 1) and 1 (the favorite channel of this secondary user is channel 1). The secondary users can recommend their favorite channels to neighbors. At the beginning time, a fraction of the secondary users have favorite channel 1. In the subsequent time, a secondary user in state 1 could be changed to state 0. The reason could be a primary user nearby emerges in channel 1 or the channel quality of channel 1 is worsened. We assume that the probability of this state change is $\mu\delta$ where μ is the change rate and δt is a sufficiently small time interval. It is also possible that a secondary user in state 0 changes to state 1, that is, it considers channel 1 as its favorite. The reasons could be:

- The secondary user receives the recommendation from a neighbor in state 1. The probability is $\lambda \delta t$, given a neighbor is in state 1.
- The secondary user finds that channel 1 is good by itself. The probability is $\phi \delta t$.

Note that the key differences between the epidemic propagation and the channel preference propagation include:

- The channel preference is propagated in a plane while the epidemic is spread in the abstract network.
- A secondary user could be changed from state 0 to 1 even if none of its neighbors is in state 1, while a susceptible person cannot be infected if none of his friends has been infected.

11.6.3.3 Dynamics of Social Behavior

To analyze how the selection of favorite channel is propagated in the cognitive radio network, we consider continuous time and denote by $x_k(t)$ the proportion of the secondary users with degree k and state 1. The dynamics of $x_k(t)$ is given by

$$\dot{x}_k(t) = -\lambda x_k(t) + \mu(1 - x_k(t))\left(\phi + \sum_{n=1}^{\infty} x_n(t)P(m|k)\right), \qquad (11.77)$$

for $k = 1, 2, \ldots$. The first term on the right hand side is the proportion of secondary users with degree k changing its preference from channel 1 to other channels while the second term is the proportion of secondary users with degree k beginning to prefer channel 1 due to the advice from other secondary users and the spontaneous discovery of the channel.

The following proposition provides a sufficient condition for the convergence of the differential equation.

Proposition 11.2 *Suppose that each secondary user has at most two neighbors and $\phi = 0$, then the differential equation in (11.77) converges to a stationary point as $t \to \infty$ if*

$$\lambda > \mu \max\{P(1|1), P(2|2)\}, \qquad (11.78)$$

and

$$\sqrt{(\lambda - \mu P(1|1))(\lambda - \mu P(2|2))} > \frac{\mu(P(2|1) + P(1|2))}{2}. \tag{11.79}$$

We assume that the differential equation in (11.77) converges, which can be demonstrated by many numerical results. At the steady state, we have $\dot{x}_k = 0$, for $k = 1, 2, \ldots$, which means that the proportions no longer change. Similarly to the discussion on the SIS model, we define

$$\theta_k = \sum_{n=1}^{\infty} x_n P(m|k), \tag{11.80}$$

which means the probability that an arbitrary neighbor of a secondary user with degree k sets channel 1 as its favorite channel. Then the steady state condition $\dot{x}_k = 0$ implies

$$x_k = \frac{\mu(\theta_k + \phi)}{\lambda + \mu(\theta_k + \phi)}, \tag{11.81}$$

as well as

$$\theta_k = \sum_{m=1}^{\infty} \frac{\mu(\theta_m + \phi)}{\lambda + \mu(\theta_m + \phi)} P(m|k). \tag{11.82}$$

Then, the steady state of the channel preference propagation is determined by Equations (11.81) and (11.82).

The following proposition provides an upper bound for the proportion of secondary users in state 1, that is, setting channel 1 as their favorite channel.

Proposition 11.3 *A solution always exists for for Equation (11.82). Moreover, the steady proportion is upper bounded by*

$$x_k \leq \frac{\mu(\theta_\infty + \phi)}{\lambda + \mu(\theta_\infty + \phi)}, \forall k, \tag{11.83}$$

where

$$\theta_\infty = \frac{-(\lambda + \mu\phi - \mu) + \sqrt{(\lambda + \mu\phi - \mu)^2 + 4\mu\phi}}{2}. \tag{11.84}$$

12

Cognitive Radio Network as Sensors

Cognitive radio network as sensors is a new initiative and tries to explore the vision of a dual-use sensing/communication system based on cognitive radio network [1331–1333]. The motivation of cognitive radio network as sensors is to push the convergence of sensing and communication systems into a unified cognitive networking system. Cognitive radio network is a cyber-physical system with the integrated capabilities of control, communication, and computing. Cognitive radio network can provide an information superhighway and a strong backbone for the next-generation intelligence, surveillance, and reconnaissance.

Multifunctional SDRs for both radar and communication have been studied in [1334–1337]. OFDM waveform is explored. OFDM is the core technology in wideband communication. OFDM is adopted by 3GPP LTE, WLAN, power line communication, cognitive radio [1263], as well as ADSL and VDSL. The OFDM waveform has also been used in the radar society [1338–1341]. The key feature of the OFDM waveform is that multiple frequencies can be exploited simultaneously and in an orthogonal way. Meanwhile, the radio resources of all frequencies in OFDM waveform can be adjusted dynamically. The advantages of using OFDM in radar have also been summarized in [1342]. Digital generation, inexpensive implementation, pulse-to-pulse shape variation, interference mitigation, noise-like waveform for low probability of intercept/detection (LPI/LPD), and so on are the benefits of the OFDM waveform [1342].

Similarly, the research about the joint OFDM-based radar and communication system has been carried about in Karlsruhe Institute of Technology, Germany [1343–1347], especially for the future intelligent transportation systems. Range estimation, angle estimation, and Doppler estimation are extensively studied. Besides, a communication waveform is proposed for radar in [1348]. OFDM waveforms can be used to solve the unambiguous radial speed in a single transmission and improve the signal-to-background contrast [1348].

Cognitive Radio Communications and Networking: Principles and Practice, First Edition.
Robert C. Qiu, Zhen Hu, Husheng Li and Michael C. Wicks.
© 2012 John Wiley & Sons, Ltd. Published 2012 by John Wiley & Sons, Ltd.

In telecommunications, direct-sequence spread spectrum (DSSS) is a modulation technique. Radar communication integration based on DSSS has been discussed in [1349]. A multifunctional RF system that integrates radar and communication can avoid mutual interference by using different pseudo random (PN) codes [1349]. Direct sequence UWB signals are also applied [1350, 1351]. Oppermann sequences are utilized to generate the weighted pulse trains for the integrated radar and communication system [1352]. The ambiguity function of weighted pulse trains with Oppermann sequences is analyzed. Oppermann sequences can facilitate both radar application and multiple-access communication.

Communication information can be embedded in the radar system through waveform diversity [1353, 1354]. Meanwhile, in the radar network, the communication message, for instance the reports on the detected targets, can be embedded into the OFDM radar waveform [1355]. A unique covert opportunistic spectrum access solution to enable the coexistence of OFDM based data communication with UWB noise radar is presented in [1356]. A multi-functional waveform has been designed, by embedding an OFDM signal into a spectrally notched UWB random noise waveform [1356]. The performance of a cognitive WiMAX system in the presence of an S-band swept pulse radar is studied in [1357]. WiMAX can still work with opportunistic transmission as long as it avoids interfering with the radar system [1357].

The USRP product family provided by Ettus Research is widely used as a hardware platform in the area of cognitive radio [1358–1361]. A major advantage of USRP is that it works with GNU Radio, an open source software with plenty of resources, which simplifies and eases the usage of USRP. The USRP product family is also exploited for sensing tasks, for example, SAR [1362], passive radar [1363–1365], multistatic radar [1366], weather radar [1367], and so on. Besides, Path Intelligence Ltd. uses USRP to track pedestrian foot traffic by receiving the control channel signal transmitted from pedestrian's cell phone. The data collected by USRP can provide information for retailer, marketing, advertising, and so on. For example, this information can help retailers to better understand customers' behaviors in their stores and measure the effectiveness of their marketing strategies.

The similar idea of cognitive radio network as sensors is called cognitive sensor networks [1368–1373]. Though cognitive radio network can provide an information superhighway and a backbone for sensing tasks, the resource limitation should always be considered. The resource limitation issues have been addressed in [1374].

Optimization theory, machine learning, real-time adaptive signal processing, and graph theory can be applied. An integrated sensing/communication cognitive network should have the capabilities of cognition, waveform diversity, network resource management, dynamic network topology, multilevel synchronization, and cyber security. In terms of cognitive radio network as sensors, the following functions should be supported:

- Interference mitigation.
- Detection and estimation.
- Classification, discrimination, and recognition.
- Tracking.
- Sensing and imaging.

12.1 Intrusion Detection by Machine Learning

Cognitive radio network is designed to utilize the radio spectrum optimally for the next generation wireless communication network. Meanwhile, cognitive radio network as sensors is also a practical trend for the future wireless sensor network as it saves infrastructure cost while taking advantage of the rich communication and computing capabilities of cognitive radio network.

Due to the embedded function of spectrum sensing in cognitive radio network. More information about the radio environment can be obtained. This valuable information can be exploited to detect, indicate, recognize, or track the target or intruder in the covered area of cognitive radio network. The data related to this kind of information are intrinsically high-dimensional and random.

Passive target intrusion detection is a very important application of cognitive radio network as sensors. The target is device free. Active sensing is performed. The intruder is detected and localized mainly through the intrusion effect on the radio environment. Angle of arrival (AOA) [1375, 1376], time of arrival (TOA) [1377, 1378], and TDOA [1379, 1380] are the commonly used methods for localization or positioning. However, for the multipath environment, it is not easy to get the accurate information of AOA, TOA, or TDOA. The performance of detection or localization will be degraded based on physical-based methods. Thus, we would like to resort to machine learning and the huge data of spectrum sensing in cognitive radio network. The passive target intrusion detection can be formulated as a multiclass classification problem. Different intruder locations correspond to different classes. Multi-class SVM can be applied. This idea is similar to location fingerprinting [1378]. Location fingerprinting refers to methods that match the fingerprint or a set of features of a signal that is location dependent [1378].

12.2 Joint Spectrum Sensing and Localization

Joint spectrum sensing and primary user localization in cognitive radio network has been discussed in [1381]. Compressed sensing has been used here with the assumption that the number of primary users is small. In order to find the location of the primary users, the area covered by the primary users has to be discretized, and the primary users are assumed to be located only at these discrete grid points. The resolution of localization is largely dependent on the discretization granularity. A sparse Bayesian learning approach is also applied to joint spectrum sensing and primary user localization [1382].

12.3 Distributed Aspect Synthetic Aperture Radar

The static cognitive radio network can be extended to a dynamic cognitive radio network. UAV can be incorporated into cognitive radio network shown in Figure 12.1. In this way, the capability of cognitive radio network as sensors can be greatly increased. The development of UAV system has gained a lot of attentions throughout the world. The importance and significance of this kind of system in aerial activities have grown continuously, especially for military and reconnaissance purposes [1383]. Also, UAV systems are greatly preferred in operations where the tasks are dangerous, tedious, and impossible for human

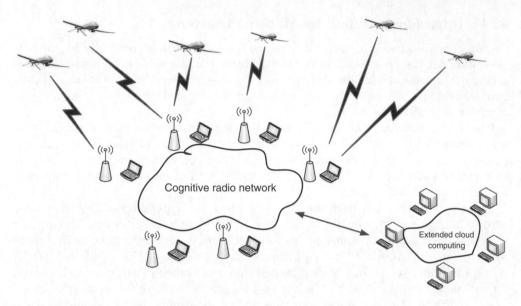

Figure 12.1 Distributed Aspect SAR.

pilots [1383]. Radio source localization by a cooperating UAV team has been presented in [1384]. Source localization is formulated as a stochastic distributed estimation problem. UAV is exploited to improve the observability in terms of the Fisher information matrix of the corresponding estimation problem [1384]. An automatic flight control algorithm that exploits network mobility and allows an autonomous UAV team to react cooperatively is developed to determine the location of a radio emitter [1384]. The problem of close target reconnaissance by a formation of three UAVs is considered in [1385]. The overall close target reconnaissance includes tasks of avoiding obstacles or no-fly-zones, avoiding inter-agent collisions, reaching a close vicinity of a specified target position, and forming an equilateral triangular formation around the target [1385]. Some search algorithms utilizing multiple mobile sensor nodes have been proposed in [1386]. The goal is to minimize the total search time in a given search area while making cooperation among these mobile sensor nodes and tolerating possible failures of one or more nodes [1386]. These search methods can be directly used in the control algorithm for UAVs if a similar search mission is required. The engineering design of a hand-launched, small UAV for ground reconnaissance has been presented in [1383]. Guidance strategy, control design, and the real data from practical flight tests are given. The main goal of this work is to implement a low-cost, portable, and reliable aerial platform for ground reconnaissance [1383]. UAVs have been used for ground thread identification in [1387]. Improved direct inference algorithm is proposed based on Bayesian network. Mobility models for UAV group reconnaissance applications have been presented in [1388]. The short-term goal is to address the specific requirements of UAV network cooperating to achieve a common mission [1388]. The long-term goal is to study the algorithms and protocols that provide optimized performance with respect to utilization of resources and exhibit robust behavior in the presence of attacks and interference [1388].

SAR [1389] is a form of imaging radar. SAR has become very popular recently because of its wide use for both military and civil applications. Most commonly, SAR is mounted on the airplane or satellite. The trajectory of these vehicles are predetermined and stale. However, if SAR is put on an UAV, some pressing challenges and interesting opportunities will appear, which leads to a lot of related research topics. Due to flexible mobility and easy control of UAVs, distributed aspect SAR based on dynamic cognitive radio network can be deployed. A fleet of UAVs can fly to anywhere at anytime in any situation to form the image of a target of interest. Meanwhile, SAR using arbitrary trajectories and UAV trajectory design for SAR are critical issues. UAV should fly along the optimized trajectory to get more useful information for SAR imaging.

PSO can be used for path planning [1390] and in-flight route replanning is needed [1391]. A multiobjective optimization under multiple constraints is applied [1391]. Image quality of each target should be maximized and route length under constraints should be minimized. Fuzzy selection is exploited to select the optimal route from the Pareto non-dominant route set with consideration of image quality, route length, and risk given by the expert system [1391]. The researchers at the Air Force Research Laboratory have presented a concept for exploiting UAV trajectories with perturbation for intelligent circular SAR applications, especially for detecting slowly moving targets [1392]. The basic concept is based on collecting subapertures of data over a given set of localized trajectories and intelligently parsing the collected data based on time-varying angle estimates between the localized UAV trajectory and subsets of a collection of moving point targets [1392]. An approximate analytic inversion method for bistatic SAR for arbitrary flight trajectories has been derived in [1393]. Performance synthesis of UAV trajectories in multistatic SAR has been analyzed in [1394]. A fleet of multiple UAVs operate in coordination over or nearby a scene of interest. A tomographic model of the SAR measurement process is used as the foundation to provide a guiding metric for trajectory optimization in the presence of measurement noise. As early as 1983, the spotlight-mode SAR was interpreted as a tomographic reconstruction problem and analyzed using the projection-slice theorem from computer-aided tomography [1395]. Bistatic SAR and multistatic SAR using UAVs lead to flexible network architecture of UAVs to perform imaging task.

Distributed aspect SAR also gives a potential to execute three-dimensional SAR imaging with complete circular apertures. Similar experimental work has been done in Air Force Research Laboratory GOTCHA project. The experiment data contains eight complete circular passes collected at an altitude of 25,000 feet and 45 degree elevation angle using an airborne fully polarimetric SAR sensor [1396]. Some three-dimensional SAR imaging methods and results are reported in [1396–1398]. These methods can be potentially used in distributed aspect SAR.

Due to atmospheric turbulence, aircraft property, and control bias, the stability of UAVs or the pre-determined flight trajectory cannot be strictly guaranteed. The motion errors can be considerably high for SAR sensors mounted on UAVs. This nonideal motion can seriously degrade SAR image quality. Thus, the motion compensation for UAV SAR should be studied. Motion compensation based on raw radar data has been shown in [1399]. The main idea is to extract necessary motion parameters, for example, forward velocity and displacement in line-of-sight direction, from radar raw data, based on an instantaneous Doppler rate estimation [1399]. Theory and application of motion compensation for linear frequency modulated (LFM) continuous wave (CW) SAR have been presented in [1400].

The effects of nonideal motion on the SAR signal are derived, and new methods for motion correction are developed, which correct for motion during the pulse [1400]. The real-time coarse compensation of motion error based on UAV SAR has been analyzed in [1401]. The algorithm is based on the relation between motion error of airborne SAR and the phase of echo [1401].

After an SAR image is obtained, we still need to extract information, knowledge, and intelligence. Sometimes, this intelligence will be more useful than the pure SAR image. One of the basic intelligence extraction strategies is change detection. SAR change detection tries to find differences by comparison of SAR images from different moments in time [1402] to indicate whether or not a change has occurred, or whether several changes might have occurred and even identify the times of any such changes. These differences can imply moving targets, terrain deformation, and so on. Hence, SAR change detection is a very important technique for homeland security, military mission, and environmental earth observation.

Generally, the change detection methods belong to two basic categories:

• coherent methods;
• noncoherent methods.

Coherent methods require phase information in the SAR imagery while noncoherent methods only use amplitude information. Two-pass SAR change detection has been reported in [1403]. The performance of coherent change detection is shown. The weakness of coherent change detection is the high false alarm rate when it is applied to an urban scenario [1403]. One potential algorithm to reduce the false alarm rate is called the clutter location, estimation, and negation (CLEAN) method [1404]. In this way, multiple classes of false alarms can be removed. Repeat-pass SAR change detection has also been studied [1405]. The detection problem is formulated as a hypothesis testing problem which leads to a new log likelihood change statistic [1405]. Complex coherence is estimated. A new statistical similarity measure for change detection has been mentioned in [1406]. Two co-registered SAR intensity images are used. The local statistics are estimated, which approximates probability density functions in the neighborhood of each pixel in the image. The degree of change of the local statistics is measured using the Kullback-Leibler divergence [1406]. A wavelet-based change-detection technique has been proposed in [1407]. Two co-registered intensity SAR images are considered. This approach exploits information at different scales which can be obtained by a proper wavelet-based multiscale decomposition of the log-ratio image [1407]. In this way, speckle reduction and preservation of geometrical details can be compromised [1407]. Entropy image instead of coherence map is explored for change detection in [1408]. The change detection performance can be improved in both low coherence areas and high coherence areas [1408]. Similar discussion can be found in [1409]. Principal component analysis has been explored for change detection in [1410]. However, the embedded feature or characteristic for no change existence has not been explicitly used.

Because of the improvement of SAR data acquisition technique and the flexibility of SAR sensor deployment, multipass SAR imageries can be easily obtained. Hence, more information, knowledge, and intelligence can be extracted from SAR imageries. In order to address the challenging issue of multipass SAR change detection, two state-of-the-art PCA-based methods can be explored. One is robust PCA and the other is template matching plus thresholding. Both methods explore the local statistics and extract some particular features for change detection. The robust PCA based approach tries to find the sparse matrix which corresponds to the potential change. The larger value of matrix Frobenius norm of the sparse matrix means the more chance of change occurrence. Traditionally, template matching is a technique used in digital image processing for finding small parts of an image which match a template image. Template matching can be used in computer-aided diagnosis [1411], image watermark [1412], mobile robot navigation [1413], and so on. For multipass SAR change detection, we should find a certain template or feature when no change exists. Template matching plus thresholding will be performed for each pixel one by one. Covariance matrix for each pixel will be calculated and the dominant eigenvector of covariance matrix will be extracted as the feature. The inner product between the feature and template will be computed. If the value of inner product is greater than the predetermined threshold, there will be no change for this pixel; otherwise change will be identified.

12.4 Wireless Tomography

As smart phones are widely used, there will be a potential large-scale communication network deployed around the world. Remote sensing can be embedded into this kind of large-scale communication network. However, communication components or communication nodes are not specifically designed for remote sensing. These components do not meet the high-accuracy requirements of remote sensing. For example, RF tomography and inverse scattering require accurate phase information to perform imaging. Accurate phase information is hard to obtain using communication components, especially in the presence of noise. Meanwhile, the nonlinear operation of noise makes noise effect more severe than expected. Thus, retrieving accurate phase information from noisy measurements is a fundamental problem.

Wireless tomography was first proposed in [1414]. Wireless tomography which combines wireless communications and RF tomography gives a novel approach to remote sensing. There are three types of tomography based on the utilization of phase information [1414],

- Incoherent tomography. Incoherent tomography uses attenuation only. Phase information is not required. Meanwhile, phase information of scattered field is not extracted. Radio tomographic imaging performed by University of Utah can be treated as incoherent tomography [1415–1417]. Without phase measurement, the equipment for incoherent tomography is cheap. Hence, the large-scale deployment of incoherent tomography is applicable. Though the resolution and performance of coherent tomography is not as good as expected, it can still be used for tracking people and vehicles in security systems, locating victims in disaster situations, and so on.
- Coherent tomography. In coherent tomography, both attenuation information and phase rotation information for the scattered field are required. The well-known diffraction tomography [1418, 1419] is coherent tomography. The three basic approaches to diffraction tomography are filtered back-propagation [1420–1422], contrast source inversion method [1423–1426], and Born iterative method [1427–1429]. Time reversal imaging is also a coherent tomography approach [1430–1433].

- Self-coherent tomography. In self-coherent tomography, we do not need to measure phase information of the total or scattered fields. However, phase information of scattered field should be reconstructed given the full data of the incident field [1434–1437]. Then, coherent tomography is performed. Meanwhile, the single step approach can also be applied to intensity only inverse scattering once the incident field is known [1436]. A subspace-based optimization method for inverse scattering problems utilizing phaseless data has been developed in [1438]. The scatterer's permittivity profile is reconstructed by using only intensity data of the total field with no phase information [1438]. The distorted Rytov iterative method with phaseless data is used for tomographic reconstruction [1439]. Phase retrieval is not required.

Different from well studied tomography technique, for example, computerized tomographic imaging [1440] or inverse scattering technique, wireless tomography starts from a system engineering's point of view [1441] and is applied to the situation with complex and dynamic harsh radio environments, for example, a low target SNR situation or severely colored interference. It is without doubt that wireless tomography will be widely and effectively used in many critical applications.

12.5 Mobile Crowdsensing

In mobile crowdsensing, individuals with sensing and computing devices collectively share data and extract information to measure and map phenomena of common interest.[1442]. These devices can be smartphones, music players, and in-vehicle sensing devices [1442]. Thus, vast amounts of data can be generated from mobile crowdsensing.

Mobile crowdsensing can be used in environmental, infrastructural, and social areas [1442]. Such applications include environmental monitoring, traffic measurement for urban transportation, human activity pattern extraction, moving target tracking, and so on [1442]. In the radar community, moving target indication has been widely studied. Moving target indication tries to to discriminate a target against clutter. Moving target indication exploits the fact that the target moves with respect to stationary clutter. Thus, the Doppler effect can be explored [1443]. Moving target indication can be extended to moving target tracking if a filtering technique is applied. As wireless communication networks, for example, 3G cellular network, Wi-Fi network, and so on, are becoming available everywhere, there will be a trend to reuse wireless communication networks to perform mobile crowdsensing. In this way, we do not need to rebuild underlying infrastructure.

There are several unique characteristics for mobile crowdsensing [1442]:

- today's mobile devices with more computing, communication, and storage resources than before;
- millions of mobile devices in use;
- dynamic architecture of mobile devices;
- data reuse;
- involvement of human intelligence;
- the incentive mechanism for human involvement.

Due to the unique characteristics of mobile crowdsensing, localized analytics, aggregate analytics, resource issue, privacy, security, load balancing, and so on should be studied within the framework of mobile crowdsensing [1442].

12.6   Integration of 3S

Cognitive radio network as sensors also reflects the concept of integration of remote sensing, geographic information system (GIS), and global positioning system (GPS) [1444]. A GIS tries to capture, store, analyze, and present all types of geographically related data [1445]. GIS become more and more useful recently with the development of computing and data storage techniques. GIS data can be stored and processed in cognitive radio network for the spatial references. Meanwhile, spatial statistics is a powerful tool to analyze the data for various sensing tasks [1446]. GPS [1447] can provide location and time information in all weather conditions based on the signals from satellites. GPS can synchronize different cognitive radios in different locations and add time reference and location reference for the sensed data. Remote sensing is one kind of sensing task which acquires information about an object or phenomenon without physical contact with the object. SAR is a remote sensing technique. The integration of 3S can be used for both military and civil applications, for example, homeland security, law enforcement, urban planning, environmental monitoring, transportation, logistics, finance, telecommunication, healthcare, and so on. Remote sensing and GIS techniques have been explored for population estimation which can be used for decisions concerning resource allocation, market area delineation, new facility/transportation development, environmental and socioeconomic assessments [1448, 1449].

12.7   The Cyber-Physical System

The term "cyber-physical systems" refers to the complex engineering system with the tight conjoining of and coordination between computational and physical resources [1450]. The materialization of the cyber-physical system can be found in aerospace, automotive, communication, civil infrastructure, energy, manufacturing, transportation, entertainment, homeland security, and so on [1450].

Within the cyber-physical system, data acquisition and data processing are equally important. Computational and data-enabled science and engineering will play an important role in the cyber-physical system. National Science Foundation has identified several research examples in computational and data-enabled science and engineering,

- Novel computational or statistical modeling for simulation, prediction, and assessment in computation-intensive and data-intensive scientific problems.
- Novel tool and theory in statistical inference and statistical learning from massive, complex, and dynamic dataset.
- Large-scale problem with particular computational difficulties, such as strong heterogeneities and anisotropies, multiphysics coupling, multiscale behavior, stochastic forcing, uncertain parameters or dynamic data, and long-time behavior.
- Mathematical and statistical challenges of uncertainty quantification.

- Large-scale data acquisition, data processing, data management, data dissemination, and data security.

Random matrix theory is a powerful tool to derive and analyze the algorithms for processing the large-scale data [18, 1451, 1452]. The applications of random matrix theory in wireless communication include detection, estimation, performance analysis of multi-antenna systems, performance analysis of multihop systems, and so on [1452]. Random matrix theory is explored for spectrum sensing in cognitive radio network [255, 258, 260], which can be easily extended to cognitive radio network as sensors for the general sensing tasks.

Moreover, random matrix theory can be applied to local failure localization of large dimensional systems. These failures include sensor failure, link failure, and so on. These failures can be easily identified through the perturbation matrix as well as its eigenvector properties. The limiting distribution of the largest eigenvector in the spiked model for Gaussian sample covariance matrices has been shown in [12, 27]. Meanwhile, the effect of matrix perturbation on singular vectors can be found in [1453].

We can also monitor the sudden parameter change in the large-scale cognitive radio network. These sudden parameter changes can be analyzed through random matrix theory. We can infer and extract information from sudden parameter change for intrusion detection, anomaly detection, moving target tracking, network tomography, and so on. For intrusion detection or moving target tracking, the perturbations of the received signal matrices are different due to the different locations of target of interest and the mobility of target. By random matrix theory, we can detect the different perturbations and extract the corresponding features which can be used to identify and locate the target. In a homogeneous network, the sudden parameter change may lead to similar amplitudes of the extreme eigenvalues [12]. Thus, leading eigenvector or leading subspace may be more sensitive to the change and perturbation than the extreme eigenvalue. Network tomography is the study of a network's internal characteristics using information derived from external observations. A cognitive radio network is a large complex system with so many nodes. Measured continuous data flows from all the nodes can be used to build a large random matrix from which we can infer the properties and traffic flows in cognitive radio network. These properties include data loss, link delay, routing state, and network fault.

12.8 Computing

Computing will always be the main issue in cognitive radio network as sensors or the cyber-physical system. Computing, including hardware and software, gives strong support to the data-enabled science, engineering, and technology. The formats of computing include:

- high performance computing;
- cloud computing [1454, 1455];
- grid computing;
- distributed computing;
- parallel computing [1456, 1457];
- cluster computing;

- mobile computing [1458];
- wireless distributed computing [1459].

Cloud storage is the new concept developed from cloud computing. Cloud storage denotes a family of popular on-line services for archiving, backup, and even primary storage of files. [1460, 1461]. Data storage similar to cloud storage is needed for cognitive radio network as sensors. The large-scale data should be kept safely in cognitive radio network. Security and access issues are two main challenges.

12.8.1 Graphics Processor Unit

Host computer with GPU can be used as computing engine. Recently, a computing enhancement technology called GPGPU appearing in the PC industry. GPGPU refers to a relatively new method by which the various cores of a GPU can be utilized for general purpose parallel computing [1462]. This idea of utilizing GPUs for nongraphical applications first became popular in 2003, but was limited by the amount of knowledge required to successfully write such programs. November 2008 saw the introduction of Nvidia's G80 architecture which brought greater versatility through support of the C computer language, and a more generalized and programmer friendly hardware structure [1463].

CUDA is both a hardware and software architecture by Nvidia, which is actually what allows GPUs to run programs that have been written using C, C++, Fortran, etc. It works by executing a kernel across several parallel threads [1463]. GPUmat allows standard MATLAB code to run on GPUs. Furthermore, CULA is a linear algebra library which has been designed to utilize the NVidia CUDA architecture for computational acceleration. This library is designed in a manner such that those with little or no GPGPU programming experience can take advantage of the parallel computing power offered by GPGPU.

The CULA library is compatible with Python, C/C++, Fortran, and MATLAB. When using C/C++, the library is designed in such a way that the user may simply replace existing functions in the program with those from the library. CULA is designed such that it automatically handles the memory allocation required for GPGPU programming. This is the most attractive feature of this software because it allows users who are not experienced with GPGPU programming to take advantage of the increased speed it offers. The code is also flexible enough so that more experienced programmers can manually adjust the memory allocation for the GPU.

12.8.2 Task Distribution and Load Balancing

Task distribution and load balancing are especially important for the computing system to deal with huge workloads. This function mainly focuses on how to allocate resources to the workload to increase the efficiency and effectiveness of the system. Load balancing is usually implemented as a scheduler plus many task pools. After the large-scale problem is divided into many subproblems, these subproblems should be processed in a parallel and distributed fashion. Meanwhile, time cost and resource cost need be taken into account when we perform load balancing. Time cost relates to the time waiting for the synchronization among different tasks and the latency for the intertask communication. Resource cost corresponds to the machine cycles for processing the task.

From a theoretical point of view, game theory is a popular approach used for load balancing [1464, 1465]. Cooperative game or noncooperative game can be exploited based on whether the tasks belong to the same user or different users. Dynamic or static game can still be applied based on whether the instantaneous system state is available or not for load balancing. In terms of implementation, dynamic load balancing on single-and multi-GPU systems has been discussed in [1466]. Experimental results show the significant performance improvement using load balancing at a fine granularity, especially for the irregular and unbalanced workload. Similarly, an efficient support for matrix computations on heterogeneous multicore and multi-GPU architectures is proposed in [1467] to achieve four objectives: a high degree of parallelism, minimized synchronization, minimized communication, and load balancing. The key idea is to developed heterogeneous rectangular-tile algorithms with two different tile sizes to cope with processor heterogeneity [1467]. The problem of designing efficient recursive algorithms on GPUs with dynamic work distribution and balancing has been addressed in [1468]. Meanwhile, implementing parallel genetic algorithm on the CUDA architecture illustrates GPU has great potential for acceleration of simple numerical function optimization [1469].

12.9 Security and Privacy

Security is as important as or even more important than any other performance of interest for both cognitive radio network and cognitive radio network as sensors. To realize a secure system, security should pervade every aspect of the system design and be integrated into every system component [1470]. Security, especially information security for cognitive radio network as sensors should include [1470, 1471]:

- data confidentiality;
- data authenticity;
- data integrity;
- data freshness;
- key infrastructure for generation and distribution;
- secure communication and routing;
- trusted computing;
- trusted storage;
- attack detection;
- robustness and attack survivability;
- privacy.

12.10 Summary

Cognitive radio network as sensors is a new initiative and tries to explore the vision of a dual use sensing/communication system based on cognitive radio network. Cognitive radio network is a cyber-physical system with the integrated capabilities of control, communication, and computing. Cognitive radio network can provide an information superhighway and a strong backbone for the next generation intelligence, surveillance, and reconnaissance. Data enabled science and engineering, computing, security, and so on are the open issues in cognitive radio network as sensors. Besides, intrusion detection by machine learning, joint spectrum sensing and localization, distributed aspect SAR, wireless tomography, mobile crowdsensing, and so on are given as potential applications in cognitive radio network as sensors.

Appendix A

Matrix Analysis

A.1 Vector Spaces and Hilbert Space

Finite-dimensional random vectors are the basic building blocks of many applications. Halmos (1958) [1472] is the standard reference. We just take the most elementary material from him.

Definition A.1 (Vector space) *A vector space is a set Ω of elements called vectors satisfying the following axioms. (A) To every pair $\mathbf{x}$ and $\mathbf{y}$, of vectors in Ω, there corresponds a vector $\mathbf{x} + \mathbf{y}$, called the sum of $\mathbf{x}$ and $\mathbf{y}$, in such a way that*

- *addition is communicative, $\mathbf{x} + \mathbf{y} = \mathbf{y} + \mathbf{x}$,*
- *addition is associative, $\mathbf{x} + (\mathbf{y} + \mathbf{z}) = (\mathbf{x} + \mathbf{y}) + \mathbf{z}$,*
- *there exists in Ω a unique vector (called the origin) such that $\mathbf{x} + \mathbf{0} = \mathbf{x}$, for every $\mathbf{x}$, and*
- *to every vector in Ω there corresponds a unique vector $-\mathbf{x}$ such that $\mathbf{x}(-\mathbf{x}) = \mathbf{0}$.*

(B) To every pair, α and $\mathbf{x}$, where α is a scalar and $\mathbf{x}$ is a vector in Ω, there corresponds a vector $\alpha\mathbf{x}$ in Ω, called the product of α and $\mathbf{x}$, in such a way that

1. *multiplication by scalars is associative, $\alpha(\beta\mathbf{x}) = (\alpha\beta\mathbf{x})$, and*
2. *$\mathbf{1}\mathbf{x} = \mathbf{x}$.*

(C)

1. *Multiplication by scalars is distributive with respect to vector addition, $\alpha(\mathbf{x} + \mathbf{y}) = \alpha\mathbf{x} + \beta\mathbf{y}$, and*
2. *Multiplication by vectors is distributive with respect to scalar addition, $(\alpha + \beta)\mathbf{x} = \alpha\mathbf{x} + \beta\mathbf{x}$.*

These axioms are not claimed to be logically independent.

Cognitive Radio Communications and Networking: Principles and Practice, First Edition.
Robert C. Qiu, Zhen Hu, Husheng Li and Michael C. Wicks.
© 2012 John Wiley & Sons, Ltd. Published 2012 by John Wiley & Sons, Ltd.

Example A.1

1. Let $C^1 (= C)$ be the set of all complex numbers; If we regard $\mathbf{x} + \mathbf{y}$ and $\alpha \mathbf{x}$ as ordinary complex numerical addition and multiplication, C^1 becomes a complex vector space.
2. Let $C^n, n = 1, 2, \ldots$, be the set of all n-tuples of complex numbers. If

$$\mathbf{x} = (\xi_1, \ldots, \xi_n) \text{ and } \mathbf{y} = (\eta_1, \ldots, \eta_n)$$

are elements of C^n, we write, by definition,

$$\mathbf{x} + \mathbf{y} = (\xi_1 + \eta_1, \ldots, \xi_n + \eta_n).$$

C^n is a vector space since all parts of our axioms are satisfied; it will be called n-dimensional complex coordinate space. □

Definition A.2 (Linear dependence and linear independence) *A finite set $\mathbf{x}_i$ of vectors is linear dependent if there exists a corresponding set α_i of scalars, not all zero, such that*

$$\sum \alpha_i \mathbf{x}_i = \mathbf{0}.$$

If, on the other hand, $\sum \alpha_i \mathbf{x}_i = \mathbf{0}$ implies that α_i for each i, the set $\mathbf{x}_i$ is linearly independent.

Theorem A.1 (Linear combination) *The set of nonzero vectors $\mathbf{x}_1, \cdots, \mathbf{x}_n$ is linearly dependent if and only if some $\mathbf{x}_k, 2 \leq k \leq n$, is a linear combination of the preceding ones.*

Definition A.3 (Finite-dimensional) *A (linear) basis (or a coordinate system) in a vector space Ω is the set Ξ of linearly independent vectors such that every vector in Ω is a linear combination of elements of Ξ. A vector space Ω is finite-dimensional if it has a finite basis.*

Recall that the basic building block in "Big Data" is a random vector which is defined in a finite-dimensional vector space. The dimension is high but still finite-dimensional. The high-dimensional data processing is critical to many modern applications.

Theorem A.2 (Basis) *If Ω is a finite-dimensional vector space and if $\{\mathbf{y}_1, \cdots, \mathbf{y}_m\}$ is any set of linearly independent vectors in Ω, then, unless the $\mathbf{y}$'s already form a basis, we can find vectors $\{\mathbf{y}_{m+1}, \cdots, \mathbf{y}_{m+p}\}$ so that the totality of the $\mathbf{y}$'s, that is, $\{\mathbf{y}_1, \cdots, \mathbf{y}_m \mathbf{y}_{m+1}, \cdots, \mathbf{y}_{m+p}\}$ is a basis. In other words, every linearly independent set can be extended to a basis.*

Theorem A.3 (Dimension) *The number of elements in any basis of a finite-dimensional vector space Ω is the same as in any other basis.*

Definition A.4 (Isomorphism) *Two vector spaces $\mathcal{U}$ and $\mathcal{V}$ (over the same field) are isomorphic, if there is a one-to-one correspondence between the vectors $\mathbf{x}$ of $\mathcal{U}$ and the vectors $\mathbf{y}$ of $\mathcal{V}$, say $\mathbf{y} = \mathbf{T}(\mathbf{x})$, such that*

$$\mathbf{T}(\alpha_1 \mathbf{x}_1 + \alpha_2 \mathbf{x}_2) = \alpha_1 \mathbf{T}\mathbf{x}_1 + \alpha_2 \mathbf{T}\mathbf{x}_2.$$

In other words, $\mathcal{U}$ and $\mathcal{V}$ are isomorphic if there is an isomorphism (such as **T***) between them, where an isomorphism is a one-to-one correspondence that preserves all linear relations.*

Theorem A.4 (Isomorphic) *Every n-dimensional vector space Ω over a field $\mathcal{F}$ is isomorphic to $\mathcal{F}^n$.*

Definition A.5 (Subspaces) *A nonempty subset Δ of a vector space Ω is a subspace or a linear manifold if along with every pair,* **x** *and* **y***, of vectors contained in Δ, every linear combination $\alpha\mathbf{x} + \beta\mathbf{y}$ is also contained in Δ.*

Theorem A.5 (Intersection of Subspace) *The intersection of any collection of subspaces is a subspace.*

Theorem A.6 (Dimension of Subspace) *A subspace Δ of an n-dimensional vector space Ω is a vector space of dimension $\leq n$.*

Definition A.6 (Linear functional) *A linear functional on a vector space Ω is a scalar-valued function y defined for every vector* **x***, with the property that (identically in the vectors $\mathbf{x}_1$ and $\mathbf{x}_2$ and the scalars α_1 and α_2)*

$$y(\alpha_1\mathbf{x}_1 + \alpha_2\mathbf{x}_2) = \alpha_1 y(\mathbf{x}_1) + \alpha_2 y(\mathbf{x}_2).$$

If y_1 and y_2 are linear functions on Ω and α_1 and α_2 are scalars, let us define the function by

$$y(\mathbf{x}) = \alpha_1 y_1(\mathbf{x}) + \alpha_2 y_2(\mathbf{x}).$$

It is easy to check that y is also a linear functional; we denote it by $\alpha_1 y_1 + \alpha_2 y_2$. With these definitions of the linear concepts (zero, addition, scalar multiplication), the set Ω' forms a vector space, the *dual space* of Ω.

A.2 Transformations

Definition A.7 (Linear transformation) *A linear transformation (or operator)* **A** *on a vector space Ω is a correspondence that assigns to every vector* **x** *in Ω a vector* **Ax** *in Ω, in such a way that*

$$\mathbf{A}(\alpha\mathbf{x} + \beta\mathbf{y}) = \alpha\mathbf{A}\mathbf{x} + \beta\mathbf{A}\mathbf{y}$$

identically in the vectors **x** *and* **y** *and the scalars α and β.*

Theorem A.7 (Linear transformation) *The set of all linear transformations on a vector space is itself a vector space.*

Linear transformations can be regarded as vectors.

Definition A.8 (Matrices) *Let Ω be an n-dimensional vector space, let $\mathcal{X} = \{\mathbf{x}_1, \ldots, \mathbf{x}_n\}$ be any basis of Ω, and let $\mathbf{A}$ be a linear transformation on Ω. Since every vector is a linear combination of the $\mathbf{x}_i$, we have in particular*

$$\mathbf{A}\mathbf{x}_j = \sum_i \alpha_{ij}\mathbf{x}_i, \, j = 1, \ldots, n.$$

The set α_{ij} of n^2 scalars, indexed with the double subscript i, j, is the matrix of $\mathbf{A}$ in the coordinate system $\mathcal{X}$; A matrix (α_{ij}) is usually written in the form of a square array:

$$[\mathbf{A}] = \begin{bmatrix} \alpha_{11} & \alpha_{12} & \cdots & \alpha_{1n} \\ \alpha_{21} & \alpha_{22} & \cdots & \alpha_{2n} \\ \vdots & \vdots & & \vdots \\ \alpha_{n1} & \alpha_{n2} & \cdots & \alpha_{nn} \end{bmatrix};$$

the scalars $(\alpha_{i1}, \cdots, \alpha_{in})$ form a column; $(\alpha_{1j}, \cdots, \alpha_{nj})$ form a row, of $\mathbf{A}$.

A.3 Trace

The trace function, $\mathrm{Tr}\mathbf{A} = \sum_i a_{ii}$, satisfies the following properties [110] for matrices $\mathbf{A}, \mathbf{B}, \mathbf{C}, \mathbf{D}, \mathbf{X}$ and scalar α:

$$\mathrm{Tr}\alpha = \alpha, \, \mathrm{Tr}(\mathbf{A} \pm \mathbf{B}) = \mathrm{Tr}\mathbf{A} \pm \mathrm{Tr}\mathbf{B}, \, \mathrm{Tr}\alpha\mathbf{A} = \alpha\mathrm{Tr}\mathbf{A}$$

$$\mathrm{Tr}\mathbf{C}\mathbf{D} = \mathrm{Tr}\mathbf{D}\mathbf{C} = \sum_{i,j} c_{ij}d_{ji},$$

$$\mathrm{Tr}\sum_{k=0}^{K-1} \mathbf{x}_k^*\mathbf{A}\mathbf{x}_k = \mathrm{Tr}(\mathbf{A}\mathbf{X}), \text{ where } \mathbf{X} = \sum_{k=0}^{K-1} \mathbf{x}_k\mathbf{x}_k^*. \tag{A.1}$$

To prove the last property, note that since $\mathbf{x}_k^*\mathbf{A}_k\mathbf{x}_k$ is a scalar, the left side of (A.1) is

$$\mathrm{Tr}\sum_{k=0}^{K-1} \mathbf{x}_k^*\mathbf{A}_k\mathbf{x}_k = \sum_{k=0}^{K-1} \mathrm{Tr}(\mathbf{x}_k^*\mathbf{A}_k\mathbf{x}_k) = \sum_{k=0}^{K-1} \mathrm{Tr}(\mathbf{A}_k\mathbf{x}_k\mathbf{x}_k^*)$$

$$= \mathrm{Tr}\mathbf{A}\mathbf{X}, \text{ where } \mathbf{X} = \sum_{k=0}^{K-1} \mathbf{x}_k\mathbf{x}_k^*, \text{ for } \mathbf{A}_k = \mathbf{A}$$

$$\mathrm{Tr}\mathbf{C}\mathbf{C}^* = \mathrm{Tr}\mathbf{C}^*\mathbf{C} = \sum_{i,j} |c_{ij}|^2.$$

A.4 Basics of C^*-Algebra

C^*-algebras (pronounced "C-star") are an important area of research in functional analysis. The prototypical example of a C^*-algebra is a complex algebra $\mathcal{A}$ of linear operators on a complex Hilbert space with two additional properties:

1. $\mathcal{A}$ is a topologically closed set in the norm topology of operators.
2. $\mathcal{A}$ is closed under the operation of taking adjoints of operators.

C*-algebras [138, 1473] are now an important tool in the theory of unitary representations of locally compact groups, and are also used in algebraic formulations of quantum mechanics. Another active area of research is the program to obtain classification, or to determine the extent of which classification is possible. It is through the latter area that its connection with our interest in hypothesis detection is made.

A.5   Noncommunicative Matrix-Valued Random Variables

We mainly follow [11] this section, but with different notations that are convenient for our context. Random variables are functions defined on a measure space, and they are often identified by their distributions in probability theory [11]. In the simplest case when the random variables are real valued, the distribution is a probability measure on the real line. In this appendix probability distributions can be represented by means of linear Hilbert space operators, as well. (In this appendix an operator is an infinite-dimensional matrix.) This observation is as old as quantum mechanics; the standard probabilistic interpretation of the quantum mechanical formalism is related.

In an algebraic generalization, elements of a typically noncommutative algebra together with a linear functional on the algebra are regarded as noncommutative random variables. The linear functional evaluated on this element is the expectation value, its use to powers of this selected element leads to the moments of the noncommunicative random variables. One does not distinguish between two random variables, when they have the same moments. A very new feature of this theory occurs when these noncommunicative random variables are truly noncommuting with each other. Then, one cannot have a joint distribution in the sense of classical probability theory, but a functional of the algebra of polynomials of noncommuting indeterminates may work as an abstract concept of joint distribution [11]. Random matrices with respect to the expectation of their trace are natural "noncommuniting" noncommutative (matrix-valued) random variables.

Random variables over a probability space form an algebra. Indeed, they are measurable functions defined on a set Ω, and so are the product and sum of two of them, that is, **AB** and **A** + **B**. As mentioned before, the expectation value $\mathbb{E}\mathbf{A}$ is a linear functional on this algebra. The algebraic approach to probability stresses this point. *An algebra over a field is a vector space* equipped with a bilinear vector product. That is to say, it is an algebraic structure consisting of a vector space together with an operation, usually called multiplication, that combines any two vectors to form a third vector; to qualify as an algebra, this multiplication must satisfy certain compatibility axioms with the given vector space structure, such as distributivity. In other words, an algebra over a field is a set together with operations of multiplication, addition, and scalar multiplication by elements of the field [1474].

If $\mathcal{A}$ is a unital algebra (a vector space defined above) over the complex numbers and ϕ is a linear functional of $\mathcal{A}$ such that

$$\varphi\mathbf{1} = 1,$$

then $(\mathcal{A}, \phi)$ will be called a noncommutative probability space and an element **A** of $\mathcal{A}$ will be called a noncommunicative random variable. of course, a random matrix is such a noncommunitative random variable. The number $\phi(\mathbf{A}^k)$ is called the n-th moment of a noncommutative random variable.

Example A.2 (Bounded operators [11])

Let $\mathcal{B}(\mathcal{H})$ denote the algebra of all bounded operators acting on a Hilbert space $\mathcal{H}$. If the linear functional $\phi : \mathcal{B}(\mathcal{H}) \to \mathbb{C}$ is defined by means of a unit vector $\mathbf{U} \in \mathcal{H}$ as

$$\varphi(\mathbf{A}) = (\mathbf{A}\mathbf{U}, \mathbf{U}),$$

then any element of $\mathcal{B}(\mathcal{H})$ is a noncommunitative random variable. $\qquad\square$

If $\mathbf{A} \in \mathcal{B}(\mathcal{H})$ is, further, self-adjoint (or Hermitian for the finite-dimensional case), then a probability measure is associated to $\mathbf{A}$ and φ, as mentioned above. The algebra used in the definition of a noncommunitative random variable is often replaced with a *-algebra. In fact, $\mathcal{B}(\mathcal{H})$ is a *-algebra, if the operation $\mathbf{A}^*$ stands for the adjoint of $\mathbf{A}$. The most familiar example of a *-algebra is the field of complex numbers C where * is just complex conjugation. Another example is the matrix algebra of $n \times n$ matrices over C with * given by the conjugate transpose. Its generalization, the Hermitian adjoint of a linear operator on a Hilbert space is also a *-algebra (or star-algebra).

A *-algebra is a unital algebra over the complex numbers which is equipped with an involution*. The revolution recalls the adjoint operation of Hilbert space operators as follows:

1. $\mathbf{A} \mapsto \mathbf{A}^*$ is conjugate linear.
2. $(\mathbf{A}\mathbf{B})^* = (\mathbf{B}\mathbf{A})^*$,
3. $\mathbf{A}^{**} = \mathbf{A}$.

When $(\mathcal{A}, \phi)$ is a noncommunitative probability space over a *-algebra $\mathcal{A}$, ϕ is always assumed to be a state on $\mathcal{A}$, that is, a linear function such that

1. $\varphi(\mathbf{1}) = 1$,
2. $\varphi(\mathbf{A}^*) = \bar{\varphi}(\mathbf{A})$ and $\varphi(\mathbf{A}^*\mathbf{A}) \geqslant 0$ for every

$$\mathbf{A} \in \mathcal{A}.$$

A matrix $\mathbf{X}$ whose entries are (classical) random variables on a (classical) probability space is called a random matrix, such as a sample covariance matrix $\mathbf{X}\mathbf{X}^{\mathbf{H}}$. Here H stands for the conjugate and transpose (Hermitian) of a complex matrix.

Random matrices form a *-algebra. For example, consider $X_{11}, X_{12}, X_{21}, X_{22}$ to be four bounded (classical) scalar random variables on a probability space. Then

$$\mathbf{X} = \begin{pmatrix} X_{11} & X_{12} \\ X_{21} & X_{22} \end{pmatrix}$$

is a bounded 2×2 random matrix. The set $\mathbf{X}$ of all such matrices has a *-algebra structure when the unit matrix operations are considered, and $(\mathcal{X}, \varphi)$ is a noncommunitative probability space when, for example,

$$\varphi(\mathbf{X}) = \mathbb{E}(X_{11}).$$

A C^*-algebra is a *-algebra $\mathcal{A}$ which is endowed with a norm such that

$$\|\mathbf{A}^*\mathbf{A}\| = \|\mathbf{A}\|^2, \|\mathbf{A}\mathbf{B}\| \leqslant \|\mathbf{A}\|\|\mathbf{B}\|, \text{ for every } \mathbf{A}, \mathbf{B} \in \mathcal{A}, \text{ and } \|\mathbf{1}\| = 1,$$

and furthermore, $\mathcal{A}$ is a Banach space with respect to this norm.

Gelfand and Naimark give two important theorems concerning the representation of C^*-algebras.

1. A communicative, unital C^*-algebra is isometrically isomorphic to the algebra of all continuous complex functions on a certain compact Hausdorff space, if the function space is equipped with the supremum norm and the involution of point-wise conjugation.
2. A general $C*$-algebra is isometrically isomorphic to the algebra of operators on a Hilbert space, if the function space is equipped with the operator norm and the involution of adjoint conjugation.

Combination of the above two theorems yields a form of the spectral theorem. For a linear functional φ of a C^*-algebra,

$$||\varphi|| = \varphi(1)$$

is equivalent to

$$\varphi(\mathbf{A}^*\mathbf{A}) \geqslant 0, \mathbf{A} \in \mathcal{A}.$$

A noncommunitative probability space $(\mathcal{A}, \varphi)$ will be called a C^*-probability space when $\mathcal{A}$ is a C^*-algebra and φ is a state on $\mathcal{A}$.

All real bounded classical scalar random variables may be considered as noncommunicative random variables.

A.6   Distances and Projections

For projections, we freely use [1475]. Let $\mathcal{B}(\mathcal{H})$ denote the algebra of linear operators acting on a finite-dimensional Hilbert space $\mathcal{H}$. The von Neumann entropy of a state ρ, i.e. that is, a positive operator of unit trace in $\mathcal{B}(\mathcal{H})$, is given by $S(\rho) = -\text{Tr}\rho \log \rho$.

For $\mathbf{A} \in M_n(\mathcal{C})$, the absolute value $|\mathbf{A}|$ is defined as $(\mathbf{A}^*\mathbf{A})^{\frac{1}{2}}$ and it is a positive matrix. The trace norm of $\mathbf{A} - \mathbf{B}$ is defined as

$$||\mathbf{A} - \mathbf{B}||_1 = \text{Tr}|\mathbf{A} - \mathbf{B}|.$$

This trace norm $||\mathbf{A} - \mathbf{B}||_1$ is a nature distance between complex $n \times n$ matrices $\mathbf{A}$ and $\mathbf{B}$, $\mathbf{A}, \mathbf{B} \in M_n(\mathbb{C})$. Similarly,

$$||\mathbf{A} - \mathbf{B}||_2 = \left(\sum_{i,j} |A_{ij} - B_{ij}|^2 \right)^{1/2}$$

is also a natural distance. We can define the p-norm as

$$||\mathbf{X}||_p = (\text{Tr}(\mathbf{X}^*\mathbf{X})^{2/p})^{1/p}, 1 \leq p, \mathbf{X} \in M_n(\mathbb{C})$$

It was Von Neumann who showed first that the Hoelder inequality remains true in the matrix setting

$$||\mathbf{AB}||_1 \leq ||\mathbf{A}||_p ||\mathbf{B}||_q, \frac{1}{p} + \frac{1}{q} = 1.$$

If **A** is a self-adjoint and written as

$$\mathbf{A} = \sum_i \lambda_i \mathbf{e}_i \mathbf{e}_i^*$$

where the vector $\mathbf{e}_i$ form an orthonormal basis, then it is defined as

$$\{\mathbf{A} \geq 0\} = \mathbf{A}_+ = \sum_{i:\lambda_i \geq 0} \lambda_i \mathbf{e}_i \mathbf{e}_i^*; \{\mathbf{A} < 0\} = \mathbf{A}_- = \sum_{i:\lambda_i < 0} \lambda_i \mathbf{e}_i \mathbf{e}_i^*.$$

Then $\mathbf{A} = \{\mathbf{A} \geq 0\} + \{\mathbf{A} < 0\} = \mathbf{A}_+ + \mathbf{A}_-$ and $|\mathbf{A}| = \{\mathbf{A} \geq 0\} - \{\mathbf{A} < 0\} = \mathbf{A}_+ - \mathbf{A}_-$. The decomposition is called the Jordan decomposition of **A**. Corresponding definitions apply for the other spectral projections $\{\mathbf{A} < 0\}$, $\{\mathbf{A} > 0\}$, and $\{\mathbf{A} \leq 0\}$. For two operators, $\{\mathbf{A} < \mathbf{B}\}$, $\{\mathbf{A} > \mathbf{B}\}$, and $\{\mathbf{A} \leq \mathbf{B}\}$.

For self-adjoint operators $\mathbf{A}, \mathbf{B}$ and any positive operator $0 \leq \mathbf{P} \leq \mathbf{I}$ we have

$$\text{Tr}[\mathbf{P}(\mathbf{A} - \mathbf{B})] \leq \text{Tr}[\{\mathbf{A} \geq \mathbf{B}\}(\mathbf{A} - \mathbf{B})]$$
$$\text{Tr}[\mathbf{P}(\mathbf{A} - \mathbf{B})] \geq \text{Tr}[\{\mathbf{A} \leq \mathbf{B}\}(\mathbf{A} - \mathbf{B})].$$

Identical conditions hold for strict inequalities in the spectral projections $\{\mathbf{A} < \mathbf{B}\}$ and $\{\mathbf{A} > \mathbf{B}\}$.

The trace distance between operators **A** and **B** is given by

$$|\mathbf{A} - \mathbf{B}||_1 = \text{Tr}[\{\mathbf{A} \geq \mathbf{B}\}(\mathbf{A} - \mathbf{B})] - \text{Tr}[\{\mathbf{A} < \mathbf{B}\}(\mathbf{A} - \mathbf{B})].$$

The fidelity of states ρ and ρ' is defined as

$$F(\rho, \rho') = \text{Tr}\sqrt{\rho^{\frac{1}{2}} \rho' \rho^{\frac{1}{2}}}.$$

The trace distance between two states is related to the fidelity as follows:

$$\frac{1}{2}||\rho - \rho'||_1 \leq \sqrt{1 - F(\rho, \rho')^2} \leq \sqrt{2(1 - F(\rho, \rho')^2)}.$$

For self-adjoint operators $\mathbf{A}, \mathbf{B}$ and any positive operator $0 \leq \mathbf{P} \leq \mathbf{I}$, the inequality

$$||\mathbf{A} - \mathbf{B}||_1 \leq \varepsilon$$

for any $\varepsilon > 0$, implies that

$$\text{Tr}[\mathbf{P}(\mathbf{A} - \mathbf{B})] \leq \varepsilon.$$

The "gentle measurement" lemma is given here: For a state ρ and any positive operator $0 \leq \mathbf{P} \leq \mathbf{I}$, if $\text{Tr}(\rho\mathbf{P}) \geq 1 - \delta$, then

$$||\rho - \sqrt{\mathbf{P}}\rho\sqrt{\mathbf{P}}||_1 \leq 2\sqrt{\delta}.$$

The same holds if ρ is only a subnormalized density operator, that is, $\text{Tr}\rho \leq 1$.

If ρ is a state and **P** is a projection operator such that $\text{Tr}(\mathbf{P}\rho) > 1 - \delta$ for a given $\delta > 0$, then

$$\tilde{\rho} = \sqrt{\mathbf{P}}\rho\sqrt{\mathbf{P}} \in B^\varepsilon(\rho),$$

where

$$B^\varepsilon(\rho) = \{\tilde{\rho} \geq 0 : ||\rho - \tilde{\rho}||_1 \leq \varepsilon, \mathrm{Tr}\tilde{\rho} \leq \mathrm{Tr}\rho\},$$

and $\varepsilon = 2\sqrt{\delta}$.

Consider a state ρ and a positive operator $\sigma \in B^\varepsilon(\rho)$, for some $\varepsilon > 0$. If π_σ denote the projection onto the support of σ, then

$$\mathrm{Tr}(\pi_\sigma \rho) \geq 1 - 2\varepsilon.$$

Lemma A.1 (Hoffman-Wielandt) *[16, 34] Let* **A** *and* **B** *be* $N \times N$ *self-adjoint matrices, with eigenvalues* $\lambda_1^A \leq \lambda_2^A \leq \cdots \leq \lambda_N^A$ *and* $\lambda_1^B \leq \lambda_2^B \leq \cdots \leq \lambda_N^B$. *Then,*

$$\sum_{i=1}^{N} |\lambda_i^A - \lambda_i^B| \leq \mathrm{Tr}(\mathbf{A} - \mathbf{B})^2.$$

The singular values of a matrix $\mathbf{A} \in M_n$ are the eigenvalues of its absolute value $|\mathbf{A}| = (\mathbf{A} * \mathbf{A})^{\frac{1}{2}}$,, we have fixed the notation $s(\mathbf{A}) = (s_1(\mathbf{A}), \ldots, s_n(\mathbf{A}))$ with $s_1(\mathbf{A}) \geq \ldots \geq s_n(\mathbf{A})$. Singular values are closely related to the unitary invariant norm. Singular values inequalities are weaker than Löwner partial order inequalities and stronger than unitarily invariant norm inequalities in the following sense [133]:

$$|\mathbf{A}| \leq |\mathbf{B}| \Rightarrow s_i(\mathbf{A}) \leq s_i(\mathbf{B}) \Rightarrow ||\mathbf{A}|| \leq ||\mathbf{B}||$$

for all unitarily invariant norms. The norm $||\mathbf{A}||_1 = \mathrm{Tr}|\mathbf{A}|$ is unitarily invariant. Singular values are unitarily invariant: $s(\mathbf{UAV}) = s(\mathbf{A})$ for every $\mathbf{A}$ and all unitary $\mathbf{U}, \mathbf{V}$. A norm $|| \cdot ||$ is called unitarily invariant if

$$||\mathbf{UAV}|| = ||\mathbf{A}|| \qquad (A.2)$$

for all $\mathbf{A} \in M_n$ and all unitary $\mathbf{U}, \mathbf{V} \in M_n$. $||\mathbf{A}|| \leq ||\mathbf{B}||$ for all unitarily invariant norms if and only if $s(\mathbf{A}) \prec_w s(\mathbf{B})$, that is,

$$s(\mathbf{A}) \prec_w s(\mathbf{B}) \Rightarrow ||\mathbf{A}|| \leq ||\mathbf{B}||. \qquad (A.3)$$

The differences of two positive semidefinite matrices $\mathbf{A}, \mathbf{B} \in M_n$ are often encountered. Denote the block diagonal matrix

$$\mathbf{A} \oplus \mathbf{B} \overset{Def}{=} \begin{pmatrix} \mathbf{A} & 0 \\ 0 & \mathbf{B} \end{pmatrix}$$

by the notation $\mathbf{A} \oplus \mathbf{B}$. Then [133]

1. $s_i(\mathbf{A} - \mathbf{B}) \leq s_i(\mathbf{A} \oplus \mathbf{B})$, $i = 1, 2, \ldots, n$,
2. $||\mathbf{A} - \mathbf{B}|| \leq ||\mathbf{A} \oplus \mathbf{B}||$ for all unitarily invariant norms,
3. $s(\mathbf{A} - |z|\mathbf{B}) \prec_{wlog} s(\mathbf{A} + z\mathbf{B}) \prec_{wlog} s(\mathbf{A} + |z|\mathbf{B})$,
4. $||\mathbf{A} - |z|\mathbf{B}|| \leq ||\mathbf{A} + z\mathbf{B}|| \leq ||\mathbf{A} + |z|\mathbf{B}||$, for any complex number z.

Note that the weak log majorization $\prec_{wlog}$ is stronger than the weak majorization $\prec_w$.

A.6.1 Matrix Inequalities

Let $\alpha : \mathcal{B}(\mathbb{H}) \to \mathcal{B}(\mathbb{K})$ be a linear mapping from finite-dimensional Hilbert spaces $\mathbb{H}$ and $\mathbb{K}$. α is called positive if it sends positive (semidefinite) operators to positive (semidefinite) operators. Let $\alpha : M_n(\mathbb{C}) \to M_k(\mathbb{C})$ be a positive, unital linear mapping and $f : \mathbb{R} \to \mathbb{R}$ be a convex function. Then, it follow [34, p. 189] that

$$\mathrm{Tr}\, f(\alpha(\mathbf{A})) \le \mathrm{Tr}\, \alpha(f(\mathbf{A})) \tag{A.4}$$

for every $\mathbf{A} \in M_n(\mathbb{C})^{\mathrm{sa}}$. Here sa denotes the self-adjoint case.

Let $\mathbf{A}$ and $\mathbf{B}$ be positive operators, then for $0 \le s \le 1$,

$$\mathrm{Tr}(\mathbf{A}^s \mathbf{B}^{1-s}) \ge \mathrm{Tr}(\mathbf{A} + \mathbf{B} - |\mathbf{A} - \mathbf{B}|)/2.$$

The triangle inequality for the matrix $\mathbf{A}^* \mathbf{A}^{\frac{1}{2}}$ is [114, p. 237]

$$|\mathbf{A} + \mathbf{B}| \le \mathbf{U}^* |\mathbf{A}| \mathbf{U} + \mathbf{V}^* |\mathbf{B}| \mathbf{V}, \tag{A.5}$$

where $\mathbf{A}$ and $\mathbf{B}$ are any square complex matrices of the same size, and $\mathbf{U}$ and $\mathbf{V}$ are unitary matrices. Taking the trace of (A.5) leads to the following

$$\mathrm{Tr}|\mathbf{A} + \mathbf{B}| \le \mathrm{Tr}|\mathbf{A}| + \mathrm{Tr}|\mathbf{B}|. \tag{A.6}$$

Replacing $\mathbf{B}$ in (A.6) with $\mathbf{B} + C$ leads to

$$\mathrm{Tr}|\mathbf{A} + \mathbf{B} + C| \le \mathrm{Tr}|\mathbf{A}| + \mathrm{Tr}|\mathbf{B} + C| \le \mathrm{Tr}|\mathbf{A}| + \mathrm{Tr}|\mathbf{B}| + \mathrm{Tr}|C|.$$

Similarly, we have

$$\mathrm{Tr}|\mathbf{A}_1 + \mathbf{A}_2 + \cdots + \mathbf{A}_K| \le \mathrm{Tr}|\mathbf{A}_1| + \mathrm{Tr}|\mathbf{A}_2| + \cdots + \mathrm{Tr}|\mathbf{A}_K|. \tag{A.7}$$

For positive operators $\mathbf{A}$ and $\mathbf{B}$,

$$\|\mathbf{A} - \mathbf{B}\|_1^2 + 4(\mathrm{Tr}(\mathbf{A}^{1/2} \mathbf{B}^{1/2}))^2 \le (\mathrm{Tr}(\mathbf{A} + \mathbf{B}))^2.$$

The n-tuples of the coefficients of real numbers may be regarded as diagonal matrices and the majorization can be extended to self-adjoint matrices. Suppose that $\mathbf{A}, \mathbf{B} \in M_n$ are so. Then $\mathbf{A} \prec \mathbf{B}$ means that the n-tuple of eigenvalues of $\mathbf{A}$ is majorized by the n- tuple of eigenvalues of $\mathbf{B}$; similarly for the weak majorization. Since the majorization depends only on the spectrums, $\mathbf{A} \prec \mathbf{B}$ holds if and only if $\mathbf{U}\mathbf{A}\mathbf{U}^* \prec V\mathbf{B}V^*$ for some unitaries $\mathbf{U}$ and V. It follows from Birkhoff's theorem [34] that $\mathbf{A} \prec \mathbf{B}$ implies that

$$\mathbf{A} = \sum_{i=1}^{n} p_i \mathbf{U}_i \mathbf{B} \mathbf{U}_i^*$$

for some $p_i > 0$ with $\sum_i p_i = 1$ and for some unitaries.

Theorem A.8 *Let ρ_1 and ρ_2 be states. Then the following statements are equivalent.*

1. $\rho_1 \prec \rho_2$.
2. ρ_1 is more mixed than ρ_2.

3. $\rho_1 = \sum_{i=1}^{n} \lambda_i \mathbf{U}_i \rho_2 \mathbf{U}_i^*$ *for some convex combination* λ_i *and for some unitaries* $\mathbf{U}_i$.

4. $\mathrm{Tr} f(\rho_1) \leq \mathrm{Tr} f(\rho_2)$ *for any convex function* $f : \mathcal{R} \to \mathcal{R}$.

A.6.2 Partial Ordering of Positive Semidefinite Matrices

Let $\mathbf{A} \geq 0$ and $\mathbf{B} \geq 0$ be of the same size. Then

1. $\mathbf{A} + \mathbf{B} \geq \mathbf{B}$,
2. $\mathbf{A}^{\frac{1}{2}} \mathbf{B} \mathbf{A}^{\frac{1}{2}} \geq 0$,
3. $\mathrm{Tr}(\mathbf{AB}) \leq \mathrm{Tr}(\mathbf{A})\mathrm{Tr}(\mathbf{B})$,
4. $(\det(\mathbf{A}+\mathbf{B}))^{\frac{1}{n}} \geq (\det(\mathbf{A}))^{\frac{1}{n}} + (\det(\mathbf{B}))^{\frac{1}{n}}$, when $n > 1$,
5. the eigenvalues of $\mathbf{AB}$ are all nonnegative, $\lambda_i(\mathbf{AB}) \geq 0$.
6. $\mathbf{AB}$ is positive semidefinite, $\mathbf{AB} \geq 0$, if and only if $\mathbf{AB} = \mathbf{BA}$. $\mathbf{AB}$ may not be even Hermitian.

If $\mathbf{A} \geq \mathbf{B} \geq 0$, then

1. $\mathrm{rank}(\mathbf{A}) \geq \mathrm{rank}(\mathbf{B})$,
2. $\det \mathbf{A} \geq \det \mathbf{B}$,
3. $\mathbf{B}^{-1} \geq \mathbf{A}^{-1}$ if $\mathbf{A}$ and $\mathbf{B}$ are nonsingular,
4. $\mathrm{Tr}\mathbf{A} \geq \mathrm{Tr}\mathbf{B}$.

Let $\mathbf{A}, \mathbf{B} \in M_n$ be positive semidefinite. Then for any complex number and any unitarily invariant norm [133],

$$||\mathbf{A} - |z|\mathbf{B}|| \leq ||\mathbf{A} + z\mathbf{B}|| \leq ||\mathbf{A} + |z|\mathbf{B}||.$$

A.6.3 Partial Ordering of Hermitian Matrices

We follow [126, p. 273] for a short review. [115] has the most exhaustive collection. Positive definite and semi-definite matrices are important since covariance matrix and sample covariance matrix (used in practice) are semi-definite. A Hermitian matrix $\mathbf{A} \in C^{n \times n}$ is called positive definite if

$$\mathbf{x}^H \mathbf{A} \mathbf{x} > 0 \text{ for all nonzero } x \in C^{n \times n}$$

and positive semidefinite if the weaker condition $\mathbf{x}^H \mathbf{A} \mathbf{x} \geq 0$ holds. A Hermitian matrix is positive definite if and only if all of its eigenvalues are positive, and positive semidefinite if and only if all of its eigenvalues are nonnegative. For $\mathbf{A}, \mathbf{B} \in C^{n \times n}$, we write $\mathbf{A} > \mathbf{B}$ when $\mathbf{A} - \mathbf{B}$ is positive definite, and $\mathbf{A} \leq \mathbf{B}$ if $\mathbf{B} - \mathbf{A}$ is positive definite. This is a partial ordering of the set of $n \times n$ Hermitian matrices. It is partial because we may have $\mathbf{A} \not\geq \mathbf{B}$ and $\mathbf{B} \not\geq \mathbf{A}$.

Let $\mathbf{A} \geq 0$ and $\mathbf{B} \geq 0$ be of the same size and C is nonsingular. We have

1. $\mathbf{C}^H \mathbf{A} \mathbf{C} > 0$,
2. $\mathbf{C}^H \mathbf{B} \mathbf{C} > 0$,
3. $\mathbf{A}^{-1} > 0$,

4. $\mathbf{A} + \mathbf{B} \geq \mathbf{B}$,
5. $\mathbf{A}^{\frac{1}{2}} \mathbf{B} \mathbf{A}^{\frac{1}{2}} \geq 0$,
6. $\mathrm{Tr}(\mathbf{AB}) \leq \mathrm{Tr}\mathbf{A}\mathrm{Tr}\mathbf{B}$,
7. $\mathbf{A} \geq \mathbf{B} \Rightarrow \lambda_i(\mathbf{A}) \geq \lambda_i(\mathbf{B})$, where λ_i are the eigenvalues (sorted in decreasing order),
8. $\det(\mathbf{A}) \geq \det(\mathbf{B})$ and $\mathrm{Tr}\mathbf{A} \geq \mathrm{Tr}\mathbf{B}$.
9. the eigenvalues of $\mathbf{AB}$ are all nonnegative. Furthermore, $\mathbf{AB}$ is positive semidefinite if and only if $\mathbf{AB} = \mathbf{BA}$.

The partitioned Hermitian matrix

$$\mathcal{A} = \begin{pmatrix} \mathbf{A} & \mathbf{B} \\ \mathbf{B}^* & \mathbf{D} \end{pmatrix}$$

with square blocks $\mathbf{A}$ and $\mathbf{D}$, is positive definite if and only if $\mathbf{A} > 0$ and its Schur complement $\mathbf{D} - \mathbf{B}^H \mathbf{A}^{-1} \mathbf{B} > 0$, or $\mathbf{D} > \mathbf{B}^H \mathbf{A}^{-1} \mathbf{B}$.

The Hadamard determinant inequality for a positive semidefinite $\mathbf{A} \in \mathcal{C}^{n \times n}$ is

$$\det \mathcal{A} \leq \det \mathbf{A} \det \mathbf{D}.$$

The Minkowski determinant inequality for positive definite $\mathbf{A}, \mathbf{B} \in \mathcal{C}^{n \times n}$ is

$$\det^{1/n}(\mathbf{A} + \mathbf{B}) \geq \det^{1/n} \mathbf{A} + \det^{1/n} \mathbf{B}$$

with equality if and only if $\mathbf{B} = c\mathbf{A}$ for some constant c.

If f is convex then

$$f(x) - f(y) - (x - y)f'(y) \geq 0$$

and

$$\mathrm{Tr} f(\mathbf{B}) \geq \mathrm{Tr} f(\mathbf{A}) + \mathrm{Tr}(\mathbf{B} - \mathbf{A}) f'(\mathbf{B}). \tag{A.8}$$

References

1. J. Manyika, M. Chui, B. Brown, *et al.* (2011) Big data: The next frontier for innovation, competition and productivity. http://www.mckinsey.com, June 2011. McKinsey Global Institute, Research Report.
2. R. Qiu, N. Guo, H. Li, *et al.* (2009) A unified multi-functional dynamic spectrum access framework: tutorial, theory and multi-ghz wideband testbed, *Sensors* **9**(8): 6530–6603.
3. S. Haykin (2005) Cognitive radio: Brain-empowered wireless communications, *IEEE Journal on Selected Areas in Communications* **23**(2): 201–20.
4. S. Haykin (2006) Cognitive radar, *IEEE Signal Processing Magazine,* Jan.: 30–40.
5. IEEE (2011) IEEE Std 802.22-2011, IEEE Standard for information technology—telecommunications and information exchange between systems—wireless regional area networks (WRAN)—specific requirements—Part 22: cognitive wireless ran medium access control (MAC) and physical layer (phy) specifications: policies and procedures for operation in the TV bands, IEEE Standard, July 2011, 672 pp.
6. M. Murroni, R. Prasad, P. Marques, *et al.* (2011) IEEE 1900.6: Spectrum sensing interfaces and data structures for dynamic spectrum access and other advanced radio communication systems standard: technical aspects and future outlook, *Communications Magazine, IEEE* **49**(12): 118–27.
7. R. Qiu, Z. Hu, Z. Chen, *et al.* (2011) Cognitive radio network for the smart grid: Experimental system architecture, control algorithms, security, and microgrid testbed, *IEEE Transactions on Smart Grid* **99**: 1–18.
8. S. Boyd and L. Vandenberghe (2004) *Convex Optimization.* Cambridge: Cambridge University Press.
9. N. El Karoui (2008) Spectrum estimation for large dimensional covariance matrices using random matrix theory, *The Annals of Statistics* **36**(6): 2757–90.
10. Z. Bai (1999) Methodologies in spectral analysis of large-dimensional random matrices, a review, *Statistica Sinica* **9**(3): 611–77.
11. F. Hiai and D. Petz (2000) *The Semicircle Law, Free Random Variables, and Entropy.* Los Angeles: American Mathematical Society.
12. R. Couillet and M. Debbah (2011) *Random Matrix Methods for Wireless Communications.* Cambridge: Cambridge University Press, 2011.
13. A. Tulino and S. Verdu (2004) *Random Matrix Theory and Wireless Communications.* Hanover, MA: Now Publishers Inc.
14. Z. Bai and J. Silverstein (2010) *Spectral Analysis of Large Dimensional Random Matrices,* 2nd edn. Berlin: Springer Verlag.
15. P.J. Forrester (2010) *Log-gases and Random Matrices,* No. 34. New Jersey: Princeton University Press.
16. G.W. Anderson, A. Guionnet, and O. Zeitouni (2010) *An Introduction to Random Matrices.* Cambridge: Cambridge University Press.
17. V. Girko (1998) *An Introduction to Statistical Analysis of Random Arrays.* The Netherlands: VSP.
18. A. Edelman and N. Rao (2005) Random matrix theory, *Acta Numerica* **14**(233–97): 139.

Cognitive Radio Communications and Networking: Principles and Practice, First Edition.
Robert C. Qiu, Zhen Hu, Husheng Li and Michael C. Wicks.
© 2012 John Wiley & Sons, Ltd. Published 2012 by John Wiley & Sons, Ltd.

19. I. Johnstone (2006) High dimensional statistical inference and random matrices, Arxiv preprint math/0611589.
20. O. Ledoit and M. Wolf (2004) A well-conditioned estimator for large-dimensional covariance matrices, *Journal of Multivariate Analysis* **88**(2): 365–411.
21. G. Akemann, J. Baik, and P. Di Francesco (eds) (2011) *The Oxford Handbook of Random Matrix Theory*. Oxford: Oxford University Press.
22. I. Johnstone (2001) On the distribution of the largest eigenvalue in principal components analysis, *The Annals of Statistics* **29**(2): 295–327.
23. K. Johansson (2000) Shape fluctuations and random matrices, *Communications in Mathematical Physics* **209**(2): 437–76.
24. C. Tracy and H. Widom (1996) On orthogonal and symplectie matrix ensembles, *Communications in Mathematical Physics* **177**: 727–54.
25. C. Tracy and H. Widom (1994) Level-spacing distributions and the airy kernel, *Communications in Mathematical Physics* **159**(1): 151–74.
26. J. Baik and J. Silverstein (2006) Eigenvalues of large sample covariance matrices of spiked population models, *Journal of Multivariate Analysis* **97**(6): 1382–1408.
27. D. Paul (2007) Asymptotics of sample eigenstructure for a large dimensional spiked covariance model, *Statistica Sinica* **17**(4): 1617.
28. A. Soshnikov (2002) A note on universality of the distribution of the largest eigenvalues in certain sample covariance matrices, *Journal of Statistical Physics* **108**(5): 1033–56.
29. B. Nadler (2008) Finite sample approximation results for principal component analysis: A matrix perturbation approach, *The Annals of Statistics* **36**(6): 2791–2817.
30. D. Hoyle and M. Rattray (2007) Statistical mechanics of learning multiple orthogonal signals: asymptotic theory and fluctuation effects, *Physical Review E* **75**(1): 016101.
31. D. Hoyle and M. Rattray (2004) Principal-component-analysis eigenvalue spectra from data with symmetry-breaking structure, *Physical Review E* **69**(2): 026124.
32. D. Hoyle and M. Rattray (2003) Limiting form of the sample covariance eigenspectrum in PCA and kernel PCA, in *Proceedings of Neural Information Processing Systems* (NIPS, 16).
33. I. Dhillon and J. Tropp (2007) Matrix nearness problems with Bregman divergences, *SIAM Journal on Matrix Analysis and Applications* **29**(4): 1120–46.
34. D. Petz (2010) *Quantum Information Theory and Quantum Statistics*. New York: Springer.
35. N.J. Higham (2008) *Functions of Matrices: Theory and Computation*. Society for Industrial and Applied Mathematics.
36. Y. Zeng, Y.C. Liang, E. Peh, and A.T. Hoang (2009) Cooperative covariance and eigenvalue based detections for robust sensing, in *Global Telecommunications Conference, Globecom*, IEEE, pp. 1–6.
37. M. Naraghi-Pour and T. Ikuma, Autocorrelation-based spectrum sensing for cognitive radios, *IEEE Transactions on Vehicular Technology* **59**(2): 718–33.
38. H. Urkowitz (1967) Energy detection of unknown deterministic signals, *Proceedings of the IEEE* **55**(4): 523–31.
39. J. Ma, G.Y. Li, and B.H. Juang (2009) Signal processing in cognitive radio, *Proceedings of the IEEE* **97**(5): 805–23.
40. V.I. Kostylev (2002) Energy detection of a signal with random amplitude, in *IEEE International Conference on Communications*, ICC'02, vol. 3, pp. 1606–10.
41. F.F. Digham, M.S. Alouini, and M.K. Simon (2003) On the energy detection of unknown signals over fading channels, in *IEEE International Conference on Communications*, ICC'03. vol. 5, pp. 3575–9.
42. F.F. Digham, M.S. Alouini, and M.K. Simon (2007) On the energy detection of unknown signals over fading channels, *IEEE Transactions on Communications* **55**(1): 21–4.
43. J.G. Proakis (2001) *Digital Communications,* 4th edn. New York: McGraw Hill.
44. C.E. Shannon (1949) Communication in the presence of noise, *Proceedings of the IRE* **37**(1): 10–21.
45. M. Abramowitz and I. Stegun (eds.) (1965) *Handbook of Mathematical Functions*. National Bureau of Standards.
46. A. Hald (1952) *Statistical Tables and Formulas*. New York: John Wiley & Sons, Inc.
47. R.A. Fisher and F. Yates (1953) *Statistical Tables for Agricultural, Biological and Medical Research,* Edinburgh: Oliver & Boyd, Ltd.
48. I. Gradshteyn and I. Ryzhik (eds.) (1994) *Tables of Integral, Series, and Products*. New York: Academic Press.

49. A.H. Nuttall (1972) Some integrals involving the QM-function, Technical report, Naval Underwater Systems Center (NUSC).

50. W.B. Davenport and W.L. Root (1958) *An Introduction to the Theory of Random Signals and Noise,* vol. 11. New York: McGraw-Hill.

51. C.W. Helstrom (1960) *Statistical Theory of Signal Detection.* London: Pergamon.

52. E.J. Kelly, I.S. Reed, and W.L. Root (1960) Detection of radar echoes in noise, I, *Journal of the Society for Industrial and Applied Mathematics* **8**(2): 309–41.

53. E.J. Kelly, I.S. Reed, and W.L. Root (1960) Detection of radar echoes in noise, II, *Journal of the Society for Industrial and Applied Mathematics* **8**(3): 481–507.

54. H.L. Van Trees (1968) *Detection, Estimation, and Modulation Theory.* New York: John Wiley & Sons, Inc.

55. H. Stark and J.W. Woods (2002) *Probability and Random Processes with Applications to Signal Processing,* vol. 3. Upper Saddle River, NJ: Prentice Hall.

56. H.J. Landau (1967) Necessary density conditions for sampling and interpolation of certain entire functions, *Acta Mathematica* **117**(1): 37–52.

57. H.J. Landau (1965) The eigenvalue behavior of certain convolution equations, *Transactions of the American Mathematics Society* **115**: 242–56.

58. H.J. Landau and H.O. Pollak (1962) Prolate spheroidal wave functions, Fourier analysis and uncertainty. III. The dimension of the space of essentially time- and band-limited signals, *Bell System Technical Journal* **41**(4): 1295–1336.

59. H.J. Landau and H.O. Pollak (1961) Prolate spheroidal wave functions, Fourier analysis and uncertainty II, *Bell System Technical Journal* **40**(1): 65–84.

60. H.J. Landau and H. Widom (1980) Eigenvalue distribution of time and frequency limiting, *Journal of Mathematical Analysisand Applications* **77**(2): 469–81.

61. D. Slepian (1983) Some comments on Fourier analysis, uncertainty and modeling, *SIAM Review,* 379–93.

62. D. Slepian (1978) Prolate spheroidal wave functions, Fourier analysis, and uncertainty. V—The discrete case, *Bell System Technical Journal* **57**: 1371–1430.

63. D. Slepian (1976) On bandwidth, *Proceedings of the IEEE* **64**(3): 292–300.

64. D. Slepian (1962) Prolate spheroidal wave functions, Fourier analysis and uncertainty. IV: Extensions to many dimensions; generalized prolate spheroidal functions, *Bell System Technical Journal* **43**: 3009–57.

65. D. Slepian (1954) Estimation of signal parameters in the presence of noise, *Transactions of IRE* **3**(1): 68–89.

66. H.J. Landau and H.O. Pollak (1961) Prolate spheroidal wave functions, Fourier analysis and uncertainty II, *Bell System Technical Journal* **40**(1): 65–84.

67. R. Brunelli and T. Poggio (1993) Face recognition: Features versus templates, *IEEE Transactions on Pattern Analysis and Machine Intelligence* **15**(10): 1042–52.

68. A. Goldsmith (2005) *Wireless Communications.* Cambridge: Cambridge University Press.

69. S.M. Kay (1998) *Fundamentals of Statistical Signal Processing, Volume II: Detection Theory,* vol. 7. New Jersey: Prentice Hall.

70. Z. Quan, S.J. Shellhammer, W. Zhang, and A.H. Sayed (2009) Spectrum sensing by cognitive radio at very low SNR, in *Globecom'09*, pp. 1–6.

71. A. Leon-Garcia (2008) *Probability, Statistics, and Random Processes for Electrical Engineering,* 3rd edn. Upper Saddle River, New Jersey: Prentice Hall.

72. C.W. Therrien (1992) *Discrete Random Signals and Statistical Signal Processing.* New Jersey: Prentice Hall PTR.

73. S.M. Kay (1988) *Modern Spectral Estimation: Theory and Application,* vol. 1. Englewood Cliffs, NJ: Prentice Hall.

74. L. Marple (1987) *Digital Analysis with Applications.* Englewood Cliffs, NJ: Prentice-Hall.

75. P. Stoica and R.L. Moses (1997) *Introduction to Spectral Analysis,* vol. 51. Upper Saddle River, NJ: Prentice Hall.

76. Y. Zeng, Y. Liang, A. Hoang, and R. Zhang (2011) A Review on Spectrum Sensing Techniques for Cognitive Radio: Challenges and Solutions, *EURASIP Journal on Advances in Signal Processing*, Hindawi Publishing Corporation, vol. 2010, Article ID 381465, 1–15.

77. T. Lim, R. Zhang, Y. Liang, and Y. Zeng (2008) GLRT-based spectrum sensing for cognitive radio, in *IEEE Globecom'08*, New Orleans, LA, pp. 1–5.

78. H.V. Poor (1994) *An Introduction to Signal Detection and Estimation*. New York: Springer.
79. D.S. Bernstein (2009) Matrix Mathematics: Theory, Facts, and Formulas. New Jersey: Princeton University Press.
80. R.M. Gray (2006) *Toeplitz and Circulant Matrices: A Review*. Hanover, MA: Now Publishers Inc.
81. I. Selin (1965) *Detection Theory*. New Jersey: Princeton University Press.
82. G. Berkooz, P. Holmes, and J. Lumley (1993) The proper orthogonal decomposition in the analysis of turbulent flows, *Annual Review of Fluid Mechanics* **25**(1): 539–75.
83. P. Holmes, J. Lumley, G. Berkooz, J. Mattingly, and R. Wittenberg (1997) Low-dimensional models of coherent structures in turbulence, *Physics Reports* **287**(4): 337–84.
84. P. Zhang, R. Qiu, and N. Guo (2011) Demonstration of spectrum sensing with blindly learned features, *Communications Letters, IEEE* **15**(99): 548–50.
85. S. Hou and R. Qiu (2011) Spectrum sensing for cognitive radio using kernel-based learning, Arxiv preprint arXiv:1105.2978, submitted to *IEEE Transactions on Communications*.
86. B. Scholkopf, A. Smola, and K. Muller (1998) Nonlinear component analysis as a kernel eigenvalue problem, *Neural Computation* **10**(5): 1299–1319.
87. V. Tawil (2006) 51 captured DTV signal. http://grouper.ieee.org/groups/802/22/Meeting documents/2006 May/Informal Documents.
88. J. Ma, G.Y. Li, and B.H. Juang (2009) Signal processing in cognitive radio, *Proceedings of the IEEE* **97**(5): 805–23.
89. W.A. Gardner, A. Napolitano, and L. Paura (2006) Cyclostationarity: Half a century of research, *Signal Processing*. **86**(4): 639–97.
90. W.A. Gardner (1991) Exploitation of spectral redundancy in cyclostationary signals, *Signal Processing Magazine, IEEE* **8**(2): 14–36.
91. W.A. Gardner (1991) Two alternative philosophies for estimation of the parameters of time-series, *IEEE Transactions on Information Theory* **37**(1): 216–18.
92. A. Sahai and D. Cabric (2005) Cyclostationary feature detection. Tutorial presented at the IEEE DySPAN 2005.
93. W.A. Gardner and C.M. Spooner (1992) Signal interception: performance advantages of cyclic-feature detectors, *IEEE Transactions on Communications* **40**(1): 149–59.
94. M. Derakhshani, M. Nasiri-Kenari, and T. Le-Ngoc (2010) Cooperative cyclostationary spectrum sensing in cognitive radios at low SNR regimes, in *IEEE International Conference on Communications* (ICC), 2010.
95. A. Fehske, J. Gaeddert, and J.H. Reed (2005) A new approach to signal classification using spectral correlation and neural networks, in *New Frontiers in Dynamic Spectrum Access Networks*, DySPAN 2005, IEEE, pp. 144–50.
96. H. Hu, Y. Wang, and J. Song (2008) Signal classification based on spectral correlation analysis and SVM in cognitive radio, in *Proceedings of the 22nd International Conference on Advanced Information Networking and Applications*, IEEE Computer Society, pp. 883–7.
97. A. Tani and R. Fantacci (2010) A low-complexity cyclostationary-based spectrum sensing for UWB and WiMAX coexistence with noise uncertainty, *IEEE Transactions on Vehicular Technology* **59**(6): 2940–50.
98. J. Lunden, V. Koivunen, A. Huttunen, and H.V. Poor (2009) Collaborative cyclostationary spectrum sensing for cognitive radio systems, *IEEE Transactions on Signal Processing* **57**(11): 4182–95.
99. K.W. Choi, W.S. Jeon, and D.G. Jeong (2009) Sequential detection of cyclostationary signal for cognitive radio systems, *IEEE Transactions on Wireless Communications* **8**(9): 4480–5.
100. R.W. Heath Jr and G.B. Giannakis (1999) Exploiting input cyclostationarity for blind channel identification in OFDM systems, *IEEE Transactions on Signal Processing* **47**(3): 848–56.
101. H. Bolcskei (2001) Blind estimation of symbol timing and carrier frequency offset in wireless OFDM systems, *IEEE Transactions on Communications* **49**(6): 988–99.
102. P.D. Sutton, K.E. Nolan, and L.E. Doyle (2008) Cyclostationary signatures in practical cognitive radio applications, *IEEE Journal on Selected Areas in Communications* **26**(1): 13–24.
103. K. Abed-Meraim, Y. Xiang, J.H. Manton, and Y. Hua (2001) Blind source-separation using second-order cyclostationary statistics, *IEEE Transactions on Signal Processing* **49**(4): 694–701.
104. R.S. Prendergast and T.Q. Nguyen (2006) Minimum mean-squared error reconstruction for generalized undersampling of cyclostationary processes, *IEEE Transactions on Signal Processing* **54**(8): 3237–42.

105. H. Yan and H.H. Fan (2007) Signal-selective DOA tracking for wideband cyclostationary sources, *IEEE Transactions on Signal Processing* **55**(5).

106. Z. Huang, Y. Zhou, and W. Jiang (2008) TDOA and Doppler estimation for cyclostationary sgnals based on multi-cycle frequencies, *IEEE Transactions on Aerospace and Electronic Systems* **44**(4): 1251–64.

107. R. Vershynin (2011) Introduction to the non-asymptotic analysis of random matrices, Arxiv preprint arXiv:1011.3027v5, July.

108. C.W. Therrien (1992) *Discrete Random Signals and Statistical Signal Processing*. New Jersey: Prentice-Hall.

109. R. Bhatia (2007) *Positive Definite Matrices*. New Jersey: Princeton University Press.

110. K. Mardia, J. Kent, and J. Bibby (1979) *Multivariate Analysis*. New York: Academic Press.

111. V.L. Girko (1990) *Theory of Random Determinant*. Dordrecht: Kluwer Academic Publishers.

112. K. Abadir and J. Magnus (2005) *Matrix Algebra*. Cambridge: Cambridge University Press.

113. J. Steele (2004) *The Cauthy-Schwarz Master Class*. Cambridge: Cambridge University Press.

114. F. Zhang (1999) *Matrix Theory*. New York: Springer.

115. D.S. Bernstein (2009) *Matrix Mathematics: Theory, Facts, and Formulas*. New Jersey: Princeton University Press.

116. J.A. Tropp (2011) User-friendly Tail Bounds for Sums of Random Matrices. In *Foundations of Computational Mathematics,* New York: Springer, pp. 1–46.

117. H. Nagaoka and M. Hayashi (2007) An information-spectrum approach to classical and quantum hypothesis testing for simple hypotheses, *IEEE Transactions on Information Theory* **53**(2): 534–49.

118. S.M. Kay (1998) *Fundamentals of Statistical Signal Processing*. New Jersey: Prentice Hall.

119. Y. Zeng and Y. Liang (2007) Covariance based signal detections for cognitive radio, in *New Frontiers in Dynamic Spectrum Access Networks*, DySPAN 2007, pp. 202–7.

120. T.J. Lim, R. Zhang, Y. Liang, and Y. Zeng (2008) Glrt-based spectrum sensing for cognitive radio, in *Global Telecommunications Conference, Globecom'08*, IEEE, pp. 1–5.

121. P. Zhang, R. Qiu, and N. Guo (2011) Demonstration of spectrum sensing with blindly learned feature, *IEEE Communications Letters* **15**: 548–50.

122. F. Lin, R.C. Qiu, Z. Hu, S. Hou, J.P. Browning, and M.C. Wicks (2012) Generalized FMD detection for spectrum sensing under low signal-to-noise ratio. Accepted by *IEEE Communications Letters*.

123. L.L. Scharf (1991) *Statistical Signal Processing*. Boston, MA: Addison-Wesley.

124. B. Levy (2008) *Principles of Signal Detection and Parameter Estimation*. Berlin: Springer Verlag.

125. P. Schreier (2008) A unifying discussion of correlation analysis for complex random vectors, IEEE Transactions on Signal Processing **56**(4): 1327–36.

126. P.J. Schreiner and L.L. Scharf (2010) *Statistical Signal Processing of Complex-Valued Data: The Theory of Improper and Noncircular Signals*. Cambridge: Cambridge University Press.

127. G. Jaeger (2007) *Quantum Information: An Overview*. New York: Springer.

128. M.A. Nielsen and I.L. Chuang (2010) *Quantum Computation and Quantum Information,* 10th edn. Cambridge: Cambridge University Press.

129. M. Hayashi (2006) *Quantum Information: An Introduction*. New York: Springer.

130. R. Bhatia (1997) *Matrix Analysis*. New York: Springer.

131. A.W. Marshall, I. Olkin, and B.C. Arnold (2011) *Inequalities: Theory of Majorization and Its Applications*. New York: Springer.

132. M. Shaked and J.G. Shanthikumar (2007) *Stochastic Order*. New York: Springer.

133. X.Z. Zhan (2002) *Matrix Inequality*. Berlin: Springer Verlag.

134. M. Loss and M.B. Ruskai (eds.) (2002) *Inequalities: Selecta of Elliott H. Lieb*. Berlin: Springer Verlag.

135. T. Kosem (2006) Inequalities between kf(A + B)k and kf(A) + f(B)k, *Linear Algebra and Its Applications* **418**(1): 153–60.

136. J. Bourin and M. Uchiyama (2007) A matrix subadditivity inequality for f (a + b) and f (a) + f (b), Arxiv preprint math/0702475.

137. R. Ahlswede and A. Winter (2002) Strong converse for identification via quantum channels, *IEEE Transactions on Information Theory* **48**(3): 569–79.

138. O. Bratteli and D.W. Robinson (1979) *Operator Algebras amd Quantum Statistical Mechanics I*. Berlin: Springer Verlag.

139. J. Lawson and Y. Lim (2001) The geometric mean, matrices, metrics, and more, *The American Mathematical Monthly* **108**(9): 797–812.

140. A.S. Holevo (1972) An analogue of the theory of statistical decisions in non-commutative probability theory, *Transactions of the Moscow Mathematics Society* (English translation) **26**: 133–49. Trudy Moskov. Mat. Obc (in Russian).

141. C.W. Helstrom (1976) *Quantum Detection and Estimation Theory*. New York: Academic Press.

142. K.M.R. Audenaert, J. Calsamiglia, R. Munoz-Tapia, *et al.* (2007) Discriminating states: The quantum Chernoff bound, *Physical Review Letters* **98**: 160501, Apr.

143. M. Nussbaum and A. Szkola (2009) The Chernoff lower bound for symmetric quantum hypothesis testing, *Annals of Statistics* **37**(2): 1040–57.

144. K. Audenaert, M. Nussbaum, A. Szkoaa, and F. Verstraete (2008) Asymptotic error rates in quantum hypothesis testing. *Communications in Mathematical Physics* **279**, pp. 251–83, 2008. 10.1007/s00220-008-0417-5.

145. K.M.R. Audenaert (2007) Quantum hypothesis testing non-commutative Chernoff and Hoeffding bounds, February 2007. Presentation Slides.

146. D. Bacon, I. Chuang, and A. Harrow (2006) Efficient quantum circuits for Schur and Clebsch-Gordan transforms, *Physical Review Letters* **97**(17): 1705021–1705024.

147. S. Barnett and S. Croke (2009) On the conditions for discrimination between quantum states with minimum error, *Journal of Physics A: Mathematical and Theoretical* **42**: 0620011–0620014.

148. S. Barnett and S. Croke (2008) Quantum state discrimination, Arxiv preprint arXiv:0810.1970.

149. V. Belavkin (1975) Optimal multiple quantum statistical hypothesis testing, *Stochastics* **1**: 315–45.

150. I. Bjelakovic (2006) Limit theorems for quantum entropies and applications, in *Turbo Codes & Related Topics: 6th International ITG-Conference on Source and Channel Coding (TURBOCODING)*, pp. 1–6.

151. I. Bjelakovic and H. Boche (2009) Classical capacities of compound and averaged quantum channels, *IEEE Transactions on Information Theory* **55**(7): 3360–74.

152. I. Bjelakovic and H. Boche (2008) Classical capacities of compound quantum channels, in *Information Theory Workshop*, ITW'08. IEEE, pp. 373–7.

153. I. Bjelakovic, J. Deuschel, T. Kruger, R. Seiler, R. Siegmund-Schultze, and A. Szkola (2005) A quantum version of Sanov's theorem, *Communications in Mathematical Physics* **260**(3): 659–71.

154. I. Bjelakovic, T. Kruger, R. Siegmund-Schultze, and A. Szkola (2004) The Shannon-McMillan theorem for ergodic quantum lattice systems, *Inventiones Mathematicae* **155**(1): 203–22.

155. S. Broadbent (1955) Quantum hypotheses, *Biometrika* **42**(1/2): 45–57.

156. J. Calsamiglia, R. Munoz-Tapia, L. Masanes, A. Acin, and E. Bagan (2008) Quantum Chernoff bound as a measure of distinguishability between density matrices: Application to Qubit and Gaussian states, *Physical Review A* **77**(3): 0323111–03231115.

157. G. Cariolaro and G. Pierobon (2010) Performance of quantum data transmission systems in the presence of thermal noise, *IEEE Transactions on Communications* **58**(2): 623–30.

158. G. Cariolaro and G. Pierobon (2010) Theory of quantum pulse position modulation and related numerical problems, *IEEE Transactions on Communications* **58**(4): 1213–22.

159. A. Chefles (2000) Quantum state discrimination, *Contemporary Physics* **41**(6): 401–24.

160. A. Chefles (1998) Unambiguous discrimination between linearly independent quantum states, *Physics Letters A* **239**(6): 339–47.

161. H. Chernoff (1952) A measure of asymptotic efficiency for tests of a hypothesis based on the sum of observations, *The Annals of Mathematical Statistics,* 493–507.

162. N. Datta (2009) Min-and max-relative entropies and a new entanglement monotone, *IEEE Transactions on Information Theory* **55**(6): 2816–26.

163. Y. Eldar (2003) Mixed-quantum-state detection with inconclusive results, *Physical Review A* **67**(4): 0423091–04230914.

164. Y. Eldar (2003) A semidefinite programming approach to optimal unambiguous discrimination of quantum states, *IEEE Transactions on Information Theory* **49**(2): 446–56.

165. Y. Eldar and G. Forney Jr (2001) On quantum detection and the square-root measurement, *IEEE Transactions on Information Theory* **47**(3): 858–72.

166. Y. Eldar, A. Megretski, and G. Verghese (2004) Optimal detection of symmetric mixed quantum states, *IEEE Transactions on Information Theory* **50**(6): 1198–1207.

167. Y. Eldar, A. Megretski, and G. Verghese (2003) Designing optimal quantum detectors via semidefinite programming, *IEEE Transactions on Information Theory* **49**(4): 1007–12.

168. Y. Eldar, M. Stojnic, and B. Hassibi (2004) Optimal quantum detectors for unambiguous detection of mixed states, *Physical Review A* **69**(6): 0623181–0623189.

169. J. Fiurášek and M. Ježek (2003) Optimal discrimination of mixed quantum states involving inconclusive results, *Physical Review A* **67**(1): 0123211–0123215.

170. R. Gill (2001) Asymptotics in quantum statistics, Lecture Notes. Monograph Series, pp. 255–85.

171. R. Gill and S. Massar (2000) State estimation for large ensembles, *Physical Review A* **61**(4): 0423121–04231216.

172. P. Hausladen and W. Wootters (1994) A "pretty good" measurement for distinguishing quantum states, *Journal of Modern Optics* **41**: 2385–90.

173. M. Hayashi (2009) Discrimination of two channels by adaptive methods and its application to quantum system, *IEEE Transactions on Information Theory* **55**(8): 3807–20.

174. M. Hayashi (2007) Error exponent in asymmetric quantum hypothesis testing and its application to classical-quantum channel coding, *Physical Review A* **76**(6): 0623011–0623014.

175. M. Hayashi (2002) Optimal sequence of quantum measurements in the sense of Stein's lemma in quantum hypothesis testing, *Journal of Physics A: Mathematical and General* **35**.

176. M. Hayashi (2001) Asymptotics of quantum relative entropy from a representation theoretical viewpoint, *Journal of Physics A: Mathematical and General* **34**.

177. M. Hayashi (2001) Optimal sequence of povms in the sense of Stein's lemma in quantum hypothesis testing, Arxiv preprint quant-ph/0107004.

178. M. Hayashi and K. Matsumoto (2004) Asymptotic performance of optimal state estimation in quantum two level system, Arxiv preprint quant-ph/0411073.

179. M. Hayashi and H. Nagaoka (2003) General formulas for capacity of classical-quantum channels, *IEEE Transactions on Information Theory* **49**(7): 1753–68.

180. C. Helstrom (1982) Bayes-cost reduction algorithm in quantum hypothesis testing (corresp.), *IEEE Transactions on Information Theory* **28**(2): 359–66.

181. C. Helstrom (1973) Resolution of point sources of light as analyzed by quantum detection theory, *IEEE Transactions on Information Theory* **19**(4): 389–98.

182. C. Helstrom (1969) Quantum detection and estimation theory, *Journal of Statistical Physics* **1**(2): 231–52.

183. C. Helstrom (1967) Detection theory and quantum mechanics, *Information and Control* **10**(3): 254–91.

184. U. Herzog and J. Bergou (2005) Optimum unambiguous discrimination of two mixed quantum states, *Physical Review A* **71**(5): 0503011–0503014.

185. U. Herzog and J. Bergou (2004) Distinguishing mixed quantum states: Minimum-error discrimination versus optimum unambiguous discrimination, *Physical Review A* **70**(2): 0223021–0223026.

186. F. Hiai, M. Mosonyi, and M. Hayashi (2009) Quantum hypothesis testing with group symmetry, *Journal of Mathematical Physics* **50**: 1033041–10330431.

187. W. Hoeffding (1965) Asymptotically optimal tests for multinomial distributions, *The Annals of Mathematical Statistics*, 369–401.

188. A. Holevo (1973) Statistical decision theory for quantum systems, *Journal of Multivariate Analysis* **3**(4): 337–94.

189. V. Kargin (2005) On the Chernoff bound for efficiency of quantum hypothesis testing, *Annals of Statistics*, 959–76.

190. K. Kato and O. Hirota (2003) Square-root measurement for quantum symmetric mixed state signals, *IEEE Transactions on Information Theory* **49**(12): 3312–17.

191. K. Kato, M. Osaki, M. Sasaki, and O. Hirota (1999) Quantum detection and mutual information for QAM and PSK signals, *IEEE Transactions on Communications* **47**(2): 248–54.

192. G. Kimura, T. Miyadera, and H. Imai (2009) Optimal state discrimination in general probabilistic theories, *Physical Review A* **79**(6): 0623061–0623069.

193. R. Konig, R. Renner, and C. Schaffner (2009) The operational meaning of min-and max-entropy, *IEEE Transactions on Information Theory* **55**(9): 4337–47.

194. M. Mohseni, A. Steinberg, and J. Bergou (2004) Optical realization of optimal unambiguous discrimination for pure and mixed quantum states, *Physical Review Letters* **93**(20): 2004031–2004034.

195. M. Mosonyi (2009) Hypothesis testing for Gaussian states on Bosonic lattices, *Journal of Mathematical Physics* **50**: 0321051–03210517.

196. M. Mosonyi and N. Datta (2009) Generalized relative entropies and the capacity of classical-quantum channels, *Journal of Mathematical Physics* **50**: 0721041–07210414.

197. H. Nagaoka (2006) The converse part of the theorem for quantum Hoeffding bound, Arxiv preprint quant-ph/0611289.

198. M. Nussbaum and A. Szkola (2008) A lower bound of Chernoff type for symmetric quantum hypothesis testing, Arxiv preprint quant-ph/0607216.
199. M. Nussbaum and A. Szkola (2010) Exponential error rates in multiple state discrimination on a quantum spin chain, *Journal of Mathematical Physics* **51**: 0722031–07220311.
200. M. Nussbaum and A. Szkola (2011) Asymptotically optimal discrimination between pure quantum states, *Theory of Quantum Computation, Communication, and Cryptography*, pp. 1–8.
201. T. Ogawa and M. Hayashi (2004) On error exponents in quantum hypothesis testing, *IEEE Transactions on Information Theory* **50**(6): 1368–72.
202. T. Ogawa and H. Nagaoka (2007) Making good codes for classical-quantum channel coding via quantum hypothesis testing, *IEEE Transactions on Information Theory* **53**(6): 2261–6.
203. T. Ogawa and H. Nagaoka (2002) A new proof of the channel coding theorem via hypothesis testing in quantum information theory, *IEEE International Symposium on Information Theory, 2002*, p. 73.
204. T. Ogawa and H. Nagaoka (2000) Strong converse and Stein's lemma in quantum hypothesis testing, *IEEE Transactions on Information Theory* **46**(7): 2428–33.
205. T. Ogawa and H. Nagaoka (1999) Strong converse theorems in the quantum information theory, in *Information Theory and Networking Workshop*, IEEE, p. 54.
206. T. Ogawa and H. Nagaoka (1999) Strong converse to the quantum channel coding theorem, *IEEE Transactions on Information Theory* **45**(7): 2486–9.
207. A. Peres and D. Terno (1998) Optimal distinction between non-orthogonal quantum states, *Journal of Physics A: Mathematical and General* **31**.
208. R. Renner (2005) Security of quantum key distribution, Arxiv preprint quant-ph/0512258.
209. E. Robinson (1982) A historical perspective of spectrum estimation, *Proceedings of the IEEE* **70**(9): 885–907.
210. B. Roysam and M. Miller (1989) A unified approach for hierarchical imaging based on joint hypothesis testing and parameter estimation, in *International Conference on Acoustics, Speech, and Signal Processing*, ICASSP-89, pp. 1779–82.
211. N. Salikhov (1999) On one generalization of Chernov's distance, *Theory of Probability and its Applications*, vol. 43, Society for Industrial and Applied Mathematics, pp. 239–55.
212. R. Stratonovich (1965) Information capacity of a quantum communication channel, ii, *Radiophysics and Quantum Electronics* **8**(1): 92–101.
213. R. Stratonovich (1965) Information capacity of a quantum communications channel. i., *Radiophysics and Quantum Electronics* **8**(1): 82–91.
214. M. Takeoka, M. Sasaki, and M. Ban (2003) Design of an optimal quantum receiver for interferometric sensing devices, in *Quantum Electronics Conference*, EQEC'03, p. 375.
215. L. Wang and R. Renner (2010) One-shot classical-quantum capacity and hypothesis testing, Arxiv preprint arXiv:1007.5456.
216. H. Yuen, R. Kennedy, and M. Lax (1975) Optimum testing of multiple hypotheses in quantum detection theory, *IEEE Transactions on Information Theory* **21**(2): 125–34.
217. S. Zhao, F. Gao, X. Dong, and B. Zheng (2010) Detection scheme for quantum multiple access channel with noisy coherent state, in *International Conference on Wireless Communications and Signal Processing (WCSP), 2010*, IEEE, pp. 1–4.
218. C.A. Fuchs (1995) Distinguishability and Accessible Information in Quantum Theory. PhD thesis, Albuquerque, NM: University of New Mexico.
219. A. Peres (1995) *Quantum Theory: Concepts and Methods*. Dordrecht: Kluwer Academic Publishers.
220. C.H. Bennett, D.P. DiVincenzo, J.A. Smolin, and W.K. Wootters (1996) Mixed-state entanglement and quantum error correction, *Physical Review A* **54**(5): 3824–51.
221. B. Schumacher and M.A. Nielsen (1996) Quantum data processing and error correction, Arxiv preprint quant-ph/9604022.
222. R. Derka, V. Buzek, and A.K. Ekert (1998) Universal algorithm for optimal estimation of quantum states from finite ensembles via realizable generalized measurement, *Physical Review Letters* **80**(8): 1571–5.
223. D. Loss and D.P. DiVincenzo (1998) Quantum computation with quantum dots, *Physical Review A* **57**(1): 120–6.
224. M.A. Nielsen (1998) Quantum Information Theory. PhD thesis, Albuquerque, New Mexico: The University of New Mexico.

225. M.A. Nielsen, C.M. Caves, B. Schumacher, and H. Barnum (1998) Information-theoretic approach to quantum error correction and reversible measurement. *Proceedings of the Royal Society of London. Series A: Mathematical, Physical and Engineering Sciences* **454**(1969): 277–304.

226. A. Steane (1998) Quantum computing. *Reports on Progress in Physics* **61**.

227. K. Banaszek, G.M. DAriano, M.G.A. Paris, and M.F. Sacchi (1999) Maximum-likelihood estimation of the density atrix, *Physical Review A* **61**(1): 0103041–0103044.

228. H.E. Brandt (1999) Positive operator valued measure in quantum information processing, *American Journal of Physics* **67**: 434–9.

229. C.M. Caves (1999) Quantum error correction and reversible operations, *Journal of Superconductivity* **12**(6): 707–18.

230. A. Imamoglu, D.D. Awschalom, G. Burkard, *et al.* (1999) Quantum information processing using quantum dot spins and cavity qed, *Physical Review Letters* **83**(20): 4204–7.

231. E. Knill, R. Laamme, and G.J. Milburn (1999) A scheme for efficient quantum computation with linear optics, *Clusters of Galaxies* **320**: 296–9.

232. G.G. Amosov, A.S. Holevo, and R.F. Werner (2000) On some additivity problems in quantum information theory, Arxiv preprint math-ph/0003002.

233. C.H. Bennett and D.P. DiVincenzo (2000) Quantum information and computation, *Nature* **404**(6775): 247–55.

234. G.M. D'Ariano and P. Lo Presti (2000) Tomography of quantum operations, arXiv:quant-ph/0012071v1.

235. S.B. Zheng and G.C. Guo (2000) Efficient scheme for two-atom entanglement and quantum information processing in cavity qed, *Physical Review Letters* **85**(11): 2392–5.

236. D. Bouwmeester, A.K. Ekert, A. Zeilinger, *et al.*, *The Physics of Quantum Information*. Berlin: Springer.

237. A.S. Holevo (2001) *Statistical Structure of Quantum Theory*. New York: Springer, 2001.

238. M.D. Lukin, M. Fleischhauer, R. Cote, *et al.* (2001) Dipole blockade and quantum information processing in mesoscopic atomic ensembles, *Physical Review Letters* **87**(3): 379011–379014.

239. A.I. Lvovsky, H. Hansen, T. Aichele, O. Benson, J. Mlynek, and S. Schiller (2001) Quantum state reconstruction of the single-photon fock state, *Physical Review Letters* **87**(5): 504021–504024.

240. M. Keyl (2002) Fundamentals of quantum information theory, *Physics Reports* **369**(5): 431–548.

241. M.A. Nielsen, I. Chuang, and L.K. Grover (2002) Quantum computation and quantum information, *American Journal of Physics* **70**: 558–60.

242. V. Vedral (2002) The role of relative entropy in quantum information theory, *Reviews of Modern Physics* **74**(1): 197–234, 2002. [110]

243. G. D'Ariano, M. Paris, and M. Sacchi (2003) Quantum tomography, *Advances in Imaging and Electron Physics* **128**: 205–308.

244. F. De Martini, A. Mazzei, M. Ricci, G.M. D Ariano, *et al.* (2003) Exploiting quantum parallelism of entanglement for a complete experimental quantum characterization of a single-qubit device, *Physical Review-Series A* **67**(6): 62307.

245. I. Chuang (2004) Quantum information joining the foundations of physics and computer science, *MIT Physics Annual*.

246. N.K. Langford, R.B. Dalton, M.D. Harvey, *et al.* (2004) Measuring entangled qutrits and their use for quantum bit commitment, *Physical Review Letters* **93**(5): 536011–536014.

247. J.M. Renes, R. Blume-Kohout, A.J. Scott, and C.M. Caves (2004) Symmetric informationally complete quantum measurements, *Journal of Mathematical Physics* **45**: 2171–80.

248. P.W. Shor (2004) Equivalence of additivity questions in quantum information theory, *Communications in Mathematical Physics* **246**(3): 453–72.

249. S.E. Ahnert and M.C. Payne (2005) General implementation of all possible positive-operator-value measurements of single photon polarization states, *Physical Review A* **71**(1): 0123301–0123304.

250. C.J. O'Loan (2010) Topics in estimation of quantum channels, Arxiv preprint arXiv:1001.3971.

251. V. Marchenko and L. Pastur (1967) Distributions of eigenvalues for some sets of random matrices, *Mathematics of the USSR-Sbornik* **1**: 457–83.

252. A. Guionnet (2009) *Large Random Matrices: Lectures on Macroscopic Asymptotics*. Berlin: Springer Verlag.

253. F. Penna and R. Garello (2009) Theoretical performance analysis of eigenvalue-based detection, Arxiv preprint arXiv:0907.1523,

254. N. Rao and A. Edelman (2006) Free probability, sample covariance matrices, and signal processing, in *IEEE International Conference on Acoustics, Speech and Signal Processing, ICASSP 2006*, vol. 5, p. V.

255. L. Cardoso, M. Debbah, P. Bianchi, and J. Najim (2008) Cooperative spectrum sensing using random matrix theory, in *3rd International Symposium on Wireless Pervasive Computing, ISWPC 2008*, pp. 334–8.

256. L. Wang, B. Zheng, J. Cui, S. Tang, and H. Dou (2009) Cooperative spectrum sensing using free probability theory, in *Global Telecommunications Conference, Globecom'09*, IEEE, pp. 1–5.

257. Y. Zeng and Y. Liang (2007) Maximum-minimum eigenvalue detection for cognitive radio, in *IEEE 18th International Symposium on Personal, Indoor and Mobile Radio Communications* (PIMRC) 2007, pp. 1–5.

258. Y. Zeng and Y. Liang (2009) Eigenvalue-based spectrum sensing algorithms for cognitive radio, *IEEE Transactions on Communications* **57**(6): 1784–93.

259. J. Baik, G. Ben Arous, and S. Péché (2005) Phase transition of the largest eigenvalue for nonnull complex sample covariance matrices, *The Annals of Probability* **33**(5): 1643–97.

260. F. Penna, R. Garello, and M. Spirito (2009) Cooperative spectrum sensing based on the limiting eigenvalue ratio distribution in Wishart matrices, *Communications Letters, IEEE* **13**(7): 507–9.

261. O. Feldheim and S. Sodin (2010) A universality result for the smallest eigenvalues of certain sample covariance matrices, *Geometric And Functional Analysis* **20**(1): 88–123.

262. O. Ryan and M. Debbah (2007) Free deconvolution for signal processing applications, *IEEE Transactions on Information Theory* **1**: 1–15.

263. R. Dozier and J. Silverstein (2007) On the empirical distribution of eigenvalues of large dimensional information-plus-noise type matrices, *Journal of Multivariate Analysis* **98**(4): 678–94.

264. Z. Bai and Y. Yin (1988) Convergence to the semicircle law, *The Annals of Probability* **16**(2): 863–75.

265. U. Grenander and J. Silverstein (1977) Spectral analysis of networks with random topologies, *SIAM Journal on Applied Mathematics*, 499–519.

266. D. Jonsson (1982) Some limit theorems for the eigenvalues of a sample covariance matrix, *Journal of Multivariate Analysis* **12**(1): 1–38.

267. K. Wachter (1978) The strong limits of random matrix spectra for sample matrices of independent elements, The Annals of Probability, pp. 1–18.

268. Yin, Y. (1986) Limiting spectral distribution for a class of random matrices, *Journal of Multivariate Analysis* **20**(1): 50–68.

269. Z. Bai and Y. Yin (1986) Limiting behavior of the norm of products of random matrices and two problems of Geman-Hwang, *Probability Theory and Related Fields*. **73**(4): 555–69.

270. S. Geman (1980) A limit theorem for the norm of random matrices, *The Annals of Probability* **8**(2): 252–61.

271. Z. Bai and Y. Yin (1993) Limit of the smallest eigenvalue of a large dimensional sample covariance matrix, *The Annals of Probability*, pp. 1275–94.

272. T. Tao and V. Vu (2011) Random matrices: Universality of eigenvectors, Arxiv preprint arXiv: 1103.2801.

273. S. Geman (1986) The spectral radius of large random matrices, *The Annals of Probability* **14**(4): 1318–28.

274. Z.D. Bai (1997) Circular law, *The Annals of Probability* **25**: 494–529.

275. Z.D. Bai (1993) Convergence rate of expected spectral distributions of large random matrices. Part II. Sample covariance matrices, *The Annals of Probability*, pp. 649–72.

276. Z. Bai, B. Miao, and J. Tsay (1997) A note on the convergence rate of the spectral distributions of large random matrices, *Statistics & Probability Letters* **34**(1): 95–101.

277. Z. Bai, J. Hu, and W. Zhou (2012) Convergence rates to the Marchenko–Pastur type distribution, *Stochastic Processes and Their Applications* **122**: 68–92.

278. R. Müller (2003) Applications of large random matrices in communications engineering, in *Proceedings of the International Conference on Advances in Internet, Processing, Systems, and Interdisciplinary Research (IPSI)*, Sveti Stefan, Montenegro.

279. W. Hachem, P. Loubaton, and J. Najim (2011) Applications of large random matrices to digital communications and statistical signal processing. *EUSIPCO*, September. Presentation (133 slides).

280. L. Pastur (2005) A simple approach to the global regime of Gaussian ensembles of random matrices, *Ukrainian Mathematical Journal* **57**(6): 936–6.

281. W. Hachem, P. Loubaton, and J. Najim (2007) Deterministic equivalents for certain functionals of large random matrices, *The Annals of Applied Probability* **17**(3): 875–930.

282. P. Vallet (2011) Random matrix theory and applications to statistical signal processing. PhD Dissertation, Université Paris-Est.

283. J. Silverstein and Z. Bai (1995) On the empirical distribution of eigenvalues of a class of large dimensional random matrices, *Journal of Multivariate Analysis* **54**(2): 175–92.

284. G. Pan (2010) Strong convergence of the empirical distribution of eigenvalues of sample covariance matrices with a perturbation matrix, *Journal of Multivariate Analysis* **101**(6): 1330–8.

285. S. Kritchman and B. Nadler (2009) Non-parametric detection of the number of signals: hypothesis testing and random matrix theory, *IEEE Transactions on Signal Processing* **57**(10): 3930–41.

286. A. Gittens and J. Tropp (2011) Tail bounds for all eigenvalues of a sum of random matrices, Arxiv preprint arXiv:1104.4513.

287. W. Bryc, A. Dembo, and T. Jiang (2006) Spectral measure of large random Hankel, Markov and Toeplitz matrices, *The Annals of Probability* **34**(1): 1–38.

288. A. Bose, S. Chatterjee, and S. Gangopadhyay (2003) Limiting spectral distrbution of large dimensional random matrices, *Journal of the Indian Statistical Association* **41**: 221–59.

289. A. Bose and A. Dey (2010) The wonderful world of eigenvalues, in C.S. Yogananda (ed.), *Math Unlimited: Essays in Mathematics,* Enfield, New Hampshire: Science Publishers.

290. A. Bose, S. Gangopadhyay, and A. Sen (2010) Limiting spectral distribution of xx'matrices, in *Annales de l'Institut Henri Poincaré, Probabilités et Statistiques,* vol. 46, pp. 677–707, Institut Henri Poincaré.

291. A. Bose, R. Hazra, and K. Saha (2010) Patterned random matrices and method of moments, in *Proceedings of the International Congress of Mathematicians*, Hyderabad, pp. 2203–30.

292. A. Bose and J. Mitra (2002) Limiting spectral distribution of a special circulant, *Statistics & Probability Letters* **60**(1): 111–20.

293. A. Bose and A. Sen (2008) Another look at the moment method for large dimensional random matrices, *Electronic Journal of Probability* **13**(21): 588–628.

294. A. Bose and A. Sen (2007) Spectral norm of random large dimensional noncentral Toeplitz and Hankel matrices, *Electronic Journal of Probability* **12**: 29–35.

295. C. Hammond and S. Miller (2005) Distribution of eigenvalues for the ensemble of real symmetric Toeplitz matrices, *Journal of Theoretical Probability* **18**(3): 537–66.

296. A. Massey, S. Miller, and J. Sinsheimer (2007) Distribution of eigenvalues of real symmetric palindromic Toeplitz matrices and circulant matrices, *Journal of Theoretical Probability* **20**(3): 637–62.

297. L. Pastur and V. Vasilchuk (2000) On the law of addition of random matrices, *Communications in Mathematical Physics* **214**(2): 249–86.

298. V. Girko (2001) *Theory of Stochastic Canonical Equations*. Dordrecht: Kluwer Academic Publishers.

299. W. Hachem, P. Loubaton, J. Najim, and P. Vallet (2010) On bilinear forms based on the resolvent of large random matrices, Arxiv preprint arXiv:1004.3848.

300. R. Dozier and J. Silverstein (2007) Analysis of the limiting spectral distribution of large dimensional information-plus-noise type matrices, *Journal of Multivariate Analysis* **98**(6): 1099–1122.

301. W. Hachem, P. Loubaton, X. Mestre, J. Najim, and P. Vallet (2011) A subspace estimator for fixed rank perturbations of large random matrices, Arxiv preprint arXiv:1106.1497.

302. P. Bianchi, M. Debbah, M. Maïda, and J. Najim (2011) Performance of statistical tests for single-source detection using random matrix theory, *IEEE Transactions on Information Theory* **57**(4): 2400–19.

303. T. Anderson (1963) Asymptotic theory for principal component analysis, *The Annals of Mathematical Statistics* **34**(1): 122–48.

304. M. Wax and T. Kailath (1985) Detection of signals by information theoretic criteria, *IEEE Transactions on Speech and Signal Processing (SSSP)* **33**: 387–92.

305. B. Nadler (2011) On the distribution of the ratio of the largest eigenvalue to the trace of a Wishart matrix, *Journal of Multivariate Analysis* **102**(2): 363–71.

306. A. Bejan (2005) Largest eigenvalues and sample covariance matrices. Tracy-Widom and Painlevé. II: Computational aspects and realization in S-plus with applications, Preprint.

307. F. Bornemann (2010) Asymptotic independence of the extreme eigenvalues of Gaussian unitary ensemble, *Journal of Mathematical Physics* **51**: 023514.

308. C. Tracy and H. Widom (2002) Distribution functions for largest eigenvalues and their applications, Arxiv preprint mathph/0210034.

309. C. Tracy and H. Widom (1998) Correlation functions, cluster functions, and spacing distributions for random matrices, *Journal of Statistical Physics* **92**(5): 809–35.

310. C. Tracy and H. Widom (1994) Level spacing distributions and the Bessel kernel, *Communications in Mathematical Physics* **161**(2): 289–309.
311. C. Tracy and H. Widom (1993) Introduction to random matrices, *Geometric and Quantum Aspects of Integrable Systems,* pp. 103–30.
312. I. Johnstone and A. Lu (2009) Sparse principal components analysis, Arxiv preprint arXiv:0901.4392.
313. I. Johnstone and Z. Ma (2011) Fast approach to the Tracy-Widom law at the edge of goe and gue, Arxiv preprint arXiv:1110.0108.
314. N. El Karoui (2010) The spectrum of kernel random matrices, *The Annals of Statistics* **38**(1): 1–50.
315. N. El Karoui (2009) Concentration of measure and spectra of random matrices: applications to correlation matrices, elliptical distributions and beyond, *The Annals of Applied Probability* **19**(6): 2362–2405.
316. N. El Karoui (2007) On spectral properties of large dimensional correlation matrices and covariance matrices computed from elliptically distributed data. Technical report from Department of Statistics, University of California, Berkeley.
317. N. El Karoui (2007) Tracy–Widom limit for the largest eigenvalue of a large class of complex sample covariance matrices, *The Annals of Probability,* **35**(2): 663–714.
318. N. El Karoui (2006) A rate of convergence result for the largest eigenvalue of complex white Wishart matrices, *The Annals of Probability* **34**(6): 2077–2117.
319. P. Bianchi, J. Najim, M. Maida, and M. Debbah (2009) Performance analysis of some eigen-based hypothesis tests for collaborative sensing, in *IEEE/SP 15th Workshop on Statistical Signal Processing, SSP'09*, pp. 5–8.
320. Y. Yin, Z. Bai, and P. Krishnaiah (1988) On the limit of the largest eigenvalue of the large dimensional sample covariance matrix, *Probability Theory and Related Fields* **78**(4): 509–21.
321. A. Van der Vaart (1998) *Asymptotic Statistics*. Cambridge: Cambridge University Press.
322. R. Nadakuditi and A. Edelman (2008) Sample eigenvalue based detection of high-dimensional signals in white noise using relatively few samples, IEEE Transactions on Signal Processing **56**(7): 2625–38.
323. R. Nadakuditi (2006) Applied stochastic eigen-analysis. PhD dissertation, Massachusetts Institute of Technology.
324. B. Nadler and I. Johnstone (2011) Detection performance of Roy's largest root test when the noise covariance matrix is arbitrary, in *Statistical Signal Processing Workshop (SSP)*, IEEE, pp. 681–4.
325. I. Johnstone (2008) Multivariate analysis and Jacobi ensembles: Largest eigenvalue, Tracy–Widom limits and rates of convergence, Annals of Statistics **36**(6): 2638.
326. A. James (1964) Distributions of matrix variates and latent roots derived from normal samples, The Annals of Mathematical Statistics, 475–501.
327. I. Johnstone (2009) Approximate null distribution of the largest root in multivariate analysis, The Annals of Applied Statistics **3**(4): 1616.
328. R. Muirhead (2005) *Aspects of Mutivariate Statistical Theory*. New York: John Wiley & Sons, Inc.
329. M. Debbah (2008) Random matrix theory and free probability, *WP 2.1–Paradigms Collection and Foundations*, pp. 161–90. http://www.bionets.eu/docs/BIONETS_D2_1_1.pdf
330. R. Couillet, J. Silverstein, Z. Bai, and M. Debbah (2011) Eigen-inference for energy estimation of multiple sources, *IEEE Transactions on Information Theory* **57**(4): 2420–39.
331. J. Dumont, W. Hachem, S. Lasaulce, P. Loubaton, and J. Najim (2010) On the capacity achieving covariance matrix for Rician mimo channels: an asymptotic approach, *IEEE Transactions on Information Theory* **56**(3): 1048–69.
332. M. Dieng (2005) Distribution functions for edge eigenvalues in orthogonal and symplectic ensembles: Painlevé representations, *International Mathematics Research Notices* **2005**(37): 2263–87.
333. P. Vallet, P. Loubaton, and X. Mestre (2010) Improved subspace estimation for multivariate observations of high dimension: the deterministic signals case, Arxiv preprint arXiv:1002.3234.
334. B. Nadler (2010) Nonparametric detection of signals by information theoretic criteria: performance analysis and an improved estimator, *IEEE Transactions on Signal Processing* **58**(5): 2746–56.
335. F. Benaych-Georges and R. Nadakuditi (2011) The eigenvalues and eigenvectors of finite, low rank perturbations of large random matrices, *Advances in Mathematics*.
336. J. Garnier (2011) Use of random matrix theory for target detection, localization, and reconstruction, *Contemporary Mathematics* **548** pp. 1–19.
337. F. Pasqualetti, R. Carli, and F. Bullo (2011) A distributed method for state estimation and false data detection in power networks. *IEEE International Conference on Smart Grid Communications (Smart-GridComm)*, pp. 469–74.

338. P. Vallet, W. Hachem, P. Loubaton, X. Mestre, and J. Najim (2011) An improved music algorithm based on low rank perturbation of large random matrices, in *Statistical Signal Processing Workshop (SSP)*, IEEE, pp. 689–92.

339. P. Vallet, P. Loubaton, and X. Mestre (2009) Improved subspace doa estimation methods with large arrays: The deterministic signals case, in *IEEE International Conference on Acoustics, Speech and Signal Processing*, ICASSP 2009, pp. 2137–40.

340. H. Li and H. Poor (2009) Large system spectral analysis of covariance matrix estimation, *IEEE Transactions on Information Theory* **55**(3): 1395–1422.

341. Z. Bai and J. Silverstein (2004) Clt for linear spectral statistics of large-dimensional sample covariance matrices, *The Annals of Probability* **32**(1A): 553–605.

342. W. Hachem, P. Loubaton, and J. Najim (2008) A clt for information-theoretic statistics of gram random matrices with a given variance profile, *The Annals of Applied Probability* **18**(6): 2071–30.

343. W. Hachem, M. Kharouf, J. Najim, and J. Silverstein (2011) A clt for information-theoretic statistics of non-centered gram random matrices, Arxiv preprint arXiv:1107.0145.

344. G. Anderson and O. Zeitouni (2008) A clt for regularized sample covariance matrices, *The Annals of Statistics* **36**(6): 2553–76.

345. R. Couillet and W. Hachem (2011) Local failure detection and diagnosis in large sensor networks, Arxiv preprint arXiv:1107.1409.

346. P. Bickel and E. Levina (2008) Regularized estimation of large covariance matrices, *The Annals of Statistics* **36**(1): 199–227.

347. P. Bickel and E. Levina (2008) Covariance regularization by thresholding. *The Annals of Statistics* **36**(6): 2577–2604.

348. V. Marčenko and L. Pastur (1967) Distribution of eigenvalues for some sets of random matrices, *Mathematics of the USSRSbornik* **1**, p. 457.

349. P. Bickel and E. Levina (2004) Some theory for Fisher's linear discriminant function, 'naive Bayes', and some alternatives when there are many more variables than observations, Bernoulli **10**(6): 989–1010.

350. U. Grenander and G. Szego (1984) *Toeplitz Forms and Their Applications*. New York: Chelsea Publishing Company.

351. T. Cai, C. Zhang, and H. Zhou (2010) Optimal rates of convergence for covariance matrix estimation, *The Annals of Statistics* **38**(4): 2118–44.

352. W. Wu and M. Pourahmadi (2003) Nonparametric estimation of large covariance matrices of longitudinal data, *Biometrika* **90**(4): 831–44.

353. W. Wu and M. Pourahmadi (2009) Banding sample autocovariance matrices of stationary processes, *Statistica Sinica* **19**(4): 1755–68.

354. E. Hannan and M. Deistler (1988) *The Statistical Theory of Linear Systems*. New York: John Wiley & Sons, Inc.

355. W. Wu (2005) Nonlinear system theory: Another look at dependence, *Proceedings of the National Academy of Sciences of the United States of America* **102**(40): 14150.

356. D. Voiculescu, K. Dykema, and A. Nica (1992) Free random variables. *American Mathematical Society*.

357. P. Biane (1998) Free probability for probabilists, *Quantum Probability Communications* **11**: 55–71.

358. D. Tse and S. Hanly (1999) Linear multiuser receivers: Effective interference, effective bandwidth and user capacity, *IEEE Transactions on Information Theory* **45**(2): 641–57.

359. D. Tse (1999) Multiuser receivers, random matrices and free probability, in *Proceedings of the Annual Allerton Conference on Communication Control and Computing* **37**: 1055–64.

360. R. Speicher (1997) Free probability theory and non-crossing partitions. *39 Seminaire Lotharingien de Combinatoire*.

361. N. Rao and A. Edelman (2008) The polynomial method for random matrices, *Foundations of Computational Mathematics* **8**(6): 649–702.

362. N. Rao (2006) Rmtool: A random matrix and free probability calculator in Matlab. Users Guide from Department of EECS, Massachusetts Institute of Technology.

363. N. Rao (2007) The analytic computability of the Shannon transform for a large class of random matrix channels, Arxiv preprint arXiv:0712.0305.

364. N. Rao, J. Mingo, R. Speicher, and A. Edelman (2008) Statistical eigen-inference from large Wishart matrices, *The Annals of Statistics* **36**(6): 2850–85.

365. O. Ryan and M. Debbah (2009) Asymptotic behavior of random Vandermonde matrices with entries on the unit circle, *IEEE Transactions on Information Theory* **55**(7): 3115–47.

366. O. Ryan and M. Debbah (2008) Channel capacity estimation using free-probability theory, *IEEE Transactions on Signal Processing* **56**(11): 5654–67.

367. M. McClure, R.C. Qiu, and L. Carin (1997) On the superresolution of observables from swept-frequency scattering data, *IEEE Transactions on Antenna Propagation* **45**: 631–41.

368. R.C. Qiu and I.T. Lu (1999) Multipath resolving with frequency dependence for broadband wireless channel modeling, *IEEE Transactions on Vehicular Technology* **48**: 273–85.

369. R.C. Qiu (2002) A study of the ultra-wideband wireless propagation channel and optimum UWB receiver design (Part 1), *IEEE Journal of Selected Areas in Commun. (JSAC), the 1st JASC special issue on UWB Radio* **20**: 1628–37.

370. R.C. Qiu (2004) A generalized time domain multipath channel and its application in ultra-wideband (UWB) wireless optimal receiver design: Part 2 Wave-based system analysis, *IEEE Transactions on Wireless Communications* **3**: 2312–24.

371. R.C. Qiu (2006) A generalized time domain multipath channel and its application in ultra-wideband (UWB) wireless optimal receiver design: Part 3 System performance analysis, *IEEE Transactions on Wireless Communications* **5**(10): 2685–95.

372. C. Qiu, C. Zhou, and Q. Liu (2005) Physics-based pulse distortion for ultra-wideband signals,, *IEEE Transactions on Vehicular Technology* **54**: 1546–54.

373. C.M. Zhou and R.C. Qiu (2007) Pulse distortion caused by cylinder diffraction and its impact on UWB communications,, *IEEE Transactions on Vehicular Technology* **56**: 2384–91.

374. A. Nordio, C. Chiasserini, and E. Viterbo (2008) Reconstruction of multidimensional signals from irregular noisy samples, *IEEE Transactions on Signal Processing* **56**(9): 4274–85.

375. C. Bordenave (2008) Eigenvalues of Euclidean random matrices, *Random Structures & Algorithms* **33**(4): 515–32.

376. G. Tucci and P. Whiting (2011) Eigenvalue results for large scale random Vandermonde matrices with unit complex entries, *IEEE Transactions on Information Theory* **57**(6): 3938–54.

377. O. Ryan and M. Debbah (2011) Convolution operations arising from Vandermonde matrices, *IEEE Transactions on Information Theory* **57**(7): 4647–59.

378. X. Mestre (2008) Improved estimation of eigenvalues and eigenvectors of covariance matrices using their sample estimates, *IEEE Transactions on Information Theory* **54**(11): 5113–29.

379. O. Ryan, A. Masucci, S. Yang, and M. Debbah (2011) Finite dimensional statistical inference, *IEEE Transactions on Information Theory* **57**(4): 2457–73.

380. D. Voiculescu (1986) Addition of certain non-commuting random variables, *Journal of Functional Analysis* **66**(3): 323–46.

381. F. Benaych-Georges and M. Debbah (2008) Free deconvolution: from theory to practice, submitted to *IEEE Transactions on Information Theory*.

382. O. Ryan (2007) Tools for convolution with finite Gaussian matrices. http://folk.uio.no/oyvindry/finitegaussian/

383. Z. Burda, J. Jurkiewicz, and B. Waclaw (2004) Spectral moments of correlated wishart matrices, Arxiv preprint condmat/0405263.

384. S. Ji, Y. Xue, and L. Carin (2008) Bayesian compressive sensing, *IEEE Transactions on Signal Processing* **56**(6): 2346–56.

385. S. Ji, D. Dunson, and L. Carin (2009) Multi-task compressive sensing, *IEEE Transactions on Signal Processing* **57**(1): 92–106.

386. L.C. Potter, E. Ertin, J.T. Parker, and M. Çetin (2010) Sparsity and compressed sensing in radar imaging, *Proceedings of the IEEE* **98**(6): 1006–20.

387. J.M. Duarte-Carvajalino and G. Sapiro (2009) Learning to sense sparse signals: Simultaneous sensing matrix and sparsifying dictionary optimization, *IEEE Transactions on Image Processing* **18**(7): 1395–1408.

388. G.R.G. Lanckriet, N. Cristianini, P. Bartlett, L.E. Ghaoui, and M.I. Jordan (2004) Learning the kernel matrix with semidefinite programming, *The Journal of Machine Learning Research* **5**: 27–72.

389. M.J. Choi, V. Chandrasekaran, and A.S. Willsky (2009) Exploiting sparse Markov and covariance structure in multiresolution models, in *Proceedings of the 26th Annual International Conference on Machine Learning* (Montreal, QC, Canada), ACM, pp. 177–84.

390. D.S. Bernstein (2009) *Matrix Mathematics: Theory, Facts, and Formulas*. New Jersey: Princeton University Press.

391. L. Vandenberghe and S. Boyd (1996) Semidefinite programming, *SIAM Review* **38**(1): 49–95.

392. M. Chiang (2005) *Geometric Programming for Communication Systems*. Hanover, MA: Now Publishers Inc.
393. Wikipedia, "Interior point method–Wikipedia, The Free Encyclopedia." http://en.wikipedia.org/wiki/ Interior point method.
394. Z. Hu, R. Qiu, and D. Singh (2009) Spectral efficiency for MIMO UWB channel in rectangular metal cavity, *Journal of Networks* **4**: 42–52.
395. Wikipedia, Particle swarm optimization—Wikipedia, The Free Encyclopedia. http://en.wikipedia.org/ wiki/Particle swarm optimization.
396. A. Ben-Tal, L. El Ghaoui, and A. Nemirovski, *Robust Optimization*. New Jersey: Princeton University Press.
397. P. Setoodeh and S. Haykin (2009) Robust transmit power control for cognitive radio, *Proceedings of IEEE* **97**(5): 915–39.
398. A.D. Maio and A. Farina (2009) *Waveform Diversity: Past, Present, and Future*. A Lecture Series on Waveform Diversity for Advanced Radar Systems, July.
399. N. Guo, J.Q. Zhang, P. Zhang, Z. Hu, Y. Song, and R.C. Qiu (2008) UWB real-time testbed with waveform-based precoding, in *IEEE Military Conference* (San Diego, USA).
400. N. Guo, Z. Hu, A.S. Saini, and R.C. Qiu (2009) Waveform-level Precoding with Simple Energy Detector Receiver for Wideband Communication, in *IEEE Southeastern Symposium on System Theory* (Tullahoma, USA).
401. Z. Hu, N. Guo, and R.C. Qiu (2009) Wideband waveform optimization for energy detector receiver with practical considerations, in *IEEE ICUWB* (Vancouver, Canada).
402. M.C. Wicks (2009) *History of Waveform Diversity and Future Benefits to Military Systems*. A Lecture Series onWaveform Diversity for Advanced Radar Systems.
403. Y. Song, Z. Hu, N. Guo, and R.C. Qiu (2010) Real-time MISO UWB radio testbed and waveform design, in *IEEE SoutheastCon*, pp. 204–9.
404. F. Barbaresco (2009) *New Agile Waveforms Based on Mathematics and Resources management of Waveform Diversity*. A Lecture Series on Waveform Diversity for Advanced Radar Systems.
405. A.D. Maio and A. Farina (2009) *New Trends in Coded Waveform Design for Radar Applications*. A Lecture Series on Waveform Diversity for Advanced Radar Systems.
406. S. Haykin (2006) Cognitive radar: A way of the future. *IEEE Signal Processing Magazine* **23**(1): 30–40.
407. N.A. Goodman, P.R. Venkata, and M.A. Neifeld (2007) Adaptive waveform design and sequential hypothesis testing for target recognition with active sensors, *IEEE Journal of Selected Topics in Signal Processing* **1**(1): 105–13.
408. A. Ben-Tal and A. Nemirovski (2000) Robust solutions of linear programming problems contaminated with uncertain data, *Mathematical Programming* **88**(3): 411–24.
409. A. Ben-Tal and A. Nemirovski (2002) Robust optimization–methodology and applications, *Mathematical Programming* **92**(3): 453–80.
410. L. El Ghaoui and H. Lebret (1997) Robust solutions to least-squares problems with uncertain data, *SIAM Journal on Matrix Analysis and Applications* **18**: 1035–64.
411. Y.C. Eldar (2008) Rethinking biased estimation: Improving maximum likelihood and the Cramér–Rao bound, *Foundations and Trends in Signal Processing* **1**(4): 305–449.
412. Y.C. Eldar and N. Merhav (2004) A competitive minimax approach to robust estimation of random parameters, *IEEE Transactions on Signal Processing* **52**(7): 1931–46.
413. Y.C. Eldar, A. Ben-Tal, and A. Nemirovski (2004) Linear minimax regret estimation of deterministic parameters with bounded data uncertainties, *IEEE Transactions on Signal Processing* **52**(8): 2177–88.
414. Y.C. Eldar (2006) Minimax estimation of deterministic parameters in linear models with a random model matrix, *IEEE Transactions on Signal Processing* **54**(2): 601–12.
415. Y.C. Eldar (2006) Minimax MSE estimation of deterministic parameters with noise covariance uncertainties, *IEEE Transactions on Signal Processing* **54**(1): 138–45.
416. A. Beck, A. Ben-Tal, and Y.C. Eldar (2006) Robust mean-squared error estimation of multiple signals in linear systems affected by model and noise uncertainties, *Mathematical Programming* **107**(1): 155–87.
417. A. Beck, Y.C. Eldar, and A. Ben-Tal (2008) Mean-squared error estimation for linear systems with block circulant uncertainty, *SIAM Journal on Matrix Analysis and Applications* **29**(3): 712–30.
418. L. El Ghaoui, F. Oustry, and H. Lebret (1998) Robust solutions to uncertain semidefinite programs, *SIAM Journal of Optimization* **9**: 33–52.

419. D. Bertsimas, D. Pachamanova, and M. Sim (2004) Robust linear optimization under general norms, *Operations Research Letters* **32**(6): 510–16.
420. K.M. Teo (2007) Nonconvex Robust Optimization. PhD thesis, Massachusetts Institute of Technology.
421. E.A. Gharavol, Y.C. Liang, and K. Mouthaan (2010) Robust downlink beamforming in multiuser MISO cognitive radio networks with imperfect channel-state information, *IEEE Transactions on Vehicular Technology* **59**: 2852–60.
422. L. Zhang, Y.C. Liang, Y. Xin, and H.V. Poor (2009) Robust cognitive beamforming with partial channel state information, *IEEE Transactions on Wireless Communications* **8**: 4143–53.
423. Y. Yang and R.S. Blum (2007) Minimax robust MIMO radar waveform design, *IEEE Journal of Selected Topics in Signal Processing* **1**(1): 147–55.
424. T. Naghibi, M. Namvar, and F. Behnia (2010) Optimal and robust waveform design for MIMO radars in the presence of clutter, *Signal Processing* **90**: 1103–17.
425. S.A. Vorobyov, A.B. Gershman, and Z. Luo (2003) Robust adaptive beamforming using worst-case performance optimization: A solution to the signal mismatch problem, *IEEE Transactions on Signal Processing* **51**(2): 313–24.
426. S.J. Kim, A. Magnani, A. Mutapcic, S.P. Boyd, and Z. Luo (2008) Robust beamforming via worst-case SINR maximization, *IEEE Transactions on Signal Processing* **56**(4): 1539–47.
427. S.A. Vorobyov, H. Chen, and A.B. Gershman (2008) On the relationship between robust minimum variance beamformers with probabilistic and worst-case distortionless response constraints, *IEEE Transactions on Signal Processing* **56**(11): 5719–24.
428. K. Slavakis and I. Yamada (2007) Robust wideband beamforming by the hybrid steepest descent method, *IEEE Transactions on Signal Processing* **55**(9): 4511–22.
429. J. Lampinen (2000) Multiobjective nonlinear pareto-optimization, Tech. Rep., Lappeenranta University of Technology, Laboratory of Information Processing, Lapperanta, Finland.
430. R.T. Marler and J.S. Arora (2004) Survey of multi-objective optimization methods for engineering, *Structural and Multidisciplinary Optimization* **26**(6): 369–95.
431. N. Srinivas and K. Deb (1994) Multiobjective optimization using nondominated sorting in genetic algorithms, *Evolutionary Computation* **2**(3): 221–48.
432. E. Zitzler, M. Laumanns, and S. Bleuler (2004) A tutorial on evolutionary multiobjective optimization, *Metaheuristics for Multiobjective Optimisation* **535**: 3–37.
433. K.C. Tan, T.H. Lee, and E.F. Khor (2002) Evolutionary algorithms for multi-objective optimization: Performance assessments and comparisons, *Artificial Intelligence Review* **17**(4): 251–90.
434. S.F. Adra and P.J. Fleming (2011) Diversity management in evolutionary many-objective optimization, *IEEE Transactions on Evolutionary Computation* **15**: 183–95.
435. T. Aittokoski and K. Miettinen (2008) Efficient evolutionary method to approximate the Pareto optimal set in multiobjective optimization, in *International Conference on Engineering Optimization* (Rio de Janeiro, Brazil).
436. V.J. Amuso and J. Enslin (2007) The Strength Pareto Evolutionary Algorithm 2 (SPEA2) applied to simultaneous multimission waveform design, in *Waveform Diversity and Design Conference* (Pisa, Italy), IEEE, pp. 407–17.
437. S. Sen, G. Tang, and A. Nehorai (2011) Multiobjective optimization of OFDM radar waveform for target detection, *IEEE Transactions on Signal Processing* **59**(2): 639–52.
438. R. Shonkwiler (1993) Parallel genetic algorithms, in *Proceedings of the 5th International Conference on Genetic Algorithms*, Citeseer, pp. 199–205.
439. B.A. Fette (ed.) (2006) *Cognitive Radio Technology*. New York: Elsevier Inc.
440. S. Chen and A.M. Wyglinski (2008) Distributed optimization of cognitive radios employed in dynamic spectrum access networks, in *5th IEEE Annual Communications Society Conference on Sensor, Mesh and Ad Hoc Communications and Networks Workshops*, IEEE, pp. 1–4.
441. S. Chen and A.M. Wyglinski (2009) Cognitive radio-enabled distributed cross-layer optimization via genetic algorithms, in *4th International Conference on Cognitive Radio Oriented Wireless Networks and Communications*, IEEE, pp. 1–6.
442. X. Zhang, Y. Huang, H. Jiang, and Y. Liu (2009) Design of cognitive radio node engine based on genetic algorithm, in *WASE International Conference on Information Engineering*, IEEE, pp. 22–5.
443. J. Ma, H. Jiang, Y. Hu, and Y. Bao (2009) Optimal design of cognitive radio wireless parameters based on non-dominated neighbor distribution Genetic algorithm, in *8th IEEE/ACIS International Conference on Computer and Information Science*, IEEE, pp. 97–101.

444. Y. Bao, H. Jiang, Y. Huang, and R. Hu (2009) Multi-objective optimization of power control and resource allocation for cognitive wireless networks, in *8th IEEE/ACIS International Conference on Computer and Information Science*, IEEE, pp. 70–4.

445. A.A. El-Saleh, M. Ismail, M.A.M. Ali, and J. Ng (2009) Development of a cognitive radio decision engine using multiobjective hybrid genetic algorithm, in *9th Malaysia International Conference on Communications (MICC)*, IEEE, pp. 343–7.

446. D.Q. Zhou (2009) Immune genetic algorithm-based parameters optimization of cognitive radios, in International Conference on Microwave Technology and Computational Electromagnetics (ICMTCE), IET, pp. 468–70.

447. H. Wang and L. Guo (2010) Multi-objective optimization of cognitive radio in clonal selection quantum genetic algorithm, in *International Conference on Measuring Technology and Mechatronics Automation*, IEEE, pp. 740–3.

448. A.A. El-Saleh, M. Ismail, and M.A.M. Ali (2010) Pragmatic trellis coded modulation for adaptive multi-objective genetic algorithm-based cognitive radio systems, in *16th Asia-Pacific Conference on Communications (APCC)*, Auckland, pp. 429–34.

449. C.K. Huynh and W.C. Lee (2011) Multicarrier cognitive radio system configuration based on interference analysis by two dimensional genetic algorithm, in *International Conference on Advanced Technologies for Communications (ATC)*, IEEE, pp. 85–8.

450. B. Zhang, K. Hu, and Y. Zhu (2010) Spectrum allocation in cognitive radio networks using swarm intelligence, in *2nd International Conference on Communication Software and Networks* (ICCSN'10), IEEE, pp. 8–12.

451. K.S. Huang, C.H. Lin, and P.A. Hsiung (2010) A space-efficient and multi-objective case-based reasoning in cognitive radio, in *IET International Conference on Frontier Computing. Theory, Technologies and Applications*, pp. 25–30.

452. J. Zander (1997) Radio resource management in future wireless networks: Requirements and limitations, *IEEE Communications Magazine* **35**(8): 30–6.

453. J. Zander and O. Queseth (2001) *Radio Resource Management for Wireless Networks*. Artech House, Inc.

454. A. Hills and B. Friday (2004) Radio resource management in wireless LANs, *IEEE Communications Magazine,* **42**(12): S9–14.

455. J. Pérez-Romero, O. Sallent, R. Agusti, and M. Diaz-Guerra (2005) *Radio Resource Management Strategies in UMTS*. New York: Wiley Online Library.

456. D. Niyato and E. Hossain (2008) A noncooperative game-theoretic framework for radio resource management in 4G heterogeneous wireless access networks, IEEE Transactions on Mobile Computing **7**(3): 332–45.

457. Wikipedia, Radio resource management–Wikipedia, The Free Encyclopedia. http://en.wikipedia.org/wiki/Radio resource management.

458. Z.Q. Luo and W. Yu (2006) An introduction to convex optimization for communications and signal processing, *IEEE Journal on Selected Areas in Communications* **24**(8): 1426–38.

459. J.O.D. Neel (2006) Analysis and Design of Cognitive Radio Networks and Distributed Radio Resource Management Algorithms. PhD thesis, Virginia Polytechnic Institute and State University.

460. I.F. Akyildiz, W.Y. Lee, M.C. Vuran, and S. Mohanty (2008) A survey on spectrum management in cognitive radio networks, *Communications Magazine, IEEE* **46**(4): 40–8.

461. F. Wang, M. Krunz, and S. Cui (2008) Price-based spectrum management in cognitive radio networks, *IEEE Journal of Selected Topics in Signal Processing* **2**(1): 74–87.

462. J. Zander (1992) Performance of optimum transmitter power control in cellular radio systems, *IEEE Transactions on Vehicular Technology* **41**(1): 57–62.

463. G.J. Foschini and Z. Miljanic (1993) A simple distributed autonomous power control algorithm and its convergence, *IEEE Transactions on Vehicular Technology* **42**(4): 641–6.

464. R. Knopp and P.A. Humblet (1995) Information capacity and power control in single-cell multiuser communications, in *IEEE International Conference on* Communications (ICC'95 Seattle, 'Gateway to Globalization'), vol. 1, pp. 331–5.

465. R.D. Yates (1995) A framework for uplink power control in cellular radio systems, *IEEE Journal on Selected Areas in Communications* **13**(7): 1341–7.

466. G. Caire, G. Taricco, and E. Biglieri (1999) Optimum power control over fading channels, *IEEE Transactions on Information Theory* **45**(5): 1468–89.

467. S.W. Kim and A.J. Goldsmith (2000) Truncated power control in code-division multiple-access communications, *IEEE Transactions on Vehicular Technology* **49**(3): 965–72.

468. Q. Liu, S. Zhou, and G.B. Giannakis (2004) Cross-layer combining of adaptive modulation and coding with truncated ARQ over wireless links, *IEEE Transactions on Wireless Communications* **3**(5): 1746–55.

469. Q. Liu, S. Zhou, and G.B. Giannakis (2005) Queuing with adaptive modulation and coding over wireless links: Cross-layer analysis and design, *IEEE Transactions on Wireless Communications* **4**(3): 1142–53.

470. K.B. Song, A. Ekbal, S. Chung, and J.M. Cioffi (2006) Adaptive modulation and coding (AMC) for bit-interleaved coded OFDM (BIC-OFDM), *IEEE Transactions on Wireless Communications* **5**(7): 1685–94.

471. X. Lin and N.B. Shroff (2004) Joint rate control and scheduling in multihop wireless networks, in *43rd IEEE Conference on Decision and Control*, CDC., vol. 2, pp. 1484–9.

472. R.W. Heath Jr, S. Sandhu, and A. Paulraj (2001) Antenna selection for spatial multiplexing systems with linear receivers, *Communications Letters, IEEE* **5**(4): 142–4.

473. D.A. Gore and A.J. Paulraj (2002) MIMO antenna subset selection with space-time coding, *IEEE Transactions on Signal Processing* **50**(10): 2580–8.

474. S. Sanayei and A. Nosratinia (2004) Antenna selection in MIMO systems, *Communications Magazine, IEEE* **42**(10): 68–73.

475. A.F. Molisch and M.Z. Win (2004) MIMO systems with antenna selection, *Microwave Magazine, IEEE* **5**(1): 46–56.

476. S. Lu, V. Bharghavan, and R. Srikant (1999) Fair scheduling in wireless packet networks, *IEEE/ACM Transactions on Networking (TON)* **7**(4): 473–89.

477. X. Liu, E.K.P. Chong, and N.B. Shroff (2001) Opportunistic transmission scheduling with resource-sharing constraints in wireless networks, *IEEE Journal on Selected Areas in Communications* **19**(10): 2053–64.

478. H. Fattah and C. Leung (2002) An overview of scheduling algorithms in wireless multimedia networks, *Wireless Communications, IEEE* **9**(5): 76–83.

479. N. Vaidya, A. Dugar, S. Gupta, and P. Bahl (2005) Distributed fair scheduling in a wireless LAN, *IEEE Transactions on Mobile Computing* **4**(6): 616–29.

480. S. Tekinay and B. Jabbari (1991) Handover and channel assignment in mobile cellular networks, *Communications Magazine, IEEE* **29**(11): 42–6.

481. E. Del Re, R. Fantacci, and G. Giambene (1995) Handover and dynamic channel allocation techniques in mobile cellular networks, *IEEE Transactions on Vehicular Technology* **44**(2): 229–37.

482. G.P. Pollini (1996) Trends in handover design, *Communications Magazine, IEEE* **34**(3): 82–90.

483. M. Naghshineh and M. Schwartz (1996) Distributed call admission control in mobile/wireless networks, *IEEE Journal on Selected Areas in Communications* **14**(4): 711–17.

484. T.K. Liu and J.A. Silvester (1998) Joint admission/congestion control for wireless CDMA systems supporting integrated services, *IEEE Journal on Selected Areas in Communications* **16**(6): 845–57.

485. B.M. Epstein and M. Schwartz (2000) Predictive QoS-based admission control for multiclass traffic in cellular wireless networks, *IEEE Journal on Selected Areas in Communications* **18**(3): 523–34.

486. Y. Xiao, P. Chen, and Y. Wang (2001) Optimal admission control for multi-class of wireless adaptive multimedia services, *IEICE Transactions on Communications* **84**(4): 795–804.

487. Y. Fang and Y. Zhang (2002) Call admission control schemes and performance analysis in wireless mobile networks, IEEE Transactions on Vehicular Technology **51**(2): 371–82.

488. D. Gao, J. Cai, and K.N. Ngan (2005) Admission control in IEEE 802.11e wireless LANs, *Network, IEEE* **19**(4): 6–13.

489. D. Niyato and E. Hossain (2005) Call admission control for QoS provisioning in 4G wireless networks: Issues and approaches, *Network, IEEE* **19**(5): 5–11.

490. S. Kunniyur and R. Srikant (2003) End-to-end congestion control schemes: Utility functions, random losses and ECN marks, *IEEE/ACM Transactions on Networking (TON)* **11**(5): 702.

491. M. Chiang (2005) Balancing transport and physical layers in wireless multihop networks: Jointly optimal congestion control and power control, *IEEE Journal on Selected Areas in Communications* **23**(1).

492. A. Eryilmaz and R. Srikant (2006) Joint congestion control, routing, and MAC for stability and fairness in wireless networks, *IEEE Journal on Selected Areas in Communications* **24**(8): 1514–24.

493. X. Lin and N.B. Shroff (2006) The impact of imperfect scheduling on cross-layer congestion control in wireless networks, *IEEE/ACM Transactions on Networking* **14**(2): 302–15.

494. Y. Yi and S. Shakkottai (2007) Hop-by-hop congestion control over a wireless multi-hop network, *IEEE/ACM Transactions on Networking* **15**(1): 133–44.

495. Y. Bejerano, S.J. Han, and L. Li (2007) Fairness and load balancing in wireless LANs using association control, *IEEE/ACM Transactions on Networking* **15**(3): 560–73.

496. B. Karp and H.T. Kung (2000) GPSR: Greedy perimeter stateless routing for wireless networks, in *Proceedings of the 6th Annual International Conference on Mobile Computing and Networking*, ACM, pp. 243–54.

497. J.H. Chang and L. Tassiulas (2000) Energy conserving routing in wireless ad-hoc networks, in *19th Annual Joint Conference of the IEEE Computer and Communications Societies, Infocom'00)*, vol. 1, pp. 22–31.

498. D.S.J.D. Couto, D. Aguayo, J. Bicket, and R. Morris (2005) A high-throughput path metric for multi-hop wireless routing, *Wireless Networks* **11**(4): 419–34.

499. M.S. Alouini and A.J. Goldsmith (1999) Capacity of Rayleigh fading channels under different adaptive transmission and diversity-combining techniques, *IEEE Transactions on Vehicular Technology* **48**(4): 1165–81.

500. L. Li and A. Goldsmith (2000) Capacity and optimal resource allocation for fading broadcast channels: Part II: outage capacity, *IEEE Transactions on Information Theory* **47**(3): 120–45.

501. A.J. Goldsmith and M. Effros (2001) The capacity region of broadcast channels with intersymbol interference and colored Gaussian noise, *IEEE Transactions on Information Theory* **47**(1): 219–40.

502. C. Zeng, L.M.C. Hoo, and J.M. Cioffi (2001) Optimal water-filling algorithms for a Gaussian multiaccess channel with intersymbol interference, in *IEEE International Conference on Communications (ICC 2001)*, vol. 8, pp. 2421–7.

503. L. Li and A.J. Goldsmith (2001) Capacity and optimal resource allocation for fading broadcast channels. I. Ergodic capacity, *IEEE Transactions on Information Theory* **47**(3): 1083–1102.

504. S. Toumpis and A.J. Goldsmith (2003) Capacity regions for wireless ad hoc networks, *IEEE Transactions on Wireless Communications* **2**(4): 736–48.

505. R.K. Mallik, M.Z. Win, J.W. Shao, M.S. Alouini, and A.J. Goldsmith (2004) Channel capacity of adaptive transmission with maximal ratio combining in correlated Rayleigh fading, *IEEE Transactions on Wireless Communications* **3**(4): 1124–33.

506. W. Yu and R. Lui (2006) Dual methods for nonconvex spectrum optimization of multicarrier systems, *IEEE Transactions on Communications* **54**(7): 1310–22.

507. W. Yu and T. Lan (2007) Transmitter optimization for the multi-antenna downlink with per-antenna power constraints, *IEEE Transactions on Signal Processing* **55**(6): 2646–60.

508. A. Paulraj, R. Nabar, and D. Gore (2003) *Introduction to Space-Time Wireless Communications*. Cambridge: Cambridge University Press.

509. Z. Hu, D. Singh, and R.C. Qiu (2008) MIMO capacity for UWB channel in rectangular metal cavity, in *IEEE Southeastcon*, pp. 129–35.

510. W. Yu and J.M. Cioffi (2006) Constant-power waterfilling: Performance bound and low-complexity implementation, *IEEE Transactions on Communications* **54**(1): 23–8.

511. D. Singh, Z. Hu, and R.C. Qiu (2008) UWB channel sounding and channel characteristics in rectangular metal cavity, in *IEEE Southeastcon* (Huntsville, USA).

512. R.C. Qiu, C. Zhou, J.Q. Zhang, and N. Guo (2007) Channel reciprocity and time-reversed propagation for ultra-wideband communications, in *IEEE AP-S International Symposium on Antennas and Propagation*, vol. 1, pp. 29–32.

513. M.C. Grant (2010) CVX: Matlab software for disciplined convex programming. http://cvxr.com/. Date accessed: Dec. 2010.

514. A. d'Aspremont (2007) Semidefinite optimization with applications in sparse multivariate statistics. *BIRS Workshop on Mathematical Programming in Data Mining and Machine Learning*.

515. Z. Hu, N. Guo, and R.C. Qiu (2010) Wideband waveform optimization for multiple input signle output cognitive radio with practical considerations, in *Proceedings of IEEE Military Communications Conference, (San Jose, CA)*.

516. T.I. Laakso, V. Valimaki, M. Karjalainen, and U.K. Laine (1996) Splitting the unit delay [FIR/all pass filters design], *IEEE Signal Processing Magazine* **13**(1): 30–60.

517. J. Li, Y. Xie, P. Stoica, X. Zheng, and J. Ward (2007) Beampattern synthesis via a matrix approach for signal power estimation, *IEEE Transactions on Signal Processing* **55**(12): 5643–7.

518. F. Wang, V. Balakrishnan, P. Zhou, *et al.* (2003) Optimal array pattern synthesis using semidefinite programming, *IEEE Transactions on Signal Processing* **51**(5): 1172–83.

519. M. Bengtsson and B. Ottersten (1999) Optimal downlink beamforming using semidefinite optimization, in *Proceedings of the Annual Allerton Conference on Communication Control and Computing*, vol. 37, Citeseer, pp. 987–96.

520. G. San Antonio and D. Fuhrmann (2005) Beampattern synthesis for wideband MIMO radar systems, in *1st IEEE International Workshop on Computational Advances in Multi-Sensor Adaptive Processing*, pp. 105–8.

521. S.F. Yan and Y.L. Ma (2005) Design of FIR beamformer with frequency invariant patterns via jointly optimizing spatial and frequency responses, in *IEEE International Conference on Acoustics, Speech, and Signal Processing*, pp. 789–92.

522. J. Sturm (1999) Using SeDuMi 1.02, a MATLAB toolbox for optimization over symmetric cones, *Optimization Methods and Software* **11**(1): 625–53.

523. Z. Hu, N. Guo, R.C. Qiu, *et al.* (2010) Design of look-up table based architecture for wideband beamforming, in *International Waveform Diversity and Design Conference (Niagara Falls, Canada)*.

524. Y. Zhao, W. Liu, and R. Langley (2009) A least squares approach to the design of frequency invariant beamformers, in *17th European Signal Processing Conference, (Glasgow, Scotland)*.

525. P. Lassila and A. Penttinen (2010) Survey on performance analysis of cognitive radio networks. http://www.netlab.tkk.fi/tutkimus/abi/publ/cogsurvey2.pdf, 2008. Date accessed: Dec. 2010.

526. Wikipedia, OSI model–Wikipedia, The Free Encyclopedia. http://en.wikipedia.org/wiki/OSI model.

527. W. Su and T. Lim (2006) Cross-layer design and optimization for wireless sensor networks, in *Proceedings of the 7th ACIS International Conference on Software Engineering, Artificial Intelligence, Networking, and Parallel/Distributed Computing*, pp. 278–84.

528. M. Chiang, S. Low, A. Calderbank, and J. Doyle (2007) Layering as optimization decomposition: A mathematical theory of network architectures, *Proceedings of the IEEE* **95**(1): 255–312.

529. L. Chen, S. Low, M. Chiang, and J. Doyle (2006) Cross-layer congestion control, routing and scheduling design in ad hoc wireless networks, in *Proceedings of the IEEE Infocom'06*, Citeseer, vol. 6.

530. A. Tang, J. Wang, S. Low, and M. Chiang (2007) Equilibrium of heterogeneous congestion control: Existence and uniqueness, *IEEE/ACM Transactions on Networking (TON)* **15**(4): 824–37.

531. S. Low (20043) A duality model of TCP and queue management algorithms, *IEEE/ACM Transactions on Networking (TON)* **11**(4): 525–36.

532. J. Lee, A. Tang, J. Huang, M. Chiang, and A. Calderbank (2007) Reverse-engineering MAC: A noncooperative game model, *IEEE Journal on Selected Areas in Communications* **25**(6): 1135–47.

533. D. Palomar and M. Chiang (2007) Alternative distributed algorithms for network utility maximization: Framework and applications, *IEEE Transactions on Automatic Control* **52**(12): 2254–69.

534. Y. Vardi (1996) Network tomography: Estimating source-destination traffic intensities from link data, *Journal of the American Statistical Association* **91**(433): 365–77.

535. R. Castro, M. Coates, G. Liang, R. Nowak, and B. Yu (2004) Network tomography: Recent developments, *Statistical Science* **19**(3): 499–517.

536. C. Fortuna and M. Mohorcic (2009) Trends in the development of communication networks: Cognitive networks, *Computer Networks* **53**(9): 1354–76.

537. L. Ding, T. Melodia, S. Batalama, J. Matyjas, and M. Medley (2010) Cross-layer routing and dynamic spectrum allocation in cognitive radio ad hoc networks, *IEEE Transactions on Vehicular Technology* **59**, 1969–79.

538. W. Lou, W. Liu, and Y. Zhang (2006) Performance optimization using multipath routing in mobile ad hoc and wireless sensor networks, *Combinatorial Optimization in Communication Networks*. New York: Springer, pp. 117–46.

539. R. Rajbanshi, A.M. Wyglinski, and G.J. Minden (2006) An efficient implementation of NC-OFDM transceivers for cognitive radios, in *1st International Conference on Cognitive Radio Oriented Wireless Networks and Communications*, (Mykonos Island), pp. 1–5.

540. A. Dutta, D. Saha, D. Grunwald, and D. Sicker (2010) Practical implementation of blind synchronization in NC-OFDM based cognitive radio networks, in *CORONET 2010* (Illinois, Chicago).

541. V.R. Cadambe and S.A. Jafar (2008) Interference alignment and spatial degrees of freedom for the k user interference channel, in *IEEE International Conference on Communications, ICC'08*, (Beijing, China), pp. 971–75.

542. R. Choudhury and N. Vaidya (2007) Mac-layer capture: A problem in wireless mesh networks using beamforming antennas, in *4th Annual IEEE Communications Society Conference on Sensor, Mesh and Ad Hoc Communications and Networks, SECON'07*, pp. 401–10.

543. Z. Quan, S. Cui, and A.H. Sayed (2008) Optimal linear cooperation for spectrum sensing in cognitive radio networks, *IEEE Journal of Selected Topics in Signal Processing* **2**(1): 28–40.

544. E.C.Y. Peh, Y.C. Liang, Y.L. Guan, and Y. Zeng (2009) Optimization of cooperative sensing in cognitive radio networks: A sensing-throughput tradeoff view, *IEEE Transactions on Vehicular Technology* **58**(9): 5294–9.

545. Q. Peng, P.C. Cosman, and L.B. Milstein (2010) Optimal sensing disruption for a cognitive radio adversary, *IEEE Transactions on Vehicular Technology* **59**(4): 1801–10.

546. N.J. Nilsson (1982) *Principles of Artificial Intelligence*. Berlin: Springer Verlag.

547. R.S. Michalski, J.G. Carbonell, and T.M. Mitchell (1985) *Machine Learning: An Artificial Intelligence Approach,* vol. 1. Morgan Kaufmann.

548. S.J. Russell, P. Norvig, J.F. Canny, J.M. Malik, and D.D. Edwards (1955) *Artificial Intelligence: A Modern Approach,* vol. 74. Englewood Cliffs, NJ: Prentice Hall.

549. D. Kortenkamp, R.P. Bonasso, and R. Murphy (1998) *Artificial Intelligence and Mobile Robots: Case Studies of Successful Robot Systems*. AAAI Press.

550. J. Ferber (1999) *Multi-agent Systems: An Introduction to Distributed Artificial Intelligence,* vol. 222. London: Addison-Wesley.

551. G.F. Luger (2005) *Artificial Intelligence: Structures and Strategies for Complex Problem Solving*. Longman: Addison-Wesley.

552. H. Henderson (2007) *Artificial Intelligence: Mirrors for the Mind*. Chelsea House.

553. D.L. Poole and A.K. Mackworth (2010) *Artificial Intelligence: Foundations of Computational Agents*. Cambridge: Cambridge University Press.

554. Wikipedia, Artificial intelligence–Wikipedia, The Free Encyclopedia. http://en.wikipedia.org/wiki/Artificial intelligence.

555. L.J. Guan and M. Duckham (2011) Decentralized reasoning about gradual changes of topological relationships between continuously evolving regions, in *Conference on Spatial Information Theory* (Belfast, Maine).

556. L.I. Perlovsky (2001) *Neural Networks and Intellect: Using Model-based Concepts*. Oxford University Press.

557. L.I. Perlovsky (2007) Neural networks, fuzzy models and dynamic logic, *Aspects of Automatic Text Analysis,* 363–86.

558. R.S. Michalski and G. Tecuci (1994) *Machine Learning: A Multistrategy Approach,* vol. 4. Morgan Kaufmann.

559. E. Alpaydin (2004) *Introduction to Machine Learning*. Cambridge, MA: The MIT Press.

560. T. Jebara (2004) *Machine Learning: Discriminative and Generative,* vol. 755. Springer Netherlands.

561. I.H. Witten and E. Frank (2005) *Data Mining: Practical Machine Learning Tools and Techniques*. Morgan Kaufmann.

562. S. Marsland (2009) *Machine Learning: An Algorithmic Perspective*. Chapman & Hall/CRC.

563. Wikipedia, Machine learning–Wikipedia, The Free Encyclopedia. http://en.wikipedia.org/wiki/Machine learning.

564. Wikipedia, Generative model–Wikipedia, The Free Encyclopedia. http://en.wikipedia.org/wiki/Generative model.

565. Wikipedia, Discriminative model–Wikipedia, The Free Encyclopedia. http://en.wikipedia.org/wiki/Discriminative model.

566. O. Yakhnenko (2009) Learning from text and images: Generative and discriminative models for partially labeled data. PhD thesis, Iowa State University.

567. J. Mitola III and G.Q. Maguire Jr (1999) Cognitive radio: making software radios more personal, *Personal Communications, IEEE* **6**(4): 13–18.

568. S. Haykin (2005) Cognitive radio: Brain-empowered wireless communications, *IEEE Journal on Selected Areas in Communications* **23**(2): 201–20.

569. C. Clancy, J. Hecker, E. Stuntebeck, and T. O'Shea (2007) Applications of machine learning to cognitive radio networks, *IEEE Wireless Communications* **14**(4): 47–52.

570. K.E. Nolan, P. Sutton, and L.E. Doyle (2006) An encapsulation for reasoning, learning, knowledge representation, and reconfiguration cognitive radio elements, in *1st International Conference on Cognitive Radio Oriented Wireless Networks and Communications,* IEEE, pp. 1–5.

571. Y. Su and M. van der Schaar (2008) Learning for cognitive wireless users, in *3rd IEEE Symposium on New Frontiers in Dynamic Spectrum Access Networks, DySPAN 2008*.

572. C. Tekin, S. Hong, and W. Stark (2009) Enhancing cognitive radio dynamic spectrum sensing through adaptive learning, in *Military Communications Conference, Milcom'09*. IEEE.

573. Y. Gai, B. Krishnamachari, and R. Jain (2010) Learning multiuser channel allocations in cognitive radio networks: A combinatorial multi-armed bandit formulation, in *IEEE Symposium on New Frontiers in Dynamic Spectrum*, IEEE.

574. T.A. Tuan, L.C. Tong, and A.B. Premkumar (2010) An adaptive learning automata algorithm for channel selection in cognitive radio network, in *International Conference on Communications and Mobile Computing (CMC)*, vol. 2, IEEE, pp. 159–63.

575. R. Zhang, F. Gao, and Y.C. Liang (2010) Cognitive beamforming made practical: Effective interference channel and learning-throughput tradeoff, *IEEE Transactions on Communications* **58**(2): 706–18.

576. F. Gao, R. Zhang, Y.C. Liang, and X. Wang (2010) Design of learning-based MIMO cognitive radio systems, *IEEE Transactions on Vehicular Technology* **59**(4): 1707–20.

577. N. Moayeri and H. Guo (2010) How often and how long should a cognitive radio sense the spectrum?, in *Symposium on New Frontiers in Dynamic Spectrum*, IEEE.

578. S. Yin, D. Chen, Q. Zhang, and S. Li (2011) Prediction-based throughput optimization for dynamic spectrum access, *IEEE Transactions on Vehicular Technology* **60**(3): 1284–9.

579. S. Thrun, W. Burgard, and D. Fox (2005) *Probabilistic Robotics-Intelligent Robotics and Autonomous Agents*. Cambridge, MA: MIT Press.

580 S. Haykin (2006) Cognitive radar, *IEEE Signal Processing Magazine* **23**(1): 30–40.

581. S. Miranda, C. Baker, K. Woodbridge, and H. Griffiths (2006) Knowledge-based resource management for multifunction radar: a look at scheduling and task prioritization, *IEEE Signal Processing Magazine* **23**(1): 66–76.

582. V. Krishnamurthy and D.V. Djonin (2009) Optimal threshold policies for multivariate POMDPs in radar resource management, *IEEE Transactions on Signal Processing* **57**(10): 3954–69.

583. T. Hanselmann, M. Morelande, B. Moran, and P. Sarunic (2008) Sensor scheduling for multiple target tracking and detection using passive measurements, in *11th International Conference on Information Fusion*, IEEE, pp. 1–8.

584. C. Rudin, D. Waltz, R. Anderson, *et al.* (2012) Machine learning for the New York City power grid, *IEEE Transactions on Pattern Analysis and Machine Intelligence* **2**: 328–45.

585. O. Wolfson (2007) Trends in computational transportation science, in *Workshop: Advanced Geographic Information Systems in Transportation*.

586. S. Winter, M. Sester, O. Wolfson, and G. Geers (2011) Towards a computational transportation science, *ACM SIGMOD Record* **39**(3): 27–32.

587. B. Ran and D.E. Boyce (1996) *Modeling Dynamic Transportation Networks: An Intelligent Transportation System Oriented Approach*. Berlin: Springer Verlag.

588. Y. Zhao and A. House (1997) *Vehicle Location and Navigation Systems: Intelligent Transportation Systems*, Artech House, pp. 221–4.

589. W. Barfield and T.A. Dingus (1998) *Human Factors in Intelligent Transportation Systems*. Lawrence Erlbaum.

590. Y. Zhao (2000) Mobile phone location determination and its impact on intelligent transportation systems, *IEEE Transactions on Intelligent Transportation Systems* **1**(1): 55–64.

591. L. Figueiredo, I. Jesus, J.A.T. Machado, J.R. Ferreira, and J.L. Martins de Carvalho (2001) Towards the development of intelligent transportation systems, in *Proceedings on Intelligent Transportation Systems*, IEEE, pp. 1206–11.

592. H.X. Ge, S.Q. Dai, L.Y. Dong, and Y. Xue (2004) Stabilization effect of traffic flow in an extended car-following model based on an intelligent transportation system application, *Physical Review E* **70**(6).

593. L.J. Guan (2007) Trip quality in peer-to-peer shared ride systems, Master's thesis, Department of Geomatics, the University of Melbourne.

594. D.J. Hand, H. Mannila, and P. Smyth (2001) *Principles of Data Mining*. Cambridge, MA: The MIT Press.

595. T. Hastie, R. Tibshirani, J. Friedman, and J. Franklin (2005) The elements of statistical learning: Data mining, inference and prediction, *The Mathematical Intelligencer* **27**(2): 83–5.

596. P.N. Tan, M. Steinbach, V. Kumar, *et al.* (2006) *Introduction to Data Mining*. Boston: Pearson Addison Wesley.

597. J. Han, M. Kamber, and J. Pei (2011) *Data Mining: Concepts and Techniques*. Morgan Kaufmann.

598. Wikipedia, Data mining–Wikipedia, The Free Encyclopedia. http://en.wikipedia.org/wiki/Data mining.

599. Wikipedia, Computer vision–Wikipedia, The Free Encyclopedia. http://en.wikipedia.org/wiki/Computer vision.

600. G. Bradski and A. Kaehler (2008) *Learning OpenCV: Computer Vision with the OpenCV Library*. O'Reilly Media,

601. Wikipedia, Robot–Wikipedia, The Free Encyclopedia. http://en.wikipedia.org/wiki/Robot.

602. Wikipedia, Web search engine–Wikipedia, The Free Encyclopedia. http://en.wikipedia.org/wiki/Web search engine.

603. Wikipedia, Human computer interaction–Wikipedia, The Free Encyclopedia. http://en.wikipedia.org/wiki/Human Computer Interaction.

604. D. Jensen and J. Neville (2003) Data mining in social networks, *in Dynamic Social Network Modeling and Analysis: Workshop Summary and Papers*, pp. 287–302.

605. S.P. Borgatti and R. Cross (2003) A relational view of information seeking and learning in social networks, Management *Science* 49(4): 432–45.

606. S. Hill, F. Provost, and C. Volinsky (2007) Learning and inference in massive social networks, in *The 5th International Workshop on Mining and Learning with Graphs*, CeDER-PP-2007-05.

607. J.M. Kleinberg (2007) Challenges in mining social network data: Processes, privacy, and paradoxes, in *Proceedings of the 13th ACM SIGKDD International Conference on Knowledge Discovery and Data Mining*, ACM, pp. 4–5.

608. J.T. Pfingsten (2007) Machine Learning for Mass Production and Industrial Engineering. PhD thesis, Universitätsbibliothek Tübingen.

609. T. Yuan and W. Kuo (2007) A model-based clustering approach to the recognition of the spatial defect patterns produced during semiconductor fabrication, *IIE Transactions* 40(2): 93–101.

610. T. Yuan and W. Kuo (2008) Spatial defect pattern recognition on semiconductor wafers using model-based clustering and Bayesian inference, *European Journal of Operational Research* 190(1): 228–40.

611. T. Yuan, S.J. Bae, and J.I. Park (2010) Bayesian spatial defect pattern recognition in semiconductor fabrication using support vector clustering, *The International Journal of Advanced Manufacturing Technology* 51(5): 671–83.

612. T. Yuan, W. Kuo, and S.J. Bae (2011) Detection of spatial defect patterns generated in semiconductor fabrication processes, *IEEE Transactions on Semiconductor Manufacturing* 24(3): 392–403.

613. Wikipedia, Biological engineering–Wikipedia, The Free Encyclopedia. http://en.wikipedia.org/wiki/Bioengineering.

614. H.B. Barlow (1989) Unsupervised learning, *Neural Computation* 1(3): 295–311.

615. M. Weber, M. Welling, and P. Perona (2000) Unsupervised learning of models for recognition, *Computer Vision-ECCV*, pp. 18–32.

616. Wikipedia, Unsupervised learning–Wikipedia, The Free Encyclopedia. http://en.wikipedia.org/wiki/Unsupervised learning.

617. Wikipedia, Cluster analysis–Wikipedia, The Free Encyclopedia. http://en.wikipedia.org/wiki/Data clustering.

618. J.A. Hartigan and M.A. Wong (1979) Algorithm AS 136: A k-means clustering algorithm, *Journal of the Royal Statistical Society. Series C (Applied Statistics)* 28(1): 100–8.

619. K. Alsabti, S. Ranka, and V. Singh (1998) An efficient k-means clustering algorithm, in *1st Workshop on High-Performance Data Mining,* Citeseer.

620. K. Wagstaff, C. Cardie, S. Rogers, and S. Schrödl (2001) Constrained k-means clustering with background knowledge, in *Proceedings of the Eighteenth International Conference on Machine Learning* 577, Citeseer.

621. T. Kanungo, D.M. Mount, N.S. Netanyahu, C.D. Piatko, R. Silverman, and A.Y. Wu (2002) A local search approximation algorithm for k-means clustering, in *Proceedings of the Eighteenth Annual Symposium on Computational Geometry*, ACM, pp. 10–18.

622. N. Roussopoulos, S. Kelley, and F. Vincent (1995) Nearest neighbor queries, in *Proceedings of the 1995 ACM SIGMOD International Conference on Management of Data*, ACM.

623. T. Seidl and H.P. Kriegel (1998) Optimal multi-step k-nearest neighbor search, in *Proceedings of the 1998 ACM SIGMOD International Conference on Management of Data*, ACM.

624. Wikipedia, Blind signal separation–Wikipedia, The Free Encyclopedia. http://en.wikipedia.org/wiki/Blind signal separation.

625. S. Wold, K. Esbensen, and P. Geladi (1987) Principal component analysis, *Chemometrics and Intelligent Laboratory Systems* 2: 37–52.

626. I. Jolliffe (2002) *Principal Component Analysis*. Berlin: Springer Verlag.
627. G.H. Golub and C. Reinsch (1970) Singular value decomposition and least squares solutions, *Numerische Mathematik* **14**(5): 403–20.
628. P. Comon (1994) Independent component analysis, a new concept?, *Signal Processing* **36**(3): 287–314.
629. A. Hyvärinen and E. Oja (2000) Independent component analysis: Algorithms and applications, *Neural Networks* **13**(4–5): 411–30.
630. A. Hyvärinen, J. Karhunen, and E. Oja (2001) *Independent Component Analysis*, vol. 26. Wiley-Interscience.
631. D.D. Lee and H.S. Seung (1999) Learning the parts of objects by non-negative matrix factorization, *Nature* **401**(6755): 788–91.
632. D.D. Lee and H.S. Seung (2001) Algorithms for non-negative matrix factorization, *Advances in Neural Information Processing Systems,* **13**: 556–62.
633. A. Cichocki, R. Zdunek, and A.H. Phan (2009) *Nonnegative Matrix and Tensor Factorizations: Applications to exploratory Multi-way Data Analysis and Blind Source Separation*. New York: John Wiley & Sons, Inc.
634. T.C. Clancy, A. Khawar, and T.R. Newman (2011) Robust signal classification using unsupervised learning, *IEEE Transactions on Wireless Communications* **10**(4): 1289–99.
635. Wikipedia, k-means clustering–Wikipedia, The Free Encyclopedia. http://en.wikipedia.org/wiki/K-means clustering.
636. M.W. Aslam, Z. Zhu, and A.K. Nandi (2010) Automatic digital modulation classification using genetic programming with k-nearest neighbor, in *Military Communications Conference, Milcom'10*, IEEE, pp. 1731–6.
637. Wikipedia, Principal component analysis–Wikipedia, The Free Encyclopedia. http://en.wikipedia.org/wiki/Principal component analysis.
638. P. Georgiev, F. Theis, A. Cichocki, and H. Bakardjian (2007) Sparse component analysis: A new tool for data mining, *Data Mining in Biomedicine*, 91–116.
639. E. Candes, X. Li, Y. Ma, and J. Wright (2009) Robust principal component analysis, preprint.
640. Z. Zhou, X. Li, J. Wright, E. Candes, and Y. Ma (2010) Stable principal component pursuit, in *International Symposium on Information Theory* (Austin, TX).
641. B. Recht, M. Fazel, and P.A. Parrilo (2010) Guaranteed minimum-rank solutions of linear matrix equations via nuclear norm minimization, *SIAM Review* **52**: 471–501.
642. B. Recht, W. Xu, and B. Hassibi (2009) Necessary and sufficient conditions for success of the nuclear norm heuristic for rank minimization, in *47th IEEE Conference on Decision and Control, CDC 2008* (Cancun, Mexico), pp. 3065–70.
643. E.J. Candes and T. Tao (2010) The power of convex relaxation: Near-optimal matrix completion, *IEEE Transactions on Information Theory* **56**(5): 2053–80.
644. Wikipedia, Independent component analysis–Wikipedia, The Free Encyclopedia. http://en.wikipedia.org/wiki/Independent component analysis.
645. J.L. Abell, J. Lee, Q. Zhao, H. Szu, and Y. Zhao (2012) Differentiating intrinsic SERS spectra from a mixture by sampling induced composition gradient and independent component analysis, *Analyst*.
646. Wikipedia, Non-negative matrix factorization–Wikipedia, The Free Encyclopedia. http://en.wikipedia.org/wiki/Nonnegative matrix factorization.
647. P.O. Hoyer (2004) Non-negative matrix factorization with sparseness constraints, *The Journal of Machine Learning Research* **5**: 1457–69.
648. A. Shashua and T. Hazan (2005) Non-negative tensor factorization with applications to statistics and computer vision, in *Proceedings of the 22nd International Conference on Machine Learning*, ACM, pp. 792–9.
649. A. Cichocki and R. Zdunek (2007) Regularized alternating least squares algorithms for non-negative matrix/tensor factorization, *Advances in Neural Networks–ISNN 2007*, pp. 793–802.
650. I. Dhillon and S. Sra (2006) Generalized nonnegative matrix approximations with bregman divergences, *Advances in Neural Information Processing Systems* **18**: 283–90.
651. L. Zhang, Z. Chen, M. Zheng, and X. He (2011) Robust non-negative matrix factorization, *Frontiers of Electrical and Electronic Engineering in China*, pp. 1–9.
652. N. Halko, P.G. Martinsson, and J.A. Tropp (2011) Finding structure with randomness: Probabilistic algorithms for constructing approximate matrix decompositions, *SIAM Review* **53**(2): 217–88.
653. T. Kohonen (1990) The self-organizing map, *Proceedings of the IEEE* **78**(9): 1464–80.

654. T. Kohonen, E. Oja, O. Simula, A. Visa, and J. Kangas (1996) Engineering applications of the self-organizing map, *Proceedings of the IEEE* **84**(10): 1358–84.

655. J. Vesanto and E. Alhoniemi (2000) Clustering of the self-organizing map, *IEEE Transactions on Neural Networks* **11**(3): 586–600.

656. S.C. Ahalt, A.K. Krishnamurthy, P. Chen, and D.E. Melton (1990) Competitive learning algorithms for vector quantization, *Neural Networks* **3**(3): 277–90.

657. R. Gray (1984) Vector quantization, *ASSP Magazine, IEEE* **1**(2): 4–29.

658. N. Farvardin (1990) A study of vector quantization for noisy channels, *IEEE Transactions on Information Theory* **36**(4): 799–809.

659. A. Gersho and R.M. Gray (1992) *Vector Quantization and Signal Compression,* vol. 159. Springer Netherlands.

660. Wikipedia, Supervised learning–Wikipedia, The Free Encyclopedia. http://en.wikipedia.org/wiki/Supervised learning.

661. D.C. Montgomery, E.A. Peck, G.G. Vining, and J. Vining (2001) *Introduction to Linear Regression Analysis,* vol. 3. New York: John Wiley & Sons, Inc.

662. R.A. Johnson and D.W. Wichern (2002) *Applied Multivariate Statistical Analysis,* vol. 4, Upper Saddle River, NJ: Prentice Hall.

663. M.H. Kutner, C. Nachtsheim, and J. Neter (2004) *Applied Linear Regression Models.* New York: McGraw-Hill.

664. S. Weisberg (2005) *Applied Linear Regression,* vol. 528. New York: John Wiley & Sons, Inc.

665. D.W. Hosmer and S. Lemeshow (2000) *Applied Logistic Regression,* vol. 354. Wiley-Interscience.

666. S.W. Menard (2002) *Applied Logistic Regression Analysis,* vol. 106. Sage Publications, Inc,

667. M.T. Hagan, H.B. Demuth, M.H. Beale, *et al.* (1996) *Neural Network Design.* Boston, MA: PWS.

668. R.J. Schalkoff (1997) *Artificial Neural Networks.* New York: McGraw-Hill.

669. T. Mitchell (1997) Decision tree learning, *Machine Learning* **414**.

670. L. Breiman (2001) Random forests, *Machine Learning* **45**(1): 5–32.

671. I. Rish (2001) An empirical study of the naive Bayes classifier, in *IJCAI 2001 Workshop on Empirical Methods in Artificial Intelligence*, pp. 41–6.

672. M.A. Hearst, S.T. Dumais, E. Osman, J. Platt, and B. Scholkopf (1998) Support vector machines, *Intelligent Systems and their Applications,* IEEE, **13**(4): 18–28.

673. T. Joachims (1998) Text categorization with support vector machines: Learning with many relevant features, *Machine Learning: ECML-98,* pp. 137–42.

674. I. Steinwart and A. Christmann (2008) *Support Vector Machines.* Berlin: Springer Verlag.

675. Wikipedia, Logistic regression–Wikipedia, The Free Encyclopedia. http://en.wikipedia.org/wiki/Logistic regression.

676. W.L. Martinez and A.R. Martinez (2002) *Computational Statistics Handbook with MATLAB.* CRC Press,

677. Wikipedia, Artificial neural network–Wikipedia, The Free Encyclopedia. http://en.wikipedia.org/wiki/Artificial neural network.

678. Wikipedia, Decision tree–Wikipedia, The Free Encyclopedia. http://en.wikipedia.org/wiki/Decision tree.

679. Wikipedia, Decision tree learning–Wikipedia, The Free Encyclopedia. http://en.wikipedia.org/wiki/Decision tree learning.

680. Wikipedia, Random forest–Wikipedia, The Free Encyclopedia. http://en.wikipedia.org/wiki/Random forest.

681. Wikipedia, Naive Bayes classifier–Wikipedia, The Free Encyclopedia. http://en.wikipedia.org/wiki/Naive Bayes classifier.

682. Wikipedia, Support vector machine–Wikipedia, The Free Encyclopedia. http://en.wikipedia.org/wiki/Support vector machine.

683. N. Cristianini and J. Shawe-Taylor (2000) *An Introduction to Support Vector Machines: and Other Kernel-based Learning Methods.* Cambridge: Cambridge University Press.

684. M. Hu, Y. Chen, and J.T.Y. Kwok, Building sparse multiple-kernel SVM classifiers, *IEEE Transactions on Neural Networks* **20**(5): 827–39.

685. J. Weston and C. Watkins (1998) Multi-class support vector machines, Technical Report CSD-TR-98-04, Department of Computer Science, Royal Holloway, University of London.

686. J. Weston and C. Watkins, Support vector machines for multi-class pattern recognition, in *Proceedings of the 7th European Symposium on Artificial Neural Networks*, pp. 219–24.

687. C.H.Q. Ding and I. Dubchak (2001) Multi-class protein fold recognition using support vector machines and neural networks, *Bioinformatics* **17**(4): 349–58.

688. C.W. Hsu and C.J. Lin (2002) A comparison of methods for multiclass support vector machines, *IEEE Transactions on Neural Networks* **13**(2): 415–25.

689. V. Franc and V. Hlaváč (2002) Multi-class support vector machine, in IEEE, *16th International Conference on Pattern Recognition*, vol. 2, pp. 236–9.

690. K. Crammer and Y. Singer (2002) On the algorithmic implementation of multiclass kernel-based vector machines, *The Journal of Machine Learning Research* **2**: 265–92.

691. S. Cheong, S.H. Oh, and S.Y. Lee (2004) Support vector machines with binary tree architecture for multi-class classification, *Neural Information Processing-Letters and Reviews* **2**(3): 47–51.

692. K.B. Duan and S. Keerthi (2005) Which is the best multiclass SVM method? An empirical study, *Multiple Classifier Systems,* pp. 732–60.

693. V. Vapnik, S.E. Golowich, and A. Smola (1996) Support vector method for function approximation, regression estimation, and signal processing, in *Advances in Neural Information Processing Systems* **9**, Citeseer, 281–7.

694. H. Drucker, C.J.C. Burges, L. Kaufman, A. Smola, and V. Vapnik (1997) Support vector regression machines, *Advances in Neural Information Processing Systems,* 155–61.

695. R. Collobert and S. Bengio (2001) SVMTorch: Support vector machines for large-scale regression problems, *The Journal of Machine Learning Research* **1**: 143–60.

696. C.C. Chuang, S.F. Su, J.T. Jeng, and C.C. Hsiao (2002) Robust support vector regression networks for function approximation with outliers, *IEEE Transactions on Neural Networks* **13**(6): 1322–30.

697. A. Smola and B. Schlkopf (2004) A tutorial on support vector regression, *Statistics and Computing* **14**(3): 199–222.

698. Y. Huang, H. Jiang, H. Hu, and Y. Yao (2009) Design of learning engine based on support vector machine in cognitive radio, in *International Conference on Computational Intelligence and Software Engineering,* IEEE.

699. Z. Yang, Y.D. Yao, S. Chen, H. He, and D. Zheng (2010) MAC protocol classification in a cognitive radio network, in *19th Annual Wireless and Optical Communications Conference (WOCC),* IEEE, pp. 1–5.

700. T. Hastie and R. Tibshirani (1998) Classification by pairwise coupling, *The Annals of Statistics* **26**(2): 451–71.

701. K. Crammer and Y. Singer (2002) On the learnability and design of output codes for multiclass problems, *Machine Learning* **47**(2): 201–33.

702. M.F. Balcan, A. Blum, and K. Yang (2005) Co-training and expansion: Towards bridging theory and practice, *Advances in Neural Information Processing Systems,* **17**: 89–96.

703. Wikipedia, Semi-supervised learning–Wikipedia, The Free Encyclopedia. http://en.wikipedia.org/wiki/Semi supervised learning.

704. O. Chapelle, B. Schölkopf, and A. Zien (2006) *Semi-supervised Learning,* vol. 2. Cambridge, MA: MIT Press.

705. N. Adams (2009) Semi-supervised learning, *Journal of the Royal Statistical Society: Series A (Statistics in Society)* **172**(2): 530.

706. N. Grira, M. Crucianu, and N. Boujemaa (2004) *Unsupervised and Semi-supervised Clustering: A Brief Survey: A Review of Machine Learning Techniques for Processing Multimedia Content*, Report of the MUSCLE European Network of Excellence (FP6).

707. S. Basu, I. Davidson, and K.L. Wagstaff (2008) *Constrained Clustering: Advances in Algorithms, Theory, and Applications*. Chapman & Hall/CRC.

708. E.P. Xing, A.Y. Ng, M.I. Jordan, and S. Russell (2002) Distance metric learning, with application to clustering with side-information, *Advances in Neural Information Processing Systems* **15**: 505–12.

709. S. Basu, A. Banerjee, and R. Mooney (2002) Semi-supervised clustering by seeding, in *Proceedings of 19th International Conference on Machine Learning*, pp. 19–26.

710. A. Blum and T. Mitchell (1998) Combining labeled and unlabeled data with co-training, in *Proceedings of the Eleventh Annual Conference on Computational Learning Theory*, ACM, pp. 92–100.

711. X. Zhu (2005) Semi-supervised learning with graphs. PhD thesis, Language Technologies Institute, School of Computer Science, Carnegie Mellon University.

712. V. Vapnik (2000) *The Nature of Statistical Learning Theory*. Berlin: Springer Verlag.

713. V. Vapnik (2006) Transductive inference and semi-supervised learning, *Semi-supervised Learning,* 454–72.

714. Wikipedia, Transduction (machine learning)–Wikipedia, The Free Encyclopedia. http://en.wikipedia.org/wiki/Transductive learning.
715. S.J. Pan and Q. Yang (2009) A survey on transfer learning. *IEEE Transactions on Knowledge and Data Engineering*, 1345–9.
716. Wikipedia, Inductive transfer–Wikipedia, The Free Encyclopedia. http://en.wikipedia.org/wiki/Transfer learning.
717. L. Mihalkova, T. Huynh, and R. J. Mooney (2007) Mapping and revising markov logic networks for transfer learning, in *Proceedings of the National Conference on Artificial Intelligence*, Menlo Park, CA; Cambridge, MA; London; AAAI Press; MIT Press.
718. A. Niculescu-Mizil and R. Caruana (2007) Inductive transfer for Bayesian network structure learning, in *Eleventh International Conference on Artificial Intelligence and Statistics (AISTATS-07)*, Citeseer.
719. Wikipedia, Multi-task learning–Wikipedia, The Free Encyclopedia. http://en.wikipedia.org/wiki/Multi-task learning.
720. B. Settles (1995) Active learning literature survey, *Science* 10(3): 237–304.
721. Wikipedia, Active learning (machine learning)–Wikipedia, The Free Encyclopedia. http://en.wikipedia.org/wiki/Active learning (machine learning).
722. L. Zhang, C. Chen, J. Bu, D. Cai, X. He, and T.S. Huang, Active learning based on locally linear reconstruction, *IEEE Transactions on Pattern Analysis and Machine Intelligence* 33(10): 2026–38.
723. L. Kaelbling, M. Littman, and A. Moore (1996) Reinforcement learning: A survey, Arxiv preprint cs/9605103.
724. R.S. Sutton and A.G. Barto (1998) *Reinforcement Learning: An Introduction,* vol. 116. Cambridge: Cambridge University Press.
725. Wikipedia, Reinforcement learning–Wikipedia, The Free Encyclopedia. http://en.wikipedia.org/wiki/Reinforcement learning.
726. R.A. Howard (1962) *Dynamic Programming and Markov Processes*. John Wiley & Sons, Inc. for The Massachusetts Institute of Technology.
727. D.P. Bertsekas (1976) *Dynamic Programming and Stochastic Control,* vol. 125. New York: Academic Press.
728. M. Sewell (2004) Exploration and exploitation in reinforcement learning. Technical report, McGill University.
729. Y. Achbany, F. Fouss, L. Yen, A. Pirotte, and M. Saerens (2005) Managing the exploration/exploitation trade-off in reinforcement learning. Technical report, Information Systems Research Unit, Université Catholique de Louvain.
730. A. Alsarhan and A. Agarwal (2009) Spectrum sharing in multi-service cognitive network using reinforcement learning, in *1st UK-India International Workshop on Cognitive Wireless Systems (UKIWCWS)*, IEEE, pp. 1–5.
731. T. Jiang, D. Grace, and P.D. Mitchell (2009) Improvement of pre-partitioning on reinforcement learning based spectrum sharing, in *IET International Communication Conference on Wireless Mobile and Computing (CCWMC)*, pp. 299–302.
732. M. Yang and D. Grace (2009) Cognitive radio with reinforcement learning applied to heterogeneous multicast terrestrial communication systems, in *4th International Conference on Cognitive Radio Oriented Wireless Networks and Communications (CROWNCOM'09)*, IEEE, pp. 1–6.
733. F. Bernardo, R. Agustí, J. Pérez-Romero, and O. Sallent (2009) Distributed spectrum management based on reinforcement learning, in *4th International Conference on Cognitive Radio Oriented Wireless Networks and Communications (CROWNCOM'09)*, IEEE, pp. 1–6.
734. A. Alsarhan and A. Agarwal (2010) Resource adaptations for revenue optimization in cognitive mesh network using reinforcement learning, in *Globecom Workshops,* IEEE, pp. 1108–12.
735. J. Oksanen, J. Lundén, and V. Koivunen (2010) Reinforcement learning method for energy efficient cooperative multiband spectrum sensing, in *International Workshop on Machine Learning for Signal Processing (MLSP)*, pp. 59–64.
736. B.F. Lo and I.F. Akyildiz (2010) Reinforcement learning-based cooperative sensing in cognitive radio ad hoc networks, in *21st International Symposium on Personal Indoor and Mobile Radio Communications (PIMRC),* IEEE pp. 2244–9.
737. P. Venkatraman, B. Hamdaoui, and M. Guizani (2010) Opportunistic bandwidth sharing through reinforcement learning, *IEEE Transactions on Vehicular Technology* 59(6): 3148–53.

738. T. Jiang, D. Grace, and Y. Liu (2011) Two-stage reinforcement-learning-based cognitive radio with exploration control, *Communications*, IET **5**(5): 644–51.

739. K.L.A. Yau, P. Komisarczuk, and P.D. Teal (2010) Enhancing network performance in distributed cognitive radio networks using single-agent and multi-agent reinforcement learning, in *35th Conference on Local Computer Networks (LCN)*, IEEE, pp. 152–9.

740. M. Di Felice, K. Chowdhury, C. Wu, L. Bononi, and W. Meleis (2010) Learning-based spectrum selection in cognitive radio ad hoc networks, *Wired/Wireless Internet Communications*, pp. 133–45.

741. N. Vucevic, I.F. Akyildiz, and J. Pérez-Romero (2010) Cooperation reliability based on reinforcement learning for cognitive radio networks, in *5th IEEE Workshop on Networking Technologies for Software Defined Radio (SDR) Networks*, IEEE, pp. 1–6.

742. C. Wu, K. Chowdhury, M. Di Felice, and W. Meleis (2010) Spectrum management of cognitive radio using multi-agent reinforcement learning, in *Proceedings of the 9th International Conference on Autonomous Agents and Multiagent Systems: Industry Track*, International Foundation for Autonomous Agents and Multiagent Systems, pp. 1705–12.

743. M. Di Felice, K.R. Chowdhury, W. Meleis, and L. Bononi (2010) To sense or to transmit: A learning-based spectrum management scheme for cognitive radio mesh networks, in *5th IEEE Workshop on Wireless Mesh Networks (WIMESH)*, pp. 1–6

744. J. Lunden, V. Koivunen, S.R. Kulkarni, and H.V. Poor (2011) Reinforcement learning based distributed, multiagent sensing policy for cognitive radio networks, in *IEEE Symposium on New Frontiers in Dynamic Spectrum Access Networks (DySPAN)*, IEEE, pp. 642–6.

745. T. Jiang, D. Grace, and P.D. Mitchell (2011) Efficient exploration in reinforcement learning-based cognitive radio spectrum sharing, *IET Communications* **5**: 1309–17.

746. W. Jouini, R. Bollenbach, M. Guillet, C. Moy, and A. Nafkha (2011) Reinforcement learning application scenario for opportunistic spectrum access, in *54th International Midwest Symposium on Circuits and Systems (MWSCAS)*, IEEE, pp. 1–4.

747. C.J.C.H. Watkins and P. Dayan (1992) Q-learning, *Machine Learning* **8**(3): 279–92.

748. Wikipedia, Q-learning–Wikipedia, The Free Encyclopedia. http://en.wikipedia.org/wiki/Q learning.

749. R. Farha, N. Abji, O. Sheikh, and A. Leon-Garcia, Market-based resource management for cognitive radios using machine learning, in *Global Telecommunications Conference, Globecom'07*, IEEE, pp. 4630–5.

750. N. Hosey, S. Bergin, I. Macaluso, and D. O'Donohue (2009) Q-learning for cognitive radios, in *Proceedings of the China-Ireland Information and Communications Technology Conference (CIICT 2009)*. National University of Ireland, Maynooth.

751. M. Li, Y. Xu, and J. Hu (2009) A Q-learning based sensing task selection scheme for cognitive radio networks, in *International Conference on Wireless Communications & Signal Processing*. IEEE, pp. 1–5.

752. H. Li (2009) Multi-agent Q-learning of channel selection in multi-user cognitive radio systems: A two by two case, in IEEE *International Conference on Systems, Man and Cybernetics*, IEEE, pp. 1893–8.

753. A. Galindo-Serrano and L. Giupponi (2009) Aggregated interference control for cognitive radio networks based on multiagent learning, in *4th International Conference on Cognitive Radio Oriented Wireless Networks and Communications*, CROWNCOM'09, IEEE, pp. 1–6.

754. K.L.A. Yau, P. Komisarczuk, and P.D. Teal (2009) Performance analysis of reinforcement learning for achieving context awareness and intelligence in cognitive radio networks, in *34th Conference on Local Computer Networks*, IEEE, pp. 1046–53.

755. B. Xia, M.H. Wahab, Y. Yang, Z. Fan, and M. Sooriyabandara (2009) Reinforcement learning based spectrum-aware routing in multi-hop cognitive radio networks, in *4th International Conference on Cognitive Radio Oriented Wireless Networks and Communications, CROWNCOM'09*, IEEE, pp. 1–5.

756. Y. Yao and Z. Feng (2010) Centralized channel and power allocation for cognitive radio networks: A Q-learning solution, in *Proceedings of Future Network and Mobile Summit 2010 Conference*, International Information Management Corporation, pp. 1–8.

757. A. Galindo-Serrano and L. Giupponi (2010) Decentralized Q-learning for aggregated interference control in completely and partially observable cognitive radio networks, in *7th IEEE Consumer Communications and Networking Conference (CCNC)*, IEEE, pp. 1–6.

758. A. Galindo-Serrano and L. Giupponi (2010) Distributed Q-learning for aggregated interference control in cognitive radio networks, *IEEE Transactions on Vehicular Technology* **59**(4): 1823–34.

759. X. Chen, Z. Zhao, and H. Zhang (2010) Green transmit power assignment for cognitive radio networks by applying multiagent Q-learning approach, in *European Wireless Technology Conference (EuWIT)*, IEEE, pp. 113–16.

760. L. Saker, S. Ben Jemaa, and S. E. Elayoubi (2010) An analytical framework for modeling distributed JRRM decision in cognitive networks, in *21st International Symposium on Personal Indoor and Mobile Radio Communications (PIMRC)*, IEEE, pp. 1654–9.

761. Y. Ren, P. Dmochowski, and P. Komisarczuk (2010) Analysis and implementation of reinforcement learning on a GNU radio cognitive radio platform, in *Proceedings of the 5th International Conference on Cognitive Radio Oriented Wireless Networks & Communications (CROWNCOM)*, IEEE, pp. 1–6.

762. P. Venkatraman and B. Hamdaoui (2011) Cooperative Q-learning for multiple secondary users in dynamic spectrum access, in *7th International Wireless Communications and Mobile Computing Conference (IWCMC)*, IEEE, pp. 238–42.

763. Wikipedia, Markov decision process–Wikipedia, The Free Encyclopedia. http://en.wikipedia.org/wiki/Markov decision process.

764. E.V. Denardo (1970) On linear programming in a Markov decision problem, *Management Science*, 281–8.

765. C. Derman (1970) *Finite State Markovian Decision Processes*. New York: Academic Press, Inc.

766. D. Bello and G. Riano (2006) Linear programming solvers for Markov decision processes, in *Systems and Information Engineering Design Symposium*, IEEE, pp. 90–5.

767. P.S. Castro and D. Precup (2007) Using linear programming for Bayesian exploration in Markov decision processes, in *Proceedings of the 20th International Joint Conference on Artificial Intelligence (IJCAI)*, pp. 2437–42.

768. W.A. Massey, K.G. Ramakrisluian, M. Aravamudan, and G. Pai (2004) Scheduling algorithms for downlink services in wireless networks: A Markov decision process approach, in *Global Telecommunications Conference, Globecom'04*, vol. 6, IEEE, pp. 4038–42.

769. D.V. Djonin, Q. Zhao, and V. Krishnamurthy (2007) Optimality and complexity of opportunistic spectrum access: A truncated Markov decision process formulation, in *International Conference on Communications, ICC'07*, IEEE, pp. 5787–92.

770. S. Yin and S. Li (2008) Optimization for time consumption on channel searching in cognitive radio system with Markov decision processes, in *4th International Conference on Wireless Communications, Networking and Mobile Computing, WiCOM'08*, IEEE, pp. 1–6.

771. S. Geirhofer, L. Tong, and B.M. Sadler (2008) Cognitive medium access: Constraining interference based on experimental models, *IEEE Journal on Selected Areas in Communications* **26**(1): 95–105.

772. R.M. Amini and Z. Dziong (2010) A framework for routing and channel allocation in cognitive wireless mesh networks, in *7th International Symposium on Wireless Communication Systems (ISWCS)*, IEEE, pp. 1017–21.

773. Y. Wu, B. Wang, and K.J.R. Liu (2010) Optimal defense against jamming attacks in cognitive radio networks using the Markov decision process approach, in *Global Telecommunications Conference, Globecom'10*, IEEE, pp. 1–5.

774. F. Wang, J. Zhu, J. Huang, and Y. Zhao (2010) Admission control and channel allocation for supporting real-time applications in cognitive radio networks, in *Global Telecommunications Conference, Globecom'10*, IEEE, pp. 1–6.

775. D. Qiang, H. Bo, S. Yan, and C. Shan-zhi (2010) An SMDP-based optimal admission control scheme in cognitive radio networks, in *6th International Conference on Wireless Communications Networking and Mobile Computing (WiCOM)*, IEEE, pp. 1–4.

776. P. Si, Y. Zhang, R. Yang, and J. Zhang (2011) Resource allocation policy of primary users in proactively-optimization cognitive radio networks, in *3rd International Conference on Computational Intelligence, Communication Systems and Networks (CICSyN)*, IEEE, pp. 343–7.

777. D. Niyato, E. Hossain, and P. Wang (2011) Optimal channel access management with QoS support for cognitive vehicular networks, *IEEE Transactions on Mobile Computing* **10**(4): 573–91.

778. X. Li, Y. Hei, and G. Yu (2011) A delay-optimal spectrum access policy for cognitive coexistence systems, in *Proceedings of 20th International Conference on Computer Communications and Networks (ICCCN)*, IEEE, pp. 1–4.

779. R. Chen and X. Liu (2011) Delay performance of threshold policies for dynamic spectrum access, *IEEE Transactions on Wireless Communications* **10**: 2283–93.

780. Y. Chen and K.J. Liu (2011) Indirect reciprocity game modelling for cooperation stimulation in cognitive networks, *IEEE Transactions on Communications* **59**(1): 159–68.
781. S. Chen and L. Tong (2011) Maximum throughput region of multiuser cognitive access of continuous time Markovian channels, *IEEE Journal on Selected Areas in Communications* **29**(10): 1959–69.
782. E. Anifantis, V. Karyotis, and S. Papavassiliou (2011) Time-based cross-layer adaptations in wireless cognitive radio ad hoc networks, in *Symposium on Computers and Communications (ISCC)*, IEEE, pp. 1044–9.
783. Wikipedia, Partially observable Markov decision process–Wikipedia, The Free Encyclopedia. http://en.wikipedia.org/wiki/POMDP. [286]
784. T.W. Yu (2002) Partially observable Markov decision processes. Presentation slides.
785. K.P. Murphy (2000) A survey of POMDP solution techniques, *Environment* **2**, 1–12.
786. E. Sondik (1971) The optimal control of partially observable Markov processes. PhD thesis, Stanford University.
787. A. Cassandra (1998) Exact and approximate algorithms for partially observable Markov decision processes. PhD thesis, Brown University.
788. M. Spaan and F. Oliehoek (2008) The multi-agent decision process toolbox: Software for decision-theoretic planning in multiagent systems, in *Proceedings of MSDM2008*, pp. 107–21.
789. T. Smith (2007) Probabilistic planning for robotic exploration. PhD thesis, Carnegie Mellon University.
790. H. Kurniawati, D. Hsu, and W.S. Lee (2008) SARSOP: Efficient point-based POMDP planning by approximating optimally reachable belief spaces, in *Proceedings of Robotics: Science and Systems*, pp. 1–8.
791. M. Spaan and N. Vlassis (2005) Perseus: Randomized point-based value iteration for POMDPs, *Journal of Artificial Intelligence Research* **24**: 195–220.
792. W. Zhao, L. Tong, A. Swami, and Y. Chen (2007) Decentralized cognitive MAC for opportunistic spectrum access in ad hoc networks: A POMDP framework, *IEEE Journal on Selected Areas in Communications* **25**(3): 589–600.
793. Q. Zhao, B. Krishnamachari, and K. Liu (2007) Low-complexity approaches to spectrum opportunity tracking, in *2nd International Conference on Cognitive Radio Oriented Wireless Networks and Communications, CrownCom 2007*, IEEE, pp. 27–35.
794. Y. Chen, Q. Zhao, and A. Swami (2008) Joint design and separation principle for opportunistic spectrum access in the presence of sensing errors, *IEEE Transactions on Information Theory* **54**(5): 2053–71.
795. T. Javidi, B. Krishnamachari, Q. Zhao, and M. Liu (2008) Optimality of myopic sensing in multi-channel opportunistic access, in *International Conference on Communications*, ICC'08, IEEE, pp. 2107–12.
796. J. Unnikrishnan and V. V. Veeravalli (2008) Dynamic spectrum access with learning for cognitive radio, in *42nd Asilomar Conference on Signals, Systems and Computers*, IEEE, pp. 103–7.
797. S. Huang, X. Liu, and Z. Ding (2008) Opportunistic spectrum access in cognitive radio networks, in *27th Conference on Computer Communications, infocom 2008*, IEEE, pp. 1427–35.
798. Y. Chen, Q. Zhao, and A. Swami (2009) Distributed spectrum sensing and access in cognitive radio networks with energy constraint, *IEEE Transactions on Signal Processing* **57**(2): 783–97.
799. R.T. Ma, Y.P. Hsu, and K.T. Feng (2009) A pomdp-based spectrum handoff protocol for partially observable cognitive radio networks, in *Wireless Communications and Networking Conference, WCNC 2009*. IEEE, pp. 1–6.
800. H. Li and Z. Han (2009) Dogfight in spectrum: Jamming and anti-jamming in multichannel cognitive radio systems, in *Global Telecommunications Conference, Globecom'09*, IEEE, pp. 1–6.
801. A.T. Hoang, Y.C. Liang, D. Wong, Y. Zeng, and R. Zhang (2009) Opportunistic spectrum access for energy-constrained cognitive radios, *IEEE Transactions on Wireless Communications* **8**(3): 1206–11.
802. S.H.A. Ahmad, M. Liu, T. Javidi, Q. Zhao, and B. Krishnamachari (2009) Optimality of myopic sensing in multichannel opportunistic access, *IEEE Transactions on Information Theory* **55**(9): 4040–50.
803. T. Zhang, Y. Wu, K. Lang, and D.H.K. Tsang (2010) Optimal scheduling of cooperative spectrum sensing in cognitive radio networks, *IEEE Systems Journal* **4**(4): 535–49.
804. J. Unnikrishnan and V.V. Veeravalli (2010) Algorithms for dynamic spectrum access with learning for cognitive radio, *IEEE Transactions on Signal Processing* **58**(2): 750–60.
805. S. Gong, P. Wang, W. Liu, and W. Yuan (2010) Maximize secondary user throughput via optimal sensing in multi-channel cognitive radio networks, in *Global Telecommunications Conference, Globecom'10*, IEEE, pp. 1–5.

806. K. Choi (2010) Adaptive sensing technique to maximize spectrum utilization in cognitive radio, *IEEE Transactions on Vehicular Technology* **59**(2): 992–8.

807. C. Luo, F.R. Yu, H. Ji, and V.C.M. Leung (2010) Cross-layer design for TCP performance improvement in cognitive radio networks, *IEEE Transactions on Vehicular Technology* **59**(5): 2485–95.

808. W. Wang, J. Cai, and A.S. Alfa (2010) Receiver-aided spectrum sensing scheme with spatial differentiation in OFDM based cognitive radio networks, in *Conference on Computer Communications Workshops, INFOCOM 2010,* IEEE, pp. 1–6.

809. S.K. Hsu, J.S. Lin, and K.T. Feng (2011) Stochastic multiple channel sensing protocol for cognitive radio networks, in *Wireless Communications and Networking Conference (WCNC),* IEEE, pp. 227–32.

810. M. Bkassiny, S.K. Jayaweera, and K.A. Avery (2011) Distributed reinforcement learning based mac protocols for autonomous cognitive secondary users, in *20th Annual Wireless and Optical Communications Conference (WOCC),* IEEE, pp. 1–6.

811. C. An and Y. Liu (2011) A matching game algorithm for spectrum allocation based on POMDP model, in *7th International Conference on Wireless Communications, Networking and Mobile Computing (WiCOM),* IEEE, pp. 1–3.

812. J.A. Han, W.S. Jeon, and D.G. Jeong (2011) Energy-efficient channel management scheme for cognitive radio sensor networks, *IEEE Transactions on Vehicular Technology* **60**(4): 1905–10.

813. K.W. Choi and E. Hossain (2011) Opportunistic access to spectrum holes between packet bursts: A learning-based approach, *IEEE Transactions on Wireless Communications* **10**(8): 2497–2509.

814. Y.L. Che, R. Zhang, and Y. Gong (2011) Opportunistic spectrum access for cognitive radio in the presence of reactive primary users, in *International Conference on Communications (ICC),* IEEE, pp. 1–5.

815. T. Zhang and D.H.K. Tsang (2011) Optimal cooperative sensing scheduling for energy-efficient cognitive radio networks, in *Proceedings, Infocom'11*, IEEE, pp. 2723–31.

816. S. Filippi, O. Cappé, and A. Garivier (2011) Optimally sensing a single channel without prior information: The tiling algorithm and regret bounds, *IEEE Journal of Selected Topics in Signal Processing* **5**(1): 68–76.

817. M. Bkassiny, S.K. Jayaweera, Y. Li, and K.A. Avery (2011) Optimal and low-complexity algorithms for dynamic spectrum access in centralized cognitive radio networks with fading channels, in *73rd Vehicular Technology Conference (VTC Spring),* IEEE, pp. 1–5.

818. Y. Li, S.K. Jayaweera, M. Bkassiny, and K.A. Avery (2011) Optimal myopic sensing and dynamic spectrum access in centralized secondary cognitive radio networks with low-complexity implementations, in *73rd Vehicular Technology Conference (VTC Spring),* IEEE, pp. 1–5.

819. K.R. Muller, S. Mika, G. Ratsch, K. Tsuda, and B. Scholkopf (2001) An introduction to kernel-based learning algorithms, *IEEE Transactions on Neural Networks* **12**(2): 181–201.

820. B. Schölkopf, A. Smola, and K. R. Müller (1997) Kernel principal component analysis, *Artificial Neural Networks ICANN'97,* pp. 583–8.

821. S. Mika, B. Scholkopf, A. Smola, K. Muller, M. Scholz, and G. Ratsch (1999) Kernel PCA and de-noising in feature spaces, *Advances in Neural Information Processing Systems* **11**(1): 536–42.

822. K.I. Kim, K. Jung, and H.J. Kim (2002) Face recognition using kernel principal component analysis, *Signal Processing Letters, IEEE* **9**(2): 40–2.

823. J.T.Y. Kwok and I.W.H. Tsang (2004) The pre-image problem in kernel methods, *IEEE Transactions on Neural Networks* **15**(6): 1517–25.

824. S. Mika, G. Ratsch, J. Weston, B. Scholkopf, and K.R. Mullers (1999) Fisher discriminant analysis with kernels, in *Proceedings of the 1999 IEEE Signal Processing Society Workshop, Neural Networks for Signal Processing IX*, pp. 41–8.

825. G. Baudat and F. Anouar (2000) Generalized discriminant analysis using a kernel approach, *Neural Computation* **12**(10): 2385–2404.

826. O. Bousquet and D.J.L. Herrmann (2003) On the complexity of learning the kernel matrix, in *Advances in Neural Information Processing Systems 15: Proceedings of the 2002 Conference*, vol. 15, Cambridge, MA: The MIT Press, pp. 415–22.

827. K.Q. Weinberger, F. Sha, and L.K. Saul (2004) Learning a kernel matrix for nonlinear dimensionality reduction, in *Proceedings of the 21st International Conference on Machine Learning*, ACM, pp. 106–13.

828. E. Hu, S. Chen, D. Zhang, and X. Yin (2010) Semisupervised kernel matrix learning by kernel propagation, *IEEE Transactions on Neural Networks* **21**(11): 1831–41.

829. B. Kulis, M.A. Sustik, and I.S. Dhillon (2009) Low-rank kernel learning with bregman matrix divergences, *Journal of Machine Learning Research* **10**: 341–76.

830. L. Wof and A. Shashua (2003) Kernel principal angles for classification machines with applications to image sequence interpretation, in *Proceedings Computer Society Conference on Computer Vision and Pattern Recognition*, IEEE, pp. 635–40.

831. L. Wolf and A. Shashua (2003) Learning over sets using kernel principal angles, *Journal of Machine Learning Research* **4**, pp. 913–31.

832. Z. Drmac (2000) On principal angles between subspaces of Euclidean space (2000) *SIAM Journal on Matrix Analysis and Applications* **22**(1): 173–94.

833. A.V. Knyazev and M.E. Argentati (2002) Principal angles between subspaces in an A-based scalar product: Algorithms and perturbation estimates, *SIAM Journal on Scientific Computing* **23**(6): 2008–40.

834. M.E. Argentati (2003) Principal angles between subspaces as related to Rayleigh quotient and Rayleigh Ritz inequalities with applications to eigenvalue accuracy and an eigenvalue solver. PhD thesis, University of Colorado at Denver.

835. P.A. Absil, A. Edelman, and P. Koev (2006) On the largest principal angle between random subspaces, *Linear Algebra and Its Applications* **414**(1): 288–94.

836. T. Gou and S.A. Jafar (2010) Degrees of freedom of the K user M × N MIMO interference channel, *IEEE Transactions on Information Theory* **56**(12): 6040–57.

837. M.A. Maddah-Ali, A.S. Motahari, and A.K. Khandani (2008) Communication over MIMO X channels: Interference alignment, decomposition, and performance analysis, *IEEE Transactions on Information Theory* **54**(8): 3457–70.

838. B. Nazer, S.A. Jafar, M. Gastpar, and S. Vishwanath (2009) Ergodic interference alignment, in *International Symposium on Information Theory, ISIT 2009 (Seoul, Korea)*, IEEE, pp. 1769–73.

839. C. Huang and S.A. Jafar (2009) Degrees of freedom of the MIMO interference channel with cooperation and cognition, *IEEE Transactions on Information Theory* **55**(9): 4211–20.

840. C.S. Vaze and V.M.K. (2010) The degrees of freedom region of the MIMO cognitive interference channel with no CSIT, in *ISIT, (Austin, Texas)*, pp. 440–4.

841. I.K. Fodor (2002) A survey of dimension reduction techniques, technical report, Center for Applied Scientific Computing, Lawrence Livermore National Laboratory.

842. Wikipedia, Dimension reduction–Wikipedia, The Free Encyclopedia. http://en.wikipedia.org/wiki/Dimension reduction.

843. J. Tenenbaum, V. Silva, and J. Langford (2000) A global geometric framework for nonlinear dimensionality reduction, *Science* **290**(5500): 2319.

844. M.H.C. Law and A.K. Jain (2006) Incremental nonlinear dimensionality reduction by manifold learning, *IEEE Transactions on Pattern Analysis and Machine Intelligence*, pp. 377–91.

845. T. Lin, H. Zha, and S. Lee (2006) Riemannian manifold learning for nonlinear dimensionality reduction, *Lecture Notes in Computer Science* **3951**: 44.

846. Z. Zhang, J. Wang, and H. Zha (2012) Adaptive manifold learning, *IEEE Transactions on Pattern Analysis and Machine Intelligence* **34**(2): 253–65.

847. L. Cayton (2005) Algorithms for manifold learning, technical report, University of California, San Diego.

848. I. Borg and P.J.F. Groenen (2005) *Modern Multidimensional Scaling: Theory and Applications*. Berlin: Springer Verlag.

849. A.M. Bronstein, M.M. Bronstein, and R. Kimmel (2006) Generalized multidimensional scaling: A framework for isometry invariant partial surface matching, *Proceedings of the National Academy of Sciences of the United States of America* **103**(5): 1168.

850. J. Venna and S. Kaski (2006) Local multidimensional scaling, *Neural Networks* **19**(6–7): 889–99.

851. M. Balasubramanian and E.L. Schwartz (2002) The isomap algorithm and topological stability, *Science* **295**(5552): 7.

852. O.C. Jenkins and M.J. Matarić (2004) A spatio-temporal extension to isomap nonlinear dimension reduction, in *Proceedings of the 21st International Conference on Machine Learning*, ACM, p. 56.

853. O. Samko, A.D. Marshall, and P.L. Rosin (2006) Selection of the optimal parameter value for the Isomap algorithm, *Pattern Recognition Letters* **27**(9): 968–79.

854. S. Roweis and L. Saul (2000) Nonlinear dimensionality reduction by locally linear embedding, *Science* **290**(5500): p. 2323.

855. D. Donoho and C. Grimes (2003) Hessian eigenmaps: Locally linear embedding techniques for high-dimensional data, *Proceedings of the National Academy of Sciences* **100**(10): 5591.

856. H. Chang and D.Y. Yeung (2006) Robust locally linear embedding, *Pattern Recognition* **39**(6): 1053–65.

857. M. Belkin and P. Niyogi (2002) Laplacian eigenmaps and spectral techniques for embedding and clustering, *Advances in Neural Information Processing Systems* **1**: 585–92.
858. M. Belkin and P. Niyogi (2003) Laplacian eigenmaps for dimensionality reduction and data representation, *Neural Computation* **15**(6): 1373–96.
859. R.R. Coifman and S. Lafon (2006) Diffusion maps, *Applied and Computational Harmonic Analysis* **21**(1): 5–30.
860. R.R. Coifman, I.G. Kevrekidis, S. Lafon, M. Maggioni, and B. Nadler (2008) Diffusion maps, reduction coordinates, and low dimensional representation of stochastic systems, *Multiscale Modelling Simulation* **7**(2): 842–64.
861. K. Weinberger and L. Saul (2006) Unsupervised learning of image manifolds by semidefinite programming, *International Journal of Computer Vision* **70**(1): 77–90.
862. K.Q. Weinberger and L.K. Saul (2006) An introduction to nonlinear dimensionality reduction by maximum variance unfolding, in *Proceedings of the National Conference on Artificial Intelligence*, Menlo Park, CA; Cambridge, MA; London; AAAI Press; ™
863. K.Q. Weinberger, F. Sha, Q. Zhu, and L.K. Saul (2007) Graph Laplacian regularization for large-scale semidefinite programming, *Advances in Neural Information Processing Systems,* vol. 19, 1489–96.
864. L. Song, A. Smola, K. Borgwardt, and A. Gretton (2008) Colored maximum variance unfolding, *Advances in Neural Information Processing Systems* **20**: 1385–92.
865. A.J. Smola and B. Schölkopf, *Learning with Kernels*. Citeseer.
866. C.K.I. Williams (2002) On a connection between kernel PCA and metric multidimensional scaling, *Machine Learning* **46**(1): 11–19.
867. Wikipedia, Multidimensional scaling–Wikipedia, The Free Encyclopedia. http://en.wikipedia.org/wiki/Multidimensional scaling.
868. L. Chen and A. Buja (2009) Local multidimensional scaling for nonlinear dimension reduction, graph drawing, and proximity analysis, *Journal of the American Statistical Association* **104**(485): 209–19.
869. Wikipedia, Isomap–Wikipedia, The Free Encyclopedia. http://en.wikipedia.org/wiki/Isomap.
870. K. Weinberger, B. Packer, and L. Saul (2005) Nonlinear dimensionality reduction by semidefinite programming and kernel matrix factorization, in *Proceedings of the 10th International Workshop on Artificial Intelligence and Statistics (Barbados)*, pp. 381–8.
871. Wikipedia, Ensemble learning–Wikipedia, The Free Encyclopedia. http://en.wikipedia.org/wiki/Ensemble learning.
872. M. Sewell (2011) Ensemble learning, technical report, Department of Computer Science, University College London, 2011.
873. A.K. Jain, R.P.W. Duin, and J. Mao (2000) Statistical pattern recognition: A review, *IEEE Transactions on Pattern Analysis and Machine Intelligence* **22**(1): 4–37.
874. T.G. Dietterich and G. Bakiri (1995) Solving multiclass learning problems via error-correcting output codes, *Journal of Artificial Intelligence Research* **2**: 263–86.
875. Wikipedia, Meta learning (computer science)–Wikipedia, The Free Encyclopedia. http://en.wikipedia.org/wiki/Meta learning (computer science).
876. C. Giraud-Carrier (2008) Metalearning-a tutorial, in *Tutorial at the 2008 International Conference on Machine Learning and Applications (ICMLA08)*, pp. 1–45.
877. W.R. Gilks, S. Richardson, and D.J. Spiegelhalter, Markov Chain Monte Carlo in Practice. Chapman & Hall/CRC, 1996. [296]
878. C. Andrieu, N. De Freitas, A. Doucet, and M.I. Jordan (2003) An introduction to MCMC for machine learning, Machine *Learning* **50**(1): 5–43.
879. Wikipedia, Markov chain Monte Carlo–Wikipedia, The Free Encyclopedia. http://en.wikipedia.org/wiki/Markov chain Monte Carlo.
880. P. Magni, R. Bellazzi, and G. De Nicolao (1998) Bayesian function learning using MCMC methods, *IEEE Transactions on Pattern Analysis and Machine Intelligence* **20**(12): 1319–31.
881. H. Zhang and S. Sheng (2004) Learning weighted naive Bayes with accurate ranking, in *4th IEEE International Conference on Data Mining, ICDM'04,* IEEE, pp. 567–70.
882. F. Bavencoff, J.M. Vanpeperstraete, and J.P. Le Cadre (2004) Constrained bearings-only target motion analysis via Monte Carlo Markov chains, in *Proceedings of the 2004 14th IEEE Signal Processing Society Workshop: Machine Learning for Signal Processing*, IEEE, pp. 153–62.
883. S. Bandyopadhyay (2005) Simulated annealing using a reversible jump Markov Chain Monte Carlo algorithm for fuzzy clustering, *IEEE Transactions on Knowledge and Data Engineering* **17**(4): 479–90.

884. D. Marinakis, G. Dudek, and D.J. Fleet (2005) Learning sensor network topology through Monte Carlo Expectation maximization, in *Proceedings of the 2005 IEEE International Conference on Robotics and Automation*, IEEE, pp. 4581–7.

885. K. Nagata and S. Watanabe (2007) Analysis of exchange ratio for exchange Monte Carlo method, in *Symposium on Foundations of Computational Intelligence*, IEEE, pp. 434–9.

886. Y. Shen, C. Archambeau, D. Cornford, M. Opper, J. Shawe-Taylor, and R. Barillec (2007) Evaluation of variational and Markov Chain Monte Carlo methods for inference in partially observed stochastic dynamic systems, in *Workshop on Machine Learning for Signal Processing*, IEEE, pp. 306–11.

887. S. Zhang and L. Liu (2008) MCMC samples selecting for online Bayesian network structure learning, in *International Conference on Machine Learning and Cybernetics*, vol. 3, IEEE, pp. 1762–7.

888. K. Nagata and S. Watanabe (2008) Exchange Monte Carlo sampling from Bayesian posterior for singular learning machines, *IEEE Transactions on Neural Networks* **19**(7): 1253–66.

889. H. Wang, Z. Li, and Y. Cheng (2008) Distributed latent Dirichlet allocation for objects-distributed cluster ensemble, in *International Conference on Natural Language Processing and Knowledge Engineering*, NLP-KE'08, IEEE, pp. 1–7.

890. X. Jun and W. Li (2009) PCMHS-Based algorithm for Bayesian networks online structure learning, in *International Forum on Computer Science-Technology and Applications, IFCSTA'09*, vol. 3, IEEE, pp. 310–14.

891. I. Saleemi, K. Shafique, and M. Shah (2009) Probabilistic modeling of scene dynamics for applications in visual surveillance, *IEEE Transactions on Pattern Analysis and Machine Intelligence* **31**(8): 1472–85.

892. W. Kim and K.M. Lee (2009) Markov chain Monte Carlo combined with deterministic methods for Markov random field optimization, in *Conference on Computer Vision and Pattern Recognition, CVPR 2009*, IEEE, pp. 1406–13.

893. C. Malagon, J.A. Barrio, and D. Nieto (2009) Automatic image classification from Cherenkov telescopes using Bayesian ensemble of neural networks, in *3rd International Workshop on Soft Computing Applications, SOFA'09*, IEEE, pp. 49–54.

894. P. McCullagh, V. Vovk, I. Nouretdinov, D. Devetyarov, and A. Gammerman (2009) Conditional prediction intervals for linear regression, in *International Conference on Machine Learning and Applications, ICMLA'09*, IEEE, pp. 131–8.

895. B. Zou, Z. Peng, H. Fan, and J. Xu (2010) Learning performance of Fisher linear discriminant based on Markov sampling, in *6th International Conference on Natural Computation (ICNC)*, vol. 3, IEEE, pp. 1114–18.

896. J. Huang, F. Wang, H. Mao, and M. Zhou (2010) A Markov chain Monte Carlo sampling relevance vector machine model for recognizing transcription start sites, in *International Conference on Artificial Intelligence and Computational Intelligence (AICI)*, IEEE, vol. 3, pp. 185–8.

897. W. Kim and K.M. Lee (2010) Continuous Markov random field optimization using fusion move driven Markov chain Monte Carlo technique, in *20th International Conference on Pattern Recognition (ICPR)*, IEEE, pp. 1364–7.

898. C. Walder (2010) Dirichlet mixtures of graph diffusions for semi supervised learning, in *International Workshop on Machine Learning for Signal Processing (MLSP)*, IEEE, pp. 421–6.

899. L. Gerencsér and S.D. Hill (2002) Discrete optimization, SPSA and Markov chain Monte Carlo methods, in *Proceedings of the 41st IEEE Conference on Decision and Control*, vol. 2, IEEE, pp. 2346–7.

900. A. Lecchini Visintini, W. Glover, J. Lygeros, and J. Maciejowski (2006) Monte Carlo optimization for conflict resolution in air traffic control, *IEEE Transactions on Intelligent Transportation Systems* **7**(4): 470–82.

901. A. Lecchini-Visintini, J. Lygeros, and J. Maciejowski (2008) On the approximate domain optimization of deterministic and expected value criteria, in *47th IEEE Conference on Decision and Control*, IEEE, pp. 4933–8.

902. F. Dabbene, P. Shcherbakov, and B. Polyak (2008) A randomized cutting plane scheme with geometric convergence: Probabilistic analysis and SDP applications, in *47th IEEE Conference on Decision and Control*, IEEE, pp. 3044–9.

903. M. Hansen, B. Hassibi, A.G. Dimakis, and W. Xu (2009) Near-optimal detection in MIMO systems using Gibbs sampling, in *Global Telecommunications Conference, Globecom'09*. IEEE, pp. 1–6.

904. B.T. Polyak and E.N. Gryazina (2010) Markov chain Monte Carlo method exploiting barrier functions with applications to control and optimization, in *International Symposium on Computer-Aided Control System Design*, IEEE, pp. 1553–7.

905. A. Lecchini-Visintini, J. Lygeros, and J.M. Maciejowski (2010) Schochastic optimization on continuous domains with finite time guarantees by Markov Chain Monte Carlo methods, *IEEE Transactions on Automatic Control* **55**(12): 2858–63.

906. S. Kimura and K. Matsumura (2011) Constrained multimodal function optimization using a simple evolutionary algorithm, in *Congress on Evolutionary Computation,* IEEE, pp. 447–54.

907. G. Zheng, Y. Zhang, C. Ji, and K. Wong (2011) A stochastic optimization approach for joint relay assignment and power allocation in orthogonal amplify-and-forward cooperative wireless networks, *IEEE Transactions on Wireless Communications* **10**: 4091–9.

908. B. Walsh (2004) Markov chain monte carlo and gibbs sampling, technical report, Massachusetts Institute of Technology.

909. S. Tascioglu and O. Ureten (2009) Bayesian wideband spectrum segmentation for cognitive radios, in *Proceedings of 18th International Conference on Computer Communications and Networks*, ICCCN 2009, IEEE, pp. 1–6.

910. X.Y. Wang, A. Wong, and P. Ho (2010) Dynamic Markov-chain Monte Carlo channel negotiation for cognitive radio, in *Conference on Computer Communications Workshops, Infocom'10,* IEEE, pp. 1–5.

911. L. Wang, B. Zheng, J. Cui, S. Tang, and H. Dou (2010) Spectrum sensing for cognitive of dm system using free probability theory, in Proceedings of the 6th International Wireless Communications and Mobile Computing Conference, ACM, pp. 575–9.

912. X.Y. Wang, A. Wong, and P.H. Ho (2011) Stochastic medium access for cognitive radio ad hoc networks, *IEEE Journal on Selected Areas in Communications* **29**(4): 770–83.

913. G. Welch and G. Bishop (2001) An introduction to the Kalman filter, *Design* **7**(1): 1–16.

914. Wikipedia, Kalman filter–Wikipedia, The Free Encyclopedia. http://en.wikipedia.org/wiki/Kalman filter.

915. S.J. Julier and J.K. Uhlmann (1997) A new extension of the Kalman filter to nonlinear systems, in *International Symposium on Aerospace/Defense Sensing, Simulation and Controls*, vol. 3, Spie Bellingham, WA, pp. 26–37.

916. E.A. Wan and R. Van Der Merwe (2001) The unscented Kalman filter, *Kalman Filtering and Neural Networks,* Wiley Online Library, pp. 221–80.

917. D. Zhang and Z. Tian (2007) Adaptive games for agile spectrum access based on extended Kalman filtering, *IEEE Journal of Selected Topics in Signal Processing* **1**(1): 79–90.

918. L. Akter, B. Natarajan, and C. Scoglio (2008) Modeling and forecasting secondary user activity in cognitive radio networks, in *Proceedings of 17th International Conference on Computer Communications and Networks, ICCCN'08*, IEEE, pp. 1–6.

919. A. Kushki (2008) A cognitive radio tracking system for indoor environments. PhD thesis, University of Toronto.

920. Z. Wen, T. Luo, W. Xiang, S. Majhi, and Y. Ma (2008) Autoregressive spectrum hole prediction model for cognitive radio systems, in *International Conference on Communications Workshops,* IEEE, pp. 154–7.

921. S.J. Kim, E. Dall'Anese, and G.B. Giannakis (2009) Sparsity-aware cooperative cognitive radio sensing using channel gain maps, in *43rd Asilomar Conference on Signals, Systems and Computers,* IEEE, pp. 518–22.

922. L. Akter and B. Natarajan (2009) Modeling and forecasting secondary user activity considering bulk arrival and bulk departure traffic model, in *70th Vehicular Technology Conference,* IEEE, pp. 1–5.

923. S.J. Kim, E. Dall'Anese, G.B. Giannakis, and S. Pupolin (2010) Collaborative channel gain map tracking for cognitive radios, in *2nd International Workshop on Cognitive Information Processing,* IEEE, pp. 338–43.

924. E. Dall'Anese, S.J. Kim, and G.B. Giannakis (2011) Channel gain map tracking via distributed kriging, *IEEE Transactions on Vehicular Technology* **60**(3): 1205–11.

925. F. Penna, J. Wang, and D. Cabric (2011) Cooperative localization of primary users by directional antennas or antenna arrays: Challenges and design issues, in *International Symposium on Antennas and Propagation (APSURSI),* IEEE, pp. 1113–15.

926. S.J. Kim, E. Dall'Anese, and G.B. Giannakis (2011) Cooperative spectrum sensing for cognitive radios using Kriged Kalman filtering, *IEEE Journal of Selected Topics in Signal Processing* **5**: 24–36.

927. H. Poveda, G. Ferré, and E. Grivel (2011) Robust frequency synchronization for an OFDMA uplink system disturbed by a cognitive radio system interference, in *International Conference on Acoustics, Speech and Signal Processing (ICASSP),* IEEE, pp. 3552–5.

928. A. Doucet, N. De Freitas, and N. Gordon (2001) *Sequential Monte Carlo Methods in Practice*. Berlin: Springer Verlag.

929. Wikipedia, Particle filter–Wikipedia, The Free Encyclopedia. http://en.wikipedia.org/wiki/Particle filter.

930. P.M. Djuric, J.H. Kotecha, J. Zhang, *et al.* (2003) Particle filtering, *Signal Processing Magazine*, **20**(5): 19–38.

931. S. Kandeepan, S. Reisenfeld, T. Aysal, D. Lowe, and R. Piesiewicz (2009) Bayesian tracking in cooperative localization for cognitive radio networks, in *69th Vehicular Technology Conference,* IEEE, pp. 1–5.

932. H. Roufarshbaf and J.K. Nelson (2009) Modulation classification in multipath environments using deterministic particle filtering, in *13th Digital Signal Processing Workshop and 5th Signal Processing Education Workshop,* IEEE, pp. 292–7.

933. S.M. Alavi, M. Mahdavi, and A.M. Doost Hosseini (2010) Prediction of state transitions in Rayleigh fading channels using particle filter, in *18th Iranian Conference on Electrical Engineering (ICEE),* IEEE, pp. 340–5.

934. Wikipedia, Collaborative filtering–Wikipedia, The Free Encyclopedia. http://en.wikipedia.org/wiki/ Collaborative filtering.

935. X. Su and T.M. Khoshgoftaar (2009) A survey of collaborative filtering techniques, *Advances in Artificial Intelligence,* pp. 1–19.

936. G. Linden, B. Smith, and J. York, Amazon.com recommendations: Item-to-item collaborative filtering, *Internet Computing* **7**(1): 76–80.

937. S. Oh (2010) Matrix completion: Fundamental limits and efficient algorithms. PhD thesis, Stanford University.

938. Y. Koren (2008) Factorization meets the neighborhood: A multifaceted collaborative filtering model, in *Proceeding of the 14th ACM SIGKDD International Conference on Knowledge Discovery and Data Mining*, ACM, pp. 426–34.

939. Y.B. Reddy and N. Gajendar (2008) Hybrid approach for spectrum bidding in wireless communications for maximizing the profit, in *5th International Conference on Information Technology: New Generations,* IEEE, pp. 341–7.

940. H. Sun, Y. Zhong, and W. Zhang (2010) Channel selection through a recommender system, in *Wireless Communications and Signal Processing (WCSP),* IEEE, pp. 1–5.

941. H. Li (2010) Learning the spectrum via collaborative filtering in cognitive radio networks, in *International Conference on New Frontiers in Dynamic Spectrum,* IEEE, pp. 1–12.

942. Y.B. Reddy (2010) Efficient spectrum allocation using case-based reasoning and collaborative filtering approaches, in *4th International Conference on Sensor Technologies and Applications (SENSORCOMM),* IEEE, pp. 375–80.

943. D. Heckerman *et al.* (1998) A tutorial on learning with Bayesian networks, *Nato Asi Series D Behavioural and Social Sciences* **89**: 301–54.

944. F.V. Jensen and T.D. Nielsen (2007) *Bayesian Networks and Decision Graphs*. Berlin: Springer Verlag.

945. K.P. Murphy (2002) Dynamic Bayesian networks: Representation, inference and learning. PhD thesis, University of California.

946. P. Demestichas, A. Katidiotis, K.A. Tsagkaris, E.F. Adamopoulou, and K.P. Demestichas (2009) Enhancing channel estimation in cognitive radio systems by means of Bayesian networks, *Wireless Personal Communications* **49**(1): 87–105.

947. H. Li and R.C. Qiu (2010) A graphical framework for spectrum modeling and decision making in cognitive radio networks, in *Global Telecommunications Conference, Globecom'10,* IEEE, pp. 1–6.

948. G. Quer, H. Meenakshisundaram, B. Tamma, B. Manoj, R. Rao, and M. Zorzi (2010) Cognitive network inference through Bayesian network analysis, in *Global Telecommunications Conference, Globecom'10,* IEEE, pp. 1–6.

949. Y. Huang, J. Wang, and H. Jiang (2010) Modeling of learning inference and decision-making engine in cognitive radio, in *2nd International Conference on Networks Security Wireless Communications and Trusted Computing (NSWCTC),* vol. 2, IEEE, pp. 258–61.

950. J. Salz (1985) Digital transmission over cross-coupled linear channels, *AT&T Technical Journal* **64**: 1147–59.

951. E. Telatar (1995) Capacity of multi-antenna gaussian channels, AT&T Bell Labs Memorandum.

952. G.J. Foschini (1996) Layered space-time architecture for wireless communication in a fading environment when using multi-element antennas, *Bell Labs Technical Journal* **1**(2): 41–59.

953. A. Paulraj, R. Nabar, and D. Gore (2003) *Introduction to Space-Time Wireless Communications*. Cambridge: Cambridge University Press.
954. Wikipedia, MIMO–Wikipedia, The Free Encyclopedia. http://en.wikipedia.org/wiki/MIMO.
955. Wikipedia, Array gain–Wikipedia, The Free Encyclopedia. http://en.wikipedia.org/wiki/Array gain.
956. Wikipedia, Diversity gain–Wikipedia, The Free Encyclopedia. http://en.wikipedia.org/wiki/Diversity gain.
957. Wikipedia, Spatial multiplexing gain–Wikipedia, The Free Encyclopedia. http://en.wikipedia.org/wiki/Spatial multiplexing gain.
958. L. Zheng and D.N.C. Tse (2003) Diversity and multiplexing: A fundamental tradeoff in multiple-antenna channels, IEEE *Transactions on Information Theory* **49**(5): 1073–96.
959. H. El Gamal, G. Caire, and M.O. Damen (2004) Lattice coding and decoding achieve the optimal diversity-multiplexing tradeoff of MIMO channels, *IEEE Transactions on Information Theory* **50**(6): 968–85.
960. Wikipedia, Spacetime code–Wikipedia, The Free Encyclopedia. http://en.wikipedia.org/wiki/Space time code.
961. Wikipedia, Spacetime block code–Wikipedia, The Free Encyclopedia. http://en.wikipedia.org/wiki/Spacetime block code.
962. S.M. Alamouti (1998) A simple transmit diversity technique for wireless communications, *IEEE Journal on Selected Areas in Communications* **16**(8): 1451–8.
963. W. Su, S.N. Batalama, and D.A. Pados (2006) On orthogonal space-time block codes and transceiver signal linearization, *Communications Letters, IEEE* **10**(2): 91–3.
964. V. Tarokh, H. Jafarkhani, and A.R. Calderbank (1999) Space-time block codes from orthogonal designs, *IEEE Transactions on Information Theory* **45**(5): 1456–7.
965. V. Tarokh, N. Seshadri, and A.R. Calderbank (1998) Space-time codes for high data rate wireless communication: Performance criterion and code construction, *IEEE Transactions on Information Theory* **44**(2): 744–65.
966. Wikipedia, Spacetime trellis code–Wikipedia, The Free Encyclopedia. http://en.wikipedia.org/wiki/Spacetime trellis code.
967. D.S. Shiu and M. Kahn (1999) Layered space-time codes for wireless communications using multiple transmit antennas, in *IEEE International Conference on Communications, ICC'99,* vol. 1, IEEE, pp. 436–40.
968. P.W. Wolniansky, G.J. Foschini, G.D. Golden, and R.A. Valenzuela (1998) V-BLAST: An architecture for realizing very high data rates over the rich-scattering wireless channel, in *URSI International Symposium on Signals, Systems, and Electronics, ISSSE 98,* IEEE, pp. 295–300.
969. G.D. Golden, C.J. Foschini, R.A. Valenzuela, and P.W. Wolniansky (1999) Detection algorithm and initial laboratory results using V-BLAST space-time communication architecture, *Electronics Letters* **35**(1): 14–16.
970. Wikipedia, Bell Laboratories Layered Space-Time–Wikipedia, The Free Encyclopedia. http://en.wikipedia.org/wiki/Bell Laboratories Layered Space-Time.
971. M. Sellathurai and S. Haykin (2002) Turbo-BLAST for wireless communications: Theory and experiments, IEEE Transactions on Signal Processing **50**(10): 2538–46.
972. M. Sellathurai, T. Ratnarajah, and P. Guinand (2007) Multirate layered space-time coding and successive interference cancellation receivers in quasi-static fading channels, *IEEE Transactions on Wireless Communications* **6**(12): 4524–33.
973. B. Hassibi (2000) An efficient square-root algorithm for BLAST, in *International Conference on Acoustics, Speech, and Signal Processing, ICASSP'00,* IEEE, vol. 2, pp. II737–II740.
974. E. Biglieri, G. Taricco, and A. Tulino (2002) Decoding space-time codes with BLAST architectures, *IEEE Transactions on Signal Processing* **50**(10): 2547–52.
975. Wikipedia, Multi-user MIMO–Wikipedia, The Free Encyclopedia. http://en.wikipedia.org/wiki/Multi-user MIMO.
976. B.D. Van Veen and K.M. Buckley (1988) Beamforming: A versatile approach to spatial filtering, *ASSP Magazine, IEEE* **5**(2): 4–24.
977. L.C. Godara (1997) Application of antenna arrays to mobile communications. II. Beam-forming and direction-of-arrival considerations, *Proceedings of the IEEE* **85**(8): 1195–1245.
978. C. Farsakh and J.A. Nossek (1998) Spatial covariance based downlink beamforming in an SDMA mobile radio system, *IEEE Transactions on Communications* **46**(11): 1497–1506.

979. J.H. Winters (1998) Smart antennas for wireless systems, *Personal Communications, IEEE* **5**(1): 23–7.
980. H. Lehne and M. Pettersen (1999) An overview of smart antenna technology for mobile communications systems, *IEEE Communications Surveys* **2**(4): 2–13.
981. P. Vandenameele, L. Van Der Perre, M.G.E. Engels, B. Gyselinckx, and H.J. De Man (2000) A combined OFDM/SDMA approach, *IEEE Journal on Selected Areas in Communications* **18**(11): 2312–21.
982. P. Viswanath, D.N.C. Tse, and R. Laroia (2002) Opportunistic beamforming using dumb antennas, *IEEE Transactions on Information Theory* **48**(6): 1277–94.
983. H. Yin and H. Liu (2002) Performance of space-division multiple-access (SDMA) with scheduling, *IEEE Transactions on Wireless Communications* **1**(4): 611–18.
984. K.K. Mukkavilli, A. Sabharwal, E. Erkip, and B. Aazhang (2003) On beamforming with finite rate feedback in multipleantenna systems, *IEEE Transactions on Information Theory* **49**(10): 2562–79.
985. A. Alexiou and M. Haardt (2004) Smart antenna technologies for future wireless systems: Trends and challenges, Communications Magazine, IEEE **42**(9): 90–7.
986. K. Huang, J.G. Andrews, and R.W. Heath (2009) Performance of orthogonal beamforming for SDMA with limited feedback, *IEEE Transactions on Vehicular Technology* **58**(1): 152–64.
987. P. Zhao, B. Daneshrad, A. Warrier, W. Zhu, and O. Takeshita (2011) Performance of a concurrent link SDMA MAC under practical PHY operating conditions, *IEEE Transactions on Vehicular Technology* **60**(3): 1301–7.
988. Y. Huang, L. Yang, M. Bengtsson, and B. Ottersten (2011) Exploiting long-term channel correlation in limited feedback SDMA through channel phase codebook, *IEEE Transactions on Signal Processing* **59**(3): 1217–28.
989. N. Jindal and A. Goldsmith (2005) Dirty-paper coding versus TDMA for MIMO broadcast channels, IEEE Transactions on Information Theory **51**(5): 1783–94.
990. N. Jindal, W. Rhee, S. Vishwanath, S.A. Jafar, and A. Goldsmith (2005) Sum power iterative water-filling for multi-antenna Gaussian broadcast channels, *IEEE Transactions on Information Theory* **51**(4): 1570–80.
991. M. Costa (1983) Writing on dirty paper, *IEEE Transactions on Information Theory* **29**(3): 439–41.
992. G. Caire and S. Shamai, On the achievable throughput of a multiantenna Gaussian broadcast channel, *IEEE Transactions on Information Theory* **49**(7): 1691–1706.
993. S. Vishwanath, N. Jindal, and A. Goldsmith (2003) Duality, achievable rates, and sum-rate capacity of Gaussian MIMO broadcast channels, *IEEE Transactions on Information Theory* **49**(10): 2658–68.
994. P. Viswanath and D.N.C. Tse (2003) Sum capacity of the vector Gaussian broadcast channel and uplink-downlink duality, *IEEE Transactions on Information Theory* **49**(8): 1912–21.
995. W. Yu and J.M. Cioffi (2004) Sum capacity of Gaussian vector broadcast channels, *IEEE Transactions on Information Theory* **50**(9): 1875–92.
996. T. Yoo and A. Goldsmith (2006) On the optimality of multiantenna broadcast scheduling using zero-forcing beamforming, *IEEE Journal on Selected Areas in Communications* **24**(3): 528–41.
997. Q.H. Spencer, A.L. Swindlehurst, and M. Haardt (2004) Zero-forcing methods for downlink spatial multiplexing in multiuser MIMO channels, *IEEE Transactions on Signal Processing* **52**(2): 461–71.
998. A. Wiesel, Y.C. Eldar, and S. Shamai (2008) Zero-forcing precoding and generalized inverses, *IEEE Transactions on Signal Processing* **56**(9): 4409–18.
999. C. Guthy, W. Utschick, and G. Dietl (2009) Low-complexity linear zero-forcing for the MIMO broadcast channel, *IEEE Journal of Selected Topics in Signal Processing* **3**(6): 1106–17.
1000. S. Wagner, R. Couillet, D. Slock, and M. Debbah (2010) Large system analysis of zero-forcing precoding in miso broadcast channels with limited feedback, in *11th International Workshop on Signal Processing Advances in Wireless Communications (SPAWC)*, IEEE, pp. 1–5.
1001. P.S. Udupa and J.S. Lehnert (2007) Optimizing zero-forcing precoders for MIMO broadcast systems, *IEEE Transactions on Communications* **55**(8): 1516–24.
1002. X. Shao, J. Yuan, and Y. Shao (2007) Error performance analysis of linear zero forcing and MMSE precoders for MIMO broadcast channels, *IET Communications* **1**(5): 1067–74.
1003. J. Duplicy and L. Vandendorpe (2007) Robust MMSE precoding for the MIMO complex Gaussian broadcast channel, in *International Conference on Acoustics, Speech and Signal Processing*, vol. 3, IEEE, pp. III–421.
1004. R.F.H. Fischer, C. Stierstorfer, and J.B. Huber (2004) Precoding for point-to-multipoint transmission over MIMO ISI channels, in *International Zurich Seminar on Communications*, IEEE, pp. 208–11.
1005. R.F.H. Fischer (2002) *Precoding and Signal Shaping for Digital Transmission*. Wiley-IEEE Press.

1006. Z. Shen, R. Chen, J.G. Andrews, R.W. Heath, and B.L. Evans (2006) Low complexity user selection algorithms for multiuser MIMO systems with block diagonalization, *IEEE Transactions on Signal Processing* **54**(9): 3658–63.

1007. Z. Shen, R. Chen, J.G. Andrews, R.W. Heath, and B.L. Evans (2007) Sum capacity of multiuser MIMO broadcast channels with block diagonalization, *IEEE Transactions on Wireless Communications* **6**(6): 2040–5.

1008. N. Ravindran and N. Jindal (2007) MIMO broadcast channels with block diagonalization and finite rate feedback, in *International Conference on Acoustics, Speech and Signal Processing,* vol. 3, IEEE, pp. III–13.

1009. W. Li and M. Latva-aho (2011) An efficient channel block diagonalization method for generalized zero forcing assisted MIMO broadcasting systems, *IEEE Transactions on Wireless Communications* **99**: 739–44.

1010. H. Sung, K.J. Lee, and I. Lee (2009) An MMSE based block diagonalization for multiuser MIMO downlink channels with other cell interference, in *70th Vehicular Technology Conference,* IEEE, pp. 1–5.

1011. J. Lee and N. Jindal (2007) High SNR analysis for MIMO broadcast channels: Dirty paper coding versus linear precoding, IEEE Transactions on Information Theory **53**(12): 4787–92.

1012. M. Sharif and B. Hassibi (2005) On the capacity of MIMO broadcast channels with partial side information, IEEE Transactions on Information Theory **51**(2): 506–22.

1013. N. Jindal (2006) MIMO broadcast channels with finite-rate feedback, *IEEE Transactions on Information Theory* **52**(11): 5045–60.

1014. M. Kountouris, R. de Francisco, D. Gesbert, D.T.M. Slock, and T. Salzer (2007) Efficient metrics for scheduling in MIMO broadcast channels with limited feedback, in *International Conference on Acoustics, Speech and Signal Processing,* vol. 3, IEEE, pp. 109–12.

1015. T. Yoo, N. Jindal, and A. Goldsmith (2007) Multi-antenna downlink channels with limited feedback and user selection, *IEEE Journal on Selected Areas in Communications* **25**(7): 1478–91.

1016. W. Zhang and K.B. Letaief (2007) MIMO broadcast scheduling with limited feedback, *IEEE Journal on Selected Areas in Communications* **25**(7): 1457–67.

1017. J. Zhang, M. Kountouris, J. Andrews, and R. Heath Jr (2009) Multi-mode transmission for the MIMO broadcast channel with imperfect channel state information, *IEEE Transactions on Communications* **99**: 1–12.

1018. M. Sharif and B. Hassibi (2007) A comparison of time-sharing, DPC, and beamforming for MIMO broadcast channels with many users, *IEEE Transactions on Communications* **55**(1): 11–15.

1019. A. Bayesteh and A. K. Khandani (2008) On the user selection for MIMO broadcast channels, *IEEE Transactions on Information Theory* **54**(3): 1086–1107.

1020. J. Dai, C. Chang, Z. Ye, and Y.S. Hung (2009) An efficient greedy scheduler for zero-forcing dirty-paper coding, *IEEE Transactions on Communications* **57**(7): 1939–43.

1021. L. Zhang, Y. Xin, and Y.C. Liang (2009) Weighted sum rate optimization for cognitive radio MIMO broadcast channels, *IEEE Transactions on Wireless Communications* **8**(6): 2950–9.

1022. K. Hamdi, W. Zhang, and K. Letaief (2009) Opportunistic spectrum sharing in cognitive MIMO wireless networks, IEEE *Transactions on Wireless Communications* **8**(8): 4098–4109.

1023. J. Tang, K. Cumanan, and S. Lambotharan (2011) Sum-rate maximization technique for spectrum-sharing MIMO OFDM broadcast channels, *IEEE Transactions on Vehicular Technology* **60**(4): 1960–4.

1024. K J. Lee, H. Sung, and I. Lee (2011) Linear precoder designs for cognitive radio multiuser MIMO downlink systems, in International Conference on Communications, IEEE, pp. 1–5.

1025. K. Lee and I. Lee (2011) MMSE based block diagonalization for cognitive radio MIMO broadcast channels, *IEEE Transactions on Wireless Communications* **99**: 1–6.

1026. H. Boche and M. Wiczanowski (2007) Optimization-theoretic analysis of stability-optimal transmission policy for multipleantenna multiple-access channel, *IEEE Transactions on Signal Processing* **55**(6): 2688–2702.

1027. W. Yu, W. Rhee, S. Boyd, and J.M. Cioffi (2004) Iterative water-filling for Gaussian vector multiple-access channels, *IEEE Transactions on Information Theory* **50**(1): 145–52.

1028. S.A. Jafar and M.J. Fakhereddin (2007) Degrees of freedom for the MIMO interference channel, *IEEE Transactions on Information Theory* **53**(7): 2637–42.

1029. F. Negro, S.P. Shenoy, I. Ghauri, and D.T.M. Slock (2010) On the MIMO interference channel, in *Information Theory and Applications Workshop* (ITA), IEEE, pp. 1–9.

1030. K. Gomadam, V.R. Cadambe, and S.A. Jafar (2008) Approaching the capacity of wireless networks through distributed interference alignment, in *Global Telecommunications Conference, Globecom'08*, IEEE, pp. 1–6.

1031. M. Amir, A. El-Keyi, and M. Nafie (2011) Constrained interference alignment and the spatial degrees of freedom of MIMO cognitive networks, *IEEE Transactions on Information Theory* **57**(5): 2994–3004.

1032. S. Perlaza, N. Fawaz, S. Lasaulce, and M. Debbah (2010) From spectrum pooling to space pooling: opportunistic interference alignment in mimo cognitive networks, *IEEE Transactions on Signal Processing* **58**(7): 3728–41.

1033. C. Shen and M.P. Fitz (2011) Opportunistic spatial orthogonalization and its application in fading cognitive radio networks, *IEEE Journal of Selected Topics in Signal Processing* **5**(1). 182–9.

1034. X. Lin, N.B. Shroff, and R. Srikant (2006) A tutorial on cross-layer optimization in wireless networks, *IEEE Journal on Selected Areas in Communications* **24**(8): 1452–63.

1035. J. Liu, Y.T. Hou, Y. Shi, and H. Sherali (2008) Cross-layer optimization for mimo-based wireless ad hoc networks: Routing, power allocation, and bandwidth allocation, *IEEE Journal on Selected Areas in Communications* **26**(6): 913–26.

1036. J. Liu and Y.T. Hou (2007) Cross-layer optimization of MIMO-based mesh networks with Gaussian vector broadcast channels, Arxiv preprint arXiv:0704.0967.

1037. Y.H. Lin, T. Javidi, R.L. Cruz, and L.B. Milstein (2006) Distributed link scheduling, power control and routing for multihop wireless MIMO networks, in *40th Asilomar Conference on Signals, Systems and Computers, ACSSC'06*, IEEE, pp. 122–6.

1038. W. Ge, J. Zhang, and G. Xue (2010) MIMO-pipe modeling and scheduling for efficient interference management in multihop MIMO networks, *IEEE Transactions on Vehicular Technology* **59**(8): 3966–78.

1039. Y. Lin and V.W.S. Wong (2009) Cross-layer design of MIMO-enabled WLANs with network utility maximization, *IEEE Transactions on Vehicular Technology* **58**(5): 2443–56.

1040. V.K.N. Lau (2009) Adaptive resource allocation for multiuser MIMO systems with transmit group MMSE, *IEEE Transactions on Wireless Communications* **8**(5): 2362–8.

1041. A.L. Toledo, X. Wang, and B. Lu (2006) A cross-layer TCP modelling framework for MIMO wireless systems, *IEEE Transactions on Wireless Communications* **5**(4): 920–9.

1042. O. Souihli and T. Ohtsuki (2010) Joint feedback and scheduling scheme for service-differentiated multiuser MIMO systems, *IEEE Transactions on Wireless Communications* **9**(2): 528–33.

1043. W. Ge, J. Zhang, and S. Shen (2007) A cross-layer design approach to multicast in wireless networks, *IEEE Transactions on Wireless Communications* **6**(3): 1063–71.

1044. S. Ma, Y. Yang, and H. Sharif (2011) Distributed MIMO technologies in cooperative wireless networks, *IEEE Communications Magazine* **49**(5): 78–82.

1045. Y. Kim and H. Liu (2008) Infrastructure relay transmission with cooperative MIMO, *IEEE Transactions on Vehicular Technology* **57**(4): 2180–8.

1046. S. Simoens, O. Muoz-Medina, J. Vidal, and A. Del Coso (2010) Compress-and-forward cooperative MIMO relaying with full channel state information, *IEEE Transactions on Signal Processing* **58**(2): 781–91.

1047. L. Dai and K. Letaief (2008) Throughput maximization of ad-hoc wireless networks using adaptive cooperative diversity and truncated ARQ, *IEEE Transactions on Communications* **56**(11): 1907–18.

1048. G. Aruma Baduge, C. Tellambura, and M. Ardakani, Performance analysis framework for transmit antenna selection strategies of cooperative MIMO AF relay networks, *IEEE Transactions on Vehicular Technology* **60**(99): 3030–44.

1049. X. He, T. Luo, and G. Yue (2010) Optimized distributed MIMO for cooperative relay networks, *IEEE Communications Letters* **14**(1): 9–11.

1050. P. Clarke and R. de Lamare (2011) Joint transmit diversity optimization and relay selection for multi-relay cooperative MIMO systems using discrete stochastic algorithms, *IEEE Communications Letters* **15**(99): 1035–7.

1051. Y. Rong (2011) Joint source and relay optimization for two-way MIMO multi-relay networks, *IEEE Communications Letters* **99**: 1329–31.

1052. X. Dong, Y. Rong, and Y. Hua (2010) Cooperative power scheduling for a network of MIMO links, *IEEE Transactions on Wireless Communications* **9**(3): 939–44.

1053. Q. Qu, L.B. Milstein, and D.R. Vaman (2010) Cooperative and constrained MIMO communications in wireless ad hoc/sensor networks, *IEEE Transactions on Wireless Communications* **9**(10): 3120–9.

1054. M.Z. Siam, M. Krunz, and O. Younis (2009) Energy-efficient clustering/routing for cooperative MIMO operation in sensor networks, in *Infocom'09,* IEEE, pp. 621–9.

1055. D. Gesbert, S. Hanly, H. Huang, S. Shamai Shitz, O. Simeone, and W. Yu (2010) Multi-cell MIMO cooperative networks: A new look at interference, *IEEE Journal on Selected Areas in Communications* **28**(9): 1380–1408.

1056. C.T.K. Ng and H. Huang (2010) Linear precoding in cooperative MIMO cellular networks with limited coordination clusters, *IEEE Journal on Selected Areas in Communications* **28**(9): 1446–54.

1057. S. Parkvall, E. Dahlman, A. Furuskar, *et al.* (2008) LTE-advanced-evolving LTE towards IMT-advanced, in *68th Vehicular Technology Conference,* IEEE, pp. 1–5.

1058. S. Brueck, L. Zhao, J. Giese, and M.A. Amin (2010) Centralized scheduling for joint transmission coordinated multi-point in LTE-Advanced, in *International ITG Workshop on Smart Antennas (WSA),* IEEE, pp. 177–84.

1059. R. Irmer, H. Droste, P. Marsch, *et al.* (2011) Coordinated multipoint: Concepts, performance, and field trial results, *IEEE Communications Magazine* **49**(2): 102–11.

1060. S.A. Ramprashad, H.C. Papadopoulos, A. Benjebbour, Y. Kishiyama, N. Jindal, and G. Caire (2011) Cooperative cellular networks using multi-user MIMO: Trade-offs, overheads, and interference control across architectures, *IEEE Communications Magazine* **49**(5): 70–7.

1061. Q. Du and X. Zhang (2011) QoS-Aware base-station selections for distributed MIMO links in broadband wireless networks, *IEEE Journal on Selected Areas in Communications* **29**(6): 1123–38.

1062. X. Ge, K. Huang, C.X. Wang, X. Hong, and X. Yang (2011) Capacity analysis of a multi-cell multi-antenna cooperative cellular network with co-channel interference, IEEE Transactions on Wireless Communications **10**(10): 3298–3309.

1063. P. Wang, H. Wang, L. Ping, and X. Lin (2011) On the capacity of MIMO cellular systems with base station cooperation, *IEEE Transactions on Wireless Communications* **10**(11): 3720–31.

1064. Z. Hu, N. Guo, and R. Qiu (2010) Wideband waveform design for relay cognitive network, in *Military Communications Conference, Milcom'10,* IEEE, pp. 749–54.

1065. H. Chen, A.B. Gershman, and S. Shahbazpanahi (2010) Filter-and-forward distributed beamforming in relay networks with frequency selective fading, *IEEE Transactions on Signal Processing* **58**(3): 1251–62.

1066. T. Al-Khasib, M. Shenouda, and L. Lampe (2011) Dynamic spectrum management for multiple-antenna cognitive radio systems: Designs with imperfect CSI, *IEEE Transactions on Wireless Communications* **9**: 2850–9.

1067. R. Zhang and Y.C. Liang (2008) Exploiting multi-antennas for opportunistic spectrum sharing in cognitive radio networks, *IEEE Journal of Selected Topics in Signal Processing* **2**(1): 88–102.

1068. M. Jung, K. Hwang, and S. Choi (2011) Interference minimization approach to precoding scheme in MIMO-based cognitive radio networks, *IEEE Communications Letters* **99**: 789–91.

1069. M. Fainan Hanif, P.J. Smith, D.P. Taylor, and P.A. Martin (2011) MIMO cognitive radios with antenna selection, *IEEE Transactions on Wireless Communications* **10**(11): 3688–99.

1070. A. Ghosh and W. Hamouda (2011) Cross-layer antenna selection and channel allocation for MIMO cognitive radios, *IEEE Transactions on Wireless Communications* **10**(11): 3666–74.

1071. S.J. Kim and G.B. Giannakis (2008) Optimal resource allocation for MIMO ad hoc cognitive radio networks, in *46th Annual Allerton Conference on Communication, Control, and Computing,* IEEE, pp. 39–45.

1072. C. Gao, Y. Shi, T. Hou, and S. Kompella (2011) On the throughput of MIMO-empowered multi-hop cognitive radio networks, *IEEE Transactions on Mobile Computing* **10**(11): 1505–19.

1073. G. Scutari, D. Palomar, and S. Barbarossa (2008) Cognitive MIMO radio, *Signal Processing Magazine,* IEEE **25**(6): 46–59.

1074. G. Scutari and D.P. Palomar (2010) MIMO cognitive radio: A game theoretical approach, *IEEE Transactions on Signal Processing* **58**(2): 761–80.

1075. J. Wang, G. Scutari, and D.P. Palomar (2011) Robust MIMO cognitive radio via game theory, *IEEE Transactions on Signal Processing* **59**(3): 1183–1201.

1076. H. Islam, Y. Liang, and A. Hoang (2008) Joint power control and beamforming for cognitive radio networks, *IEEE Transactions on Wireless Communications* **7**(7): 2415–19.

1077. L. Zhang, Y.C. Liang, and Y. Xin (2008) Joint beamforming and power allocation for multiple access channels in cognitive radio networks, *IEEE Journal on Selected Areas in Communications* **26**(1): 38–51.

1078. G. Zheng, K.K. Wong, and B. Ottersten (2009) Robust cognitive beamforming with bounded channel uncertainties, *IEEE Transactions on Signal Processing* **57**(12): 4871–81.

1079. L. Zhang, Y.C. Liang, Y. Xin, and H.V. Poor (2009) Robust cognitive beamforming with partial channel state information, *IEEE Transactions on Wireless Communications* **8**(8): 4143–53.

1080. E.A. Gharavol, Y.C. Liang, and K. Mouthaan (2010) Robust downlink beamforming in multiuser MISO cognitive radio networks with imperfect channel-state information, *IEEE Transactions on Vehicular Technology* **59**(6): 2852–60.

1081. S. Yiu, M. Vu, and V. Tarokh (2009) Interference and noise reduction by beamforming in cognitive networks, *IEEE Transactions on Communications* **57**(10): 3144–53.

1082. K. Cumanan, R. Krishna, L. Musavian, and S. Lambotharan (2010) Joint beamforming and user maximization techniques for cognitive radio networks based on branch and bound method, *IEEE Transactions on Wireless Communications* **9**(10): 3082–92.

1083. K.L. Du and W.H. Mow (2010) Affordable cyclostationarity-based spectrum sensing for cognitive radio with smart antennas, *IEEE Transactions on Vehicular Technology* **59**(4): 1877–86.

1084. A. Tajer, N. Prasad, and X. Wang (2010) Beamforming and rate allocation in MISO cognitive radio networks, *IEEE Transactions on Signal Processing* **58**(1): 362–77.

1085. G. Zheng, S. Ma, K.K. Wong, and T.S. Ng (2010) Robust beamforming in cognitive radio, *IEEE Transactions on Wireless Communications* **9**(2): 570–6.

1086. Y. Pei, Y.C. Liang, L. Zhang, K.C. Teh, and K.H. Li (2010) Secure communication over MISO cognitive radio channels, *IEEE Transactions on Wireless Communications* **9**(4): 1494–1502.

1087. K. Cumanan, L. Musavian, S. Lambotharan, and A.B. Gershman (2010) SINR balancing technique for downlink beamforming in cognitive radio networks, *Signal Processing Letters, IEEE* **17**(2): 133–6.

1088. G. Xiong and S. Kishore (2011) Cooperative spectrum sensing with beamforming in cognitive radio networks, *Communications Letters, IEEE* **15**(2): 220–2.

1089. Z. Xiong, R. Krishna, K. Cumanan, and S. Lambotharan (2011) Grassmannian beamforming and null space broadcasting protocols for cognitive radio networks, *Signal Processing, IET* **5**(5): 451–60.

1090. J. Tang and S. Lambotharan (2011) Beamforming and temporal power optimisation for an overlay cognitive radio relay network, *Signal Processing, IET* **5**(6): 582–8.

1091. R. Mochaourab and E.A. Jorswieck (2011) Optimal beamforming in interference networks with perfect local channel information, *IEEE Transactions on Signal Processing* **59**(3): 1128–41.

1092. Y. Pei, Y.C. Liang, K.C. Teh, and K.H. Li (2011) Secure communication in multiantenna cognitive radio networks with imperfect channel state information, *IEEE Transactions on Signal Processing* **59**(4): 1683–93.

1093. E.A. Gharavol, Y.C. Liang, and K. Mouthaan (2011) Robust linear transceiver design in MIMO ad hoc cognitive radio networks with imperfect channel state information, *IEEE Transactions on Wireless Communications* **10**(5): 1448–57.

1094. Y. Huang, Q. Li, W. Ma, and S. Zhang (2012) Robust multicast beamforming for spectrum sharing-based cognitive radios, *IEEE Transactions on Signal Processing* **60**(1): 527–33.

1095. R. Ramanathan (2001) On the performance of ad hoc networks with beamforming antennas, in *Proceedings of the 2nd ACM International Symposium on Mobile Ad Hoc Networking & Computing*, ACM, pp. 95–105.

1096. S. Roy, D. Saha, S. Bandyopadhyay, T. Ueda, and S. Tanaka (2003) A network-aware MAC and routing protocol for effective load balancing in ad hoc wireless networks with directional antenna, in *Proceedings of the 4th ACM International Symposium on Mobile Ad Hoc Networking & Computing*, ACM, pp. 88–97.

1097. L. Cimini Jr (1985) Analysis and simulation of a digital mobile channel using orthogonal frequency division multiplexing, *IEEE Transactions on Communications* **33**(7): 665–75.

1098. Y. Wu and W.Y. Zou (1995) Orthogonal frequency division multiplexing: A multi-carrier modulation scheme, *IEEE Transactions on Consumer Electronics* **41**(3): 392–9.

1099. Wikipedia, Orthogonal frequency division multiplexing–Wikipedia, The Free Encyclopedia. http://en.wikipedia.org/wiki/OFDM. [319]

1100. R.W. Chang (1966) Synthesis of band-limited orthogonal signals for multichannel data transmission, Bell System Technical Journal **45**: 1775–96.

1101. B. Saltzberg (1967) Performance of an efficient parallel data transmission system, *IEEE Transactions on Communication Technology* **15**(6): 805–11.

1102. S. Weinstein and P. Ebert (1971) Data transmission by frequency-division multiplexing using the discrete Fourier transform, *IEEE Transactions on Communication Technology* **19**(5): 628–34.

1103. A. Peled and A. Ruiz (1980) Frequency domain data transmission using reduced computational complexity algorithms, in *IEEE International Conference on Acoustics, Speech, and Signal Processing, ICASSP'80*, vol. 5, IEEE, pp. 964–7.

1104. T. de Couasnon, R. Monnier, and J. Bernard Rault (1994) OFDM for digital TV broadcasting, *Signal Processing* **39**(1–2): 1–32.

1105. J.S. Chow, J.C. Tu, and J.M. Cioffi (1991) A discrete multitone transceiver system for HDSL applications, *IEEE Journal on Selected Areas in Communications* **9**(6): 895–908.

1106. H. Tang, K.Y. Lau, and R.W. Brodersen (2003) Synchronization schemes for packet OFDM system, in *IEEE International Conference on Communications*, ICC'03, vol. 5, pp. 3346–50.

1107. T. Pollet, M. Van Bladel, and M. Moeneclaey (1995) BER sensitivity of OFDM systems to carrier frequency offset and Wiener phase noise, *IEEE Transactions on Communications* **43**(234): 191–3.

1108. T.M. Schmidl and D.C. Cox (1997) Robust frequency and timing synchronization for OFDM, *IEEE Transactions on Communications* **45**(12): 1613–21.

1109. L. Hanzo (2003) *OFDM and MC-CDMA for Broadband Multi-user Communications, WLANs, and Broadcasting*. Wiley-IEEE Press.

1110. H. Zhou, A. Malipatil, and Y.F. Huang (2008) OFDM carrier synchronization based on time-domain channel estimates, *IEEE Transactions on Wireless Communications* **7**(8): 2988–9.

1111. P.H. Moose (1994) A technique for orthogonal frequency division multiplexing frequency offset correction, *IEEE Transactions on Communications* **42**(10): 2908–14.

1112. J.J. Van de Beek, M. Sandell, and P.O. Borjesson (1997) ML estimation of time and frequency offset in OFDM systems, *IEEE Transactions on Signal Processing* **45**(7): 1800–5.

1113. N. Lashkarian and S. Kiaei (2000) Class of cyclic-based estimators for frequency-offset estimation of OFDM systems, *IEEE Transactions on Communications* **48**(12): 2139–49.

1114. U. Tureli, H. Liu, and M.D. Zoltowski (2000) OFDM blind carrier offset estimation: ESPRIT, *IEEE Transactions on Communications* **48**(9): 1459–61.

1115. B. Chen and H. Wang (2004) Blind estimation of OFDM carrier frequency offset via oversampling, *IEEE Transactions on Signal Processing* **52**(7): 2047–57.

1116. B. Park, H. Cheon, E. Ko, C. Kang, and D. Hong (2004) A blind OFDM synchronization algorithm based on cyclic correlation, *Signal Processing Letters, IEEE* **11**(2): 83–5.

1117. W.L. Chin and S.G. Chen (2009) A blind synchronizer for OFDM systems based on SINR maximization in multipath fading channels, *IEEE Transactions on Vehicular Technology* **58**(2): 625–35.

1118. H. Liu and U. Tureli (1998) A high-efficiency carrier estimator for OFDM communications, *Communications Letters, IEEE* **2**(4): 104–6.

1119. U. Tureli, D. Kivanc, and H. Liu (2001) Experimental and analytical studies on a high-resolution OFDM carrier frequency offset estimator, *IEEE Transactions on Vehicular Technology* **50**(2): 629–43.

1120. B. Chen and H. Wang (2002) Maximum likelihood estimation of OFDM carrier frequency offset, in *International Conference on Communications*, vol. 1, IEEE, pp. 49–53.

1121. M. Sliskovic (2001) Sampling frequency offset estimation and correction in OFDM systems, in *8th IEEE International Conference on Electronics, Circuits and Systems*, vol. 1, IEEE, pp. 437–40.

1122. Y.H. You, S.T. Kim, K.T. Lee, and H.K. Song (2008) An improved sampling frequency offset estimator for OFDM-based digital radio mondiale systems, *IEEE Transactions on Broadcasting* **54**(2): 283–6.

1123. G. Liu, B. Li, and H. Pang (2009) Blind sampling clock offset estimation algorithm for OFDM system, in *2nd International Congress on Image and Signal Processing*, CISP'09, IEEE, pp. 1–4.

1124. W.L. Chin, S.G. Chen, and C.L. Chen (2007) A joint synchronization algorithm for OFDM systems, in *18th International Symposium on Personal, Indoor and Mobile Radio Communications*, IEEE, pp. 1–5.

1125. E. del Castillo-Sanchez, F.J. Lopez-Martinez, E. Martos-Naya, and J.T. Entrambasaguas (2009) Joint time, frequency and sampling clock synchronization for OFDM-based systems, in *Wireless Communications and Networking Conference*, IEEE, pp. 1–6.

1126. Y.H. Kim and J.H. Lee (2011) Joint maximum likelihood estimation of carrier and sampling frequency offsets for OFDM systems, *IEEE Transactions on Broadcasting* **57**(2): 277–83.

1127. M. Morelli and M. Moretti (2008) Robust frequency synchronization for OFDM-based cognitive radio systems, *IEEE Transactions on Wireless Communications* **7**(12): 5346–55.

1128. Y.J. Kou, W.S. Lu, and A. Antoniou (2005) Application of sphere decoding in intercarrier-interference reduction for OFDM systems, in *Pacific Rim Conference on Communications, Computers and Signal Processing*, IEEE, pp. 360–3.

1129. T. Wang, J.G. Proakis, and J.R. Zeidler (2005) Techniques for suppression of intercarrier interference in OFDM systems, in *Wireless Communications and Networking Conference*, vol. 1, IEEE, pp. 39–44.

1130. A.F. Molisch, M. Toeltsch, and S. Vermani (2007) Iterative methods for cancellation of intercarrier interference in OFDM systems, *IEEE Transactions on Vehicular Technology* **56**(4): 2158–67.

1131. L. Favalli, P. Savazzi, and A. Vizziello (2008) Frequency domain estimation and compensation of inter-carrier interference in OFDM systems, in *10th International Symposium on Spread Spectrum Techniques and Application*, Bologna.

1132. H. Hijazi and L. Ros (2009) Polynomial estimation of time-varying multipath gains with intercarrier interference tigation in OFDM systems, *IEEE Transactions on Vehicular Technology* **58**(1): 140–51.

1133. Y. Zhao, J.D. Leclercq, and S.G. Haggman (1998) Intercarrier interference compression in OFDM communication systems by using correlative coding, *Communications Letters, IEEE* **2**(8): 214–16.

1134. J. Armstrong (1999) Analysis of new and existing methods of reducing intercarrier interference due to carrier frequency offset in OFDM, IEEE Transactions on Communications **47**(3): 365–9.

1135. Y. Zhao and S.G. Haggman (2001) Intercarrier interference self-cancellation scheme for OFDM mobile communication systems, *IEEE Transactions on Communications* **49**(7): 1185–91.

1136. Y. Fu and C.C. Ko (2002) A new ICI self-cancellation scheme for OFDM systems based on a generalized signal mapper, in *5th International Symposium on Wireless Personal Multimedia Communications*, vol. 3, IEEE, pp. 995–9.

1137. H. Zhang and Y. Li (2003) Optimum frequency-domain partial response encoding in OFDM system, in *IEEE International Conference on Communications, ICC'03*, vol. 3, IEEE, pp. 2025–9.

1138. J.Y. Yun and Y.H. Lee (2004) A bandwidth efficient precode to reduce intercarrier interference in OFDM, in *59th Vehicular Technology Conference*, vol. 2, IEEE, pp. 944–6.

1139. Y.H. Peng, Y.C. Kuo, G.R. Lee, and J.H. Wen (2007) Performance analysis of a new ICI-self-cancellation-scheme in OFDM systems, *IEEE Transactions on Consumer Electronics* **53**(4): 1333–8.

1140. H.G. Yeh and C.C. Wang (2004) New parallel algorithm for mitigating the frequency offset of OFDM systems, in *60th Vehicular Technology Conference*, vol. 3, IEEE, pp. 2087–91.

1141. H.G. Yeh and Y.K. Chang (2004) A conjugate operation for mitigating intercarrier interference of OFDM systems, in *60th Vehicular Technology Conference*, vol. 6, IEEE, pp. 3965–9.

1142. H.G. Yeh, Y.K. Chang, and B. Hassibi (2007) A scheme for cancelling intercarrier interference using conjugate transmission in multicarrier communication systems, *IEEE Transactions on Wireless Communications* **6**(1): 3–7.

1143. C.L. Wang and Y.C. Huang (2010) Intercarrier interference cancellation using general phase rotated conjugate transmission for OFDM systems, *IEEE Transactions on Communications* **58**(3): 812–19.

1144. K. Sathananthan and C. Tellambura (2002) Reducing intercarrier interference in OFDM systems by partial transmit sequence and selected mapping, in *Proceedings of the International Symposium on DSP for Communication Systems, Manly-Sydney, Australia*, pp. 234–8.

1145. K. Sathananthan and C. Tellambura (2002) Partial transmit sequence arid selected mapping schemes to reduce ICI in OFDM systems, *Communications Letters, IEEE* **6**(8): 313–15.

1146. Q. Shi (2010) ICI mitigation for OFDM using PEKF, *Signal Processing Letters, IEEE* **17**(12): 981–4.

1147. M. Sandell, D. McNamara, and S. Parker (2006) Analysis of frequency-offset tracking in MIMO OFDM systems, *IEEE Transactions on Communications* **54**(8): 1481–9.

1148. D. Petrovic, W. Rave, and G. Fettweis (2004) Intercarrier interference due to phase noise in OFDM-estimation and suppression, in *60th Vehicular Technology Conference*, pp. 2191–5.

1149. K. Nikitopoulos, S. Stefanatos, and A.K. Katsaggelos (2009) Decision-aided compensation of severe phase-impairmentinduced inter-carrier interference in frequency-selective OFDM, *IEEE Transactions on Wireless Communications* **8**(4): 1614–19.

1150. P. Rabiei, W. Namgoong, and N. Al-Dhahir (2010) A non-iterative technique for phase noise ICI mitigation in packet-based OFDM systems, *IEEE Transactions on Signal Processing* **58**(11): 5945–50.

1151. M. Mousa Pasandi and D. Plant (2011) Non-iterative interpolation-based partial phase noise ICI mitigation for CO-OFDM transport systems, *Photonics Technology Letters, IEEE* **23**(21): 1594–6.

1152. Y. Shen and E. Martinez (2006) Channel estimation in OFDM systems, Application Note, Freescale Semiconductor.

1153. J. Oh, J. Kim, and J. Lim (2011) On the design of pilot symbols for OFDM systems over doubly-selective channels, *Communications Letters, IEEE* **12**: 1335–7.

1154. M.H. Hsieh and C.H. Wei (1998) Channel estimation for OFDM systems based on comb-type pilot arrangement in frequency selective fading channels, *IEEE Transactions on Consumer Electronics* **44**(1): 217–25.

1155. K.M.Z. Islam, T.Y. Al-Naffouri, and N. Al-Dhahir (2011) On optimum pilot design for comb-type OFDM transmission over doubly-selective channels, *IEEE Transactions on Communications* **59**(4): 930–5.

1156. J.J. Van de Beek, O. Edfors, M. Sandell, S. Wilson, and P.O. Borjesson (1995) On channel estimation in OFDM systems, in *45th Vehicular Technology Conference,* vol. 2, IEEE, pp. 815–19.

1157. M. Morelli and U. Mengali (2001) A comparison of pilot-aided channel estimation methods for OFDM systems, *IEEE Transactions on Signal Processing* **49**(12), pp. 3065–73.

1158. B. Yang, K.B. Letaief, R.S. Cheng, and Z. Cao (2001) Channel estimation for OFDM transmission in multipath fading channels based on parametric channel modeling, *IEEE Transactions on Communications* **49**(3): 467–79.

1159. D. Hu, L. He, and X. Wang (2011) An efficient pilot design method for OFDM-based cognitive radio systems, *IEEE Transactions on Wireless Communications* **10**(4): 1252–9.

1160. I. Rashad, I. Budiarjo, and H. Nikookar (2007) Efficient pilot pattern for ofdm-based cognitive radio channel estimation-part 1, in *14th IEEE Symposium on Communications and Vehicular Technology in the Benelux,* IEEE, pp. 1–5.

1161. I. Budiarjo, I. Rashad, and H. Nikookar (2007) Efficient pilot pattern for OFDM-based cognitive radio channel estimation—Part 2, in *14th IEEE Symposium on Communications and Vehicular Technology in the Benelux,* IEEE, pp. 1–5.

1162. J. Liu, S. Feng, and H. Wang (2009) Comb-type pilot aided channel estimation in non-contiguous OFDM systems for cognitive radio, in *5th International Conference Wireless Communications, Networking and Mobile Computing,* WiCom'09, IEEE, pp. 1–4.

1163. B. Hamilton, X. Ma, J. Kleider, and R. Baxley (2011) OFDM pilot design for channel estimation with null edge subcarriers, *IEEE Transactions on Wireless Communications* **10**(10): 3145–50.

1164. S. Roy and C. Li (2002) A subspace blind channel estimation method for OFDM systems without cyclic prefix, *IEEE Transactions on Wireless Communications* **1**(4): 572–9.

1165. C. Li and S. Roy (2003) Subspace-based blind channel estimation for OFDM by exploiting virtual carriers, *IEEE Transactions on Wireless Communications* **2**(1): 141–50.

1166. M.C. Necker and G.L. Stuber (2004) Totally blind channel estimation for OFDM on fast varying mobile radio channels, *IEEE Transactions on Wireless Communications* **3**(5): 1514–25.

1167. L. Mazet, V. Buzenac-Settineri, M. De Courville, and P. Duhamel (2002) An EM based semi-blind channel estimation algorithm designed for OFDM systems, in *36th Asilomar Conference on Signals, Systems and Computers,* vol. 2, IEEE, pp. 1642–6.

1168. B. Muquet, M. De Courville, and P. Duhamel (2002) Subspace-based blind and semi-blind channel estimation for OFDM systems, *IEEE Transactions on Signal Processing* **50**(7): 1699–1712.

1169. Y. Zeng and T.S. Ng (2004) A semi-blind channel estimation method for multiuser multiantenna OFDM systems, *IEEE Transactions on Signal Processing* **52**(5): 1419–29.

1170. W.C. Lim, B. Kannan, and T.T. Tjhung (2004) Joint channel estimation and OFDM synchronization in multipath fading, in International Conference on Communications, vol. 2, IEEE, pp. 983–7.

1171. M.M. Freda, J.F. Weng, and T. Le-Ngoc (2004) Joint channel estimation and synchronization for OFDM systems, in *60th Vehicular Technology Conference,* vol. 3, IEEE, pp. 1673–7.

1172. J.H. Lee, J.C. Han, and S.C. Kim (2006) Joint carrier frequency synchronization and channel estimation for OFDM systems via the EM algorithm, *IEEE Transactions on Vehicular Technology* **55**(1): 167–72.

1173. D.D. Lin, R.A. Pacheco, T.J. Lim, and D. Hatzinakos (2006) Joint estimation of channel response, frequency offset, and phase noise in OFDM, *IEEE Transactions on Signal Processing* **54**(9): 3542–54.

1174. H. Nguyen-Le, T. Le-Ngoc, and C.C. Ko (2007) Joint channel estimation and synchronization with inter-carrier interference reduction for OFDM, in *IEEE International Conference on Communications, ICC'07,* IEEE, pp. 2841–6.

1175. H. Nguyen-Le, T. Le-Ngoc, and C.C. Ko (2009) RLS-based joint estimation and tracking of channel response, sampling, and carrier frequency offsets for OFDM, *IEEE Transactions on Broadcasting* **55**(1): 84–94.

1176. K.G. Paterson and V. Tarokh (2000) On the existence and construction of good codes with low peak-to-average power ratios, *IEEE Transactions on Information Theory* **46**(6): 1974–87.

1177. J. Tellado (2000) *Multicarrier Modulation with Low PAR: Applications to DSL and Wireless.* Dordrecht: Kluwer Academic Publishers,

1178. T. Jiang and Y. Wu (2008) An overview: Peak-to-average power ratio reduction techniques for OFDM signals, *IEEE Transactions on Broadcasting* **54**(2): 257–68.
1179. H. Ochiai and H. Imai (2000) Performance of the deliberate clipping with adaptive symbol selection for strictly band-limited OFDM systems, *IEEE Journal on Selected Areas in Communications* **18**(11): 2270–7.
1180. G. Ren, H. Zhang, and Y. Chang (2003) A complementary clipping transform technique for the reduction of peak-to-average power ratio of OFDM system, *IEEE Transactions on Consumer Electronics* **49**(4): 922–6.
1181. D.W. Lim, J.S. No, C.W. Lim, and H. Chung (2005) A new SLM OFDM scheme with low complexity for PAPR reduction, *Signal Processing Letters, IEEE* **12**(2): 93–6.
1182. C.L. Wang and Y. Ouyang (2005) Low-complexity selected mapping schemes for peak-to-average power ratio reduction in OFDM systems, *IEEE Transactions on Signal Processing* **53**(12): 4652–60.
1183. R.J. Baxley and G.T. Zhou (2007) Comparing selected mapping and partial transmit sequence for PAR reduction, *IEEE Transactions on Broadcasting* **53**(4): 797–803.
1184. S.H. Muller and J.B. Huber (1997) OFDM with reduced peak-to-average power ratio by optimum combination of partial transmit sequences, *Electronics Letters* **33**(5): 368–9.
1185. S.H. Han and J.H. Lee (2004) PAPR reduction of OFDM signals using a reduced complexity PTS technique, *Signal Processing Letters, IEEE* **11**(11): 887–90.
1186. Y. Xiao, X. Lei, Q. Wen, and S. Li (2007) A class of low complexity PTS techniques for PAPR reduction in OFDM systems, *Signal Processing Letters, IEEE* **14**(10): 680–3.
1187. T. Jiang, W. Xiang, P.C. Richardson, J. Guo, and G. Zhu (2007) PAPR reduction of OFDM signals using partial transmit sequences with low computational complexity, *IEEE Transactions on Broadcasting* **53**(3): 719–24.
1188. H. Nikookar and K.S. Lidsheim (2002) Random phase updating algorithm for OFDM transmission with low PAPR, *IEEE Transactions on Broadcasting* **48**(2): 123–8.
1189. T. Jiang, W. Yao, P. Guo, Y. Song, and D. Qu (2006) Two novel nonlinear companding schemes with iterative receiver to reduce PAPR in multi-carrier modulation systems, *IEEE Transactions on Broadcasting* **52**(2): 268–73.
1190. T. Jiang and G. Zhu (2005) Complement block coding for reduction in peak-to-average power ratio of OFDM signals, *Communications Magazine, IEEE* **43**(9): S17–S22.
1191. A. Mobasher and A.K. Khandani (2006) Integer-based constellation-shaping method for PAPR reduction in OFDM systems, *IEEE Transactions on Communications* **54**(1): 119–27.
1192. Y.J. Kou, W.S. Lu, and A. Antoniou (2007) A new peak-to-average power-ratio reduction algorithm for OFDM systems via constellation extension, IEEE Transactions on Wireless Communications **6**(5): 1823–32.
1193. S.B. Slimane (2000) Peak-to-average power ratio reduction of OFDM signals using pulse shaping, in *Global Telecommunications Conference, Globecom'00,* vol. 3, IEEE, pp. 1412–16.
1194. S. Catreux, V. Erceg, D. Gesbert, and R.W. Heath Jr (2002) Adaptive modulation and MIMO coding for broadband wireless data networks, *Communications Magazine, IEEE* **40**(6): 108–15.
1195. A.J. Goldsmith and S.G. Chua (1998) Adaptive coded modulation for fading channels, *IEEE Transactions on Communications* **46**(5): 595–602.
1196. C. Mehlführer, S. Caban, and M. Rupp (2008) Experimental evaluation of adaptive modulation and coding in MIMOWiMAX with limited feedback, *EURASIP Journal on Advances in Signal Processing* no. 8, Article ID 837102, 1–12.
1197. A. Forenza, A. Pandharipande, H. Kim, and R.W. Heath Jr (2005) Adaptive MIMO transmission scheme: Exploiting the spatial selectivity of wireless channels, in *61st Vehicular Technology Conference,* vol. 5, IEEE, pp. 3188–92.
1198. S. Stiglmayr, M. Bossert, and E. Costa (2007) Adaptive coding and modulation in OFDM systems using BICM and ratecompatible punctured codes, in *Proceedimgs of the European Wireless Conference,* pp. 1999–2003.
1199. Y. Li and W.E. Ryan (2007) Mutual-information-based adaptive bit-loading algorithms for LDPC-coded OFDM, *IEEE Transactions on Wireless Communications* **6**(5): 1670–80.
1200. S. Stiglmayr, M. Bossert, and E. Costa (2008) Mutual-information-based adaptive coding and modulation in bit-interleaved OFDM systems using punctured LDPC codes, *European Transactions on Telecommunications* **19**(7): 801–11.

1201. C. Bockelmann, D. Wubben, and K.D. Kammeyer (2009) Rate enhancement of BICM-OFDM with adaptive coding and modulation via a bisection approach, in *10th Workshop on Signal Processing Advances in Wireless Communications,* SPAWC'09, IEEE, pp. 658–62.

1202. D. Hughes-Hartogs (1988) Ensemble modem structure for imperfect transmission media, US Patent 4,731,816.

1203. H. Zhang, J. Fu, and J. Song (2010) A Hughes-Hartogs algorithm based bit loading algorithm for OFDM systems, in *IEEE International Conference on Communications*, pp. 1–5.

1204. R.F.H. Fischer and J.B. Huber (1996) A new loading algorithm for discrete multitone transmission, in *Global Telecommunications Conference, Globecom'96. Communications: The Key to Global Prosperity,* vol. 1, pp. 724–8, IEEE.

1205. A. Seyedi and G. Saulnier (2004) Symbol-error rate analysis of Fischer's bit-loading algorithm, *IEEE Transactions on Communications* **52**(9): 1480–3.

1206. M. Bohge, J. Gross, A. Wolisz, and M. Meyer (2007) Dynamic resource allocation in OFDM systems: An overview of cross-layer optimization principles and techniques, *Network, IEEE* **21**(1): 53–9.

1207. H. Boostanimehr and V. Bhargava (2011) Selective subcarrier pairing and power allocation for DF OFDM relay systems with perfect and partial CSI, *IEEE Transactions on Wireless Communications* **10**(12): 4057–67.

1208. H.A. Mahmoud and H. Arslan (2008) Spectrum shaping of OFDM-based cognitive radio signals, in *Radio and Wireless Symposium,* IEEE, pp. 113–16.

1209. H.A. Mahmoud and H. Arslan (2008) Sidelobe suppression in OFDM-based spectrum sharing systems using adaptive symbol transition, *IEEE Communications Letters* **12**(2): 133–5.

1210. S. Brandes, I. Cosovic, and M. Schnell (2006) Reduction of out-of-band radiation in OFDM systems by insertion of cancellation carriers, *Communications Letters, IEEE* **10**(6): 420–2.

1211. Z. Wang, D. Qu, T. Jiang, and Y. He (2008) Spectral sculpting for OFDM based opportunistic spectrum access by extended active interference cancellation, in *Global Telecommunications Conference, Globecom'08,* IEEE, pp. 1–5.

1212. S.G. Huang and C.H. Hwang (2009) Improvement of active interference cancellation: Avoidance technique for OFDM cognitive radio, *IEEE Transactions on Wireless Communications* **8**(12), pp. 5928–37.

1213. D. Qu, Z. Wang, and T. Jiang (2010) Extended active interference cancellation for sidelobe suppression in cognitive radio of dm systems with cyclic prefix, *IEEE Transactions on Vehicular Technology* **59**(4): 1689–95.

1214. D. Rammoorthy, P.P.A. Murugesa, and S. Srikanth (2010) A low complexity active interference cancellation method for OFDM based cognitive radios, in *National Conference on Communications (NCC),* IEEE, pp. 1–4.

1215. D.R. Joshi, D.C. Popescu, and O.A. Dobre (2009) Dynamic spectral shaping in cognitive radios with quality of service constraints, in *43rd Asilomar Conference on Signals, Systems and Computers,* IEEE, pp. 539–43.

1216. I.G. Jang, Z.Y. Piao, Z.H. Dong, J.G. Chung, and K.Y. Lee (2011) Low-power FFT design for NC-OFDM in cognitive radio systems, in *International Symposium on Circuits and Systems (ISCAS),* IEEE, pp. 2449–52.

1217. J.W. Mwangoka, K. Ben Letaief, and Z. Cao (2008) Robust end-to-end QoS maintenance in non-contiguous OFDM based cognitive radios, in *International Conference on Communications, ICC'08,* IEEE, pp. 2905–9.

1218. Wikipedia, Orthogonal frequency division multiple access–Wikipedia, The Free Encyclopedia. http://en.wikipedia.org/wiki/OFDMA.

1219. B. Bai, W. Chen, K. Ben Letaief, and Z. Cao (2011) Diversity-multiplexing tradeoff in OFDMA systems: An H-matching approach, *IEEE Transactions on Wireless Communications* **10**(11): 3675–87.

1220. C. Xiong, G.Y. Li, S. Zhang, Y. Chen, and S. Xu (2011) Energy-and spectral-efficiency tradeoff in downlink OFDMA networks, in *International Conference on Communications (ICC),* IEEE, pp. 1–5.

1221. L. Sanguinetti, M. Morelli, and H.V. Poor (2010) Uplink synchronization in OFDMA spectrum-sharing systems, *IEEE Transactions on Signal Processing* **58**(5): 2771–82.

1222. K. Seong, M. Mohseni, and J.M. Cioffi (2006) Optimal resource allocation for OFDMA downlink systems, in *International Symposium on Information Theory,* IEEE, pp. 1394–8.

1223. M.Z. Bocus, J.P. Coon, C.N. Canagarajah *et al.* (2011) Resource allocation for OFDMA-based cognitive radio networks with application to H. 264 scalable video transmission, *EURASIP Journal on Wireless Communications and Networking,* Article ID 245673, 1–10.

1224. F.S. Chu and K.C. Chen (2007) Radio resource allocation in OFDMA cognitive radio systems, in *18th International Symposium on Personal, Indoor and Mobile Radio Communications,* IEEE, pp. 1–5.

1225. D.T. Ngo, C. Tellambura, and H.H. Nguyen (2010) Resource allocation for OFDMA-based cognitive radio multicast networks with primary user activity consideration, *IEEE Transactions on Vehicular Technology* **59**(4): 1668–79.

1226. R. Wang, V.K.N. Lau, L. Lv, and B. Chen (2009) Joint cross-layer scheduling and spectrum sensing for OFDMA cognitive radio systems, *IEEE Transactions on Wireless Communications* **8**(5): 2410–16.

1227. Y. Pan, A. Nix, and M. Beach (2011) Distributed resource allocation for OFDMA-based relay networks, *IEEE Transactions on Vehicular Technology* **60**(3): 919–31.

1228. K. Choi, E. Hossain, and D. Kim (2011) Downlink subchannel and power allocation in multi-cell OFDMA cognitive radio networks, IEEE Transactions on Wireless Communications **99**: 1–13.

1229. Y. Ma, D.I. Kim, and Z. Wu (2010) Optimization of ofdma-based cellular cognitive radio networks, *IEEE Transactions on Communications* **58**(8): 2265–76.

1230. S. Almalfouh and G. Stuber (2011) Interference-aware radio resource allocation in OFDMA-based cognitive radio networks, *IEEE Transactions on Vehicular Technology* **99**: 1.

1231. D. Ng, E. Lo, and R. Schober (2011) Secure resource allocation and scheduling for OFDMA decode-and-forward relay networks, *IEEE Transactions on Wireless Communications* **10**(10): 3528–40.

1232. S. Gao, L. Qian, and D. Vaman (2009) Distributed energy efficient spectrum access in cognitive radio wireless ad hoc networks, *IEEE Transactions on Wireless Communications* **8**(10): 5202–13.

1233. D. Ngo and T. Le-Ngoc (2011) Distributed resource allocation for cognitiveradio networks with spectrum-sharing constraints, *IEEE Transactions on Vehicular Technology* **60**(7): 3436–49.

1234. B. Da, R. Zhang, and C.C. Ko (2010) Spectrum trading in OFDMA-based cognitive radio systems, in *12th IEEE International Conference on Communication Technology (ICCT),* pp. 33–5.

1235. M.B. Ghorbel, A. Goldsmith, and M.S. Alouini (2011) Joint pricing and resource allocation for OFDMA-based cognitive radio systems, in *Workshop on Cognitive and Cooperative Networks. Infocom'11,* pp. 30–4.

1236. H. Xu and B. Li (2010) Efficient resource allocation with flexible channel cooperation in ofdma cognitive radio networks, in *Infocom'10 Proceedings,* IEEE, pp. 1–9.

1237. H. Sampath, S. Talwar, J. Tellado, V. Erceg, and A. Paulraj (2002) A fourth-generation MIMO-OFDM broadband wireless system: Design, performance, and field trial results, *Communications Magazine, IEEE* **40**(9): 143–9.

1238. G.L. Stuber, J. Barry, S.W. McLaughlin, Y. Li, M.A. Ingram, and T.G. Pratt (2004) Broadband MIMO-OFDM wireless communications, *Proceedings of the IEEE* **92**(2): 271–94.

1239. C. Dubuc, D. Starks, T. Creasy, and Y. Hou (2004) A MIMO-OFDM prototype for next-generation wireless WANs, *Communications Magazine, IEEE* **42**(12): 82–7.

1240. H. Yang (2005) A road to future broadband wireless access: MIMO-OFDM-based air interface, *Communications Magazine, IEEE* **43**(1): 53–60.

1241. H. Bolcskei (2006) MIMO-OFDM wireless systems: Basics, perspectives, and challenges, *Wireless Communications, IEEE* **13**(4): 31–7.

1242. A.N. Mody and G.L. Stuber (2001) Synchronization for MIMO OFDM systems, in *Global Telecommunications Conference, Globecom'01,* vol. 1, IEEE, pp. 509–13.

1243. C. Oberli and B. Daneshrad (2004) Maximum likelihood tracking algorithms for MIMO-OFDM, in *International Conference on Communications,* vol. 4, IEEE, pp. 2468–72.

1244. E. Zhou, X. Zhang, H. Zhao, and W. Wang (2005) Synchronization algorithms for MIMO OFDM systems, in *Wireless Communications and Networking Conference,* vol. 1, IEEE, pp. 18–22.

1245. I. Barhumi, G. Leus, and M. Moonen (2003) Optimal training design for MIMO OFDM systems in mobile wireless channels, *IEEE Transactions on Signal Processing* **51**(6): 1615–24.

1246. M. Shin, H. Lee, and C. Lee (2004) Enhanced channel-estimation technique for MIMO-OFDM systems, *IEEE Transactions on Vehicular Technology* **53**(1): 261–5.

1247. M.S. Baek, M.J. Kim, Y.H. You, and H.K. Song (2004) Semi-blind channel estimation and PAR reduction for MIMOOFDMsystem with multiple antennas, *IEEE Transactions on Broadcasting* **50**(4): 414–24.

1248. H. Miao and M.J. Juntti (2005) Space-time channel estimation and performance analysis for wireless MIMO-OFDM systems with spatial correlation, *IEEE Transactions on Vehicular Technology* **54**(6): 2003–16.

1249. Y. Qiao, S. Yu, P. Su, and L. Zhang (2005) Research on an iterative algorithm of LS channel estimation in MIMO OFDM systems, *IEEE Transactions on Broadcasting* **51**(1): 149–53.

1250. H. Minn and N. Al-Dhahir (2006) Optimal training signals for MIMO OFDM channel estimation, *IEEE Transactions on Wireless Communications* **5**(5): 1158–68.

1251. D. Hu, L. Yang, Y. Shi, and L. He (2006) Optimal pilot sequence design for channel estimation in MIMO OFDM systems, *Communications Letters, IEEE* **10**(1): 1–3.

1252. C. Shin, R.W. Heath, and E.J. Powers (2007) Blind channel estimation for MIMO-OFDM systems, *IEEE Transactions on Vehicular Technology* **56**(2): 670–85.

1253. P. Xia, S. Zhou, and G.B. Giannakis (2004) Adaptive MIMO-OFDM based on partial channel state information, *IEEE Transactions on Signal Processing* **52**(1): 202–13.

1254. R.Y. Mesleh, H. Haas, S. Sinanovic, C.W. Ahn, and S. Yun (2008) Spatial modulation, *IEEE Transactions on Vehicular Technology* **57**(4): 2228–41.

1255. W.L. Huang, K. Letaief, and Y.J. Zhang (2008) Cross-layer multi-packet reception based medium access control and resource allocation for space-time coded MIMO/OFDM, *IEEE Transactions on Wireless Communications* **7**(9): 3372–84.

1256. J.M. Choi, J.S. Kwak, H.S. Kim, and J.H. Lee (2004) Adaptive subcarrier, bit, and power allocation algorithm for MIMOOFDMA system, in *59th Vehicular Technology Conference*, vol. 3, IEEE, pp. 1801–5.

1257. J. Xu, J. Kim, W. Paik, and J.S. Seo (2006) Adaptive resource allocation algorithm with fairness for MIMO-OFDMA system, in *63rd Vehicular Technology Conference*, vol. 4, IEEE, pp. 1585–9.

1258. M.S. Maw and I. Sasase (2007) Resource allocation scheme in MIMO-OFDMA system for user's different data throughput requirements, in *Wireless Communications and Networking Conference*, IEEE, pp. 1706–10.

1259. E.S. Lo, P.W.C. Chan, V.K.N. Lau, *et al.* (2007) Adaptive resource allocation and capacity comparison of downlink multiuser MIMO-MC-CDMA and MIMO-OFDMA, *IEEE Transactions on Wireless Communications* **6**(3): 1083–93.

1260. Y. Peng, S.M.D. Armour, and J.P. McGeehan (2007) An investigation of dynamic subcarrier allocation in MIMO–OFDMA systems, *IEEE Transactions on Vehicular Technology* **56**(5): 2990–3005.

1261. N. Hassan and M. Assaad (2009) Low complexity margin adaptive resource allocation in downlink MIMO-OFDMA system, *IEEE Transactions on Wireless Communications* **8**(7): 3365–71.

1262. C.M. Yen, C.J. Chang, and L.C. Wang (2010) A utility-based TMCR scheduling scheme for downlink multiuser MIMOOFDMA systems, *IEEE Transactions on Vehicular Technology* **59**(8): 4105–15.

1263. B. Farhang-Boroujeny and R. Kempter (2008) Multicarrier communication techniques for spectrum sensing and communication in cognitive radios, *Communications Magazine, IEEE* **46**(4): 80–5.

1264. H. Mahmoud, T. Yucek, and H. Arslan (2009) OFDM for cognitive radio: Merits and challenges, *Wireless Communications, IEEE* **16**(2): 6–15.

1265. G. Bansal, M.J. Hossain, and V.K. Bhargava (2008) Optimal and suboptimal power allocation schemes for OFDM-based cognitive radio systems, *IEEE Transactions on Wireless Communications* **7**(11): 4710–18.

1266. G. Bansal, M. Hossain, and V.K. Bhargava (2011) Adaptive power loading for OFDM-based cognitive radio systems with statistical interference constraint, *IEEE Transactions on Wireless Communications* **99**, pp. 1–6.

1267. Y. Zhang and C. Leung (2010) An efficient power-loading scheme for OFDM-based cognitive radio systems, *IEEE Transactions on Vehicular Technology* **59**(4): 1858–64.

1268. Z. Hasan, G. Bansal, E. Hossain, and V. Bhargava (2009) Energy-efficient power allocation in OFDM-based cognitive radio systems: A risk-return model, *IEEE Transactions on Wireless Communications* **8**(12): 6078–88.

1269. S. Wang, F. Huang, and Z. Zhou (2011) Fast power allocation algorithm for cognitive radio networks, *Communications Letters, IEEE* **15**(99): 845–7.

1270. Y. Zhang and C. Leung (2009) Resource allocation in an OFDM-based cognitive radio system, *IEEE Transactions on Communications* **57**(7): 1928–31.

1271. Y. Zhang and C. Leung (2009) Cross-layer resource allocation for mixed services in multiuser OFDM-based cognitive radio systems, *IEEE Transactions on Vehicular Technology* **58**(8): 4605–19.

1272. Y. Zhang and C. Leung (2011) A distributed algorithm for resource allocation in OFDM cognitive radio systems, *IEEE Transactions on Vehicular Technology* **60**(2): 546–54.

1273. L.B. Le, P. Mitran, and C. Rosenberg (2009) Queue-aware subchannel and power allocation for downlink OFDM-based cognitive radio networks, in *Wireless Communications and Networking Conference*, IEEE, pp. 1–6.

1274. P. Mitran, L.B. Le, and C. Rosenberg (2010) Queue-aware resource allocation for downlink OFDMA cognitive radio networks, *IEEE Transactions on Wireless Communications* **9**(10): 3100–11.

1275. X. Kang, H. Garg, Y. Liang, and R. Zhang (2010) Optimal power allocation for OFDM-based cognitive radio with new primary transmission protection criteria, *IEEE Transactions on Wireless Communications* **9**(6): 2066–75.

1276. D. Bharadia, G. Bansal, P. Kaligineedi, and V.K. Bhargava (2011) Relay and power allocation schemes for OFDM-based cognitive radio systems, *IEEE Transactions on Wireless Communications* **10**(9): 2812–17.

1277. W. Zhu, B. Daneshrad, J. Bhatia, *et al.* (2006) A real time MIMO OFDM testbed for cognitive radio & networking research, in *Proceedings of the 1st International Workshop on Wireless Network Testbeds, Experimental Evaluation & Characterization*, ACM, pp. 115–16.

1278. H. Kim, J. Kim, S. Yang, *et al.* (2007) An effective MIMO-OFDM transmission scheme for IEEE 802.22 WRAN systems, in *2nd International Conference on Cognitive Radio Oriented Wireless Networks and Communications,* IEEE, pp. 394–9.

1279. Y. Rahulamathavan, K. Cumanan, R. Krishna, and S. Lambotharan (2009) Adaptive subcarrier and bit allocation techniques for MIMO-OFDMA based uplink cognitive radio networks, in *1st UK-India International Workshop on Cognitive Wireless Systems (UKIWCWS),* IEEE, pp. 1–5.

1280. Y. Rahulamathavan, K. Cumanan, and S. Lambotharan (2010) Optimal resource allocation techniques for MIMO-OFDMA based cognitive radio networks using integer linear programming, in *11th International Workshop on Signal Processing Advances in Wireless Communications (SPAWC),* IEEE, pp. 1–5.

1281. H.S. Shahrokh and K. Mohamed-Pour (2010) Sub-optimal power allocation in MIMO-OFDM based cognitive radio networks, in *6th International Conference on Wireless Communications Networking and Mobile Computing (WiCOM),* IEEE, pp. 1–5.

1282. H.S. Shahraki, K. Mohamed-Pour, and L. Vangelista (2011) Efficient resource allocation for MIMO-OFDMA based cognitive radio networks, in *Wireless Telecommunications Symposium (WTS),* IEEE, pp. 1–6.

1283. H.S. Shahraki and K. Mohamed-Pour (2011) Power allocation in multiple-input multipleoutput orthogonal frequency division multiplexing-based cognitive radio networks, *Communications, IET* **5**(3): 362–70.

1284. J. Von Neumann and O. Morgenstern (2007) *Theory of Games and Economic Behavior*. New Jersey: Princeton University Press.

1285. R.B. Myerson (1991) *Game Theory: Analysis of Conflict*. Harvard University Press.

1286. N. Nisan, T. Roughgarden, E. Tardos, and V.V. Vazirani (2007) *Algorithmic Game Theory*. Cambridge: Cambridge University Press.

1287. T. Başar and G.J. Olsder (1999) *Dynamic Noncooperative Game Theory,* vol. 23. Society for Industrial Mathematics.

1288. J.A. Filar and K. Vrieze (1997) *Competitive Markov Decision Processes*. Berlin: Springer Verlag.

1289. S. Nasar (1998) *A Beautiful Mind*. New York: Simon & Schuster.

1290. L.S. Shapley (1953) Stochastic games, *Proceedings of the National Academy of Sciences of the United States of America* **39**(10): 1095.

1291. R. Chen, J. Park, and J. Reed (2008) Defense against primary user emulation attacks in cognitive radio networks, *IEEE Journal on Selected Areas in Communications* **26**(1): 25–37.

1292. R. Chen, J.M. Park, and J.H. Reed (2008) Defense against primary user emulation attacks in cognitive radio networks, *IEEE Journal on Selected Areas in Communications* **26**(1): 25–37.

1293. Z. Jin, S. Anand, and K.P. Subbalakshmi (2009) Detecting primary user emulation attacks in dynamic spectrum access networks, in *International Conference on Communications,* IEEE, pp. 1–5.

1294. H. Li and Z. Han (2009) Dogfight in spectrum: Jamming and anti-jamming in multichannel cognitive radio systems, in *Global Telecommunications Conference, Globecom'09,* IEEE, pp. 1–6.

1295. H. Li and Z. Han (2010) Blind dogfight in spectrum: combating primary user emulation attacks in cognitive radio systems with unknown channel statistics, in *International Conference on Communications (ICC),* IEEE, pp. 1–6.

1296. L. Tassiulas and A. Ephremides (1992) Stability properties of constrained queueing systems and scheduling policies for maximum throughput in multihop radio networks, *IEEE Transactions on Automatic Control* **37**(12): 1936–48.

1297. R. Urgaonkar and M.J. Neely (2009) Opportunistic scheduling with reliability guarantees in cognitive radio networks, *IEEE Transactions on Mobile Computing* **8**(6): 766–77.

1298. Neely, M.J. (2010) *Stochastic Network Optimization with Application to Communication and Queuing Systems*. Morgan & Claypool Publishers.

1299. X. Wu and R. Srikant (2005) Regulated maximal matching: A distributed scheduling algorithm for multi-hop wireless networks with node-exclusive spectrum sharing, in *44th IEEE Conference on Decision and Control: European Control Conference*. CDC-ECC'05, IEEE, pp. 5342–7.

1300. H. Li (2010) Band synchronization in the control channel of cognitive radio systems: A collaboration game, in *International Conference on Communications (ICC)*, IEEE, pp. 1–5.

1301. L.A. DaSilva and I. Guerreiro (2008) Sequence-based rendezvous for dynamic spectrum access, in *3rd IEEE Symposium on New Frontiers in Dynamic Spectrum Access Networks, DySPAN*, IEEE, pp. 1–7.

1302. B. Horine and D. Turgut (2007) Link rendezvous protocol for cognitive radio networks, in *2nd IEEE International Symposium on New Frontiers in Dynamic Spectrum Access Networks, DySPAN*, IEEE, pp. 444–7.

1303. H. Li and Z. Han (2010) Collaborative spectrum sensing with a stranger: Trust, or not to trust?, in *Wireless Communications and Networking Conference (WCNC)*, IEEE, pp. 1–6.

1304. R. Chen, J.M. Park, and K. Bian (2008) Robust distributed spectrum sensing in cognitive radio networks, in *27th Conference on Computer Communications, Infocom'08*, IEEE, pp. 1876–84.

1305. T.C. Clancy and N. Goergen (2008) Security in cognitive radio networks: Threats and mitigation, in *3rd International Conference on Cognitive Radio Oriented Wireless Networks and Communications, Crown-Com 2008*, IEEE, pp. 1–8.

1306. W. Wang, H. Li, Y. Sun, and Z. Han (2009) Catchit: detect malicious nodes in collaborative spectrum sensing, in *Global Telecommunications Conference, Globecom'09*, IEEE, pp. 1–6.

1307. W. Wang, M. Chatterjee, and K. Kwiat (2009) Coexistence with malicious nodes: A game theoretic approach, in *International Conference on Game Theory for Networks, GameNets' 09*, IEEE, pp. 277–86.

1308. D. Fudenberg and J. Tirole (1991) *Game Theory*, Cambridge, MA: The MIT Press.

1309. A.S. Tanenbaum (2003) *Computer Networks*. Prentice-Hall.

1310. R. Knopp and P.A. Humblet (1995) Information capacity and power control in single-cell multiuser communications, in *International Conference on Communications: Gateway to Globalization*, vol. 1, IEEE, pp. 331–5.

1311. M. Franceschetti and R. Meester (2007) *Random Networks for Communication: From Statistical Physics to Information Systems*. Cambridge: Cambridge University Press.

1312. A. Barrat, M. Barthlemy, and A. Vespignani (2008) *Dynamical Processes on Complex Networks*. Cambridge: Cambridge University Press.

1313. F. Dorfler and F. Bullo (2010) Spectral analysis of synchronization in a lossless structure-preserving power network model in *1st IEEE International Conference on Smart Grid Communications (SmartGrid-Comm)*, IEEE, pp. 179–84.

1314. M. Barahona and L.M. Pecora (2002) Synchronization in small-world systems, *Physical Review Letters* **89**(5).

1315. I. Blekhman (1988) *Synchronization in Science and Technology*. American Society of Mechanical Engineers.

1316. L. Le and E. Hossain (2008) Resource allocation for spectrum underlay in cognitive radio networks, *IEEE Transactions on Wireless Communications* **7**(12): 5306–15.

1317. Q. Zhao and B.M. Sadler (2007) A survey of dynamic spectrum access, *IEEE Signal Processing Magazine* **24**: 79–89.

1318. F. Hou and J. Huang (2010) Dynamic channel selection in cognitive radio network with channel heterogeneity, in *IEEE Global Telecommunications Conference, Globecom'10*, pp. 1–6.

1319. R. Urgaonkar and M.J. Neely (2008) Opportunistic scheduling with reliability guarantees in cognitive radio networks, in *27th Conference on Computer Communications, Infocom'08*, IEEE, pp. 1301–9.

1320. M.J. Neely, E. Modiano, and C.E. Rohrs (2005) Dynamic power allocation and routing for time-varying wireless networks, *IEEE Journal on Selected Areas in Communications* **23**(1): 89–103.

1321. K.R. Chowdhury and I.F. Akyildiz (2011) CRP: A routing protocol for cognitive radio ad hoc networks, *IEEE Journal on Selected Areas in Communications* **29**(4): 794–804.

1322. K.R. Chowdhury, M. Di Felice, and I.F. Akyildiz (2009) Tp-crahn: A transport protocol for cognitive radio ad-hoc networks, in *Infocom'09*, IEEE, pp. 2482–90.

1323. D.E. Comer (2006) *Internet Working with TCP/IP: Principles, Protocols and Architectures*, 5th edn. Upper Saddle River, NJ: Prentice Hall.

1324. M. Bando, K. Hasebe, A. Nakayama, A. Shibata, and Y. Sugiyama (1995) Dynamical model of traffic congestion and numerical simulation, *Physical Review. E, Statistical Physics, Plasmas, Fluids, and Related Interdisciplinary Topics* **51**(2): 1035–42.

1325. H. Li (2011) Impact of primary user interruptions on data traffic in cognitive radio networks: Phantom jam on highway, in *Proceedings of IEEE Conference on Global Communications, Globecom'11*, pp. 1–5.

1326. D. Helbing (2001) Traffic and related self-driven many-particle systems, *Reviews of Modern Physics* **73**(4): 1067–1141.

1327. R.N. Mantegna and H.E. Stanley (2000) An introduction to econophysics: correlations and complexity in finance. Cambridge: Cambridge University Press.

1328. P.A. Samuelson (1965) Proof that properly anticipated prices fluctuate randomly, *Industrial Management Review* **6**(2): 965.

1329. A. Eydeland and K. Wolyniec (2003) *Energy and Power Risk Management: New Developments in Modeling, Pricing, and Hedging,* vol. 97. New York: John Wiley & Sons Inc.

1330. R. Merton (1976) Option pricing when underlying stock returns are discontinuous, *Journal of Financial Economics* **3**(1–2): 125–44.

1331. R.C. Qiu (2011) Cognitive radio network as sensors, technical report, Tennessee Technological University, Air Force Summer Faculty Fellowship Program Report.

1332. J. Yu, C. Zhang, Z. Hu, *et al.* (2011) Cognitive radio network as wireless sensor network (I): Architecture, testbed, and experiment, in *IEEE National Aerospace and Electronics Conference* (Fairborn, OH), pp. 1–7.

1333. F. Lin, Z. Hu, S. Hou, *et al.* (2011) Cognitive radio network as wireless sensor network (II): Security consideration, in *IEEE National Aerospace and Electronics Conference* (Fairborn, OH), pp. 1–5.

1334. D. Garmatyuk, J. Schuerger, Y.T. Morton, K. Binns, M. Durbin, and J. Kimani (2007) Feasibility study of a multi-carrier dual-use imaging radar and communication system, in European Radar Conference, EuRAD 2007 (Munich, Germany), IEEE, pp. 194–7.

1335. D. Garmatyuk and J. Schuerger (2008) Conceptual design of a dual-use radar/communication system based on OFDM, in *Military Communications Conference, Milcom'08* (San Diego, CA), IEEE, pp. 1–7.

1336. D. Garmatyuk and K. Kauffman (2009) Radar and data communication fusion with uwb-ofdm software-defined system, in *International Conference on Ultra-Wideband, ICUWB,* IEEE, pp. 454–8.

1337. D. Garmatyuk, J. Schuerger, and K. Kauffman (2011) Multifunctional software-defined radar sensor and data communication system, *Sensors Journal, IEEE* **11**(1): 99–106.

1338. S. Sen and A. Nehorai (2011) Sparsity-based multi-target tracking using OFDM radar, *IEEE Transactions on Signal Processing* **59**: 1902–6.

1339. S. Sen and A. Nehorai (2011) Adaptive OFDM radar for target detection in multipath scenarios, *IEEE Transactions on Signal Processing* **59**: 78–90.

1340. S. Sen and A. Nehorai (2010) OFDM MIMO radar with mutual-information waveform design for low-grazing angle tracking, *IEEE Transactions on Signal Processing* **58**(6): 3152–62.

1341. S. Sen and A. Nehorai (2009) Target detection in clutter using adaptive OFDM radar, *Signal Processing Letters, IEEE* **16**(7): 592–5.

1342. D. Garmatyuk (2006) High-resolution ultrawideband SAR based on OFDM architecture. White paper.

1343. C. Sturm, T. Zwick, and W. Wiesbeck (2009) An OFDM system concept for joint radar and communications operations, in *69th Vehicular Technology Conference,* IEEE, pp. 1–5.

1344. M. Braun, C. Sturm, A. Niethammer, and F.K. Jondral (2009) Parametrization of joint ofdm-based radar and communication systems for vehicular applications, in *20th International Symposium on Personal, Indoor and Mobile Radio Communications,* IEEE, pp. 3020–4.

1345. C. Sturm, M. Braun, and W. Wiesbeck (2010) Deterministic propagation modeling for joint radar and communication systems, in *URSI International Symposium on Electromagnetic Theory (EMTS),* pp. 942–5.

1346. Y.L. Sit, C. Sturm, and T. Zwick (2011) Doppler estimation in an OFDM joint radar and communication system, in *Microwave Conference (GeMIC),* IEEE, pp. 1–4.

1347. Y.L. Sit, L. Reichardt, C. Sturm, and T. Zwick (2011) Extension of the OFDM joint radar-communication system for a multipath, multiuser scenario, in *IEEE Radar Conference*, pp. 718–23.

1348. P. van Genderen (2010) A communication waveform for radar, in *8th International Conference on Communications (COMM),* IEEE, pp. 289–92.

1349. X. Shaojian, C. Bing, and Z. Ping (2006) Radar-communication integration based on DSSS techniques, in *8th International Conference on Signal* Processing, vol. 4, IEEE, pp. 1–4.

1350. S.J. Xu, Y. Chen, and P. Zhang (2006) Integrated radar and communication based on ds-uwb, in *3rd International Conference on Ultrawideband and Ultrashort Impulse Signals,* IEEE, pp. 142–4.

1351. Z. Lin and P. Wei (2006) Pulse amplitude modulation direct sequence ultra wideband sharing signal for communication and radar systems, in *7th International Symposium on Antennas, Propagation & EM Theory*, ISAPE'06, IEEE, pp. 1–5.

1352. M. Jamil, H. Zepernick, and M.I. Pettersson (2008) On integrated radar and communication systems using Oppermann sequences, in *Military Communications Conference, Milcom'08*, IEEE, pp. 1–6.

1353. S.D. Blunt, M.R. Cook, and J. Stiles (2010) Embedding information into radar emissions via waveform implementation, in *International Waveform Diversity and Design Conference (WDD)*, IEEE, pp. 195–9.

1354. S.D. Blunt and P. Yantham (2007) Waveform design for radar-embedded communications, in *International Waveform Diversity and Design Conference*, IEEE, pp. 214–18.

1355. G. Lellouch and H. Nikookar (2007) On the capability of a radar network to support communications, in *14th IEEE Symposium on Communications and Vehicular Technology in the Benelux*, IEEE, pp. 1–5.

1356. S.C. Surender, R.M. Narayanan, and C.R. Das (2010) Performance analysis of communications and radar coexistence in a covert UWB OSA system, in *Globecom'10*, IEEE, pp. 1–5.

1357. L. Wang, J. McGeehan, C. Williams, and A. Doufexi (2008) Radar spectrum opportunities for cognitive communications transmission, in *3rd International Conference on Cognitive Radio Oriented Wireless Networks and Communications, CrownCom 2008*, IEEE, pp. 1–6.

1358. Z. Yan, Z. Ma, H. Cao, G. Li, and W. Wang (2008) Spectrum sensing, access and coexistence testbed for cognitive radio using USRP, in *4th IEEE International Conference on Circuits and Systems for Communications*, IEEE, pp. 270–4.

1359. J. Jia, J. Zhang, and Q. Zhang (2009) Cooperative relay for cognitive radio networks, in *Infocom'09*, IEEE, pp. 2304–12.

1360. S. Wenmiao (2009) Configure cognitive radio using GNU radio and USRP, in *3rd IEEE International Symposium on Microwave, Antenna, Propagation and EMC Technologies for Wireless Communications*, IEEE, pp. 1123–6.

1361. K.R. Chowdhury and T. Melodia (2010) Platforms and testbeds for experimental evaluation of cognitive ad hoc networks, *Communications Magazine, IEEE* **48**(9): 96–104.

1362. D.M. Chinnam, J. Madhusudhan, C. Nandhini, *et al.* (2010) Implementation of a low cost synthetic aperture radar using software defined radio, in *International Conference on Computing Communication and Networking Technologies (ICCCNT)*, IEEE, pp. 1–7.

1363. F. Berizzi, M. Martorella, D. Petri, M. Conti, and A. Capria (2010) USRP technology for multiband passive radar, in *Radar Conference*, IEEE, pp. 225–9.

1364. B. Szlachetko and A. Lewandowski (2010) Signal receiving and processing platform of the experimental passive radar for intelligent surveillance system using software defined radio approach, *Knowledge-Based and Intelligent Information and Engineering Systems*, pp. 311–20.

1365. B. Szlachetko, A. Lewandowski, and G. Haza (2010) Universal software radio peripheral as a receiver and DSP platform for a passive radar, in *Proceedings of SPIE*, vol. 7745, pp. 1–6.

1366. B. Szlachetko and A. Lewandowski (2009) Application of the GNU radio platform in the multistatic radar, in *Proceedings of SPIE*, vol. 7502, pp. 1–6.

1367. A. Prabaswara, A. Munir, and A.B. Suksmono (2011) GNU radio based software-defined FMCW radar for weather surveillance application, in *6th International Conference on Telecommunication Systems, Services, and Applications (TSSA)*, IEEE, pp. 227–30.

1368. K. Shenai and S. Mukhopadhyay (2008) Cognitive sensor networks, in *26th International Conference on Microelectronics, MIEL*, IEEE, pp. 315–20.

1369. L. Bixio, L. Ciardelli, M. Ottonello, and C.S. Regazzoni (2009) Distributed cognitive sensor network approach for surveillance applications, in *6th IEEE International Conference on Advanced Video and Signal Based Surveillance*, AVSS'09, IEEE, pp. 232–7.

1370. O. Akan, O. Karli, and O. Ergul (2009) Cognitive radio sensor networks, *Network, IEEE* **23**(4): 34–40.

1371. Z. Liang and D. Zhao (2010) Quality of service performance of a cognitive radio sensor network, in *International Conference on Communications (ICC)*, IEEE, pp. 1–5.

1372. R. Yu, Y. Zhang, W. Yao, L. Song, and S. Xie (2010) Spectrum-aware routing for reliable end-to-end communications in cognitive sensor network, in Global Telecommunications Conference, *Globecom'10*, IEEE, pp. 1–5.

1373. S. Maleki, A. Pandharipande, and G. Leus (2011) Energy-efficient distributed spectrum sensing for cognitive sensor networks, *Sensors Journal, IEEE* **11**(3): 565–73.

1374. J.L. Williams, J.W. Fisher, and A.S. Willsky (2007) Approximate dynamic programming for communication-constrained sensor network management, *IEEE Transactions on Signal Processing* **55**(8): 4300–11.

1375. P. Rong and M.L. Sichitiu (2006) Angle of arrival localization for wireless sensor networks, in *3rd Annual IEEE Communications Society on Sensor and Ad Hoc Communications and Networks, SECON'06,* vol. 1, IEEE, pp. 374–82.

1376. J. Friedman, Z. Charbiwala, T. Schmid, Y. Cho, and M. Srivastava (2008) Angle-of-arrival assisted radio interferometry (ARI) target localization, in *Military Communications Conference, Milcom'08,* IEEE, pp. 1–7.

1377. G. Marchetti, G. Picchi, and L. Verrazzani, Detection and time-of-arrival estimation of received pulses in radar beacon systems, *IEEE Transactions on Aerospace and Electronic Systems* **16**(3): 294–304.

1378. H. Liu, H. Darabi, P. Banerjee, and J. Liu (2007) Survey of wireless indoor positioning techniques and systems, *IEEE Transactions on Systems, Man, and Cybernetics, Part C: Applications and Reviews* **37**(6): 1067–80.

1379. A. Dersan and Y. Tanik (2002) Passive radar localization by time difference of arrival, in *Proceedings, Milcom'02*, vol. 2, IEEE, pp. 1251–7.

1380. F. Gustafsson and F. Gunnarsson (2003) Positioning using time-difference of arrival measurements, in *International Conference on Acoustics, Speech, and Signal Processing, ICASSP'03,* vol. 6, IEEE, pp. 553–6.

1381. X. Li, Q. Han, V. Chakravarthy, and Z. Wu (2010) Joint spectrum sensing and primary user localization for cognitive radio via compressed sensing, in *Military Communications Conference, Milcom'10,* IEEE, pp. 329–34.

1382. D.T. Huang, S.H. Wu, and P.H. Wang (2010) Cooperative spectrum sensing and locationing: A sparse Bayesian learning approach, in *Global Telecommunications Conference, Globecom'10,* IEEE, pp. 1–5.

1383. P. Iscold and G.A.S. Pereira (2010) Development of a hand-launched small UAV for ground reconnaissance, *IEEE Transactions on Aerospace and Electronic Systems* **46**(1): 335–48.

1384. E. Frew, C. Dixon, B. Argrow, and T. Brown (2005) Radio source localization by a cooperative UAV team, in *Infotech@ Aerospace* (Arlington, VA), September 2005.

1385. I. Shames, B. Fidan, and B. Anderson (2008) Close target reconnaissance using autonomous uav formations, in *47th IEEE Conference on Decision and Control, CDC'08* (Cancun, Mexico), IEEE, pp. 1729–34.

1386. S. Yoon and C. Qiao (2006) A novel approach to reconnaissance using cooperative mobile sensor nodes, in *Military Communications Conference, Milcom'06,* IEEE, pp. 1–7.

1387. S. Yang, X. Gao, and H. Chen (2010) Ground thread identification of the reconnaissance and strike integrated UAV based on improved direct Inference algorithm, in *3rd International Conference on Advanced Computer Theory and Engineering (ICACTE),* vol. 1, IEEE, pp. V1–577.

1388. E. Kuiper and S. Nadjm-Tehrani (2006) Mobility models for UAV group reconnaissance applications, in *International Conference on Wireless and Mobile Communications, ICWMC '06,* IEEE Computer Society, pp. 33-1–33-7.

1389. Wikipedia, Synthetic aperture radar–Wikipedia, The Free Encyclopedia. http://en.wikipedia.org/wiki/ Synthetic aperture radar.

1390. Y. Bao, X. Fu, and X. Gao (2010) Path planning for reconnaissance UAV based on particle swarm optimization, in *2nd International Conference on Computational Intelligence and Natural Computing Proceedings, CINC'10*, IEEE, vol. 2, pp. 28–32.

1391. Y. Zhang and J. Gao (2008) In-flight route re-planning for endurance reconnaissance unmanned aerial vehicles, in *2nd International Symposium on Systems and Control in Aerospace and Astronautics, ISSCAA'08,* IEEE, pp. 1–5.

1392. A.K. Mitra, T. Lewis, and C.L. Willemsen (2007) Exploitation of UAV trajectories with perturbation for intelligent circular SAR applications, in *Proceedings of SPIE, Algorithms for Synthetic Aperture Radar Imagery XIV* (Orlando, Florida), p. 656805.

1393. C.E. Yarman, B. Yazici, and M. Cheney (2007) Bistatic synthetic aperture inversion for arbitrary flight trajectories, in *Proceedings of SPIE, Algorithms for Synthetic Aperture Radar Imagery XIV* (Orlando, Florida)), p. 656807.

1394. J.D. Coker and A.H. Tewfik (2011) Performance synthesis of UAV trajectories in multistatic SAR, *IEEE Transactions on Aerospace and Electronic Systems* **47**: 848–61.

1395. D.C. Munson, J.D. O'Brien, and W.K. Jenkins (1983) A tomographic formulation of spotlight-mode synthetic aperture radar, *Proceedings of the IEEE* **71**: 917–25.

1396. E. Ertin, C.D. Austin, S. Sharma, R.L. Moses, and L.C. Potter (2007) GOTCHA experiment report: Three-dimensional SAR imaging with complete circular apertures, in *Proceedings of SPIE, Algorithms for Synthetic Aperture Radar Imagery* XIV), p. 656802.

1397. L.J. Moore and L.C. Potter (2007) Three-dimensional resolution for circular synthetic aperture radar, in *Proceedings of SPIE, Algorithms for Synthetic Aperture Radar Imagery* XIV), p. 656804.

1398. C.D. Austin, E. Ertin, and R.L. Moses (2009) Sparse multipass 3D SAR imaging: Applications to the GOTCHA data set, in *Proceedings of SPIE, Algorithms for Synthetic Aperture Radar Imagery* XVI, Citeseer., vol. 7337, 03-01–03–12.

1399. M. Xing, X. Jiang, R. Wu, F. Zhou, and Z. Bao (2009) Motion compensation for UAV SAR based on raw radar data, *IEEE Transactions on Geoscience and Remote Sensing* 47: 2870–83.

1400. E.C. Zaugg and D.G. Long (2008) Theory and application of motion compensation for LFM-CW SAR, *IEEE Transactions on Geoscience and Remote Sensing* 46(10): 2990–8.

1401. J. Huanq, W. Lei, W. Yu, and H. Jian (2006) The real-time coarse compensation of motion error based on UAV SAR, in *International Conference on Radar, CIE'06 (Shanghai, China)*, IEEE, pp. 1–4.

1402. R.J. Dekker (2005) SAR change detection techniques and applications, in *25th EARSeL Symposium on Global Developments in Environmental Earth Observation from Space (Porto, Portugal)*, pp. 63–9.

1403. S.M. Scarborough, L.R. Gorham, M.J. Minardi, *et al.* (2010) A challenge problem for SAR change detection and data compression, in *Proceedings of SPIE*, pp. 1–5.

1404. R.D. Phillips (2011) CLEAN: A false alarm reduction method for SAR CCD, in *International Conference on Acoustics, Speech and Signal Processing (ICASSP) (Prague)*, IEEE, pp. 1365–8.

1405. M. Priess, D. Gray, and N. Stacy (2003) A change detection technique for repeat pass interferometric SAR, in *International Geoscience and Remote Sensing Symposium, IGARSS'03*, vol. 2, IEEE, pp. 938–40.

1406. J. Inglada and G. Mercier (2007) A new statistical similarity measure for change detection in multitemporal SAR images and its extension to multiscale change analysis, *IEEE Transactions on Geoscience and Remote Sensing* 45(5): 1432–45.

1407. F. Bovolo and L. Bruzzone (2005) A wavelet-based change-detection technique for multitemporal SAR images, in *International Workshop on the Analysis of Multi-Temporal Remote Sensing Images*, IEEE, pp. 85–9.

1408. R.Z. Schneider and D. Fernandes (2003) Entropy among a sequency of SAR images for change detection, in *International Geoscience and Remote Sensing Symposium, IGARSS'03*, vol. 2, IEEE, pp. 1389–91.

1409. M. He, X.F. He, and H.B. Luo (2007) Detection of information change on SAR images based on entropy theory, in 1st *Asian and Pacific Conference on Synthetic Aperture Radar, APSAR'07*, pp. 775–8.

1410. S. Baronti, R. Carla, S. Sigismondi, and L. Alparone (1994) Principal component analysis for change detection on polarimetric multitemporal SAR data, in *International Geoscience and Remote Sensing Symposium, IGARSS'94: Surface and Atmospheric Remote Sensing: Technologies, Data Analysis and Interpretation*, vol. 4, IEEE, pp. 2152–4.

1411. Y. Lee, T. Hara, H. Fujita, S. Itoh, and T. Ishigaki (2001) Automated detection of pulmonary nodules in helical CT images based on an improved template-matching technique, *IEEE Transactions on Medical Imaging* 20(7): 595–604.

1412. S. Pereira and T. Pun (2000) Fast robust template matching for affine resistant image watermarks, in *Information Hiding*, New York: Springer, pp. 199–210.

1413. G.N. DeSouza and A.C. Kak (2002) Vision for mobile robot navigation: A survey, *IEEE Transactions on Pattern Analysis and Machine Intelligence* 24(2): 237–67.

1414. R.C. Qiu, M.C. Wicks, L. Li, Z. Hu, and S.J. Hou (2010) Wireless tomography, Part I: A novel approach to remote sensing, in *5th International Waveform Diversity & Design Conference (Niagara Falls, Canada)*, pp. 244–56.

1415. J. Wilson and N. Patwari (2010) Radio tomographic imaging with wireless networks, *IEEE Transactions on Mobile Computing* 9(5): 621–32.

1416. J. Wilson and N. Patwari (2011) See through walls: Motion tracking using variance-based radio tomography networks, *IEEE Transactions on Mobile Computing* 99: 1.

1417. Y. Zhao and N. Patwari (2011) Noise reduction for variance-based device-free localization and tracking, in *Proceedings of the 8th IEEE Conference on Sensor, Mesh and Ad Hoc Communications and Networks (SECON'11), Salt Lake City, Utah*.

1418. L. Lo Monte, D. Erricolo, F. Soldovieri, and M.C. Wicks (2010) Radio frequency tomography for tunnel detection, IEEE *Transactions on Geoscience and Remote Sensing* 48(3): 1128–37.

1419. L. Lo Monte, D. Erricolo, F. Soldovieri, and M.C. Wicks (2010) RF tomography for below-ground imaging of extended areas and close-in sensing, *Geoscience and Remote Sensing Letters*, IEEE **7**(3): 496–500.

1420. A.J. Devaney (1982) A filtered backpropagation algorithm for diffraction tomography, *Ultrasonic Imaging* **4**(4): 336–50.

1421. A.J. Devaney (1984) Geophysical diffraction tomography, *IEEE Transactions on Geoscience and Remote Sensing* **GE-22**(1): 3–13.

1422. A.T. Vouldis, C.N. Kechribaris, T.A. Maniatis, K.S. Nikita, and N.K. Uzunoglu (2006) Three-dimensional diffraction tomography using filtered backpropagation and multiple illumination planes, *IEEE Transactions on Instrumentation and Measurement* **55**(6): 1975–84.

1423. P.M. Berg and R.E. Kleinman (1997) A contrast source inversion method, *Inverse Problems* **13**: 1607.

1424. P.M. Berg, A.L. Broekhoven, and A. Abubakar (1999) Extended contrast source inversion, *Inverse Problems* **15**: 1325.

1425. A. Abubakar, P.M. Van den Berg, and J.J. Mallorqui (2002) Imaging of biomedical data using a multiplicative regularized contrast source inversion method, *IEEE Transactions on Microwave Theory and Techniques* **50**(7): 1761–71.

1426. A. Abubakar, T.M. Habashy, and P.M. Van den Berg (2006) Nonlinear inversion of multi-frequency microwave fresnel data using the multiplicative regularized contrast source inversion, *Progress in Electromagnetics Research* **62**: 193–201.

1427. W.C. Chew and Y.M. Wang (1990) Reconstruction of two-dimensional permittivity distribution using the distorted born iterative method, *IEEE Transactions on Medical Imaging* **9**(2): 218–25.

1428. O.S. Haddadin, S.D. Lucas, and E.S. Ebbini (1995) Solution to the inverse scattering problem using a modified distorted Born iterative algorithm, in *Ultrasonics Symposium,* vol. 2, IEEE, pp. 1411–14.

1429. T.J. Cui, W.C. Chew, A.A. Aydiner, and S. Chen (2001) Inverse scattering of two-dimensional dielectric objects buried in a lossy earth using the distorted born iterative method, *IEEE Transactions on Geoscience and Remote Sensing* **39**(2): 339–46.

1430. F.K. Gruber, E.A. Marengo, and A.J. Devaney, Time-reversal imaging with multiple signal classification considering multiple scattering between the targets, *Journal of the Acoustical Society of America* **115**: 3042–7.

1431. A.J. Devaney, E.A. Marengo, and F.K. Gruber (2005) Time-reversal-based imaging and inverse scattering of multiply scattering point targets, *Journal of the Acoustical Society of America* **118**: 3129–38.

1432. E.A. Marengo, F.K. Gruber, and F. Simonetti (2007) Time-reversal MUSIC imaging of extended targets, *IEEE Transactions on Image Processing* **16**(8): 1967–84.

1433. J.M.F. Moura and Y. Jin (2008) Time reversal imaging by adaptive interference canceling, *IEEE Transactions on Signal Processing* **56**(1): 233–47.

1434. O. Bucci, L. Crocco, M. D'Urso, and T. Isernia (2006) Inverse scattering from phaseless measurements of the total field on open lines, *JOSA A* **23**(10): 2566–77.

1435. L. Crocco, M. DUrso, and T. Isernia (2004) Inverse scattering from phaseless measurements of the total field on a closed curve, *JOSA A* **21**(4): 622–31.

1436. M. D'Urso, K. Belkebir, L. Crocco, T. Isernia, and A. Litman (2008) Phaseless imaging with experimental data: Facts and challenges, *JOSA A.* **25**(1): 271–81.

1437. W.J. Zhang, L.L. Li, and F. Li (2009) Inverse scattering from phaseless data in the freespace, *Science in China Series F: Information Sciences* **52**(8): 1389–98.

1438. L. Pan, Y. Zhong, X. Chen, and S.P. Yeo (2011) Subspace-based optimization method for inverse scattering problems utilizing phaseless data, *IEEE Transactions on Geoscience and Remote Sensing* **49**(3): 981–7.

1439. L. Li, W. Zhang, and F. Li (2008) Tomographic reconstruction using the distorted rytov iterative method with phaseless data, *Geoscience and Remote Sensing Letters*, IEEE **5**(3): 479–83.

1440. A.C. Kak and M. Slaney (1988) *Principles of Computerized Tomographic Imaging.* IEEE Service Center, Piscataway, NJ.

1441. R.C. Qiu, Z. Hu, M. Wicks, L. Li, S.J. Hou, and L. Gary (2010) Wireless tomography, Part II: A system engineering approach, in *5th International Waveform Diversity & Design Conference (Niagara Falls, Canada),* pp. 277–82.

1442. R.K. Ganti, F. Ye, and H. Lei (2011) Mobile crowdsensing: Current state and future challenges, *IEEE Communications Magazine* **49**(11): 32–9.

1443. R. Mahafza (2000) *Radar Systems Analysis and Design Using MATLAB*. CRC Press.

1444. J. Gao (2002) Integration of GPS with remote sensing and GIS: Reality and prospect, *Photogrammetric Engineering and Remote Sensing* **68**(5): 447–54.

1445. Wikipedia, Geographic information system–Wikipedia, The Free Encyclopedia. http://en.wikipedia.org/wiki/Geographic information system.

1446. B.D. Ripley (1981) *Spatial Statistics*. Wiley Online Library.

1447. Wikipedia, Global Positioning System–Wikipedia, The Free Encyclopedia. http://en.wikipedia.org/wiki/Gps.

1448. S. Wu, X. Qiu, and L. Wang (2005) Population estimation methods in GIS and remote sensing: A review, *GIScience & Remote Sensing* **42**(1): 80–96.

1449. L. Wang and C. Wu (2010) Population estimation using remote sensing and GIS technologies, *International Journal of Remote Sensing* **31**(21): 5569–70.

1450. Wikipedia, Cyber physical system–Wikipedia, The Free Encyclopedia. http://en.wikipedia.org/wiki/Cyberphysical system.

1451. A.M. Tulino and S. Verdú (2004) *Random Matrix Theory and Wireless Communications,* vol. 1. Hanover, MA: Now Publishers Inc.

1452. R. Couillet and M. Debbah (2011) *Random Matrix Methods for Wireless Communications*. Cambridge: Cambridge University Press.

1453. V. Vu (2011) Singular vectors under random perturbation, *Random Structures & Algorithms* **39**(4): 526–38.

1454. M.A. Vouk (2004) Cloud computing—issues, research and implementations, *Journal of Computing and Information Technology* **16**(4): 235–46.

1455. M. Armbrust, A. Fox, R. Griffith, *et al.* (2009) Above the clouds: A Berkeley view of cloud computing, Technical Report UCB/EECS-2009-28, EECS Department, University of California, Berkeley.

1456. F.T. Leighton (1992) *Introduction to Parallel Algorithms and Architectures*. Kaufmann.

1457. V. Kumar (2002) *Introduction to Parallel Computing*. Addison-Wesley Longman Publishing Co., Inc.

1458. T. Imielinski and H. Korth (1996) Introduction to mobile computing, *Mobile Computing,* pp. 1–43.

1459. D. Datla, X. Chen, T. Tsou, *et al.* (2012) Wireless distributed computing: a survey of research challenges, *Communications Magazine, IEEE* **50**(1): 144–52.

1460. K.D. Bowers, A. Juels, and A. Oprea (2009) HAIL: A high-availability and integrity layer for cloud storage, in *Proceedings of the 16th ACM Conference on Computer and Communications Security*, ACM, pp. 187–98.

1461. C. Wang, S.S.M. Chow, Q. Wang, K. Ren, and W. Lou (2010) Privacy preserving public auditing for secure cloud storage, *IEE Transactions on Computers,, Infocom'10*, IEEE, pp. 1–14.

1462. J. Glenn-Anderson (2008) *Gpu-based Desktop Supercomputing*. enParallel Inc.

1463. Nvidia (2009) Nvidia's Next Generation CUDA Compute Architecture: Fermi, 2009. available online: http://www.nvidia.com/content/PDF/fermi white papers/NVIDIA Fermi Compute Architecture Whitepaper.pdf.

1464. S. Penmatsa (2007) Game Theory Based Job Allocation/Load Balancing in Distributed Systems with Applications to Grid Computing. PhD thesis, The University of Texas at San Antonio.

1465. R. Subrata, A.Y. Zomaya, and B. Landfeldt (2008) Game-theoretic approach for load balancing in computational grids, *IEEE Transactions on Parallel and Distributed Systems* **19**(1): 66–76.

1466. L. Chen, O. Villa, S. Krishnamoorthy, and G.R. Gao (2010) Dynamic load balancing on single-and multi-GPU systems, in *International Symposium on Parallel & Distributed Processing (IPDPS)*. IEEE, pp. 1–12.

1467. F. Song, S. Tomov, and J. Dongarra (2011) fficient support for matrix computations on heterogeneous multi-core and multi-GPU architectures, technical report, EECS Department, University of Tennessee.

1468. C. Lauterback, Q. Mo, and D. Manocha (2011) Work distribution methods on GPUs, tech. rep., Department of Computer Science, University of North Carolina.

1469. P. Pospíchal, J. Jaros, and J. Schwarz (2010) Parallel genetic algorithm on the CUDA architecture, *Applications of Evolutionary Computation,* pp. 442–51.

1470. A. Perrig, J. Stankovic, and D. Wagner (2004) Security in wireless sensor networks, *Communications of the ACM* **47**(6): 53–7.

1471. M. Saraogi (2005) Security in wireless sensor networks, technical report, Department of Computer Science, University of Tennessee, Knoxville.

1472. P. Halmos (1958) *Finite-Dimensional Vector Spaces*. New York: Springer.

1473. W. Arveson (1974) *An Invitation to C*-Algebra*. New York: Springer.

1474. M. Hazewinkel, N. Gubareni, and V. Kirichenko (2004) *Algebras, Rings and Modules*. New York: Springer.

1475. N. Datta and R. Renner (2009) Smooth entropies and the quantum information spectrum, *IEEE Transactions on Information Theory* **55**(6): 2807–15.

Index

Cognitive Radio Communications and Networking: Principles and Practice, First Edition.
Robert C. Qiu, Zhen Hu, Husheng Li and Michael C. Wicks.
© 2012 John Wiley & Sons, Ltd. Published 2012 by John Wiley & Sons, Ltd.